Q-tables — 220
Fourier transform — 81 & 82

FUNDAMENTALS OF
COMMUNICATION SYSTEMS

FUNDAMENTALS OF COMMUNICATION SYSTEMS

John G. Proakis
Masoud Salehi

PEARSON
Prentice Hall

Upper Saddle River, New Jersey 07458

Library of Congress Cataloging-in-Publication Data on File

Vice President and Editorial Director, ECS: *Marcia J. Horton*
Vice President and Director of Production and Manufacturing, ESM: *David W. Riccardi*
Executive Managing Editor: *Vince O'Brien*
Managing Editor: *David A. George*
Production Editors: *Daniel Sandin and Wendy Kopf*
Director of Creative Services: *Paul Belfanti*
Creative Director: *Jayne Conte*
Cover Designer: *Bruce Kenselaar*
Art Editor: *Greg Dulles*
Manufacturing Buyer: *Lisa McDowell*
Senior Marketing Manager: *Holly Stark*

© 2005 Pearson Education, Inc.
Pearson Prentice Hall
Pearson Education, Inc.
Upper Saddle River, New Jersey 07458

Pearson Prentice Hall® is a trademark of Pearson Education, Inc.

The author and publisher of this book have used their best efforts in preparing this book. These efforts include the development, research, and testing of the theories and programs to determine their effectiveness. The author and publisher make no warranty of any kind, expressed or implied, with regard to these programs or the documentation contained in this book. The author and publisher shall not be liable in any event for incidental or consequential damages in connection with, or arising out of, the furnishing, performance, or use of these programs.

Printed in the United States of America

10 9 8 7 6 5

ISBN 0-13-147135-X

Pearson Education Ltd., *London*
Pearson Education Australia Pty, Ltd., *Sydney*
Pearson Education Singapore, Pte. Ltd.
Pearson Education North Asia Ltd., *Hong Kong*
Pearson Education Canada, Inc., *Toronto*
Pearson Educación de Mexico, S.A. de C.V.
Pearson Education—Japan, *Tokyo*
Pearson Education Malaysia, Pte. Ltd.
Pearson Education, Inc., Upper Saddle River, *New Jersey*

To Felia, George, and Elena.
  −John G. Proakis
To Fariba, Omid, Sina, and my parents.
  −Masoud Salehi

Contents

Preface

This book is intended as a senior level undergraduate textbook on communication systems for Electrical Engineering majors. Its primary objective is to introduce the basic techniques used in modern communication systems and to provide fundamental tools and methodologies used in the analysis and design of these systems. Although the book is mainly written as an undergraduate level textbook, it can be equally be useful to the practicing engineer, or as a self study tool.

The emphasis of the book is on digital communication systems, which are treated in detail in Chapters 7 through 13. These systems are the backbone of modern communication systems, including new generations of wireless communication systems, satellite communications, and data transmission networks. Traditional analog communication systems are also covered with due detail in Chapters 3, 4, and 6. In addition, the book provides detailed coverage of the background required for the course in two chapters, one on linear system analysis with emphasis on the frequency domain approach and Fourier techniques, and one on probability, random variables, and random processes. Although these topics are now covered in separate courses in the majority of electrical engineering colloquia, it is the experience of the authors that the students frequently need to review these topics in a course on communications, and therefore it is essential to have quick access the relevant material from these courses.

It is assumed that the students taking this course have background in calculus, linear algebra, basic electronic circuits, linear system theory, and probability and random variables. These latter two topics are reviewed in two chapters of the book.

ORGANIZATION OF THE BOOK

The book starts with a brief review of communication systems in Chapter 1 followed by methods of signal representation and system analysis in both time and frequency domains in Chapter 2. Emphasis is placed on the Fourier series and the Fourier transform representation of signals and the use of transforms in linear systems analysis.

Chapters 3 and 4 cover the modulation and demodulation of analog signals. In Chapter 3 amplitude modulation (AM), and in Chapter 4 frequency modulation (FM), and phase modulation (PM) are covered. Radio and television broadcasting and analog mobile radio cellular communication systems are also treated in these chapters.

In Chapter 5, we present a review of the basic definitions and concepts in probability and random processes. Special emphasis is placed on Gaussian random processes, which provide mathematically tractable models for additive noise disturbances. Both time domain and frequency domain representations of random signals are presented.

Chapter 6 covers the effects of additive noise in the demodulation of amplitude modulated (AM) and angle modulated (FM,PM) analog signals and a comparison of these analog signal modulations in terms of their signal-to-noise ratio performance. Also discussed in this chapter is the problem of estimating the carrier phase using a phase-locked loop (PLL). Finally, we describe the characterization of thermal noise and the effect of transmission losses in analog communication systems.

Chapter 7 is devoted to analog-to-digital conversion. Sampling theorem and quantization techniques are treated first, followed by waveform encoding methods including PCM, DPCM, and DM. This chapter concludes with brief discussion of LPC speech coding and the JPEG standard for image compression.

Chapter 8 treats modulation methods for baseband AWGN channels. Various types of binary and non-binary modulation methods are described based on a geometric representation of signals and their performance is evaluated in terms of the probability of error. The final topic of this chapter is focused on signal synchronization methods for digital communication systems.

In Chapter 9, we consider the problem of digital communication through bandlimited, AWGN channels. The effect of channel distortion on the transmitted signals is characterized in terms of intersymbol interference (ISI) and the design of adaptive equalizers for suppressing ISI is described.

Digital signal transmission via carrier modulation is described in Chapter 10. The carrier modulation methods treated in this chapter are pulse amplitude modulation (PAM), phase-shift keying (PSK), quadrature-amplitude modulation (QAM), frequency-shift keying (FSK), and continuous-phase frequency-shift keying (CPFSK).

A number of selected topics in digital communications are treated in Chapter 11. Topics include digital communication in fading multipath channels, multicarrier modulation (orthogonal frequency-division multiplexing), spread spectrum signals and systems, a brief description of the GSM and IS95 digital cellular communication systems, and link budget analysis in free space (line-of-sight) channels.

Chapter 12 is focused on the basic limits on communication of information, including the information content of memoryless sources and the capacity of the additive white Gaussian noise channel. Two widely used algorithms for encoding the output of digital sources, namely, the Huffman coding algorithm and the Lempel-Ziv algorithm are also described in this chapter.

Chapter 13, the last chapter in the book, treats channel coding and decoding. Linear block codes and convolutional codes are described for enhancing the performance of a digital communication system in the presence of additive, white Gaussian noise. Both hard-decision and soft-decision decoding of block and convolutional codes are also treated. Coding for bandwidth limited channels (trellis coded modulation), turbo codes, and low density parity check codes are also treated in this chapter.

Throughout the book many worked examples are provided to emphasize the use of the techniques developed in theory. Each chapter follows with a large number of problems at different levels of difficulty. The problems in each chapter are followed by a selection of computer problems which usually ask for simulation of various algorithms developed in that chapter using MATLAB. The solutions to the MATLAB problems are made available on the PH website for the book.

COURSE OPTIONS

This book can serve as a text in either a one-semester or a two-semester course in communication systems. An important consideration in the design of the course is whether or not the students have had a prior course in probability and random processes. Another important consideration is whether or not analog modulation and demodulation techniques are to be covered. Below, we outline three scenarios. Others are certainly possible.

1. A one-term course in analog and digital communication: Selected review sections from Chapters 2 and 5, all of Chapters 3, 4, 6, 7, and 8, and selections from Chapters 7–13.

2. A one-term course in digital communication: Selected review sections from Chapters 2 and 5, and Chapters 7–13.

3. A two-term course sequence on analog and digital communications:

   - Chapters 2–6 for the first course.
   - Chapters 7–13 for the second course.

Acknowledgments

We wish to thank the reviewers of the manuscript (Stephen Wilson, University of Virginia; Dennis Goeckel, University of Massachusettes, Amherst; Costas N. Georghiades, Texas A&M University; Selin Aviyente, Michigan State University; and Robert Morelos-Zaragoza, San Jose State University) for their comments and recommendations. Their suggestions have resulted in significant improvements to our treatment

of several topics. We also thank Gloria Doukakis for her assistance in the preparation
of the manuscript.

John G. Proakis
Adjunct Professor,
University of California at San Diego
and Professor Emeritus,
Northeastern University,
Masoud Salehi
Northeastern University.

1

Introduction

Every day, in our work and in our leisure time, we use and come in contact with a variety of modern communication systems and communication media, the most common being the telephone, radio, and television. Through these media, we are able to communicate (nearly) instantaneously with people on different continents, transact our daily business, and receive information about various developments and noteworthy events that occur all around the world. Electronic mail and facsimile transmission have made it possible to rapidly communicate written messages across great distances.

Can you imagine a world without telephones, radios, and televisions? Yet, when you think about it, most of these modern communication systems were invented and developed during the past century. Here, we present a brief historical review of major developments within the last two hundred years that have had a major role in the development of modern communication systems.

1.1 HISTORICAL REVIEW

Telegraphy and Telephony

One of the earliest inventions of major significance to communications was the invention of the electric battery by Alessandro Volta in 1799. This invention made it possible for Samuel Morse to develop the electric telegraph, which he demonstrated in 1837. The first telegraph line linked Washington with Baltimore and became operational in May 1844. Morse devised the variable-length binary code given in Table 1.1, in which letters of the English alphabet were represented by a sequence of dots and dashes (code words). In this code, more frequently occurring letters are represented by short code words, while less frequently occurring letters are represented by longer code words.

TABLE 1.1   THE MORSE CODE

A	$\cdot$ —	N	— $\cdot$		
B	— $\cdot\,\cdot\,\cdot$	O	— — —		
C	— $\cdot$ — $\cdot$	P	$\cdot$ — — $\cdot$		
D	— $\cdot\,\cdot$	Q	— — $\cdot$ —	1	$\cdot$ — — — —
E	$\cdot$	R	$\cdot$ — $\cdot$	2	$\cdot\,\cdot$ — — —
F	$\cdot\,\cdot$ — $\cdot$	S	$\cdot\,\cdot\,\cdot$	3	$\cdot\,\cdot\,\cdot$ — —
G	— — $\cdot$	T	—	4	$\cdot\,\cdot\,\cdot\,\cdot$ —
H	$\cdot\,\cdot\,\cdot\,\cdot$	U	$\cdot\,\cdot$ —	5	$\cdot\,\cdot\,\cdot\,\cdot\,\cdot$
I	$\cdot\,\cdot$	V	$\cdot\,\cdot\,\cdot$ —	6	— $\cdot\,\cdot\,\cdot\,\cdot$
J	$\cdot$ — — —	W	$\cdot$ — —	7	— — $\cdot\,\cdot\,\cdot$
K	— $\cdot$ —	X	— $\cdot\,\cdot$ —	8	— — — $\cdot\,\cdot$
L	$\cdot$ — $\cdot\,\cdot$	Y	— $\cdot$ — —	9	— — — — $\cdot$
M	— —	Z	— — $\cdot\,\cdot$	0	— — — — —

(a) Letters	(b) Numbers

Period (.)	$\cdot$ — $\cdot$ — $\cdot$ —	Wait sign (AS)	$\cdot$ — $\cdot\,\cdot\,\cdot$
Comma (,)	— — $\cdot\,\cdot$ — —	Double dash (break)	— $\cdot\,\cdot\,\cdot$ —
Interrogation (?)	$\cdot\,\cdot$ — — $\cdot\,\cdot$	Error sign	$\cdot\,\cdot\,\cdot\,\cdot\,\cdot\,\cdot\,\cdot$
Quotation Mark (")	$\cdot$ — $\cdot\,\cdot$ — $\cdot$	Fraction bar (/)	— $\cdot\,\cdot$ — $\cdot$
Colon (:)	— — — $\cdot\,\cdot\,\cdot$	End of message (AR)	$\cdot$ — $\cdot$ — $\cdot$
Semicolon (;)	— $\cdot$ — $\cdot$ — $\cdot$	End of transmission (SK)	$\cdot\,\cdot\,\cdot$ — $\cdot$ —
Parenthesis ( )	— $\cdot$ — — $\cdot$ —		

(c) Punctuation and special characters

The *Morse code* was the precursor to the variable-length source coding method that is described in Chapter 12. It is remarkable that the earliest form of electrical communications that was developed by Morse, namely *telegraphy*, was a binary digital communication system in which the letters of the English alphabet were efficiently encoded into corresponding variable-length code words with binary elements.

Nearly forty years later, in 1875, Emile Baudot developed a code for telegraphy in which each letter was encoded into fixed-length binary code words of length 5. In the *Baudot code*, the binary code elements have equal length and are designated as mark and space.

An important milestone in telegraphy was the installation of the first transatlantic cable that linked the United States and Europe in 1858. This cable failed after about four weeks of operation. A second cable was laid a few years later and became operational in July 1866.

Telephony came into being with the invention of the telephone in the 1870s. Alexander Graham Bell patented his invention of the telephone in 1876; in 1877, established the Bell Telephone Company. Early versions of telephone communication systems were relatively simple and provided service over several hundred miles. Significant advances in the quality and range of service during the first two decades of the twentieth century resulted from the invention of the carbon microphone and the induction coil.

In 1906, the invention of the triode amplifier by Lee DeForest made it possible to introduce signal amplification in telephone communication systems and, thus, to allow for telephone signal transmission over great distances. For example, transcontinental telephone transmission became operational in 1915.

Two world wars and the Great Depression during the 1930s must have been a deterrent to the establishment of transatlantic telephone service. It was not until 1953, when the first transatlantic cable was laid, that telephone service became available between the United States and Europe.

Automatic switching was another important advance in the development of telephony. The first automatic switch, developed by Strowger in 1897, was an electromechanical step-by-step switch. This type of switch was used for several decades. With the invention of the transistor, electronic (digital) switching became economically feasible. After several years of development at the Bell Telephone Laboratories, a digital switch was placed in service in Illinois in June 1960.

During the past forty years, there have been significant advances in telephone communications. Fiber optic cables are rapidly replacing copper wire in the telephone plant, and electronic switches have replaced the old electromechanical systems.

Wireless Communications.   The development of wireless communications stems from the works of Oersted, Faraday, Gauss, Maxwell, and Hertz during the nineteenth century. In 1820, Oersted demonstrated that an electric current produces a magnetic field. On August 29, 1831, Michael Faraday showed that an induced current is produced by moving a magnet in the vicinity of a conductor. Thus, he demonstrated that a changing magnetic field produces an electric field. With this early work as background, James C. Maxwell in 1864 predicted the existence of electromagnetic radiation and formulated the basic theory that has been in use for over a century. Maxwell's theory was verified experimentally by Hertz in 1887.

In 1894, a sensitive device that could detect radio signals, called the *coherer*, was used by its inventor, Oliver Lodge, to demonstrate wireless communication over a distance of 150 yards in Oxford, England. Guglialmo Marconi is credited with the development of *wireless telegraphy*. In 1895, Marconi demonstrated the transmission of radio signals at a distance of approximately 2 kilometers. Two years later, in 1897, he patented a radio telegraph system and established the Wireless Telegraph and Signal Company. On December 12, 1901, Marconi received a radio signal at Signal Hill in Newfoundland; this signal was transmitted from Cornwall, England, a city located about 1700 miles away.

The invention of the vacuum tube was especially instrumental in the development of radio communication systems. The vacuum diode was invented by Fleming in 1904, and the vacuum triode amplifier was invented by DeForest in 1906, as previously indicated. In the early part of the twentieth century, the invention of the triode made radio broadcast possible. The AM (*amplitude modulation*) broadcast was initiated in 1920 when radio station KDKA, Pittsburgh, went on the air. From that date, AM-radio broadcasting grew very rapidly across the country and around the world. The *superheterodyne AM radio receiver*, as we know it today, was invented by Edwin Armstrong during World War I. Another significant development in radio communications was the invention of FM (*frequency modulation*), also by Armstrong. In 1933, Armstrong built and demonstrated the first FM communication system. However, the use of FM was developed more slowly than the use of AM broadcast. It was not until the end of World War II that FM broadcast gained in popularity and developed commercially.

The first television system was built in the United States by V. K. Zworykin and demonstrated in 1929. Commercial-television broadcasting was initiated in London in 1936 by the British Broadcasting Corporation (BBC). Five years later, the Federal Communications Commission (FCC) authorized television broadcasting in the United States.

The Past 50 Years.   The growth in communications services over the past 50 years has been phenomenal. Significant achievements include the invention of the transistor in 1947 by Walter Brattain, John Bardeen, and William Shockley; the integrated circuit in 1958 by Jack Kilby and Robert Noyce; and the laser in 1958 by Townes and Schawlow. These inventions have made possible the development of small size, low power, low weight, and high speed electronic circuits that are used in the construction of satellite communication systems, wideband microwave radio systems, and lightwave communication systems using fiber optic cables. A satellite named Telstar I was launched in 1962 and used to relay TV signals between Europe and the United States. Commercial satellite communication services began in 1965 with the launching of the Early Bird satellite.

Currently, most of the wireline communication systems are being replaced by fiber optic cables, which provide extremely high bandwidth and make possible the transmission of a wide variety of information sources, including voice, data, and video. Cellular radio has been developed to provide telephone service to people in automobiles, buses, and trains. High-speed communication networks link computers and a variety of peripheral devices, literally around the world.

Today, we are witnessing a significant growth in the introduction and use of personal communications services, including voice, data, and video transmission. Satellite and fiber optic networks provide high speed communication services around the world. Indeed, this is the dawn of the modern telecommunications era.

There are several historical treatments in the development of radio and telecommunications covering the past century. We cite the books by McMahon, entitled *The Making of a Profession—A Century of Electrical Engineering in America* (IEEE Press, 1984); Ryder and Fink, entitled *Engineers and Electronics* (IEEE Press, 1984); and S. Millman, Editor, entitled *A History of Engineering and Science in the Bell System—Communications Sciences (1925–1980)* (AT&T Bell Laboratories, 1984).

1.2 ELEMENTS OF AN ELECTRICAL COMMUNICATION SYSTEM

Electrical communication systems are designed to send messages or information from a source that generates the messages to one or more destinations. In general, a communication system can be represented by the functional block diagram shown in Figure 1.1. The information generated by the source may be of the form of voice (speech source), a picture (image source), or plain text in some particular language, such as English, Japanese, German, French, etc. An essential feature of any source that generates information is that its output is described in probabilistic terms, i.e., the output of a source is not deterministic. Otherwise, there would be no need to transmit the message.

A transducer is usually required to convert the output of a source into an electrical signal that is suitable for transmission. For example, a microphone serves as the transducer that converts an acoustic speech signal into an electrical signal, and a video camera converts an image into an electrical signal. At the destination, a similar transducer is required to convert the electrical signals that are received into a form that is suitable for the user, e.g., acoustic signals, images, etc.

The heart of the communication system consists of three basic parts, namely, the *transmitter*, the *channel*, and the *receiver*. The functions performed by these three elements are described next.

The Transmitter.   The transmitter converts the electrical signal into a form that is suitable for transmission through the physical channel or transmission medium. For example, in radio and TV broadcast, the Federal Communications Commission (FCC) specifies the frequency range for each transmitting station. Hence, the transmitter must translate the outgoing information signal into the appropriate frequency range that matches the frequency allocation assigned to the transmitter. Thus, signals transmitted by multiple radio stations do not interfere with one another. Similar functions are performed in telephone communication systems where the electrical speech signals from many users are transmitted over the same wire.

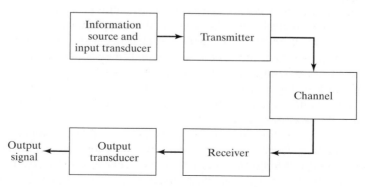

Figure 1.1   Functional diagram of a communication system.

In general, the transmitter matches the message signal to the channel via a process called *modulation*. Usually, modulation involves the use of the information signal to systematically vary either the amplitude or the frequency or the phase of a sinusoidal carrier. For example, in AM radio broadcast, the information signal that is transmitted is contained in the amplitude variations of the sinusoidal carrier, which is the center frequency in the frequency band allocated to the radio transmitting station. This is an example of *amplitude modulation*. In an FM radio broadcast, the information signal that is transmitted is contained in the frequency variations of the sinusoidal carrier. This is an example of *frequency modulation. Phase modulation (PM)* is yet a third method for impressing the information signal on a sinusoidal carrier.

In general, carrier modulation such as AM, FM, and PM is performed at the transmitter, as previously indicated, to convert the information signal to a form that matches the characteristics of the channel. Thus, through the process of modulation, the information signal is translated in frequency to match the allocation of the channel. The choice of the type of modulation is based on several factors, such as the amount of bandwidth allocated, the types of noise and interference that the signal encounters in transmission over the channel, and the electronic devices that are available for signal amplification prior to transmission. In any case, the modulation process makes it possible to accommodate the transmission of multiple messages from many users over the same physical channel.

In addition to modulation, other functions that are usually performed at the transmitter are filtering of the information-bearing signal, amplification of the modulated signal and, in the case of wireless transmission, radiation of the signal by means of a transmitting antenna.

The Channel.   The communications channel is the physical medium that is used to send the signal from the transmitter to the receiver. In wireless transmission, the channel is usually the atmosphere (free space). On the other hand, telephone channels usually employ a variety of physical media, including wirelines, fiber optic cables, and wireless (microwave radio). Whatever the physical medium for signal transmission, the essential feature is that the transmitted signal is corrupted in a random manner by a variety of possible mechanisms. The most common form of signal degradation comes in the form of additive noise, which is generated at the front end of the receiver, where signal amplification is performed. This noise is often called *thermal noise*. In wireless transmission, additional additive disturbances are man-made noise and atmospheric noise picked up by a receiving antenna. Automobile ignition noise is an example of man-made noise, and electrical lightning discharges from thunderstorms is an example of atmospheric noise. Interference from other users of the channel is another form of additive noise that often arises in both wireless and wireline communication systems.

In some radio communication channels, such as the ionospheric channel that is used for long-range, short-wave radio transmission, another form of signal degradation is multipath propagation. Such signal distortion is characterized as a non-additive signal disturbance that manifests itself as time variations in the signal amplitude, usually called fading. This phenomenon is described in more detail in Section 1.3.

Both additive and non-additive signal distortions are usually characterized as random phenomena and described in statistical terms. The effect of these signal distortions must be considered in the design of the communication system.

In the design of a communication system, the system designer works with mathematical models that statistically characterize the signal distortion encountered on physical channels. Often, the statistical description that is used in a mathematical model is a result of actual empirical measurements obtained from experiments involving signal transmission over such channels. In such cases, there is a physical justification for the mathematical model used in the design of communication systems. On the other hand, in some communication system designs, the statistical characteristics of the channel may vary significantly with time. In such cases, the system designer may design a communication system that is robust to the variety of signal distortions. This can be accomplished by having the system adapt some of its parameters to the channel distortion encountered.

The Receiver.   The function of the receiver is to recover the message signal contained in the received signal. If the message signal is transmitted by carrier modulation, the receiver performs *carrier demodulation* to extract the message from the sinusoidal carrier. Since the signal demodulation is performed in the presence of additive noise and possibly other signal distortions, the demodulated message signal is generally degraded to some extent by the presence of these distortions in the received signal. As we shall see, the fidelity of the received message signal is a function of the type of modulation and the strength of the additive noise.

Besides performing the primary function of signal demodulation, the receiver also performs a number of peripheral functions, including signal filtering and noise suppression.

1.2.1 Digital Communication System

Up to this point, we have described an electrical communication system in rather broad terms based on the implicit assumption that the message signal is a continuous time-varying waveform. We refer to such continuous-time signal waveforms as *analog signals* and to the corresponding information sources that produce such signals as *analog sources*. Analog signals can be transmitted directly over the communication channel via carrier modulation and demodulated accordingly at the receiver. We call such a communication system an *analog communication system*.

Alternatively, an analog source output may be converted into a digital form, and the message can be transmitted via digital modulation and demodulated as a digital signal at the receiver. There are some potential advantages to transmitting an analog signal by means of digital modulation. The most important reason is that signal fidelity is better controlled through digital transmission than through analog transmission. In particular, digital transmission allows us to regenerate the digital signal in long distance transmission, thus eliminating effects of noise at each regeneration point. In contrast, the noise added in analog transmission is amplified along with the signal when amplifiers are periodically used to boost the signal level in long distance

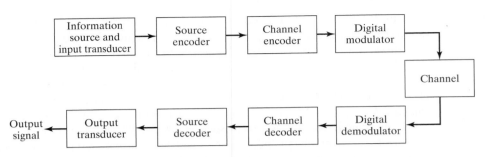

Figure 1.2   Basic elements of a digital communication system.

transmission. Another reason for choosing digital transmission over analog is that the analog message signal may be highly redundant. With digital processing, redundancy may be removed prior to modulation, thus conserving channel bandwidth. Yet a third reason may be that digital communication systems are often cheaper to implement.

In some applications, the information to be transmitted is inherently digital, e.g., in the form of English text, computer data, etc. In such cases, the information source that generates the data is called a *discrete (digital) source*.

In a digital communication system, the functional operations performed at the transmitter and receiver must be expanded to include message signal discretization at the transmitter and message signal synthesis or interpolation at the receiver. Additional functions include redundancy removal, as well as channel coding and decoding.

Figure 1.2 illustrates the functional diagram and the basic elements of a digital communication system. The source output may be either an analog signal, such as an audio or video signal, or a digital signal, such as computer output, which is discrete in time and has a finite number of output characters. In a digital communication system, the messages produced by the source are usually converted into a sequence of binary digits. Ideally, we would like to represent the source output (message) with as few binary digits as possible. In other words, we seek an efficient representation of the source output that results in little or no redundancy. The process of efficiently converting the output of either an analog or a digital source into a sequence of binary digits is called *source encoding* or *data compression*. We shall describe source-encoding methods in Chapter 12.

The source encoder outputs a sequence of binary digits, which we call the *information sequence*; this is passed to the *channel encoder*. The purpose of the channel encoder is to introduce, in a controlled manner, some redundancy in the binary information sequence that can be used at the receiver to overcome the effects of noise and interference encountered in the transmission of the signal through the channel. Thus, the added redundancy serves to increase the reliability of the received data and improves the fidelity of the received signal. In effect, redundancy in the information sequence aids the receiver in decoding the desired information sequence. For example, a (trivial) form of encoding of the binary information sequence is simply to repeat each binary digit m times, where m is some positive integer. More sophisticated (nontrivial) encoding involves taking k information bits at a time and mapping each k-bit sequence into a unique n-bit sequence, called a *code word*. The amount of redundancy introduced by encoding the data in this manner is measured by the ratio n/k. The reciprocal of this ratio, namely, k/n, is called the rate of the code or, simply, the *code rate*. Channel coding and decoding is discussed in Chapter 13.

The binary sequence at the output of the channel encoder is passed to the *digital modulator*, which serves as the interface to the communications channel. Since nearly all of the communication channels encountered in practice are capable of transmitting electrical signals (waveforms), the primary purpose of the digital modulator is to map the binary information sequence into signal waveforms. To elaborate on this point, let us suppose that the coded information sequence is to be transmitted one bit at a time at some uniform rate R bits/s. The digital modulator may simply map the binary digit 0 into a waveform $s_0(t)$ and the binary digit 1 into a waveform $s_1(t)$. In this manner, each bit from the channel encoder is transmitted separately. We call this *binary modulation*. Alternatively, the modulator may transmit b coded information bits at a time by using $M = 2^k$ distinct waveforms $s_i(t)$, $i = 0, 1, \ldots, M - 1$. This provides one waveform for each of the 2^k possible k-bit sequences. We call this *M-ary modulation* $(M > 2)$. Note that a new k-bit sequence enters the modulator every k/R seconds. Hence, when the channel bit rate R is fixed, the amount of time available to transmit one of the M waveforms corresponding to a k-bit sequence is k times the time period in a system that uses binary modulation.

At the receiving end of a digital communications system, the *digital demodulator* processes the channel-corrupted transmitted waveform and reduces each waveform to a single number that represents an estimate of the transmitted data symbol (binary or M-ary). For example, when binary modulation is used, the demodulator may process the received waveform and decide whether the transmitted bit is a 0 or a 1. In such a case, we say the *demodulator has made a binary decision* or *hard decision*. As one alternative, the demodulator may make a ternary decision; that is, it decides that the transmitted bit is either a 0 or 1 or it makes no decision at all, depending on the apparent quality of the received signal. When no decision is made on a particular bit, we say that *the demodulator has inserted an erasure in the demodulated data*. Using the redundancy in the transmitted data, the decoder attempts to fill in the positions where erasures occurred. Viewing the decision process performed by the demodulator as a form of quantization, we observe that binary and ternary decisions are special cases of a demodulator that quantizes to Q levels, where $Q \geq 2$. In general, if the digital communication system employs M-ary modulation, where $m = 0, 1, \ldots, M - 1$ represents the M possible transmitted symbols, each corresponding to $k = \log_2 M$ bits, the demodulator may make a Q-ary decision, where $Q \geq M$. In the extreme case where no quantization is performed, $Q = \infty$.

When there is no redundancy in the transmitted information, the demodulator must decide which of the M waveforms was transmitted in any given time interval. Consequently $Q = M$, and since there is no redundancy in the transmitted information, no discrete channel decoder is used following the demodulator. On the other hand, when there is redundancy introduced by a discrete channel encoder at the transmitter, the Q-ary output from the demodulator occurring every k/R seconds is fed to the decoder, which attempts to reconstruct the original information sequence from knowledge of the code used by the channel encoder and the redundancy contained in the received data.

A measure of how well the demodulator and decoder perform is the frequency with which errors occur in the decoded sequence. More precisely, the average probability of a bit error at the output of the decoder is a measure of the performance of the demodulator–decoder combination. In general, the probability of error is a function of the code characteristics, the types of waveforms used to transmit the information over the channel, the transmitter power, the characteristics of the channel, i.e., the amount of noise, and the method of demodulation and decoding.

These items and their effect on performance will be discussed in detail in Chapters 8 through 10.

As a final step, when an analog output is desired, the source decoder accepts the output sequence from the channel decoder and, from knowledge of the source encoding method used, attempts to reconstruct the original signal from the source. Due to channel decoding errors and possible distortion introduced by the source encoder and, perhaps, the source decoder, the signal at the output of the source decoder is an approximation to the original source output. The difference or some function of the difference between the original signal and the reconstructed signal is a measure of the distortion introduced by the digital communications system.

1.2.2 Early Work in Digital Communications

Although Morse is responsible for the development of the first electrical digital communication system (telegraphy), the beginnings of what we now regard as modern digital communications stem from the work of Nyquist (1924), who investigated the problem of determining the maximum signaling rate that can be used over a telegraph channel of a given bandwidth without intersymbol interference. He formulated a model of a telegraph system in which a transmitted signal has the general form

$$s(t) = \sum_n a_n g(t - nT),$$

where $g(t)$ represents a basic pulse shape and $\{a_n\}$ is the binary data sequence of $\{\pm 1\}$ transmitted at a rate of $1/T$ bits/sec. Nyquist set out to determine the optimum pulse shape that was bandlimited to W Hz and maximized the bit rate $1/T$ under the constraint that the pulse caused no intersymbol interference at the sampling times $k/T, k = 0, \pm 1, \pm 2, \dots$. His studies led him to conclude that the maximum pulse rate $1/T$ is $2W$ pulses/sec. This rate is now called the *Nyquist rate*. Moreover, this pulse rate can be achieved by using the pulses $g(t) = (\sin 2\pi Wt)/2\pi Wt$. This pulse shape allows the recovery of the data without intersymbol interference at the sampling instants. Nyquist's result is equivalent to a version of the sampling theorem for bandlimited signals, which was later stated precisely by Shannon (1948). The sampling theorem states that a signal of bandwidth W can be reconstructed from samples taken at the Nyquist rate of $2W$ samples/sec using the interpolation formula

$$s(t) = \sum_n s\left(\frac{n}{2W}\right) \frac{\sin 2\pi W(t - n/2W)}{2\pi W(t - n/2W)}.$$

In light of Nyquist's work, Hartley (1928) considered the amount of data that can be reliably transmitted over a bandlimited channel when multiple amplitude levels are used. Due to the presence of noise and other interference, Hartley postulated that the receiver can reliably estimate the received signal amplitude to some accuracy, say A_δ. This investigation led Hartley to conclude that there is a maximum data rate that can be communicated reliably over a bandlimited channel, when the maximum signal amplitude is limited to $A_{\max}$ (fixed power constraint) and the amplitude resolution is A_δ.

Another significant advance in the development of communications was the work of Wiener (1942), who considered the problem of estimating a desired signal waveform $s(t)$ in the presence of additive noise $n(t)$ based on observation of the received signal $r(t) = s(t) + n(t)$. This problem arises in signal demodulation. Wiener determined the linear filter whose output is the best mean-square approximation to the desired signal $s(t)$. The resulting filter is called the *optimum linear (Wiener) filter*.

Hartley's and Nyquist's results on the maximum transmission rate of digital information were precursors to the work of Shannon (1948a,b), who established the mathematical foundations for information theory and derived the fundamental limits for digital communication systems. In his pioneering work, Shannon formulated the basic problem of reliable transmission of information in statistical terms, using probabilistic models for information sources and communication channels. Based on this statistical formulation, he adopted a logarithmic measure for the information content of a source. He also demonstrated that the effect of a transmitter power constraint, a bandwidth constraint, and additive noise can be associated with the channel and incorporated into a single parameter, called the *channel capacity*. For example, in the case of an additive white (spectrally flat) Gaussian noise interference, an ideal bandlimited channel of bandwidth W has a capacity C given by

$$C = W \log_2\left(1 + \frac{P}{W N_0}\right) \text{ bits/s,}$$

where P is the average transmitted power and N_0 is the power-spectral density of the additive noise. The significance of the channel capacity is as follows: If the information rate R from the source is less than C ($R < C$), then it is theoretically possible to achieve reliable (error-free) transmission through the channel by appropriate coding. On the other hand, if $R > C$, reliable transmission is not possible regardless of the amount of signal processing performed at the transmitter and receiver. Thus, Shannon established basic limits on communication of information and gave birth to a new field that is now called *information theory*.

Initially, the fundamental work of Shannon had a relatively small impact on the design and development of new digital communications systems. In part, this was due to the small demand for digital information transmission during the 1950s. Another reason was the relatively large complexity and, hence, the high cost of digital hardware required to achieve the high efficiency and the high reliability predicted by Shannon's theory.

Another important contribution to the field of digital communications is the work of Kotelnikov (1947). His work provided a coherent analysis of the various digital communication systems, based on a geometrical approach. Kotelnikov's approach was later expanded by Wozencraft and Jacobs (1965).

The increase in the demand for data transmission during the last three decades, coupled with the development of more sophisticated integrated circuits, has led to the development of very efficient and more reliable digital communications systems. In the course of these developments, Shannon's original results and the generalization of his results on maximum transmission limits over a channel and on bounds on the performance achieved, have served as benchmarks relative to which any given communications system design is compared. The theoretical limits, derived by Shannon and other researchers that contributed to the development of information theory, serve as an ultimate goal in the continuing efforts to design and develop more efficient digital communications systems.

Following Shannon's publications came the classic work of Hamming (1950), which used error-detecting and error-correcting codes to combat the detrimental effects of channel noise. Hamming's work stimulated many researchers in the years that followed, and a variety of new and powerful codes were discovered, many of which are used today in the implementation of modern communication systems.

1.3 COMMUNICATION CHANNELS AND THEIR CHARACTERISTICS

As indicated in our preceding discussion, the communication channel provides the connection between the transmitter and the receiver. The physical channel may be a pair of wires that carry the electrical signal, or an optical fiber that carries the information on a modulated light beam, or an underwater ocean channel in which the information is transmitted acoustically, or free space over which the information-bearing signal is radiated by use of an antenna. Other media that can be characterized as communication channels are data storage media, such as magnetic tape, magnetic disks, and optical disks.

One common problem in signal transmission through any channel is additive noise. In general, additive noise is generated internally by components, such as resistors and solid-state devices, used to implement the communication system. This is sometimes called *thermal noise*. Other sources of noise and interference may arise externally to the system, such as interference from other users of the channel. When such noise and interference occupy the same frequency band as the desired signal, the effect can be minimized by proper design of the transmitted signal and the demodulator at the receiver. Other types of signal degradations may be encountered in transmission over the channel, such as signal attenuation, amplitude and phase distortion, and multipath distortion.

The effects of noise may be minimized by increasing the power in the transmitted signal. However, equipment and other practical constraints limit the power level in the transmitted signal. Another basic limitation is the available channel bandwidth. A bandwidth constraint is usually due to the physical limitations of the medium and the electronic components used to implement the transmitter and the receiver. These two limitations result in constraining the amount of data that can be transmitted reliably over any communications channel.

Next, we describe some of the important characteristics of several communication channels.

Wireline Channels

The telephone network makes extensive use of wirelines for voice signal transmission, as well as data and video transmission. Twisted-pair wirelines and coaxial cable are basically guided electromagnetic channels that provide relatively modest bandwidths. Telephone wire generally used to connect a customer to a central office has a bandwidth of several hundred kilohertz (kHz). On the other hand, coaxial cable has a usable bandwidth of several megahertz (MHz). Figure 1.3 illustrates the frequency range of guided electromagnetic channels, which include waveguides and optical fibers.

Signals transmitted through such channels are distorted in both amplitude and phase, and they are further corrupted by additive noise. Twisted-pair wireline channels are also prone to crosstalk interference from physically adjacent channels. Because wireline channels carry a large percentage of our daily communications around the country and the world, much research has been performed on the characterization of their transmission properties and on methods for mitigating the amplitude and phase distortion encountered in signal transmission. In Chapter 9, we describe methods for designing optimum transmitted signals and their demodulation, including the design of *channel equalizers* that compensate for amplitude and phase distortion.

Fiber Optic Channels.   Optical fibers offer the communications system designer a channel bandwidth that is several orders of magnitude larger than coaxial cable channels. During the past decade, researchers have developed optical fiber cables, which have a relatively low signal attenuation, and highly reliable photonic devices, which improves signal generation and signal detection. These technological advances have resulted in a rapid deployment of optical fiber channels both in domestic telecommunication systems as well as for transatlantic and transpacific communications. With the large bandwidth available on fiber optic channels, it is possible for the telephone companies to offer subscribers a wide array of telecommunication services, including voice, data, facsimile, and video.

The transmitter or modulator in a fiber-optic communication system is a light source, either a light-emitting diode (LED) or a laser. Information is transmitted by varying (modulating) the intensity of the light source with the message signal. The light propagates through the fiber as a light wave and is amplified periodically (in the case of digital transmission, it is detected and regenerated by repeaters) along

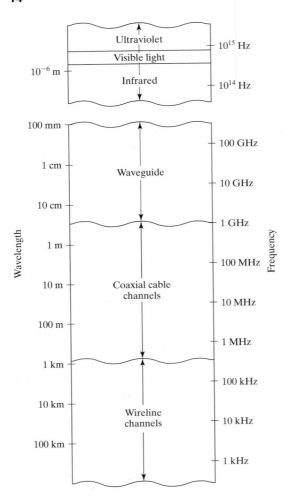

Figure 1.3 Frequency range for guided wireline channels.

the transmission path to compensate for signal attenuation. At the receiver, the light intensity is detected by a photodiode, whose output is an electrical signal that varies in direct proportion to the power of the light impinging on the photodiode.

It is envisioned that optical fiber channels will replace nearly all wireline channels in the telephone network in the next few years.

Wireless Electromagnetic Channels. In radio communication systems, electromagnetic energy is coupled to the propagation medium by an *antenna*, which serves as the radiator. The physical size and the configuration of the antenna depend primarily on the frequency of operation. To obtain efficient radiation of electromagnetic energy, the antenna must be longer than $1/10$ of the wavelength. Consequently, a radio station transmitting in the AM frequency band, say at 1 MHz (corresponding to a wavelength of $\lambda = c/f_c = 300$ m), requires an antenna of at least 30 meters.

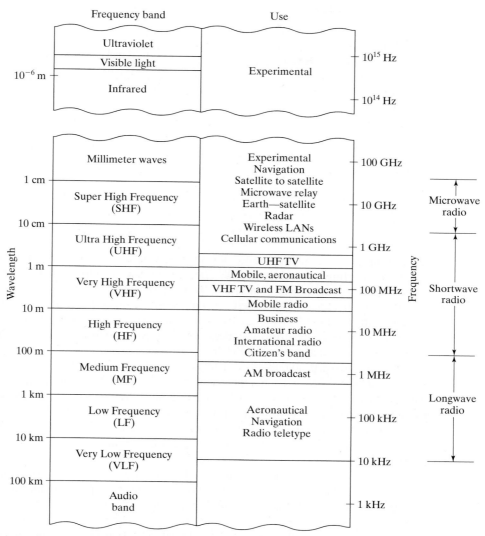

Figure 1.4 Frequency range for wireless electromagnetic channels. (Adapted from Carlson, Sec. Ed.; ©1975 McGraw-Hill. Reprinted with permission of the publisher.)

Figure 1.4 illustrates the various frequency bands of the electromagnetic spectrum. The mode of propagation of electromagnetic waves in the atmosphere and in free space may be subdivided into three categories, namely, ground-wave propagation, sky-wave propagation, and line-of-sight (LOS) propagation. In the VLF and ELF frequency bands where the wavelengths exceed 10 km, the earth and the ionosphere act as a waveguide for electromagnetic wave propagation. In these frequency ranges, communication signals practically propagate around the globe. For

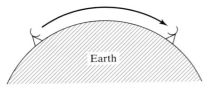

Figure 1.5   Illustration of ground-wave propagation.

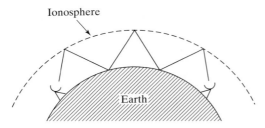

Figure 1.6   Illustration of sky-wave propagation.

this reason, these frequency bands are primarily used to provide navigational aids from shore to ships around the world. The channel bandwidths available in these frequency bands are relatively small (usually 1%–10% of the center frequency); hence, the information that is transmitted through these channels is relatively slow speed and generally confined to digital transmission. A dominant type of noise at these frequencies is generated from thunderstorm activity around the globe, especially in tropical regions. Interference results from the many users of these frequency bands.

Ground-wave propagation, illustrated in Figure 1.5, is the dominant mode of propagation for frequencies in the MF band (0.3–3 MHz). This is the frequency band used for AM broadcasting and maritime radio broadcasting. In AM broadcast, ground-wave propagation limits the range of even the most powerful radio stations to about 100 miles. Atmospheric noise, man-made noise, and thermal noise from electronic components at the receiver are dominant disturbances for signal transmission at MF.

Sky-wave propagation, as illustrated in Figure 1.6, results from transmitted signals being reflected (bent or refracted) from the ionosphere, which consists of several layers of charged particles ranging in altitude from 30 to 250 miles above the surface of the earth. During the daytime hours, the heating of the lower atmosphere by the sun causes the formation of the lower layers at altitudes below 75 miles. These lower layers, especially the D-layer, absorb frequencies below 2 MHz; thus, they severely limit sky-wave propagation of AM radio broadcast. However, during the nighttime hours, the electron density in the lower layers of the ionosphere drops sharply and the frequency absorption that occurs during the day is significantly reduced. As a consequence, powerful AM radio broadcast stations can propagate over large distances via sky-wave over the F-layer of the ionosphere, which ranges from 90 miles to 250 miles above the surface of the earth.

A common problem with electromagnetic wave propagation via sky-wave in the HF frequency range is *signal multipath*. Signal multipath occurs when the transmitted signal arrives at the receiver via multiple propagation paths at different delays. Signal multipath generally results in intersymbol interference in a digital communication system. Moreover, the signal components arriving via different propagation paths may add destructively, resulting in a phenomenon called *signal fading*. Most people have experienced this phenomenon when listening to a distant radio station at night, when sky-wave is the dominant propagation mode. Additive noise at HF is a combination of atmospheric noise and thermal noise.

Sky-wave ionospheric propagation ceases to exist at frequencies above approximately 30 MHz, which is the end of the HF band. However, it is possible to have ionospheric scatter propagation at frequencies in the range of 30–60 MHz; this is a result of signal scattering from the lower ionosphere. It is also possible to communicate over distances of several hundred miles using tropospheric scattering at frequencies in the range of 40–300 MHz. Troposcatter results from signal scattering due to particles in the atmosphere at altitudes of 10 miles or less. Generally, ionospheric scatter and tropospheric scatter involve large signal propagation losses and require a large amount of transmitter power and relatively large antennas.

Frequencies above 30 MHz propagate through the ionosphere with relatively little loss and make satellite and extraterrestrial communications possible. Hence, at frequencies in the VHF band and higher, the dominant mode of electromagnetic propagation is line-of-sight (LOS) propagation. For terrestrial communication systems, this means that the transmitter and receiver antennas must be in direct LOS with relatively little or no obstruction. For this reason, television stations transmitting in the VHF and UHF frequency bands mount their antennas on high towers in order to achieve a broad coverage area.

In general, the coverage area for LOS propagation is limited by the curvature of the earth. If the transmitting antenna is mounted at a height h feet above the surface of the earth, the distance to the radio horizon is approximately $d = \sqrt{2h}$ miles (assuming no physical obstructions such as a mountain). For example, a TV antenna mounted on a tower of 1000 feet in height provides a coverage of approximately 50 miles. As another example, microwave radio relay systems used extensively for telephone and video transmission at frequencies above 1 GHz have antennas mounted on tall towers or on the top of tall buildings.

The dominant noise limiting the performance of communication systems in the VHF and UHF frequency ranges is thermal noise generated in the front end of the receiver and cosmic noise picked up by the antenna. At frequencies above 10 GHz in the SHF band, atmospheric conditions play a major role in signal propagation. Figure 1.7 illustrates the signal attenuation in dB/mile due to precipitation for frequencies in the range of 10–100 GHz. We observe that heavy rain introduces extremely high propagation losses that can result in service outages (total breakdown in the communication system).

At frequencies above the EHF band, we have the infrared and visible light regions of the electromagnetic spectrum, which can be used to provide LOS optical communication in free space. To date, these frequency bands have been used in experimental communication systems, such as satellite to satellite links.

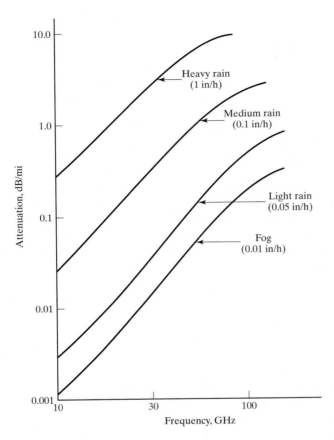

Figure 1.7 Signal attenuation due to precipitation. [From Ziemer and Tranter (1990); ©Houghton Mifflin. Reprinted with permission of the publisher.]

Underwater Acoustic Channels. Over the past few decades, ocean exploration activity has been steadily increasing. Coupled with this increase in ocean exploration is the need to transmit data, which is collected by sensors placed under-water, to the surface of the ocean. From there, it is possible to relay the data via a satellite to a data collection center.

Electromagnetic waves do not propagate over long distances underwater, except at extremely low frequencies. However, the transmission of signals at such low frequencies is prohibitively expensive because of the large and powerful transmitters required. The attenuation of electromagnetic waves in water can be expressed in terms of the *skin depth*, which is the distance a signal is attenuated by $1/e$. For sea water, the skin depth $\delta = 250/\sqrt{f}$, where f is expressed in Hertz and δ is in meters. For example, at 10 kHz, the skin depth is 2.5 meters. In contrast, acoustic signals propagate over distances of tens and even hundreds of kilometers.

A shallow-water acoustic channel is characterized as a multipath channel due to signal reflections from the surface and the bottom of the sea. Due to wave motion, the signal multipath components undergo time-varying propagation delays that result in signal fading. In addition, there is frequency dependent attenuation, which is approximately proportional to the square of the signal frequency.

Ambient ocean acoustic noise is caused by shrimp, fish, and various mammals. Additionally, man-made acoustic noise exists near harbors.

In spite of this hostile environment, it is possible to design and implement efficient and highly reliable underwater acoustic communication systems for transmitting digital signals over large distances.

Storage Channels.   Information storage and retrieval systems constitute a significant part of our data-handling activities on a daily basis. Magnetic tape (including digital audio tape and video tape), magnetic disks (used for storing large amounts of computer data), and optical disks (used for computer data storage, music, and video) are examples of data storage systems that can be characterized as communication channels. The process of storing data on a magnetic tape, magnetic disk, or optical disk is equivalent to transmitting a signal over a telephone or a radio channel. The readback process and the signal processing used to recover the stored information is equivalent to the functions performed by a telephone receiver or radio communication system to recover the transmitted information.

Additive noise generated by the electronic components and interference from adjacent tracks is generally present in the readback signal of a storage system.

The amount of data that can be stored is generally limited by the size of the disk or tape and the density (number of bits stored per square inch) that can be achieved by the write/read electronic systems and heads. For example, a packing density of 10^9 bits/sq in has been achieved in magnetic disk storage systems. The speed at which data can be written on a disk or tape and the speed at which it can be read back is also limited by the associated mechanical and electrical subsystems that constitute an information storage system.

Channel coding and modulation are essential components of a well-designed digital magnetic or optical storage system. In the readback process, the signal is demodulated and the added redundancy introduced by the channel encoder is used to correct errors in the readback signal.

1.4 MATHEMATICAL MODELS FOR COMMUNICATION CHANNELS

While designing communication systems to transmit information through physical channels, we find it convenient to construct mathematical models that reflect the most important characteristics of the transmission medium. Then the mathematical model for the channel is used in the design of the channel encoder and modulator at the transmitter and the demodulator and channel decoder at the receiver. Next, we provide a brief description of three channel models that are frequently used to characterize many of the physical channels that we encounter in practice.

The Additive Noise Channel.   The simplest mathematical model for a communication channel is the additive noise channel, illustrated in Figure 1.8. In this

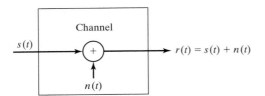

Figure 1.8 The additive noise channel.

model, the transmitted signal $s(t)$ is corrupted by the additive random-noise process $n(t)$. Physically, the additive noise process may arise from electronic components and amplifiers at the receiver of the communication system, or from interference encountered in transmission, as in the case of radio signal transmission.

If the noise is introduced primarily by electronic components and amplifiers at the receiver, it may be characterized as thermal noise. This type of noise is characterized statistically as a *Gaussian noise process*. Hence, the resulting mathematical model for the channel is usually called the *additive Gaussian noise channel*. Because this channel model applies to a broad class of physical communication channels and because it has mathematical tractability, this is the predominant channel model used in the analysis and design of our communication system. Channel attenuation is easily incorporated into the model. When the signal undergoes attenuation in transmission through the channel, the received signal is

$$r(t) = as(t) + n(t), \tag{1.4.1}$$

where a represents the attenuation factor.

The Linear Filter Channel. In some physical channels, such as wireline telephone channels, filters are used to ensure that the transmitted signals do not exceed specified bandwidth limitations; thus, they do not interfere with one another. Such channels are generally characterized mathematically as linear filter channels with additive noise, as illustrated in Figure 1.9. Hence, if the channel input is the signal $s(t)$, the channel output is the signal

$$r(t) = s(t) \star h(t) + n(t)$$

$$= \int_0^\infty h(\tau)s(t - \tau)\, d\tau + n(t), \tag{1.4.2}$$

where $h(t)$ is the impulse response of the linear filter and $\star$ denotes convolution.

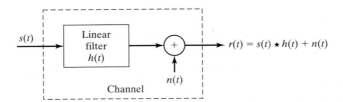

Figure 1.9 The linear filter channel with additive noise.

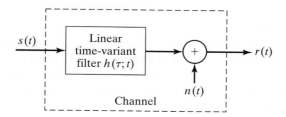

Figure 1.10   Linear time-variant filter channel with additive noise.

The Linear Time-Variant Filter Channel.   Physical channels, such as underwater acoustic channels and ionospheric radio channels, which result in time-variant multipath propagation of the transmitted signal, may be characterized mathematically as time-variant linear filters. Such linear filters are characterized by the time-variant channel impulse response $h(\tau; t)$, where $h(\tau; t)$ is the response of the channel at time t, due to an impulse applied at time $t - \tau$. Thus, τ represents the "age" (elapsed time) variable. The linear time-variant filter channel with additive noise is illustrated in Figure 1.10. For an input signal $s(t)$, the channel output signal is

$$r(t) = s(t) \star h(\tau; t) + n(t)$$

$$= \int_{-\infty}^{\infty} h(\tau; t)s(t - \tau)\, d\tau + n(t). \tag{1.4.3}$$

Consider a multipath signal propagation through physical channels, such as the ionosphere (at frequencies below 30 MHz) and mobile cellular radio channels. For such channels, a good model for the time-variant impulse response has the form

$$h(\tau; t) = \sum_{k=1}^{L} a_k(t)\delta(\tau - \tau_k), \tag{1.4.4}$$

where the $\{a_k(t)\}$ represents the possibly time-variant attenuation factor for the L multipath propagation paths. If Equation (1.4.4) is substituted into Equation (1.4.3), the received signal has the form

$$r(t) = \sum_{k=1}^{L} a_k(t)s(t - \tau_k) + n(t). \tag{1.4.5}$$

Hence, the received signal consists of L multipath components, where each component is attenuated by $\{a_k\}$ and delayed by $\{\tau_k\}$.

The three mathematical models previously described characterize a majority of physical channels encountered in practice. These three channel models are used in this text for the analysis and design of communication systems.

1.5 FURTHER READING

We have already cited several historical books on radio and telecommunications of the past century. These include the books by McMahon (1984), Ryder and Fink (1984), and Millman (1984). In addition, the classical works of Nyquist (1924), Hartley (1928), Kotelnikov (1947), Shannon (1948), and Hamming (1950) are particularly important because they lay the groundwork of modern communication systems engineering.

2

Signals and Linear Systems

In this chapter, we will review the basics of *signals* and *linear systems*. The motivation for studying these fundamental concepts stems from the basic role they play in modeling various types of communication systems. In particular, signals are used to transmit information over a communication channel. Such signals are usually called *information-bearing signals*. Speech signals, video signals, and the output of an ASCII terminal are examples of information-bearing signals.

When an information-bearing signal is transmitted over a communication channel, the shape of the signal is changed, or *distorted*, by the channel. In other words, the output of the communication channel, which is called the *received signal*, is not an exact replica of the channel input due to many factors, including channel distortion. The communication channel is an example of a system, i.e., an entity that produces an output signal when excited by an input signal. A large number of communication channels can be modeled closely by a subclass of systems called *linear systems*. Linear systems arise naturally in many practical applications and are rather easy to analyze.

2.1 BASIC CONCEPTS

Recall that a signal is defined as any function that carries information. However, we will only consider signals with time as the independent variable. Therefore, we can say that a signal is any function of time. Examples include audio signals (speech and music), video signals, and data signals. In our calculations, a signal is represented by the symbol used for mathematical functions, e.g., $x(t)$ or $f(t)$.

Another example of a signal is a speech signal. A *sample waveform* of speech is shown in Figure 2.1.

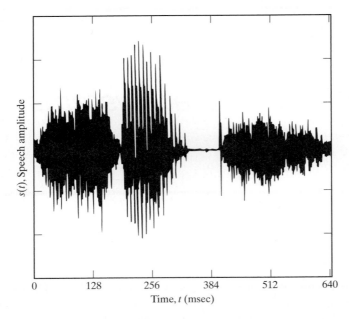

Figure 2.1 A sample speech waveform.

2.1.1 Basic Operations on Signals

Basic operations on signals involve shifting, time-reversal (flipping), and time-scaling. In this section, we describe the effect of these operations on signals.

Time Shifting. Shifting, or delaying, a signal $x(t)$ by a given constant time t_0 results in the signal $x(t - t_0)$. If t_0 is positive, this action is equivalent to a delay of t_0; thus, the result is a shifted version of $x(t)$ by t_0 to the right. If t_0 is negative, then the result is a shift to the left by an amount equal to $|t_0|$. This is equivalent to the prediction of the signal by a time amount equal to $|t_0|$. A plot of a signal shift for positive t_0 is shown in Figure 2.2.

Time Reversal. Time-reversal, or flipping, of a signal results in flipping the signal around the vertical axis, or creating the mirror image of the plot with respect to the vertical axis. We can visualize this flipping of a signal as playing an

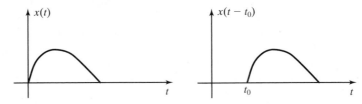

Figure 2.2 Time shifting of a signal.

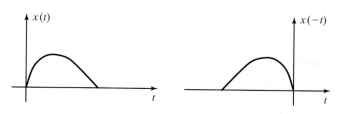

Figure 2.3   Time reversal of a signal.

audio tape in reverse. As a result, positive times are mapped as negative times and vice versa. In mathematical terms, time reversal of $x(t)$ results in $x(-t)$. Figure 2.3 shows this operation.

Time Scaling.   Time scaling of a signal results in a change in the time unit against which the signal is plotted. Time scaling results in either an expanded version of the signal (if the new time unit is a fraction of the original time unit) or a contracted version of the original signal (if the new time unit is a multiple of the original time unit). In general, time scaling is expressed as $x(at)$ for some $a > 0$. If $a < 1$, then the result is an expanded version of the original signal (such as a tape which is played at a slower speed than it was recorded). If $a > 1$, the result is a contracted form of the original signal (such as a tape that is played at a higher speed than it was recorded). The case of $a > 1$ is shown in Figure 2.4.

In general, we may have a combination of these operations. For instance, $x(-2t)$ is a combination of flipping the signal and then contracting it by a factor of 2. Also, $x(2t - 3)$ is equal to $x[2(t - 1.5)]$, which is equivalent to contracting the signal by a factor of 2 and then shifting it to the right by 1.5.

2.1.2 Classification of Signals

The classification of signals makes their study easier. Depending on the point of view, signals can be classified in a variety of ways. In this section we present the most important ways to classify signals.

Continuous-Time and Discrete-Time Signals.   Based on the range of the independent variable, signals can be divided into two classes: *continuous-time* signals and *discrete-time* signals. A *continuous-time* signal is a signal $x(t)$ for which the independent variable t takes real numbers. A *discrete-time* signal, denoted by $x[n]$, is a signal for which the independent variable n takes its values in the set of integers.

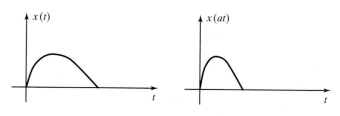

Figure 2.4   Time scaling of a signal.

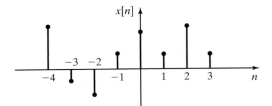

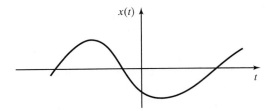

Figure 2.5 Examples of discrete-time and continuous-time signals.

By sampling a continuous-time signal $x(t)$ at time instants separated by T_0, we can define the discrete-time signal $x[n] = x(nT_0)$. Figure 2.5 shows examples of discrete-time and continuous-time signals.

Example 2.1.1
Let

$$x(t) = A\cos(2\pi f_0 t + \theta).$$

This is an example of a continuous-time signal called a *sinusoidal* signal. A sketch of this signal is given in Figure 2.6.   ∎

Example 2.1.2
Let

$$x[n] = A\cos(2\pi f_0 n + \theta),$$

where $n \in \mathbb{Z}$ ($\mathbb{Z}$ is the set of integers). A sketch of this discrete-time signal is given in Figure 2.7.   ∎

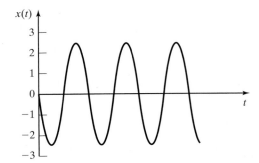

Figure 2.6 Sinusoidal signal.

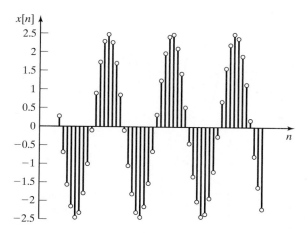

Figure 2.7   Discrete-time sinusoidal signal.

Real and Complex Signals.   Signals are functions, and functions at a given value of their independent variable are just numbers, which can be either real or complex. A *real signal* takes its values in the set of real numbers, i.e., $x(t) \in \mathbb{R}$. A *complex signal* takes its values in the set of complex numbers, i.e., $x(t) \in \mathbb{C}$.

In communications, complex signals are usually used to model signals that convey amplitude and phase information. Like complex numbers, a complex signal can be represented by two real signals. These two real signals can be either the real and imaginary parts or the absolute value (or *modulus* or *magnitude*) and phase. A graph of a complex signal can be given by graphs in either of these representations. However, the magnitude and phase graphs are more widely used.

Example 2.1.3

The signal

$$x(t) = Ae^{j(2\pi f_0 t + \theta)},$$

is a complex signal. Its real part is

$$x_r(t) = A\cos(2\pi f_0 t + \theta)$$

and its imaginary part is

$$x_i(t) = A\sin(2\pi f_0 t + \theta),$$

where we have used Euler's relation $e^{j\phi} = \cos\phi + j\sin\phi$. We could equivalently describe this signal in terms of its modulus and phase. The absolute value of $x(t)$ is

$$|x(t)| = \sqrt{x_r^2(t) + x_i^2(t)} = |A|,$$

and its phase is

$$\angle x(t) = 2\pi f_0 t + \theta.$$

Graphs of these functions are given in Figure 2.8.   ∎

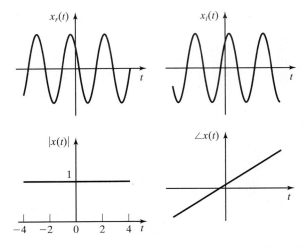

Figure 2.8 Real–imaginary and magnitude–phase graphs of the complex exponential signal in Example 2.1.3.

The real and complex components, as well as the modulus and phase of any complex signal, are represented by the following relations:

$$x_r(t) = |x(t)| \cos(\angle x(t)) ; \tag{2.1.1}$$

$$x_i(t) = |x(t)| \sin(\angle x(t)) ; \tag{2.1.2}$$

$$|x(t)| = \sqrt{x_r^2(t) + x_i^2(t)}; \tag{2.1.3}$$

$$\angle x(t) = \arctan \frac{x_i(t)}{x_r(t)}. \tag{2.1.4}$$

Deterministic and Random Signals. In a deterministic signal at any time instant t, the value of $x(t)$ is given as a real or a complex number. In a random (or stochastic) signal at any given time instant t, $x(t)$ is a random variable, i.e., it is defined by a probability density function.

All of our previous examples were deterministic signals. Random signals are discussed in Chapter 5.

Periodic and Nonperiodic Signals. A periodic signal repeats in time; hence, it is sufficient to specify the signal in a basic interval called the *period*. More formally, a periodic signal is a signal $x(t)$ that satisfies the property

$$x(t + T_0) = x(t)$$

for all t, and some positive real number T_0 (called the *period* of the signal). For discrete-time periodic signals, we have

$$x[n + N_0] = x[n]$$

for all integers n, and a positive integer N_0 (called the *period*). A signal that does not satisfy the conditions of periodicity is called nonperiodic.

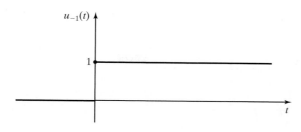

Figure 2.9   The unit-step signal.

Example 2.1.4

The signals

$$x(t) = A\cos(2\pi f_0 t + \theta)$$

and

$$x(t) = A e^{j(2\pi f_0 t + \theta)}$$

are examples of real and complex periodic signals. The period of both signals is $T_0 = \frac{1}{f_0}$. The signal

$$u_{-1}(t) = \begin{cases} 1 & t \geq 0 \\ 0 & t < 0 \end{cases} \tag{2.1.5}$$

illustrated in Figure 2.9 is an example of a nonperiodic signal. This signal is known as the *unit-step* signal.   ∎

Example 2.1.5

The discrete-time sinusoidal signal shown in Figure 2.7 is not periodic for all values of f_0. For this signal to be periodic, we must have

$$2\pi f_0(n + N_0) + \theta = 2\pi f_0 n + \theta + 2m\pi \tag{2.1.6}$$

for all integers n, some positive integer N_0, and some integer m. Thus, we conclude that

$$2\pi f_0 N_0 = 2\pi m$$

or

$$f_0 = \frac{m}{N_0},$$

i.e., the discrete sinusoidal signal is periodic *only for rational values of* f_0. For instance, $A\cos(3\pi n + \theta)$ is periodic, but $A\cos(\sqrt{2}\pi n + \theta)$ is not periodic.   ∎

Causal and Noncausal Signals.   Causality is an important concept in classifying *systems*. This concept has a close relationship to the realizability of a system. We will cover discuss this issue later, during our discussion on various types of systems. Now we define the concept of causal signals, which is closely related to the concept of causal systems. A signal $x(t)$ is called *causal* if for all $t < 0$, we have $x(t) = 0$; otherwise, the signal is noncausal. Equivalently, a discrete-time signal is a causal signal if it is identically equal to zero for $n < 0$.

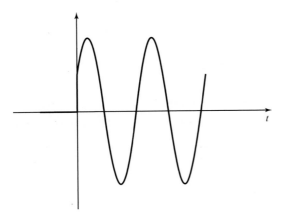

Figure 2.10   An example of a causal signal.

Example 2.1.6
The signal

$$x(t) = \begin{cases} A\cos(2\pi f_0 t + \theta) & \text{for } t \geq 0 \\ 0 & \text{otherwise,} \end{cases}$$

is a causal signal. Its graph is shown in Figure 2.10.   ∎

Similarly, we can define *anticausal* signals as signals whose time inverse is causal. Therefore, an anticausal signal is identically equal to zero for $t > 0$.

Even and Odd Signals.   Evenness and oddness are expressions of various types of symmetry present in signals. A signal $x(t)$ is *even* if it has mirror symmetry with respect to the vertical axis. A signal is *odd* if it is symmetric with respect to the origin.

The signal $x(t)$ is even if and only if, for all t,

$$x(-t) = x(t),$$

and is odd if and only if, for all t,

$$x(-t) = -x(t).$$

Figure 2.11 shows graphs of even and odd signals.

In general, any signal $x(t)$ can be written as the sum of its even and odd parts as

$$x(t) = x_e(t) + x_o(t), \tag{2.1.7}$$

where

$$x_e(t) = \frac{x(t) + x(-t)}{2}, \tag{2.1.8}$$

$$x_o(t) = \frac{x(t) - x(-t)}{2}. \tag{2.1.9}$$

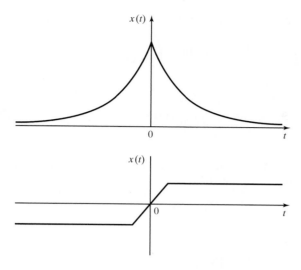

Figure 2.11   Examples of even and odd signals.

Example 2.1.7

The sinusoidal signal $x(t) = A\cos(2\pi f_0 t + \theta)$ is generally neither even nor odd. However, the special cases $\theta = 0$ and $\theta = \pm\frac{\pi}{2}$ correspond to even and odd signals, respectively. In general,

$$x(t) = \frac{A}{2}\cos(\theta)\cos(2\pi f_0 t) - \frac{A}{2}\sin(\theta)\sin(2\pi f_0 t).$$

Since $\cos(2\pi f_0 t)$ is even and $\sin(2\pi f_0 t)$ is odd, we conclude that

$$x_e(t) = \frac{A}{2}\cos(\theta)\cos(2\pi f_0 t)$$

and

$$x_o(t) = -\frac{A}{2}\sin(\theta)\sin(2\pi f_0 t). \qquad \blacksquare$$

Example 2.1.8

From Figure 2.8, we can see that for $\theta = 0$ and $x(t) = Ae^{j2\pi f_0 t}$, the real part and the magnitude are even and the imaginary part and the phase are odd. $\qquad \blacksquare$

Hermitian Symmetry for Complex Signals.   For complex signals, another form of symmetry, called Hermitian symmetry, is also defined. A complex signal $x(t)$ is called *Hermitian* if its real part is even and its imaginary part is odd. In addition, we can easily show that its magnitude is even and its phase is odd. The signal $x(t) = Ae^{j2\pi f_0 t}$ is an example of a Hermitian signal.

Energy-Type and Power-Type Signals.   This classification deals with the energy content and the power content of signals. Before classifying these signals, we need to define the energy content (or simply the energy) and the power content (or power).

For any signal $x(t)$, the *energy content* of the signal is defined by[1]

$$\mathcal{E}_x = \int_{-\infty}^{+\infty} |x(t)|^2 dt = \lim_{T \to \infty} \int_{-T/2}^{T/2} |x(t)|^2 \, dt. \tag{2.1.10}$$

The *power content* is defined by

$$P_x = \lim_{T \to \infty} \frac{1}{T} \int_{-T/2}^{T/2} |x(t)|^2 dt. \tag{2.1.11}$$

For real signals, $|x(t)|^2$ is replaced by $x^2(t)$.

A signal $x(t)$ is an *energy-type* signal if and only if $\mathcal{E}_x$ is finite. A signal is a *power-type* signal if and only if P_x satisfies

$$0 < P_x < \infty.$$

Example 2.1.9

Find the energy in the signal described by

$$x(t) = \begin{cases} 3 & |x| < 3 \\ 0 & \text{otherwise.} \end{cases}$$

Solution   We have

$$\mathcal{E}_x = \int_{-\infty}^{+\infty} |x(t)|^2 dt = \int_{-3}^{3} 9 \, dt = 54.$$

Therefore, this signal is an energy-type signal.   ∎

Example 2.1.10

The energy content of $A \cos(2\pi f_0 t + \theta)$ is

$$\mathcal{E}_x = \lim_{T \to \infty} \int_{-T/2}^{T/2} A^2 \cos^2(2\pi f_0 t + \theta) \, dt = \infty.$$

Therefore, this signal is not an energy-type signal. However, the power of this signal is

$$P_x = \lim_{T \to \infty} \frac{1}{T} \int_{-T/2}^{T/2} A^2 \cos^2(2\pi f_0 t + \theta) \, dt$$

$$= \lim_{T \to \infty} \frac{1}{T} \int_{-T/2}^{T/2} \frac{A^2}{2} \left[1 + \cos(4\pi f_0 t + 2\theta) \right] dt$$

$$= \lim_{T \to \infty} \left[\frac{A^2 T}{2T} + \left[\frac{A^2}{8\pi f_0 T} \sin(4\pi f_0 t + 2\theta) \right]_{-T/2}^{T/2} \right]$$

$$= \frac{A^2}{2} < \infty. \tag{2.1.12}$$

Hence, $x(t)$ is a power-type signal and its power is $\frac{A^2}{2}$.   ∎

---

[1] If $x(t)$ indicates the voltage across a $1 \, \Omega$ resistor, then the current flowing through the resistor is $x(t)$ and the instantaneous power is $x^2(t)$. We can justify Equation (2.1.10) by noting that energy is the integral of power.

Example 2.1.11

For any periodic signal with period T_0, the energy is

$$\mathcal{E}_x = \lim_{T \to \infty} \int_{-T/2}^{T/2} |x(t)|^2 dt$$

$$= \lim_{n \to \infty} \int_{-\frac{nT_0}{2}}^{+\frac{nT_0}{2}} |x(t)|^2 dt$$

$$= \lim_{n \to \infty} n \int_{-\frac{T_0}{2}}^{+\frac{T_0}{2}} |x(t)|^2 dt$$

$$= \infty. \tag{2.1.13}$$

Therefore, *periodic signals are not typically energy type*. The power content of any periodic signal is

$$P_x = \lim_{T \to \infty} \frac{1}{T} \int_{-T/2}^{T/2} |x(t)|^2 dt$$

$$= \lim_{n \to \infty} \frac{1}{nT_0} \int_{-\frac{nT_0}{2}}^{+\frac{nT_0}{2}} |x(t)|^2 dt$$

$$= \lim_{n \to \infty} \frac{n}{nT_0} \int_{-\frac{T_0}{2}}^{+\frac{T_0}{2}} |x(t)|^2 dt$$

$$= \frac{1}{T_0} \int_{-\frac{T_0}{2}}^{+\frac{T_0}{2}} |x(t)|^2 dt. \tag{2.1.14}$$

This means that the power content of a periodic signal is equal to the average power in one period.   ∎

2.1.3 Some Important Signals and Their Properties

In our study of communication systems, certain signals appear frequently. In this section, we briefly introduce these signals and describe some of their properties.

The Sinusoidal Signal.   The sinusoidal signal is defined by

$$x(t) = A \cos(2\pi f_0 t + \theta),$$

where the parameters A, f_0, and θ are the *amplitude, frequency,* and *phase* of the signal. A sinusoidal signal is periodic with the period $T_0 = 1/f_0$. For a graph of this signal, see Figure 2.6.

The Complex Exponential Signal.   The complex exponential signal is defined by $x(t) = A e^{j(2\pi f_0 t + \theta)}$. Again A, f_0, and θ are the amplitude, frequency, and phase of the signal. This signal is shown in Figure 2.8.

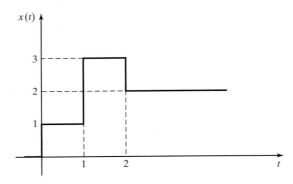

Figure 2.12   The signal
$u_{-1}(t) + 2u_{-1}(t - 1) - u_{-1}(t - 2)$.

The Unit-Step Signal.   The unit-step signal, which is defined in Section 2.1.2, is another frequently encountered signal. The unit step multiplied by any signal produces a "causal version" of the signal. The unit-step signal is shown in Figure 2.9. Note that for positive a, we have $u_{-1}(at) = u_{-1}(t)$.

Example 2.1.12

To plot the signal $u_{-1}(t) + 2u_{-1}(t - 1) - u_{-1}(t - 2)$, we note that this is a result of time shifting the unit-step function. The plot is shown in Figure 2.12.   ∎

The Rectangular Pulse.   This signal is defined as

$$\Pi(t) = \begin{cases} 1 & -\frac{1}{2} \leq t \leq \frac{1}{2} \\ 0 & \text{otherwise} \end{cases} . \tag{2.1.15}$$

The graph of the rectangular pulse is shown in Figure 2.13.

Example 2.1.13

To plot the signal $2\Pi\left(\frac{t-3}{6}\right) - \Pi\left(\frac{t-3}{4}\right)$, we note that this signal is the difference between two time shifted and scaled versions of $\Pi(t)$. Its plot is shown in Figure 2.14.   ∎

The Triangular Signal.   This signal is defined as

$$\Lambda(t) = \begin{cases} t + 1 & -1 \leq t \leq 0 \\ -t + 1 & 0 \leq t \leq 1 \\ 0 & \text{otherwise} \end{cases} . \tag{2.1.16}$$

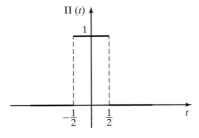

Figure 2.13   The rectangular pulse.

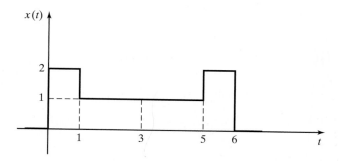

Figure 2.14   The signal $2\Pi\left(\frac{t-3}{6}\right) - \Pi\left(\frac{t-3}{4}\right)$.

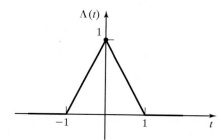

Figure 2.15   The triangular signal.

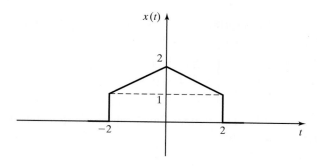

Figure 2.16   The signal $\Pi\left(\frac{t}{4}\right) + \Lambda\left(\frac{t}{2}\right)$.

Its plot is shown in Figure 2.15. It is not difficult to verify that[2]

$$\Lambda(t) = \Pi(t) \star \Pi(t). \tag{2.1.17}$$

Example 2.1.14

To plot $\Pi\left(\frac{t}{4}\right) + \Lambda\left(\frac{t}{2}\right)$, we use the time scaling of signals. The result is shown in Figure 2.16.   ∎

---

[2]Note that $x(t) \star y(t)$ denotes the convolution of two signals, which is defined by

$$x(t) \star y(t) = \int_{-\infty}^{+\infty} x(\tau)y(t-\tau)\,d\tau = \int_{-\infty}^{+\infty} x(t-\tau)y(\tau)\,d\tau = y(t) \star x(t).$$

For more details, see Section 2.1.5.

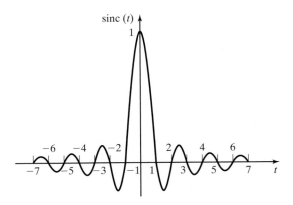

Figure 2.17 The sinc signal.

The Sinc Signal.   The sinc signal is defined as

$$\text{sinc}(t) = \begin{cases} \frac{\sin(\pi t)}{\pi t} & t \neq 0 \\ 1 & t = 0 \end{cases}. \tag{2.1.18}$$

The waveform corresponding to this signal is shown in Figure 2.17. From this figure, we can see that the sinc signal achieves its maximum of 1 at $t = 0$. The zeros of the sinc signal are at $t = \pm 1, \pm 2, \pm 3, \ldots$.

The Sign or the Signum Signal.   The sign or the signum signal denotes the sign of the independent variable t and is defined by

$$\text{sgn}(t) = \begin{cases} 1 & t > 0 \\ -1 & t < 0 \\ 0 & t = 0 \end{cases}. \tag{2.1.19}$$

This signal is shown in Figure 2.18. The signum signal can be expressed as the limit of the signal $x_n(t)$, which is defined by

$$x_n(t) = \begin{cases} e^{-\frac{t}{n}} & t > 0 \\ -e^{\frac{t}{n}} & t < 0 \\ 0 & t = 0 \end{cases}, \tag{2.1.20}$$

when $n \rightarrow \infty$. We will later use this definition of the signum signal to find its Fourier transform. This limiting behavior is also shown in Figure 2.18.

The Impulse or Delta Signal.   The impulse or delta signal is a mathematical model for representing physical phenomena that occur in a very small time duration, are so small that they are beyond the resolution of the measuring instruments involved, and for all practical purposes, have a duration that can be assumed to be equal to zero. Examples of such phenomena are a hammer blow, a very narrow voltage or current pulse, etc. In the precise mathematical sense, the impulse signal $\delta(t)$ is not a function (or signal)—it is a *distribution* or a *generalized function*. A

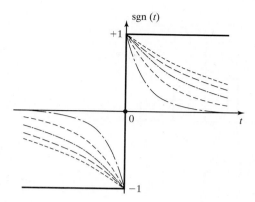

Figure 2.18 The signum signal as the limit of $x_n(t)$.

distribution is defined in terms of its effect on another function (usually called the "test function") under the integral sign. The impulse distribution (or signal) can be defined by the relation

$$\int_{-\infty}^{+\infty} \phi(t)\delta(t)\,dt = \phi(0),\qquad(2.1.21)$$

which expresses the effect of the impulse distribution on the "test function" $\phi(t)$, assumed to be continuous at the origin. This property is called the *sifting property* of the impulse signal. In other words, the effect of the impulse signal on the "test function" $\phi(t)$ under the integral sign is to extract or *sift* its value at the origin. As shown, $\delta(t)$ is defined in terms of its action on $\phi(t)$; it is not defined in terms of its value for different values of t.

Sometimes it is helpful to visualize $\delta(t)$ as the limit of certain known signals. The most commonly used forms are

$$\delta(t) = \lim_{\epsilon \downarrow 0} \frac{1}{\epsilon}\Pi\left(\frac{t}{\epsilon}\right)\qquad(2.1.22)$$

and

$$\delta(t) = \lim_{\epsilon \downarrow 0} \frac{1}{\epsilon}\operatorname{sinc}\left(\frac{t}{\epsilon}\right).\qquad(2.1.23)$$

Figure 2.19 shows graphs of these signals. (The symbol $\epsilon \downarrow 0$ means that ϵ tends to zero from above, i.e., it remains positive.)

The following properties are derived from the definition of the impulse signal:

1. $\delta(t) = 0$ for all $t \neq 0$ and $\delta(0) = \infty$.
2. $x(t)\delta(t - t_0) = x(t_0)\delta(t - t_0)$.
3. For any $\phi(t)$ continuous at t_0,

$$\int_{-\infty}^{\infty} \phi(t)\delta(t - t_0)\,dt = \phi(t_0).\qquad(2.1.24)$$

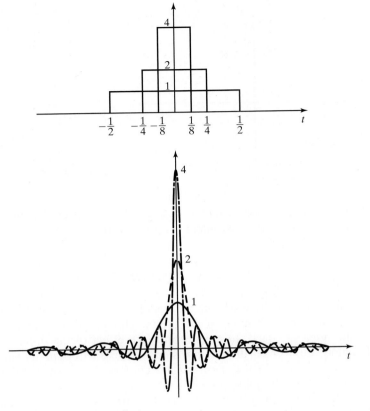

Figure 2.19 The impulse signal as a limit.

4. For any $\phi(t)$ continuous at t_0,

$$\int_{-\infty}^{\infty} \phi(t + t_0)\delta(t)\, dt = \phi(t_0). \qquad (2.1.25)$$

5. For all $a \neq 0$,

$$\delta(at) = \frac{1}{|a|}\delta(t). \qquad (2.1.26)$$

6. The result of the convolution of any signal with the impulse signal is the signal itself:

$$x(t) \star \delta(t) = x(t). \qquad (2.1.27)$$

Also,

$$x(t) \star \delta(t - t_0) = x(t - t_0). \qquad (2.1.28)$$

7. The unit-step signal is the integral of the impulse signal, and the impulse signal is the *generalized* derivative of the unit-step signal, i.e.,

$$u_{-1}(t) = \int_{-\infty}^{t} \delta(\tau)\,d\tau \tag{2.1.29}$$

and

$$\delta(t) = \frac{d}{dt}u_{-1}(t). \tag{2.1.30}$$

8. Similar to the way we defined $\delta(t)$ we can define $\delta'(t)$, $\delta''(t)$, ..., $\delta^{(n)}(t)$, the *generalized derivatives* of $\delta(t)$, by the following equation:

$$\int_{-\infty}^{+\infty} \delta^{(n)}(t)\phi(t)\,dt = (-1)^{n}\frac{d^{n}}{dt^{n}}\phi(t)\Big|_{t=0}. \tag{2.1.31}$$

We can generalize this result to

$$\int_{-\infty}^{+\infty} \delta^{(n)}(t - t_0)\phi(t)\,dt = (-1)^{n}\frac{d^{n}}{dt^{n}}\phi(t)\Big|_{t=t_0}. \tag{2.1.32}$$

9. The result of the convolution of any signal with nth derivative of $x(t)$ is the nth derivative of $x(t)$, i.e.,

$$x(t) \star \delta^{(n)}(t) = x^{(n)}(t) \tag{2.1.33}$$

and in particular

$$x(t) \star \delta'(t) = x'(t). \tag{2.1.34}$$

10. The result of the convolution of any signal $x(t)$ with the unit-step signal is the integral of the signal $x(t)$, i.e.,

$$x(t) \star u_{-1}(t) = \int_{-\infty}^{t} x(\tau)\,d\tau. \tag{2.1.35}$$

11. For even values of n, $\delta^{(n)}(t)$ is even; for odd values of n, it is odd. In particular, $\delta(t)$ is even and $\delta'(t)$ is odd.

A schematic representation of the impulse signal is given in Figure 2.20.

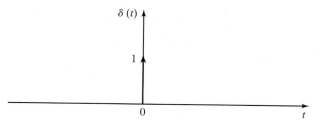

Figure 2.20   The impulse signal.

Example 2.1.15
Determine $(\cos t)\delta(t)$, $(\cos t)\delta(2t - 3)$, and $\int_{-\infty}^{\infty} e^{-t}\delta'(t - 1)\,dt$.

Solution   To determine $(\cos t)\delta(t)$, we can use Property 2:

$$(\cos t)\delta(t) = (\cos 0)\delta(t) = \delta(t).$$

To determine $(\cos t)\delta(2t - 3)$, we can use Property 5:

$$\delta(2t - 3) = \frac{1}{2}\delta\left(t - \frac{3}{2}\right).$$

Then, from Property 1, we have

$$(\cos t)\delta(2t - 3) = \frac{1}{2}(\cos t)\delta\left(t - \frac{3}{2}\right) = \frac{\cos 1.5}{2}\delta\left(t - \frac{3}{2}\right) \approx 0.035\delta\left(t - \frac{3}{2}\right).$$

Finally, to determine $\int_{-\infty}^{\infty} e^{-t}\delta'(t - 1)\,dt$, we use Property 8 to obtain

$$\int_{-\infty}^{\infty} e^{-t}\delta'(t - 1)\,dt = (-1)\frac{d}{dt}e^{-t}\Big|_{t=1} = e^{-1}. \qquad \blacksquare$$

2.1.4 Classification of Systems

A system is an interconnection of various elements or devices that, from a certain viewpoint, behave as a whole. From a communication point of view, a system is an entity that is excited by an *input signal* and, as a result of this excitation, produces an *output signal*. From a communication engineer's point of view, a system is a law that assigns output signals to various input signals. For example, an electric circuit with some voltage source as the input and some current in a certain branch is a system. The most important point in the definition of a system is that *its output must be uniquely defined for any legitimate input*. This definition can be written mathematically as

$$y(t) = \mathcal{T}[x(t)], \qquad (2.1.36)$$

where $x(t)$ is the input, $y(t)$ is the output, and $\mathcal{T}$ is the operation performed by the system. Figure 2.21 shows a pictorial representation of a system.

Example 2.1.16
The input–output relationship $y(t) = 3x(t) + 3x^2(t)$ defines a system. For any input $x(t)$, the output $y(t)$ is uniquely determined. $\blacksquare$

A system is defined by two characteristics: 1) the operation that describes the system and 2) the set of legitimate input signals. In this text, we will use the operator $\mathcal{T}$ to denote the operation that describes the system. We will use $\mathcal{X}$ to denote the space of legitimate inputs to the system.

Figure 2.21   A system with an input and output.

Example 2.1.17

The system described by the input–output relationship

$$y(t) = \mathscr{T}[x(t)] = \frac{d}{dt}x(t), \qquad (2.1.37)$$

for which $\mathscr{X}$ is the space of all differentiable signals, describes a system. This system is referred to as the *differentiator*. ∎

The space $\mathscr{X}$ is usually defined by the system operation; therefore, it is not usually given explicitly.

Systems, like signals, can be classified according to their properties. Based on this point of view, various classifications are possible. We will now briefly introduce some of the fundamental system classifications.

Discrete-Time and Continuous-Time Systems.

Systems are defined by the operation that the input signals use to produce the corresponding output signal. Systems can accept either discrete-time or continuous-time signals as their inputs and outputs. This is the basis of system classification into continuous-time and discrete-time systems.

A *discrete-time system* accepts discrete-time signals as the input and produces discrete-time signals at the output. For a *continuous-time system*, both input and output signals are continuous-time signals.

Example 2.1.18

The systems described in the two previous examples are both continuous-time systems. An example of a discrete-time system is the system described by

$$y[n] = x[n] - x[n-1]. \qquad (2.1.38)$$

This system is a *discrete-time differentiator*. ∎

Linear and Nonlinear Systems.

Linear systems are systems for which the *superposition* property is satisfied, i.e., the system's response to a linear combination of the inputs is the linear combination of the responses to the corresponding inputs.

A system $\mathscr{T}$ is *linear* if and only if, for any two input signals $x_1(t)$ and $x_2(t)$ and for any two scalars α and β, we have

$$\mathscr{T}[\alpha x_1(t) + \beta x_2(t)] = \alpha \mathscr{T}[x_1(t)] + \beta \mathscr{T}[x_2(t)]. \qquad (2.1.39)$$

A system that does not satisfy this relationship is called *nonlinear*.

Linearity can also be defined in terms of the following two properties:

$$\begin{cases} \mathscr{T}[x_1(t) + x_2(t)] &= \mathscr{T}[x_1(t)] + \mathscr{T}[x_2(t)] \\ \mathscr{T}[\alpha x(t)] &= \alpha \mathscr{T}[x(t)]. \end{cases} \qquad (2.1.40)$$

A system that satisfies the first property is called *additive*, and a system that satisfies the second property is called *homogeneous*. From the second property, it is obvious that $\mathcal{T}[0] = 0$ in a linear system. In other words, the response of a linear system to a zero input is always zero (for linearity, this is a necessary condition but not a sufficient condition).

Linearity is a very important property. In a linear system, we can decompose the input into a linear combination of some fundamental signals whose output can be derived easily, and then we can find the linear combination of the corresponding outputs. We denote the operation of linear systems by $\mathcal{L}$, rather than $\mathcal{T}$.

Example 2.1.19

The differentiator described earlier is an example of a linear system. This is true because if $x_1(t)$ and $x_2(t)$ are differentiable, $\alpha x_1(t) + \beta x_2(t)$ must also be differentiable for any choice of α and β, and

$$\frac{d}{dt}[\alpha x_1(t) + \beta x_2(t)] = \alpha x_1'(t) + \beta x_2'(t).$$

The system described by

$$y(t) = ax^2(t)$$

is nonlinear because its response to $2x(t)$ is

$$\mathcal{T}[2x(t)] = 4x^2(t) \neq 2x^2(t) = 2\mathcal{T}[x(t)];$$

therefore, the system is not homogeneous.   ■

Example 2.1.20

A delay system is defined by $y(t) = x(t - \Delta)$, that is, the output is a delayed version of the input. (See Figure 2.22.) If $x(t) = \alpha x_1(t) + \beta x_2(t)$, then the response of the system is obviously $\alpha x_1(t - \Delta) + \beta x_2(t - \Delta)$. Therefore, the system is linear.   ■

Time-Invariant and Time-Varying Systems.   A system is called time invariant if its input–output relationship does not change with time. This means that a delayed version of an input results in a delayed version of the output.

A system is *time invariant* if and only if, for all $x(t)$ and all values of t_0, its response to $x(t - t_0)$ is $y(t - t_0)$, where $y(t)$ is the response of the system to $x(t)$. (See Figure 2.23.)

Example 2.1.21

The differentiator is a time-invariant system, since

$$\frac{d}{dt}x(t - t_0) = x'(t)\big|_{t=t-t_0}.$$ ■

Example 2.1.22

The modulator, defined by $y(t) = x(t)\cos 2\pi f_0 t$, is an example of a time-varying system. The response of this system to $x(t - t_0)$ is

$$x(t - t_0)\cos(2\pi f_0 t),$$

which is not equal to $y(t - t_0)$.   ■

$x(t) \longrightarrow \boxed{\Delta} \longrightarrow y(t) = x(t - \Delta)$ **Figure 2.22**   The input–output relation for the delay system.

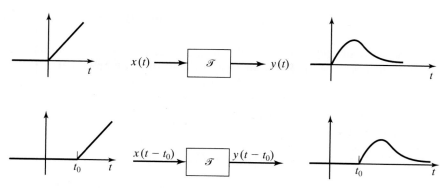

Figure 2.23   A time-invariant system.

The class of *linear time-invariant* (LTI) systems is particularly important. The response of these systems to inputs can be derived simply by finding the convolution of the input and the impulse response of the system. We will discuss this property in more detail in Section 2.1.5.

Causal and Noncausal Systems.   Causality deals with the physical realizability of systems. Since no physical system can predict the values that its input signal will assume in the future, we can assume that in a physically realizable system, the output at any time depends only on the values of the input signal up to that time and does not depend on the future values of the input.

A system is *causal* if its output at any time t_0 depends on the input at times prior to t_0, i.e.,

$$y(t_0) = \mathscr{T}[x(t) : t \leq t_0].$$

A necessary and sufficient condition for an LTI system to be causal is that its impulse response $h(t)$ (i.e., the output when the input is $\delta(t)$, see Section 2.1.5) must be a causal signal, i.e., for $t < 0$, we must have $h(t) = 0$. For noncausal systems, the value of the output at t_0 also depends on the values of the input at times after t_0. Noncausal systems are encountered in situations where signals are not processed in real time.[3]

Example 2.1.23

A differentiator is an example of a causal system since it is LTI and its impulse response, $h(t) = \delta'(t)$, is zero for $t < 0$. A modulator is a causal but time-varying system since its output, $y(t) = x(t)\cos(2\pi f_0 t)$, at time t depends on the value of the input at time t and not on future values of $x(t)$. The delay system defined in Example 2.1.20 is causal for $\Delta \geq 0$ and noncausal for $\Delta < 0$. (Why?), since its impulse response $\delta(t - \Delta)$ is zero for $t < 0$ if $\Delta > 0$ and nonzero if $\Delta < 0$.   ∎

[3]For instance, when the entire signal is recorded and then processed. In such a case when processing the signal at time t_0 we have access to its future values.

2.1.5 Analysis of LTI Systems in the Time Domain

The class of LTI systems plays an important role both in communication and system theory. For this class of systems, the input–output relationship is particularly simple and can be expressed in terms of the convolution integral. To develop this relationship, we first introduce the concept of the *impulse response* of a system.

The *impulse response* $h(t)$ of a system is the response of the system to a unit impulse input $\delta(t)$:

$$h(t) = \mathcal{T}[\delta(t)].$$

The response of the system to a unit impulse applied at time τ, i.e., $\delta(t - \tau)$, is denoted by $h(t, \tau)$. Obviously, for time-invariant systems, $h(t, \tau) = h(t - \tau)$.[4]

The Convolution Integral. Now, we will derive the output $y(t)$ of an LTI system to any input signal $x(t)$. We will show that $y(t)$ can be expressed in terms of the input $x(t)$ and the impulse response $h(t)$ of the system.

Section 2.1.3 showed that, for any signal $x(t)$, we have

$$x(t) = x(t) \star \delta(t) = \int_{-\infty}^{+\infty} x(\tau)\delta(t - \tau)\, d\tau. \tag{2.1.41}$$

Now, if we denote the response of the LTI system to the input $x(t)$ by $y(t)$, we can write

$$
\begin{aligned}
y(t) &= \mathcal{L}[x(t)] \\
&= \mathcal{L}\left[\int_{-\infty}^{+\infty} x(\tau)\delta(t - \tau)\, d\tau \right] \\
&\overset{a}{=} \int_{-\infty}^{+\infty} x(\tau)\mathcal{L}[\delta(t - \tau)]\, d\tau \\
&\overset{b}{=} \int_{-\infty}^{+\infty} x(\tau)h(t - \tau)\, d\tau \\
&= x(t) \star h(t),
\end{aligned}
\tag{2.1.42}
$$

where (a) follows from the linearity of the system (note that an integral is basically the limit of a sum) and (b) follows from the time-invariance. This shows that the response to $x(t)$ is the convolution of $x(t)$ and the impulse response $h(t)$. Therefore, for the class of LTI systems, the impulse response completely characterizes the system. This means that the impulse response contains all the information we need to describe the system behavior.

---

[4]Note that this notation is sloppy. It uses h to denote two different functions, $h(t, \tau)$, which is a function of two variables, and $h(t)$, which is a function of one variable.

Example 2.1.24

The system described by

$$y(t) = \int_{-\infty}^{t} x(\tau)\,d\tau \qquad (2.1.43)$$

is called an integrator. Since integration is linear, this system is linear. Also, the response to $x(t - t_0)$ is

$$
\begin{aligned}
y_1(t) &= \int_{-\infty}^{t} x(\tau - t_0)\,d\tau \\
&= \int_{-\infty}^{t - t_0} x(u)\,du \\
&= y(t - t_0), \qquad (2.1.44)
\end{aligned}
$$

where we have used the change of variables $u = \tau - t_0$. We can see that the system is LTI. The impulse response is obtained by applying an impulse at the input, i.e.,

$$h(t) = \int_{-\infty}^{t} \delta(\tau)\,d\tau = u_{-1}(t). \qquad \blacksquare$$

Example 2.1.25

Let a linear time-invariant system have the impulse response $h(t)$. Assume this system has a complex exponential signal as input, i.e., $x(t) = Ae^{j(2\pi f_0 t + \theta)}$. The response to this input can be obtained by

$$
\begin{aligned}
y(t) &= \int_{-\infty}^{+\infty} h(\tau) A e^{j(2\pi f_0 (t - \tau) + \theta)}\,d\tau \\
&= A e^{j\theta} e^{j2\pi f_0 t} \int_{-\infty}^{+\infty} h(\tau) e^{-j2\pi f_0 \tau}\,d\tau \\
&= A |H(f_0)| e^{j\left(2\pi f_0 t + \theta + \angle H(f_0)\right)}, \qquad (2.1.45)
\end{aligned}
$$

where

$$H(f_0) = |H(f_0)| e^{j\angle H(f_0)} = \int_{-\infty}^{+\infty} h(\tau) e^{-j2\pi f_0 \tau}\,d\tau. \qquad (2.1.46)$$

This shows that *the response of an LTI system to the complex exponential with frequency f_0 is a complex exponential with the same frequency.* The amplitude of the response can be obtained by multiplying the amplitude of the input by $|H(f_0)|$, and its phase is obtained by adding $\angle H(f_0)$ to the input phase. Note that $H(f_0)$ is a function of the impulse response and the input frequency. Because of this property, complex exponentials are called *eigenfunctions* of the class of linear time-invariant systems. The eigenfunctions of a system are the set of inputs for which the output is a scaling of the input. Because of this important property, finding the response of LTI systems to the class of complex exponential signals is particularly simple; therefore, it is desirable to find ways to express arbitrary signals in terms of complex exponentials. We will later explore ways to do this.

$\blacksquare$

2.2 FOURIER SERIES

A large number of building blocks in a communication system can be modeled by linear time-invariant systems. LTI systems provide good and accurate models for a large class of communication channels. Also, some basic components of transmitters and receivers, such as filters, amplifiers, and equalizers, are LTI systems.

Our main objective is to develop methods and tools necessary to analyze linear time-invariant systems. When analyzing a system, we want to determine the output corresponding to a given input and, at the same time, provide insight into the behavior of the system. We have already seen that the input and output of an LTI system are related by the convolution integral, given by

$$y(t) = \int_{-\infty}^{+\infty} h(\tau)x(t-\tau)\,d\tau = \int_{-\infty}^{+\infty} h(t-\tau)x(\tau)\,d\tau, \qquad (2.2.1)$$

where $h(t)$ denotes the impulse response of the system. The convolution integral provides the basic tool for analyzing LTI systems. However, there are major drawbacks in the direct application of the convolution integral. First, using the convolution integral to find the response of an LTI system may be straightforward, but is not always an easy task. Second, even when the convolution integral can be performed with reasonable effort, it does not provide good insight into what the system *really does*.

In the next two sections we will develop another approach to analyzing LTI systems. The basic idea is *to expand the input as a linear combination of some basic signals whose output can be easily obtained, and then to employ the linearity properties of the system to obtain the corresponding output*. This approach is much easier than a direct computation of the convolution integral; at the same time, it provides better insight into the behavior of LTI systems. This method is based on the close similarity between the expansion of signals in terms of a basic signal set and the expansion of vectors in Euclidean space in terms of unit vectors.

2.2.1 Fourier Series and Its Properties

The set of complex exponentials are the eigenfunctions of LTI systems. The response of an LTI system to a complex exponential is a complex exponential with the same frequency *with a change in amplitude and phase*. Example 2.1.25 showed that the change in phase and amplitude are functions of the frequency of the complex exponential and the impulse response of the LTI system. So, which signals can be expanded in terms of complex exponentials? To answer this question, we will give the conditions for a periodic signal to be expandable in terms of complex exponentials. The expansion of nonperiodic signals will be discussed later.

Let the signal $x(t)$ be a periodic signal with period T_0. First, we need to determine whether the following Dirichlet conditions are satisfied:

1. $x(t)$ is absolutely integrable over its period, i.e.,

$$\int_0^{T_0} |x(t)|\,dt < \infty,$$

2. The number of maxima and minima of $x(t)$ in each period is finite,
3. The number of discontinuities of $x(t)$ in each period is finite.

If these conditions are met, then $x(t)$ can be expanded in terms of the complex exponential signals $\{e^{j2\pi \frac{n}{T_0}t}\}_{n=-\infty}^{+\infty}$ as

$$x(t) = \sum_{n=-\infty}^{+\infty} x_n e^{j2\pi \frac{n}{T_0}t}, \tag{2.2.2}$$

where

$$x_n = \frac{1}{T_0} \int_{\alpha}^{\alpha+T_0} x(t) e^{-j2\pi \frac{n}{T_0}t} dt \tag{2.2.3}$$

for some arbitrary α.

Some observations concerning this theorem are in order:

- The coefficients x_n are called the Fourier-series coefficients of the signal $x(t)$. These are generally complex numbers (even when $x(t)$ is a real signal).
- The parameter α in the limits of the integral is arbitrary. It can be chosen to simplify the computation of the integral. Usually $\alpha = 0$ or $\alpha = -T_0/2$ are good choices.
- The Dirichlet conditions are only *sufficient* conditions for the existence of the Fourier series expansion. For some signals that do not satisfy these conditions, we can still find the Fourier-series expansion.
- The quantity $f_0 = \frac{1}{T_0}$ is called the *fundamental frequency* of the signal $x(t)$. We observe that the frequencies of the complex exponential signals are multiples of this fundamental frequency. The nth multiple of f_0 is called the nth *harmonic*.
- Conceptually, this is a very important result. It states that the periodic signal $x(t)$ can be described by the period T_0 (or the fundamental frequency f_0) and the sequence of complex numbers $\{x_n\}$. Thus, to describe $x(t)$, we may specify a *countable* set of complex numbers. This considerably reduces the complexity of describing $x(t)$, since to define $x(t)$ for all values of t, we have to specify its values on an *uncountable* set of points.
- The Fourier-series expansion can be expressed in terms of the angular frequency $\omega_0 = 2\pi f_0$ by

$$x_n = \frac{\omega_0}{2\pi} \int_{\alpha}^{\alpha+\frac{2\pi}{\omega_0}} x(t) e^{-jn\omega_0 t} dt \tag{2.2.4}$$

and

$$x(t) = \sum_{n=-\infty}^{+\infty} x_n e^{jn\omega_0 t}. \tag{2.2.5}$$

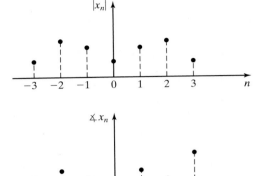

Figure 2.24   The discrete spectrum of $x(t)$.

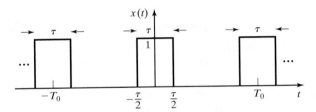

Figure 2.25   Periodic signal $x(t)$ in Equation (2.2.6).

- In general, $x_n = |x_n|e^{j\angle x_n}$. Thus, $|x_n|$ gives the magnitude of the nth harmonic and $\angle x_n$ gives its phase. Figure 2.24 shows a graph of the magnitude and phase of various harmonics in $x(t)$. This type of graph is called the *discrete spectrum* of the periodic signal $x(t)$.

Example 2.2.1

Let $x(t)$ denote the periodic signal depicted in Figure 2.25 and described analytically by

$$x(t) = \sum_{n=-\infty}^{+\infty} \Pi\left(\frac{t - nT_0}{\tau}\right), \tag{2.2.6}$$

where τ is a given positive constant (pulse width). Determine the Fourier-series expansion for this signal.

Solution   We first observe that the period of the signal is T_0 and

$$x_n = \frac{1}{T_0} \int_{-\frac{T_0}{2}}^{+\frac{T_0}{2}} x(t)e^{-jn\frac{2\pi t}{T_0}} dt$$

$$= \frac{1}{T_0} \int_{-\frac{\tau}{2}}^{+\frac{\tau}{2}} 1 \, e^{-jn\frac{2\pi t}{T_0}} \, dt$$

$$= \frac{1}{T_0} \frac{T_0}{-jn2\pi} \left[e^{-jn\frac{\pi\tau}{T_0}} - e^{+jn\frac{\pi\tau}{T_0}} \right] \quad n \neq 0$$

$$= \frac{1}{\pi n} \sin\left(\frac{n\pi\tau}{T_0} \right) \quad n \neq 0$$

$$= \frac{\tau}{T_0} \text{sinc}\left(\frac{n\tau}{T_0} \right) \quad n \neq 0, \tag{2.2.7}$$

where we have used the relation $\sin\phi = \frac{e^{j\phi} - e^{-j\phi}}{2j}$. For $n = 0$, the integration is very simple and yields $x_0 = \frac{\tau}{T_0}$. Therefore,

$$x(t) = \sum_{n=-\infty}^{+\infty} \frac{\tau}{T_0} \text{sinc}\left(\frac{n\tau}{T_0} \right) e^{jn\frac{2\pi t}{T_0}}. \tag{2.2.8}$$

A graph of these Fourier-series coefficients is shown in Figure 2.26.   ∎

Example 2.2.2

Determine the Fourier-series expansion for the signal $x(t)$ shown in Figure 2.27 and described by

$$x(t) = \sum_{n=-\infty}^{+\infty} (-1)^n \Pi(t - n). \tag{2.2.9}$$

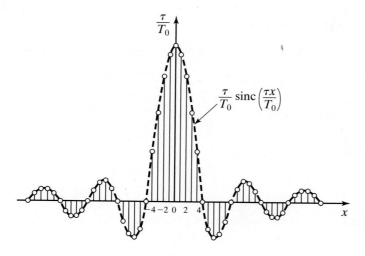

Figure 2.26   The discrete spectrum of the rectangular-pulse train.

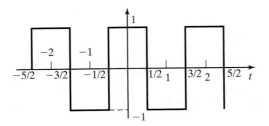

Figure 2.27   Signal $x(t)$ in Equation (2.2.9)

Solution   Since $T_0 = 2$, it is convenient to choose $\alpha = -\frac{1}{2}$. First, we note that for $n = 0$, we can easily find the integral to be zero; therefore $x_0 = 0$. For $n \neq 0$, we have

$$x_n = \frac{1}{2} \int_{-\frac{1}{2}}^{\frac{3}{2}} x(t) e^{-jn\pi t} \, dt$$

$$= \frac{1}{2} \int_{-\frac{1}{2}}^{\frac{1}{2}} e^{-jn\pi t} \, dt - \frac{1}{2} \int_{\frac{1}{2}}^{\frac{3}{2}} e^{-jn\pi t} \, dt$$

$$= -\frac{1}{j2\pi n} \left[e^{-j\frac{n\pi}{2}} - e^{j\frac{n\pi}{2}} \right] - \frac{1}{-j2\pi n} \left[e^{-jn\frac{3\pi}{2}} - e^{-jn\frac{\pi}{2}} \right]$$

$$= \frac{1}{n\pi} \sin\left(\frac{n\pi}{2}\right) - \frac{1}{n\pi} e^{-jn\pi} \sin\left(\frac{n\pi}{2}\right)$$

$$= \frac{1}{n\pi} (1 - \cos(n\pi)) \sin\left(\frac{n\pi}{2}\right)$$

$$= \begin{cases} \frac{2}{n\pi} & n = 4k + 1 \\ -\frac{2}{n\pi} & n = 4k + 3 \\ 0 & n \text{ even.} \end{cases} \tag{2.2.10}$$

From these values of x_n, we have the following Fourier-series expansion for $x(t)$:

$$x(t) = \frac{2}{\pi}(e^{j\pi t} + e^{-j\pi t}) - \frac{2}{3\pi}(e^{j3\pi t} + e^{-j3\pi t}) + \frac{2}{5\pi}(e^{j5\pi t} + e^{-j5\pi t}) - \cdots$$

$$= \frac{4}{\pi} \cos(\pi t) - \frac{4}{3\pi} \cos(3\pi t) + \frac{4}{5\pi} \cos(5\pi t) - \cdots$$

$$= \frac{4}{\pi} \sum_{k=0}^{\infty} \frac{(-1)^k}{2k + 1} \cos[(2k + 1)\pi t]. \tag{2.2.11}$$

∎

Example 2.2.3

Determine the Fourier-series representation of an *impulse train* denoted by

$$x(t) = \sum_{n=-\infty}^{+\infty} \delta(t - nT_0) \tag{2.2.12}$$

and shown in Figure 2.28.

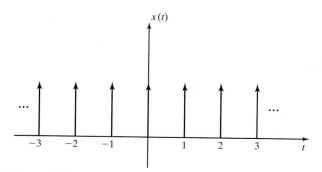

Figure 2.28   An impulse train.

Solution   We have

$$x_n = \frac{1}{T_0} \int_{-\frac{T_0}{2}}^{+\frac{T_0}{2}} x(t) e^{-j2\pi \frac{n}{T_0} t} dt$$

$$= \frac{1}{T_0} \int_{-\frac{T_0}{2}}^{+\frac{T_0}{2}} \delta(t) e^{-j2\pi \frac{n}{T_0} t} dt$$

$$= \frac{1}{T_0}. \tag{2.2.13}$$

With these coefficients, we have the following expansion:

$$\sum_{n=-\infty}^{+\infty} \delta(t - nT_0) = \frac{1}{T_0} \sum_{n=-\infty}^{+\infty} e^{j2\pi \frac{n}{T_0} t}. \tag{2.2.14}$$

This is a very useful relation, and we will employ it frequently.   ∎

Positive and Negative Frequencies.   We have seen that the Fourier-series expansion of a periodic signal $x(t)$ is expressed as

$$x(t) = \sum_{n=-\infty}^{\infty} x_n e^{j2\pi \frac{n}{T_0} t},$$

in which all positive and negative multiples of the fundamental frequency $\frac{1}{T_0}$ are present. A positive frequency corresponds to a term of the form $e^{j\omega t}$ (for a positive ω), and a negative frequency corresponds to $e^{-j\omega t}$. The term $e^{j\omega t}$ corresponds to a phasor rotating counterclockwise at an angular frequency of ω, and $e^{-j\omega t}$ corresponds to a phasor rotating clockwise at the same angular frequency. A plot of these two phasors is shown in Figure 2.29.

Note that if the two signals $e^{j\omega t}$ and $e^{-j\omega t}$ are added, their sum is $2\cos \omega t$, which is a real signal with two frequency components at $\pm\frac{\omega}{2\pi}$. We will soon see that this property holds for all real signals; i.e., in real signals, frequencies appear in positive and negative pairs with amplitudes that are conjugates.

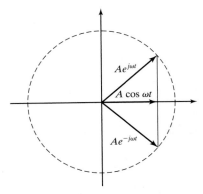

Figure 2.29   Phasors representing positive and negative frequencies.

Fourier Series for Real Signals.   If the signal $x(t)$ is a real signal satisfying the conditions of the Fourier-series theorem, then there must be alternative ways to expand the signal. For real $x(t)$, we have

$$
x_{-n} = \frac{1}{T_0} \int_\alpha^{\alpha+T_0} x(t) e^{j2\pi \frac{n}{T_0} t} \, dt
$$

$$
= \left[\frac{1}{T_0} \int_\alpha^{\alpha+T_0} x(t) e^{-j2\pi \frac{n}{T_0} t} \, dt \right]^*
$$

$$
= x_n^*. \tag{2.2.15}
$$

This means that for real $x(t)$, the positive and negative coefficients are conjugates. Hence, $|x_n|$ has even symmetry ($|x_n| = |x - n|$) and $\angle x_n$ has odd symmetry ($\angle x_n = -\angle x_{-n}$) with respect to the $n = 0$ axis. An example of the discrete spectrum for a real signal is shown in Figure 2.30.

From $x_{-n} = x_n^*$, it follows that if we denote

$$
x_n = \frac{a_n - jb_n}{2},
$$

then

$$
x_{-n} = \frac{a_n + jb_n}{2};
$$

therefore, for $n \geq 1$,

$$
x_n e^{j2\pi \frac{n}{T_0} t} + x_{-n} e^{-j2\pi \frac{n}{T_0} t} = \frac{a_n - jb_n}{2} e^{j2\pi \frac{n}{T_0} t} + \frac{a_n + jb_n}{2} e^{-j2\pi \frac{n}{T_0} t}
$$

$$
= a_n \cos\left(2\pi \frac{n}{T_0} t \right) + b_n \sin\left(2\pi \frac{n}{T_0} t \right).
$$

Since x_0 is real and given as $x_0 = \frac{a_0}{2}$, we conclude that

$$
x(t) = \frac{a_0}{2} + \sum_{n=1}^{\infty} \left[a_n \cos\left(2\pi \frac{n}{T_0} t \right) + b_n \sin(2\pi \frac{n}{T_0} t) \right]. \tag{2.2.16}
$$

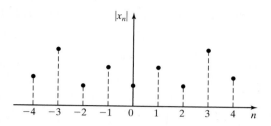

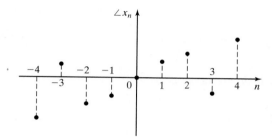

Figure 2.30 Discrete spectrum of a real-valued signal.

This relation, *which only holds for real periodic signals*, is called the *trigonometric Fourier-series expansion*. To obtain a_n and b_n, we have

$$x_n = \frac{a_n - jb_n}{2} = \frac{1}{T_0} \int_\alpha^{\alpha+T_0} x(t)e^{-j2\pi \frac{n}{T_0}t}\, dt;$$

therefore,

$$\frac{a_n - jb_n}{2} = \frac{1}{T_0} \int_\alpha^{\alpha+T_0} x(t)\cos\left(2\pi \frac{n}{T_0}t\right) dt - \frac{j}{T_0} \int_\alpha^{\alpha+T_0} x(t)\sin\left(2\pi \frac{n}{T_0}t\right) dt.$$

Thus, we obtain

$$a_n = \frac{2}{T_0} \int_\alpha^{\alpha+T_0} x(t)\cos\left(2\pi \frac{n}{T_0}t\right) dt \tag{2.2.17}$$

$$b_n = \frac{2}{T_0} \int_\alpha^{\alpha+T_0} x(t)\sin\left(2\pi \frac{n}{T_0}t\right) dt. \tag{2.2.18}$$

A third way exists to represent the Fourier-series expansion of a real signal. Note that

$$x_n e^{j2\pi \frac{n}{T_0}t} + x_{-n} e^{-j2\pi \frac{n}{T_0}t} = 2|x_n|\cos\left(2\pi \frac{n}{T_0}t + \angle x_n\right). \tag{2.2.19}$$

Substituting (2.2.19) in (2.2.2), we have

$$x(t) = x_0 + 2\sum_{n=1}^{\infty} |x_n|\cos\left(2\pi \frac{n}{T_0}t + \angle x_n\right). \tag{2.2.20}$$

In summary, for a real periodic signal $x(t)$, we have three alternatives to represent the Fourier-series expansion

$$x(t) = \sum_{n=-\infty}^{+\infty} x_n e^{j2\pi \frac{n}{T_0} t} \tag{2.2.21}$$

$$= \frac{a_0}{2} + \sum_{n=1}^{\infty} \left[a_n \cos \left(2\pi \frac{n}{T_0} t \right) + b_n \sin \left(2\pi \frac{n}{T_0} t \right) \right] \tag{2.2.22}$$

$$= x_0 + 2 \sum_{n=1}^{\infty} |x_n| \cos \left(2\pi \frac{n}{T_0} t + \angle x_n \right), \tag{2.2.23}$$

where the corresponding coefficients are obtained from

$$x_n = \frac{1}{T_0} \int_{\alpha}^{\alpha+T_0} x(t) e^{-j2\pi \frac{n}{T_0} t} dt = \frac{a_n}{2} - j\frac{b_n}{2} \tag{2.2.24}$$

$$a_n = \frac{2}{T_0} \int_{\alpha}^{\alpha+T_0} x(t) \cos \left(2\pi \frac{n}{T_0} t \right) dt \tag{2.2.25}$$

$$b_n = \frac{2}{T_0} \int_{\alpha}^{\alpha+T_0} x(t) \sin \left(2\pi \frac{n}{T_0} t \right) dt \tag{2.2.26}$$

$$|x_n| = \frac{1}{2} \sqrt{a_n^2 + b_n^2} \tag{2.2.27}$$

$$\angle x_n = -\arctan \left(\frac{b_n}{a_n} \right). \tag{2.2.28}$$

Example 2.2.4

Determine the sine and cosine coefficients in Example 2.2.1.

Solution   We have previously seen that

$$x_n = \frac{\tau}{T_0} \operatorname{sinc} \left(\frac{n\tau}{T_0} \right) = \frac{a_n - jb_n}{2}.$$

Therefore,

$$\begin{cases} a_n = \frac{2\tau}{T_0} \operatorname{sinc} \left(\frac{n\tau}{T_0} \right) \\ b_n = 0 \end{cases}$$

and

$$\begin{cases} |x_n| = \left| \frac{2\tau}{T_0} \operatorname{sinc} \left(\frac{n\tau}{T_0} \right) \right| \\ \angle x_n = 0 \text{ or } \pi \end{cases}$$

∎

Fourier-Series Expansion for Even and Odd Signals.   If, in addition to being real, a signal is either even or odd, then the Fourier-series expansion can be further simplified. For even $x(t)$, we have

$$b_n = \frac{2}{T_0} \int_{-\frac{T_0}{2}}^{\frac{T_0}{2}} x(t) \sin\left(2\pi \frac{n}{T_0}t\right) dt = 0. \qquad (2.2.29)$$

Since $x(t) \sin(2\pi \frac{n}{T_0}t)$ is the product of an even and an odd signal, it will be odd and its integral will be zero. Therefore, for even signals, the Fourier-series expansion has only cosine terms, i.e., we have

$$x(t) = \frac{a_0}{2} + \sum_{n=1}^{\infty} a_n \cos\left(2\pi \frac{n}{T_0}t\right). \qquad (2.2.30)$$

Equivalently, since $x_n = \frac{a_n - jb_n}{2}$, we conclude that for an even signal, every x_n is real (or all phases are either 0 or π, depending on the sign of x_n).

For odd signals, we can conclude in a similar way that every a_n vanishes; therefore, the Fourier-series expansion only contains the sine terms, or equivalently, every x_n is imaginary. In this case, we have

$$x(t) = \sum_{n=1}^{\infty} b_n \sin\left(2\pi \frac{n}{T_0}t\right). \qquad (2.2.31)$$

Example 2.2.5

Assuming $T_0 = 2$, determine the Fourier-series expansion of the signal shown in Figure 2.31.

Solution   For $n = 0$, we can easily verify that $x_0 = 0$. For $n \neq 0$, we have

$$x_n = \frac{1}{T_0} \int_{-\frac{T_0}{2}}^{\frac{T_0}{2}} x(t) e^{-j2\pi \frac{n}{T_0}} \, dt$$

$$= \frac{1}{2} \int_{-1}^{1} x(t) e^{-j\pi nt} \, dt$$

$$= \frac{1}{2} \left[\int_{-1}^{0} (-t - 1) e^{-j\pi nt} dt + \int_{0}^{+1} t e^{-j\pi nt} dt \right].$$

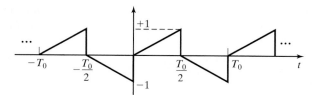

Figure 2.31   An example of an odd-harmonic signal.

Using

$$\int te^{\alpha t}\, dt = \frac{1}{\alpha} te^{\alpha t} - \frac{1}{\alpha^2} e^{\alpha t},$$

we have

$$\int te^{-j\pi nt}\, dt = \frac{j}{\pi n} te^{-j\pi nt} + \frac{1}{\pi^2 n^2} e^{-j\pi nt};$$

therefore,

$$x_n = -\frac{j}{2\pi n} te^{-j\pi nt}\Big]_{-1}^{0} - \frac{1}{2\pi^2 n^2} e^{-j\pi nt}\Big]_{-1}^{0}$$

$$- \frac{j}{2\pi n} e^{-j\pi nt}\Big]_{-1}^{0} + \frac{j}{2\pi n} te^{-j\pi nt}\Big]_{0}^{1} + \frac{1}{2\pi^2 n^2} e^{-j\pi nt}\Big]_{0}^{1}$$

$$= (\cos(\pi n) - 1)\left(\frac{1}{\pi^2 n^2} + \frac{j}{2\pi n}\right)$$

$$= \begin{cases} -\frac{2}{\pi^2 n^2} - \frac{j}{\pi n} & n \text{ odd} \\ 0 & n \text{ even} \end{cases}.$$

Noting that $x_n = \frac{a_n - jb_n}{2}$, we have

$$a_n = \begin{cases} -\frac{4}{\pi^2 n^2} & n \text{ odd} \\ 0 & n \text{ even} \end{cases}$$

and

$$b_n = \begin{cases} \frac{2}{\pi n} & n \text{ odd} \\ 0 & n \text{ even}. \end{cases}$$

The Fourier-series expansion is

$$x(t) = \sum_{n=0}^{\infty} \left[\frac{2}{\pi(2n+1)} \sin(2n+1)\pi t - \frac{4}{\pi^2(2n+1)^2} \cos(2n+1)\pi t \right].$$

Obviously, this signal only contains the odd harmonics. No even harmonics are present in the Fourier-series expansion of this signal. ∎

2.2.2 Response of LTI Systems to Periodic Signals

As we have already observed, the response of an LTI system to a complex exponential is a complex exponential with the same frequency and a change in amplitude and phase. In particular, if $h(t)$ is the impulse response of the system, then from Example 2.1.25, we know that the response to the exponential $e^{j2\pi f_0 t}$ is $H(f_0)e^{j2\pi f_0 t}$, where

$$H(f) = \int_{-\infty}^{+\infty} h(t)e^{-j2\pi ft}\, dt.$$

Now let us assume that $x(t)$, the input to the LTI system, is periodic with period T_0 and has a Fourier-series representation

$$x(t) = \sum_{n=-\infty}^{+\infty} x_n e^{j2\pi \frac{n}{T_0} t}.$$

Then we have

$$y(t) = \mathcal{L}[x(t)]$$

$$= \mathcal{L}\left[\sum_{-\infty}^{+\infty} x_n e^{j2\pi \frac{n}{T_0} t}\right]$$

$$= \sum_{-\infty}^{+\infty} x_n \mathcal{L}\left[e^{j2\pi \frac{n}{T_0} t}\right]$$

$$= \sum_{-\infty}^{+\infty} x_n H\left(\frac{n}{T_0}\right) e^{j2\pi \frac{n}{T_0} t}, \qquad (2.2.32)$$

where

$$H(f) = \int_{-\infty}^{+\infty} h(t) e^{-j2\pi f t} dt.$$

From this relation, we can draw the following conclusions:

* If the input to an LTI system is periodic with period T_0, then the output is also periodic. (What is the period of the output?) The output has a Fourier-series expansion given by

$$y(t) = \sum_{n=-\infty}^{+\infty} y_n e^{j2\pi \frac{n}{T_0} t},$$

where

$$y_n = x_n H\left(\frac{n}{T_0}\right).$$

This is equivalent to

$$|y_n| = |x_n| \cdot \left| H\left(\frac{n}{T_0}\right) \right|$$

and

$$\angle y_n = \angle x_n + \angle H\left(\frac{n}{T_0}\right).$$

- Only the frequency components that are present at the input can be present at the output. This means that *an LTI system cannot introduce new frequency components in the output, if these component are different from those already present at the input*. In other words, all systems capable of introducing new frequency components are either nonlinear and/or time varying.

- The amount of change in amplitude $\left| H\left(\frac{n}{T_0}\right) \right|$ and phase $\angle H\left(\frac{n}{T_0}\right)$ are functions of n, the harmonic order, and $h(t)$, the impulse response of the system. The function

$$H(f) = \int_{-\infty}^{+\infty} h(t) e^{-j2\pi f t} dt \qquad (2.2.33)$$

is called the *frequency response* or *frequency characteristics* of the LTI system. In general, $H(f)$ is a complex function that can be described by its magnitude $|H(f)|$ and phase $\angle H(f)$. The function $H(f)$, or equivalently $h(t)$, is the only information needed to find the output of an LTI system for a given periodic input.

Example 2.2.6

Let $x(t)$ denote the signal shown in Figure 2.27, but set the period equal to $T_0 = 10^{-5}$ seconds. This signal is passed through a filter with the frequency response depicted in Figure 2.32. Determine the output of the filter.

Solution   We first start with the Fourier-series expansion of the input. This can be easily obtained as

$$
\begin{aligned}
x(t) &= \frac{4}{\pi} \sum_{n=0}^{\infty} \frac{(-1)^n}{2n+1} \cos\left(2\pi(2n+1)10^5 t\right) \\
&= \frac{2}{\pi} e^{j2\pi 10^5 t} + \frac{2}{\pi} e^{-j2\pi 10^5 t} \\
&\quad - \frac{2}{3\pi} e^{j6\pi 10^5 t} - \frac{2}{3\pi} e^{-j6\pi 10^5 t} \\
&\quad + \frac{2}{5\pi} e^{j10\pi 10^5 t} + \frac{2}{5\pi} e^{-j10\pi 10^5 t} - \cdots .
\end{aligned} \qquad (2.2.34)
$$

To find the output corresponding to each frequency, we must multiply the coefficient of each frequency component by the $H(f)$ corresponding to that frequency. Therefore, we have

$$
\begin{aligned}
H(10^5) &= 1 e^{j\frac{\pi}{2}} = e^{j\frac{\pi}{2}} \\
H(-10^5) &= 1 e^{-j\frac{\pi}{2}} = e^{-j\frac{\pi}{2}} \\
H(3 \times 10^5) &= 3 e^{j\frac{\pi}{2}} \\
H(-3 \times 10^5) &= 3 e^{-j\frac{\pi}{2}} \\
H(5 \times 10^5) &= 5 e^{j\frac{\pi}{2}} \\
H(-5 \times 10^5) &= 5 e^{-j\frac{\pi}{2}} .
\end{aligned} \qquad (2.2.35)
$$

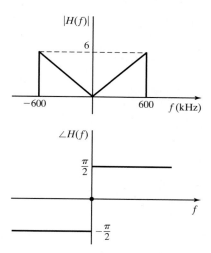

Figure 2.32 Frequency response of the filter.

For higher frequencies, $H(f) = 0$. Therefore, we have

$$y(t) = \frac{2}{\pi} e^{j(2\pi 10^5 t + \frac{\pi}{2})} + \frac{2}{\pi} e^{j(-2\pi 10^5 t - \frac{\pi}{2})}$$

$$- \frac{2}{\pi} e^{j(6\pi 10^5 t + \frac{\pi}{2})} - \frac{2}{\pi} e^{j(-6\pi 10^5 t - \frac{\pi}{2})}$$

$$+ \frac{2}{\pi} e^{j(10\pi 10^5 t + \frac{\pi}{2})} + \frac{2}{\pi} e^{j(-10\pi 10^5 t - \frac{\pi}{2})}$$

or, equivalently,

$$y(t) = -\frac{4}{\pi} \sin(2\pi 10^5 t) + \frac{4}{\pi} \sin(6\pi 10^5 t) - \frac{4}{\pi} \sin(10\pi 10^5 t). \qquad (2.2.36)$$

∎

2.2.3 Parseval's Relation

Parseval's relation says that the power content of a periodic signal is the sum of the power contents of its components in the Fourier-series representation of that signal. This relation is a consequence of the orthogonality of the basis used for the Fourier-series expansion, that is, the exponential signals.

Let us assume that the Fourier-series representation of the periodic signal $x(t)$ is given by

$$x(t) = \sum_{n=-\infty}^{+\infty} x_n e^{j2\pi \frac{n}{T_0} t}.$$

Then, the complex conjugate of both sides of this relation is

$$x^*(t) = \sum_{n=-\infty}^{+\infty} x_n^* e^{-j2\pi \frac{n}{T_0} t}.$$

By multiplying the two equations, we obtain

$$|x(t)|^2 = \sum_{n=-\infty}^{+\infty} \sum_{m=-\infty}^{+\infty} x_n x_m^* e^{j2\pi \frac{n-m}{T_0} t}.$$

Next, we integrate both sides over one period and note

$$\int_{\alpha}^{\alpha+T_0} e^{j2\pi \frac{n-m}{T_0} t} dt = T_0 \delta_{mn} = \begin{cases} T_0 & n = m \\ 0 & n \neq m \end{cases}.$$

Thus,

$$\int_{\alpha}^{\alpha+T_0} |x(t)|^2 dt = \sum_{n=-\infty}^{+\infty} \sum_{m=-\infty}^{+\infty} x_n x_m^* T_0 \delta_{mn}$$

$$= T_0 \sum_{n=-\infty}^{+\infty} |x_n|^2 \qquad (2.2.37)$$

or, finally,

$$\frac{1}{T_0} \int_{\alpha}^{\alpha+T_0} |x(t)|^2 dt = \sum_{n=-\infty}^{+\infty} |x_n|^2. \qquad (2.2.38)$$

This is the formal statement of Parseval's relation. Note that by Equation (2.1.14) the left-hand side of this relation is P_x, the power content of the signal $x(t)$, and $|x_n|^2$ is the power content of $x_n e^{j2\pi \frac{n}{T_0}}$, the nth harmonic. Therefore, Parseval's relation says that the power content of the periodic signal is the sum of the power contents of its harmonics.

If we substitute $x_n = \frac{a_n - jb_n}{2}$ in Parseval's relation, we obtain

$$P_x = \frac{1}{T_0} \int_{\alpha}^{\alpha+T_0} |x(t)|^2 dt = \frac{a_0^2}{4} + \frac{1}{2} \sum_{n=1}^{\infty} (a_n^2 + b_n^2). \qquad (2.2.39)$$

Since the power content of $a_n \cos\left(2\pi \frac{n}{T_0}\right)$ and $b_n \sin\left(2\pi \frac{n}{T_0}\right)$ are $\frac{a_n^2}{2}$ and $\frac{b_n^2}{2}$, respectively, we see that the power content of $x(t)$ is the sum of the power contents of its harmonics.

Example 2.2.7
    Determine the power contents of the input and output signals in Example 2.2.6.
    Solution   We have

$$P_x = \frac{1}{2} \int_{-\frac{1}{2}}^{\frac{3}{2}} |x(t)|^2 dt = \frac{1}{2} \left[\int_{-\frac{1}{2}}^{\frac{1}{2}} 1 dt + \int_{\frac{1}{2}}^{\frac{3}{2}} 1 dt \right] = 1.$$

We could employ Parseval's relation to obtain the same result. To do this, we have

$$P_x = \sum_{n=1}^{\infty} \frac{a_n^2}{2} = \frac{8}{\pi^2} \left(1 + \frac{1}{9} + \frac{1}{25} + \frac{1}{49} + \cdots \right).$$

Equating the two relations, we obtain

$$\sum_{k=0}^{\infty} \frac{1}{(2k+1)^2} = \frac{\pi^2}{8}.$$

To find the output power, we have

$$P_y = \frac{1}{2} \int_{\alpha}^{\alpha+2} |y(t)|^2 dt = \sum_{n=1}^{\infty} \frac{b_n^2}{2} = \frac{1}{2}\left[\frac{16}{\pi^2} + \frac{16}{\pi^2} + \frac{16}{\pi^2}\right] = \frac{24}{\pi^2}. \qquad \blacksquare$$

2.3 FOURIER TRANSFORM

2.3.1 From Fourier Series to Fourier Transforms

As we have observed, the Fourier series may be used to represent a periodic signal in terms of complex exponentials. This Fourier-series representation considerably decreases the complexity of the description of the periodic signal. Simultaneously, the series representation in terms of complex exponentials is particularly useful when analyzing LTI systems. This is because the complex exponentials are the eigenfunctions of LTI systems. Equivalent to any periodic signal is the sequence $\{x_n\}$ of the Fourier-series expansion coefficients which, together with the fundamental frequency f_0, completely describe the signal.

In this section, we will apply the Fourier series representation to nonperiodic signals. We will see that it is still possible to expand a nonperiodic signal in terms of complex exponentials. However, the resulting spectrum will no longer be discrete. In other words, the spectrum of nonperiodic signals covers a continuous range of frequencies. This result is the well-known Fourier transform given next.

First, the signal $x(t)$ must satisfy the following Dirichlet conditions:

1. $x(t)$ is absolutely integrable on the real line; that is,

$$\int_{-\infty}^{+\infty} |x(t)| dt < \infty.$$

2. The number of maxima and minima of $x(t)$ in any finite interval on the real line is finite.
3. The number of discontinuities of $x(t)$ in any finite interval on the real line is finite.

Then the Fourier transform (or Fourier integral) of $x(t)$, defined by

$$X(f) = \int_{-\infty}^{+\infty} x(t)e^{-j2\pi ft} dt, \qquad (2.3.1)$$

exists and the original signal can be obtained from its Fourier transform by

$$x(t) = \int_{-\infty}^{+\infty} X(f)e^{j2\pi ft}df. \tag{2.3.2}$$

We make the following observations concerning the Fourier transform:

- $X(f)$ is generally a complex function. Its magnitude $|X(f)|$ and phase $\angle X(f)$ represent the amplitude and phase of various frequency components in $x(t)$. The function $X(f)$ is sometimes referred to as the *spectrum*[5] of the signal $x(t)$.
- To denote that $X(f)$ is the Fourier transform of $x(t)$, we frequently employ the following notation:

$$X(f) = \mathscr{F}[x(t)].$$

To denote that $x(t)$ is the *inverse Fourier transform* of $X(f)$, we use the following notation:

$$x(t) = \mathscr{F}^{-1}[X(f)].$$

Sometimes we use a shorthand for both relations:

$$x(t) \Leftrightarrow X(f).$$

- If the variable in the Fourier transform is ω rather than f, then we have

$$X(\omega) = \int_{-\infty}^{+\infty} x(t)e^{-j\omega t}\, dt$$

and

$$x(t) = \frac{1}{2\pi} \int_{-\infty}^{+\infty} X(\omega)e^{j\omega t}\, d\omega.$$

- The Fourier-transform and the inverse Fourier-transform relations can be written as

$$x(t) = \int_{-\infty}^{+\infty} \overbrace{\left[\int_{-\infty}^{+\infty} x(\tau)e^{-j2\pi f\tau}\, d\tau \right]}^{X(f)} e^{j2\pi ft}df$$

$$= \int_{-\infty}^{+\infty} \left[\int_{-\infty}^{+\infty} e^{j2\pi f(t-\tau)}\, df \right] x(\tau)\, d\tau. \tag{2.3.3}$$

where in the last step we have exchanged the order of integration. On the other hand,

$$x(t) = \int_{-\infty}^{+\infty} \delta(t - \tau)x(\tau)\, d\tau. \tag{2.3.4}$$

---

[5]Sometimes $X(f)$ is referred to as a *voltage spectrum*, as opposed to a *power spectrum*, which will be defined later.

Comparing Equation (2.3.3) with Equation (2.3.4), we obtain

$$\delta(t - \tau) = \int_{-\infty}^{+\infty} e^{j2\pi f(t-\tau)}\,df,\qquad(2.3.5)$$

or

$$\delta(t) = \int_{-\infty}^{+\infty} e^{j2\pi ft}\,df.\qquad(2.3.6)$$

• Equation (2.3.1) is similar to Equation (2.2.3) and Equation (2.3.2) is similar to Equation (2.2.2). The difference is that the sum in Equation (2.2.2) is substituted by the integral.

Example 2.3.1

Determine the Fourier transform of the signal $\Pi(t)$ given in Equation (2.1.15) and shown in Figure 2.33.

Solution   We have

$$
\begin{aligned}
\mathscr{F}[\Pi(t)] &= \int_{-\infty}^{+\infty} \Pi(t)e^{-j2\pi ft}\,dt \\
&= \int_{-\frac{1}{2}}^{+\frac{1}{2}} e^{-j2\pi ft}\,dt \\
&= \frac{1}{-j2\pi f}\left[e^{-j\pi f} - e^{j\pi f}\right] \\
&= \frac{\sin \pi f}{\pi f} \\
&= \text{sinc}(f).
\end{aligned}
\qquad(2.3.7)
$$

Therefore,

$$\mathscr{F}[\Pi(t)] = \text{sinc}(f).$$

Figure 2.33 illustrates the Fourier-transform relationship for this signal.   ∎

Example 2.3.2

Determine the Fourier transform of an impulse signal $x(t) = \delta(t)$.

Solution   The Fourier transform can be obtained by

$$
\begin{aligned}
\mathscr{F}[\delta(t)] &= \int_{-\infty}^{+\infty} \delta(t)e^{-j2\pi ft}\,dt, \\
&= 1
\end{aligned}
\qquad(2.3.8)
$$

where we have used the sifting property of $\delta(t)$. This shows that all frequencies are present in the spectrum of $\delta(t)$ with unity magnitude and zero phase. The graphs of $x(t)$ and its Fourier transform are given in Figure 2.34. Similarly, from the relation

$$\int_{-\infty}^{+\infty} \delta(f)e^{j2\pi ft}\,df = 1,$$

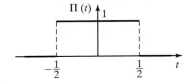

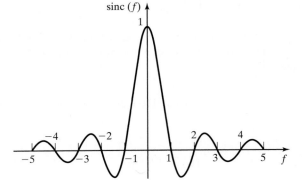

Figure 2.33 $\Pi(t)$ and its Fourier transform.

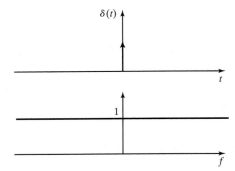

Figure 2.34 Impulse signal and its spectrum.

we conclude that

$$\mathscr{F}[1] = \delta(f). \qquad\blacksquare$$

Example 2.3.3

Determine the Fourier transform of signal sgn(t).

Solution We begin with the definition of sgn(t) as a limit of an exponential, as shown in Figure 2.18 and given by

$$x_n(t) = \begin{cases} e^{-\frac{t}{n}} & t > 0 \\ -e^{\frac{t}{n}} & t < 0 \\ 0 & t = 0. \end{cases} \qquad (2.3.9)$$

For this signal, the Fourier transform is

$$X_n(f) = \mathscr{F}[x_n(t)]$$

$$= \int_{-\infty}^{0} (-e^{\frac{t}{n}})e^{-j2\pi ft}\,dt + \int_{0}^{+\infty} e^{-\frac{t}{n}}e^{-j2\pi ft}\,dt$$

$$= -\int_{-\infty}^{0} e^{t(\frac{1}{n}-j2\pi f)}\,dt + \int_{0}^{+\infty} e^{-t(\frac{1}{n}+j2\pi f)}\,dt$$

$$= -\frac{1}{\frac{1}{n} - j2\pi f} + \frac{1}{\frac{1}{n} + j2\pi f}$$

$$= \frac{-j4\pi f}{\frac{1}{n^2} + 4\pi^2 f^2}. \qquad (2.3.10)$$

Now, letting $n \to \infty$, we obtain

$$\mathscr{F}[\text{sgn}(t)] = \lim_{n\to\infty} X_n(f)$$

$$= \lim_{n\to\infty} \frac{-j4\pi f}{\frac{1}{n^2} + 4\pi^2 f^2}$$

$$= \frac{1}{j\pi f}. \qquad (2.3.11)$$

The graphs of $\text{sgn}(t)$ and its spectrum are shown in Figure 2.35.   ∎

Fourier Transform of Real, Even, and Odd Signals.   The Fourier-transform relation can be generally written as

$$\mathscr{F}[x(t)] = \int_{-\infty}^{+\infty} x(t)e^{-j2\pi ft}\,dt$$

$$= \int_{-\infty}^{+\infty} x(t)\cos(2\pi ft)\,dt - j\int_{-\infty}^{+\infty} x(t)\sin(2\pi ft)\,dt.$$

For real $x(t)$, both integrals

$$\int_{-\infty}^{+\infty} x(t)\cos(2\pi ft)\,dt$$

and

$$\int_{-\infty}^{+\infty} x(t)\sin(2\pi ft)\,dt$$

are real; therefore, they denote the real and imaginary parts of $X(f)$, respectively. Since cos is an even function and sin is an odd function, we know that for real $x(t)$, the real part of $X(f)$ is an even function of f and the imaginary part is an odd

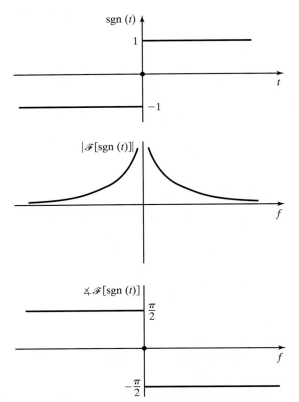

Figure 2.35 The signum signal and its spectrum.

function of f. Therefore, in general, for real $x(t)$, the transform $X(f)$ is a Hermitian function

$$X(-f) = X^*(f).$$

This is equivalent to the following relations:

$$\text{Re}[X(-f)] = \text{Re}[X(f)]$$
$$\text{Im}[X(-f)] = -\text{Im}[X(f)]$$
$$|X(-f)| = |X(f)|$$
$$\angle X(-f) = -\angle X(f).$$

Typical plots of $|X(f)|$ and $\angle X(f)$ for a real $x(t)$ are given in Figure 2.36. If, in addition to being real, $x(t)$ is an even signal, then the integral

$$\int_{-\infty}^{+\infty} x(t) \sin(2\pi f t)\, dt$$

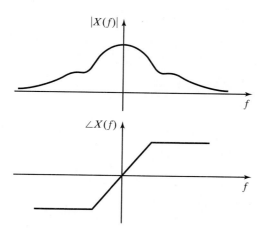

Figure 2.36 Magnitude and phase of the spectrum of a real signal.

vanishes because the integrand is the product of even and odd signals; therefore, it is odd. Hence, the Fourier transform $X(f)$ will be real and even. Similarly, if $x(t)$ is real and odd, the real part of its Fourier transform vanishes and $X(f)$ will be imaginary and odd.

Signal Bandwidth. The bandwidth of a signal represents the range of frequencies present in the signal. The higher the bandwidth, the larger the variations in the frequencies present. In general, we define the bandwidth of a real signal $x(t)$ as the range of *positive* frequencies present in the signal. In order to find the bandwidth of $x(t)$, we first find $X(f)$, which is the Fourier transform of $x(t)$; then, we find the range of positive frequencies that $X(f)$ occupies. The bandwidth is $BW = W_{max} - W_{min}$, where W_{max} is the highest positive frequency present in $X(f)$ and W_{min} is the lowest positive frequency present in $X(f)$.

2.3.2 Basic Properties of the Fourier Transform

In this section, we will develop the basic properties of the Fourier transform. With each property, we will present examples of its applications.

Linearity. The Fourier-transform operation is linear. That is, if $x_1(t)$ and $x_2(t)$ are signals possessing Fourier transforms $X_1(f)$ and $X_2(f)$ respectively, the Fourier transform of $\alpha x_1(t) + \beta x_2(t)$ is $\alpha X_1(f) + \beta X_2(f)$, where α and β are two arbitrary (real or complex) scalars. This property is a direct result of the linearity of integration.

Example 2.3.4

Determine the Fourier transform of $u_{-1}(t)$, the unit-step signal.

Solution Using the relation

$$u_{-1}(t) = \frac{1}{2} + \frac{1}{2}\text{sgn}(t) = \frac{1}{2} \times 1 + \frac{1}{2}\text{sgn}(t)$$

and the linearity theorem, we obtain

$$\mathscr{F}[u_{-1}(t)] = \mathscr{F}[\frac{1}{2} \times 1 + \frac{1}{2}\text{sgn}(t)]$$

$$= \frac{1}{2}\delta(f) + \frac{1}{j2\pi f}. \tag{2.3.12}$$

∎

Duality.   If

$$X(f) = \mathscr{F}[x(t)],$$

then

$$x(f) = \mathscr{F}[X(-t)]$$

and

$$x(-f) = \mathscr{F}[X(t)].$$

To show this property, we begin with the inverse Fourier-transform relation

$$x(t) = \int_{-\infty}^{+\infty} X(f)e^{j2\pi ft}df.$$

Then, we introduce the change of variable $u = -f$ to obtain

$$x(t) = \int_{-\infty}^{+\infty} X(-u)e^{-j2\pi ut}du.$$

Letting $t = f$, we have

$$x(f) = \int_{-\infty}^{+\infty} X(-u)e^{-j2\pi uf}du;$$

finally, substituting t for u, we get

$$x(f) = \int_{-\infty}^{+\infty} X(-t)e^{-j2\pi tf}dt$$

or

$$x(f) = \mathscr{F}[X(-t)]. \tag{2.3.13}$$

Using the same technique once more, we obtain

$$x(-f) = \mathscr{F}[X(t)]. \tag{2.3.14}$$

Example 2.3.5

Determine the Fourier transform of $\text{sinc}(t)$.

Solution   Noting that $\Pi(t)$ is an even signal and, therefore, that $\Pi(-f) = \Pi(f)$, we can use the duality theorem to obtain

$$\mathscr{F}[\text{sinc}(t)] = \Pi(-f) = \Pi(f). \tag{2.3.15}$$

∎

Example 2.3.6

Determine the Fourier transform of $\frac{1}{t}$.

Solution   Here again, we apply the duality theorem to the transform pair derived in Example 2.3.3

$$\mathscr{F}[\text{sgn}(t)] = \frac{1}{j\pi f}$$

to obtain

$$\mathscr{F}[\frac{1}{j\pi t}] = \text{sgn}(-f) = -\text{sgn}(f).$$

Using the linearity theorem, we have

$$\mathscr{F}[\frac{1}{t}] = -j\pi\,\text{sgn}(f). \qquad (2.3.16)$$

∎

Shift in Time Domain.   A shift of t_0 in the time origin causes a phase shift of $-2\pi f t_0$ in the frequency domain. In other words,

$$\mathscr{F}[x(t - t_0)] = e^{-j2\pi f t_0}\mathscr{F}[x(t)].$$

To see this, we start with the Fourier transform of $x(t - t_0)$, namely,

$$\mathscr{F}[x(t - t_0)] = \int_{-\infty}^{\infty} x(t - t_0)e^{-j2\pi ft}dt.$$

With a change of variable of $u = t - t_0$, we obtain

$$\mathscr{F}[x(t - t_0)] = \int_{-\infty}^{\infty} x(u)e^{-j2\pi f t_0}e^{-j2\pi fu}du$$

$$= e^{-j2\pi f t_0}\int_{-\infty}^{\infty} x(u)e^{-j2\pi fu}du$$

$$= e^{-j2\pi f t_0}\mathscr{F}[x(t)]. \qquad (2.3.17)$$

Note that a change in the time origin does not change the magnitude of the transform. It only introduces a phase shift linearly proportional to the time shift (or delay).

Example 2.3.7

Determine the Fourier transform of the signal shown in Figure 2.37.

Solution   We have

$$x(t) = \Pi\left(t - \frac{3}{2}\right).$$

By applying the shift theorem, we obtain

$$\mathscr{F}[x(t)] = e^{-j2\pi f \times \frac{3}{2}} = e^{-j3\pi f}\text{sinc}(f). \qquad (2.3.18)$$

∎

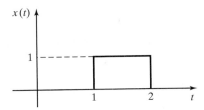

Figure 2.37 Signal $x(t)$.

Example 2.3.8
Determine the Fourier transform of the impulse train

$$\sum_{n=-\infty}^{\infty} \delta(t - nT_0).$$

Solution   Applying the shift theorem, we have

$$\mathscr{F}[\delta(t - nT_0)] = e^{-j2\pi f n T_0} \mathscr{F}[\delta(t)] = e^{-j2\pi f n T_0}.$$

Therefore,

$$\mathscr{F}[\sum_{n=-\infty}^{\infty} \delta(t - nT_0)] = \sum_{n=-\infty}^{\infty} e^{-j2\pi f n T_0}.$$

Using Equation (2.2.14)

$$\sum_{n=-\infty}^{+\infty} \delta(t - nT_0) = \frac{1}{T_0} \sum_{n=-\infty}^{+\infty} e^{j2\pi \frac{n}{T_0} t}$$

and substituting f for t and $\frac{1}{T_0}$ for T_0, we obtain

$$\sum_{n=-\infty}^{\infty} \delta\left(f - \frac{n}{T_0}\right) = T_0 \sum_{n=-\infty}^{\infty} e^{j2\pi n f T_0} = T_0 \sum_{n=-\infty}^{\infty} e^{-j2\pi n f T_0} \qquad (2.3.19)$$

or

$$\sum_{n=-\infty}^{\infty} e^{-j2\pi n f T_0} = \frac{1}{T_0} \sum_{n=-\infty}^{\infty} \delta\left(f - \frac{n}{T_0}\right). \qquad (2.3.20)$$

This relation yields,

$$\mathscr{F}[\sum_{n=-\infty}^{\infty} \delta(t - nT_0)] = \frac{1}{T_0} \sum_{n=-\infty}^{\infty} \delta\left(f - \frac{n}{T_0}\right). \qquad (2.3.21)$$

The case of $T_0 = 1$ is particularly interesting. For this case, we have

$$\mathscr{F}[\sum_{n=-\infty}^{\infty} \delta(t - n)] = \sum_{n=-\infty}^{\infty} \delta(f - n). \qquad (2.3.22)$$

That is, after substituting f for t, we find that the Fourier transform of $\sum_{n=-\infty}^{\infty} \delta(t - n)$
is itself.   ∎

Scaling.   For any real $a \neq 0$, we have

$$\mathcal{F}[x(at)] = \frac{1}{|a|} X \left(\frac{f}{a} \right). \qquad (2.3.23)$$

To see this, we note that

$$\mathcal{F}[x(at)] = \int_{-\infty}^{\infty} x(at) e^{-j2\pi f t} dt \qquad (2.3.24)$$

and make the change in variable $u = at$. Then,

$$\mathcal{F}[x(at)] = \frac{1}{|a|} \int_{-\infty}^{\infty} x(u) e^{-j2\pi f u/a} du$$

$$= \frac{1}{|a|} X \left(\frac{f}{a} \right), \qquad (2.3.25)$$

where we have treated the cases $a > 0$ and $a < 0$ separately.

Note that in the previous expression, if $a > 1$, then $x(at)$ is a contracted form of $x(t)$, whereas if $a < 1$, $x(at)$ is an expanded version of $x(t)$. Thus, if we expand a signal in the time domain, its frequency-domain representation (Fourier transform) contracts; if we contract a signal in the time domain, its frequency domain representation expands. This is exactly what we expect since contracting a signal in the time domain makes the changes in the signal more abrupt, thus increasing its frequency content.

Example 2.3.9

Determine the Fourier transform of the signal

$$x(t) = \begin{cases} 3 & 0 \leq t \leq 4 \\ 0 & \text{otherwise.} \end{cases}$$

Solution   Note that $x(t)$ is a rectangular signal amplified by a factor of 3, expanded by a factor of 4, and then shifted to the right by 2. In other words, $x(t) = 3\Pi \left(\frac{t-2}{4} \right)$. Using the linearity, time shift, and scaling properties, we have

$$\mathcal{F}[3\Pi \left(\frac{t-2}{4} \right)] = 3\mathcal{F}[\Pi \left(\frac{t-2}{4} \right)]$$

$$= 3e^{-4j\pi f} \mathcal{F}[\Pi \left(\frac{t}{4} \right)]$$

$$= 12e^{-4j\pi f} \operatorname{sinc}(4f). \qquad \blacksquare$$

Convolution.   If the signals $x(t)$ and $y(t)$ both possess Fourier transforms, then

$$\mathcal{F}[x(t) \star y(t)] = \mathcal{F}[x(t)] \cdot \mathcal{F}[y(t)] = X(f) \cdot Y(f). \qquad (2.3.26)$$

For a proof, we have

$$\mathcal{F}[x(t) \star y(t)] = \int_{-\infty}^{\infty} \left[\int_{-\infty}^{\infty} x(\tau) y(t - \tau) \, d\tau \right] e^{-j2\pi ft} dt$$

$$= \int_{-\infty}^{\infty} x(\tau) \left[\int_{-\infty}^{\infty} y(t - \tau) e^{-j2\pi f(t-\tau)} \, dt \right] e^{-j2\pi f\tau} d\tau.$$

Now with the change of variable $u = t - \tau$, we have

$$\int_{-\infty}^{\infty} y(t - \tau) e^{-j2\pi f(t-\tau)} dt = \int_{-\infty}^{\infty} y(u) e^{-j2\pi fu} du$$

$$= \mathcal{F}[y(t)]$$

$$= Y(f);$$

therefore,

$$\mathcal{F}[x(t) \star y(t)] = \int_{-\infty}^{\infty} x(\tau) Y(f) e^{-j2\pi f\tau} d\tau$$

$$= X(f) \cdot Y(f). \tag{2.3.27}$$

This theorem is very important and is a direct result of the fact that the complex exponentials are eigenfunctions of LTI systems (or, equivalently, eigenfunctions of the convolution operation). Finding the response of an LTI system to a given input is much easier in the frequency domain than it is the time domain. This theorem is the basis of the frequency-domain analysis of LTI systems.

Example 2.3.10
Determine the Fourier transform of the signal $\Lambda(t)$, shown in Figure 2.15.
Solution It is enough to note that $\Lambda(t) = \Pi(t) \star \Pi(t)$ and use the convolution theorem. Thus, we obtain

$$\mathcal{F}[\Lambda(t)] = \mathcal{F}[\Pi(t)] \cdot \mathcal{F}[\Pi(t)] = \text{sinc}^2(f). \tag{2.3.28}$$

■

Example 2.3.11
Determine the Fourier transform of the signal $x(t) = \Pi\left(\frac{t}{4}\right) + \Lambda\left(\frac{t}{2}\right)$, which is shown in Figure 2.16 and discussed in Example 2.1.14.
Solution Using scaling and linearity, we have

$$X(f) = 4\text{sinc}(4f) + 2\text{sinc}^2(2f).$$

■

Modulation. The Fourier transform of $x(t)e^{j2\pi f_0 t}$ is $X(f - f_0)$. To show this relation, we have

$$\mathcal{F}[x(t)e^{j2\pi f_0 t}] = \int_{-\infty}^{\infty} x(t) e^{j2\pi f_0 t} e^{-j2\pi ft} dt$$

$$= \int_{-\infty}^{\infty} x(t) e^{-j2\pi t(f - f_0)} dt$$

$$= X(f - f_0). \tag{2.3.29}$$

This theorem is the dual of the time-shift theorem. The time-shift theorem says that a shift in the time domain results in a multiplication by a complex exponential in the frequency domain. The modulation theorem states that a multiplication in the time domain by a complex exponential results in a shift in the frequency domain. A shift in the frequency domain is usually called modulation.

Example 2.3.12

Determine the Fourier transform of $x(t) = e^{j2\pi f_0 t}$.

Solution   Using the modulation theorem, we have

$$\mathscr{F}[e^{j2\pi f_0 t}] = \mathscr{F}[1e^{j2\pi f_0 t}]$$

$$= \delta(f - f_0).$$

Note that, since $x(t)$ is not a real signal, its Fourier transform does not have Hermitian symmetry. (The magnitude is not even and the phase is not odd.)   ∎

Example 2.3.13

Determine the Fourier transform of the signal $\cos(2\pi f_0 t)$.

Solution   Using Euler's relation, we have $\cos(2\pi f_0 t) = \frac{1}{2}e^{j2\pi f_0 t} + \frac{1}{2}e^{-j2\pi f_0 t}$. Now using the linearity property and the result of Example 2.3.12, we have

$$\mathscr{F}[\cos(2\pi f_0 t)] = \frac{1}{2}\mathscr{F}[e^{j2\pi f_0 t}] + \frac{1}{2}\mathscr{F}[e^{-j2\pi f_0 t}]$$

$$= \frac{1}{2}\delta(f - f_0) + \frac{1}{2}\delta(f + f_0).$$ ∎

Example 2.3.14

Determine the Fourier transform of the signal

$$x(t)\cos(2\pi f_0 t).$$

Solution   We have

$$\mathscr{F}[x(t)\cos(2\pi f_0 t)] = \mathscr{F}[\frac{1}{2}x(t)e^{j2\pi f_0 t} + \frac{1}{2}x(t)e^{-j2\pi f_0 t}]$$

$$= \frac{1}{2}X(f - f_0) + \frac{1}{2}X(f + f_0). \tag{2.3.30}$$

Figure 2.38 shows a graph of this relation. In Chapter 3, we will see that this relation is the basis of the operation of amplitude modulation systems.   ∎

Example 2.3.15

Determine the Fourier transform of the signal

$$x(t) = \begin{cases} \cos(\pi t) & |t| \leq \frac{1}{2} \\ 0 & \text{otherwise} \end{cases} \tag{2.3.31}$$

shown in Figure 2.39.

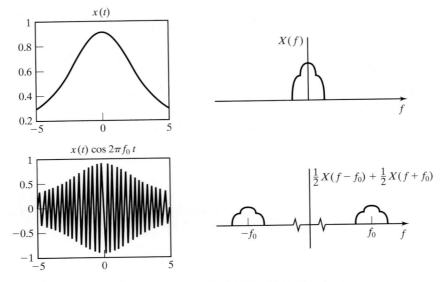

Figure 2.38 Effect of modulation in both the time and frequency domain.

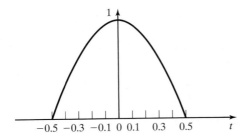

Figure 2.39 Signal $x(t)$.

Solution   Note that $x(t)$ can be expressed as

$$x(t) = \Pi(t) \cos(\pi t).$$ (2.3.32)

Therefore,

$$\mathscr{F}[\Pi(t) \cos(\pi t)] = \frac{1}{2} \text{sinc}\left(f - \frac{1}{2}\right) + \frac{1}{2} \text{sinc}\left(f + \frac{1}{2}\right),$$ (2.3.33)

where we have used the result of Example 2.3.14, with $f_0 = \frac{1}{2}$.   ∎

Parseval's Relation.   If the Fourier transforms of the signals $x(t)$ and $y(t)$ are denoted by $X(f)$ and $Y(f)$ respectively, then

$$\int_{-\infty}^{\infty} x(t)y^*(t)\, dt = \int_{-\infty}^{\infty} X(f)Y^*(f)\, df.$$ (2.3.34)

To prove Parseval's relation, we note that

$$\int_{-\infty}^{\infty} x(t)y^*(t)\,dt = \int_{-\infty}^{\infty} \left[\int_{-\infty}^{\infty} X(f)e^{j2\pi ft}df\right]\left[\int_{-\infty}^{\infty} Y(f')e^{j2\pi f't}df'\right]^* dt$$

$$= \int_{-\infty}^{\infty} \left[\int_{-\infty}^{\infty} X(f)e^{j2\pi ft}df\right]\left[\int_{-\infty}^{\infty} Y^*(f')e^{-j2\pi f't}df'\right] dt$$

$$= \int_{-\infty}^{\infty} X(f)\left[\int_{-\infty}^{\infty} Y^*(f')\left[\int_{-\infty}^{\infty} e^{j2\pi t(f-f')}\,dt\right]df'\right]df.$$

Now using Equation (2.3.5), we have

$$\int_{-\infty}^{\infty} e^{j2\pi t(f-f')}dt = \delta(f - f');$$

therefore,

$$\int_{-\infty}^{\infty} x(t)y^*(t)\,dt = \int_{-\infty}^{\infty} X(f)\left[\int_{-\infty}^{\infty} Y^*(f')\delta(f - f')\,df'\right]df$$

$$= \int_{-\infty}^{\infty} X(f)Y^*(f)\,df, \tag{2.3.35}$$

where we have employed the sifting property of the impulse signal in the last step. Note that if we let $y(t) = x(t)$, we obtain

$$\int_{-\infty}^{\infty} |x(t)|^2 dt = \int_{-\infty}^{\infty} |X(f)|^2 df. \tag{2.3.36}$$

This is known as *Rayleigh's theorem* and is similar to Parseval's relation for periodic signals.

Example 2.3.16

Using Parseval's theorem, determine the values of the integrals

$$\int_{-\infty}^{\infty} \text{sinc}^4(t)\,dt$$

and

$$\int_{-\infty}^{\infty} \text{sinc}^3(t)\,dt.$$

Solution   We have $\mathcal{F}[\text{sinc}^2(t)] = \Lambda(t)$. Therefore, using Rayleigh's theorem, we get

$$\int_{-\infty}^{\infty} \text{sinc}^4(t)\,dt = \int_{-\infty}^{\infty} \Lambda^2(f)\,df$$

$$= \int_{-1}^{0} (f + 1)^2 df + \int_{0}^{1} (-f + 1)^2 df$$

$$= \frac{2}{3}.$$

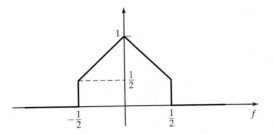

Figure 2.40   Product of Π and Λ.

For the second integral, we note that $\mathscr{F}[\text{sinc}(t)] = \Pi(f)$; therefore, by Parseval's theorem, we have

$$\int_{-\infty}^{\infty} \text{sinc}^2(t)\text{sinc}(t)\,dt = \int_{-\infty}^{\infty} \Pi(f)\Lambda(f)\,df.$$

Figure 2.40 shows the product of $\Pi(f)$ and $\Lambda(f)$. From this figure, we see that

$$\int_{-\infty}^{\infty} \Pi(f)\Lambda(f)\,df = 1 \times \frac{1}{2} + \frac{1}{2} \times \frac{1}{2};$$

therefore,

$$\int_{-\infty}^{\infty} \text{sinc}^3(t)\,dt = \frac{3}{4}. \qquad\blacksquare$$

Autocorrelation.   The *(time) autocorrelation function* of the signal $x(t)$ is denoted by $R_x(\tau)$ and is defined by

$$R_x(\tau) = \int_{-\infty}^{\infty} x(t)x^*(t-\tau)\,dt. \qquad (2.3.37)$$

The autocorrelation theorem states that

$$\mathscr{F}[R_x(\tau)] = |X(f)|^2. \qquad (2.3.38)$$

We note that $R_x(\tau) = x(\tau) \star x^*(-\tau)$. By using the convolution theorem, the auto-correlation theorem follows easily.[6]

From this theorem, we conclude that the Fourier transform of the autocorrelation of a signal is always a real-valued, positive function.

Differentiation.   The Fourier transform of the derivative of a signal can be obtained from the relation

$$\mathscr{F}[\frac{d}{dt}x(t)] = j2\pi f X(f). \qquad (2.3.39)$$

---

[6]The definition of the autocorrelation function for random signals is given in Chapter 5.

To see this, we have

$$\frac{d}{dt}x(t) = \frac{d}{dt}\int_{-\infty}^{\infty} X(f)e^{j2\pi ft}df$$

$$= \int_{-\infty}^{\infty} j2\pi f X(f)e^{j2\pi ft}df. \qquad (2.3.40)$$

Thus, we conclude that

$$\mathscr{F}^{-1}[j2\pi f X(f)] = \frac{d}{dt}x(t)$$

or

$$\mathscr{F}[\frac{d}{dt}x(t)] = j2\pi f X(f).$$

With repeated application of the differentiation theorem, we obtain the relation

$$\mathscr{F}[\frac{d^n}{dt^n}x(t)] = (j2\pi f)^n X(f). \qquad (2.3.41)$$

Example 2.3.17

Determine the Fourier transform of the signal shown in Figure 2.41.

Solution   Obviously, $x(t) = \frac{d}{dt}\Lambda(t)$. Therefore, by applying the differentiation theorem, we have

$$\mathscr{F}[x(t)] = \mathscr{F}[\frac{d}{dt}\Lambda(t)]$$

$$= j2\pi f \mathscr{F}[\Lambda(t)]$$

$$= j2\pi f \operatorname{sinc}^2(f). \qquad (2.3.42)$$

∎

Differentiation in Frequency Domain.   We begin with

$$\mathscr{F}[tx(t)] = \frac{j}{2\pi}\frac{d}{df}X(f). \qquad (2.3.43)$$

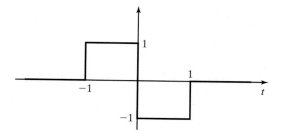

Figure 2.41   Signal $x(t)$.

The proof is basically the same as the differentiation theorem in the time domain and is left as an exercise. Repeated use of this theorem yields

$$\mathscr{F}[t^n x(t)] = \left(\frac{j}{2\pi}\right)^n \frac{d^n}{df^n} X(f). \tag{2.3.44}$$

Example 2.3.18

Determine the Fourier transform of $x(t) = t$.

Solution   We have

$$\mathscr{F}[t] = \mathscr{F}[t \times 1]$$

$$= \frac{j}{2\pi} \frac{d}{df} \mathscr{F}[1]$$

$$= \frac{j}{2\pi} \frac{d}{df} \delta(f)$$

$$= \frac{j}{2\pi} \delta'(f). \tag{2.3.45}$$

∎

Integration.   The Fourier transform of the integral of a signal can be determined from the relation

$$\mathscr{F}[\int_{-\infty}^{t} x(\tau)\, d\tau] = \frac{X(f)}{j2\pi f} + \frac{1}{2} X(0)\delta(f). \tag{2.3.46}$$

To show this, we use the result of Problem 2.15 to obtain

$$\int_{-\infty}^{t} x(\tau)\, d\tau = x(t) \star u_{-1}(t).$$

Now using the convolution theorem and the Fourier transform of $u_{-1}(t)$, we have

$$\mathscr{F}[\int_{-\infty}^{t} x(\tau)\, d\tau] = X(f)\left[\frac{1}{j2\pi f} + \frac{1}{2}\delta(f)\right]$$

$$= \frac{X(f)}{j2\pi f} + \frac{1}{2} X(0)\delta(f). \tag{2.3.47}$$

Example 2.3.19

Determine the Fourier transform of the signal shown in Figure 2.42.

Solution   Note that

$$x(t) = \int_{-\infty}^{t} \Pi(\tau)\, d\tau.$$

Therefore, by using the integration theorem, we obtain

$$\mathscr{F}[x(t)] = \frac{\text{sinc}(f)}{j2\pi f} + \frac{1}{2}\text{sinc}(0)\delta(f)$$

$$= \frac{\text{sinc}(f)}{j2\pi f} + \frac{1}{2}\delta(f). \tag{2.3.48}$$

∎

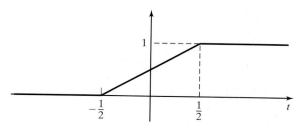

Figure 2.42   Signal $x(t)$.

Moments.   If $\mathcal{F}[x(t)] = X(f)$, then $\int_{-\infty}^{\infty} t^n x(t)\, dt$, the nth moment of $x(t)$, can be obtained from the relation

$$\int_{-\infty}^{\infty} t^n x(t)\, dt = \left(\frac{j}{2\pi}\right)^n \frac{d^n}{df^n} X(f)\Big|_{f=0}. \qquad (2.3.49)$$

This can be shown by using the differentiation in the frequency domain result. We have

$$\mathcal{F}[t^n x(t)] = \left(\frac{j}{2\pi}\right)^n \frac{d^n}{df^n} X(f). \qquad (2.3.50)$$

This means that

$$\int_{-\infty}^{\infty} t^n x(t) e^{-j2\pi ft}\, dt = \left(\frac{j}{2\pi}\right)^n \frac{d^n}{df^n} X(f).$$

Letting $f = 0$ on both sides, we obtain the desired result.

For the special case of $n = 0$, we obtain this simple relation for finding the area under a signal, i.e.,

$$\int_{-\infty}^{\infty} x(t)\, dt = X(0). \qquad (2.3.51)$$

Example 2.3.20

Let $\alpha > 0$ and $x(t) = e^{-\alpha t} u_{-1}(t)$. Determine the nth moment of $x(t)$.

Solution   First we solve for $X(f)$. We have

$$X(f) = \int_{-\infty}^{\infty} e^{-\alpha t} u_{-1}(t) e^{-j2\pi ft}\, dt$$

$$= \int_{0}^{\infty} e^{-t(\alpha + j2\pi ft)}\, dt$$

$$= -\frac{1}{\alpha + j2\pi f}(0 - 1)$$

$$= \frac{1}{\alpha + j2\pi f}. \qquad (2.3.52)$$

By differentiating n times, we obtain

$$\frac{d^n}{df^n}X(f) = \frac{(-j2\pi)^n}{(\alpha + j2\pi f)^{n+1}};$$

(2.3.53)

hence,

$$\int_{-\infty}^{\infty} t^n e^{-\alpha t} u_{-1}(t)\, dt = \left(\frac{j}{2\pi}\right)^n (-j2\pi)^n \cdot \frac{1}{\alpha^{n+1}}$$

$$= \frac{1}{\alpha^{n+1}}.$$

(2.3.54)

■

Example 2.3.21

Determine the Fourier transform of $x(t) = e^{-\alpha|t|}$, where $\alpha > 0$. (See Figure 2.43.)

Solution   We have

$$x(t) = e^{-\alpha t} u_{-1}(t) + e^{\alpha t} u_{-1}(-t) = x_1(t) + x_1(-t),$$

(2.3.55)

and we have already seen that

$$\mathcal{F}[x_1(t)] = \mathcal{F}[e^{-\alpha t} u_{-1}(t)] = \frac{1}{\alpha + j2\pi f}.$$

By using the scaling theorem with $a = -1$, we obtain

$$\mathcal{F}[x_1(-t)] = \mathcal{F}[e^{\alpha t} u_{-1}(-t)] = \frac{1}{\alpha - j2\pi f}.$$

Hence, by the linearity property, we have

$$\mathcal{F}[e^{-\alpha|t|}] = \frac{1}{\alpha + j2\pi f} + \frac{1}{\alpha - j2\pi f}$$

$$= \frac{2\alpha}{\alpha^2 + 4\pi^2 f^2}.$$

(2.3.56)

■

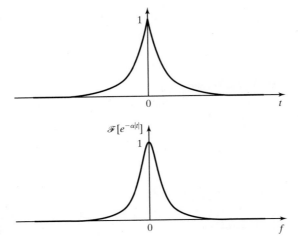

Figure 2.43 Signal $e^{-\alpha|t|}$ and its Fourier transform.

TABLE 2.1   TABLE OF FOURIER-TRANSFORM PAIRS

Time Domain	Frequency Domain
$\delta(t)$	1
1	$\delta(f)$
$\delta(t - t_0)$	$e^{-j2\pi f t_0}$
$e^{j2\pi f_0 t}$	$\delta(f - f_0)$
$\cos(2\pi f_0 t)$	$\frac{1}{2}\delta(f - f_0) + \frac{1}{2}\delta(f + f_0)$
$\sin(2\pi f_0 t)$	$-\frac{1}{2j}\delta(f + f_0) + \frac{1}{2j}\delta(f - f_0)$
$\Pi(t)$	$\text{sinc}(f)$
$\text{sinc}(t)$	$\Pi(f)$
$\Lambda(t)$	$\text{sinc}^2(f)$
$\text{sinc}^2(t)$	$\Lambda(f)$
$e^{-\alpha t}u_{-1}(t), \alpha > 0$	$\frac{1}{\alpha + j2\pi f}$
$te^{-\alpha t}u_{-1}(t), \alpha > 0$	$\frac{1}{(\alpha + j2\pi f)^2}$
$e^{-\alpha|t|}$	$\frac{2\alpha}{\alpha^2 + (2\pi f)^2}$
$e^{-\pi t^2}$	$e^{-\pi f^2}$
$\text{sgn}(t)$	$\frac{1}{j\pi f}$
$u_{-1}(t)$	$\frac{1}{2}\delta(f) + \frac{1}{j2\pi f}$
$\delta'(t)$	$j2\pi f$
$\delta^{(n)}(t)$	$(j2\pi f)^n$
$\frac{1}{t}$	$-j\pi\,\text{sgn}(f)$
$\sum_{n=-\infty}^{n=+\infty} \delta(t - nT_0)$	$\frac{1}{T_0}\sum_{n=-\infty}^{n=+\infty} \delta\left(f - \frac{n}{T_0}\right)$

Table 2.1 provides a collection of frequently used Fourier-transform pairs. Table 2.2 outlines the main properties of the Fourier transform.

2.3.3 Fourier Transform for Periodic Signals

In this section, we extend our results to develop methods for finding the Fourier transform of periodic signals. We have already obtained the Fourier transform for some periodic signals (see Table 2.1). These include $e^{j2\pi f_0 t}$, $\cos(2\pi f_0 t)$, $\sin(2\pi f_0 t)$, and $\sum_{n=-\infty}^{n=+\infty} \delta(t - nT_0)$. The Fourier transform of all these periodic signals have a

TABLE 2.2   TABLE OF FOURIER-TRANSFORM PROPERTIES

Signal	Fourier Transform
$\alpha x_1(t) + \beta x_2(t)$	$\alpha X_1(f) + \beta X_2(f)$
$X(t)$	$x(-f)$
$x(at)$	$\frac{1}{|a|} X\left(\frac{f}{a}\right)$
$x(t - t_0)$	$e^{-j2\pi f t_0} X(f)$
$e^{j2\pi f_0 t} x(t)$	$X(f - f_0)$
$x(t) \star y(t)$	$X(f)Y(f)$
$x(t)y(t)$	$X(f) \star Y(f)$
$\frac{d}{dt} x(t)$	$j2\pi f X(f)$
$\frac{d^n}{dt^n} x(t)$	$(j2\pi f)^n X(f)$
$t x(t)$	$\left(\frac{j}{2\pi}\right) \frac{d}{df} X(f)$
$t^n x(t)$	$\left(\frac{j}{2\pi}\right)^n \frac{d^n}{df^n} X(f)$
$\int_{-\infty}^{t} x(\tau)\,d\tau$	$\frac{X(f)}{j2\pi f} + \frac{1}{2} X(0)\delta(f)$

common property: the Fourier transform consists of impulse functions in the frequency domain. In this section, we will show that this property holds for all periodic signals; in fact, there exists a close relationship between the Fourier transform of a periodic signal and the Fourier-series representation of the signal.

Let $x(t)$ be a periodic signal with the period T_0. Let $\{x_n\}$ denote the Fourier-series coefficients corresponding to this signal. Then

$$x(t) = \sum_{n=-\infty}^{\infty} x_n e^{j2\pi \frac{n}{T_0} t}.$$

By taking the Fourier transform of both sides and using the fact that

$$\mathscr{F}[e^{j2\pi \frac{n}{T_0} t}] = \delta\left(f - \frac{n}{T_0}\right),$$

we obtain

$$X(f) = \sum_{n=-\infty}^{\infty} x_n \delta\left(f - \frac{n}{T_0}\right). \tag{2.3.57}$$

From this relation, we observe that the Fourier transform of a periodic signal $x(t)$ consists of a sequence of impulses in frequency at multiples of the fundamental frequency of the periodic signal. The weights of the impulses are simply the Fourier-series coefficients of the periodic signal. This relation gives a shortcut for computing Fourier-series coefficients of a signal using the properties of the Fourier transform. If we define the truncated signal $x_{T_0}(t)$ as

$$x_{T_0}(t) = \begin{cases} x(t) & -\frac{T_0}{2} < t \leq \frac{T_0}{2} \\ 0 & \text{otherwise} \end{cases}, \tag{2.3.58}$$

we see that

$$x(t) = \sum_{n=-\infty}^{\infty} x_{T_0}(t - nT_0). \tag{2.3.59}$$

Noting that $x_{T_0}(t - nT_0) = x_{T_0}(t) \star \delta(t - nT_0)$, we have

$$x(t) = x_{T_0}(t) \star \sum_{n=-\infty}^{\infty} \delta(t - nT_0). \tag{2.3.60}$$

Therefore, using the convolution theorem and Table 2.1, we obtain

$$X(f) = X_{T_0}(f) \left[\frac{1}{T_0} \sum_{n=-\infty}^{\infty} \delta\left(f - \frac{n}{T_0}\right) \right], \tag{2.3.61}$$

which simplifies to

$$X(f) = \frac{1}{T_0} \sum_{n=-\infty}^{\infty} X_{T_0}\left(\frac{n}{T_0}\right) \delta\left(f - \frac{n}{T_0}\right). \tag{2.3.62}$$

Comparing this result with

$$X(f) = \sum_{n=-\infty}^{\infty} x_n \delta\left(f - \frac{n}{T_0}\right), \tag{2.3.63}$$

we conclude that

$$x_n = \frac{1}{T_0} X_{T_0}\left(\frac{n}{T_0}\right). \tag{2.3.64}$$

This equation gives an alternative way to find the Fourier-series coefficients.

Given the periodic signal $x(t)$, we can find x_n by using the following steps:

1. First, we determine the truncated signal $x_{T_0}(t)$ using Equation (2.3.58).
2. Then, we determine the Fourier transform of the truncated signal using Table 2.1 and the Fourier-transform theorems and properties.
3. Finally, we evaluate the Fourier transform of the truncated signal at $f = \frac{n}{T_0}$ and scale it by $\frac{1}{T_0}$, as shown in Equation (2.3.64).

Example 2.3.22

Determine the Fourier-series coefficients of the signal $x(t)$, as shown in Figure 2.25.

Solution   We follow the preceding steps. The truncated signal is

$$x_{T_0}(t) = \Pi\left(\frac{t}{\tau}\right)$$

and its Fourier transform is

$$X_{T_0}(f) = \tau \operatorname{sinc}(\tau f).$$

Therefore,

$$x_n = \frac{\tau}{T_0} \operatorname{sinc}\left(\frac{n\tau}{T_0}\right). \qquad \blacksquare$$

2.3.4 Transmission over LTI Systems

The convolution theorem is the basis for the analysis of LTI systems in the frequency domain. We have seen that the output of an LTI system is equal to the convolution of the input and the impulse response of the system. If we translate this relationship in the frequency domain using the convolution theorem, then $X(f)$, $H(f)$, and $Y(f)$ are the Fourier transforms of the input, system impulse response, and the output, respectively. Thus,

$$Y(f) = X(f)H(f).$$

Hence, the input–output relation for an LTI system in the frequency domain is much simpler than the corresponding relation in the time domain. In the time domain, we have the convolution integral; however, in the frequency domain we have simple multiplication. This simplicity occurs because the signal's Fourier transformation represents it in terms of the eigenfunctions of LTI systems. Therefore, to find the output of an LTI system for a given input, we must find the Fourier transform of the input and the Fourier transform of the system impulse response. Then, we must multiply them to obtain the Fourier transform of the output. To get the time-domain representation of the output, we find the inverse Fourier transform of the result. In most cases, computing the inverse Fourier transform is not necessary. Usually, the frequency-domain representation of the output provides enough information about the output.

Example 2.3.23

Let the input to an LTI system be the signal

$$x(t) = \operatorname{sinc}(W_1 t),$$

and let the impulse response of the system be

$$h(t) = \operatorname{sinc}(W_2 t).$$

Determine the output signal.

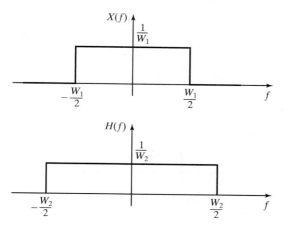

Figure 2.44 Lowpass signal and lowpass filter.

Solution First, we transform the signals to the frequency domain. Thus, we obtain

$$X(f) = \frac{1}{W_1} \Pi \left(\frac{f}{W_1} \right)$$

and

$$H(f) = \frac{1}{W_2} \Pi \left(\frac{f}{W_2} \right).$$

Figure 2.44 shows $X(f)$ and $H(f)$. To obtain the output in the frequency domain, we have

$$Y(f) = X(f)H(f)$$

$$= \frac{1}{W_1 W_2} \Pi \left(\frac{f}{W_1} \right) \Pi \left(\frac{f}{W_2} \right)$$

$$= \begin{cases} \frac{1}{W_1 W_2} \Pi \left(\frac{f}{W_1} \right) & W_1 \leq W_2 \\ \frac{1}{W_1 W_2} \Pi \left(\frac{f}{W_2} \right) & W_1 > W_2 \end{cases}. \tag{2.3.65}$$

From this result, we obtain

$$y(t) = \begin{cases} \frac{1}{W_2} \text{sinc}(W_1 t) & W_1 \leq W_2 \\ \frac{1}{W_1} \text{sinc}(W_2 t) & W_1 > W_2 \end{cases}. \qquad\blacksquare$$

Signals such as $x(t)$ in the preceding example are called *lowpass* signals. Lowpass signals are signals with a frequency domain representation that contains frequencies around the zero frequency and does not contain frequencies beyond some W_1. Similarly, an LTI system that can pass all frequencies less than some W and rejects all frequencies beyond W is called an *ideal lowpass filter*. An ideal lowpass filter will have a frequency response that is 1 for all frequencies $-W \leq f \leq W$ and is 0 outside this interval. W is the *bandwidth* of the filter. Similarly, we can have

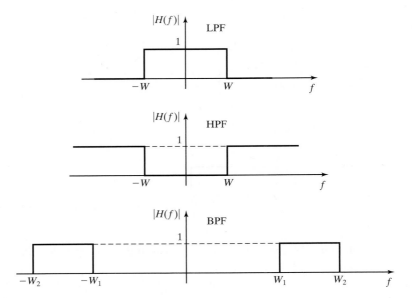

Figure 2.45 Various filter types.

ideal highpass filters. In a highpass filter, $H(f)$, there is unity outside the interval $-W \leq f \leq W$ and zero inside. *Ideal bandpass filters* have a frequency response that is unity in some interval $W_1 \leq |f| \leq W_2$ and zero otherwise. In this case, the bandwidth of the filter is $W_2 - W_1$. Figure 2.45 shows the frequency-response functions of various filter types.

For nonideal lowpass or bandpass filters, the bandwidth is usually defined as the band of frequencies at which the power-transfer ratio of the filter is at least half of the maximum power-transfer ratio. This bandwidth is usually called the 3 dB bandwidth of the filter, because reducing the power by a factor of two is equivalent to decreasing it by 3 dB on the logarithmic scale. Figure 2.46 shows the 3 dB bandwidth of filters.

It is worth emphasizing that the bandwidth of a filter is always the set of *positive frequencies* that a filter can pass.

Example 2.3.24

The magnitude of the transfer function of a filter is given by

$$|H(f)| = \frac{1}{\sqrt{1 + \left(\frac{f}{10000}\right)^2}}.$$

Determine the filter type and its 3 dB bandwidth.

Solution At $f = 0$, we have $|H(f)| = 1$ and $|H(f)|$ decreases and tends to zero as f tends to infinity. Therefore, this is a lowpass filter. Since power is proportional to

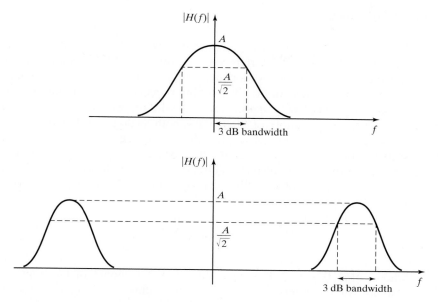

Figure 2.46   3 dB bandwidth of filters in Example 2.3.24.

the square of the amplitude, the equation must be

$$|H(f_0)|^2 = \frac{1}{1 + \left(\frac{f_0}{10000}\right)^2} = \frac{1}{2}.$$

This yields $f_0 = \pm 10,000$. Therefore, this is a lowpass filter with a 3 dB bandwidth of 10 kHz. A plot of $|H(f)|$ is shown in Figure 2.47.   ∎

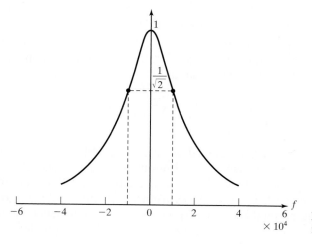

Figure 2.47   3 dB bandwidth of filter in Example 2.3.24.

2.4 FILTER DESIGN

In communication systems, filters are widely used to separate desired signals from undesired signals and interference. Usually, the desired signal characteristics are specified in the frequency domain, in terms of the desired magnitude and phase response of the filter. In the filter design process, we determine the coefficients of a causal filter that closely approximates the desired frequency response specifications.

There are a variety of filter types, both analog and digital. We are particularly interested in the design of digital filters, because they are easily implemented in software on a computer. Digital filters are generally classified as having either a finite duration impulse response (FIR) or an infinite duration impulse response (IIR). An FIR filter is an all-zero filter that is characterized in the z-domain by the system function

$$H(z) = \sum_{k=0}^{M-1} h(k)z^{-k},$$

where $\{h(k), 0 \leq k \leq M - 1\}$ is the impulse response of the filter. The frequency response of the filter is obtained by evaluating $H(z)$ on the unit circle, i.e., by substituting $z = e^{j\omega}$ in $H(z)$ to yield $H(\omega)$. In the discrete-time domain, the FIR filter is characterized by the (difference) equation

$$y(n) = \sum_{k=0}^{M-1} h(k)x(n - k),$$

where $\{x(n)\}$ is the input sequence to the filter and $y(n)$ is the output sequence.

An IIR filter has both poles and zeros, and it is generally characterized in the z-domain by the rational system function

$$H(z) = \frac{\sum_{k=0}^{M-1} b(k)z^{-k}}{1 - \sum_{k=1}^{N} a(k)z^{-k}},$$

where $\{a(k)\}$ and $\{b(k)\}$ are the filter coefficients. The frequency response $H(\omega)$ is obtained by evaluating $H(z)$ on the unit circle. In the discrete-time domain, the IIR filter is characterized by the difference equation

$$y(n) = \sum_{k=1}^{N} a(k)y(n - k) + \sum_{k=0}^{M-1} b(k)x(n - k).$$

In practice, FIR filters are employed in filtering problems where there is a requirement of a linear-phase characteristic within the passband of the filter. If there is no requirement for a linear-phase characteristic, either an IIR or an FIR filter may be employed. In general, if some phase distortion is either tolerable or unimportant, an IIR filter is preferable, primarily because the implementation involves fewer coefficients and consequently has a lower computational complexity. In the

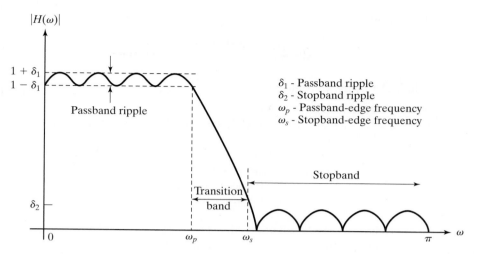

Figure 2.48 Magnitude characteristics of physically realizable filters.

communication systems that are considered in this text, phase distortion is very undesirable; hence, our focus will be on the design of linear-phase FIR filters. A typical frequency-response specification for a causal digital filter is illustrated in Figure 2.48.

The filter is a lowpass filter with the passband-edge frequency $f_p = \omega_p/2\pi$, stopband-edge frequency $f_s = \omega_s/2\pi$, passband ripple δ_1, and stopband ripple δ_2. The transition of the frequency response from passband to stopband defines the *transition band* of the filter. The width of the passband, $0 \le f \le f_p$, is usually called the bandwidth of the filter.

Often, the graph of the frequency response of a filter has a large dynamic range, and to accommodate this range, it is common practice to use a logarithmic scale for the magnitude $|H(f)|$. Consequently, the ripple in the passband is $20\log_{10}\frac{1+\delta_1}{1-\delta_1}$, and the ripple in the stopband is $20\log_{10}\delta_2$.

In a filter design problem, we usually specify several filter parameter (1) the maximum tolerable passband ripple, (2) the maximum tolerable stopband ripple, (3) the passband-edge frequency f_p, and (4) the stopband-edge frequency f_s. On the basis of these specifications, we select the filter coefficients that are closest to the desired frequency response specifications.

MATLAB provides several functions for designing digital filters based on various design methods. Our focus will be on the design and implementation of digital FIR filters (lowpass filters, Hilbert-transform filters, and differentiator filters) based on the Remez algorithm, which provides optimum equiripple, linear-phase FIR filters. The Remez algorithm requires that we specify the length of the FIR filter, the critical frequencies f_p and f_s and the ratio δ_2/δ_1. However, it is more natural in filter design to specify f_p, f_s, δ_1, and δ_2 and to determine the filter length that satisfies the specifications. Given the specifications, a simple formula for approximating the filter length M is

$$\hat{M} = \frac{-20 \log_{10}\left(\sqrt{\delta_1 \delta_2}\right) - 13}{14.6 \Delta f} + 1, \tag{2.4.1}$$

where Δf is the transition bandwidth $\Delta f = f_s - f_p$.

Example 2.4.1

Use the "remez" function to design an FIR-lowpass filter that meets the following specifications:

> Passband ripple ≤ 1 dB,
> Stopband attenuation ≥ 40 dB,
> Passband-edge frequency $= 0.2$,
> Stopband-edge frequency $= 0.35$.

1. Estimate the length of the FIR filter using Equation (2.4.1).

2. Plot the magnitude frequency response $20 \log_{10} |H(f)|$.

Solution   From the preceding data, the ripple in the passband is given by $20 \log_{10} \frac{1+\delta_1}{1-\delta_1}$. This is equal to 1 dB, so we have

$$1 = 20 \log_{10} \frac{1 + \delta_1}{1 - \delta_1}.$$

Solving for δ_1, we get

$$\delta_1 = 0.0575.$$

On the other hand, the attenuation in the stopband is measured with respect to the maximum passband gain, i.e., $1 + \delta_1$. Therefore, we have

$$20 \log_{10} \frac{\delta_2}{1 + \delta_1} = -40.$$

Substituting $\delta_1 = 0.0575$ and solving for δ_2, we obtain

$$\delta_2 = 0.0106.$$

We can also determine the value of Δf to be

$$\Delta f = f_s - f_p = 0.35 - 0.2 = 0.15.$$

Now that we know the values of δ_1, δ_2, and Δf, we can use Equation (2.4.1) to find the length of the filter $\hat{M}$:

$$\hat{M} = \frac{-20 \log_{10}\left(\sqrt{\delta_1 \delta_2}\right) - 13}{14.6 \Delta f} + 1 = 9.7491.$$

Therefore, we design a filter of length 10.

     The MATLAB command used here is the `remez` command, which has a syntax of the form

```
h=remez(M,f,m,w)
```

where M is the order of the filter (note that the order of the filter is related to the length of the filter by $\hat{M}$ through $M = \hat{M} - 1$) and f is a vector of frequencies (the first component is equal to 0 and the last component is equal to 1). The components of f are in increasing order and specify the important frequencies, such as the end of the passband and the beginning of the stopband. Vector m is a vector whose size is equal to the size of vector f and contains the desired magnitude response of the filter corresponding to the values of f. Vector w is a weighting vector; in the passband, it is equal to $\frac{\delta_2}{\delta_1}$, and in the stopband, it is equal to 1. In our example, we have

$$f = [0 \quad 0.1 \quad 0.15 \quad 1]$$

$$m = [1 \quad 1 \quad 0 \quad 0]$$

$$w = \left[\frac{0.0106}{0575} \quad 1\right] = [0.1839 \quad 1].$$

Using these results and calling

```
>> [h]=remez(M,f,m,w)
```

at the MATLAB prompt returns the impulse response of the desired filter. Once we have the impulse response, we can use the MATLAB `freqz` command to determine the frequency response of the filter. MATLAB's complete set of commands for designing this filter are as follows:

```
>> fp=0.2;
>> fs=0.35;
>> df=fs-fp;
>> Rp=1;
>> As=40;
>> delta1=(10^(Rp/20)-1)/(10^(Rp/20)+1);
>> delta2=(1+delta1)*(10^(-As/20));
>> Mhat=ceil((-20*log10(sqrt(delta1*delta2))-13)/
        (14.6*df)+1);
>> f=[0 fp fs 1];
>> m=[1 1 0 0];
>> w=[delta2/delta1 1];
>> h=remez(Mhat+20,f,m,w);
>> [H,W]=freqz(h,[1],3000);
>> plot(W/pi,20*log10(abs(H)))
```

The transfer function of the resulting filter, in decibels, is plotted in Figure 2.49.   ∎

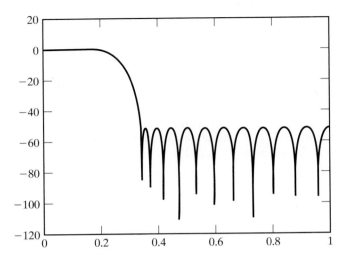

Figure 2.49 Frequency response of the designed filter.

2.5 POWER AND ENERGY

The concepts of power and energy, as well as power-type and energy-type signals, were defined in Section 2.1.2. In this section, we expand these concepts both in the time and frequency domains.

The energy and power of a signal represent the energy or power delivered by the signal when it is interpreted as a voltage or current source feeding a 1-ohm resistor. The energy content of a (generally complex-valued) signal $x(t)$ is defined as

$$\mathcal{E}_x = \int_{-\infty}^{\infty} |x(t)|^2 dt,$$

and the power content of a signal is

$$P_x = \lim_{T \to \infty} \frac{1}{T} \int_{-\frac{T}{2}}^{+\frac{T}{2}} |x(t)|^2 dt.$$

A signal is energy-type if $\mathcal{E}_x < \infty$, and it is power-type if $0 < P_x < \infty$. A signal *cannot* be both power- and energy-type because $P_x = 0$ for energy-type signals, and $\mathcal{E}_x = \infty$ for power-type signals. A signal can be neither energy-type nor power-type. An example of such a signal is given in Problem 2.9. However, most of the signals of interest are either energy-type or power-type. Practically all periodic signals are power-type and have power

$$P_x = \frac{1}{T_0} \int_{\alpha}^{\alpha+T_0} |x(t)|^2 dt,$$

where T_0 is the period and α is any arbitrary real number.

2.5.1 Energy-Type Signals

For an energy-type signal $x(t)$, we define the autocorrelation function

$$R_x(\tau) = x(\tau) \star x^*(-\tau)$$

$$= \int_{-\infty}^{\infty} x(t)x^*(t-\tau)\,dt$$

$$= \int_{-\infty}^{\infty} x(t+\tau)x^*(t)\,dt. \tag{2.5.1}$$

By setting $\tau = 0$ in the definition of the autocorrelation function of $x(t)$, we obtain its energy content, i.e.,

$$\mathscr{E}_x = \int_{-\infty}^{\infty} |x(t)|^2 dt$$

$$= R_x(0). \tag{2.5.2}$$

Using the autocorrelation property of the Fourier transform (see Section 2.3.2), we derive the Fourier transform of $R_x(\tau)$ to be $|X(f)|^2$. Using this result, or employing Rayleigh's theorem, we have

$$\mathscr{E}_x = \int_{-\infty}^{\infty} |x(t)|^2 dt$$

$$= \int_{-\infty}^{\infty} |X(f)|^2 df. \tag{2.5.3}$$

If we pass the signal $x(t)$ through a filter with the (generally complex) impulse response $h(t)$ and frequency response $H(f)$, the output will be $y(t) = x(t) \star h(t)$ and in the frequency domain $Y(f) = X(f)H(f)$. To find the energy content of the output signal $y(t)$, we have

$$\mathscr{E}_y = \int_{-\infty}^{\infty} |y(t)|^2 dt$$

$$= \int_{-\infty}^{\infty} |Y(f)|^2 df$$

$$= \int_{-\infty}^{\infty} |X(f)|^2 |H(f)|^2 df$$

$$= R_y(0), \tag{2.5.4}$$

where $R_y(\tau) = y(\tau) \star y^*(-\tau)$ is the autocorrelation function of the output. The inverse Fourier transform of $|Y(f)|^2$ is

$$R_y(\tau) = \mathscr{F}^{-1}\left[|Y(f)|^2\right]$$

$$= \mathscr{F}^{-1}\left[|X(f)|^2 |H(f)|^2\right]$$

$$\overset{a}{=} \mathscr{F}^{-1}\left[|X(f)|^2\right] \star \mathscr{F}^{-1}\left[|H(f)|^2\right]$$

$$\overset{b}{=} R_x(\tau) \star R_h(\tau), \tag{2.5.5}$$

where (a) follows from the convolution property and (b) follows from the autocorrelation property. Now let us assume that

$$H(f) = \begin{cases} 1 & W < f < W + \Delta W \\ 0 & \text{otherwise} \end{cases}.$$

Then

$$|Y(f)|^2 = \begin{cases} |X(f)|^2 & W < f < W + \Delta W \\ 0 & \text{otherwise} \end{cases}$$

and

$$\mathscr{E}_y = \int_{-\infty}^{\infty} |Y(f)|^2 df$$

$$\approx |X(W)|^2 \Delta W. \tag{2.5.6}$$

This filter passes the frequency components in a small interval around $f = W$, and rejects all the other components. Therefore, the output energy represents the amount energy located in the vicinity of $f = W$ in the input signal. This means that $|X(W)|^2 \Delta W$ is the amount of energy in $x(t)$, which is located in the bandwidth $[W, W + \Delta W]$. Thus,

$$|X(W)|^2 = \frac{\text{Energy in } [W, W + \Delta W] \text{ bandwidth}}{\Delta W}.$$

This shows why $|X(f)|^2$ is called the *energy spectral density* of a signal $x(t)$, and why it represents the amount of energy per unit bandwidth present in the signal at various frequencies. Hence, we define the energy spectral density (or energy spectrum) of the signal $x(t)$ as

$$\mathscr{G}_x(f) = |X(f)|^2$$

$$= \mathscr{F}[R_x(\tau)]. \tag{2.5.7}$$

To summarize,

1. For any energy-type signal $x(t)$, we define the autocorrelation function $R_x(\tau) = x(\tau) \star x^*(-\tau)$.
2. The energy spectral density of $x(t)$, denoted by $\mathscr{G}_x(f)$, is the Fourier transform of $R_x(\tau)$. It is equal to $|X(f)|^2$.

3. The energy content of $x(t)$, $\mathcal{E}_x$, is the value of the autocorrelation function evaluated at $\tau = 0$ or, equivalently, the integral of the energy spectral density over all frequencies, i.e.,

$$\mathcal{E}_x = R_x(0)$$
$$= \int_{-\infty}^{\infty} \mathcal{G}_x(f)\, df. \tag{2.5.8}$$

4. If $x(t)$ is passed through a filter with the impulse response $h(t)$ and the output is denoted by $y(t)$, we have

$$y(t) = x(t) \star h(t)$$
$$R_y(\tau) = R_x(\tau) \star R_h(\tau)$$
$$\mathcal{G}_y(f) = \mathcal{G}_x(f)\mathcal{G}_h(f) = |X(f)|^2 |H(f)|^2.$$

Example 2.5.1

Determine the autocorrelation function, energy spectral density, and energy content of the signal $x(t) = e^{-\alpha t} u_{-1}(t)$, $\alpha > 0$.

Solution First we find the Fourier transform of $x(t)$. From Table 2.1, we have

$$X(f) = \frac{1}{\alpha + j2\pi f}.$$

Hence,

$$\mathcal{G}_x(f) = |X(f)|^2 = \frac{1}{\alpha^2 + (2\pi f)^2}$$

and

$$R_x(\tau) = \mathcal{F}^{-1}\left[|X(f)|^2\right] = \frac{1}{2\alpha} e^{-\alpha|\tau|}.$$

To find the energy content, we can simply find the value of the autocorrelation function at zero:

$$\mathcal{E}_x = R_x(0) = \frac{1}{2\alpha}. \qquad\blacksquare$$

Example 2.5.2

If the signal in the preceding example is passed through a filter with impulse response $h(t) = e^{-\beta t} u_{-1}(t)$, $\beta > 0$, $\beta \neq \alpha$, determine the autocorrelation function, the power spectral density, and the energy content of the signal at the output.

Solution The frequency response of the filter is

$$H(f) = \frac{1}{\beta + j2\pi f}.$$

Therefore,

$$|Y(f)|^2 = |X(f)|^2 |H(f)|^2$$

$$= \frac{1}{(\alpha^2 + 4\pi^2 f^2)(\beta^2 + 4\pi^2 f^2)}$$

$$= \frac{1}{\beta^2 - \alpha^2}\left[\frac{1}{\alpha^2 + 4\pi^2 f^2} - \frac{1}{\beta^2 + 4\pi^2 f^2}\right].$$

Note that in the last step we used partial fraction expansion. From this result, and using Table 2.1, we obtain

$$R_y(\tau) = \frac{1}{\beta^2 - \alpha^2}\left[\frac{1}{2\alpha}e^{-\alpha|\tau|} - \frac{1}{2\beta}e^{-\beta|\tau|}\right]$$

and

$$\mathcal{E}_y = R_y(0)$$

$$= \frac{1}{2\alpha\beta(\alpha + \beta)}. \qquad \blacksquare$$

2.5.2 Power-Type Signals

For the class of power-type signals, a similar development is possible. In this case, we define the *time-average autocorrelation function* of the power-type signal $x(t)$ as

$$R_x(\tau) = \lim_{T \to \infty} \frac{1}{T} \int_{-\frac{T}{2}}^{+\frac{T}{2}} x(t)x^*(t - \tau)\, dt. \qquad (2.5.9)$$

Obviously, the power content of the signal can be obtained from

$$P_x = \lim_{T \to \infty} \frac{1}{T} \int_{-\frac{T}{2}}^{+\frac{T}{2}} |x(t)|^2 dt$$

$$= R_x(0). \qquad (2.5.10)$$

We define $\mathcal{S}_x(f)$, the *power-spectral density* or the *power spectrum* of the signal $x(t)$, to be the Fourier transform of the time-average autocorrelation function:

$$\mathcal{S}_x(f) = \mathcal{F}[R_x(\tau)]. \qquad (2.5.11)$$

Subsequently, we will justify this definition. Now we can express the power content of the signal $x(t)$ in terms of $\mathcal{S}_x(f)$ by noting that $R_x(0) = \int_{-\infty}^{\infty} \mathcal{S}_x(f)e^{j2\pi f\tau}df|_{\tau=0} = \int_{-\infty}^{\infty} \mathcal{S}_x(f)\, df$, i.e.,

$$P_x = R_x(0)$$

$$= \int_{-\infty}^{\infty} \mathcal{S}_x(f)\, df. \qquad (2.5.12)$$

If a power-type signal $x(t)$ is passed through a filter with impulse response $h(t)$, the output is

$$y(t) = \int_{-\infty}^{\infty} x(\tau)h(t-\tau)\,d\tau$$

and the time-average autocorrelation function for the output signal is

$$R_y(\tau) = \lim_{T\to\infty} \frac{1}{T} \int_{-\frac{T}{2}}^{+\frac{T}{2}} y(t)y^*(t-\tau)\,dt.$$

Substituting for $y(t)$, we obtain

$$R_y(\tau) = \lim_{T\to\infty} \frac{1}{T} \int_{-\frac{T}{2}}^{+\frac{T}{2}} \left[\int_{-\infty}^{\infty} h(u)x(t-u)\,du\right]\left[\int_{-\infty}^{\infty} h^*(v)x^*(t-\tau-v)\,dv\right]dt.$$

By making a change of variables $w = t - u$ and changing the order of integration, we obtain

$$R_y(\tau) = \int_{-\infty}^{\infty}\int_{-\infty}^{\infty} h(u)h^*(v)$$

$$\times \lim_{T\to\infty} \frac{1}{T} \int_{-\frac{T}{2}-u}^{\frac{T}{2}+u} \left[x(w)x^*(u+w-\tau-v)\,dw\right]du\,dv$$

$$\stackrel{a}{=} \int_{-\infty}^{\infty}\int_{-\infty}^{\infty} R_x(\tau+v-u)h(u)h^*(v)\,du\,dv$$

$$\stackrel{b}{=} \int_{-\infty}^{\infty} [R_x(\tau+v)\star h(\tau+v)]\,h^*(v)\,dv$$

$$\stackrel{c}{=} R_x(\tau)\star h(\tau)\star h^*(-\tau), \tag{2.5.13}$$

where (a) uses the definition of R_x, given in Equation (2.5.9), and (b) and (c) use the definition of the convolution integral. Taking the Fourier transform of both sides of this equation, we obtain

$$\mathcal{S}_y(f) = \mathcal{S}_x(f)H(f)H^*(f)$$

$$= \mathcal{S}_x(f)|H(f)|^2. \tag{2.5.14}$$

This relation between the input–output power-spectral densities is the same as the relation between the energy-spectral densities at the input and the output of a filter. Now we can use the same arguments used for the case of energy-spectral density; that is, we can employ an ideal filter with a very small bandwidth, pass the signal through the filter, and interpret the power at the output of this filter as the power content of the input signal in the passband of the filter. Thus, we conclude that $\mathcal{S}_x(f)$, as just defined, represents the amount of power at various frequencies. This justifies the definition of the power-spectral density as the Fourier transform of the time-average autocorrelation function.

We have already seen that periodic signals are power-type signals. For periodic signals, the time-average autocorrelation function and the power-spectral density simplify considerably. Let us assume that the signal $x(t)$ is a periodic signal with the period T_0 and has the Fourier-series coefficients $\{x_n\}$. To find the time-average autocorrelation function, we have

$$
\begin{aligned}
R_x(\tau) &= \lim_{T \to \infty} \frac{1}{T} \int_{-\frac{T}{2}}^{+\frac{T}{2}} x(t)x^*(t - \tau)\, dt \\[2mm]
&= \lim_{k \to \infty} \frac{1}{kT_0} \int_{-\frac{kT_0}{2}}^{+\frac{kT_0}{2}} x(t)x^*(t - \tau)\, dt \\[2mm]
&= \lim_{k \to \infty} \frac{k}{kT_0} \int_{-\frac{T_0}{2}}^{+\frac{T_0}{2}} x(t)x^*(t - \tau)\, dt \\[2mm]
&= \frac{1}{T_0} \int_{-\frac{T_0}{2}}^{+\frac{T_0}{2}} x(t)x^*(t - \tau)\, dt.
\end{aligned}
\tag{2.5.15}
$$

This relation gives the time-average autocorrelation function for a periodic signal. If we substitute the Fourier-series expansion of the periodic signal in this relation, we obtain

$$
R_x(\tau) = \frac{1}{T_0} \int_{-\frac{T_0}{2}}^{+\frac{T_0}{2}} \sum_{n=-\infty}^{\infty} \sum_{m=-\infty}^{+\infty} x_n x_m^* e^{j2\pi \frac{m}{T_0}\tau} e^{j2\pi \frac{n-m}{T_0}t}\, dt.
\tag{2.5.16}
$$

Now, using the fact that

$$
\frac{1}{T_0} \int_{-\frac{T_0}{2}}^{+\frac{T_0}{2}} e^{j2\pi \frac{n-m}{T_0}t}\, dt = \delta_{mn},
$$

we obtain

$$
R_x(\tau) = \sum_{n=-\infty}^{\infty} |x_n|^2 e^{j2\pi \frac{n}{T_0}\tau}.
\tag{2.5.17}
$$

From this relation, we see that the time-average autocorrelation function of a periodic signal is itself periodic; it has with the same period as the original signal, and its Fourier-series coefficients are magnitude squares of the Fourier-series coefficients of the original signal.

To determine the power-spectral density of a periodic signal, we can simply find the Fourier transform of $R_x(\tau)$. Since we are dealing with a periodic function, the Fourier transform consists of impulses in the frequency domain. We expect this result, because a periodic signal consists of a sum of sinusoidal (or exponential) signals; therefore, the power is concentrated at discrete frequencies (the harmonics). Thus, the power spectral density of a periodic signal is given by

$$S_x(f) = \sum_{n=-\infty}^{\infty} |x_n|^2 \delta \left(f - \frac{n}{T_0} \right). \tag{2.5.18}$$

To find the power content of a periodic signal, we must integrate this relation over the whole frequency spectrum. When we do this, we obtain

$$P_x = \sum_{n=-\infty}^{\infty} |x_n|^2.$$

This is the same relation we obtained in Section 2.2.3. If this periodic signal passes through an LTI system with the frequency response $H(f)$, the output will be periodic and the power spectral density of the output can be obtained by employing the relation between the power spectral densities of the input and the output of a filter. Thus,

$$S_y(f) = |H(f)|^2 \sum_{n=-\infty}^{\infty} |x_n|^2 \delta \left(f - \frac{n}{T_0} \right)$$

$$= \sum_{n=-\infty}^{\infty} |x_n|^2 \left| H \left(\frac{n}{T_0} \right) \right|^2 \delta \left(f - \frac{n}{T_0} \right) \tag{2.5.19}$$

and the power content of the output signal is

$$P_y = \sum_{n=-\infty}^{\infty} |x_n|^2 \left| H \left(\frac{n}{T_0} \right) \right|^2.$$

2.6 HILBERT TRANSFORM AND ITS PROPERTIES

In this section, we will introduce the Hilbert transform of a signal and explore some of its properties. The Hilbert transform is unlike many other transforms because it does not involve a change of domain. In contrast, Fourier, Laplace, and z-transforms start from the time-domain representation of a signal and introduce the transform as an equivalent frequency-domain (or more precisely, transform-domain) representation of the signal. The resulting two signals are equivalent representations of the *same signal* in terms of *two different arguments*, time and frequency. Strictly speaking, the Hilbert transform is not a transform in this sense. First, the result of a Hilbert transform is not equivalent to the original signal, rather it is a completely different signal. Second, the Hilbert transform does not involve a domain change, i.e., the Hilbert transform of a signal $x(t)$ is *another* signal denoted by $\hat{x}(t)$ in the same domain (i.e., time domain with *the same* argument t).

The Hilbert transform of a signal $x(t)$ is a signal $\hat{x}(t)$ whose frequency components lag the frequency components of $x(t)$ by $90°$. In other words, $\hat{x}(t)$ has exactly the same frequency components present in $x(t)$ with the same amplitude—except there is a 90° phase delay. For instance, the Hilbert transform of $x(t) = A\cos(2\pi f_0 t + \theta)$ is $A\cos(2\pi f_0 t + \theta - 90°) = A\sin(2\pi f_0 t + \theta)$.

A delay of $\frac{\pi}{2}$ at all frequencies means that $e^{j2\pi f_0 t}$ will become $e^{j2\pi f_0 t - \frac{\pi}{2}} = -je^{j2\pi f_0 t}$ and $e^{-j2\pi f_0 t}$ will become $e^{-j(2\pi f_0 t - \frac{\pi}{2})} = je^{-j2\pi f_0 t}$. In other words, at positive frequencies, the spectrum of the signal is multiplied by $-j$; at negative frequencies, it is multiplied by $+j$. This is equivalent to saying that the spectrum (Fourier transform) of the signal is multiplied by $-j\,\mathrm{sgn}(f)$. In this section, we assume that $x(t)$ is real and has no DC component, i.e., $X(f)|_{f=0} = 0$.

Therefore,

$$\mathscr{F}[\hat{x}(t)] = -j\,\mathrm{sgn}(f)X(f). \tag{2.6.1}$$

Using Table 2.1, we have

$$\mathscr{F}[-j\,\mathrm{sgn}(f)] = \frac{1}{\pi t}.$$

Hence,

$$\hat{x}(t) = \frac{1}{\pi t} \star x(t) = \frac{1}{\pi}\int_{-\infty}^{\infty} \frac{x(\tau)}{t - \tau}\,d\tau. \tag{2.6.2}$$

Thus, the operation of the Hilbert transform is equivalent to a convolution, i.e., filtering.

Example 2.6.1

Determine the Hilbert transform of the signal $x(t) = 2\mathrm{sinc}\,(2t)$.

Solution   We use the frequency-domain approach to solve this problem. Using the scaling property of the Fourier transform, we have

$$\mathscr{F}[x(t)] = 2 \times \frac{1}{2}\Pi\left(\frac{f}{2}\right) = \Pi\left(\frac{f}{2}\right) = \Pi\left(f + \frac{1}{2}\right) + \Pi\left(f - \frac{1}{2}\right).$$

In this expression, the first term contains all the negative frequencies and the second term contains all the positive frequencies.

To obtain the frequency-domain representation of the Hilbert transform of $x(t)$, we use the relation $\mathscr{F}[\hat{x}(t)] = -j\,\mathrm{sgn}(f)\mathscr{F}[x(t)]$, which results in

$$\mathscr{F}[\hat{x}(t)] = j\Pi\left(f + \frac{1}{2}\right) - j\Pi\left(f - \frac{1}{2}\right).$$

Taking the inverse Fourier transform, we have

$$\hat{x}(t) = je^{-j\pi t}\mathrm{sinc}(t) - je^{j\pi t}\mathrm{sinc}(t)$$

$$= -j\left(e^{j\pi t} - e^{-j\pi t}\right)\mathrm{sinc}(t)$$

$$= -j \times 2j\sin(\pi t)\mathrm{sinc}(t)$$

$$= 2\sin(\pi t)\,\mathrm{sinc}(t). \qquad\blacksquare$$

Obviously performing the Hilbert transform on a signal is equivalent to a $90°$ phase shift in all its frequency components. Therefore, the only change that the Hilbert transform performs on a signal is changing its phase. Most important, the amplitude of the frequency components of the signal, and, therefore the energy and power of the signal, do not change by performing the Hilbert-transform operation. On the other hand, since performing the Hilbert transform changes cosines into sines, it is not surprising that the Hilbert transform $\hat{x}(t)$ of a signal $x(t)$ is orthogonal to $x(t)$.[7] Also, since the Hilbert transform introduces a $90°$ phase shift, carrying it out twice causes a $180°$ phase shift, which can cause a sign reversal of the original signal. These are some of the most important properties of the Hilbert transform. In all the following results, we assume that $X(f)$ does not have any impulses at zero frequency.

Evenness and Oddness.   The Hilbert transform of an even signal is odd, and the Hilbert transform of an odd signal is even.

Proof. If $x(t)$ is even, then $X(f)$ is a real and even function; therefore, $-j\,\mathrm{sgn}(f)X(f)$ is an imaginary and odd function. Hence, its inverse Fourier transform $\hat{x}(t)$ will be odd. If $x(t)$ is odd, then $X(f)$ is imaginary and odd; thus $-j\,\mathrm{sgn}(f)X(f)$ is real and even. Therefore, $\hat{x}(t)$ is even.

Sign Reversal.   Applying the Hilbert-transform operation to a signal twice causes a sign reversal of the signal, i.e.,

$$\hat{\hat{x}}(t) = -x(t).   \tag{2.6.3}$$

Proof. Since

$$\mathcal{F}[\hat{\hat{x}}(t)] = [-j\,\mathrm{sgn}(f)]^2 X(f),   \tag{2.6.4}$$

it follows that

$$\mathcal{F}[\hat{\hat{x}}(t)] = -X(f),   \tag{2.6.5}$$

where $X(f)$ does not contain any impulses at the origin.

Energy.   The energy content of a signal is equal to the energy content of its Hilbert transform.

Proof. Using Rayleigh's theorem of the Fourier transform, we have

$$\mathcal{E}_x = \int_{-\infty}^{\infty} |x(t)|^2 dt = \int_{-\infty}^{\infty} |X(f)|^2 df   \tag{2.6.6}$$

and

$$\mathcal{E}_{\hat{x}} = \int_{-\infty}^{\infty} |\hat{x}(t)|^2 dt = \int_{-\infty}^{\infty} |-j\,\mathrm{sgn}(f)X(f)|^2 df.   \tag{2.6.7}$$

Using the fact that $|-j\,\mathrm{sgn}(f)|^2 = 1$ except for $f = 0$, and the fact that $X(f)$ does not contain any impulses at the origin completes the proof.

[7] $x(t)$ and $y(t)$ are orthogonal if $\int_{-\infty}^{+\infty} x(t)y^*(t)\, dt = 0$.

Orthogonality. The signal $x(t)$ and its Hilbert transform are orthogonal.

Proof. Using Parseval's theorem of the Fourier transform, we obtain

$$\int_{-\infty}^{\infty} x(t)\hat{x}(t)\,dt = \int_{-\infty}^{\infty} X(f)[-j\,\mathrm{sgn}(f)X(f)]^*df$$

$$= -j\int_{-\infty}^{0} |X(f)|^2 df + j\int_{0}^{+\infty} |X(f)|^2 df$$

$$= 0, \tag{2.6.8}$$

where, in the last step, we have used the fact that $X(f)$ is Hermitian; therefore, $|X(f)|^2$ is even.

2.7 LOWPASS AND BANDPASS SIGNALS

A *lowpass signal* is a signal in which the spectrum (frequency content) of the signal is located around the zero frequency. A bandpass signal is a signal with a spectrum far from the zero frequency. The frequency spectrum of a bandpass signal is usually located around a frequency f_c, which is much higher than the bandwidth of the signal (recall that the bandwidth of a signal is the set of the range of all positive frequencies present in the signal). Therefore, the bandwidth of the bandpass signal is usually much less than the frequency f_c, which is close to the location of the frequency content.

The extreme case of a bandpass signal is a single frequency signal whose frequency is equal to f_c. The bandwidth of this signal is zero, and generally, it can be written as

$$x(t) = A\cos(2\pi f_c t + \theta).$$

This is a sinusoidal signal that can be represented by a phasor

$$x_l = Ae^{j\theta}.$$

Note that A is assumed to be positive and the range of θ is $-\pi$ to π, as shown in Figure 2.50.

The phasor has a magnitude of A and a phase of θ. If this phasor rotates counterclockwise with an angular velocity of $\omega_c = 2\pi f_c$ (equivalent to multiplying

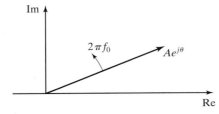

Figure 2.50 Phasor corresponding to a sinusoid signal.

it by $e^{j2\pi f_c t}$), the result would be $Ae^{j2\pi f_c t}$. Its projection on the real axis (its real part) is $x(t) = A\cos(2\pi f_c t)$.

We can expand the signal $x(t)$ as

$$x(t) = A\cos(2\pi f_c t + \theta) \qquad (2.7.1)$$

$$= A\cos(\theta)\cos(2\pi f_c t) - A\sin(\theta)\sin(2\pi f_c t) \qquad (2.7.2)$$

$$= x_c\cos(2\pi f_c t) - x_s\sin(2\pi f_c t). \qquad (2.7.3)$$

From our previous discussion, we can see that this single-frequency signal has two components. The first component is $x_c = A\cos(\theta)$, which is in the direction of $\cos(2\pi f_c t)$. This is called the *in-phase component*. The other component is $x_s = A\sin(\theta)$, which is in the direction of $-\sin(2\pi f_c t)$. This is called the *quadrature component*. Note that we can also write

$$x_l = Ae^{j\theta} = x_c + jx_s. \qquad (2.7.4)$$

Now assume that instead of the phasor shown in Figure 2.50, we have a phasor with slowly varying magnitude and phase. This is represented by

$$x_l(t) = A(t)e^{j\theta(t)}, \qquad (2.7.5)$$

where $A(t)$ and $\theta(t)$ are slowly varying signals (compared to f_c). In this case, similar to Equation (2.7.3), we have

$$x(t) = \text{Re}[A(t)e^{j(2\pi f_c t + \theta(t))}] \qquad (2.7.6)$$

$$= A(t)\cos(\theta(t))\cos(2\pi f_c t) - A\sin(\theta(t))\sin(2\pi f_c t) \qquad (2.7.7)$$

$$= x_c(t)\cos(2\pi f_c t) - x_s(t)\sin(2\pi f_c t). \qquad (2.7.8)$$

Unlike the single frequency signal previously studied, this signal contains a range of frequencies; therefore, its bandwidth is not zero. However, since the amplitude (also called the *envelope*) and the phase are slowly varying, this signal's frequency components constitute a small bandwidth around f_c. The spectra of three bandpass signals are shown in Figure 2.51.

In this case, the in-phase and quadrature components are

$$x_c(t) = A(t)\cos(\theta(t)) \qquad (2.7.9)$$

$$x_s(t) = A(t)\sin(\theta(t)) \qquad (2.7.10)$$

and we have

$$x(t) = x_c(t)\cos(2\pi f_c t) - x_s(t)\sin(2\pi f_c t). \qquad (2.7.11)$$

Note that both the in-phase and quadrature components of a bandpass signal are slowly varying signals; therefore, they are both lowpass signals.

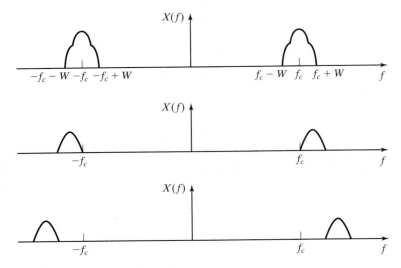

Figure 2.51   Spectra of three bandpass signals.

Equation (2.7.11) is a very useful relation; it basically says that a bandpass signal can be represented in terms of two lowpass signals, namely, its in-phase and quadrature components.

In this case, the complex lowpass signal

$$x_l(t) = x_c(t) + jx_s(t) \tag{2.7.12}$$

is called the *lowpass equivalent* of the bandpass signal $x(t)$. If we represent $x_l(t)$ in polar coordinates, we have

$$x_l(t) = \sqrt{x_c^2(t) + x_s^2(t)}\, e^{j\, \arctan \frac{x_s(t)}{x_c(t)}}.$$

Now if we define the *envelope* and the *phase* of the bandpass signal as

$$\begin{cases} |x_l(t)| = A(t) = \sqrt{x_c^2(t) + x_s^2(t)} \\ \angle x_l(t) = \theta(t) = \arctan \frac{x_s(t)}{x_c(t)}, \end{cases} \tag{2.7.13}$$

we can express $x_l(t)$ as

$$x_l(t) = A(t)e^{j\theta(t)}. \tag{2.7.14}$$

Using Equations (2.7.14) and (2.7.11), we have

$$x(t) = \mathrm{Re}[x_l(t)e^{j2\pi f_c t}] \tag{2.7.15}$$

$$= \mathrm{Re}[A(t)e^{j2\pi f_c t + \theta(t)}] \tag{2.7.16}$$

$$= A(t)\cos(2\pi f_c t + \theta(t)). \tag{2.7.17}$$

Equations (2.7.17) and (2.7.11) represent two methods for expressing a band-pass signal in terms of two lowpass signals. We can express the signal in terms of the in-phase and quadrature components or in terms of the envelope and phase of the bandpass signal.

2.8 FURTHER READING

Numerous references cover the analysis of LTI systems in both the time and frequency domain. Oppenheim, Willsky, and Young (1983) contains wide coverage of time- and frequency-domain analysis of both discrete-time and continuous-time systems. Papoulis (1962) and Bracewell (1965) provide in-depth analysis of the Fourier-series and transform techniques. A more advanced treatment of linear systems based on linear operator theory can be found in Franks (1969). Digital filters are studied in detail by Proakis and Manolakis (1996). The book by Ingle and Proakis (1997) emphasizes design techniques for digital filters using MATLAB.

PROBLEMS

2.1 Plot the following signals:

1. $x_1(t) = \Pi(2t + 5)$

2. $x_2(t) = \Pi(-2t + 8)$

3. $x_3(t) = \sum_{n=0}^{\infty} \Lambda(t - n)$

4. $x_4(t) = 2\sum_{n=-\infty}^{\infty} \Lambda\left(\frac{t-4n}{2}\right) - \sum_{n=-\infty}^{\infty} \Lambda(t - 4n)$

5. $x_5(t) = \text{sgn}(2t) - \text{sgn}(t)$

6. $x_6(t) \sum_{n=-\infty}^{\infty} (-1)^n \Lambda(t - 2n)$

7. $x_7(t) = \text{sinc}(10t)$

8. $x_8(t) = \text{sinc}\frac{t}{10}$

9. $x_9(t) = 4\Pi\left(\frac{t}{6}\right)\cos(4\pi t)$

2.2 Plot the discrete version of the given signals. Assume they are sampled at multiples of T_0, i.e., $x[n] = x(nT_0)$.

1. $x(t) = \text{sinc}(3t)$ and $T_0 = \frac{1}{9}$

2. $x(t) = \Pi\left(\frac{2t-1}{3}\right)$ and $T_0 = \frac{1}{8}$

3. $x(t) = tu_{-1}(t) - (t - 1)u_{-1}(t - 1)$ and $T_0 = \frac{1}{4}$

2.3 Two signals, $x_1(t) = 1$ and $x_2(t) = \cos 2\pi t$, are sampled at $t = 0 \pm 1, \pm 2, \ldots$. The resulting discrete-time signals are denoted by $x_1[n]$ and $x_2[n]$. Verify that $x_1[n] = x_2[n]$. What can you conclude from this observation?

2.4 Show that the sum of two discrete periodic signals is periodic, whereas the sum of two continuous periodic signals is not necessarily periodic. Under what condition is the sum of two continuous periodic signals periodic?

2.5 Determine whether the given signals are periodic. For periodic signals, determine the period.

    1. $\sin(4000\pi t) + \cos(11{,}000\pi t)$

    2. $\sin(4000\pi t) + \cos(11{,}000t)$

    3. $\sin[4000\pi n] + \cos[11{,}000\pi n]$

    4. $\sin[4000\pi n] + \cos[11{,}000n]$

2.6 Classify the signals that follow into even signals, odd signals, and signals that are neither even nor odd. In the latter case, find the even and odd parts of the signals.

    1. $x_1(t) = \begin{cases} e^{-t} & t > 0 \\ -e^t & t < 0 \\ 0 & t = 0 \end{cases}$

    2. $\cos\left(120\pi t + \frac{\pi}{3}\right)$

    3. $x_2(t) = e^{-|t|}$

    4. $x_3(t) = \begin{cases} \frac{t}{|t|} & t \neq 0 \\ 0 & t = 0 \end{cases}$

    5. $x_4(t) = \begin{cases} t & t \geq 0 \\ 0 & t < 0 \end{cases}$

    6. $x_5(t) = \sin t + \cos t$

    7. $x_6(t) = x_1(t) - x_2(t)$, where $x_1(t)$ is even and $x_2(t)$ is odd

2.7 Classify these signals into energy-type signals, power-type signals, and signals that are neither energy type nor power type signals. For energy-type and power-type signals, find the energy or the power content of the signal.

    1. $x_1(t) = \left(e^{-t}\cos t\right) u_{-1}(t)$

    2. $x_2(t) = e^{-t}\cos t$

    3. $x_3(t) = \text{sgn}(t)$

    4. $x_4(t) = A\cos 2\pi f_1 t + B\cos 2\pi f_2 t$

2.8 Classify these signals into periodic and nonperiodic:

    1. $x_1(t) = 2\sum_{n=-\infty}^{\infty} \Lambda\left(\frac{t-4n}{2}\right) - \sum_{n=-\infty}^{\infty} \Lambda(t - 4n)$

   2. $x_2(t) = \sum_{n=-\infty}^{\infty} \Lambda(t - n)$

   3. $x_3(t) = \sin t + \sin 2\pi t$

   4. $x_4[n] = \sin n$

   5. $x_5(t) = \sum_{n=-\infty}^{\infty} y(t - nT)$, where $y(t)$ is an arbitrary signal and T is an arbitrary constant

2.9 Using the definition of power-type and energy-type signals,

   1. Show that $x(t) = A e^{j(2\pi f_0 t + \theta)}$ is a power-type signal and its power content is A^2.

   2. Show that $x(t) = A \cos(2\pi f_0 t + \theta)$ is power-type and its power is $\frac{A^2}{2}$.

   3. Show that the unit-step signal $u_{-1}(t)$ is a power-type signal and find its power content.

   4. Show that the signal

$$x(t) = \begin{cases} K t^{-\frac{1}{4}} & t > 0 \\ 0 & t \le 0 \end{cases}$$

is neither an energy- nor a power-type signal.

2.10 Find the even and odd parts of the signal $x(t) = \Lambda(t) u_{-1}(t)$.

2.11 Using the definition of even and odd signals,

   1. Show that the decomposition of a signal into even and odd parts is unique.

   2. Show that the product of two even or two odd signals is even, whereas the product of an even and an odd signal is odd.

   3. Try to find an example of two signals that are neither even nor odd, but have an even product.

2.12 Plot the following signals:

   1. $x_1(t) = \Pi(t) + \Pi(-t)$

   2. $x_2(t) = \Pi(t) - \Pi(t - 1)$

   3. $x_3(t) = \Lambda(t)\Pi(t)$

   4. $x_4(t) = \sum_{n=-\infty}^{\infty} \Lambda(t - 2n)$

   5. $x_5(t) = \sum_{n=-\infty}^{\infty} (-1)^n \Lambda(t - n)$

   6. $x_6(t) = \text{sgn}(t) + \text{sgn}(1 - t)$

   7. $x_7(t) = 1 + \text{sgn}(t)$

   8. $x_8(t) = \text{sgn}^2(t)$

   9. $x_9(t) = \text{sinc}(t) \, \text{sgn}(t)$

10. $x_{10}(t) = \sum_{n=-\infty}^{\infty} (-1)^n n\delta(t-n)$

11. $x_{11}(t) = \sum_{n=1}^{\infty} \frac{1}{2^n} \Pi(\frac{t}{n})$

2.13 By using the properties of the impulse function, find the values of these expressions:

    1. $x_1(t) = \text{sinc}(t)\,\delta(t)$

    2. $x_1(t) = \text{sinc}(t)\,\delta(t-3)$

    3. $x_1(t) = \text{sinc}(t-2)\,\delta(t)$

    4. $x_2(t) = \Lambda(t) \star \sum_{n=-\infty}^{\infty} \delta(t-2n)$

    5. $x_3(t) = \Lambda(t) \star \delta'(t)$

    6. $x_4(t) = \cos t\,\delta(3t)$

    7. $x_4(t) = \cos\left(2t + \frac{\pi}{3}\right)\delta(3t)$

    8. $x_4(t) = \cos t\,\delta(3t+1)$

    9. $x_5(t) = \delta(5t) \star \delta(4t)$

    10. $x_6(t) = \delta(5t) \star \delta'(3t)$

    11. $x_7(t) = \cos t\,\delta'(t)$

    12. $x_8(t) = \int_{-\infty}^{\infty} \Pi(t)\,\delta(2t-1)\,dt$

    13. $\int_{-\infty}^{\infty} \text{sinc}(t)\,\delta(t)\,dt$

    14. $\int_{-\infty}^{\infty} \text{sinc}(t+1)\,\delta(t)\,dt$

    15. $\int_{-\infty}^{\infty} \left[\sum_{n=1}^{\infty} \frac{1}{2^n} \Pi(\frac{t}{n})\right]\delta(t)\,dt$

    16. $\int_{-\infty}^{\infty} \cos t \left[\sum_{n=1}^{\infty} \delta(2^n t)\right]dt$

2.14 Show that the impulse signal is even. What can you say about evenness or oddness of its nth derivative?

2.15 We have seen that $x(t) \star \delta(t) = x(t)$. Show that

$$x(t) \star \delta^{(n)}(t) = \frac{d^n}{dt^n} x(t)$$

and

$$x(t) \star u_{-1}(t) = \int_{-\infty}^{t} x(\tau)\,d\tau.$$

2.16 Classify these systems into linear and nonlinear:

    1. $y(t) = 2x(t) - 3$

    2. $y(t) = |x(t)|$

3. $y(t) = 0$

4. $y(t) = 2^{x(t)}$

5. $y(t) = \begin{cases} 1 & x(t) > 0 \\ 0 & x(t) \leq 0 \end{cases}$

6. $y(t) = \begin{cases} \frac{x(t)}{|x(t)|} & x(t) \neq 0 \\ 0 & x(t) = 0 \end{cases}$

7. $y(t) = \begin{cases} \frac{x^2(t)}{|x(t)|} & x(t) \neq 0 \\ 0 & \text{otherwise} \end{cases}$

8. $y(t) = e^{-t}x(t)$

9. $y(t) = x(t)u_{-1}(t)$

10. $y(t) = x(t)\delta(t)$

11. $y(t) = x(t)\sum_{n=-\infty}^{\infty} \delta(t - nT)$

12. $y(t) = \begin{cases} \frac{d}{dt}x(t) & t > 0 \\ x(t) & t < 0 \\ 0 & t = 0 \end{cases}$

13. $y(t) = x(t) + y(t - 1)$

14. $y(t) = $ Algebraic sum of jumps in $x(t)$ in the interval $(-\infty, t]$

2.17 Prove that a system is linear if and only if

1. It is *homogeneous*, i.e., for all input signals $x(t)$ and all real numbers α, we have $\mathcal{T}[\alpha x(t)] = \alpha \mathcal{T}[x(t)]$.

2. It is *additive*, i.e., for all input signals $x_1(t)$ and $x_2(t)$, we have

$$\mathcal{T}[x_1(t) + x_2(t)] = \mathcal{T}[x_1(t)] + \mathcal{T}[x_2(t)].$$

In other words, show that the two definitions of linear systems given by Equations (2.1.39) and (2.1.40) are equivalent.

2.18 Verify whether any (or both) of the conditions described in Problem 2.17 are satisfied by the systems given in Problem 2.16.

2.19 Prove that if a system satisfies the additivity property described in Problem 2.17, then it is homogeneous for all *rational* α.

2.20 Show that the system described by

$$y(t) = \begin{cases} \frac{x^2(t)}{x'(t)} & x'(t) \neq 0 \\ 0 & x'(t) = 0 \end{cases}$$

is homogeneous but nonlinear. Can you give another example of such a system?

2.21 Show that the response of a linear system to the input which is identically zero is an output which is identically zero.

2.22 The system defined by the input–output relation

$$y(t) = x(t)\cos(2\pi f_0 t),$$

where f_0 is a constant, is called a *modulator*. Is this system linear? Is it time invariant?

2.23 Are these statements true or false? Why?

    1. A system whose components are nonlinear is necessarily nonlinear.

    2. A system whose components are time variant is necessarily a time-variant system.

    3. The response of a causal system to a causal signal is itself causal.

2.24 Determine whether the following systems are time variant or time invariant:

    1. $y(t) = 2x(t) + 3$

    2. $y(t) = (t + 2)x(t)$

    3. $y(t) = x(t) + t$

    4. $y(t) = x(-t)$

    5. $y(t) = x(t)u_{-1}(t)$

    6. $y(t) = x(t)\delta(t)$

    7. $y(t) = x(t)\sum_{n=-\infty}^{\infty}\delta(t - nT)$

    8. $y(t) = \int_{-\infty}^{t}x(\tau)\,d\tau$

    9. $y(t) = x(t) + y(t - 1)$

    10. $y(t) = \begin{cases} \frac{x(t)}{|x(t)|} & x(t) \neq 0 \\ 0 & x(t) = 0 \end{cases}$

2.25 Prove that if the response of an LTI system to $x(t)$ is $y(t)$, then the response of this system to $\frac{d}{dt}x(t)$ is $\frac{d}{dt}y(t)$.

2.26 Prove that if the response of an LTI system to $x(t)$ is $y(t)$, then the response of this system to $\int_{-\infty}^{t}x(\tau)\,d\tau$ is $\int_{-\infty}^{t}y(\tau)\,d\tau$.

2.27 The response of a linear time-invariant system to the input $x(t) = e^{-\alpha t}u_{-1}(t)$ is $\delta(t)$. Using time-domain analysis and the result of Problem 2.25, determine the impulse response of this system. What is the response of the system to a general input $x(t)$?

2.28 Let a system be defined by

$$y(t) = \frac{1}{2T} \int_{t-T}^{t+T} x(\tau)\, d\tau.$$

Is this system causal?

2.29 For an LTI system to be causal, it is required that $h(t)$ be zero for $t < 0$. Give an example of a nonlinear system which is causal, but its impulse response is nonzero for $t < 0$.

2.30 Determine whether the impulse response of these LTI systems is causal:

    1. $h(t) = \text{sinc}(t)$
    2. $h(t) = \Pi\left(\frac{t-3}{6}\right)$
    3. $h(t) = \text{sinc}(t)u_{-1}(t)$

2.31 Using the convolution integral, show that the response of an LTI system to $u_{-1}(t)$ is given by $\int_{-\infty}^{t} h(\tau)\, d\tau$.

2.32 What is the impulse response of a differentiator? Find the output of this system to an arbitrary input $x(t)$ by finding the convolution of the input and the impulse response. Repeat for the delay system.

2.33 The system defined by

$$y(t) = \int_{t-T}^{t} x(\tau)\, d\tau.$$

(T is a constant) is a finite-time integrator. Is this system LTI? Find the impulse response of this system.

2.34 Compute the following convolution integrals:

    1. $e^{-t}u_{-1}(t) \star e^{-t}u_{-1}(t)$
    2. $e^{-t}u_{-1}(t) \star u_{-1}(t)$
    3. $\Pi(t) \star \Lambda(t)$
    4. $(\Lambda(t)\text{sgn}(t)) \star u_{-1}(t)$
    5. $\Lambda(t) \star \text{sgn}(t)$
    6. $(\Lambda(t)u_{-1}(t)) \star \Pi(t)$

2.35 Show that in a causal LTI system, the convolution integral reduces to

$$y(t) = \int_{0}^{+\infty} x(t-\tau)h(\tau)\, d\tau = \int_{-\infty}^{t} x(\tau)h(t-\tau)\, d\tau.$$

2.36 Show that the set of signals $\psi_n(t) = \sqrt{\frac{1}{T_0}} e^{j2\pi \frac{n}{T_0}t}$ constitutes an orthonormal set of signals on the interval $[\alpha, \alpha + T_0]$, where α is arbitrary.

2.37 In this problem, we present the proof of the Cauchy–Schwartz inequality.

  1. Show that for nonnegative $\{\alpha_i\}_{i=1}^n$ and $\{\beta_i\}_{i=1}^n$,

$$\sum_{i=1}^{n} \alpha_i \beta_i \leq \left[\sum_{i=1}^{n} \alpha_i^2\right]^{\frac{1}{2}} \left[\sum_{i=1}^{n} \beta_i^2\right]^{\frac{1}{2}}.$$

  What are the conditions for equality?

  2. Let $\{x_i\}_{i=1}^n$ and $\{y_i\}_{i=1}^n$ be complex numbers. Show that

$$\left|\sum_{i=1}^{n} x_i y_i^*\right| \leq \sum_{i=1}^{n} |x_i y_i^*| = \sum_{i=1}^{n} |x_i||y_i^*|.$$

  What are the conditions for equality?

  3. From (1) and (2), conclude that

$$\left|\sum_{i=1}^{n} x_i y_i^*\right| \leq \left[\sum_{i=1}^{n} |x_i|^2\right]^{\frac{1}{2}} \left[\sum_{i=1}^{n} |y_i|^2\right]^{\frac{1}{2}}.$$

  What are the conditions for equality?

  4. Generalize the preceding results to integrals and prove the Cauchy–Schwartz inequality

$$\left|\int_{-\infty}^{\infty} x(t) y^*(t)\, dt\right| \leq \left[\int_{-\infty}^{\infty} |x(t)|^2\, dt\right]^{\frac{1}{2}} \left[\int_{-\infty}^{\infty} |y(t)|^2\, dt\right]^{\frac{1}{2}}.$$

  What are the conditions for equality?

2.38 Let $\{\phi_i(t)\}_{i=1}^N$ be an orthogonal set on N signals, i.e.,

$$\int_0^\infty \phi_i(t)\phi_j^*(t)\, dt = \begin{cases} 1 & i \neq j \\ 0 & i = j \end{cases} \qquad 1 \leq i, j \leq N,$$

and let $x(t)$ be an arbitrary signal. Let $\hat{x}(t) = \sum_{i=1}^N \alpha_i \phi_i(t)$ be a linear approximation of $x(t)$ in terms of $\{\phi_i(t)\}_{i=1}^N$. Find α_i's such that

$$\epsilon^2 = \int_{-\infty}^{\infty} |x(t) - \hat{x}(t)|^2\, dt$$

is minimized.

1. Show that the minimizing α_i's satisfy

$$\alpha_i = \int_{-\infty}^{\infty} x(t)\phi_i^*(t)\,dt.$$

2. Show that with the preceding choice of α_i's, we have

$$\epsilon_{\min}^2 = \int_{-\infty}^{\infty} |x(t)|^2\,dt - \sum_{i=1}^{N} |\alpha_i|^2.$$

2.39 Determine the Fourier-series expansion of the following signals:

1. $x_1(t) = \cos(2\pi t) + \cos(4\pi t)$

2. $x_2(t) = \cos(2\pi t) - \cos(4\pi t + \pi/3)$

3. $x_3(t) = 2\cos(2\pi t) - \sin(4\pi t)$

4. $x_4(t) = \sum_{n=-\infty}^{\infty} \Lambda(t - 2n)$

5. $x_5(t) = \sum_{n=-\infty}^{\infty} \Lambda(t - n)$

6. $x_6(t) = e^{t-n}$ for $n \le t < n+1$

7. $x_7(t) = \cos t + \cos 2.5t$

8. $x_8(t) = \sum_{n=-\infty}^{\infty} \Lambda(t - n)u_{-1}(t - n)$

9. $x_9(t) = \sum_{n=-\infty}^{\infty} (-1)^n \delta(t - nT)$

10. $x_{10}(t) = \sum_{n=-\infty}^{\infty} \delta'(t - nT)$

11. $x_{11}(t) = |\cos 2\pi f_0 t|$ (full-wave rectifier output)

12. $x_{12}(t) = \cos 2\pi f_0 t + |\cos 2\pi f_0 t|$ (half-wave rectifier output)

2.40 Show that for a real, periodic $x(t)$, the even and odd parts are given by

$$x_e(t) = \frac{a_0}{2} + \sum_{n=1}^{\infty} a_n \cos\left(2\pi \frac{n}{T_0} t\right);$$

$$x_o(t) = \sum_{n=1}^{\infty} a_n \sin\left(2\pi \frac{n}{T_0} t\right).$$

2.41 Let x_n and y_n represent the Fourier-series coefficients of $x(t)$ and $y(t)$, respectively. Assuming the period of $x(t)$ is T_0, express y_n in terms of x_n in each of the following cases:

1. $y(t) = x(t - t_0)$

2. $y(t) = x(t)e^{j2\pi f_0 t}$

3. $y(t) = x(at)$, $a \neq 0$

4. $y(t) = \frac{d}{dt}x(t)$

2.42 Let $x(t)$ and $y(t)$ be two periodic signals with the period T_0, and let x_n and y_n denote the Fourier-series coefficients of these two signals. Show that

$$\frac{1}{T_0} \int_{\alpha}^{\alpha+T_0} x(t) y^*(t)\, dt = \sum_{n=-\infty}^{\infty} x_n y_n^*.$$

2.43 Determine the Fourier-series expansion of each of the periodic signals shown in Figure P-2.43. For each signal, also determine the trigonometric Fourier series.

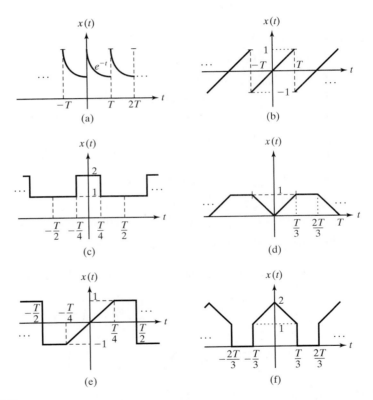

Figure P-2.43

2.44 Determine the output of each LTI system when the signals in Problem 2.43 are the inputs:

    1. $H(f) = 10\Pi\!\left(\frac{f}{4}\right)$

2.

$$H(f) = \begin{cases} -j & 0 < f \leq 4 \\ j & -4 \leq f < 0 \\ 0 & \text{otherwise} \end{cases}$$

2.45 Show that for all periodic physical signals that have finite power, the coefficients of the Fourier-series expansion x_n tend to zero as $n \to \pm\infty$.

2.46 Determine the Fourier transform of each of the following signals (α is positive):

1. $x(t) = \frac{1}{1+t^2}$
2. $\Pi(t - 3) + \Pi(t + 3)$
3. $\Lambda(2t + 3) + \Lambda(3t - 2)$
4. $4\Pi\left(\frac{t}{4}\right) \cos(2\pi f_0 t)$
5. $4\Pi\left(\frac{t-2}{4}\right) \cos(2\pi f_0 t)$
6. $\text{sinc}^3 t$
7. $t \, \text{sinc} t$
8. $t \cos 2\pi f_0 t$
9. $e^{-\alpha t} \cos(\beta t)$
10. $t e^{-\alpha t} \cos(\beta t)$

2.47 The Fourier transform of a signal is shown in Figure P-2.47. Determine and sketch the Fourier transform of the signal $x_1(t) = -x(t) + x(t) \cos(2000\pi t) + 2x(t) \cos^2(3000\pi t)$.

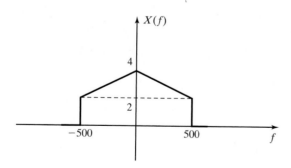

Figure P-2.47

2.48 Show that the Fourier transform of $\frac{1}{2}\delta\left(t + \frac{1}{2}\right) + \frac{1}{2}\delta\left(t - \frac{1}{2}\right)$ is $\cos(\pi f)$. Prove the following transform pairs:

$$\mathscr{F}[\cos(\pi t)] = \frac{1}{2}\delta\left(f + \frac{1}{2}\right) + \frac{1}{2}\delta\left(f - \frac{1}{2}\right)$$

and

$$\mathscr{F}[\sin(\pi t)] = \frac{j}{2}\delta\left(f + \frac{1}{2}\right) - \frac{j}{2}\delta\left(f - \frac{1}{2}\right).$$

2.49 Determine the Fourier transform of the signals shown in Figure P-2.49.

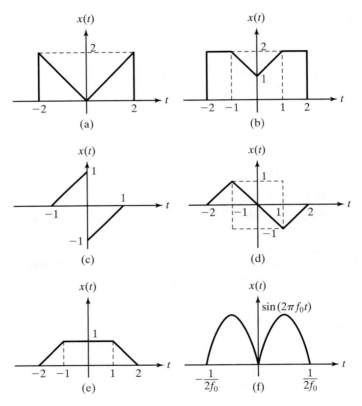

Figure P-2.49

2.50 Use the convolution theorem to show that

$$\operatorname{sinc}(t) \star \operatorname{sinc}(t) = \operatorname{sinc}(t).$$

2.51 Prove that convolution in the frequency domain is equivalent to multiplication in the time domain. That is,

$$\mathscr{F}[x(t)y(t)] = X(f) \star Y(f).$$

2.52 Let $x(t)$ be an arbitrary signal and define $x_1(t) = \sum_{n=-\infty}^{\infty} x(t - nT_0)$.

1. Show that $x_1(t)$ is a periodic signal.
2. How can you write $x_1(t)$ in terms of $x(t)$ and $\sum_{n=-\infty}^{\infty} \delta(t - nT_0)$?
3. Find the Fourier transform of $x_1(t)$ in terms of the Fourier transform of $x(t)$.

2.53 Using the properties of the Fourier transform, evaluate the following integrals (α is positive):

1. $\int_{-\infty}^{\infty} \text{sinc}^5(t)\, dt$
2. $\int_{0}^{\infty} e^{-\alpha t} \text{sinc}(t)\, dt$
3. $\int_{0}^{\infty} e^{-\alpha t} \text{sinc}^2(t)\, dt$
4. $\int_{0}^{\infty} e^{-\alpha t} \cos(\beta t)\, dt$.

2.54 A linear time-invariant system with impulse response $h(t) = e^{-\alpha t} u_{-1}(t)$ is driven by the input $x(t) = e^{-\beta t} u_{-1}(t)$. Assume that $\alpha, \beta > 0$. Using frequency-domain analysis, find the output of the system. Is the output power type or energy type? Find its power or energy.

2.55 Let $x(t)$ be periodic with period T_0, and let $0 \le \alpha < T_0$. Define

$$x_\alpha(t) = \begin{cases} x(t) & \alpha \le t < \alpha + T_0 \\ 0 & \text{otherwise} \end{cases}$$

and let $X_\alpha(f)$ denote the Fourier transform of $x_\alpha(t)$. Prove that for all n, $X_\alpha(\frac{n}{T_0})$ is independent of the choice of α.

2.56 Using the identity

$$\sum_{n=-\infty}^{\infty} \delta(t - nT_s) = \frac{1}{T_s} \sum_{n=-\infty}^{\infty} e^{jn\frac{2\pi t}{T_s}},$$

show that for any signal $x(t)$ and any T_s, the following identity holds

$$\sum_{n=-\infty}^{\infty} x(t - nT_s) = \frac{1}{T_s} \sum_{n=-\infty}^{\infty} X\left(\frac{n}{T_s}\right) e^{jn\frac{2\pi t}{T_s}},$$

From this, derive the following relation known as *Poisson's sum formula*:

$$\sum_{n=-\infty}^{\infty} x(nT_s) = \frac{1}{T_s} \sum_{n=-\infty}^{\infty} X\left(\frac{n}{T_s}\right).$$

2.57 Using the Poisson's sum formula discussed in Problem 2.56, show that

1. $\sum_{n=-\infty}^{\infty} \frac{2\alpha}{\alpha^2 + 4\pi^2 n^2} = \sum_{n=-\infty}^{\infty} e^{-\alpha|n|}$

2. $\sum_{n=-\infty}^{\infty} \text{sinc}(\frac{n}{K}) = K$ for all $K \in \{1, 2, \dots\}$

3. $\sum_{n=-\infty}^{\infty} \text{sinc}^2(\frac{n}{K}) = K$ for all $K \in \{1, 2, \dots\}$.

2.58 The response of an LTI system to $e^{-\alpha t}u_{-1}(t)$, where $(\alpha > 0)$, is $\delta(t)$. Using frequency-domain analysis techniques, find the response of the system to $x(t) = e^{-\alpha t}\cos(\beta t)u_{-1}(t)$.

2.59 Find the output of an LTI system with impulse response $h(t)$ when driven by the input $x(t)$ in each of the following cases:

1. $h(t) = \text{sinc}(t)$ $x(t) = \text{sinc}(t)$

2. $h(t) = \delta(t) + \delta'(t)$ $x(t) = e^{-\alpha|t|}$, $(\alpha > 0)$

3. $h(t) = e^{-\alpha t}u_{-1}(t)$ $x(t) = e^{-\beta t}u_{-1}(t)$, $(\alpha, \beta > 0)$
   (treat the special case $\alpha = \beta$ separately)

4. $h(t) = e^{-\alpha t}\cos(\beta t)u_{-1}(t)$ $x(t) = e^{-\beta t}u_{-1}(t)$, $(\alpha, \beta > 0)$
   (treat the special case $\alpha = \beta$ separately)

5. $h(t) = e^{-\alpha|t|}u_{-1}(t)$ $x(t) = e^{-\beta t}u_{-1}(t)$, $(\alpha, \beta > 0)$
   (treat the special case $\alpha = \beta$ separately)

6. $h(t) = \text{sinc}(t)$ $x(t) = \text{sinc}^2(t)$.

2.60 Can the response of an LTI system to $x(t) = \text{sinc}(t)$ be $y(t) = \text{sinc}^2(t)$? Justify your answer.

2.61 Let the response of an LTI system to $\Pi(t)$ be $\Lambda(t)$.

1. Can you find the response of this system to $x(t) = \cos 2\pi t$ from the preceding information?

2. Show that $h_1(t) = \Pi(t)$ and $h_2(t) = \Pi(t) + \cos 2\pi t$ can both be impulse responses of this system, and therefore having the response of a system to $\Pi(t)$ does not uniquely determine the system.

3. Does the response of an LTI system to $u_{-1}(t)$ uniquely determine the system? Does the response to $e^{-\alpha t}u_{-1}(t)$ for some $\alpha > 0$? In general, what conditions must the input $x(t)$ satisfy so that the system can be uniquely determined by knowing its corresponding output?

2.62 Show that the Hilbert transform of $A\sin(2\pi f_0 t + \theta)$ is $-A\cos(2\pi f_0 t + \theta)$.

2.63 Show that the Hilbert transform of the signal $e^{j2\pi f_0 t}$ is equal to $-j\,\text{sgn}(f_0)e^{j2\pi f_0 t}$.

2.64 Show that

$$\mathcal{F}[\widehat{\frac{d}{dt}x(t)}] = 2\pi |f| \mathcal{F}[x(t)].$$

2.65 Show that the Hilbert transform of the derivative of a signal is equal to the derivative of its Hilbert transform.

2.66 The real narrowband signal $x(t)$, whose frequency components are in the neighborhood of some f_0 (and $-f_0$), is passed through a filter with the transfer function $H(f)$, and the output is denoted by $y(t)$. The magnitude of the transfer function is denoted by $A(f)$, and its phase is denoted by $\theta(f)$. Assume that the transfer function of the filter is so smooth that in the bandwidth of the input signal, the magnitude of the transfer function is essentially constant and its phase can be approximated by its first-order Taylor-series expansion, i.e.,

$$A(f) \approx A(f_0)$$
$$\theta(f) \approx \theta(f_0) + (f - f_0)\theta'(f)|_{f=f_0}.$$

1. Show that $Y_l(f)$, the Fourier transform of the lowpass equivalent of the output, can be written as

$$Y_l(f) \approx X_l(f)A(f_0)e^{j\left(\theta(f_0) + f\theta'(f)|_{f=f_0}\right)}.$$

2. Conclude that

$$y(t) \approx A(f_0) + V_x(t - t_g)\cos(2\pi f_0 t - t_p),$$

where $V_x(t)$ is the envelope of the input $x(t)$ and

$$t_g = -\frac{1}{2\pi}\frac{d\theta(f)}{df}\bigg|_{f=f_0}$$

$$t_p = -\frac{1}{2\pi}\frac{\theta(f)}{f}\bigg|_{f=f_0}.$$

3. The quantities t_g and t_p are called *envelope delay* (or *group delay*) and *phase delay*, respectively. Can you interpret their role and justify this nomenclature?

2.67 We have seen that the Hilbert transform introduces a 90° phase shift in the components of a signal and the transfer function of a quadrature filter can be written as

$$H(f) = \begin{cases} e^{-j\frac{\pi}{2}} & f > 0 \\ 0 & f = 0 \\ e^{j\frac{\pi}{2}} & f < 0 \end{cases}.$$

We can generalize this concept to a new transform that introduces a phase shift of θ in the frequency components of a signal, by introducing

$$H_\theta(f) = \begin{cases} e^{-j\theta} & f > 0 \\ 0 & f = 0 \\ e^{j\theta} & f < 0 \end{cases}$$

and denoting the result of this transform by $x_\theta(t)$, i.e., $X_\theta(f) = X(f)H_\theta(f)$, where $X_\theta(f)$ denotes the Fourier transform of $x_\theta(t)$. Throughout this problem, assume that the signal $x(t)$ does not contain any DC components.

1. Find $h_\theta(t)$, the impulse response of the filter representing this transform.
2. Show that $x_\theta(t)$ is a linear combination of $x(t)$ and its Hilbert transform.
3. Show that if $x(t)$ is an energy-type signal, $x_\theta(t)$ will also be an energy-type signal and its energy content will be equal to the energy content of $x(t)$.

COMPUTER PROBLEMS

2.1 Fourier Series

The purpose of this problem is to evaluate and plot the Fourier-series coefficients of a periodic signal. The periodic signal $x(t)$, with period T_0, is defined as

$$x(t) = A\Pi\left(\frac{t}{2t_0}\right) = \begin{cases} A, & |t| \leq t_0 \\ 0 & \text{otherwise} \end{cases}$$

for $|t| \leq T_0/2$, where $t_0 < T_0/2$. A plot of $x(t)$ is shown in Figure CP–2.1. Let $A = 1$, $T_0 = 4$, and $t_0 = 1$.

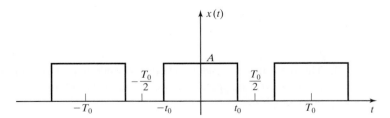

Figure CP-2.1 Signal $x(t)$.

1. Demonstrate mathematically that the Fourier-series coefficients in the expansion of $x(t)$ are given as

$$x_n = \frac{1}{2}\text{sinc}\left(\frac{n}{2}\right) = \frac{\sin(n\pi/2)}{\pi n}.$$

2. Use MATLAB to plot the original signal $x(t)$ and the Fourier-series approximation of $x(t)$ over one period for $n = 1, 3, 5, 7$, and 9. Note that as n increases, the approximation becomes closer to the original signal $x(t)$.

3. Plot the discrete magnitude spectrum $|x_n|$ and the phase spectrum $\angle x_n$ for $|n| \leq 20$.

2.2 Filtering of a Periodic Signal

The objective of this problem is to demonstrate the effect of passing a periodic signal through a linear, time-invariant system. The periodic signal $x(t)$ is a triangular pulse train with period $T_0 = 2$ and defined over a period as

$$\Lambda(t) = \begin{cases} t+1 & -1 \leq t \leq 0 \\ -t+1 & 0 \leq t \leq 1 \\ 0 & \text{otherwise.} \end{cases}$$

The frequency response characteristic of the LTI filter is

$$H(f) = \frac{1}{\sqrt{1+f^2}}.$$

1. Demonstrate mathematically that the Fourier-series coefficients in the expansion of $x(t)$ are given as

$$x_n = \frac{1}{2}\text{sinc}^2\left(\frac{n}{2}\right).$$

2. Use MATLAB to plot the spectrum $\{x_n\}$ for $0 \leq |n| \leq 10$.

3. Since the fundamental frequency of the input signal $x(t)$ is $F_0 = 1/T_0 = 1/2$, the Fourier-series coefficients of the filter output are $y_n = x_n H(n/2)$. Plot the discrete spectrum $\{y_n\}$ of the filter output.

4. On a single graph, plot the Fourier-series approximation of the filter output signal for $n = 1, 3, \ldots, 9$. Comment on the results.

2.3 Fourier Transform

Plot the magnitude and the phase spectra of the two signals shown in Figure CP–2.3. Note that the signal $x_2(t)$ is a time-shifted version of $x_1(t)$. Therefore, we expect the two signals to have identical magnitude spectra.

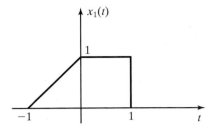

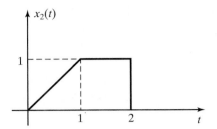

Figure CP-2.3   Signals $x_1(t)$ and $x_2(t)$.

2.4 LTI System Analysis in the Frequency Domain

The signal

$$x(t) = \begin{cases} e^{-t} & t \geq 0 \\ 0 & t < 0 \end{cases}$$

is passed through a lowpass filter, whose frequency response is specified as

$$H(f) = \begin{cases} \cos(\pi f/3) & |f| \leq 1.5 \text{ Hz} \\ 0 & \text{otherwise} \end{cases}.$$

1. Determine the magnitude and phase spectra of the input signal $x(t)$ and plot these on two separate graphs.

2. Determine and plot the magnitude and phase spectra of the output signal $y(t)$.

3. Use MATLAB to compute the inverse Fourier transform of the output signal and plot it.

2.5 Bandpass to Lowpass Transformation

The signal

$$x(t) = \text{sinc}(100t) \cos(400\pi t)$$

is a bandpass signal whose frequency content is centered at $f_0 = 200$ Hz.

1. Plot $x(t)$ and the magnitude spectrum $|X(f)|$.

2. With $f_0 = 200$ Hz, determine the equivalent lowpass signal and plot its magnitude spectrum. Also, plot the in-phase and quadrature components of $x(t)$ and the envelope of $x(t)$.

2.6 FIR Filter Design

Use the `remez` function in MATLAB to design the lowest order (smallest length) FIR lowpass filter that satisfies the following specifications:

Passband: $0 \le \omega \le 0.4\pi$
Stopband: $0.5\pi \le \omega \le \pi$
Passband ripple: 0.5 dB
Stopband attenuation: 40 dB.

1. Plot the impulse response coefficients of the FIR filter.

2. Plot the magnitude and phase of the frequency response of the filter.

2.7 FIR Hilbert Transform Filter

Use the `remez` function to design an FIR Hilbert-transform filter that satisfies the following specifications:

Passband: $0.1\pi \le |\omega| \le 0.5\pi$
Stopband: $0.6\pi \le |\omega| \le \pi$
Passband ripple: $\delta_1 = 0.01$
Stopband attenuation: $\delta_2 = 0.01$.

1. Plot the impulse-response coefficients of the FIR filter.

2. Plot the magnitude of the frequency response of the filter.

2.8 FIR Differentiator

Use the `remez` function to design a 25-tap FIR differentiator that has a passband in the range $0.1\pi \le \omega \le 0.6\pi$.

1. Plot the impulse response of the filter.

2. Plot the magnitude of the frequency response.

3. Generate 100 samples of the sinusoid

$$x(n) = 5 \sin\left(\frac{\pi}{4}n\right), \qquad n = 0, 1, \ldots$$

and pass them through the differentiator. Plot the filter output $y(n)$ and compare $y(n)$ with $x(n)$. Note that there is a 12-sample delay in the FIR filter.

3

Amplitude Modulation

A large number of information sources produce analog signals. Analog signals can be modulated and transmitted directly, or they can be converted into digital data and transmitted using digital-modulation techniques. The notion of analog-to-digital conversion will be examined in detail in Chapter 7.

Speech, music, images, and video are examples of analog signals. Each of these signals is characterized by its bandwidth, dynamic range, and the nature of the signal. For instance, in the case of audio and black-and-white video, the signal has just one component, which measures air pressure or light intensity. But in the case of color video, the signal has four components, namely, the red, green, and blue color components, plus a fourth component for the intensity. In addition to the four video signals, an audio signal carries the audio information in color-TV broadcasting. Various analog signals exhibit a large diversity in terms of signal bandwidth. For instance, speech signals have a bandwidth of up to 4 kHz, music signals typically have a bandwidth of 20 kHz, and video signals have a much higher bandwidth, about 6 MHz.

In spite of the general trend toward the digital transmission of analog signals, we still have a significant amount of analog signal transmission, especially in audio and video broadcast. In Chapters 3 and 4, we treat the transmission of analog signals by carrier modulation. (The treatment of the performance of these systems in the presence of noise is deferred to Chapter 6.) We consider the transmission of an analog signal by impressing it on the amplitude, the phase, or the frequency of a sinusoidal carrier. Methods for demodulation of the carrier-modulated signal to recover the analog information signal are also described. This chapter is devoted to amplitude-modulation systems, where the message signals change the amplitude of the carrier. Chapter 4 is devoted to phase- and frequency-modulation systems, in which either the phase or frequency of the carrier is changed according to variations in the message signal.

3.1 INTRODUCTION TO MODULATION

The analog signal to be transmitted is denoted by $m(t)$, which is assumed to be a lowpass signal of bandwidth W; in other words $M(f) \equiv 0$, for $|f| > W$. The power content of this signal is denoted by

$$P_m = \lim_{T \to \infty} \frac{1}{T} \int_{-T/2}^{T/2} |m(t)|^2 \, dt.$$

The message signal $m(t)$ is transmitted through the communication channel by impressing it on a *carrier* signal of the form

$$c(t) = A_c \cos(2\pi f_c t + \phi_c), \tag{3.1.1}$$

where A_c is the carrier amplitude, f_c is the carrier frequency, and ϕ_c is the carrier phase. The value of ϕ_c depends on the choice of the time origin. Without loss of generality, we assume that the time origin is chosen such that $\phi_c = 0$. We say that the message signal $m(t)$ modulates the carrier signal $c(t)$ in either amplitude, frequency, or phase if after modulation, the amplitude, frequency, or phase of the signal become functions of the message signal. In effect, modulation converts the message signal $m(t)$ from lowpass to bandpass, in the neighborhood of the carrier frequency f_c.

Modulation of the carrier $c(t)$ by the message signal $m(t)$ is performed to achieve one or more of the following objectives:

(1) To translate the frequency of the lowpass signal to the passband of the channel so that the spectrum of the transmitted bandpass signal will match the passband characteristics of the channel. For instance, in transmission of speech over microwave links in telephony transmission, the transmission frequencies must be increased to the gigahertz range for transmission over the channel. This means that modulation, or a combination of various modulation techniques, must be used to translate the speech signal from the low-frequency range (up to 4 kHz) to the gigahertz range.

(2) To simplify the structure of the transmitter by employing higher frequencies. For instance, in the transmission of information using electromagnetic waves, transmission of the signal at low frequencies requires huge antennas. Modulation helps translate the frequency band to higher frequencies, thus requiring smaller antennas. This simplifies the structure of the transmitter (and the receiver).

(3) To accommodate for the simultaneous transmission of signals from several message sources, by means of frequency-division multiplexing. (See Section 3.4.)

(4) To expand the bandwidth of the transmitted signal in order to increase its noise and interference immunity in transmission over a noisy channel, as we will see in our discussion of angle-modulation in Chapter 6.

Objectives (1), (2), and (3) are met by all of the modulation methods described in this chapter. Objective (4) is met by employing angle modulation to spread the signal $m(t)$ over a larger bandwidth, as discussed in Chapter 4.

In the following sections, we consider the transmission and reception of analog signals by carrier-amplitude modulation (AM). We will compare these modulation methods on the basis of their bandwidth requirements and their implementation complexity. Their performance in the presence of additive noise disturbances and their power efficiency will be discussed in Chapter 6.

3.2 AMPLITUDE MODULATION (AM)

In amplitude modulation, the message signal $m(t)$ is impressed on the amplitude of the carrier signal $c(t) = A_c \cos(2\pi f_c t)$. This results in a sinusoidal signal whose amplitude is a function of the message signal $m(t)$. There are several different ways of amplitude modulating the carrier signal by $m(t)$; each results in different spectral characteristics for the transmitted signal. We will describe these methods, which are called (a) double sideband, suppressed-carrier AM, (b) conventional double-sideband AM, (c) single-sideband AM, and (d) vestigial-sideband AM.

3.2.1 Double-Sideband Suppressed-Carrier AM

A double-sideband, suppressed-carrier (DSB-SC) AM signal is obtained by multiplying the message signal $m(t)$ with the carrier signal $c(t) = A_c \cos(2\pi f_c t)$. Thus, we have the amplitude-modulated signal

$$u(t) = m(t)c(t)$$
$$= A_c m(t) \cos(2\pi f_c t).$$

An example of the message signal $m(t)$, the carrier $c(t)$, and the modulated signal $u(t)$ are shown in Figure 3.1. This figure shows that a relatively slowly varying message signal $m(t)$ is changed into a rapidly varying modulated signal $u(t)$, and due to its rapid changes with time, it contains higher frequency components. At the same time, the modulated signal retains the main characteristics of the message signal; therefore, it can be used to retrieve the message signal at the receiver.

Spectrum of the DSB-SC AM Signal.   The spectrum of the modulated signal can be obtained by taking the Fourier transform of $u(t)$ and using the result of Example (2.3.14). Thus, we obtain

$$U(f) = \frac{A_c}{2} \left[M(f - f_c) + M(f + f_c) \right].$$

Figure 3.2 illustrates the magnitude and phase spectra for $M(f)$ and $U(f)$.

The magnitude of the spectrum of the message signal $m(t)$ has been translated or shifted in frequency by an amount f_c. Furthermore, the bandwidth occupancy of

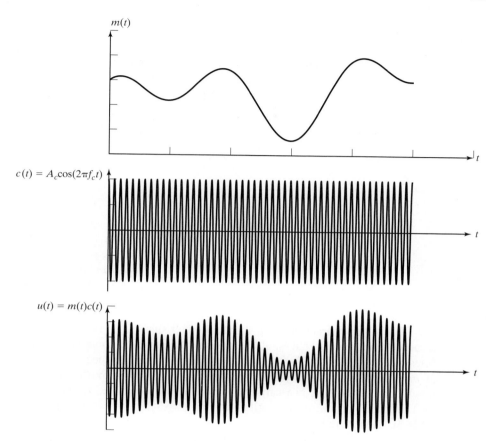

Figure 3.1   An example of message, carrier, and DSB-SC modulated signals.

the amplitude-modulated signal is $2W$, whereas the bandwidth of the message signal $m(t)$ is W. Therefore, the channel bandwidth required to transmit the modulated signal $u(t)$ is $B_c = 2W$.

The frequency content of the modulated signal $u(t)$ in the frequency band $|f| > f_c$ is called the *upper sideband* of $U(f)$, and the frequency content in the frequency band $|f| < f_c$ is called the *lower sideband* of $U(f)$. It is important to note that either one of the sidebands of $U(f)$ contains all the frequencies that are in $M(f)$. That is, the frequency content of $U(f)$ for $f > f_c$ corresponds to the frequency content of $M(f)$ for $f > 0$, and the frequency content of $U(f)$ for $f < -f_c$ corresponds to the frequency content of $M(f)$ for $f < 0$. Hence, the upper sideband of $U(f)$ contains all the frequencies in $M(f)$. A similar statement applies to the lower sideband of $U(f)$. Therefore, the lower sideband of $U(f)$ contains all the frequency content of the message signal $M(f)$. Since $U(f)$ contains both the upper and the lower sidebands, it is called a *double-sideband (DSB) AM signal*.

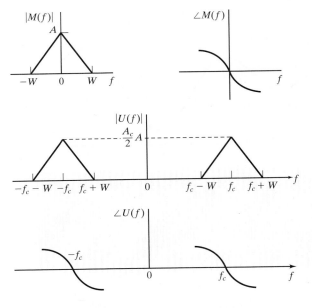

Figure 3.2   Magnitude and phase spectra of the message signal $m(t)$ and the DSB-AM modulated signal $u(t)$.

The other characteristic of the modulated signal $u(t)$ is that it does not contain a carrier component. That is, all the transmitted power is contained in the modulating (message) signal $m(t)$. This is evident from observing the spectrum of $U(f)$. As long as $m(t)$ does not have any DC component, there is no impulse in $U(f)$ at $f = f_c$; this would be the case if a carrier component was contained in the modulated signal $u(t)$. For this reason, $u(t)$ is called a *suppressed-carrier signal*. Therefore, $u(t)$ is a DSB-SC AM signal.

Example 3.2.1

Suppose that the modulating signal $m(t)$ is a sinusoid of the form

$$m(t) = a \cos 2\pi f_m t \qquad f_m \ll f_c.$$

Determine the DSB-SC AM signal and its upper and lower sidebands.

Solution   The DSB-SC AM is expressed in the time domain as

$$u(t) = m(t)c(t) = A_c a \cos(2\pi f_m t)\cos(2\pi f_c t)$$

$$= \frac{A_c a}{2}\cos[2\pi(f_c - f_m)t] + \frac{A_c a}{2}\cos[2\pi(f_c + f_m)t].$$

Taking the Fourier transform, the modulated signal in the frequency domain will have the following form:

$$U(f) = \frac{A_c a}{4}\left[\delta(f - f_c + f_m) + \delta(f + f_c - f_m)\right]$$

$$+ \frac{A_c a}{4}\left[\delta(f - f_c - f_m) + \delta(f + f_c + f_m)\right].$$

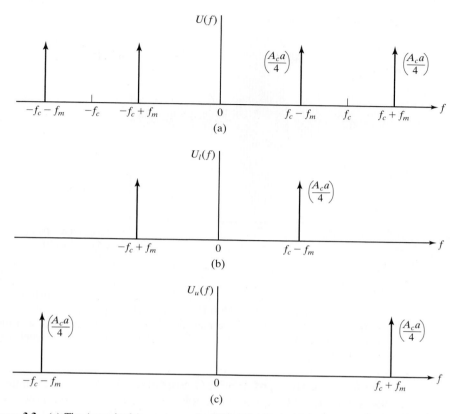

Figure 3.3 (a) The (magnitude) spectrum of a DSB-SC AM signal for a sinusoidal message signal and (b) its lower and (c) upper sidebands.

This spectrum is shown in Figure 3.3(a). The lower sideband of $u(t)$ is the signal

$$u_\ell(t) = \frac{A_c a}{2} \cos[2\pi(f_c - f_m)t],$$

and its spectrum is illustrated in Figure 3.3(b). Finally, the upper sideband of $u(t)$ is the signal

$$u_u(t) = \frac{A_c a}{2} \cos[2\pi(f_c + f_m)t],$$

and its spectrum is illustrated in Figure 3.3(c).   ∎

Example 3.2.2

Let the message signal be $m(t) = \text{sinc}(10^4 t)$. Determine the DSB-SC modulated signal and its bandwidth when the carrier is a sinusoid with a frequency of 1 MHz.

Solution   In this example, $c(t) = \cos(2\pi \times 10^6 t)$. Therefore, $u(t) = \text{sinc}(10^4 t) \cos(2\pi \times 10^6 t)$. A plot of $u(t)$ is shown in Figure 3.4. To obtain the bandwidth of

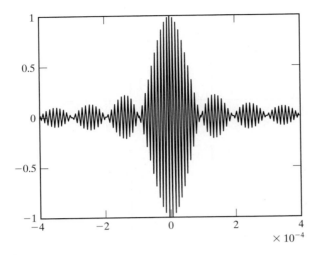

Figure 3.4   Plot of
$u(t) = \text{sinc}(10^4 t) \cos(2\pi \times 10^6 t)$.

the modulated signal, we first need to have the bandwidth of the message signal. We
have $M(f) = \mathcal{F}\,\text{sinc}(10^4 t) = 10^{-4}\Pi(10^{-4} f)$. The Fourier transform is constant in
the frequency range from -5000 Hz to 5000 Hz, and it is zero at other frequencies.
Therefore, the bandwidth of the message signal is $W = 5000$ Hz, and the bandwidth
of the modulated signal is twice the bandwidth of the message signal, i.e., 10000 Hz
or 10 kHz.   ■

Power Content of DSB-SC Signals.   In order to compute the power
content of the DSB-SC signal, we employ the definition of the power content of a
signal given in Equation (2.1.11). Thus,

$$P_u = \lim_{T \to \infty} \frac{1}{T} \int_{-T/2}^{T/2} u^2(t)\,dt$$

$$= \lim_{T \to \infty} \frac{1}{T} \int_{-T/2}^{T/2} A_c^2 m^2(t) \cos^2(2\pi f_c t)\,dt$$

$$= \frac{A_c^2}{2} \lim_{T \to \infty} \frac{1}{T} \int_{-T/2}^{T/2} m^2(t)[1 + \cos(4\pi f_c t)]\,dt \qquad (3.2.1)$$

$$= \frac{A_c^2}{2} P_m, \qquad (3.2.2)$$

where P_m indicates the power in the message signal $m(t)$. The last step follows from
the fact that $m^2(t)$ is a slowly varying signal and when multiplied by $\cos(4\pi f_c t)$,
which is a high frequency sinusoid, the result is a high-frequency sinusoid with a
slowly varying envelope, as shown in Figure 3.5.

Since the envelope is slowly varying, the positive and the negative halves of
each cycle have almost the same amplitude. Hence, when they are integrated, they

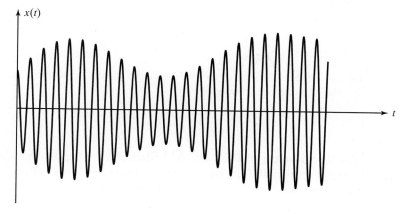

Figure 3.5   Plot of $m^2(t)\cos(4\pi f_c t)$.

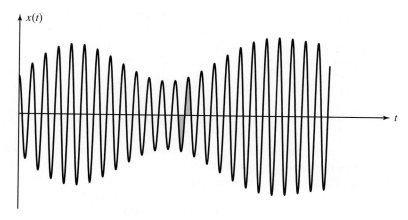

Figure 3.6   This figure shows why the second term in Equation (3.2.1) is zero.

cancel each other. Thus, the overall integral of $m^2(t)\cos(4\pi f_c t)$ is almost zero. This is depicted in Figure 3.6. Since the result of the integral is divided by T, and T becomes very large, the second term in Equation (3.2.1) is zero.

Example 3.2.3

In Example 3.2.1, determine the power in the modulated signal and the power in each of the sidebands.

Solution   The message signal is $m(t) = a\cos 2\pi f_m t$. Its power was obtained in Example 2.1.10 and Equation (2.1.12) as

$$P_m = \frac{a^2}{2}.$$

Therefore,

$$P_u = \frac{A_c^2}{2} P_m = \frac{A_c^2 a^2}{4}.$$

Because of the symmetry of the sidebands, the powers in the upper and lower sidebands, P_{us} and P_{ls}, are equal and given by

$$P_{us} = P_{ls} = \frac{A_c^2 a^2}{8}. \qquad \blacksquare$$

Demodulation of DSB-SC AM Signals.   Suppose that the DSB-SC AM signal $u(t)$ is transmitted through an ideal channel (with no channel distortion and no noise). Then the received signal is equal to the modulated signal, i.e.,

$$r(t) = u(t)$$

$$= A_c m(t) \cos(2\pi f_c t). \qquad (3.2.3)$$

Suppose we demodulate the received signal by first multiplying $r(t)$ by a locally generated sinusoid $\cos(2\pi f_c t + \phi)$, where ϕ is the phase of the sinusoid. Then, we pass the product signal through an ideal lowpass filter with the bandwidth W. The multiplication of $r(t)$ with $\cos(2\pi f_c t + \phi)$ yields

$$r(t) \cos(2\pi f_c t + \phi) = A_c m(t) \cos(2\pi f_c t) \cos(2\pi f_c t + \phi)$$

$$= \frac{1}{2} A_c m(t) \cos(\phi) + \frac{1}{2} A_c m(t) \cos(4\pi f_c t + \phi).$$

The spectrum of the signal is illustrated in Figure 3.7. Since the frequency content of the message signal $m(t)$ is limited to W Hz, where $W \ll f_c$, the lowpass filter can be designed to eliminate the signal components centered at frequency $2f_c$ and to pass the signal components centered at frequency $f = 0$ without experiencing distortion. An ideal lowpass filter that accomplishes this objective is also illustrated in Figure 3.7. Consequently, the output of the ideal lowpass filter is

$$y_\ell(t) = \frac{1}{2} A_c m(t) \cos(\phi). \qquad (3.2.4)$$

Note that $m(t)$ is multiplied by $\cos(\phi)$; therefore, the power in the demodulated signal is decreased by a factor of $\cos^2 \phi$. Thus, the desired signal is scaled in amplitude by a factor that depends on the phase ϕ of the locally generated sinusoid.

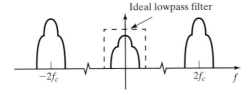

Figure 3.7   Frequency-domain representation of the DSB-SC AM demodulation.

When $\phi \neq 0$, the amplitude of the desired signal is reduced by the factor $\cos(\phi)$. If $\phi = 45°$, the amplitude of the desired signal is reduced by $\sqrt{2}$ and the signal power is reduced by a factor of two. If $\phi = 90°$, the desired signal component vanishes.

The preceding discussion demonstrates the need for a *phase-coherent or synchronous demodulator* for recovering the message signal $m(t)$ from the received signal. That is, the phase ϕ of the locally generated sinusoid should ideally be equal to 0 (the phase of the received-carrier signal).

A sinusoid that is phase-locked to the phase of the received carrier can be generated at the receiver in one of two ways. One method is to add a carrier component into the transmitted signal, as illustrated in Figure 3.8. We call such a carrier component "a pilot tone." Its amplitude A_p and its power $A_p^2/2$ are selected to be significantly smaller than those of the modulated signal $u(t)$. Thus, the transmitted signal is a double-sideband, but it is no longer a suppressed carrier signal. At the receiver, a narrowband filter tuned to frequency f_c filters out the pilot signal component; its output is used to multiply the received signal, as shown in Figure 3.9. We may show that the presence of the pilot signal results in a *DC* component in the demodulated signal; this must be subtracted out in order to recover $m(t)$.

Adding a pilot tone to the transmitted signal has a disadvantage: it requires that a certain portion of the transmitted signal power must be allocated to the transmission of the pilot. As an alternative, we may generate a phase-locked sinusoidal carrier from the received signal $r(t)$ without the need of a pilot signal. This can be accomplished by the use of a *phase-locked loop*, as described in Section 6.4.

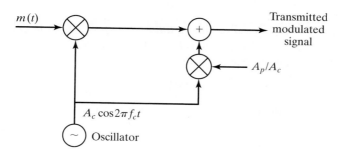

Figure 3.8   Addition of a pilot tone to a DSB-AM signal.

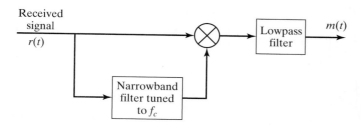

Figure 3.9   Use of a pilot tone to demodulate a DSB-AM signal.

3.2.2 Conventional Amplitude Modulation

A conventional AM signal consists of a large carrier component, in addition to the double-sideband AM modulated signal. The transmitted signal is expressed mathematically as

$$u(t) = A_c[1 + m(t)] \cos(2\pi f_c t), \tag{3.2.5}$$

where the message waveform is constrained to satisfy the condition that $|m(t)| \leq 1$. We observe that $A_c m(t) \cos(2\pi f_c t)$ is a double-sideband AM signal and $A_c \cos(2\pi f_c t)$ is the carrier component. Figure 3.10 illustrates an AM signal in the time domain. As we will see later in this chapter, the existence of this extra carrier results in a very simple structure for the demodulator. That is why commercial AM broadcasting generally employs this type of modulation.

As long as $|m(t)| \leq 1$, the amplitude $A_c[1 + m(t)]$ is always positive. This is the desired condition for conventional DSB AM that makes it easy to demodulate, as we will describe. On the other hand, if $m(t) < -1$ for some t, the AM signal is *overmodulated* and its demodulation is rendered more complex. In practice, $m(t)$ is scaled so that its magnitude is always less than unity.

It is sometimes convenient to express $m(t)$ as

$$m(t) = am_n(t),$$

where $m_n(t)$ is normalized such that its minimum value is -1. This can be done, for example, by defining

$$m_n(t) = \frac{m(t)}{\max |m(t)|}.$$

In this case, the scale factor a is called the *modulation index*, which is generally a constant less than 1. Since $|m_n(t)| \leq 1$ and $0 < a < 1$, we have $1 + am_n(t) > 0$, and the modulated signal can be expressed as

$$u(t) = A_c [1 + am_n(t)] \cos 2\pi f_c t, \tag{3.2.6}$$

which will never be overmodulated.

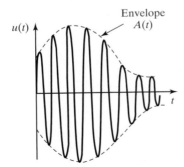

Figure 3.10   A conventional AM signal in the time domain.

Spectrum of the Conventional AM Signal.   If $m(t)$ is a message signal with Fourier transform (spectrum) $M(f)$, the spectrum of the amplitude-modulated signal $u(t)$ is

$$U(f) = \mathscr{F}[A_c a m_n(t) \cos(2\pi f_c t)] + \mathscr{F}[A_c \cos(2\pi f_c t)]$$

$$= \frac{A_c a}{2} \left[M_n(f - f_c) + M_n(f + f_c)\right] + \frac{A_c}{2} \left[\delta(f - f_c) + \delta(f + f_c)\right].$$

A message signal $m(t)$, its spectrum $M(f)$, the corresponding modulated signal $u(t)$, and its spectrum $U(f)$ are shown in Figure 3.11. Obviously, the spectrum of a conventional AM signal occupies a bandwidth twice the bandwidth of the message signal.

Example 3.2.4

Suppose that the modulating signal $m(t)$ is a sinusoid of the form

$$m(t) = \cos 2\pi f_m t \quad f_m \ll f_c.$$

Determine the DSB-AM signal, its upper and lower sidebands, and its spectrum, assuming a modulation index of a.

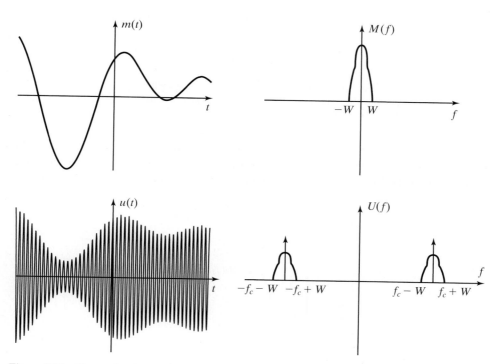

Figure 3.11   Conventional AM in both the time and frequency domain.

Solution　From Equation (3.2.6), the DSB-AM signal is expressed as

$$u(t) = A_c[1 + a \cos 2\pi f_m t] \cos(2\pi f_c t)$$

$$= A_c \cos(2\pi f_c t) + \frac{A_c a}{2} \cos[2\pi(f_c - f_m)t]$$

$$+ \frac{A_c a}{2} \cos[2\pi(f_c + f_m)t].$$

The lower sideband component is

$$u_\ell(t) = \frac{A_c a}{2} \cos[2\pi(f_c - f_m)t],$$

while the upper sideband component is

$$u_u(t) = \frac{A_c a}{2} \cos[2\pi(f_c + f_m)t].$$

The spectrum of the DSB-AM signal $u(t)$ is

$$U(f) = \frac{A_c}{2} \left[\delta(f - f_c) + \delta(f + f_c)\right]$$

$$+ \frac{A_c a}{4} \left[\delta(f - f_c + f_m) + \delta(f + f_c - f_m)\right]$$

$$+ \frac{A_c a}{4} \left[\delta(f - f_c - f_m) + \delta(f + f_c + f_m)\right].$$

The magnitude spectrum $|U(f)|$ is shown in Figure 3.12. It is interesting to note that the power of the carrier component, which is $A_c^2/2$, exceeds the total power $\left(A_c^2 a^2/4\right)$ of the two sidebands because $a < 1$.　■

Power for the Conventional AM Signal.　A conventional AM signal is similar to a DSB when $m(t)$ is substituted with $1 + m_n(t)$. As we have already seen in the DSB-SC case, the power in the modulated signal is (see Equation 3.2.2)

$$P_u = \frac{A_c^2}{2} P_m,$$

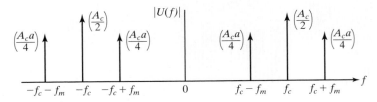

Figure 3.12　Spectrum of a DSB-AM signal in Example 3.2.4.

where P_m denotes the power in the message signal. For the conventional AM,

$$P_m = \lim_{T\to\infty} \frac{1}{T} \int_{-T/2}^{T/2} [1 + am_n(t)]^2 \, dt$$

$$= \lim_{T\to\infty} \frac{1}{T} \int_{-T/2}^{T/2} [1 + a^2 m_n^2(t)] \, dt,$$

where we have assumed that the average of $m_n(t)$ is zero. This is a valid assumption for many signals, including audio signals. Therefore, for conventional AM,

$$P_m = 1 + a^2 P_{m_n};$$

hence,

$$P_u = \frac{A_c^2}{2} + \frac{A_c^2}{2} a^2 P_{m_n}.$$

The first component in the preceding relation applies to the existence of the carrier, and this component does not carry any information. The second component is the information-carrying component. Note that the second component is usually much smaller than the first component ($a < 1$, $|m_n(t)| < 1$, and for signals with a large dynamic range, $P_{m_n} \ll 1$). This shows that the conventional AM systems are far less power efficient than the DSB-SC systems. The advantage of conventional AM is that it is easily demodulated.

Example 3.2.5

The signal $m(t) = 3\cos(200\pi t) + \sin(600\pi t)$ is used to modulate the carrier $c(t) = \cos(2 \times 10^5 t)$. The modulation index is $a = 0.85$. Determine the power in the carrier component and in the sideband components of the modulated signal.

Solution   The message signal is shown in Figure 3.13. First, we determine $m_n(t)$, the normalized message signal. In order to find $m_n(t)$, we have to determine $\max |m(t)|$.

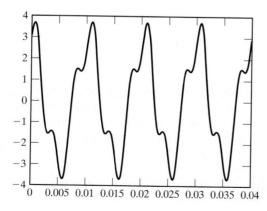

Figure 3.13   The message signal in Example 3.2.5.

To determine the extrema of $m(t)$, we find its derivative and make it equal to zero. We then have

$$m'(t) = -600\pi \sin(200\pi t) + 600\pi \cos(600\pi t)$$

$$= 0,$$

which results in

$$\cos(600\pi t) = \sin(200\pi t)$$

$$= \cos\left(\frac{\pi}{2} - 200\pi t\right).$$

One solution of this equation is $800\pi t = \frac{\pi}{2}$, or $t = \frac{1}{1600}$. Substituting this value into $m(t)$, we obtain

$$m\left(\frac{1}{1600}\right) = 3.6955,$$

which is the maximum value of the signal $m(t)$. Therefore,

$$m_n(t) = \frac{3\cos(200\pi t) + \sin(600\pi t)}{3.6955}$$

$$= 0.8118 \cos(200\pi t) + 0.2706 \sin(600\pi t).$$

The power in the sum of two sinusoids with different frequencies is the sum of powers in them. Therefore,

$$P_{m_n} = \frac{1}{2}[0.8118^2 + 0.2706^2]$$

$$= 0.3661.$$

The power in the carrier component of the modulated signal is

$$\frac{A_c^2}{2} = 0.5,$$

and the power in the sidebands is

$$\frac{A_c^2}{2} a^2 P_{m_n} = \frac{1}{2} \times 0.85^2 \times 0.3661$$

$$= 0.1323. \qquad\blacksquare$$

Demodulation of Conventional DSB-AM Signals.   The major advantage of conventional AM signal transmission is the ease in which the signal can be demodulated. There is no need for a synchronous demodulator. Since the message signal $m(t)$ satisfies the condition $|m(t)| < 1$, the envelope (amplitude) $1+m(t) > 0$. If we rectify the received signal, we eliminate the negative values without affecting the message signal, as shown in Figure 3.14.

The rectified signal is equal to $u(t)$ when $u(t) > 0$, and it is equal to zero when $u(t) < 0$. The message signal is recovered by passing the rectified signal through a

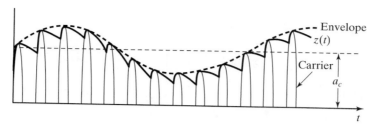

Figure 3.14 Envelope detection of a conventional AM signal.

lowpass filter whose bandwidth matches that of the message signal. The combination of the rectifier and the lowpass filter is called an *envelope detector.*

Ideally, the output of the envelope detector is of the form

$$d(t) = g_1 + g_2 m(t), \tag{3.2.7}$$

where g_1 represents a DC component and g_2 is a gain factor due to the signal demodulator. The DC component can be eliminated by passing $d(t)$ through a transformer, whose output is $g_2 m(t)$.

The simplicity of the demodulator has made conventional DSB-AM a practical choice for AM-radio broadcasting. Since there are literally billions of radio receivers, an inexpensive implementation of the demodulator is extremely important. The power inefficiency of conventional AM is justified by the fact that there are few broadcast transmitters relative to the number of receivers. Consequently, it is cost-effective to construct powerful transmitters and sacrifice power efficiency in order to simplify the signal demodulation at the receivers.

3.2.3 Single-Sideband AM

In Section 3.2.1, we showed that a DSB-SC AM signal required a channel bandwidth of $B_c = 2W$ Hz for transmission, where W is the bandwidth of the message signal. However, the two sidebands are redundant. We will demonstrate that the transmission of either sideband is sufficient to reconstruct the message signal $m(t)$ at the receiver. Thus, we reduce the bandwidth of the transmitted signal to that of the baseband message signal $m(t)$.

In the appendix at the end of this chapter, we will demonstrate that a single-sideband (SSB) AM signal is represented mathematically as

$$u(t) = A_c m(t) \cos 2\pi f_c t \mp A_c \hat{m}(t) \sin 2\pi f_c t, \tag{3.2.8}$$

where $\hat{m}(t)$ is the Hilbert transform of $m(t)$ that was introduced in Section 2.6, and the plus or minus sign determines which sideband we obtain. The plus sign indicates the lower sideband, and the minus sign indicates the upper sideband. Recall that the Hilbert transform may be viewed as a linear filter with impulse response $h(t) = 1/\pi t$

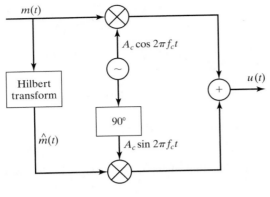

Figure 3.15   Generation of a lower single-sideband AM signal.

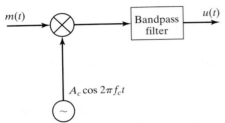

Figure 3.16   Generation of a single-sideband AM signal by filtering one of the sidebands of a DSB-SC AM signal.

and frequency response

$$H(f) = \begin{cases} -j, & f > 0 \\ j, & f < 0 \\ 0, & f = 0 \end{cases}. \tag{3.2.9}$$

Therefore, the SSB-AM signal $u(t)$ may be generated by using the system configuration shown in Figure 3.15.

The method shown in Figure 3.15 employs a Hilbert-transform filter. Another method, illustrated in Figure 3.16, generates a DSB-SC AM signal and then employs a filter that selects either the upper sideband or the lower sideband of the double-sideband AM signal.

Example 3.2.6

Suppose that the modulating signal is a sinusoid of the form

$$m(t) = \cos 2\pi f_m t, \quad f_m \ll f_c.$$

Determine the two possible SSB-AM signals.

Solution   The Hilbert transform of $m(t)$ is

$$\hat{m}(t) = \sin 2\pi f_m t. \tag{3.2.10}$$

Hence,

$$u(t) = A_c \cos 2\pi f_m t \cos 2\pi f_c t \mp A_c \sin 2\pi f_m t \sin 2\pi f_c t. \tag{3.2.11}$$

If we take the upper ($-$) sign, we obtain the upper-sideband signal

$$u_u(t) = A_c \cos 2\pi(f_c + f_m)t.$$

On the other hand, if we take the lower ($+$) sign in Equation (3.2.11), we obtain the lower-sideband signal

$$u_\ell(t) = A_c \cos 2\pi(f_c - f_m)t.$$

The spectra of $u_u(t)$ and $u_\ell(t)$ were previously given in Figure 3.3.   ∎

Demodulation of SSB-AM Signals.   To recover the message signal $m(t)$ in the received SSB-AM signal, we require a phase-coherent or synchronous demodulator, as was the case for DSB-SC AM signals. Thus, for the USSB signal given in Equation (3A.7), we have

$$r(t)\cos(2\pi f_c t + \phi) = u(t)\cos(2\pi f_c t + \phi)$$

$$= \frac{1}{2}A_c m(t)\cos\phi + \frac{1}{2}A_c \hat{m}(t)\sin\phi$$

$$+ \text{ double frequency terms.} \qquad (3.2.12)$$

By passing the product signal in Equation (3.2.12) through an ideal lowpass filter, the double-frequency components are eliminated, leaving us with

$$y_\ell(t) = \frac{1}{2}A_c m(t)\cos\phi + \frac{1}{2}A_c \hat{m}(t)\sin\phi. \qquad (3.2.13)$$

Note that the phase offset not only reduces the amplitude of the desired signal $m(t)$ by $\cos\phi$, but it also results in an undesirable sideband signal due to the presence of $\hat{m}(t)$ in $y_\ell(t)$. The latter component was not present in the demodulation of a DSB-SC signal. However, it is a factor that contributes to the distortion of the demodulated SSB signal.

The transmission of a pilot tone at the carrier frequency is a very effective method for providing a phase-coherent reference signal for performing synchronous demodulation at the receiver. Thus, the undesirable sideband-signal component is eliminated. However, this means that a portion of the transmitted power must be allocated to the transmission of the carrier.

The spectral efficiency of SSB AM makes this modulation method very attractive for use in voice communications over telephone channels (wirelines and cables). In this application, a pilot tone is transmitted for synchronous demodulation and shared among several channels.

The filter method shown in Figure 3.16, which selects one of the two signal sidebands for transmission, is particularly difficult to implement when the message signal $m(t)$ has a large power concentrated in the vicinity of $f = 0$. In such a case, the sideband filter must have an extremely sharp cutoff in the vicinity of the carrier in order to reject the second sideband. Such filter characteristics are very difficult to implement in practice.

3.2.4 Vestigial-Sideband AM

The stringent-frequency response requirements on the sideband filter in an SSB-AM system can be relaxed by allowing *vestige*, which is a portion of the unwanted sideband, to appear at the output of the modulator. Thus, we simplify the design of the sideband filter at the cost of a modest increase in the channel bandwidth required to transmit the signal. The resulting signal is called *vestigial-sideband* (VSB) AM. This type of modulation is appropriate for signals that have a strong low-frequency component, such as video signals. That is why this type of modulation is used in standard TV broadcasting.

To generate a VSB-AM signal, we begin by generating a DSB-SC AM signal and passing it through a sideband filter with the frequency response $H(f)$, as shown in Figure 3.17. In the time domain, the VSB signal may be expressed as

$$u(t) = [A_c m(t) \cos 2\pi f_c t] \star h(t), \tag{3.2.14}$$

where $h(t)$ is the impulse response of the VSB filter. In the frequency domain, the corresponding expression is

$$U(f) = \frac{A_c}{2}[M(f - f_c) + M(f + f_c)]H(f). \tag{3.2.15}$$

To determine the frequency-response characteristics of the filter, we will consider the demodulation of the VSB signal $u(t)$. We multiply $u(t)$ by the carrier component $\cos 2\pi f_c t$ and pass the result through an ideal lowpass filter, as shown in Figure 3.18. Thus, the product signal is

$$v(t) = u(t) \cos 2\pi f_c t,$$

or equivalently,

$$V(f) = \frac{1}{2}[U(f - f_c) + U(f + f_c)]. \tag{3.2.16}$$

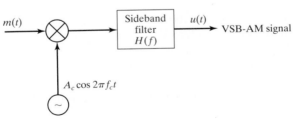

Figure 3.17 Generation of vestigial-sideband AM signal.

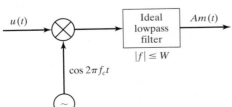

Figure 3.18 Demodulation of VSB signal.

If we substitute $U(f)$ from Equation (3.2.15) into Equation (3.2.16), we obtain

$$V(f) = \frac{A_c}{4}[M(f - 2f_c) + M(f)]H(f - f_c)$$

$$+ \frac{A_c}{4}[M(f) + M(f + 2f_c)]H(f + f_c). \tag{3.2.17}$$

The lowpass filter rejects the double-frequency terms and passes only the components in the frequency range $|f| \le W$. Hence, the signal spectrum at the output of the ideal lowpass filter is

$$V_\ell(f) = \frac{A_c}{4}M(f)[H(f - f_c) + H(f + f_c)]. \tag{3.2.18}$$

The message signal at the output of the lowpass filter must be undistorted. Hence, the VSB-filter characteristic must satisfy the condition

$$H(f - f_c) + H(f + f_c) = \text{constant} \quad |f| \le W. \tag{3.2.19}$$

This condition is satisfied by a filter that has the frequency-response characteristic shown in Figure 3.19.

We note that $H(f)$ selects the upper sideband and a vestige of the lower sideband. It has odd symmetry about the carrier frequency f_c in the frequency range $f_c - f_a < f < f_c + f_a$, where f_a is a conveniently selected frequency that is some small fraction of W, i.e., $f_a \ll W$. Thus, we obtain an undistorted version of the transmitted signal. Figure 3.20 illustrates the frequency response of a VSB filter that selects the lower sideband and a vestige of the upper sideband.

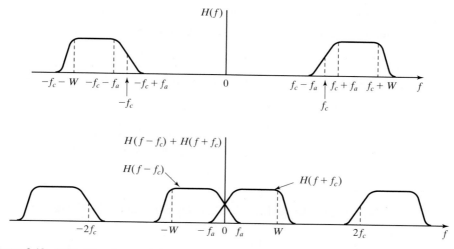

Figure 3.19 VSB-filter characteristics.

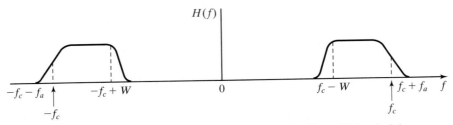

Figure 3.20 Frequency response of the VSB filter for selecting the lower sideband of the message signals.

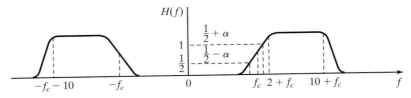

Figure 3.21 Frequency-response characteristics of the VSB filter in Example 3.2.7.

In practice, the VSB filter is designed to have some specified phase characteristic. To avoid distortion of the message signal, the VSB filter should have a linear phase over its passband $f_c - f_a \leq |f| \leq f_c + W$.

Example 3.2.7

Suppose that the message signal is given as

$$m(t) = 10 + 4\cos 2\pi t + 8\cos 4\pi t + 10\cos 20\pi t.$$

Specify both the frequency-response characteristic of a VSB filter that passes the upper sideband and the first frequency component of the lower sideband.

Solution The spectrum of the DSB-SC AM signal $u(t) = m(t)\cos 2\pi f_c t$ is

$$U(f) = 5[\delta(f - f_c) + \delta(f + f_c)] + 2[\delta(f - f_c - 1) + \delta(f + f_c + 1)]$$
$$+ 4[\delta(f - f_c - 2) + \delta(f + f_c + 2)] + 5[\delta(f - f_c - 10) + \delta(f + f_c + 10)].$$

The VSB filter can be designed to have unity gain in the range $2 \leq |f - f_c| \leq 10$, a gain of $1/2$ at $f = f_c$, a gain of $1/2 + \alpha$ at $f = f_c + 1$, and a gain of $1/2 - \alpha$ at $f = f_c - 1$, where α is some conveniently selected parameter that satisfies the condition $0 < \alpha < 1/2$. Figure 3.21 illustrates the frequency-response characteristic of the VSB filter. ∎

3.3 IMPLEMENTATION OF AM MODULATORS AND DEMODULATORS

There are several different methods for generating AM-modulated signals. In this section, we shall describe the methods most commonly used in practice. Since

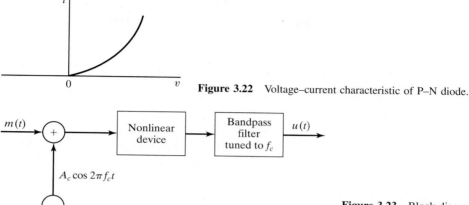

Figure 3.22   Voltage–current characteristic of P–N diode.

Figure 3.23   Block diagram of power-law AM modulator.

the process of modulation involves the generation of new frequency components, modulators are generally characterized as nonlinear and/or time-variant systems.

Power-Law Modulation.   Let us consider the use of a nonlinear device such as a P–N diode, which has voltage–current characteristic shown in Figure 3.22.

Suppose that the voltage input to such a device is the sum of the message signal $m(t)$ and the carrier $A_c \cos 2\pi f_c t$, as illustrated in Figure 3.23. The nonlinearity will generate a product of the message $m(t)$ with the carrier, plus additional terms. The desired modulated signal can be filtered out by passing the output of the nonlinear device through a bandpass filter.

To elaborate, suppose that the nonlinear device has an input–output (square-law) characteristic of the form

$$v_0(t) = a_1 v_i(t) + a_2 v_i^2(t), \tag{3.3.1}$$

where $v_i(t)$ is the input signal, $v_0(t)$ is the output signal, and the parameters (a_1, a_2) are constants. Then, if the input to the nonlinear device is

$$v_i(t) = m(t) + A_c \cos 2\pi f_c t, \tag{3.3.2}$$

its output is

$$
\begin{aligned}
v_0(t) &= a_1[m(t) + A_c \cos 2\pi f_c t] \\
&\quad + a_2[m(t) + A_c \cos \cos 2\pi f_c t]^2 \\
&= a_1 m(t) + a_2 m^2(t) + a_2 A_c^2 \cos^2 2\pi f_c t \\
&\quad + A_c a_1 \left[1 + \frac{2a_2}{a_1} m(t) \right] \cos 2\pi f_c t.
\end{aligned}
\tag{3.3.3}
$$

The output of the bandpass filter with a bandwidth $2W$ centered at $f = f_c$ yields

$$u(t) = A_c a_1 \left[1 + \frac{2a_2}{a_1} m(t)\right] \cos 2\pi f_c t, \qquad (3.3.4)$$

where $2a_2|m(t)|/a_1 < 1$ by design. Thus, the signal generated by this method is a conventional AM signal.

Switching Modulator.   Another method for generating an AM-modulated signal is by means of a switching modulator. Such a modulator can be implemented by the system illustrated in Figure 3.24(a). The sum of the message signal and the carrier $v_i(t)$, which is given by Equation (3.3.2), are applied to a diode that has the input–output voltage characteristic shown in Figure 3.24(b), where $A_c \gg m(t)$. The

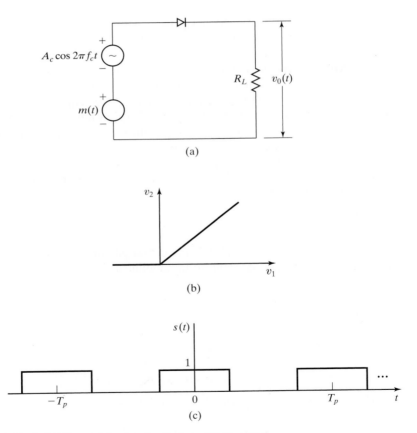

Figure 3.24   Switching modulator and periodic switching signal.

output across the load resistor is simply

$$v_0(t) = \begin{cases} v_i(t), & c(t) > 0 \\ 0, & c(t) < 0. \end{cases} \tag{3.3.5}$$

This switching operation may be viewed mathematically as a multiplication of the input $v_i(t)$ with the switching function $s(t)$, i.e.,

$$v_0(t) = [m(t) + A_c \cos 2\pi f_c t] s(t), \tag{3.3.6}$$

where $s(t)$ is shown in Figure 3.24(c).

Since $s(t)$ is a periodic function, it is represented in the Fourier series as

$$s(t) = \frac{1}{2} + \frac{2}{\pi} \sum_{n=1}^{\infty} \frac{(-1)^{n-1}}{2n-1} \cos[2\pi f_c t(2n-1)]. \tag{3.3.7}$$

This is similar to Equation (2.2.11).

Hence,

$$\begin{aligned} v_0(t) &= [m(t) + A_c \cos 2\pi f_c t] s(t) \\ &= \frac{A_c}{2}\left[1 + \frac{4}{\pi A_c} m(t)\right] \cos 2\pi f_c t + \text{other terms.} \end{aligned} \tag{3.3.8}$$

The desired AM-modulated signal is obtained by passing $v_0(t)$ through a bandpass filter with the center frequency $f = f_c$ and the bandwidth $2W$. At its output, we have the desired conventional AM signal

$$u(t) = \frac{A_c}{2}\left[1 + \frac{4}{\pi A_c} m(t)\right] \cos 2\pi f_c t. \tag{3.3.9}$$

Balanced Modulator.   A relatively simple method for generating a DSB-SC AM signal is to use two conventional-AM modulators arranged in the configuration illustrated in Figure 3.25.

For example, we may use two square-law AM modulators as previously described. Care must be taken to select modulators with approximately identical characteristics so that the carrier component cancels out at the summing junction.

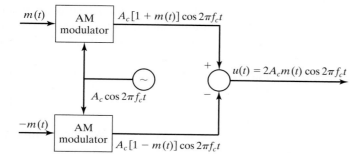

Figure 3.25   Block diagram of a balanced modulator.

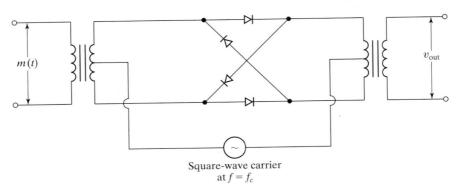

Square-wave carrier
at $f = f_c$

Figure 3.26 Ring modulator for generating a DSB-SC AM signal.

Ring Modulator. Another type of modulator for generating a DSB-SC AM signal is the *ring modulator* illustrated in Figure 3.26.

The switching of the diodes is controlled by a square wave of frequency f_c, denoted as $c(t)$, which is applied to the center taps of the two transformers. When $c(t) > 0$, the top and bottom diodes conduct, while the two diodes in the crossarms are off. In this case, the message signal $m(t)$ is multiplied by $+1$. When $c(t) < 0$, the diodes in the crossarms of the ring conduct, while the other two diodes are switched off. In this case, the message signal $m(t)$ is multiplied by -1. Consequently, the operation of the ring modulator may be described mathematically as a multiplier of $m(t)$ by the square-wave carrier $c(t)$, i.e.,

$$v_0(t) = m(t)c(t), \tag{3.3.10}$$

as shown in Figure 3.26.

Since $c(t)$ is a periodic function, it is represented by the Fourier series

$$c(t) = \frac{4}{\pi} \sum_{n=1}^{\infty} \frac{(-1)^{n-1}}{2n-1} \cos[2\pi f_c(2n-1)t]. \tag{3.3.11}$$

Again, this equation is similar to Equation (2.2.11). Hence, the desired DSB-SC AM signal $u(t)$ is obtained by passing $v_0(t)$ through a bandpass filter with the center frequency f_c and the bandwidth $2W$.

From the preceding discussion, we observe that the balanced modulator and the ring modulator systems, in effect, multiply the message signal $m(t)$ with the carrier to produce a DSB-SC AM signal. The multiplication of $m(t)$ with $A_c \cos w_c t$ is called a mixing operation. Hence, a *mixer* is basically a balanced modulator.

The method shown in Figure 3.15 for generating an SSB signal requires two mixers, i.e., two balanced modulators, in addition to the Hilbert transformer. On the other hand, the filter method illustrated in Figure 3.16 for generating an SSB signal requires a single balanced modulator and a sideband filter.

Let us now consider the demodulation of AM signals. We begin with a description of the envelope detector.

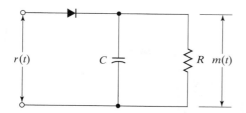

Figure 3.27   An envelope detector.

Envelope Detector.   As previously indicated, conventional DSB-AM signals are easily demodulated by an envelope detector. A circuit diagram for an envelope detector is shown in Figure 3.27. It consists of a diode and an RC circuit, which is basically a simple lowpass filter.

During the positive half-cycle of the input signal, the diode conducts and the capacitor charges up to the peak value of the input signal. When the input falls below the voltage on the capacitor, the diode becomes reverse-biased and the input disconnects from the output. During this period, the capacitor discharges slowly through the load resistor R. On the next cycle of the carrier, the diode again conducts when the input signal exceeds the voltage across the capacitor. The capacitor again charges up to the peak value of the input signal and the process is repeated.

The time constant RC must be selected to follow the variations in the envelope of the carrier-modulated signal. If RC is too small, then the output of the filter falls very rapidly after each peak and will not follow the envelope of the modulated signal closely. This corresponds to the case where the bandwidth of the lowpass filter is too large. If RC is too large, then the discharge of the capacitor is too slow and again the output will not follow the envelope of the modulated signal. This corresponds to the case where the bandwidth of the lowpass filter is too small. The effect of large and small RC values is shown in Figure 3.28.

In effect, for good performance of the envelope detector, we should have

$$\frac{1}{f_c} \ll RC \ll \frac{1}{W}.$$

In such a case, the capacitor discharges slowly through the resistor; thus, the output of the envelope detector, which we denote as $\tilde{m}(t)$, closely follows the message signal.

Example 3.3.1

An audio signal of bandwidth $W = 5$ kHz is modulated on a carrier of frequency 1 MHz using conventional AM modulation. Determine the range of values of RC for successful demodulation of this signal using an envelope detector.

Solution   We must have $\frac{1}{f_c} \ll RC \ll \frac{1}{W}$; therefore, $10^{-6} \ll RC \ll 2 \times 10^{-4}$. In this case, $RC = 10^{-5}$ is an appropriate choice.   ∎

Demodulation of DSB-SC AM Signals.   As previously indicated, the demodulation of a DSB-SC AM signal requires a synchronous demodulator. That is, the demodulator must use a coherent phase reference, which is usually generated by

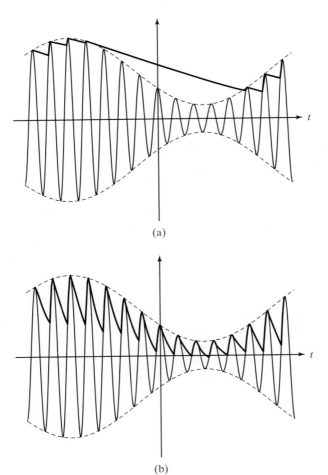

(a)

(b)

Figure 3.28 Effect of (a) large and (b) small RC values on the performance of the envelope detector.

means of a phase-locked loop (PLL) (see Section 6.4), to demodulate the received signal.

The general configuration is shown in Figure 3.29. A PLL is used to generate a phase-coherent carrier signal that is mixed with the received signal in a balanced modulator. The output of the balanced modulator is passed through a lowpass filter of bandwidth W that passes the desired signal and rejects all signal and noise components above W Hz. The characteristics and operation of the PLL are described in Section 6.4.

Demodulation of SSB Signals. The demodulation of SSB-AM signals also requires the use of a phase-coherent reference. In the case of signals, such as speech, that have relatively little or no power content at DC, it is simple to generate the SSB signal, as shown in Figure 3.16. Then, we can insert a small

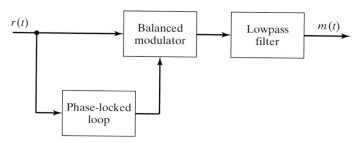

Figure 3.29 Demodulator for a DSB-SC signal.

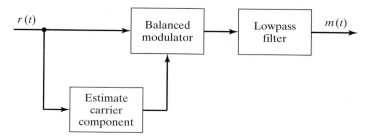

Figure 3.30 Demodulation of SSB-AM signal containing a carrier component.

carrier component that is transmitted along with the message. In such a case, we may use the configuration shown in Figure 3.30 to demodulate the SSB signal. We observe that a balanced modulator is used to convert the frequency of the bandpass signal to lowpass or baseband.

Demodulation of VSB Signals.   In VSB, a carrier component is generally transmitted along with the message sidebands. The existence of the carrier component makes it possible to extract a phase-coherent reference for demodulation in a balanced modulator, as shown in Figure 3.30.

In applications such as a TV broadcast, a large carrier component is transmitted along with the message in the VSB signal. In such a case, it is possible to recover the message by passing the received VSB signal through an envelope detector.

3.4 SIGNAL MULTIPLEXING

When we use a message signal $m(t)$ to modulate the amplitude of a sinusoidal carrier, we translate the message signal by an amount equal to the carrier frequency f_c. If we have two or more message signals to transmit simultaneously over the communication channel, we can have each message signal modulate a carrier of a different frequency, where the minimum separation between two adjacent carriers is either $2W$ (for DSB AM) or W (for SSB AM), where W is the bandwidth of each of the message signals. Thus, the various message signals occupy separate frequency bands of the channel and do not interfere with one another during transmission.

Combining separate message signals into a composite signal for transmission over a common channel is called *multiplexing*. There are two commonly used methods for signal multiplexing: (1) time-division multiplexing and (2) frequency-division multiplexing. Time-division multiplexing is usually used to transmit digital information; this will be described in a subsequent chapter. Frequency-division multiplexing (FDM) may be used with either analog or digital signal transmission.

3.4.1 Frequency-Division Multiplexing

In FDM, the message signals are separated in frequency, as previously described. A typical configuration of an FDM system is shown in Figure 3.31. This figure illustrates the frequency-division multiplexing of K message signals at the transmitter and their demodulation at the receiver. The lowpass filters at the transmitter ensure that the bandwidth of the message signals is limited to W Hz. Each signal modulates a separate carrier; hence, K modulators are required. Then, the signals from the K modulators are summed and transmitted over the channel. For SSB and VSB modulation, the modulator outputs are filtered prior to summing the modulated signals.

At the receiver of an FDM system, the signals are usually separated by passing through a parallel bank of bandpass filters. There, each filter is tuned to one of the carrier frequencies and has a bandwidth that is wide enough to pass the desired signal. The output of each bandpass filter is demodulated, and each demodulated signal is

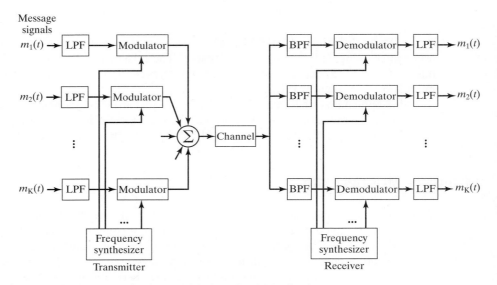

Figure 3.31   Frequency-division multiplexing of multiple signals.

fed to a lowpass filter that passes the baseband message signal and eliminates the double-frequency components.

FDM is widely used in radio and telephone communications. In telephone communications, each voice-message signal occupies a nominal bandwidth of 4 kHz. The message signal is single-sideband modulated for bandwidth-efficient transmission. In the first level of multiplexing, 12 signals are stacked in frequency, with a frequency separation of 4 kHz between adjacent carriers. Thus, a composite 48 kHz channel, called a *group channel*, transmits the 12 voice-band signals simultaneously. In the next level of FDM, a number of group channels (typically five or six) are stacked together in frequency to form a *supergroup channel*. Then the composite signal is transmitted over the channel. Higher-order multiplexing is obtained by combining several supergroup channels. Thus, an FDM hierarchy is employed in telephone communication systems.

3.4.2 Quadrature-Carrier Multiplexing

Another type of multiplexing allows us to transmit two message signals on the same carrier frequency. This type of multiplexing uses two quadrature carriers, $A_c \cos 2\pi f_c t$ and $A_c \sin 2\pi f_c t$. To elaborate, suppose that $m_1(t)$ and $m_2(t)$ are two separate message signals to be transmitted over the channel. The signal $m_1(t)$ amplitude modulates the carrier $A_c \cos 2\pi f_c t$, and the signal $m_2(t)$ amplitude modulates the quadrature carrier $A_c \sin 2\pi f_c t$. The two signals are added together and transmitted over the channel. Hence, the transmitted signal is

$$u(t) = A_c m_1(t) \cos 2\pi f_c t + A_c m_2(t) \sin 2\pi f_c t. \qquad (3.4.1)$$

Therefore, each message signal is transmitted by DSB-SC AM. This type of signal multiplexing is called *quadrature-carrier multiplexing*. Quadrature-carrier multiplexing results in a bandwidth-efficient communication system that is comparable in bandwidth efficiency to SSB AM. Figure 3.32 illustrates the modulation and demodulation of the quadrature-carrier multiplexed signals. As shown, a synchronous demodulator is required at the receiver to separate and recover the quadrature-carrier modulated signals.

Demodulation of $m_1(t)$ is done by multiplying $u(t)$ by $\cos 2\pi f_c t$ and then passing the result through a lowpass filter. We have

$$u(t) \cos 2\pi f_c t = A_c m_1(t) \cos^2 2\pi f_c t + A_c m_2(t) \cos 2\pi f_c t \sin 2\pi f_c t$$

$$= \frac{A_c}{2} m_1(t) + \frac{A_c}{2} m_1(t) \cos 4\pi f_c t + \frac{A_c}{2} m_2(t) \sin 4\pi f_c t.$$

This signal has a lowpass component $\frac{A_c}{2} m_1(t)$ and two high-frequency components. The lowpass component can be separated using a lowpass filter. Similarly, to demodule $m_2(t)$, we can multiply $u(t)$ by $\sin 2\pi f_c t$ and then pass the product through a lowpass filter.

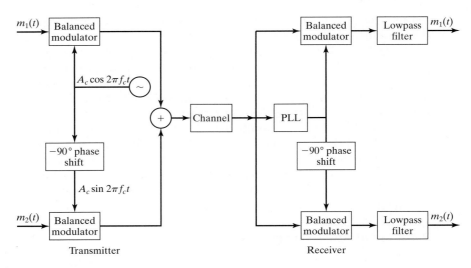

Figure 3.32   Quadrature-carrier multiplexing.

3.5 AM-RADIO BROADCASTING

AM-radio broadcasting is a familiar form of communication via analog-signal trans-mission. Commercial AM-radio broadcasting utilizes the frequency band 535–1605 kHz for the transmission of voice and music. The carrier-frequency allocations range from 540–1600 kHz with 10 kHz spacing.

Radio stations employ conventional AM for signal transmission. The baseband-message signal $m(t)$ is limited to a bandwidth of approximately 5 kHz. Since there are billions of receivers and relatively few radio transmitters, the use of conventional AM for broadcast is justified from an economic standpoint. The major objective is to reduce the cost of implementing the receiver.

The receiver most commonly used in AM-radio broadcast is the so-called *super-heterodyne receiver* shown in Figure 3.33. It consists of a radio-frequency (RF) tuned amplifier, a mixer, a local oscillator, an intermediate-frequency (IF) amplifier, an envelope detector, an audio-frequency amplifier, and a loudspeaker. Tuning for the desired radio frequency is provided by a variable capacitor, which simultaneously tunes the RF amplifier and the frequency of the local oscillator.

In the superheterodyne receiver, every AM-radio signal is converted to a com-mon IF frequency of $f_{IF} = 455$ kHz. This conversion allows the use of a single-tuned IF amplifier for signals from any radio station in the frequency band. The IF ampli-fier is designed to have a bandwidth of 10 kHz, which matches the bandwidth of the transmitted signal.

The frequency conversion to IF is performed by the combination of the RF amplifier and the mixer. The frequency of the local oscillator is

$$f_{LO} = f_c + f_{IF},$$

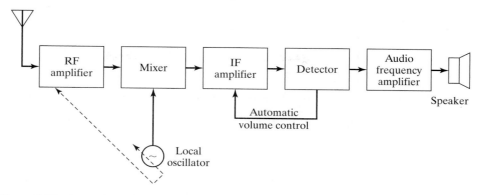

Figure 3.33   A superheterodyne receiver.

where f_c is the carrier frequency of the desired AM-radio signal. The tuning range of the local oscillator is 995–2055 kHz. By tuning the RF amplifier to the frequency f_c and mixing its output with the local oscillator frequency $f_{L0} = f_c + f_{IF}$, we obtain two signal components; one is centered at the difference frequency f_{IF}, and the second is centered at the sum frequency $2f_c + f_{IF}$. Only the first component is passed by the IF amplifier.

At the input to the RF amplifier, we have signals that are picked up by the antenna from all radio stations. By limiting the bandwidth of the RF amplifier to the range $B_c < B_{RF} < 2f_{IF}$, where B_c is the bandwidth of the AM-radio signal (10 kHz), we can reject the radio signal transmitted at the so-called *image frequency* $f_c' = f_{L0} + f_{IF}$. When we mix the local oscillator output $\cos 2\pi f_{L0}t$ with the received signals

$$r_1(t) = A_c[1 + m_1(t)] \cos 2\pi f_c t$$
$$r_2(t) = A_c[1 + m_2(t)] \cos 2\pi f_c' t,$$

where $f_c = f_{L0} - f_{IF}$ and $f_c' = f_{L0} + f_{IF}$, the mixer output consists of the two signals

$$y_1(t) = A_c[1 + m_1(t)] \cos 2\pi f_{IF}(t) + \text{double frequency term}$$
$$y_2(t) = A_c[1 + m_2(t)] \cos 2\pi f_{IF}(t) + \text{double frequency term},$$

where $m_1(t)$ represents the desired signal and $m_2(t)$ is the signal sent by the radio station transmitting at the carrier frequency $f_c' = f_{L0} + f_{IF}$. To prevent the signal $r_2(t)$ from interfering with the demodulation of the desired signal $r_1(t)$, the RF-amplifier bandwidth is sufficiently narrow so the image-frequency signal is rejected. Hence, $B_{RF} < 2f_{IF}$ is the upper limit on the bandwidth of the RF amplifier. In spite of this constraint, the bandwidth of the RF amplifier is still considerably wider than the bandwidth of the IF amplifier. Thus, the IF amplifier, with its narrow bandwidth, provides signal rejection from adjacent channels, and the RF amplifier provides signal rejection from image channels. Figure 3.34 illustrates the bandwidths of both the RF and IF amplifiers and the requirement for rejecting the image-frequency signal.

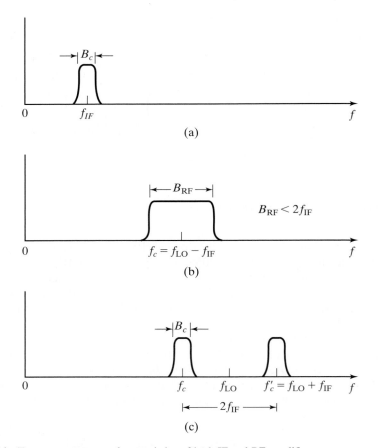

Figure 3.34   Frequency response characteristics of both IF and RF amplifiers.

The output of the IF amplifier is passed through an envelope detector, which produces the desired audio-band message signal $m(t)$. Finally, the output of the envelope detector is amplified, and this amplified signal drives a loudspeaker. Automatic volume control (AVC) is provided by a feedback-control loop, which adjusts the gain of the IF amplifier based on the power level of the signal at the envelope detector.

APPENDIX 3A: DERIVATION OF THE EXPRESSION FOR SSB-AM SIGNALS

Let $m(t)$ be a signal with the Fourier transform (spectrum) $M(f)$. An upper single-sideband amplitude-modulated signal (USSB AM) is obtained by eliminating the lower sideband of a DSB amplitude-modulated signal. Suppose we eliminate the lower sideband of the DSB AM signal, $u_{\text{DSB}}(t) = 2A_c m(t) \cos 2\pi f_c t$, by passing it through a highpass filter whose transfer function is given by

$$H(f) = \begin{cases} 1, & |f| > f_c \\ 0, & \text{otherwise,} \end{cases}$$

as shown in Figure 3.16. Obviously, $H(f)$ can be written as

$$H(f) = u_{-1}(f - f_c) + u_{-1}(-f - f_c),$$

where $u_{-1}(\cdot)$ represents the unit-step function. Therefore, the spectrum of the USSB-AM signal is given by

$$U_u(f) = A_c M(f - f_c)u_{-1}(f - f_c) + A_c M(f + f_c)u_{-1}(-f - f_c),$$

or equivalently,

$$U_u(f) = A_c M(f)u_{-1}(f)|_{f=f-f_c} + A_c M(f)u_{-1}(-f)|_{f=f+f_c}. \tag{3A.1}$$

Taking the inverse Fourier transform of both sides of Equation (3A.1) and using the modulation and convolution properties of the Fourier transform, as shown in Example 2.3.14 and Equation (2.3.26), we obtain

$$u_u(t) = A_c m(t) \star \mathscr{F}^{-1}\left[u_{-1}(f)\right] e^{j2\pi f_c t} +$$
$$A_c m(t) \star \mathscr{F}^{-1}\left[u_{-1}(-f)\right] e^{-j2\pi f_c t}. \tag{3A.2}$$

Next, we note that

$$\mathscr{F}\left[\frac{1}{2}\delta(t) + \frac{j}{2\pi t}\right] = u_{-1}(f)$$

$$\mathscr{F}\left[\frac{1}{2}\delta(t) - \frac{j}{2\pi t}\right] = u_{-1}(-f), \tag{3A.3}$$

which follows from Equation (2.3.12) and the duality theorem of the Fourier transform. Substituting Equation (3A.3) in Equation (3A.2), we obtain

$$u_u(t) = A_c m(t) \star \left[\frac{1}{2}\delta(t) + \frac{j}{2\pi t}\right] e^{j2\pi f_c t}$$

$$+ A_c m(t) \star \left[\frac{1}{2}\delta(t) - \frac{j}{2\pi t}\right] e^{-j2\pi f_c t}$$

$$= \frac{A_c}{2}\left[m(t) + j\hat{m}(t)\right] e^{j2\pi f_c t} +$$

$$\frac{A_c}{2}\left[m(t) - j\hat{m}(t)\right] e^{-j2\pi f_c t}, \tag{3A.4}$$

where we have used the identities

$$m(t) \star \delta(t) = m(t)$$

$$m(t) \star \frac{1}{\pi t} = \hat{m}(t).$$

Using Euler's relations in Equation (3A.4), we obtain

$$u_u(t) = A_c m(t) \cos 2\pi f_c t - A_c \hat{m}(t) \sin 2\pi f_c t, \qquad (3A.5)$$

which is the time-domain representation of a USSB-AM signal. The expression for the LSSB-AM signal can be derived by noting that

$$u_u(t) + u_\ell(t) = u_{\text{DSB}}(t)$$

or

$$A_c m(t) \cos 2\pi f_c t - A_c \hat{m}(t) \sin 2\pi f_c t + u_\ell(t) = 2 A_c m(t) \cos 2\pi f_c t.$$

Therefore,

$$u_\ell(t) = A_c m(t) \cos 2\pi f_c t + A_c \hat{m}(t) \sin 2\pi f_c t. \qquad (3A.6)$$

Thus, the time-domain representation of a SSB-AM signal can generally be expressed as

$$u_{\text{SSB}}(t) = A_c m(t) \cos 2\pi f_c t \mp A_c \hat{m}(t) \sin 2\pi f_c t, \qquad (3A.7)$$

where the minus sign corresponds to the USSB-AM signal, and the plus sign corresponds to the LSSB-AM signal.

PROBLEMS

3.1 The message signal $m(t) = 2 \cos 400t + 4 \sin(500t + \frac{\pi}{3})$ modulates the carrier signal $c(t) = A \cos(8000\pi t)$, using DSB-amplitude modulation. Find the time domain and frequency domain representation of the modulated signal and plot the spectrum (Fourier transform) of the modulated signal. What is the power content of the modulated signal?

3.2 In a DSB system, the carrier is $c(t) = A \cos 2\pi f_c t$ and the message signal is given by $m(t) = \text{sinc}(t) + \text{sinc}^2(t)$. Find the frequency-domain representation and the bandwidth of the modulated signal.

3.3 The two signals (a) and (b), shown in Figure P-3.3, DSB modulate a carrier signal $c(t) = A \cos 2\pi f_0 t$. Precisely plot the resulting modulated signals as a function of time and discuss their differences and similarities.

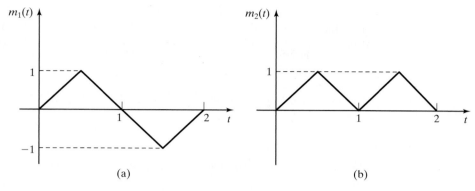

Figure P-3.3

3.4 Suppose the signal $x(t) = m(t) + \cos 2\pi f_c t$ is applied to a nonlinear system whose output is $y(t) = x(t) + \frac{1}{2}x^2(t)$.

1. Determine and sketch the spectrum of $y(t)$ when $M(f)$ is as shown in Figure P-3.4 and $W \ll f_c$.

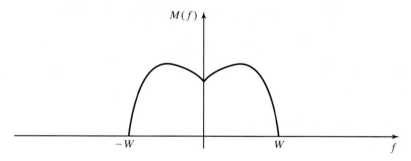

Figure P-3.4

3.5 The modulating signal

$$m(t) = 2\cos 4000\pi t + 5\cos 6000\pi t$$

is multiplied by the carrier

$$c(t) = 100\cos 2\pi f_c t,$$

where $f_c = 50$ kHz. Determine and sketch the spectrum of the DSB signal.

3.6 A DSB-modulated signal $u(t) = Am(t)\cos 2\pi f_c t$ is mixed (multiplied) with a local carrier $x_L(t) = \cos(2\pi f_c t + \theta)$, and the output is passed through an

LPF with a bandwidth equal to the bandwidth of the message $m(t)$. The signal power at the output of the lowpass filter is denoted by P_{out}. The modulated signal power is denoted by P_U. Plot $\frac{P_{out}}{P_U}$ as a function of θ for $0 \le \theta \le \pi$.

3.7 An AM signal has the form

$$u(t) = [20 + 2\cos 3000\pi t + 10\cos 6000\pi t]\cos 2\pi f_c t,$$

where $f_c = 10^5$ Hz.

    1. Sketch the (voltage) spectrum of $u(t)$.

    2. Determine the power in each of the frequency components.

    3. Determine the modulation index.

    4. Determine the sidebands' power, the total power, and the ratio of the sidebands' power to the total power.

3.8 A message signal $m(t) = \cos 2000\pi t + 2\cos 4000\pi t$ modulates the carrier $c(t) = 100\cos 2\pi f_c t$, where $f_c = 1$ MHz to produce the DSB signal $m(t)c(t)$.

    1. Determine the expression for the upper-sideband (USB) signal.

    2. Determine and sketch the spectrum of the USB signal.

3.9 A DSB-SC signal is generated by multiplying the message signal $m(t)$ with the periodic rectangular waveform (shown in Figure P-3.6), then filtering the product with a bandpass filter tuned to the reciprocal of the period T_p, with the bandwidth $2W$, where W is the bandwidth of the message signal. Demonstrate that the output $u(t)$ of the BPF is the desired DSB-SC AM signal

$$u(t) = m(t)\sin 2\pi f_c t,$$

where $f_c = 1/T_p$.

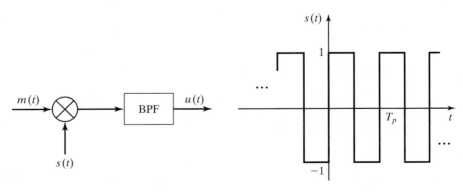

Figure P-3.9

3.10 Show that while generating a DSB-SC signal in Problem 3.9, it is not necessary for the periodic signal to be rectangular. This means that any periodic signal with the period T_p can substitute for the rectangular signal in Figure P-3.9.

3.11 The message signal $m(t)$ has the Fourier transform shown in Figure P-3.11(a). This signal is applied to the system shown in Figure P-3.11(b) to generate the signal $y(t)$.

    1. Plot $Y(f)$, the Fourier transform of $y(t)$.

    2. Show that if $y(t)$ is transmitted, the receiver can pass it through a replica of the system shown in Figure 3.11(b) to obtain $m(t)$ back. This means that this system can be used as a simple scrambler to enhance communication privacy.

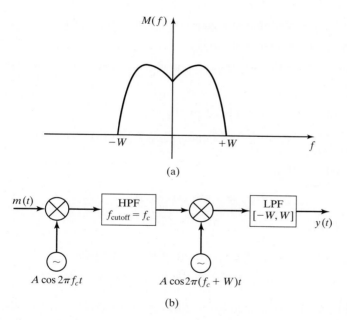

(a)

(b)

Figure P-3.11

3.12 Show that in a DSB-modulated signal, the envelope of the resulting bandpass signal is proportional to the *absolute value* of the message signal. This means that an envelope detector can be employed as a DSB demodulator if we know that the message signal is always positive.

3.13 An AM signal is generated by modulating the carrier $f_c = 800$ kHz by the signal

$$m(t) = \sin 2000\pi t + 5\cos 4000\pi t.$$

The AM signal
$$u(t) = 100 \, [1 + m \, (t)] \cos 2\pi f_c t$$

is fed to a 50-Ω load.

1. Determine and sketch the spectrum of the AM signal.
2. Determine the average power in the carrier and in the sidebands.
3. What is the modulation index?
4. What is the peak power delivered to the load?

3.14 The output signal from an AM modulator is

$$u \, (t) = 5 \cos 1800\pi t + 20 \cos 2000\pi t + 5 \cos 2200\pi t.$$

1. Determine the modulating signal $m \, (t)$ and the carrier $c \, (t)$.
2. Determine the modulation index.
3. Determine the ratio of the power in the sidebands to the power in the carrier.

3.15 A DSB-SC AM signal is modulated by the signal

$$m \, (t) = 2 \cos 2000\pi t + \cos 6000\pi t.$$

The modulated signal is

$$u(t) = 100m \, (t) \cos 2\pi f_c t,$$

where $f_c = 1$ MHz.

1. Determine and sketch the spectrum of the AM signal.
2. Determine the average power in the frequency components.

3.16 An SSB-AM signal is generated by modulating an 800 kHz carrier by the signal $m \, (t) = \cos 2000\pi t + 2 \sin 2000\pi t$. The amplitude of the carrier is $A_c = 100$.

1. Determine the signal $\hat{m} \, (t)$.
2. Determine the (time-domain) expression for the lower sideband of the SSB-AM signal.
3. Determine the magnitude spectrum of the lower-sideband-SSB signal.

3.17 Weaver's SSB modulator is illustrated in Figure P-3.17. By taking the input signal as $m \, (t) = \cos 2\pi f_m t$ where $f_m < W$, demonstrate that by proper choice of f_1 and f_2, the output is a SSB signal.

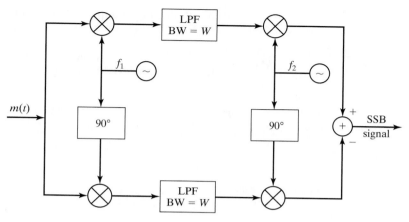

Figure P-3.17

3.18 The message signal $m(t)$, whose spectrum is shown in Figure P-3.18, is passed through the system shown in that figure.

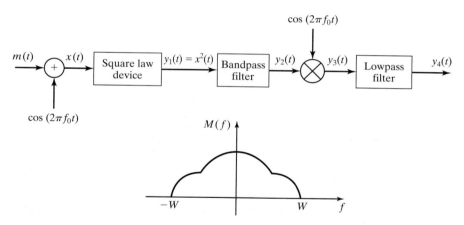

Figure P-3.18

The bandpass filter has a bandwidth of $2W$ centered at f_0, and the lowpass filter has a bandwidth of W. Plot the spectra of the signals $x(t)$, $y_1(t)$, $y_2(t)$, $y_3(t)$, and $y_4(t)$. What are the bandwidths of these signals?

3.19 The system shown in Figure P-3.19 is used to generate an AM signal. The modulating signal $m(t)$ has zero mean and its maximum (absolute) value is $A_m = \max |m(t)|$. The nonlinear device has the input–output characteristic

$$y(t) = ax(t) + bx^2(t).$$

1. Express $y(t)$ in terms of the modulating signal $m(t)$ and the carrier $c(t) = \cos 2\pi f_c t$.

2. What is the modulation index?

3. Specify the filter characteristics that yield an AM signal at its output.

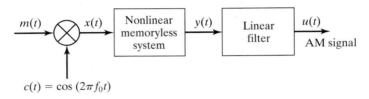

$$c(t) = \cos(2\pi f_0 t)$$

Figure P-3.19

3.20 The signal $m(t)$, whose Fourier transform $M(f)$ is shown in Figure P-3.20, is to be transmitted from point A to point B. We know that the signal is normalized, meaning that $-1 \le m(t) \le 1$.

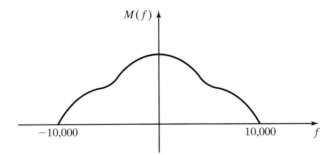

Figure P-3.20

1. If USSB is employed, what is the bandwidth of the modulated signal?

2. If DSB is employed, what is the bandwidth of the modulated signal?

3. If an AM-modulation scheme with $a = 0.8$ is used, what is the bandwidth of the modulated signal?

3.21 A vestigial-sideband modulation system is shown in Figure P-3.21. The bandwidth of the message signal $m(t)$ is W, and the transfer function of the bandpass filter is shown in the figure.

1. Determine $h_l(t)$, which is the lowpass equivalent of $h(t)$, where $h(t)$ represents the impulse response of the bandpass filter.

2. Derive an expression for the modulated signal $u(t)$.

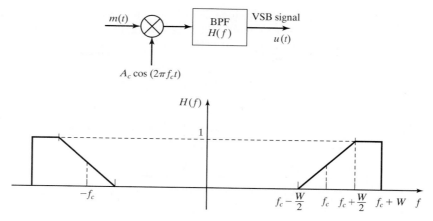

Figure P-3.21

3.22 Find expressions for the in-phase and quadrature components, $x_c(t)$ and $x_s(t)$, as well as the envelope and phase, $V(t)$ and $\Theta(t)$, for DSB, SSB, conventional AM, USSB, and LSSB.

3.23 The normalized signal $m_n(t)$ has a bandwidth of 10,000 Hz, and its power content is 0.5 Watts. The carrier $A \cos 2\pi f_0 t$ has a power content of 200 Watts.

    1. If $m_n(t)$ modulates the carrier using SSB-amplitude modulation, what will be the bandwidth and the power content of the modulated signal?

    2. If the modulation scheme is DSB SC, what is the answer to Part 1?

    3. If the modulation scheme is AM with a modulation index of 0.6, what is the answer to Part 1?

3.24 We wish to transmit 60 voice-band signals by SSB (upper-sideband) modulation and frequency-division multiplexing (FDM). Each of the 60 signals has a spectrum as shown in Figure P-3.24. Note that the voiceband signal is band limited to 3 kHz. If each signal is frequency translated separately, we require a frequency synthesizer that produces 60 carrier frequencies to perform the frequency-division multiplexing. On the other hand, if we subdivide the channels into L groups of K subchannels each, such that $LK = 60$, we may reduce the number of frequencies from the synthesizer to $L + K$.

    1. Illustrate the spectrum of the SSB signals in a group of K subchannels. Assume that a 1 kHz guard band separates the signals in adjacent frequency subchannels and that the carrier frequencies are $f_{c_1} = 10$ kHz, $f_{c_2} = 14$ kHz, ..., etc.

    2. Sketch L and K such that $LK = 60$ and $L + K$ is a minimum.

3. Determine the frequencies of the carriers if the 60-FDM signals occupy the frequency band 300 kHz–540 kHz, and each group of K signals occupies the band 10 kHz to $(10 + 4K)$ kHz.

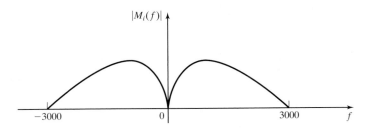

Figure P-3.24

COMPUTER PROBLEMS

3.1 Double-sideband (DSB) AM

The message signal $m(t)$ is given by

$$m(t) = \begin{cases} \text{sinc } (100t), & 0 \le t \le t_0 \\ 0, & \text{otherwise} \end{cases},$$

where $t_0 = 0.1$. The message signal modulates the carrier $c(t) = \cos 2\pi f_c t$, where $f_c = 250$ Hz, to produce a DSB-AM signal $u(t)$.

1. By selecting the sampling interval $t_s = 0.0001$, generate samples of $m(t)$ and $u(t)$ for $0 \le t \le t_0$ and plot them.
2. Determine and plot the spectra of $m(t)$ and $u(t)$.
3. Repeat Parts 1 and 2 when $t_0 = 0.4$, and comment on the results between $t_0 = 0.1$ and $t_0 = 0.4$.

3.2 Conventional AM

The message signal $m(t)$, which is given in Problem CP-3.1, modulates the carrier $c(t) = \cos 2\pi f_c t$ using conventional AM. The carrier frequency is $f_c = 250$ Hz and the modulation index is $a = 0.80$.

1. Plot the message signal $m(t)$ and the modulated signal $u(t)$, using a sampling interval $t_s = 0.0001$.
2. Determine and plot the spectra of the message signal $m(t)$ and the modulated signal $u(t)$.
3. Repeat Parts 1 and 2 when $t_0 = 0.4$, and comment on the results between $t_0 = 0.1$ and $t_0 = 0.4$.

3.3 Single-sideband AM

The message signal $m(t)$, which is given in Problem CP-3.1, modulates the carrier $c(t) = \cos 2\pi f_c t$ and produces the lower SSB signal $u(t)$. The carrier frequency is $f_c = 250$ Hz.

1. Plot the message signal $m(t)$, its Hilbert transform $\hat{m}(t)$, and the modulated LSSB signal $u(t)$.

2. Determine and plot the spectra of the message signal $m(t)$ and the modulated LSSB signal $u(t)$.

3. Repeat Parts 1 and 2 when $t_0 = 0.4$, and comment on the results between $t_0 = 0.1$ and $t_0 = 0.4$.

3.4 Demodulation of the DSB-AM Signal

The message signal $m(t)$, which is given in Problem CP-3.1, modulates the carrier $c(t) = \cos 2\pi f_c t$ and results in the DSB-AM signal $u(t) = m(t)c(t)$. The carrier frequency $f_c = 250$ Hz and $t_0 = 0.1$.

1. By selecting the sampling interval $t_s = 0.0001$, generate 1000 samples of the message signal $m(t)$ and the modulated signal $u(t)$, and plot both signals.

2. Demodulate the sampled DSB-AM signal $u(t)$ generated in Part 1 by using the demodulator shown in Figure CP-3.4. Perform the demodulation for $\phi = 0$, $\pi/8$, $\pi/4$, and $\pi/2$, and plot the received message signal $m_r(t)$. The lowpass filter is a linear-phase-FIR filter having 31 taps, a cutoff frequency (-3 dB) of 100 Hz, and a stopband attenuation of at least 30 dB.

3. Comment on the results obtained in Part 2.

4. Instead of using a time-domain-lowpass filter to reject the frequency components centered at $2f_c$, compute the discrete-Fourier transform (DFT) of 1000 samples of the mixer output, set to zero those frequency components centered at $2f_c$ and compute the inverse DFT to obtain the time-domain signal. Compare the results of this frequency-domain filtering with the time-domain filtering in Part 2.

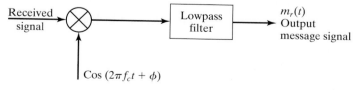

Figure CP-3.4   Demodulation for DSB-SC AM signal.

3.5 Demodulation of the SSB-AM signal

The message signal $m(t)$, which is given in Problem CP-3.1, modulates the carrier $c(t) = \cos 2\pi f_c t$ and produces the lower SSB signal $u(t)$. The carrier frequency is $f_c = 250$ Hz.

1. By selecting the sampling interval $t_s = 0.0001$, generate 1000 samples of the message signal $m(t)$, its Hilbert transform $\hat{m}(t)$, and the modulated LSSB signal $u(t)$. Plot these three signals.

2. Demodulate the sampled LSSB signal $u(t)$ generated in Part 1 by using the demodulator shown in Figure CP-3.4. Perform the demodulation for $\phi = 0$, $\pi/8$, $\pi/4$, and $\pi/2$, and plot the demodulated received message signal $m_r(t)$. The characteristics of the lowpass filter to be designed are given in Part 2 of Problem CP-3.4.

3. Comment on the results obtained in Part 2.

4. Instead of using the time-domain filter in Part 2, suppose the filtering of the frequency components centered at $2f_c$ is performed in the frequency domain by using the DFT as described in Part 4 of Problem CP-3.4. Perform the filtering in the frequency domain and compare the demodulated signal with that obtained by time-domain filtering.

3.6 Demodulation of Conventional AM

The message signal $m(t)$, which is given in Problem CP-3.1, modulates the carrier $c(t) = \cos 2\pi f_c t$ to produce a conventional-AM signal. The carrier frequency is $f_c = 250$ Hz and the modulation index is $a = 0.80$.

1. By selecting the sampling interval $t_s = 0.0001$, generate 1000 samples of the message signal $m(t)$ and the modulated conventional AM signal $u(t)$. Plot these two signals.

2. Demodulate the sampled conventional-AM signal $u(t)$ generated in Part 1 by computing the envelope of $u(t)$, i.e., by computing

$$e(t) = \sqrt{[1 + am(t)]^2} = |1 + am(t)|$$

and subtracting the DC value term to obtain the demodulated signal $m_r(t)$. Plot the demodulated received-message signal $m_r(t)$.

3. Comment on the results obtained in Part 2.

4

Angle Modulation

In Chapter 3, we considered amplitude modulation of the carrier as a means for transmitting the message signal. Amplitude-modulation methods are also called *linear-modulation methods*, although conventional AM is not linear in the strict sense.

Another class of modulation methods include frequency and phase modulation, which are described in this chapter. In frequency-modulation (FM) systems, the frequency of the carrier f_c is changed by the message signal; in phase modulation (PM) systems, the phase of the carrier is changed according to the variations in the message signal. Frequency and phase modulation are nonlinear, and often they are jointly called *angle-modulation methods*. In the following sections, we will show that, angle modulation, due to its inherent nonlinearity, is more complex to implement and much more difficult to analyze. In many cases, only an approximate analysis can be done. Another property of angle modulation is its bandwidth-expansion property. Frequency and phase modulation systems generally expand the bandwidth such that the effective bandwidth of the modulated signal is usually many times the bandwidth of the message signal.[1] With a higher implementation complexity and a higher bandwidth occupancy, we would naturally question the usefulness of these systems. As our analysis in Chapter 6 will show, the major benefit of these systems is their high degree of noise immunity. In fact, these systems sacrifice bandwidth for high-noise immunity. That is the reason that FM systems are widely used in high-fidelity music broadcasting and point-to-point communication systems, where the transmitter power is quite limited. Another advantage of angle-modulated signals is their constant envelope, which is beneficial when amplified by nonlinear amplifiers.

---

[1] Strictly speaking, the bandwidth of the modulated signal is infinite. That is why we talk about the *effective bandwidth*.

4.1 REPRESENTATION OF FM AND PM SIGNALS

An angle-modulated signal generally can be written as

$$u(t) = A_c \cos(2\pi f_c t + \phi(t)), \qquad (4.1.1)$$

where f_c denotes the carrier frequency and $\phi(t)$ denotes a time-varying phase. The instantaneous frequency of this signal is given by

$$f_i(t) = f_c + \frac{1}{2\pi} \frac{d}{dt} \phi(t). \qquad (4.1.2)$$

If $m(t)$ is the message signal, then in a PM system, the phase is proportional to the message, i.e.,

$$\phi(t) = k_p m(t), \qquad (4.1.3)$$

and in an FM system, the instantaneous frequency deviation from the carrier frequency is proportional with the message signal, i.e.,

$$f_i(t) - f_c = k_f m(t) = \frac{1}{2\pi} \frac{d}{dt} \phi(t), \qquad (4.1.4)$$

where k_p and k_f are phase and frequency *deviation constants*. From the preceding relationships, we have

$$\phi(t) = \begin{cases} k_p m(t), & \text{PM} \\ 2\pi k_f \int_{-\infty}^{t} m(\tau) d\tau, & \text{FM} \end{cases}. \qquad (4.1.5)$$

The foregoing expression shows a close and interesting relationship between FM and PM systems. This close relationship allows us to analyze these systems in parallel and only emphasize their main differences. First, note that if we phase modulate the carrier with the integral of a message, it is equivalent to the frequency modulation of the carrier with the original message. On the other hand, this relation can be expressed as

$$\frac{d}{dt} \phi(t) = \begin{cases} k_p \frac{d}{dt} m(t), & \text{PM,} \\ 2\pi k_f m(t), & \text{FM} \end{cases} \qquad (4.1.6)$$

which shows that if we frequency modulate the carrier with the derivative of a message, the result is equivalent to the phase modulation of the carrier with the message itself. Figure 4.1 shows the above relation between FM and PM. Figure 4.2 illustrates a square-wave signal and its integral, a sawtooth signal, and their corresponding FM and PM signals.

The demodulation of an FM signal involves finding the instantaneous frequency of the modulated signal and then subtracting the carrier frequency from it. In the demodulation of PM, the demodulation process is done by finding the phase of the signal and then recovering $m(t)$. The maximum phase deviation in a PM system is given by

$$\Delta\phi_{\max} = k_p \max[|m(t)|], \qquad (4.1.7)$$

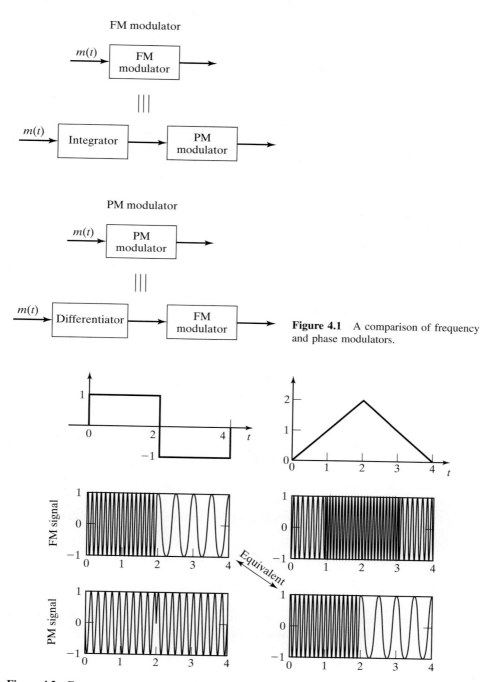

Figure 4.1   A comparison of frequency and phase modulators.

Figure 4.2   Frequency and phase modulation of square and sawtooth waves.

and the maximum frequency deviation in an FM system is given by

$$\Delta f_{\max} = k_f \max[|m(t)|]. \tag{4.1.8}$$

Example 4.1.1

The message signal

$$m(t) = a\cos(2\pi f_m t)$$

is used to either frequency modulate or phase modulate the carrier $A_c \cos(2\pi f_c t)$. Find the modulated signal in each case.

Solution   In PM, we have

$$\phi(t) = k_p m(t) = k_p a \cos(2\pi f_m t), \tag{4.1.9}$$

and in FM, we have

$$\phi(t) = 2\pi k_f \int_{-\infty}^{t} m(\tau)\, d\tau = \frac{k_f a}{f_m} \sin(2\pi f_m t). \tag{4.1.10}$$

Therefore, the modulated signals will be

$$u(t) = \begin{cases} A_c \cos\left(2\pi f_c t + k_p a \cos(2\pi f_m t)\right), & \text{PM} \\ A_c \cos\left(2\pi f_c t + \frac{k_f a}{f_m} \sin(2\pi f_m t)\right), & \text{FM} \end{cases}. \tag{4.1.11}$$

By defining

$$\beta_p = k_p a \tag{4.1.12}$$

and

$$\beta_f = \frac{k_f a}{f_m}, \tag{4.1.13}$$

we have

$$u(t) = \begin{cases} A_c \cos\left(2\pi f_c t + \beta_p \cos(2\pi f_m t)\right), & \text{PM} \\ A_c \cos\left(2\pi f_c t + \beta_f \sin(2\pi f_m t)\right), & \text{FM} \end{cases}. \tag{4.1.14}$$

The parameters β_p and β_f are called the *modulation indices* of the PM and FM systems, respectively. ∎

We can extend the definition of the modulation index for a general nonsinusoidal signal $m(t)$ as

$$\beta_p = k_p \max[|m(t)|]; \tag{4.1.15}$$

$$\beta_f = \frac{k_f \max[|m(t)|]}{W}, \tag{4.1.16}$$

where W denotes the bandwidth of the message signal $m(t)$. In terms of the maximum phase and frequency deviation $\Delta\phi_{\max}$ and $\Delta f_{\max}$, we have

$$\beta_p = \Delta\phi_{\max}; \tag{4.1.17}$$

$$\beta_f = \frac{\Delta f_{\max}}{W}. \tag{4.1.18}$$

Narrowband Angle Modulation[2]

Consider an angle modulation system in which the deviation constants k_p and k_f and the message signal $m(t)$ are such that for all t, we have $\phi(t) \ll 1$. Then we can use a simple approximation to expand $u(t)$ in Equation (4.1.1) as

$$u(t) = A_c \cos 2\pi f_c t \cos \phi(t) - A_c \sin 2\pi f_c t \sin \phi(t)$$

$$\approx A_c \cos 2\pi f_c t - A_c \phi(t) \sin 2\pi f_c t, \qquad (4.1.19)$$

where we have used the approximations $\cos \phi(t) \approx 1$ and $\sin \phi(t) \approx \phi(t)$ for $\phi(t) \ll 1$. Equation (4.1.19) shows that in this case, the modulated signal is very similar to a conventional-AM signal given in Equation (3.2.5). The only difference is that the message signal $m(t)$ is modulated on a sine carrier rather than a cosine carrier. The bandwidth of this signal is similar to the bandwidth of a conventional AM signal, which is twice the bandwidth of the message signal. Of course, this bandwidth is only an approximation of the real bandwidth of the FM signal. A phasor diagram for this signal and the comparable conventional-AM signal are given in Figure 4.3. Compared to conventional AM, the narrowband angle-modulation scheme has far less amplitude variations. Of course, the angle-modulation system has constant amplitude and, hence, there should be no amplitude variations in the phasor-diagram representation of the system. These slight variations are due to the first-order approximation that we have used for the expansions of $\sin(\phi(t))$ and $\cos(\phi(t))$. As we will see later, the narrowband angle-modulation method does not provide better noise immunity than a conventional AM system. Therefore, narrowband angle-modulation is seldom used in practice for communication purposes. However, these systems can be used as an intermediate stage for the generation of wideband angle-modulated signals, as we will discuss in Section 4.3.

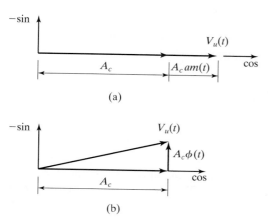

(a)

(b)

Figure 4.3 Phasor diagram for the conventional AM and narrowband angle modulation.

---

[2]Also known as *low-index angle modulation.*

4.2 SPECTRAL CHARACTERISTICS OF ANGLE-MODULATED SIGNALS

Due to the inherent nonlinearity of angle modulation systems, the precise characterization of their spectral properties, even for simple message signals, is mathematically intractable. Therefore, the derivation of the spectral characteristics of these signals usually involves the study of simple modulating signals and certain approximations. Then the results are generalized to the more complicated messages. We will study the spectral characteristics of an angle-modulated signal when the modulating signal is a sinusoidal signal.

4.2.1 Angle Modulation by a Sinusoidal Signal

Consider the case where the message signal is a sinusoidal signal (to be more precise, sine in PM and cosine in FM). As we have seen in Example 4.1.1, in this case for both FM and PM we have

$$u(t) = A_c \cos(2\pi f_c t + \beta \sin 2\pi f_m t), \qquad (4.2.1)$$

where β is the modulation index that can be either β_p or β_f and in PM $\sin 2\pi f_m t$ is substituted by $\cos 2\pi f_m t$. Using Euler's relation, the modulated signal can be written as

$$u(t) = \text{Re}\left(A_c e^{j2\pi f_c t} e^{j\beta \sin 2\pi f_m t}\right). \qquad (4.2.2)$$

Since $\sin 2\pi f_m t$ is periodic with period $T_m = \frac{1}{f_m}$, the same is true for the complex exponential signal

$$e^{j\beta \sin 2\pi f_m t}.$$

Therefore, it can be expanded in a Fourier-series representation. The Fourier-series coefficients are obtained from the integral

$$
\begin{aligned}
c_n &= f_m \int_0^{\frac{1}{f_m}} e^{j\beta \sin 2\pi f_m t} e^{-jn2\pi f_m t} \, dt \\
&\overset{u=2\pi f_m t}{=} \frac{1}{2\pi} \int_0^{2\pi} e^{j(\beta \sin u - nu)} \, du.
\end{aligned}
\qquad (4.2.3)
$$

This latter expression is a well-known integral called the *Bessel function of the first kind of order n* and is denoted by $J_n(\beta)$. Therefore, we have the Fourier series for the complex exponential as

$$e^{j\beta \sin 2\pi f_m t} = \sum_{n=-\infty}^{\infty} J_n(\beta) e^{j2\pi n f_m t}. \qquad (4.2.4)$$

By substituting Equation (4.2.4) in to Equation (4.2.2), we obtain

$$u(t) = \text{Re}\left(A_c \sum_{n=-\infty}^{\infty} J_n(\beta) e^{j2\pi n f_m t} e^{j2\pi f_c t}\right)$$

$$= \sum_{n=-\infty}^{\infty} A_c J_n(\beta) \cos\left(2\pi(f_c + n f_m)t\right). \tag{4.2.5}$$

The preceding relation shows that, even in this very simple case where the modulating signal is a sinusoid of frequency f_m, the angle-modulated signal contains all frequencies of the form $f_c + n f_m$ for $n = 0, \pm 1, \pm 2, \dots$. Therefore, the actual bandwidth of the modulated signal is infinite. However, the amplitude of the sinusoidal components of frequencies $f_c \pm n f_m$ for large n is very small. Hence, we can define a finite *effective bandwidth* for the modulated signal. For small β, we can use the approximation

$$J_n(\beta) \approx \frac{\beta^n}{2^n n!}. \tag{4.2.6}$$

Thus, for a small modulation index β, only the first sideband corresponding to $n = 1$ is important. Also, we can easily verify the following symmetry properties of the Bessel function:

$$J_{-n}(\beta) = \begin{cases} J_n(\beta), & n \text{ even} \\ -J_n(\beta), & n \text{ odd} \end{cases}. \tag{4.2.7}$$

Plots of $J_n(\beta)$ for various values of n are given in Figure 4.4. The values of the Bessel function are given in Table 4.1.

Example 4.2.1

Let the carrier be given by $c(t) = 10\cos(2\pi f_c t)$, and let the message signal be $\cos(20\pi t)$. Further assume that the message is used to frequency modulate the carrier with $k_f = 50$. Find the expression for the modulated signal and determine how many harmonics should be selected to contain 99% of the modulated-signal power.

Solution The power content of the carrier signal is given by

$$P_c = \frac{A_c^2}{2} = \frac{100}{2} = 50. \tag{4.2.8}$$

The modulated signal is represented by

$$u(t) = 10\cos\left(2\pi f_c t + 2\pi k_f \int_{-\infty}^{t} \cos(20\pi\tau)\,d\tau\right)$$

$$= 10\cos\left(2\pi f_c t + \frac{50}{10}\sin(20\pi t)\right)$$

$$= 10\cos(2\pi f_c t + 5\sin(20\pi t)). \tag{4.2.9}$$

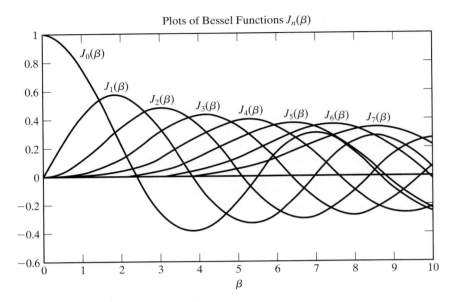

Figure 4.4   Bessel functions for various values of n.

TABLE 4.1   TABLE OF BESSEL FUNCTION VALUES

n	$\beta = 0.1$	$\beta = 0.2$	$\beta = 0.5$	$\beta = 1$	$\beta = 2$	$\beta = 5$	$\beta = 8$	$\beta = 10$
0	0.997	0.990	0.938	0.765	0.224	−0.178	0.172	−0.246
1	0.050	0.100	0.242	0.440	0.577	−0.328	0.235	0.043
2	0.001	0.005	0.031	0.115	0.353	0.047	−0.113	0.255
3				0.020	0.129	0.365	−0.291	0.058
4				0.002	0.034	0.391	−0.105	−0.220
5					0.007	0.261	0.186	−0.234
6					0.001	0.131	0.338	−0.014
7						0.053	0.321	0.217
8						0.018	0.223	0.318
9						0.006	0.126	0.292
10						0.001	0.061	0.207
11							0.026	0.123
12							0.010	0.063
13							0.003	0.029
14							0.001	0.012
15								0.004
16								0.001

(From Ziemer and Tranter; © 1990 Houghton Mifflin, reprinted with permission.)
Single and double underlines indicate the number of harmonics containing 70% and 98% of total power, respectively.

The modulation index is given by Equation (4.1.16) as

$$\beta = k_f \frac{\max[|m(t)|]}{f_m} = 5; \tag{4.2.10}$$

therefore, the FM-modulated signal is

$$u(t) = \sum_{n=-\infty}^{\infty} A_c J_n(\beta) \cos(2\pi(f_c + nf_m)t)$$

$$= \sum_{n=-\infty}^{\infty} 10 J_n(5) \cos(2\pi(f_c + 10n)t). \tag{4.2.11}$$

The frequency content of the modulated signal is concentrated at frequencies of the form $f_c + 10n$ for various n. To make sure that at least 99% of the total power is within the effective bandwidth, we must choose a k large enough such that

$$\sum_{n=-k}^{n=k} \frac{100 J_n^2(5)}{2} \geq 0.99 \times 50. \tag{4.2.12}$$

This is a nonlinear equation and its solution (for k) can be found by trial-and-error and by using tables of the Bessel functions. In finding the solution to this equation, we must employ the symmetry properties of the Bessel function given in Equation (4.2.7). Using these properties, we have

$$50 \left[J_0^2(5) + 2 \sum_{n=1}^{k} J_n^2(5) \right] \geq 49.5. \tag{4.2.13}$$

Starting with small values of k and increasing it, we see that the smallest value of k for which the left-hand side exceeds the right-hand side is $k = 6$. Therefore, taking frequencies $f_c \pm 10k$ for $0 \leq k \leq 6$ guarantees that 99% of the power of the modulated signal has been included and only 1% has been left out. This means that if the modulated signal is passed through an ideal bandpass filter centered at f_c with a bandwidth of at least 120 Hz, only 1% of the signal power will be eliminated. This gives us a practical way to define the *effective bandwidth* of the angle-modulated signal as 120 Hz. Figure 4.5 shows the frequencies present in the effective bandwidth of the modulated signal. ∎

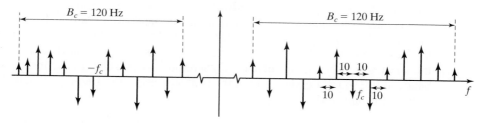

Figure 4.5   The harmonics present inside the effective bandwidth of Example 4.2.1.

In general, the effective bandwidth of an angle-modulated signal, which contains at least 98% of the signal power, is given by the relation

$$B_c = 2(\beta + 1)f_m, \tag{4.2.14}$$

where β is the modulation index and f_m is the frequency of the sinusoidal message signal. It is instructive to study the effect of the amplitude and frequency of the sinusoidal message signal on the bandwidth and the number of harmonics in the modulated signal. Let the message signal be given by

$$m(t) = a\cos(2\pi f_m t). \tag{4.2.15}$$

Using Equations (4.2.14), (4.1.12), and (4.1.13), the bandwidth[3] of the modulated signal is given by

$$B_c = 2(\beta + 1)f_m = \begin{cases} 2(k_p a + 1)f_m, & \text{PM} \\ 2\left(\dfrac{k_f a}{f_m} + 1\right)f_m, & \text{FM} \end{cases} \tag{4.2.16}$$

or

$$B_c = \begin{cases} 2(k_p a + 1)f_m, & \text{PM} \\ 2(k_f a + f_m), & \text{FM} \end{cases}. \tag{4.2.17}$$

The preceding relation shows that increasing a, the amplitude of the modulating signal, in PM and FM has almost the same effect on increasing the bandwidth B_c. On the other hand, increasing f_m, the frequency of the message signal, has a more profound effect in increasing the bandwidth of a PM signal as compared to an FM signal. In both PM and FM, the bandwidth B_c increases by increasing f_m; but in PM, this increase is a proportional increase, and in FM, this is only an additive increase which usually (for large β) is not substantial. Now if we look at the number of harmonics in the bandwidth (including the carrier) and denote it by M_c, we have

$$M_c = 2(\lfloor \beta \rfloor + 1) + 1 = 2\lfloor \beta \rfloor + 3 = \begin{cases} 2\lfloor k_p a \rfloor + 3, & \text{PM} \\ 2\left\lfloor \dfrac{k_f a}{f_m} \right\rfloor + 3, & \text{FM} \end{cases}. \tag{4.2.18}$$

In both cases, increasing the amplitude a increases the number of harmonics in the bandwidth of the modulated signal. However, increasing f_m has no effect on the number of harmonics in the bandwidth of the PM signal, and it almost linearly decreases the number of harmonics in the FM signal. This explains the relative insensitivity of the FM-signal bandwidth to the message frequency. First, increasing f_m decreases the number of harmonics in the bandwidth, and at the same time, it increases the spacing between the harmonics. The net effect is a slight increase in the bandwidth. In PM, however, the number of harmonics remains constant and only the spacing between them increases. Therefore, the net effect is a linear increase in bandwidth. Figure 4.6 shows the effect of increasing the frequency of the message in both FM and PM.

---

[3]For the remainder of the text, "bandwidth" refers to the effective bandwidth, unless otherwise stated.

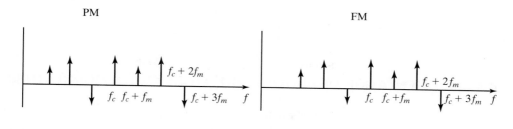

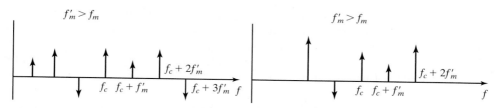

Figure 4.6   The effect of doubling the bandwidth of the message in FM and PM.

4.2.2 Angle Modulation by an Arbitrary Message Signal

The spectral characteristics of an angle-modulated signal for a general message signal $m(t)$ is quite involved due to the nonlinear nature of the modulation process. However, there exists an approximate relation for the effective bandwidth of the modulated signal. This is known as *Carson's rule* and is given by

$$B_c = 2(\beta + 1)W, \tag{4.2.19}$$

where β is the modulation index defined as

$$\beta = \begin{cases} k_p \max[|m(t)|], & \text{PM} \\ \dfrac{k_f \max[|m(t)|]}{W}, & \text{FM} \end{cases}, \tag{4.2.20}$$

and W is the bandwidth of the message signal $m(t)$. Since wideband FM has a β with a value that is usually around 5 or more, the bandwidth of an angle-modulated signal is much greater than the bandwidth of various amplitude-modulation schemes. This bandwidth is either W (in SSB) or $2W$ (in DSB or conventional AM).

Example 4.2.2

Assuming that $m(t) = 10 \operatorname{sinc}(10^4 t)$, determine the transmission bandwidth of an FM-modulated signal with $k_f = 4000$.

Solution   For FM, we have $B_c = 2(\beta + 1)W$. To find W, we have to find the spectrum of $m(t)$. We have $M(f) = 10^{-3} \Pi(10^{-4} f)$, which shows that $m(t)$ has a bandwidth of 5000 Hz. Since the maximum amplitude of $m(t)$ is 10, we have

$$\beta = \frac{k_f \max[|m(t)|]}{W} = \frac{4000 \times 10}{5000} = 8$$

and

$$B_c = 2(8+1) \times 5000 = 90000 \text{ Hz} = 90 \text{ kHz}. \qquad \blacksquare$$

4.3 IMPLEMENTATION OF ANGLE MODULATORS AND DEMODULATORS

Any modulation and demodulation process involves the generation of new frequencies that were not present in the input signal. This is true for both amplitude and angle-modulation systems. Thus, consider a modulator system with the message signal $m(t)$ as the input and with the modulated signal $u(t)$ as the output; this system has frequencies in its output that were not present in the input. Therefore, a modulator (and demodulator) cannot be modeled as a linear time-invariant system, because a linear time-invariant system cannot produce any frequency components in the output that are not present in the input signal.

Angle Modulators.   Angle modulators are generally time-varying and nonlinear systems. One method for directly generating an FM signal is to design an oscillator whose frequency changes with the input voltage. When the input voltage is zero, the oscillator generates a sinusoid with frequency f_c; when the input voltage changes, this frequency changes accordingly. There are two approaches to designing such an oscillator, usually called a VCO or *voltage-controlled oscillator*. One approach is to use a *varactor diode*. A varactor diode is a capacitor whose capacitance changes with the applied voltage. Therefore, if this capacitor is used in the tuned circuit of the oscillator and the message signal is applied to it, the frequency of the tuned circuit and the oscillator will change in accordance with the message signal. Let us assume that the inductance of the inductor in the tuned circuit of Figure 4.7 is L_0 and the capacitance of the varactor diode is given by

$$C(t) = C_0 + k_0 m(t). \qquad (4.3.1)$$

When $m(t) = 0$, the frequency of the tuned circuit is given by $f_c = \frac{1}{2\pi \sqrt{L_0 C_0}}$. In general, for nonzero $m(t)$, we have

$$
\begin{aligned}
f_i(t) &= \frac{1}{\pi \sqrt{L_0(C_0 + k_0 m(t))}} \\
&= \frac{1}{2\pi \sqrt{L_0 C_0}} \frac{1}{\sqrt{1 + \frac{k_0}{C_0} m(t)}} \\
&= f_c \frac{1}{\sqrt{1 + \frac{k_0}{C_0} m(t)}}. \qquad (4.3.2)
\end{aligned}
$$

Assuming that

$$\epsilon = \frac{k_0}{C_0} m(t) \ll 1$$

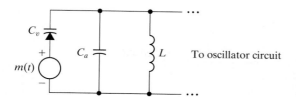

To oscillator circuit

Figure 4.7   Varactor-diode implementation of an angle modulator.

and using the approximations

$$\sqrt{1 + \epsilon} \approx 1 + \frac{\epsilon}{2} \qquad (4.3.3)$$

and

$$\frac{1}{1 + \epsilon} \approx 1 - \epsilon, \qquad (4.3.4)$$

we obtain

$$\frac{1}{\sqrt{1 + \epsilon}} \approx 1 - \frac{\epsilon}{2}.$$

Hence,

$$f_i(t) \approx f_c \left(1 - \frac{k_0}{2C_0} m(t)\right), \qquad (4.3.5)$$

which is the relation for a frequency-modulated signal.

A second approach for generating an FM signal is to use a *reactance tube*. In the reactance-tube implementation, an inductor whose inductance varies with the applied voltage is employed; the analysis is very similar to the analysis presented for the varactor diode. Although we described these methods for the generation of FM signals, basically the same methods can be applied for the generation of PM signals (see Figure 4.1), due to the close relation between FM and PM signals.

Another approach for generating an angle-modulated signal is to generate a narrowband angle-modulated signal and then change it to a wideband signal. This method is usually known as the *indirect method* for the generation of FM and PM signals. Due to the similarity of conventional AM signals, the generation of narrowband angle-modulated signals is straightforward. In fact, any modulator for conventional AM generation can be easily modified to generate a narrowband angle-modulated signal. Figure 4.8 shows the block diagram of a narrowband angle modulator. Next, we use the narrowband angle-modulated signal to generate a wideband angle-modulated signal. Figure 4.9 shows the block diagram of such a system. The first stage of this system is to create a narrowband angle modulator, such as the one shown in Figure 4.8. The narrowband angle-modulated signal enters a frequency multiplier which multiplies the instantaneous frequency of the input by some constant n. This is usually done by applying the input signal to a nonlinear element and then passing its output through a bandpass filter tuned to the desired central frequency. If the narrowband modulated signal is represented by

$$u_n(t) = A_c \cos(2\pi f_c t + \phi(t)), \qquad (4.3.6)$$

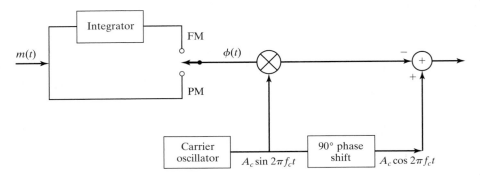

Figure 4.8 Generation of a narrowband angle-modulated signal.

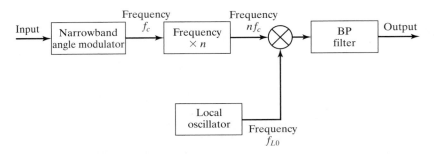

Figure 4.9 Indirect generation of angle-modulated signals.

the output of the frequency multiplier (which is the output of the bandpass filter) is given by

$$y(t) = A_c \cos(2\pi n f_c t + n\phi(t)). \qquad (4.3.7)$$

In general, this is a wideband angle-modulated signal. However, there is no guarantee that the carrier frequency of this signal, nf_c, will be the desired carrier frequency. In the last stage, the modulator performs an up/down conversion to shift the modulated signal to the desired center frequency. This stage consists of a mixer and a bandpass filter. If the frequency of the local oscillator of the mixer is f_{LO} and we are using a down converter, the final wideband angle-modulated signal is given by

$$u(t) = A_c \cos\left(2\pi (nf_c - f_{\text{LO}})t + n\phi(t)\right). \qquad (4.3.8)$$

Since we can freely choose n and f_{LO}, we can generate any modulation index at any desired carrier frequency using this method.

FM demodulators are implemented by generating an AM signal, whose amplitude is proportional to the instantaneous frequency of the FM signal, and then using an AM demodulator to recover the message signal. To implement the first step, i.e., to transform the FM signal into an AM signal, it is enough to pass the FM signal through an LTI system, whose frequency response is approximately a straight line in the frequency band of the FM signal. If the frequency response of such a system is given by

$$|H(f)| = V_0 + k(f - f_c) \qquad \text{for } |f - f_c| < \frac{B_c}{2} \qquad (4.3.9)$$

and if the input to the system is

$$u(t) = A_c \cos\left(2\pi f_c t + 2\pi k_f \int_{-\infty}^{t} m(\tau)\, d\tau\right), \qquad (4.3.10)$$

then the output will be the signal

$$v_o(t) = A_c(V_0 + kk_f m(t)) \cos\left(2\pi f_c t + 2\pi k_f \int_{-\infty}^{t} m(\tau)\, d\tau\right). \qquad (4.3.11)$$

The next step is to demodulate this signal to obtain $A_c(V_0 + kk_f m(t))$, from which the message $m(t)$ can be recovered. Figure 4.10 shows a block diagram of these two steps.

Many circuits can be used to implement the first stage of an FM demodulator, i.e., FM to AM conversion. One such candidate is a simple differentiator with

$$|H(f)| = 2\pi f. \qquad (4.3.12)$$

Another candidate is the rising half of the frequency characteristics of a tuned circuit, as shown in Figure 4.11. Such a circuit can be easily implemented, but usually the linear region of the frequency characteristic may not be wide enough. To obtain linear characteristics over a wide range of frequencies, usually two circuits tuned at two frequencies f_1 and f_2 are connected in a configuration, which is known as a *balanced discriminator*. A balanced discriminator with the corresponding frequency characteristics is shown in Figure 4.12.

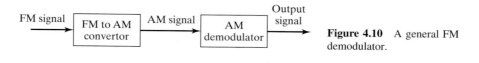

Figure 4.10   A general FM demodulator.

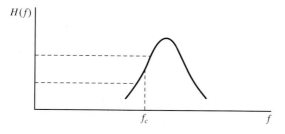

Figure 4.11   A tuned circuit used in an FM demodulator.

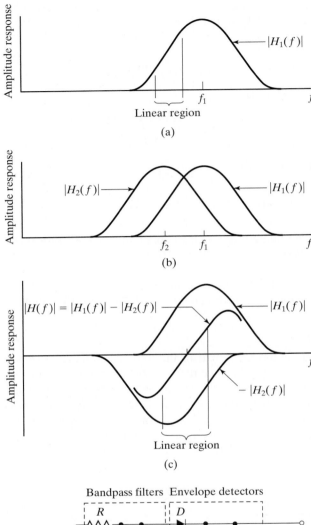

Figure 4.12 A balanced discriminator and the corresponding frequency response.

These FM-demodulation methods, which transform the FM signal into an AM signal, have a bandwidth equal to the channel bandwidth B_c occupied by the FM signal. Consequently, the noise that is passed by the demodulator is the noise contained within B_c.

A different approach to FM-signal demodulation is to use feedback in the FM demodulator to narrow the bandwidth of the FM detector and, as will be seen in Chapter 6, to reduce the noise power at the output of the demodulator. Figure 4.13 illustrates such a system. In this figure, the FM discriminator is placed in the feedback branch of a feedback system that employs a voltage-controlled oscillator (VCO) path.

The bandwidth of the discriminator and the subsequent lowpass filter are designed to match the bandwidth of the message signal $m(t)$. The output of the lowpass filter is the desired message signal. This type of FM demodulator is called an FM demodulator with feedback (FMFB). An alternative to the FMFB demodulator is the use of a phase-locked loop (PLL), as shown in Figure 4.14 (phase-locked loops are studied in detail in Section 6.4).

The input to the PLL is the angle-modulated signal (we will neglect the presence of noise in this discussion)

$$u(t) = A_c \cos[2\pi f_c t + \phi(t)], \qquad (4.3.13)$$

where, for FM,

$$\phi(t) = 2\pi k_f \int_{-\infty}^{t} m(\tau)\, d\tau. \qquad (4.3.14)$$

The VCO generates a sinusoid of a fixed frequency; in this case, it generates the carrier frequency f_c, in the absence of an input control voltage.

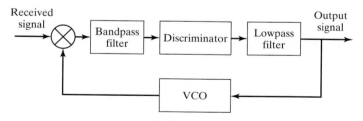

Figure 4.13   Block diagram of an FMFB demodulator.

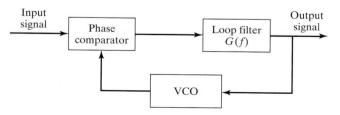

Figure 4.14   Block diagram of PLL-FM demodulator.

Now, suppose that the control voltage to the VCO is the loop filter's output, denoted as $v(t)$. Then, the instantaneous frequency of the VCO is

$$f_v(t) = f_c + k_v v(t), \tag{4.3.15}$$

where k_v is a deviation constant with units of Hz/volt. Consequently, the VCO output may be expressed as

$$y_v(t) = A_v \sin[2\pi f_c t + \phi_v(t)], \tag{4.3.16}$$

where

$$\phi_v(t) = 2\pi k_v \int_0^t v(\tau)\, d\tau. \tag{4.3.17}$$

The phase comparator is basically a multiplier and a filter that rejects the signal component centered at $2f_c$. Hence, its output may be expressed as

$$e(t) = \frac{1}{2} A_v A_c \sin[\phi(t) - \phi_v(t)], \tag{4.3.18}$$

where the difference $\phi(t) - \phi_v(t) \equiv \phi_e(t)$ constitutes the phase error. The signal $e(t)$ is the input to the loop filter.

Let us assume that the PLL is in lock position, so the phase error is small. Then,

$$\sin[\phi(t) - \phi_v(t)] \approx \phi(t) - \phi_v(t) = \phi_e(t) \tag{4.3.19}$$

under this condition, so we may deal with the linearized model of the PLL, shown in Figure 4.15.

We may express the phase error as

$$\phi_e(t) = \phi(t) - 2\pi k_v \int_0^t v(\tau)\, d\tau, \tag{4.3.20}$$

or equivalently, either as

$$\frac{d}{dt}\phi_e(t) + 2\pi k_v v(t) = \frac{d}{dt}\phi(t) \tag{4.3.21}$$

or as

$$\frac{d}{dt}\phi_e(t) + 2\pi k_v \int_0^\infty \phi_e(\tau) g(t - \tau)\, d\tau = \frac{d}{dt}\phi(t). \tag{4.3.22}$$

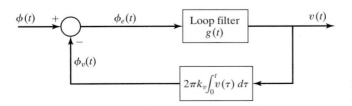

Figure 4.15   Linearized PLL.

The Fourier transform of the integro-differential equation in Equation (4.3.22) is

$$(j2\pi f)\Phi_e(f) + 2\pi k_v \Phi_e(f)G(f) = (j2\pi f)\Phi(f); \qquad (4.3.23)$$

hence,

$$\Phi_e(f) = \frac{1}{1 + \left(\frac{k_v}{jf}\right)G(f)}\Phi(f). \qquad (4.3.24)$$

The corresponding equation for the control voltage to the VCO is

$$V(f) = \Phi_e(f)G(f)$$

$$= \frac{G(f)}{1 + \left(\frac{k_v}{jf}\right)G(f)}\Phi(f). \qquad (4.3.25)$$

Now, suppose that we design $G(f)$ such that

$$\left| k_v \frac{G(f)}{jf} \right| \gg 1 \qquad (4.3.26)$$

in the frequency band $|f| < W$ of the message signal. Then, from Equation (4.3.25), we have

$$V(f) = \frac{j2\pi f}{2\pi k_v}\Phi(f), \qquad (4.3.27)$$

or equivalently,

$$v(t) = \frac{1}{2\pi k_v}\frac{d}{dt}\phi(t)$$

$$= \frac{k_f}{k_v}m(t). \qquad (4.3.28)$$

Since the control voltage of the VCO is proportional to the message signal, $v(t)$ is the demodulated signal.

We observe that the output of the loop filter with the frequency response $G(f)$ is the desired message signal. Hence, the bandwidth of $G(f)$ should be the same as the bandwidth W of the message signal. Consequently, the noise at the output of the loop filter is also limited to the bandwidth W. On the other hand, the output from the VCO is a wideband FM signal with an instantaneous frequency that follows the instantaneous frequency of the received FM signal.

The major benefit of using feedback in FM-signal demodulation is to reduce the threshold effect that occurs when the input signal-to-noise-ratio to the FM demodulator drops below a critical value. The threshold effect is treated in Chapter 6.

4.4 FM-RADIO AND TELEVISION BROADCASTING

In this section, we describe two types of broadcasting systems that employ angle modulation or a combination of amplitude- and angle-modulation techniques. These systems are FM-radio broadcasting and television broadcasting.

4.4.1 FM-Radio Broadcasting

Commercial FM-radio broadcasting utilizes the frequency band 88–108 MHz for the transmission of voice and music signals. The carrier frequencies are separated by 200 kHz and the peak frequency deviation is fixed at 75 kHz. Preemphasis is generally used, as described in Chapter 6, to improve the demodulator performance in the presence of noise in the received signal.

The receiver most commonly used in FM-radio broadcast is a superheterodyne type. The block diagram of such a receiver is shown in Figure 4.16.

As in AM-radio reception, common tuning between the RF amplifier and the local oscillator allows the mixer to bring all FM-radio signals to a common IF bandwidth of 200 kHz, centered at $f_{IF} = 10.7$ MHz. Since the message signal $m(t)$ is embedded in the frequency of the carrier, any amplitude variations in the received signal are a result of additive noise and interference. The amplitude limiter removes any amplitude variations in the received signal at the output of the IF amplifier by bandlimiting the signal. A bandpass filter, which is centered at $f_{IF} = 10.7$ MHz with

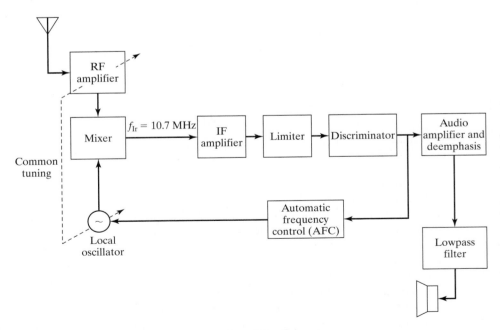

Figure 4.16 Block diagram of a superheterodyne FM-radio receiver.

a bandwidth of 200 kHz, is included in the limiter to remove higher-order frequency components introduced by the nonlinearity inherent in the hard limiter.

A balanced frequency discriminator is used for frequency demodulation. The resulting message signal is then passed to the audio-frequency amplifier, which performs the functions of deemphasis and amplification. The output of the audio amplifier is further filtered by a lowpass filter to remove out-of-band noise, and this output is used to drive a loudspeaker.

FM-Stereo Broadcasting.   Many FM-radio stations transmit music programs in stereo by using the outputs of two microphones placed on two different parts of the stage. Figure 4.17 shows a block diagram of an FM-stereo transmitter. The signals from the left and right microphones, $m_\ell(t)$ and $m_r(t)$, are added and subtracted as shown. The sum signal $m_\ell(t) + m_r(t)$ is left unchanged and occupies the frequency band 0–15 kHz. The difference signal $m_\ell(t) - m_r(t)$ is used to AM modulate (DSB-SC) a 38 kHz carrier that is generated from a 19-kHz oscillator. A pilot tone at the frequency of 19 kHz is added to the signal for the purpose of demodulating the DSB-SC AM signal. We place the pilot tone at 19 kHz instead of 38 kHz because the pilot is more easily separated from the composite signal at the receiver. The combined signal is used to frequency modulate a carrier.

By configuring the baseband signal as an FDM signal, a monophonic FM receiver can recover the sum signal $m_\ell(t) + m_r(t)$ by using a conventional FM demodulator. Hence, FM-stereo broadcasting is compatible with conventional FM. In addition, the resulting FM signal does not exceed the allocated 200-kHz bandwidth.

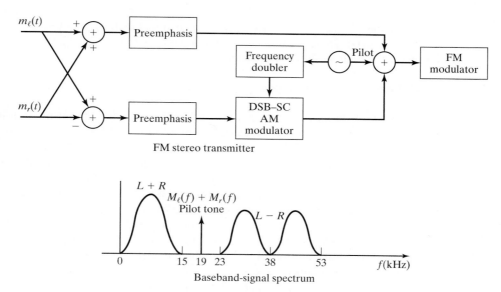

Figure 4.17 FM-stereo transmitter and signal spacing.

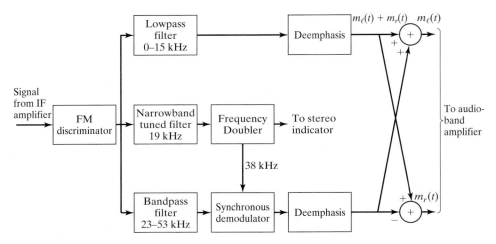

Figure 4.18   FM-stereo receiver.

The FM demodulator for FM stereo is basically the same as a conventional FM demodulator down to the limiter/discriminator. Thus, the received signal is converted to baseband. Following the discriminator, the baseband message signal is separated into the two signals, $m_\ell(t) + m_r(t)$ and $m_\ell(t) - m_r(t)$, and passed through de-emphasis filters, as shown in Figure 4.18. The difference signal is obtained from the DSB-SC signal via a synchronous demodulator using the pilot tone. By taking the sum and difference of the two composite signals, we recover the two signals, $m_\ell(t)$ and $m_r(t)$. These audio signals are amplified by audio-band amplifiers, and the two outputs drive dual loudspeakers. As indicated, an FM receiver that is not configured to receive the FM stereo sees only the baseband signal $m_\ell(t) + m_r(t)$ in the frequency range 0–15 kHz. Thus, it produces a monophonic output signal that consists of the sum of the signals at the two microphones.

4.4.2 Television Broadcasting

Commercial TV broadcasting began as black-and-white picture transmission in London in 1936 by the British Broadcasting Corporation (BBC). Color television was demonstrated a few years later, but commercial TV stations were slow to develop the transmission of color-TV signals. To a large extent, this was due to the high cost of color-TV receivers. With the development of the transistor and microelectronic components, the cost of color TV receivers decreased significantly. By the middle 1960s, color TV broadcasting was widely used by the industry.

The frequencies allocated for TV broadcasting fall in the VHF and UHF bands. Table 4.2 lists the TV channels allocated in the United States. The channel bandwidth allocated for the transmission of TV signals is 6 MHz.

In contrast to radio broadcasting, standards for television-signal transmission vary from country to country. The US standard was set by the National Television Systems Committee (NTSC).

TABLE 4.2 ALLOCATION OF VHF AND UHF
FREQUENCIES FOR COMMERCIAL TELEVISION

Channel	Frequency Band (6 MHz bandwidth/station)
VHF 2–4	54–72 MHz
VHF 5–6	76–88 MHz
VHF 7–13	174–216 MHz
UHF 14–69	470–896 MHz

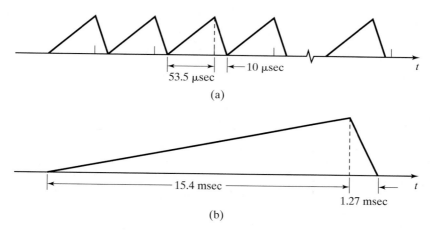

Figure 4.19 Signal waveforms applied to horizontal and vertical deflection plates.

Black-and-White TV Signals. The first step in TV-signal transmission is to convert a visual image into an electrical signal. The two-dimensional image is converted to a one-dimensional electrical signal by sequentially scanning the image and producing an electrical signal that is proportional to the brightness level of the image. The scanning is performed in a TV camera, which optically focuses the image on a photocathode tube, which consists of a photosensitive surface.

The image is scanned by an electron beam that produces an output current or voltage that is proportional to the brightness of the image. The resulting electrical signal is called a *video signal*.

The scanning of the electron beam is controlled by two voltages applied across the horizontal and vertical deflection plates. These two voltages are shown in Figure 4.19. In this scanning method, the image is divided into 525 lines that define a *frame*, as illustrated in Figure 4.20. The resulting signal is transmitted in 1/30 of a second. The number of lines determines the picture resolution and, in combination with the rate of transmission, determines the channel bandwidth required for transmission of the image.

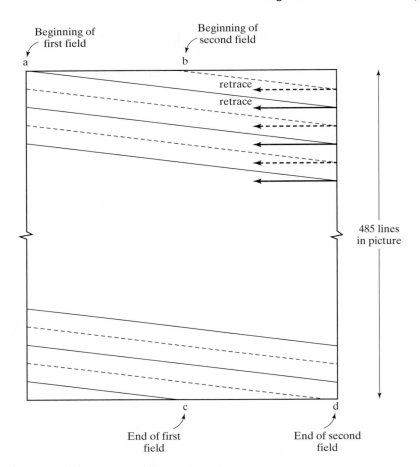

Figure 4.20   Interlaced scanning pattern.

Transmitting a complete image in 1/30 of a second is generally not fast enough to avoid the flicker that is annoying to the eyes of the average viewer. To overcome flicker, the image scanning is performed in an *interlaced pattern*, as shown in Figure 4.20. The interlaced pattern consists of two fields, each consisting of 262.5 lines. Each field is transmitted in 1/60 of a second, which exceeds the flicker rate that is observed by the average eye. The first field begins at point "a" and terminates at point "c." The second field begins at point "b" and terminates at point "d."

A horizontal line is scanned in 53.5 μsec, as indicated by the sawtooth signal waveform applied to the horizontal deflection plates. The beam has 10 μsec to move to the next line. During this interval, a blanking pulse is inserted to avoid the appearance of retrace lines across the TV receiver. A 5 μsec pulse is added to the blanking pulse to provide synchronization for the horizontal-sweep circuit at the receiver. A typical video signal is shown in Figure 4.21.

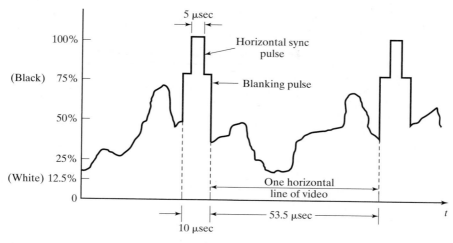

Figure 4.21   Typical video signal for one horizontal sweep.

After the transmission of one interlaced field, the vertical sawtooth signal wave-form applied to the vertical deflection plates is reset to zero. The retrace interval of 1.27 msec, corresponding to 20 line scans, allows the beam to move from the bottom to the top of the picture. A vertical blanking pulse is inserted during the interval to avoid the appearance of retrace lines at the TV receiver. When we allow for the vertical retrace (twice per frame), the actual number of horizontal lines in the image is 485.

The bandwidth of the video signal can be estimated by viewing the image as a rectangular array of 485 rows by $485 \times \frac{4}{3}$ columns, where 4/3 is the *aspect ratio* (the ratio of the width to height of the image). Thus, we have 313,633 picture elements *(pixels)* per frame, which are transmitted in 1/30 of a second. This is equivalent to a sampling rate of 10.5 MHz, which can represent a signal as large as 5.25 MHz. However, the light intensity of adjacent pixels in an image is highly correlated. Hence, the bandwidth of the video signal is less than 5.25 MHz. In commercial TV broadcasting, the bandwidth of the video signal is limited to $W = 4.2$ MHz.

Since the allocated channel bandwidth for commercial TV is 6 MHz, it is clear that DSB transmission is not possible. The large low-frequency content of the video signal also rules out SSB as a practical modulation method. Hence, VSB is the only viable alternative. By transmitting a large carrier component, the received VSB signal can be simply demodulated via an envelope detector. This type of detection significantly simplifies the implementation of the receiver.

The range of frequencies occupied by the transmitted video signal is shown in Figure 4.22. We note that the full upper sideband (4.2 MHz) of the video signal is transmitted along with a portion (1.25 MHz) of the lower sideband. Unlike the conventional VSB-spectral shaping described in Section 3.2.4, the lower sideband signal in the frequency range f_c and $f_c - 0.75$ MHz is transmitted without attenuation. The frequencies in the range $f_c - 1.25$ MHz to $f_c - 0.75$ MHz are attenuated, as

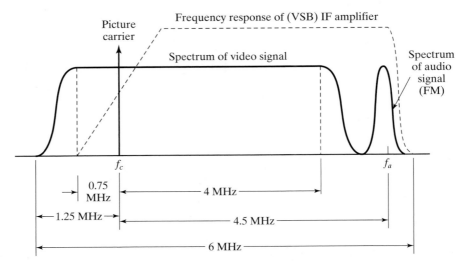

Figure 4.22   Spectral characteristics of a black-and-white television signal.

shown in Figure 4.22, and all frequency components below $f_c - 1.25$ MHz are blocked. VSB spectral shaping is performed at the IF amplifier of the receiver.

In addition to the video signal, the audio portion of the TV signal is transmitted by frequency modulating a carrier at $f_c + 4.5$ MHz. The audio-signal bandwidth is limited to $W = 10$ kHz. The frequency deviation in the FM-modulated signal is selected as 25 kHz, and the FM-signal bandwidth is 70 kHz. Hence, the total channel bandwidth required to transmit the video and audio signals is 5.785 MHz.

Figure 4.23 shows the block diagram of a black-and-white TV transmitter. The corresponding receiver is shown in Figure 4.24. It is a heterodyne receiver. There are two separate tuners, one for the UHF band and one for the VHF band. The TV signals in the UHF band are brought down to the VHF band by a UHF mixer. This frequency conversion makes it possible to use a common RF amplifier for the two frequency bands. Then, the video signal selected by the tuner is translated to a common IF-frequency band of 41–47 MHz. The IF amplifier also provides the VSB shaping required prior to signal detection. The output of the IF amplifier is envelope detected to produce the baseband signal.

The audio portion of the signal centered at 4.5 MHz is filtered out via an IF-filter–amplifier and is passed to the FM demodulator. The demodulated audio-band signal is then amplified by an audio amplifier and its output drives the speaker.

The video component of the baseband signal is passed through a video amplifier, which passes frequency components in the range 0–4.2 MHz. Its output is passed to the DC restorer, which clamps the blanking pulses and sets the correct DC level. The DC-restored video signal is then fed to the picture tube. The synchronizing pulses contained in the received video signal are separated and applied to the horizontal and vertical sweep generators.

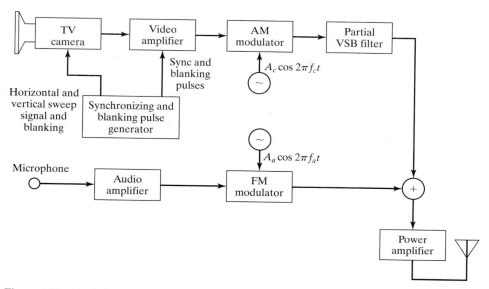

Figure 4.23 Block diagram of a black-and-white TV transmitter.

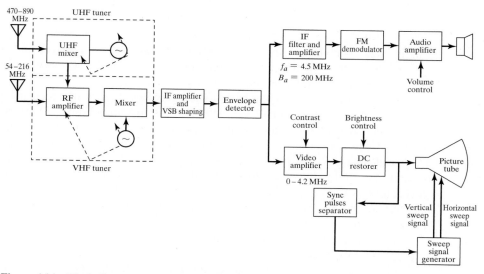

Figure 4.24 Block diagram of a black-and-white TV receiver.

Compatible Color Television. The transmission of color information contained in an image can be accomplished by decomposing the pixel colors into primary colors and transmitting the electrical signals corresponding to these colors. In general, all natural colors are well approximated by appropriate mixing of three primary colors—blue, green, and red. Consequently, if we employ three cameras, one with a blue filter, one with a green filter, and one with a red filter, and transmit the electrical signals $m_b(t), m_g(t)$, and $m_r(t)$, which are generated by the three color cameras that view the color image, the received signals can be combined to produce a replica of the original color image.

Such a transmission scheme has two major disadvantages. First, it requires three times the channel bandwidth of black-and-white television. Secondly, the transmitted color-TV signal cannot be received by a black-and-white (monochrome) TV receiver.

The NTSC standard, which was adopted in 1953 in the United States, avoids these two problems by transmitting a mixture of the three primary color signals. Specifically, the three signals transmitted in the standard color-TV system are the following three linearly independent combinations:

$$m_L(t) = 0.11m_b(t) + 0.59m_g(t) + 0.30m_r(t);$$

$$m_I(t) = -0.32m_b(t) - 0.28m_g(t) + 0.60m_r(t); \qquad (4.4.1)$$

$$m_Q(t) = 0.31m_b(t) - 0.52m_g(t) + 0.21m_r(t).$$

The transformation matrix

$$M = \begin{bmatrix} 0.11 & 0.59 & 0.30 \\ -0.32 & -0.28 & 0.60 \\ 0.31 & -0.52 & 0.21 \end{bmatrix} \qquad (4.4.2)$$

that is used to construct the new transmitted signals $m_L(t), m_I(t)$ and $m_Q(t)$ is nonsingular and is inverted at the receiver to recover the primary color signals $m_b(t), m_g(t)$, and $m_r(t)$ from $m_L(t), m_I(t)$, and $m_Q(t)$.

The signal $m_L(t)$ is called the *luminance signal*. It is assigned a bandwidth of 4.2 MHz and is transmitted via VSB-AM, as in monochrome TV transmission. When this signal is received by a monochrome receiver, the result is a conventional black-and-white version of the color image. Thus, compatibility with monochrome TV broadcasting is achieved by transmitting $m_L(t)$. There remains the problem of transmitting the additional information that can be used by a color-TV receiver to reconstruct the color image. It is remarkable that the two composite color signals, $m_I(t)$ and $m_Q(t)$, can be transmitted in the same bandwidth as $m_L(t)$ without interfering with $m_L(t)$.

The signals, $m_I(t)$ and $m_Q(t)$, are called *chrominance signals*; they are related to the hue and saturation of colors. It has been determined experimentally, through subjective tests, that human vision cannot discriminate changes in $m_I(t)$ and $m_Q(t)$ over short time intervals and, hence, over small areas of the image. This implies that the high-frequency content in the signals $m_I(t)$ and $m_Q(t)$ can be eliminated without significantly compromising the quality of the reconstructed image. The end result is that $m_I(t)$ is limited in bandwidth to 1.6 MHz, and $m_Q(t)$ is limited to 0.6 MHz prior to transmission. These two signals are quadrature-carrier multiplexed on a subcarrier frequency $f_{sc} = f_c + 3.579545$ MHz, as illustrated in Figure 4.25. The signal $m_I(t)$ is passed through a VSB filter, which removes the part of the upper sideband that is

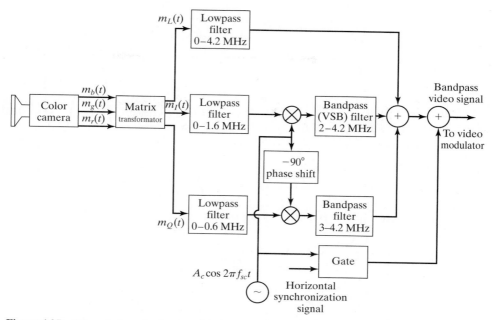

Figure 4.25 Transmission of primary color signals and multiplexing of chrominance and luminance signals.

above 4.2 MHz. The signal $m_Q(t)$ is transmitted by DSB-SC amplitude modulation. Therefore, the composite video signal may be expressed as

$$m(t) = m_L(t) + m_Q(t) \sin 2\pi f_{sc}t + m_I(t) \cos 2\pi f_{sc}t$$
$$+ \hat{m}'_I(t) \sin 2\pi f_{sc}t. \tag{4.4.3}$$

The last two terms in Equation (4.4.3) that involve $m_I(t)$ and $\hat{m}'_I(t)$, constitute the VSB-AM signal for the chrominance $m_I(t)$. The composite signal $m(t)$ is transmitted by VSB-plus carrier in a 6 MHz bandwidth, as shown in Figure 4.26.

The spectrum of the luminance signal $m_L(t)$ has periodic gaps between the harmonics of the horizontal-sweep frequency f_h. In color TV f_h is 4.5 MHz/286. The subcarrier frequency, $f_{sc} = 3.579545$ for transmission of the chrominance signals, was chosen because it corresponds to one of these gaps in the spectrum of $m_L(t)$. Specifically, it falls between the 227 and 228 harmonics of f_h. Thus, the chrominance signals are interlaced in the frequency domain with the luminance signal, as illustrated in Figure 4.27. As a consequence, the effect of the chrominance signal on the luminance signal $m_L(t)$ is not perceived by the human eye, due to the persistence of human vision. Therefore, the chrominance signals do not interfere with the demodulation of the luminance signal in either a monochrome TV receiver or a color-TV receiver.

Horizontal and vertical synchronization pulses are added to $m(t)$ at the transmitter. In addition, eight cycles of the color subcarrier $A_c \cos 2\pi f_{sc}t$, called a "*color*

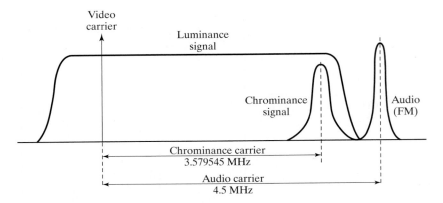

Figure 4.26   Spectral characteristics of color TV signal.

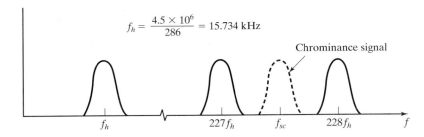

Figure 4.27   Interlacing of chrominance signals with luminance signals.

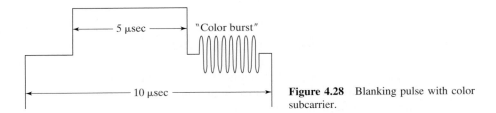

Figure 4.28   Blanking pulse with color subcarrier.

burst," are superimposed on the trailing edge of the blanking pulses, as shown in Figure 4.28. This provides a signal for subcarrier phase synchronization at the receiver.

The front end of the color TV receiver is similar to that of a monochrome receiver, down to the envelope-detector that converts the 6-MHz VSB signal to baseband. The remaining demultiplexing operations in the color TV receiver are shown in Figure 4.29. We note that a lowpass filter with the bandwidth 4.2 MHz is used to recover the luminance signal $m_L(t)$. The chrominance signals are stripped off by bandpass filtering and demodulated by the quadrature-carrier demodulator using the output of a VCO that is phase locked to the received color-carrier frequency

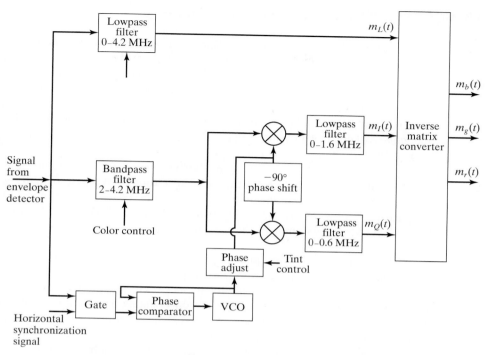

Figure 4.29 Demultiplexing and demodulation of luminance and chrominance signals in a color TV receiver.

burst transmitted in each horizontal sweep. The demodulated chrominance signals are lowpass filtered and, along with the luminance signal, are passed to the "inverse matrix" converter that reconstructs the three color signals $m_b(t), m_g(t)$, and $m_r(t)$, i.e.,

$$\begin{bmatrix} m_b(t) \\ m_g(t) \\ m_r(t) \end{bmatrix} = \begin{bmatrix} 1.00 & -1.10 & 1.70 \\ 1.00 & -0.28 & -0.64 \\ 1.00 & -0.96 & 0.62 \end{bmatrix} \begin{bmatrix} m_L(t) \\ m_I(t) \\ m_Q(t) \end{bmatrix}. \qquad (4.4.4)$$

The resulting color signals control the three electron guns that strike corresponding blue, green, and red picture elements in a color picture tube. Although color picture tubes are constructed in many different ways, the color-mask tube is commonly used in practice. The face of the picture tube contains a matrix of phosphor dots. This matrix contains the three primary colors with three such dots in each group. Behind each dot color group, there is a mask with holes; there is one hole for each group. The three electron guns are aligned so that each gun can excite one of the three types of color dots. Thus, the three types of color dots are excited simultaneously in different intensities to generate color images.

4.5 MOBILE WIRELESS TELEPHONE SYSTEMS

The demand to provide telephone service for people traveling in automobiles, buses, trains, and airplanes has been steadily increasing over the past three to four decades. To meet this demand, radio-transmission systems have been developed to link the mobile-telephone user to the terrestrial-telephone network. Today, radio-based systems make it possible for people to communicate via telephone while traveling on airplanes and motor vehicles. In this section, we will briefly describe the cellular telephone system that provides telephone service to people with handheld portable telephones and automobile telephones.

A major problem with the establishment of any radio communication system is the availability of a portion of the radio spectrum. In the case of radio telephone service, the Federal Communications Commission (FCC) in the United States has assigned parts of the UHF frequency band in the range 806–890 MHz for this use. Similar frequency assignments in the UHF band have been made in Europe and Japan.

The cellular-radio concept was adopted as a method for the efficient utilization of the available frequency spectrum, especially in highly populated metropolitan areas, where the demand for mobile telephone services is the greatest. A geographic area is subdivided into cells, each of which contains a *base station*, as illustrated in Figure 4.30. Each base station is connected via telephone lines to a *mobile-telephone-switching office* (MTSO) which, in turn, is connected via telephone lines to a telephone central office (CO) of the terrestrial-telephone network.

A mobile user communicates via radio with the base station within the cell. The base station routes the call through the MTSO to another base station (if the called party is located in another cell) or to the central office of the terrestrial-telephone network (if the called party is not a mobile). Each mobile telephone is identified by its telephone number and the telephone serial number assigned by the manufacturer. These numbers are automatically transmitted to the MTSO during the initialization of the call for authentication and billing purposes.

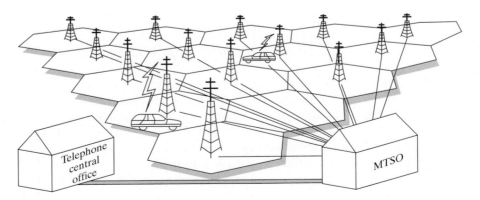

Figure 4.30   Mobile-radio base station.

A mobile user initiates a telephone call in the usual manner by keying in the desired telephone number and pressing the "send" button. The MTSO checks the authentication of the mobile user and assigns an available frequency channel for the radio transmission of the voice signal from the mobile to the base station. The frequency assignment is sent to the mobile telephone via a supervisory control channel. A second frequency is assigned for the radio transmission from the base station to the mobile user. The simultaneous transmission between the two parties is called *full-duplex operation*. The MTSO interfaces with the central office of the telephone network to complete the connection to the called party. All telephone communications between the MTSO and the telephone network are via wideband trunk lines that carry speech signals from many users. Upon completion of the telephone call, the two parties hang up and the radio channel becomes available for other users.

During the phone call, the MTSO monitors the signal strength of the radio transmission from the mobile user to the base station. If the signal strength drops below a preset threshold, the MTSO views this as an indication that the mobile user is moving out of the initial cell into a neighboring cell. By communicating with the base stations of neighboring cells, the MTSO finds a neighboring cell that receives a stronger signal and automatically switches or hands-off the mobile user to the base station of the adjacent cell. The switching is performed in a fraction of a second and is generally transparent to the two parties.

In the analog transmission of voiceband audio signals via radio, which are between the base station and the mobile user, the 3 kHz-wide audio signal is transmitted via FM using a channel bandwidth of 30 kHz. This represents a bandwidth expansion of approximately a factor of 10. Such a large bandwidth expansion is necessary to obtain a sufficiently large SNR at the output of the FM demodulator. However, the use of FM is highly wasteful of the radio frequency spectrum. The new generation of cellular telephone systems use digital transmission of digitized compressed speech (at bit rates of about 10,000 bps), based on LPC encoding and vector quantization of the speech-model parameters, as described in Chapter 12. With digital transmission, the cellular telephone system can accommodate a 4-fold to 10-fold increase in the number of simultaneous users with the same available channel bandwidth.

The cellular radio telephone system is designed such that the transmitter powers of the base station and the mobile users are sufficiently small, so that signals do not propagate beyond immediately adjacent cells. This allows frequencies to be reused in other cells outside of the adjacent cells. Consequently, by making the cells smaller and reducing the radiated power, it is possible to increase frequency reuse and, thus, to increase the bandwidth efficiency and the number of mobile users. Current cellular systems employ cells with a radius in the range of 5–18 km. The base station normally transmits at a power level of 35 watts or less, and the mobile users transmit at a power level of approximately 3 watts. Digital transmission systems are capable of communicating reliably at lower power levels.

The cellular radio concept is being extended to different types of personal communication services using low-power, hand-held radio transmitters and receivers.

These emerging communication services are made possible by rapid advances in the fabrication of small and powerful integrated circuits that consume very little power and are relatively inexpensive. As a consequence, we will continue to experience exciting new developments in the telecommunications industry well into the 21st century.

4.6 FURTHER READING

Analog communication systems are treated in numerous books on basic communication theory, including Sakrison (1968), Shanmugam (1979), Carlson (1986), Stremler (1990), Ziemer and Tranter (2002), Couch (1993), Gibson (1993), Ziemer and Tranter (1988), and Haykin (2000). Implementation of analog communication-systems are discussed in depth in Clarke and Hess (1971).

PROBLEMS

4.1 The message signal $m(t) = 10 \, \text{sinc}(400t)$ frequency modulates the carrier $c(t) = 100 \cos 2\pi f_c t$. The modulation index is 6.

1. Write an expression for the modulated signal $u(t)$.
2. What is the maximum frequency deviation of the modulated signal?
3. What is the power content of the modulated signal?
4. Find the bandwidth of the modulated signal.

4.2 Signal $m(t)$ is shown in Figure P-4.2; this signal is used once to frequency modulate a carrier and once to phase modulate the same carrier.

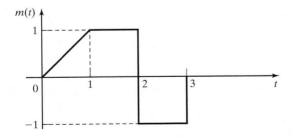

Figure P-4.2

1. Find a relation between k_p and k_f such that the maximum phase of the modulated signals in both cases are equal.
2. If $k_p = f_d = 1$, what is the maximum instantaneous frequency in each case?

4.3 Determine the in-phase and quadrature components as well as the envelope and the phase of FM- and PM-modulated signals.

4.4 An angle-modulated signal has the form

$$u(t) = 100 \cos \left[2\pi f_c t + 4 \sin 2000\pi t \right],$$

where $f_c = 10$ MHz.

1. Determine the average transmitted power.
2. Determine the peak-phase deviation.
3. Determine the peak-frequency deviation.
4. Is this an FM or a PM signal? Explain.

4.5 Find the smallest value of the modulation index in an FM system that guarantees that all the modulated-signal power is contained in the sidebands and no power is transmitted at the carrier frequency when the modulating signal is sinusoidal.

4.6 To generate wideband FM, we can first generate a narrowband FM signal, and then use frequency multiplication to spread the signal bandwidth. Figure P-4.6 illustrates such a scheme, which is called an Armstrong-type FM modulator. The narrowband FM signal has a maximum angular deviation of 0.10 radians to keep distortion under control.

1. If the message signal has a bandwidth of 15 kHz and the output frequency from the oscillator is 100 kHz, determine the frequency multiplication that is necessary to generate an FM signal at a carrier frequency of $f_c = 104$ MHz and a frequency deviation of $f = 75$ kHz.

2. If the carrier frequency for the wideband FM signal is to be within ± 2 Hz, determine the maximum allowable drift of the 100 kHz oscillator.

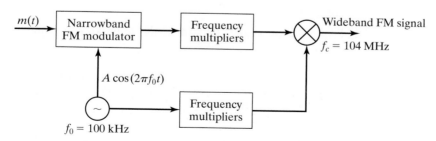

Figure P-4.6   Armstrong type FM modulator.

4.7 Determine the amplitude and phase of the various frequency components of a PM signal with $k_p = 1$ and with $m(t)$ a periodic signal given by

$$m(t) = \begin{cases} 1, & 0 \le t < \frac{T_m}{2} \\ -1, & \frac{T_m}{2} \le t \le T_m \end{cases} \tag{4.6.1}$$

in one period.

4.8 An FM signal is given as

$$u(t) = 100 \cos \left[2\pi f_c t + 100 \int_{-\infty}^{t} m(\tau) \, d\tau \right],$$

where $m(t)$ is shown in Figure P-4.8.

 1. Sketch the instantaneous frequency as a function of time.

 2. Determine the peak-frequency deviation.

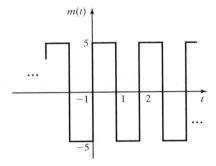

Figure P-4.8

4.9 The carrier $c(t) = 100 \cos 2\pi f_c t$ is frequency modulated by the signal $m(t) = 5 \cos 20,000\pi t$, where $f_c = 10^8$ Hz. The peak-frequency deviation is 20 kHz.

 1. Determine the amplitude and frequency of all signal components that have a power level of at least 10% of the power of the unmodulated carrier component.

 2. From Carson's rule, determine the approximate bandwidth of the FM signal.

4.10 The carrier $c(t) = A \cos 2\pi 10^6 t$ is angle modulated (PM or FM) by the sinusoid signal $m(t) = 2 \cos 2000\pi t$. The deviation constants are $k_p = 1.5$ rad/v and $k_f = 3000$ Hz/v.

 1. Determine β_f and β_p.

 2. Determine the bandwidth in each case using Carson's rule.

3. Plot the spectrum of the modulated signal in each case. (Plot only those frequency components that lie within the bandwidth derived in Part 2.)

4. If the amplitude of $m(t)$ is decreased by a factor of two, how would your answers to Parts 1–3 change?

5. If the frequency of $m(t)$ is increased by a factor of two, how would your answers to Parts 1–3 change?

4.11 The carrier $c(t) = 100 \cos 2\pi f_c t$ is phase modulated by the signal $m(t) = 5 \cos 2000\pi t$. The PM signal has a peak-phase deviation of $\pi/2$. The carrier frequency is $f_c = 10^8$ Hz.

1. Determine the magnitude spectrum of the sinusoidal components and sketch the results.

2. Using Carson's rule, determine the approximate bandwidth of the PM signal and compare the result with the analytical result in Part 1.

4.12 An angle-modulated signal has the form

$$u(t) = 100 \cos \left[2\pi f_c t + 4 \sin 2\pi f_m t \right],$$

where $f_c = 10$ MHz and $f_m = 1000$ Hz.

1. Assuming that this is an FM signal, determine the modulation index and the transmitted-signal bandwidth.

2. Repeat Part 1 if f_m is doubled.

3. Assuming that this is a PM signal, determine the modulation index and the transmitted-signal bandwidth.

4. Repeat Part 3 if f_m is doubled.

4.13 It is easy to demonstrate that amplitude modulation satisfies the superposition principle, whereas angle modulation does not. To be specific, let $m_1(t)$ and $m_2(t)$ represent two message signals and let $u_1(t)$ and $u_2(t)$ represent the corresponding modulated versions.

1. Show that when the combined message signal $m_1(t) + m_2(t)$ DSB modulates a carrier $A_c \cos 2\pi f_c t$, the result is the sum of the two DSB amplitude-modulated signals $u_1(t) + u_2(t)$.

2. Show that if $m_1(t) + m_2(t)$ frequency modulates a carrier, the modulated signal is not equal to $u_1(t) + u_2(t)$.

4.14 An FM discriminator is shown in Figure P-4.14. The envelope detector is assumed to be ideal and has an infinite input impedance. Select the values for L and C if the discriminator is to be used to demodulate an FM signal with a carrier $f_c = 80$ MHz and a peak frequency deviation of 6 MHz.

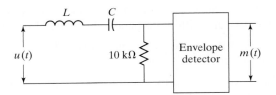

Figure P-4.14

4.15 An angle-modulated signal is given as

$$u(t) = 100 \cos\left[2000\pi t + \phi(t)\right],$$

where (a) $\phi(t) = 5 \sin 20\pi t$ and (b) $\phi(t) = 5 \cos 20\pi t$. Determine and sketch the amplitude and phase spectra for (a) and (b), and compare the results.

4.16 The message signal $m(t)$ into an FM modulator with a peak frequency deviation $f_d = 25$ Hz/v is shown in Figure P-4.16. Plot the frequency deviation in Hz and the phase-deviation in radians.

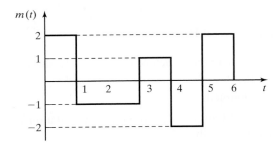

Figure P-4.16

4.17 A message signal $m(t)$ has a bandwidth of 10 kHz and a peak magnitude $|m(t)|$ of 1 volt. Estimate the bandwidth of the signal $u(t)$ obtained when the $m(t)$ frequency modulates a carrier with a peak-frequency deviation of (a) $f_d = 10$ Hz/v, (b) 100 Hz/v and (c) 1000 Hz/v.

4.18 The modulating signal that is the input into an FM modulator is

$$m(t) = 10 \cos 16\pi t.$$

The output of the FM modulator is

$$u(t) = 10 \cos\left[4000\pi t + 2\pi k_f \int_{-\infty}^{t} m(\tau)\, d\tau\right],$$

where $k_f = 10$. (See Figure P-4.18.) If the output of the FM modulator is passed through an ideal BPF centered at $f_c = 2000$ with a bandwidth of 62 Hz, determine the power of the frequency components at the output of the filter. What percentage of the transmitter power appears at the output of the BPF?

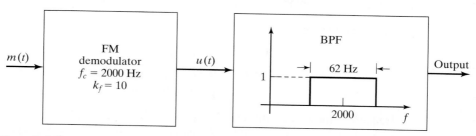

Figure P-4.18

4.19 The message signal $m_1(t)$ is shown in Figure P-4.19 and the message signal $m_2(t) = \text{sinc}(2 \times 10^4 t)$, volts.

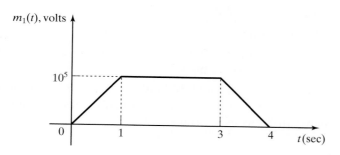

Figure P-4.19

1. If $m_1(t)$ is frequency modulated on a carrier with frequency 10^6 Hz and a frequency-deviation constant (k_f) equal to 5 Hz/v, what is the maximum instantaneous frequency of the modulated signal?

2. If $m_1(t)$ is phase modulated with a phase-deviation constant $k_p = 3$ rad/v, what is the maximum instantaneous frequency of the modulated signal? What is the minimum instantaneous frequency of the modulated signal?

3. If $m_2(t)$ is frequency modulated with $k_f = 10^3$ Hz/v, what is the maximum instantaneous frequency of the modulated signal? What is the bandwidth of the modulated signal?

4.20 A superheterodyne FM receiver operates in the frequency range of 88–108 MHz. The IF and local-oscillator frequencies are chosen such that $f_{IF} < f_{LO}$. We require that the image frequency f_c' fall outside of the 88–108 MHz region. Determine the minimum required f_{IF} and the range of variation in f_{LO}.

COMPUTER PROBLEMS

4.1 Frequency Modulation

The message signal

$$m(t) = \begin{cases} 1 & 0 \le t < t_0/3 \\ -2 & t_0/3 \le t < 2t_0/3 \\ 0 & \text{otherwise} \end{cases}$$

frequency modulates the carrier $c(t) = \cos 2\pi f_c t$, where $f_c = 200$ Hz and $t_0 = 0.15$ sec. The frequency-deviation constant is $k_f = 50$. Therefore, the frequency-modulated signal is

$$u(t) = \cos\left(2\pi f_c t + 2\pi k_f \int_{-\infty}^{t} m(\tau)\,d\tau\right).$$

1. Plot the message signal $m(t)$ and its integral on two separate graphs. The sampling interval is $t_s = 0.0001$.

2. Plot the FM signal $u(t)$.

3. Use MATLAB's Fourier-transform routine to compute and plot the spectra of $m(t)$ and $u(t)$ on separate plots.

4. Suppose we define the bandwidth W of $m(t)$ as the width of the main lobe of its spectrum. Determine the modulation index β and the modulation-signal bandwidth B_c using Carson's rule.

4.2 Frequency Modulation

The message signal

$$m(t) = \begin{cases} \text{sinc}(100t) & |t| \le t_0 \\ 0 & \text{otherwise} \end{cases}$$

frequency modulates the carrier $c(t) = \cos(2\pi f_c t)$, when $f_c = 250$ Hz and $t_0 = 0.1$. The frequency-deviation constant is $k_f = 100$. Therefore, the frequency-modulated signal is

$$u(t) = \cos\left(2\pi f_c t + 2\pi k_f \int_{-\infty}^{t} m(\tau)\,d\tau\right).$$

1. Plot the message signal and its integral on two separate graphs. The sampling interval is $t_s = 0.0001$.

2. Plot the FM signal $u(t)$.

3. Use MATLAB's Fourier-transform routine to compute and plot the spectra of $m(t)$ and $u(t)$ on separate graphs.

4. Demodulate the FM signal $u(t)$ to obtain the message signal and compare the result with the original message signal. The FM signal can be demodulated by first finding the phase of $u(t)$, i.e., the integral of $m(t)$, which can be differentiated and divided by $2\pi k_f$ to yield $m(t)$. Use MATLAB function `unwrap.m` to undo the effect of 2π-phase foldings. Comment on how well the demodulated message signal matches the original message signal $m(t)$.

4.3 Frequency Modulation

The message signal

$$m(t) = \begin{cases} t & 0 \le t < 1 \\ -t + 2 & 1 \le t < 2 \\ 0 & \text{otherwise} \end{cases}$$

frequency modulates the carrier $c(t) = \cos(2\pi f_c t)$, when $f_c = 1000$ Hz. The frequency-deviation constant is $k_f = 25$.

1. Plot the message signal and its integral on two separate graphs.

2. Plot the FM signal

$$u(t) = \cos\left(2\pi f_c t + 2\pi k_f \int_{-\infty}^{t} m(\tau) \, d\tau\right).$$

3. Use MATLAB's Fourier-transform routine to compute and plot the spectra of $m(t)$ and $u(t)$ on separate graphs.

4. Determine the modulation index, the bandwidth, and the range of the instantaneous frequency of $u(t)$.

5. Demodulate the FM signal $u(t)$ to obtain the message signal and compare the result with the original message signal. The FM signal can be demodulated by first finding the phase of $u(t)$, i.e., the integral of $m(t)$, which can be differentiated and divided by $2\pi k_f$ to yield $m(t)$. Use the MATLAB function `unwrap.m` to undo the effect of 2π phase foldings. Comment on how well the demodulated message signal matches the original message signal $m(t)$.

5

Probability and Random Processes

This chapter is devoted to a review of the basics of probability and the study of random processes and their properties. Random processes provide good models for both information sources and noise. When a signal is transmitted through a communication channel, there are two types of imperfections that cause the received signal to be different from the transmitted signal. One class of imperfections is deterministic in nature, such as linear and nonlinear distortion, intersymbol interference, etc. The second class is nondeterministic, such as the addition of noise, multipath fading, etc. For a quantitative study of these phenomena, we model them as random processes.

The information that is to be transmitted is best modeled as a random process. This is because any signal that conveys information must have some uncertainty in it, otherwise its transmission is of no interest. We will explore this aspect in greater detail in Chapter 12. In this chapter, we will briefly review the basics of probability theory and random variables, then we will introduce the concept of a random process and the basic tools used in analysis of random processes.

5.1 REVIEW OF PROBABILITY AND RANDOM VARIABLES

In this section, we will briefly review some basics of probability theory that are needed for our treatment of random processes. Throughout the text, we assume that you have already been exposed to probability theory elsewhere; therefore, our treatment in this section will be brief.

5.1.1 Sample Space, Events, and Probability

The fundamental concept in any probabilistic model is the concept of a *random experiment*, which is any experiment whose outcome cannot be predicted with certainty. Flipping a coin, throwing a die, and drawing a card from a deck of cards are examples of random experiments. What is common in all these cases is that the result (or outcome) of the experiment is uncertain. A random experiment has certain *outcomes*, which are the elementary results of the experiment. In flipping of a coin, "head" and "tail" are the possible outcomes. In throwing a die, 1,2,3,4,5, and 6 are the possible outcomes. The set of all possible outcomes is called the *sample space* and is denoted by Ω. Outcomes are denoted by ω's, and each ω lies in Ω, i.e., $\omega \in \Omega$.

A sample space is *discrete* if the number of its elements are finite or *countably infinite*[1], otherwise it is a *non-discrete* sample space. All the random experiments given above have discrete sample spaces. If we randomly choose a number between 0 and 1, then the sample space corresponding to this random experiment is the set of all numbers between 0 and 1, which is infinite and uncountable. Such a sample space is nondiscrete.

Events are subsets of the sample space; in other words, an event is a collection of outcomes. For instance, in throwing a die, the event "the outcome is odd" consists of outcomes 1, 3, and 5; the event "the outcome is greater than 3" consists of outcomes 4, 5, and 6; and the event "the outcome divides 4" consists of the single outcome 4. For the experiment of picking a number between 0 and 1, we can define an event as "the outcome is less than 0.7," or "the outcome is between 0.2 and 0.5," or "the outcome is 0.5." Events are *disjoint* if their intersection is empty. For instance, in throwing a die, the events "the outcome is odd" and "the outcome divides 4" are disjoint.

We define a *probability* P as a set function assigning nonnegative values to all events E such that the following conditions are satisfied:

1. $0 \leq P(E) \leq 1$ for all events.
2. $P(\Omega) = 1$.
3. For disjoint events E_1, E_2, E_3, $\cdots$ (i.e., events for which $E_i \cap E_j = \emptyset$ for all $i \neq j$, where $\emptyset$ is the empty set), we have $P(\cup_{i=1}^{\infty} E_i) = \sum_{i=1}^{\infty} P(E_i)$.

Some basic properties of probability follow easily from the set theoretical properties of events combined with the basic properties of probability. Some of the most important properties are as follows:

1. $P(E^c) = 1 - P(E)$, where E^c denotes the complement of E.
2. $P(\emptyset) = 0$.
3. $P(E_1 \cup E_2) = P(E_1) + P(E_2) - P(E_1 \cap E_2)$.
4. If $E_1 \subset E_2$ then $P(E_1) \leq P(E_2)$.

5.1.2 Conditional Probability

Let us assume that the two events, E_1 and E_2, have probabilities $P(E_1)$ and $P(E_2)$. If an observer knows that the event E_2 has occurred, then the probability that event E_1

---

[1]Countably infinite means infinite but enumerable, that is, the number of outcomes is infinite but it can be put in one-to-one correspondence with the set of natural numbers, i.e., it can be counted.

will occur will not be $P(E_1)$ anymore. In fact, the information that the observer receives changes the probabilities of various events; thus, new probabilities, called *conditional probabilities*, are defined. The conditional probability of the event E_1, given the event E_2, is defined by

$$P(E_1|E_2) = \begin{cases} \frac{P(E_1 \cap E_2)}{P(E_2)}, & P(E_2) \neq 0 \\ 0, & \text{otherwise} \end{cases} . \tag{5.1.1}$$

If it happens that $P(E_1|E_2) = P(E_1)$, then the knowledge of E_2 does not change the probability of the occurrence of E_1. In this case, the events E_1 and E_2 are said to be *independent*. For independent events, $P(E_1 \cap E_2) = P(E_1)P(E_2)$.

Example 5.1.1

In throwing a fair die, the probability of

$$A = \{\text{The outcome is greater than 3}\}$$

is

$$P(A) = P(4) + P(5) + P(6) = \frac{1}{2}.$$

The probability of

$$B = \{\text{The outcome is even}\}$$

is

$$P(B) = P(2) + P(4) + P(6) = \frac{1}{2}.$$

In this case,

$$P(A|B) = \frac{P(A \cap B)}{P(B)} = \frac{P(4) + P(6)}{\frac{1}{2}} = \frac{2}{3}. \qquad \blacksquare$$

If the events $\{E_i\}_{i=1}^{n}$ are disjoint and their union makes the entire sample space, then they make a *partition* of the sample space Ω. Then, if for an event A, we have the conditional probabilities $\{P(A|E_i)\}_{i=1}^{n}$, $P(A)$ can be obtained by applying the *total probability theorem* stated as

$$P(A) = \sum_{i=1}^{n} P(E_i)P(A|E_i). \tag{5.1.2}$$

Bayes's rule gives the conditional probabilities $P(E_i|A)$ by the following relation

$$P(E_i|A) = \frac{P(E_i)P(A|E_i)}{\sum_{j=1}^{n} P(E_j)P(A|E_j)}. \tag{5.1.3}$$

Example 5.1.2

In a certain city, 50% of the population drive to work, 30% take the subway, and 20% take the bus. The probability of being late for those who drive is 10%, for those who take the subway is 3%, and for those who take the bus is 5%.

1. What is the probability that an individual in this city will be late for work?

2. If an individual is late for work, what is the probability that he drove to work?

Solution   Let D, S, and B denote the events of an individual driving, taking the subway, or taking the bus. Then $P(D) = 0.5$, $P(S) = 0.3$, and $P(B) = 0.2$. If L denotes the event of being late, then, from the assumptions, we have

$$P(L|D) = 0.1;$$

$$P(L|S) = 0.03;$$

$$P(L|B) = 0.05.$$

1. From the total probability theorem,

$$P(L) = P(D)P(L|D) + P(S)P(L|S) + P(B)P(L|B)$$
$$= 0.5 \times 0.1 + 0.3 \times 0.03 + 0.2 \times 0.05$$
$$= 0.069.$$

2. Applying Bayes's rule, we have

$$P(D|L) = \frac{P(D)P(L|D)}{P(D)P(L|D) + P(S)P(L|S) + P(B)P(L|B)}$$
$$= \frac{0.05}{0.069}$$
$$\approx 0.725.$$ ∎

Example 5.1.3

In a binary communication system, the input bits transmitted over the channel are either 0 or 1 with probabilities 0.3 and 0.7, respectively. When a bit is transmitted over the channel, it can be either received correctly or incorrectly (due to channel noise). Let us assume that if a 0 is transmitted, the probability of it being received in error (i.e., being received as 1) is 0.01, and if a 1 is transmitted, the probability of it being received in error (i.e., being received as 0) is 0.1.

1. What is the probability that the output of this channel is 1?

2. Assuming we have observed a 1 at the output of this channel, what is the probability that the input to the channel was a 1?

Solution   Let X denote the input and Y denote the output. From the problem assumptions, we have

$$P(X = 0) = 0.3; \qquad\qquad P(X = 1) = 0.7;$$
$$P(Y = 0|X = 0) = 0.99; \qquad P(Y = 1|X = 0) = 0.01;$$
$$P(Y = 0|X = 1) = 0.1; \qquad P(Y = 1|X = 1) = 0.9.$$

1. From the total probability theorem, we have

$$P(Y = 1) = P(Y = 1, X = 0) + P(Y = 1, X = 1)$$
$$= P(X = 0)P(Y = 1|X = 0) + P(X = 1)P(Y = 1|X = 1)$$
$$= 0.3 \times 0.01 + 0.7 \times 0.9$$
$$= 0.003 + 0.63$$
$$= 0.633.$$

2. From the Bayes rule, we have

$$P(X = 1|Y = 1) = \frac{P(X = 1)P(Y = 1|X = 1)}{P(X = 0)P(Y = 1|X = 0) + P(X = 1)P(Y = 1|X = 1)}$$
$$= \frac{0.7 \times 0.9}{0.3 \times 0.01 + 0.7 \times 0.9}$$
$$= \frac{0.63}{0.633}$$
$$= 0.995. \qquad \blacksquare$$

5.1.3 Random Variables

A *random variable* is a mapping from the sample space Ω to the set of real numbers. In other words, a random variable is an assignment of real numbers to the outcomes of a random experiment. A schematic diagram representing a random variable is given in Figure 5.1.

Example 5.1.4

In throwing dice, if the player wins the amount that the die shows if the result is even and loses that amount if it is odd, then the random variable is

$$X(\omega) = \begin{cases} \omega & \omega = 2, 4, 6 \\ -\omega & \omega = 1, 3, 5 \end{cases}. \qquad \blacksquare$$

Random variables are denoted by capital letters, i.e., X, Y, etc., and individual values of the random variable X are $X(\omega)$. A random variable is *discrete* if the

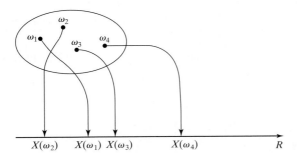

Figure 5.1 A random variable as a mapping from Ω to $\mathbb{R}$.

range of its values is either finite or countably infinite. This range is usually denoted by $\{x_i\}$. A *continuous* random variable is one in which the range of values is a continuum.

The *cumulative distribution function* or CDF of a random variable X is defined as

$$F_X(x) = P\{\omega \in \Omega : X(\omega) \le x\},$$

which can be simply written as

$$F_X(x) = P(X \le x)$$

and has the following properties:

1. $0 \le F_X(x) \le 1$.
2. $F_X(x)$ is nondecreasing.
3. $\lim_{x \to -\infty} F_X(x) = 0$ and $\lim_{x \to +\infty} F_X(x) = 1$.
4. $F_X(x)$ is continuous from the right, i.e., $\lim_{\epsilon \downarrow 0} F(x + \epsilon) = F(x)$.
5. $P(a < X \le b) = F_X(b) - F_X(a)$.
6. $P(X = a) = F_X(a) - F_X(a^-)$.

For discrete random variables, $F_X(x)$ is a staircase function. A random variable is *continuous* if $F_X(x)$ is a continuous function. A random variable is *mixed* if it is neither discrete nor continuous. Examples of CDFs for discrete, continuous, and mixed random variables are shown in Figures 5.2, 5.3, and 5.4, respectively.

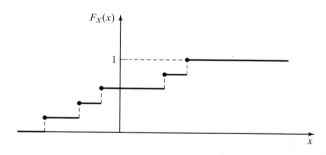

Figure 5.2 The CDF for a discrete random variable.

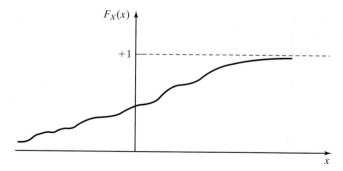

Figure 5.3 CDF for a continuous random variable.

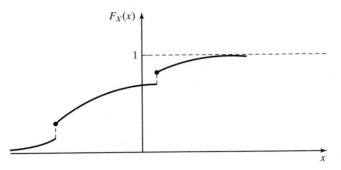

Figure 5.4 CDF for a mixed random variable.

The *probability density function*, or PDF, of a continuous random variable X is defined as the derivative of its CDF. It is denoted by $f_X(x)$, i.e.,

$$f_X(x) = \frac{d}{dx} F_X(x). \tag{5.1.4}$$

The basic properties of PDF are as follows:

1. $f_X(x) \geq 0$.
2. $\int_{-\infty}^{\infty} f_X(x)\, dx = 1$.
3. $\int_a^b f_X(x)\, dx = P(a < X \leq b)$.
4. In general, $P(X \in A) = \int_A f_X(x)\, dx$.
5. $F_X(x) = \int_{-\infty}^{x^+} f_X(u)\, du$.

For discrete random variables, we generally define the *probability mass function*, or PMF, which is defined as $\{p_i\}$, where $p_i = P(X = x_i)$. Obviously, for all i, we have $p_i \geq 0$ and $\sum_i p_i = 1$.

Important Random Variables.   In communications, the most commonly used random variables are the following:

Bernoulli random variable.   This is a discrete random variable taking two values, 1 and 0, with probabilities p and $1 - p$. A Bernoulli random variable is a good model for a binary-data generator. When binary data is transmitted over a communication channel, some bits are received in error. We can model an error by modulo-2 addition of a 1 to the input bit; thus, we change a 0 into a 1 and a 1 into a 0. Therefore, a Bernoulli random variable can be employed to model the channel errors.

Binomial random variable.   This is a discrete random variable giving the number of 1's in a sequence of n-independent Bernoulli trials. The PMF is given by

$$P(X = k) = \begin{cases} \binom{n}{k} p^k (1-p)^{n-k}, & 0 \leq k \leq n \\ 0, & \text{otherwise} \end{cases}. \tag{5.1.5}$$

This random variable models, for example, the total number of bits received in error when a sequence of n bits is transmitted over a channel with a bit–error probability of p.

Example 5.1.5

Assume 10,000 bits are transmitted over a channel in which the error probability is 10^{-3}. What is the probability that the total number of errors is less than 3?

Solution   In this example, $n = 10,000$, $p = 0.001$, and we are looking for $P(X < 3)$. We have

$$P(X < 3) = P(X = 0) + P(X = 1) + P(X = 2)$$

$$= \binom{10,000}{0} 0.001^0 (1 - 0.001)^{10,000}$$

$$+ \binom{10,000}{1} 0.001^1 (1 - 0.001)^{10,000-1}$$

$$+ \binom{10,000}{2} 0.001^2 (1 - 0.001)^{10,000-2}$$

$$= 0.0028. \qquad\blacksquare$$

Uniform random variable.   This is a continuous random variable taking values between a and b with equal probabilities for intervals of equal length. The density function is given by

$$f_X(x) = \begin{cases} \frac{1}{b-a}, & a < x < b \\ 0, & \text{otherwise} \end{cases}.$$

This is a model for continuous random variables whose range is known, but nothing else is known about the likelihood of various values that the random variable can assume. For example, when the phase of a sinusoid is random, it is usually modeled as a uniform random variable between 0 and 2π.

Gaussian or normal random variable.   The Gaussian, or normal, random variable is a continuous random variable described by the density function

$$f_X(x) = \frac{1}{\sqrt{2\pi}\sigma} e^{-\frac{(x-m)^2}{2\sigma^2}}. \tag{5.1.6}$$

There are two parameters involved in the definition of the Gaussian random variable. The parameter m is called the *mean* and can assume any finite value. The parameter σ is called the *standard deviation* and can assume any finite and positive value. The square of the standard deviation, i.e., σ^2, is called the *variance*. A Gaussian random variable with mean m and variance σ^2 is denoted by $\mathcal{N}(m, \sigma^2)$. The random variable $\mathcal{N}(0, 1)$ is usually called *standard normal*.

The Gaussian random variable is the most important and frequently encountered random variable in communications. The reason is that thermal noise, which is the major source of noise in communication systems, has a Gaussian distribution. The properties of Gaussian noise will be investigated in more detail later in this chapter.

We have graphed the PDF and PMF of the above random variables. They are given in Figures 5.5–5.8.

Assuming that X is a standard normal random variable, we define the function $Q(x)$ as $P(X > x)$. The Q-function is given by the relation

$$Q(x) = P(X > x) = \int_x^\infty \frac{1}{\sqrt{2\pi}} e^{-\frac{t^2}{2}} \, dt. \tag{5.1.7}$$

This function represents the area under the tail of a standard normal random variable, as shown in Figure 5.9. From this figure, it is clear that the Q-function is a decreasing function. This function is well tabulated and frequently used in analyzing the

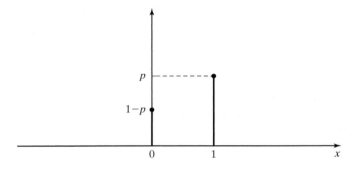

Figure 5.5 The PMF for the Bernoulli random variable.

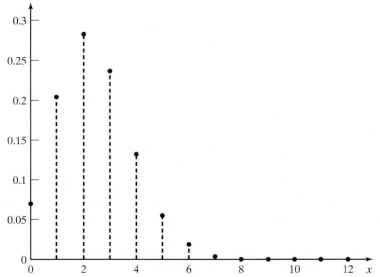

Figure 5.6 The PMF for the binomial random variable.

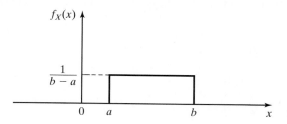

Figure 5.7 The PDF for the uniform random variable.

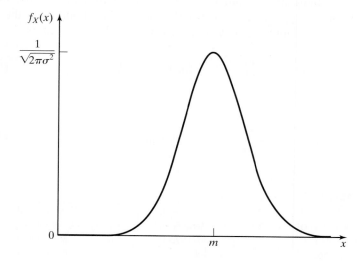

Figure 5.8 The PDF for the Gaussian random variable.

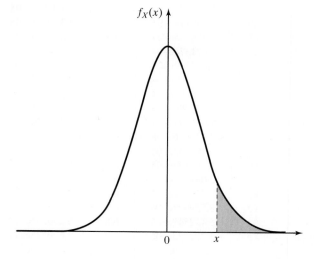

Figure 5.9 The Q-function as the area under the tail of a standard normal random variable.

performance of communication systems. It is easy to see that $Q(x)$ satisfies the following relations:

$$Q(-x) = 1 - Q(x);$$

$$Q(0) = \frac{1}{2};$$

$$Q(\infty) = 0.$$

Table 5.1 gives the values of this function for various values of x. Two important upper bounds on the Q function are widely used to find bounds on the error probability of various communication systems. These bounds are given as

$$Q(x) \leq \frac{1}{2}e^{-\frac{x^2}{2}} \qquad \text{for all } x \geq 0, \qquad (5.1.8)$$

and

$$Q(x) < \frac{1}{\sqrt{2\pi}x}e^{-\frac{x^2}{2}} \qquad \text{for all } x \geq 0. \qquad (5.1.9)$$

TABLE 5.1 TABLE OF THE Q FUNCTION

0	5.000000e–01	2.4	8.197534e–03	4.8	7.933274e–07
0.1	4.601722e–01	2.5	6.209665e–03	4.9	4.791830e–07
0.2	4.207403e–01	2.6	4.661189e–03	5.0	2.866516e–07
0.3	3.820886e–01	2.7	3.466973e–03	5.1	1.698268e–07
0.4	3.445783e–01	2.8	2.555131e–03	5.2	9.964437e–06
0.5	3.085375e–01	2.9	1.865812e–03	5.3	5.790128e–08
0.6	2.742531e–01	3.0	1.349898e–03	5.4	3.332043e–08
0.7	2.419637e–01	3.1	9.676035e–04	5.5	1.898956e–08
0.8	2.118554e–01	3.2	6.871378e–04	5.6	1.071760e–08
0.9	1.840601e–01	3.3	4.834242e–04	5.7	5.990378e–09
1.0	1.586553e–01	3.4	3.369291e–04	5.8	3.315742e–09
1.1	1.356661e–01	3.5	2.326291e–04	5.9	1.817507e–09
1.2	1.150697e–01	3.6	1.591086e–04	6.0	9.865876e–10
1.3	9.680049e–02	3.7	1.077997e–04	6.1	5.303426e–10
1.4	8.075666e–02	3.8	7.234806e–05	6.2	2.823161e–10
1.5	6.680720e–02	3.9	4.809633e–05	6.3	1.488226e–10
1.6	5.479929e–02	4.0	3.167124e–05	6.4	7.768843e–11
1.7	4.456546e–02	4.1	2.065752e–05	6.5	4.016001e–11
1.8	3.593032e–02	4.2	1.334576e–05	6.6	2.055790e–11
1.9	2.871656e–02	4.3	8.539898e–06	6.7	1.042099e–11
2.0	2.275013e–02	4.4	5.412542e–06	6.8	5.230951e–12
2.1	1.786442e–02	4.5	3.397673e–06	6.9	2.600125e–12
2.2	1.390345e–02	4.6	2.112456e–06	7.0	1.279813e–12
2.3	1.072411e–02	4.7	1.300809e–06		

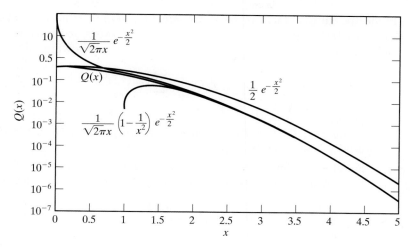

Figure 5.10   Bounds on the Q-function.

A frequently used lower bound is

$$Q(x) > \frac{1}{\sqrt{2\pi}x}\left(1 - \frac{1}{x^2}\right)e^{-\frac{x^2}{2}} \qquad \text{for all } x \geq 0. \qquad (5.1.10)$$

A plot of $Q(x)$ and its bounds is given in Figure 5.10.

For an $\mathcal{N}(m, \sigma^2)$ random variable, a simple change of variable in the integral that computes $P(X > x)$ results in $P(X > x) = Q\left(\frac{x-m}{\sigma}\right)$. This gives the so-called *tail* probability in a Gaussian random variable.

Example 5.1.6

X is a Gaussian random variable with mean 1 and variance 4. Find the probability that X is between 5 and 7.

Solution   We have $m = 1$ and $\sigma = \sqrt{4} = 2$. Thus,

$$P(5 < X < 7) = P(X > 5) - P(X > 7)$$

$$= Q\left(\frac{5-1}{2}\right) - Q\left(\frac{7-1}{2}\right)$$

$$= Q(2) - Q(3)$$

$$\approx 0.0214.$$ ∎

5.1.4 Functions of a Random Variable

A function of a random variable $Y = g(X)$ is itself a random variable. In general, to find the CDF of $Y = g(X)$, we can use the definition of the CDF to obtain

$$F_Y(y) = P\{\omega \in \Omega : g(X(\omega)) \leq x\}.$$

In the special case that, for all y, the equation $g(x) = y$ has a countable number of solutions $\{x_i\}$, and for all these solutions, $g'(x_i)$ exists and is nonzero, we can find the PDF of the random variable $Y = g(X)$ with the following relation:

$$f_Y(y) = \sum_i \frac{f_X(x_i)}{|g'(x_i)|}. \tag{5.1,11}$$

Example 5.1.7

Assuming X is a Gaussian random variable with $m = 0$ and $\sigma = 1$, find the probability-density function of the random variable Y given by $Y = aX + b$.

Solution In this case, $g(x) = ax + b$; therefore, $g'(x) = a$. The equation $ax + b = y$ has only one solution, which is given by $x_1 = \frac{y-b}{a}$. Using these results, we obtain

$$f_Y(y) = \frac{f_X(\frac{y-b}{a})}{|a|}$$

$$= \frac{1}{\sqrt{2\pi a^2}} e^{-\frac{(y-b)^2}{2a^2}} \tag{5.1.12}$$

It is observed that Y is a Gaussian random variable $\mathcal{N}(b, a^2)$. ∎

Using an approach similar to the one used in the preceding example, we can show that if X is $\mathcal{N}(m, \sigma^2)$, then $Y = aX + b$ is also a Gaussian random variable of the form $\mathcal{N}(am + b, a^2\sigma^2)$.

Example 5.1.8

Assume X is a $\mathcal{N}(3, 6)$ random variable. Find the density function of $Y = -2X + 3$.

Solution We know Y is a Gaussian random variable with the mean $m = -2 \times 3 + 3 = -3$ and variance $\sigma^2 = 4 \times 6 = 24$. Therefore, Y is a $\mathcal{N}(-3, 24)$ random variable and

$$f(y) = \frac{1}{\sqrt{48\pi}} e^{-\frac{(y+3)^2}{48}}.$$ ∎

From this example, we arrive at the important conclusion that *a linear function of a Gaussian random variable is itself a Gaussian random variable.*

Statistical Averages. The *mean, expected value,* or *expectation* of the random variable X is defined as

$$E(X) = \int_{-\infty}^{\infty} x f_X(x)\, dx \tag{5.1.13}$$

and is also denoted by m_X. The expected value of a random variable is a measure of the average of the value s that the random variable takes in a large number of experiments. Note that $E(X)$ is just a real number. In general, the nth moment of a random variable X is defined as

$$m_X^{(n)} \overset{\text{def}}{=} \int_{-\infty}^{\infty} x^n f_X(x)\, dx. \tag{5.1.14}$$

The expected value of $Y = g(X)$ is

$$E(g(X)) = \int_{-\infty}^{\infty} g(x) f_X(x)\,dx. \tag{5.1.15}$$

For discrete random variables, these equations become

$$E(X) = \sum_i x_i P(X = x_i) \tag{5.1.16}$$

and

$$E(g(X)) = \sum_i g(x_i) P(X = x_i). \tag{5.1.17}$$

In the special case where $g(X) = (X - E(X))^2$, $E(Y)$ is called the *variance* of X, which is a measure of the spread of the density function of X. If the variance is small, it indicates that the random variable is very concentrated around its mean, and in a sense is "less random." However, if the variance is large, then the random variable is highly spread; hence, it is less predictable. The variance is denoted by σ_X^2 and its square root, σ_X, is called the *standard deviation*. The relation for the variance can be written as

$$\sigma_X^2 = E(X^2) - (E(X))^2.$$

For any constant c, the following relations hold:

1. $E(cX) = cE(X)$.
2. $E(c) = c$.
3. $E(X + c) = E(X) + c$.

It is also easy to verify that the variance has the following properties:

1. $\text{VAR}(cX) = c^2 \text{VAR}(X)$.
2. $\text{VAR}(c) = 0$.
3. $\text{VAR}(X + c) = \text{VAR}(X)$.

For the important random variables we introduced earlier, we have the following relations for the mean and the variance:

      Bernoulli Random Variable. $E(X) = p$ $\text{VAR}(X) = p(1 - p)$.

      Binomial Random Variable. $E(X) = np$ $\text{VAR}(X) = np(1 - p)$.

      Uniform Random Variable. $E(X) = \frac{a+b}{2}$ $\text{VAR}(X) = \frac{(b-a)^2}{12}$.

      Gaussian Random Variable. $E(X) = m$ $\text{VAR}(X) = \sigma^2$.

5.1.5 Multiple Random Variables

Let X and Y represent two random variables defined on the same sample space Ω. For these two random variables, we can define the *joint CDF* as

$$F_{X,Y}(x, y) = P\{\omega \in \Omega : X(\omega) \leq x, Y(\omega) \leq y\},$$

or simply as

$$F_{X,Y}(x, y) = P(X \leq x, Y \leq y).$$

The *joint PDF* denoted as $f_{X,Y}(x, y)$ is defined by

$$f_{XY}(x, y) = \frac{\partial^2}{\partial x \partial y} F_{XY}(x, y). \tag{5.1.18}$$

The basic properties of the joint and *marginal* CDF's and PDF's can be summarized with the following relations:

1. $F_X(x) = F_{X,Y}(x, \infty)$.
2. $F_Y(y) = F_{X,Y}(\infty, y)$.
3. $f_X(x) = \int_{-\infty}^{\infty} f_{X,Y}(x, y)\, dy$.
4. $f_Y(y) = \int_{-\infty}^{\infty} f_{X,Y}(x, y)\, dx$.
5. $\int_{-\infty}^{\infty} \int_{-\infty}^{\infty} f_{X,Y}(x, y)\, dx\, dy = 1$.
6. $P((X, Y) \in A) = \iint_{(x,y) \in A} f_{X,Y}(x, y)\, dx\, dy$.
7. $F_{X,Y}(x, y) = \int_{-\infty}^{x} \int_{-\infty}^{y} f_{X,Y}(u, v)\, du\, dv$.

The *conditional probability density function* of the random variable Y, given that the value of the random variable X is equal to x, is denoted by $f_{Y|X}(y|x)$ and defined as

$$f_{Y|X}(y|x) = \begin{cases} \frac{f_{X,Y}(x,y)}{f_X(x)}, & f_Y(x) \neq 0 \\ 0, & \text{otherwise.} \end{cases} \tag{5.1.19}$$

If the density function after the knowledge of X is the same as the density function before the knowledge of X, then the random variables are said to be *statistically independent*. For statistically independent random variables,

$$f_{X,Y}(x, y) = f_X(x) f_Y(y). \tag{5.1.20}$$

Example 5.1.9
     Let X be $\mathcal{N}(3, 4)$ and Y be $\mathcal{N}(-2, 6)$. Assuming X and Y are independent, determine $f_{X,Y}(x, y)$.

Solution   We have

$$f(x, y) = f_X(x) f_Y(y)$$

$$= \frac{1}{\sqrt{8\pi}} e^{-\frac{(x-3)^2}{8}} \times \frac{1}{\sqrt{12\pi}} e^{-\frac{(y+2)^2}{12}}$$

$$= \frac{1}{4\pi\sqrt{6}} e^{-\frac{(x-3)^2}{8} - \frac{(y+2)^2}{12}}.$$

$\blacksquare$

The expected value of $g(X, Y)$, where $g(X, Y)$ is any function of X and Y, is obtained from

$$E(g(X, Y)) = \int_{-\infty}^{\infty} \int_{-\infty}^{\infty} g(x, y) f_{X,Y}(x, y)\, dx\, dy. \qquad (5.1.21)$$

$E(XY)$ is called the *correlation* of X and Y. Note that if X and Y are independent, then $E(XY) = E(X)E(Y)$. The *covariance* of X and Y is defined as

$$\text{COV}(X, Y) = E(XY) - E(X)E(Y).$$

If $\text{COV}(X, Y) = 0$, i.e., if $E(XY) = E(X)E(Y)$, then X and Y are called *uncorrelated* random variables. The normalized version of the covariance, called the *correlation coefficient*, is denoted by $\rho_{X,Y}$ and is defined as

$$\rho_{X,Y} = \frac{\text{COV}(X, Y)}{\sigma_X \sigma_Y}.$$

Using the Cauchy–Schwartz inequality (see Problem 2.37), we can see that $|\rho_{X,Y}| \leq 1$ and $\rho = \pm 1$ indicates a first-order relationship between X and Y, i.e., a relation of the form $Y = aX + b$. The case $\rho = 1$ corresponds to a positive a, and $\rho = -1$ corresponds to a negative a.

It is obvious that if X and Y are independent, then $\text{COV}(X, Y) = 0$ and X and Y will be uncorrelated. In other words, independence implies lack of correlation. It should be noted that lack of correlation *does not* generally imply independence. That is, the covariance (or ρ) might be zero, but the random variables may still be statistically *dependent*.

Some properties of the expected value and variance applied to multiple random variables are as follows:

1. $E(\sum_i c_i X_i) = \sum_i c_i E(X_i)$.
2. $\text{VAR}(\sum_i c_i X_i) = \sum_i c_i^2 \text{VAR}(X_i) + \sum_i \sum_{j \neq i} c_i c_j \text{COV}(X_i, X_j)$.
3. $\text{VAR}(\sum_i c_i X_i) = \sum_i c_i^2 \text{VAR}(X_i)$, if X_i and X_j are uncorrelated for $i \neq j$.

In each of these properties, the c_i's are constants.

Example 5.1.10

Assuming that X is $\mathcal{N}(3, 4)$, Y is $\mathcal{N}(-1, 2)$, and X and Y are independent, determine the covariance of the two random variables $Z = X - Y$ and $W = 2X + 3Y$.

Solution   We have

$$E(Z) = E(X) - E(Y) = 3 + 1 = 4,$$

$$E(W) = 2E(X) + 3E(Y) = 6 - 3 = 3,$$

$$E(X^2) = VAR(X) + (E(X))^2 = 4 + 9 = 13,$$

$$E(Y^2) = VAR(Y) + (E(Y))^2 = 2 + 1 = 3,$$

and

$$E(XY) = E(X)E(Y) = -3.$$

Therefore,

$$COV(W, Z) = E(WZ) - E(W)E(Z)$$

$$= E(2X^2 - 3Y^2 + XY) - E(Z)E(W)$$

$$= 2 \times 13 - 3 \times 3 - 3 - 4 \times 3$$

$$= 2 \qquad \blacksquare$$

Multiple Functions of Multiple Random Variables.   If we define two functions of the random variables X and Y by

$$\begin{cases} Z = g(X, Y) \\ W = h(X, Y) \end{cases},$$

then the joint CDF and PDF of Z and W can be obtained directly by applying the definition of the CDF. However, if it happens that for all z and w, the set of equations

$$\begin{cases} g(x, y) = z \\ h(x, y) = w \end{cases},$$

has a countable number of solutions $\{(x_i, y_i)\}$, and if at these points the determinant of the Jacobian matrix

$$J(x, y) = \begin{bmatrix} \frac{\partial z}{\partial x} & \frac{\partial z}{\partial y} \\ \frac{\partial w}{\partial x} & \frac{\partial w}{\partial y} \end{bmatrix}$$

is nonzero, then we have

$$f_{Z,W}(z, w) = \sum_i \frac{f(x_i, y_i)}{|\det J(x_i, y_i)|}, \qquad (5.1.22)$$

where $\det J$ denotes the determinant of the matrix J.

Example 5.1.11

The two random variables X and Y are independent and identically distributed, each with a Gaussian density function with the mean equal to zero and the variance equal to σ^2. If these two random variables denote the coordinates of a point in the plane, find the probability density function of the magnitude and the phase of that point in polar coordinates.

Solution   First, we have to find the joint probability density function of X and Y. Since X and Y are independent, their joint probability density function is the product of their marginal probability density functions; that is,

$$f_{X,Y}(x, y) = f_X(x)f_Y(y)$$

$$= \frac{1}{2\pi\sigma^2}e^{-\frac{x^2+y^2}{2\sigma^2}}. \tag{5.1.23}$$

The magnitude of the point with coordinates (X, Y) in the polar plane is given by $V = \sqrt{X^2 + Y^2}$, and its phase is given by $\Theta = \arctan\frac{Y}{X}$. We begin by deriving the joint probability density function of V and Θ. In this case, $g(X, Y) = \sqrt{X^2 + Y^2}$ and $h(X, Y) = \arctan\frac{Y}{X}$. The Jacobian matrix is given by

$$J(x, y) = \begin{bmatrix} \frac{x}{\sqrt{x^2+y^2}} & \frac{y}{\sqrt{x^2+y^2}} \\ -\frac{y}{x^2+y^2} & \frac{x}{x^2+y^2} \end{bmatrix}. \tag{5.1.24}$$

The determinant of the Jacobian matrix can be easily determined to be equal to

$$|\det J(x, y)| = \frac{1}{\sqrt{x^2 + y^2}}$$

$$= \frac{1}{v}. \tag{5.1.25}$$

The set of equations

$$\begin{cases} \sqrt{x^2 + y^2} = v \\ \arctan\frac{y}{x} = \theta \end{cases} \tag{5.1.26}$$

has only one solution, which is given by

$$\begin{cases} x = v\cos\theta \\ y = v\sin\theta \end{cases}. \tag{5.1.27}$$

Substituting these results into Equation (5.1.22), we obtain the joint probability density function of the magnitude and the phase as

$$f_{V,\Theta}(v, \theta) = vf_{X,Y}(v\cos\theta, v\sin\theta)$$

$$= \frac{v}{2\pi\sigma^2}e^{-\frac{v^2}{2\sigma^2}}. \tag{5.1.28}$$

To derive the marginal probability density functions for the magnitude and the phase, we have to integrate the joint probability density function. To obtain the probability density function of the phase, we have

$$f_{\Theta}(\theta) = \int_0^{\infty} f_{V,\Theta}(v, \theta)\, dv$$

$$= \frac{1}{2\pi}\int_0^{\infty} \frac{v}{\sigma^2}e^{-\frac{v^2}{2\sigma^2}}\, dv$$

$$= \frac{1}{2\pi}\left[-e^{-\frac{v^2}{2\sigma^2}}\right]_0^\infty$$

$$= \frac{1}{2\pi}. \tag{5.1.29}$$

Hence, the phase *is uniformly distributed* on $[0, 2\pi]$. To obtain the marginal probability density function for the magnitude, we have

$$f_V(v) = \int_0^{2\pi} f_{V,\Theta}(v, \theta)\, d\theta$$

$$= \frac{v}{\sigma^2} e^{-\frac{v^2}{2\sigma^2}}. \tag{5.1.30}$$

This relation holds only for positive v. For negative v, $f_V(v) = 0$. Therefore,

$$f_V(v) = \begin{cases} \frac{v}{\sigma^2}e^{-\frac{v^2}{2\sigma^2}}, & v \geq 0 \\ 0, & v < 0 \end{cases}. \tag{5.1.31}$$

This probability density function is known as the *Rayleigh probability density function*, and it has widespread applications in the study of the fading communication channels. It is also interesting to note that in the preceding example

$$f_{V,\Theta}(v, \theta) = f_V(v) f_\Theta(\theta); \tag{5.1.32}$$

therefore, the magnitude and the phase are independent random variables.   ∎

Jointly Gaussian Random Variables.   *Jointly Gaussian* or *binormal* random variables X and Y are distributed according to a joint PDF of the form

$$f_{X,Y}(x, y) = \frac{1}{2\pi\sigma_1\sigma_2\sqrt{1-\rho^2}} \exp\left\{-\frac{1}{2(1-\rho^2)}\right.$$

$$\left. \times \left[\frac{(x-m_1)^2}{\sigma_1^2} + \frac{(y-m_2)^2}{\sigma_2^2} - \frac{2\rho(x-m_1)(y-m_2)}{\sigma_1\sigma_2}\right]\right\},$$

where m_1, m_2, σ_1^2, and σ_2^2 are the mean and variances of X and Y, respectively, and ρ is their correlation coefficient. When two random variables X and Y are distributed according to a binormal distribution, it can be shown that X and Y are normal random variables and the conditional densities $f(x|y)$ and $f(y|x)$ are also Gaussian. This property shows the main difference between jointly Gaussian random variables and two random variables that each have a Gaussian distribution.

The definition of two jointly Gaussian random variables can be extended to more random variables. For instance, X_1, X_2, and X_3 are jointly Gaussian if any pair of them are jointly Gaussian, and the conditional density function of any pair given the third one is also jointly Gaussian.

Here are the main properties of jointly Gaussian random variables:

1. If n random variables are jointly Gaussian, any subset of them is also distributed according to a jointly Gaussian distribution of the appropriate size. In particular, all individual random variables are Gaussian.

2. Jointly Gaussian random variables are completely characterized by the means of all random variables $m_1, m_2, \ldots, m_n$ and the set of all covariance $COV(X_i, X_j)$ for all $1 \leq i \leq n$ and $1 \leq j \leq n$. These so-called *second-order* properties completely describe the random variables.

3. Any set of linear combinations of $(X_1, X_2, \cdots, X_n)$ are themselves jointly Gaussian. In particular, any linear combination of X_i's is a Gaussian random variable.

4. Two uncorrelated jointly Gaussian random variables are independent. Therefore, *for jointly Gaussian random variables, independence and uncorrelatedness are equivalent.* As previously stated, this is not true in general for nonjointly Gaussian random variables.

5.1.6 Sums of Random Variables

If we have a sequence of random variables $(X_1, X_2, \cdots, X_n)$ with basically the same properties, then the behavior of their average $Y = \frac{1}{n} \sum_{i=1}^n X_i$ is expected to be "less random" than each X_i. The *law of large numbers* and the *central limit theorem* are statements of this intuitive fact.

The *law of large numbers* (LLN) states that if the sequence of random variables $X_1, X_2, \cdots, X_n$ are uncorrelated with the same mean m_X and variance $\sigma_X^2 < \infty$, then for any $\epsilon > 0$, $\lim_{n \to \infty} P(|Y - m_X| > \epsilon) = 0$, where $Y = \frac{1}{n} \sum_{i=1}^n X_i$. This means that the average converges (in probability) to the expected value.

The *central limit theorem* not only states the convergence of the average to the mean, but it also gives some insight into the distribution of the average. This theorem states that if X_i's are i.i.d. (*independent and identically distributed*) random variables which each have a mean m and variance σ^2, then $Y = \frac{1}{n} \sum_{i=1}^n X_i$ converges to a $\mathcal{N}\left(m, \frac{\sigma^2}{n}\right)$. We can generally say that the central limit theorem states that the sum of many i.i.d. random variables converges to a Gaussian random variable. As we will see later, this theorem explains why thermal noise follows a Gaussian distribution.

This concludes our brief review of the basics of the probability theory. References at the end of this chapter provide sources for further study.

5.2 RANDOM PROCESSES: BASIC CONCEPTS

A random process is the natural extension of random variables when dealing with signals. In analyzing communication systems, we basically deal with time-varying signals. In our development so far, we have assumed that all the signals are deterministic.

In many situations, the deterministic assumption on time-varying signals is not a valid assumption, and it is more appropriate to model signals as random rather than deterministic functions. One such example is the case of thermal noise in electronic circuits. This type of noise is due to the random movement of electrons as a result of thermal agitation, therefore, the resulting current and voltage can only be described statistically. Another example is the reflection of radio waves from different layers of the ionosphere; they make long range broadcasting of short-wave radio possible. Due to the randomness of these reflections, the received signal can again be modeled as a random signal. These two examples show that random signals can describe certain phenomena in signal transmission.

Another situation where modeling by random processes proves useful is in the characterization of information sources. An information source, such as a speech source, generates time-varying signals whose contents are not known in advance. Otherwise there would be no need to transmit them. Therefore, random processes also provide a natural way to model information sources.

A *random process*, or a *random signal*, can be viewed as a set of possible realizations of signal waveforms. The realization of one from the set of possible signals is governed by some probabilistic law. This is similar to the definition of random variables where one from a set of possible *values* is realized according to some probabilistic law. The difference is that in random processes, we have signals (functions) instead of values (numbers).

Example 5.2.1

Assume that we have a signal generator that can generate one of the six possible sinusoids. The amplitude of all sinusoids is one, and the phase for all of them is zero, but the frequencies can be 100, 200, ... , 600 Hz. We throw a die, and depending on its outcome, which we denote by F, we generate a sinusoid whose frequency is 100 times what the die shows ($100F$). This means that each of the six possible signals will be realized with equal probability. This is an example of a random process. This random process can be defined as $X(t) = \cos(2\pi \times 100Ft)$. ∎

Example 5.2.2

Assume that we uniformly choose a phase Θ between 0 and 2π and generate a sinusoid with a fixed amplitude and frequency but with a random phase Θ. In this case, the random process is $X(t) = A\cos(2\pi f_0 t + \Theta)$, where A and f_0 denote the fixed amplitude and frequency and Θ denotes the random phase. Some sample functions for this random process are shown in Figure 5.11. ∎

Example 5.2.3

The process $X(t)$ is defined by $X(t) = X$, where X is a random variable uniformly distributed on $[-1, 1]$. In this case, an analytic description of the random process is given. For this random process, each sample is a constant signal. Sample functions of this process are shown in Figure 5.12. ∎

From the preceding examples, we see that corresponding to each outcome ω_i in a sample space Ω, there exists a signal $x(t; \omega_i)$. This description is similar to the description of random variables in which a real number is assigned to each outcome ω_i. Figure 5.13 depicts this general view of random processes. Thus, for each ω_i,

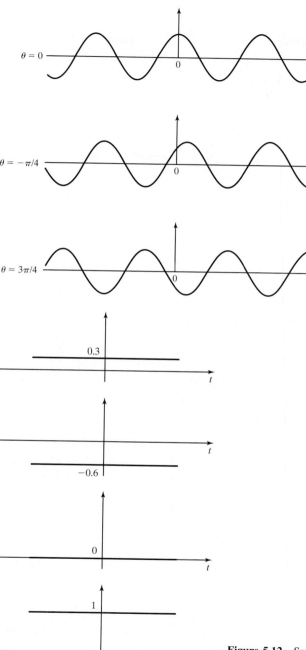

Figure 5.11 Sample functions of the random process given in Example 5.2.2.

Figure 5.12 Sample functions of the random process given in Example 5.2.3.

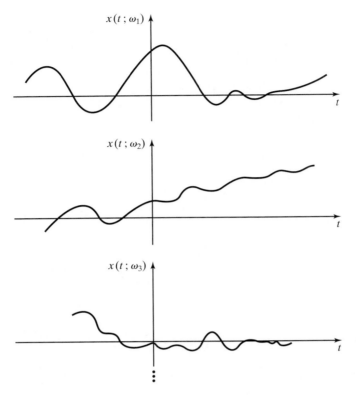

Figure 5.13 Sample functions of a random process.

there exists a deterministic time function $x(t; \omega_i)$, which is called a *sample function* or a *realization* of the random process. At each time instant t_0 and for each $\omega_i \in \Omega$, we have the number $x(t_0; \omega_i)$. For the different outcomes (ω_i's) at a fixed time t_0, the numbers $x(t_0; \omega_i)$ constitute a random variable denoted by $X(t_0)$. After all, a random variable is nothing but an assignment of real numbers to the outcomes of a random experiment. This is a very important observation and a bridge that connects the concept of a random process to the more familiar concept of a random variable. In other words, *at any time instant, the value of a random process is a random variable.*

Example 5.2.4
In Example 5.2.1, determine the values of the random variable $X(0.001)$.

Solution   The possible values are $\cos(0.2\pi)$, $\cos(0.4\pi)$, ... , $\cos(1.2\pi)$ and each has a probability $\frac{1}{6}$. ∎

Example 5.2.5
Let Ω denote the sample space corresponding to the random experiment of throwing a die. Obviously, in this case $\Omega = \{1, 2, 3, 4, 5, 6\}$. For all ω_i, let $x(t; \omega_i) = \omega_i e^{-t} u_{-1}(t)$ denote a random process. Then $X(1)$ is a random variable taking values $e^{-1}, 2e^{-1}, \ldots, 6e^{-1}$ and each has probability $\frac{1}{6}$. Sample functions of this random process are shown in Figure 5.14. ∎

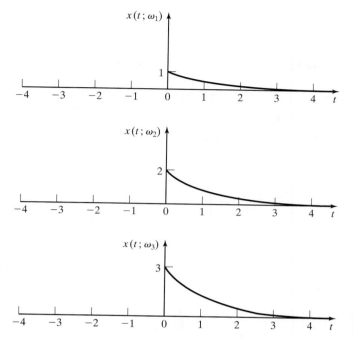

Figure 5.14 Sample functions of Example 5.2.5.

Example 5.2.6

We can have discrete-time random processes, which are similar to continuous-time random processes. For instance, let ω_i denote the outcome of a random experiment consisting of independent drawings from a Gaussian random variable distributed according to $\mathcal{N}(0, 1)$. Let the discrete-time random process $\{X_n\}_{n=0}^{\infty}$ be defined by $X_0 = 0$ and $X_n = X_{n-1} + \omega_n$ for all $n \geq 1$. This is an example of a discrete-time random process, which is nothing but a sequence of random variables.   ■

5.2.1 Statistical Averages

At any given time, the random process defines a random variable; at any given set of times, it defines a random vector. This fact enables us to define various statistical averages for the process via statistical averages of the corresponding random variables. For instance, we know that, at any time instance t_0, the random process at that time, i.e., $X(t_0)$, is an ordinary random variable; it has a density function and we can find its mean and its variance at that point. Obviously, both mean and variance are ordinary deterministic numbers, but they depend on the time t_0. That is, at time t_1, the density function, and thus the mean and the variance, of $X(t_1)$ will generally be different from those of $X(t_0)$.

Definition 5.2.1.   The *mean*, or *expectation*, of the random process $X(t)$ is a deterministic function of time denoted by $m_X(t)$ that at each time instant t_0 equals the mean of the random variable $X(t_0)$. That is, $m_X(t) = E[X(t)]$ for all t.   ■

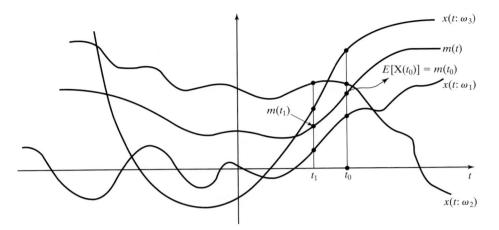

Figure 5.15 The mean of a random process.

Since at any t_0 the random variable $X(t_0)$ is well defined with a probability density function $f_{X(t_0)}(x)$, we have

$$E[X(t_0)] = m_X(t_0) = \int_{-\infty}^{\infty} x f_{X(t_0)}(x)dx. \qquad (5.2.1)$$

Figure 5.15 gives a pictorial description of this definition.

Example 5.2.7

The mean of the random process in Example 5.2.2 is obtained by noting that

$$f_\Theta(\theta) = \begin{cases} \frac{1}{2\pi} & 0 \le \theta < 2\pi \\ 0 & \text{otherwise} \end{cases}.$$

Hence,

$$E[X(t)] = \int_0^{2\pi} A \cos(2\pi f_0 t + \theta) \frac{1}{2\pi} d\theta = 0.$$

We observe that, in this case, $m_X(t)$ is independent of t. ∎

Another statistical average that plays a very important role in our study of random processes is the *autocorrelation function*. The autocorrelation function is especially important because it completely describes the power spectral density and the power content of a large class of random processes.

Definition 5.2.2. The *autocorrelation function* of the random process $X(t)$, denoted by $R_X(t_1, t_2)$, is defined by $R_X(t_1, t_2) = E[X(t_1)X(t_2)]$. ∎

From this definition, it is clear that $R_X(t_1, t_2)$ is a deterministic function of two variables t_1 and t_2 given by

$$R_X(t_1, t_2) = \int_{-\infty}^{\infty} \int_{-\infty}^{\infty} x_1 x_2 f_{X(t_1), X(t_2)}(x_1, x_2) \, dx_1 \, dx_2. \qquad (5.2.2)$$

Example 5.2.8

The autocorrelation function of the random process in Example 5.2.2 is

$$R_X(t_1, t_2) = E[A\cos(2\pi f_0 t_1 + \Theta)A\cos(2\pi f_0 t_2 + \Theta)]$$

$$= A^2 E[\frac{1}{2}\cos 2\pi f_0(t_1 - t_2) + \frac{1}{2}\cos(2\pi f_0(t_1 + t_2) + 2\Theta)]$$

$$= \frac{A^2}{2}\cos 2\pi f_0(t_1 - t_2),$$

where we have used

$$E[\cos(2\pi f_0(t_1 + t_2) + 2\Theta)] = \int_0^{2\pi} \cos[2\pi f_0(t_1 + t_2) + 2\theta]\frac{1}{2\pi}d\theta = 0. \quad\blacksquare$$

Example 5.2.9

For the random process given in Example 5.2.3, we have

$$R_X(t_1, t_2) = E(X^2) = \int_{-1}^{+1} \frac{x^2}{2}dx = \frac{1}{3}. \quad\blacksquare$$

5.2.2 Wide-Sense Stationary Processes

As we have seen, a random process observed at any given time is just a random variable, and the properties of this random variable depend on the time at which the random process is observed. It can happen that some of the properties of this random variable are independent of time. Depending on what properties are independent of time, different notions of stationarity can be defined. One of the most useful notions of stationarity is the notion of *wide-sense* stationary random processes. A process is wide-sense stationary if its mean and autocorrelation do not depend on the choice of the time origin. A formal definition is given in Definition 5.2.3.

Definition 5.2.3. A process $X(t)$ is *wide-sense stationary* (WSS) if the following conditions are satisfied:

1. $m_X(t) = E[X(t)]$ is independent of t.
2. $R_X(t_1, t_2)$ depends only on the time difference $\tau = t_1 - t_2$ and not on t_1 and t_2 individually.

   $\blacksquare$

Hereafter, we will use the term *stationary* as a shorthand for wide-sense stationary processes, and their mean and autocorrelation will be denoted by m_X and $R_X(\tau)$.

Example 5.2.10

For the random process in Exercise 5.2.2, we have already seen that $m_X = 0$ and $R_X(t_1, t_2) = \frac{A^2}{2}\cos 2\pi f_0(t_1 - t_2)$. Therefore, the process is WSS. $\blacksquare$

Example 5.2.11

Let the random process $Y(t)$ be similar to the random process $X(t)$ defined in Exercise 5.2.2, but assume that Θ is uniformly distributed between 0 and π. In this case,

$$m_Y(t) = E[A\cos(2\pi f_0 t + \Theta)]$$

$$= A\int_0^\pi \frac{1}{\pi}\cos(2\pi f_0 t + \theta)\,d\theta$$

$$= \frac{A}{\pi}\left[\sin(2\pi f_0 t + \theta)\right]_0^\pi$$

$$= \frac{A}{\pi}(-2\sin(2\pi f_0 t))$$

$$= -\frac{2A}{\pi}\sin(2\pi f_0 t).$$

Since $m_Y(t)$ is not independent of t, the process $Y(t)$ is not stationary. ∎

From the definition of the autocorrelation function, it follows that $R_X(t_1, t_2) = R_X(t_2, t_1)$. This means that if the process is stationary, we have $R_X(\tau) = R_X(-\tau)$, i.e., the autocorrelation function is an even function in stationary processes.

5.2.3 Multiple Random Processes

Multiple random processes arise naturally when dealing with two or more random processes. For example, take the case where we are dealing with a random process $X(t)$ and we pass it through a linear time-invariant system. For each sample function input $x(t; \omega_i)$, we have a sample function output defined by $y(t; \omega_i) = x(t; \omega_i) \star h(t)$, where $h(t)$ denotes the impulse response of the system. We can see that for each $\omega_i \in \Omega$, we have the two signals $x(t; \omega_i)$ and $y(t; \omega_i)$. Therefore, we are dealing with two random processes, $X(t)$ and $Y(t)$. When dealing with two random processes, we naturally question the dependence between the random processes under consideration. To this end, we define the independence of two random processes.

Definition 5.2.4. Two random processes $X(t)$ and $Y(t)$ are *independent* if for all positive integers m, n and for all $t_1, t_2, \ldots, t_n$ and $\tau_1, \tau_2, \ldots, \tau_m$, the random vectors $(X(t_1), X(t_2), \ldots, X(t_n))$ and $(Y(\tau_1), Y(\tau_2), \ldots, Y(\tau_m))$ are independent. Similarly $X(t)$ and $Y(t)$ are *uncorrelated* if the two random vectors are uncorrelated. ∎

From the properties of random variables, we know that the independence of random processes implies that they are uncorrelated, whereas uncorrelatedness generally does not imply independence, except for the important class of Gaussian processes (to be defined in Section 5.3) for which the two properties are equivalent. Next, we define the correlation function between two random processes.

Definition 5.2.5. The *cross correlation* between two random processes $X(t)$ and $Y(t)$ is defined as

$$R_{XY}(t_1, t_2) = E[X(t_1)Y(t_2)]. \tag{5.2.3}$$

∎

From the preceding definition, in general,

$$R_{XY}(t_1, t_2) = R_{YX}(t_2, t_1). \tag{5.2.4}$$

The concept of stationarity can also be generalized to joint stationarity for the case of two random processes.

Definition 5.2.6.   Two random processes $X(t)$ and $Y(t)$ are *jointly wide-sense stationary*, or simply *jointly stationary*, if both $X(t)$ and $Y(t)$ are individually stationary and the cross correlation $R_{XY}(t_1, t_2)$ depends only on $\tau = t_1 - t_2$.   ∎

Note that for jointly stationary random processes, from the definition and Equation 5.2.4, it follows that

$$R_{XY}(\tau) = R_{YX}(-\tau). \tag{5.2.5}$$

Example 5.2.12

Assuming that the two random processes $X(t)$ and $Y(t)$ are jointly stationary, determine the autocorrelation of the process $Z(t) = X(t) + Y(t)$.

Solution   By definition,

$$
\begin{aligned}
R_Z(t + \tau, t) &= E[Z(t + \tau)Z(t)] \\
&= E[(X(t + \tau) + Y(t + \tau))(X(t) + Y(t))] \\
&= R_X(\tau) + R_Y(\tau) + R_{XY}(\tau) + R_{XY}(-\tau).
\end{aligned}
$$ ∎

5.2.4 Random Processes and Linear Systems

In the section on multiple random processes, we saw that when a random process passes through a linear time-invariant system, the output is also a random process defined on the original probability space. In this section, we will study the properties of the output process based on the knowledge of the input process. We are assuming that a stationary process $X(t)$ is the input to a linear time-invariant system with the impulse response $h(t)$ and the output process is denoted by $Y(t)$, as shown in Figure 5.16.

We are interested in the following questions: Under what conditions will the output process be stationary? Under what conditions will the input and output processes be jointly stationary? How can we obtain the mean and the autocorrelation of the output process, as well as the cross correlation between the input and output processes?

We next demonstrate that if a stationary process $X(t)$ with mean m_X and autocorrelation function $R_X(\tau)$ is passed through a linear time-invariant system with

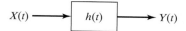

$X(t) \longrightarrow \boxed{h(t)} \longrightarrow Y(t)$ **Figure 5.16**   A random process passing through a linear time-invariant system.

impulse response $h(t)$, the input and output processes $X(t)$ and $Y(t)$ will be jointly stationary with

$$m_Y = m_X \int_{-\infty}^{\infty} h(t)\,dt; \tag{5.2.6}$$

$$R_{XY}(\tau) = R_X(\tau) \star h(-\tau); \tag{5.2.7}$$

$$R_Y(\tau) = R_X(\tau) \star h(\tau) \star h(-\tau). \tag{5.2.8}$$

By using the convolution integral to relate the output $Y(t)$ to the input $X(t)$, i.e., $Y(t) = \int_{-\infty}^{\infty} X(\tau)h(t-\tau)\,d\tau$, we have

$$
\begin{aligned}
E[Y(t)] &= E\left[\int_{-\infty}^{\infty} X(\tau)h(t-\tau)\,d\tau\right] \\
&= \int_{-\infty}^{\infty} E[X(\tau)]h(t-\tau)\,d\tau \\
&= \int_{-\infty}^{\infty} m_X h(t-\tau)\,d\tau \\
&\overset{u=t-\tau}{=} m_X \int_{-\infty}^{\infty} h(u)\,du \equiv m_Y.
\end{aligned}
$$

This proves that m_Y is independent of t.

The cross correlation function between the output and the input is

$$
\begin{aligned}
E[X(t_1)Y(t_2)] &= E[X(t_1)Y(t_2)] \\
&= E\left[X(t_1)\int_{-\infty}^{\infty} X(s)h(t_2-s)\,ds\right] \\
&= \int_{-\infty}^{\infty} E[X(t_1)X(s)]h(t_2-s)\,ds \\
&= \int_{-\infty}^{\infty} R_X(t_1-s)h(t_2-s)\,ds \\
&\overset{u=s-t_2}{=} \int_{-\infty}^{\infty} R_X(t_1-t_2-u)h(-u)\,du \\
&= \int_{-\infty}^{\infty} R_X(\tau-u)h(-u)\,du \\
&= R_X(\tau) \star h(-\tau) \equiv R_{XY}(\tau).
\end{aligned}
$$

This shows that $R_{XY}(t_1, t_2)$ depends only on $\tau = t_1 - t_2$.

The autocorrelation function of the output is

$$
\begin{aligned}
E[Y(t_1)Y(t_2)] &= E[Y(t_1)Y(t_2)] \\
&= E\left[\left(\int_{-\infty}^{\infty} X(s)h(t_1-s)\,ds\right)Y(t_2)\right] \\
&= \int_{-\infty}^{\infty} R_{XY}(s-t_2)h(t_1-s)\,ds \\
&\overset{u=s-t_2}{=} \int_{-\infty}^{\infty} R_{XY}(u)h(t_1-t_2-u)\,du \\
&= R_{XY}(\tau)\star h(\tau) \\
&= R_X(\tau)\star h(-\tau)\star h(\tau) \equiv R_Y(\tau).
\end{aligned}
$$

In the last step, we have used the result of the preceding step. This shows that R_Y and R_{XY} depend only on $\tau = t_1 - t_2$ and, hence, the output process is stationary. Therefore, the input and output processes are jointly stationary.

Example 5.2.13

Assume a stationary process passes through a differentiator. What are the mean and autocorrelation functions of the output? What is the cross correlation between the input and output?

Solution In a differentiator, $h(t) = \delta'(t)$. Since $\delta'(t)$ is odd, it follows that

$$
m_Y = m_X \int_{-\infty}^{\infty} \delta'(t)\,dt
$$
$$
= 0.
$$

The cross correlation function between output and input is

$$
R_{XY}(\tau) = R_X(\tau)\star\delta'(-\tau) = -R_X(\tau)\star\delta'(\tau) = -\frac{d}{d\tau}R_X(\tau),
$$

and the autocorrelation function of the output is

$$
R_Y(\tau) = -\frac{d}{d\tau}R_X(\tau)\star\delta'(\tau) = -\frac{d^2}{d\tau^2}R_X(\tau),
$$

where we have used Equation 2.1.34.   ∎

Example 5.2.14

Repeat the previous example for the case where the LTI system is a quadrature filter defined by $h(t) = \frac{1}{\pi t}$; therefore, $H(f) = -j\,\mathrm{sgn}(f)$. In this case, the output of the filter is the Hilbert transform of the input (see Section 2.6).

Solution We have

$$
m_Y = m_X \int_{-\infty}^{\infty} \frac{1}{\pi t}\,dt = 0
$$

because $\frac{1}{\pi t}$ is an odd function. The cross correlation function is

$$R_{XY}(\tau) = R_X(\tau) \star \frac{1}{-\pi \tau} = -\hat{R}_X(\tau),$$

and the autocorrelation function of the output is

$$R_Y(\tau) = R_{XY}(\tau) \star \frac{1}{\pi \tau} = -\widehat{\hat{R}_X}(\tau) = R_X(\tau),$$

where we have used the fact that $\hat{\hat{x}}(t) = -x(t)$ and assumed that the $R_X(\tau)$ has no DC component.   ∎

5.2.5 Power Spectral Density of Stationary Processes

A random process is a collection of signals, and the spectral characteristics of these signals determine the spectral characteristics of the random process. If the signals of the random process are slowly varying, then the random process will mainly contain low frequencies and its power will be mostly concentrated at low frequencies. On the other hand, if the signals change very fast, then most of the power in the random process will be at the high frequency components.

A useful function that determines the distribution of the power of the random process at different frequencies is the *power spectral density* or *power spectrum* of the random process. The power spectral density of a random process $X(t)$ is denoted by $S_X(f)$, and denotes the strength of the power in the random process as a function of frequency. The unit for power spectral density is Watts/Hz.

For stationary random processes, we use a very useful theorem that relates the power spectrum of the random process to its autocorrelation function. This theorem is known as the *Wiener–Khinchin* theorem.

Theorem [Wiener–Khinchin] For a stationary random process $X(t)$, the power spectral density is the Fourier transform of the autocorrelation function, i.e.,

$$S_X(f) = \mathscr{F}[R_X(\tau)]. \tag{5.2.9}$$

Example 5.2.15

For the stationary random process in Example 5.2.2, we had

$$R_X(\tau) = \frac{A^2}{2} \cos(2\pi f_0 \tau).$$

Hence,

$$S_X(f) = \frac{A^2}{4}[\delta(f - f_0) + \delta(f + f_0)].$$

The power spectral density is shown in Figure 5.17. All the power content of the process is located at f_0 and $-f_0$. This is expected because the sample functions of this process are sinusoidals with their power at those frequencies.   ∎

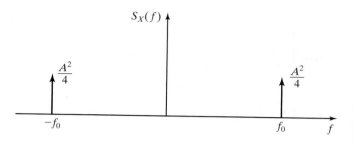

$S_X(f)$

$\dfrac{A^2}{4}$

$\dfrac{A^2}{4}$

$-f_0$

f_0 f

Figure 5.17   Power spectral density of the random process of Example 5.2.15.

Example 5.2.16

The process of Example 5.2.3 is stationary. In this case,

$$R_X(\tau) = E[X(t + \tau)X(t)] = E[X^2] = \frac{1}{3}.$$

Hence,

$$S_X(f) = \mathscr{F}\left[\frac{1}{3}\right] = \frac{1}{3}\delta(f)$$

Obviously, in this case, for each realization of the process, we have a different power spectrum. ∎

The power content, or simply the *power*, of a random process is the sum of the powers at all frequencies in that random process. In order to find the total power, we have to integrate the power spectral density over all frequencies. This means that the power in the random process $X(t)$, denoted by P_X, is obtained using the relation

$$P_X = \int_{-\infty}^{\infty} S_X(f)\, df. \tag{5.2.10}$$

Since $S_X(f)$ is the Fourier transform of $R_X(\tau)$, then $R_X(\tau)$ will be the inverse Fourier transform of $S_X(f)$. Therefore, we can write

$$R_X(\tau) = \int_{-\infty}^{\infty} S_X(f)e^{j2\pi f\tau}\, d\tau. \tag{5.2.11}$$

Substituting $\tau = 0$ into this relation yields

$$R_X(0) = \int_{-\infty}^{\infty} S_X(f)\, df. \tag{5.2.12}$$

Comparing this with Equation (5.2.10), we conclude that

$$P_X = R_X(0). \tag{5.2.13}$$

From Equations (5.2.10) and (5.2.13), we see that the power in a stationary random process can be found either by integrating its power spectral density (adding all power components) or substituting $\tau = 0$ in the autocorrelation function of the process.

Example 5.2.17

Find the power in the process given in Example 5.2.15

Solution   We can use either the relation

$$P_X = \int_{-\infty}^{\infty} S_X(f) \, df$$

$$= \int_{-\infty}^{\infty} \left[\frac{A^2}{4} [\delta(f - f_0) + \delta(f + f_0)] \right] df$$

$$= 2 \times \frac{A^2}{4}$$

$$= \frac{A^2}{2}$$

or the relation

$$P_X = R_X(0)$$

$$= \left. \frac{A^2}{2} \cos(2\pi f_0 \tau) \right|_{\tau=0}$$

$$= \frac{A^2}{2}. \qquad \blacksquare$$

Power Spectra in LTI Systems.   We have seen that when a stationary random process with mean m_x and autocorrelation function $R_X(\tau)$ passes through a linear time-invariant system with the impulse response $h(t)$, the output process will be also stationary with mean

$$m_Y = m_X \int_{-\infty}^{\infty} h(t) \, dt$$

and autocorrelation

$$R_Y(\tau) = R_X(\tau) \star h(\tau) \star h(-\tau).$$

We have also seen that $X(t)$ and $Y(t)$ will be jointly stationary with the cross-correlation function

$$R_{XY}(\tau) = R_X(\tau) \star h(-\tau).$$

Translation of these relations into the frequency domain is straightforward. By noting that $\mathscr{F}[h(-\tau)] = H^*(f)$ and $\int_{-\infty}^{\infty} h(t) \, dt = H(0)$, we can compute the Fourier transform of both sides of these relations to obtain

$$m_Y = m_X H(0) \tag{5.2.14}$$

$$S_Y(f) = S_X(f)|H(f)|^2. \tag{5.2.15}$$

The first equation says that since the mean of a random process is basically its DC value, i.e., the mean value of the response of the system only depends on the value of $H(f)$ at $f = 0$ (DC response). The second equation says that when dealing with the power spectrum, the phase of $H(f)$ is irrelevant; only the magnitude of $H(f)$ affects the output power spectrum. This is also intuitive since power depends on the amplitude and not the phase of the signal. For instance, if a random process passes through a differentiator, we have $H(f) = j2\pi f$; hence,

$$m_Y = m_X H(0) = 0 \tag{5.2.16}$$

$$S_Y(f) = 4\pi^2 f^2 S_X(f). \tag{5.2.17}$$

We can also define a frequency-domain relation for the cross-correlation function. Let us define the *cross spectral density* $S_{XY}(f)$ as

$$S_{XY}(f) \overset{\text{def}}{=} \mathscr{F}[R_{XY}(\tau)]. \tag{5.2.18}$$

Then

$$S_{XY}(f) = S_X(f) H^*(f), \tag{5.2.19}$$

and since $R_{YX}(\tau) = R_{XY}(-\tau)$, we have

$$S_{YX}(f) = S_{XY}^*(f) = S_X(f) H(f). \tag{5.2.20}$$

Note that although $S_X(f)$ and $S_Y(f)$ are real non-negative functions, $S_{XY}(f)$ and $S_{YX}(f)$ can generally be complex functions. Figure 5.18 shows how these quantities are related.

Example 5.2.18

If the process in Example 5.2.2 passes through a differentiator, we have $H(f) = j2\pi f$; therefore,

$$S_Y(f) = 4\pi^2 f^2 \left[\frac{A^2}{4} (\delta(f - f_0) + \delta(f + f_0)) \right] = A^2 \pi^2 f_0^2 [\delta(f - f_0) + \delta(f + f_0)]$$

and

$$S_{XY}(f) = (-j2\pi f)S_X(f) = \frac{jA^2\pi f_0}{2} \left[\delta(f + f_0) - \delta(f - f_0) \right]. \qquad \blacksquare$$

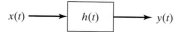

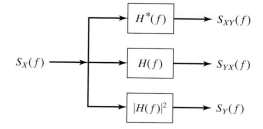

Figure 5.18 Input–output relations for the power spectral density and the cross spectral density.

Example 5.2.19

Passing the process in Example 5.2.3 through a differentiator results in

$$S_Y(f) = 4\pi^2 f^2 \left(\frac{1}{3}\delta(f) \right) = 0$$

$$S_{XY}(f) = (-j2\pi f) \left(\frac{1}{3}\delta(f) \right) = 0,$$

where we have used the basic properties of the impulse signal in both relations. These results are intuitive because the sample functions of this process are constant and differentiating them results in zero output.   ∎

5.2.6 Power Spectral Density of a Sum Process

In practice, we often encounter the sum of two random processes. For example, in the case of communication over a channel with additive noise, the noise process is added to the signal process. Next, we determine the power spectral density for the sum of two jointly stationary processes.

Let us assume that $Z(t) = X(t) + Y(t)$, where $X(t)$ and $Y(t)$ are jointly stationary random processes. We already know that $Z(t)$ is a stationary process with

$$R_Z(\tau) = R_X(\tau) + R_Y(\tau) + R_{XY}(\tau) + R_{YX}(\tau). \tag{5.2.21}$$

Taking the Fourier transform of both sides of this equation and using the result of Problem 5.39, we obtain

$$S_Z(f) = S_X(f) + S_Y(f) + S_{XY}(f) + \underbrace{S_{YX}(f)}_{S_{XY}^*(f)}$$

$$= S_X(f) + S_Y(f) + 2\operatorname{Re}\left[S_{XY}(f)\right]. \tag{5.2.22}$$

The preceding relation shows that the power spectral density of the sum process is the sum of the power spectra of the individual processes plus a third term, which depends on the cross correlation between the two processes.

If the two processes $X(t)$ and $Y(t)$ are uncorrelated, then

$$R_{XY}(\tau) = m_X \, m_Y.$$

Now if at least one of the processes is zero mean, we will have $R_{XY}(\tau) = 0$ and

$$S_Z(f) = S_X(f) + S_Y(f). \tag{5.2.23}$$

Example 5.2.20

Let $X(t)$ represent the process in Example 5.2.2, and let $Z(t) = X(t) + \frac{d}{dt}X(t)$. Then

$$S_{XY}(f) = \frac{jA^2\pi f_0}{2}\left[\delta(f + f_0) - \delta(f - f_0)\right];$$

therefore,

$$\mathrm{Re}[S_{XY}(f)] = 0.$$

Hence,

$$S_Z(f) = S_X(f) + S_Y(f) = A^2 \left(\frac{1}{4} + \pi^2 f_0^2 \right) [\delta(f - f_0) + \delta(f + f_0)]. \qquad \blacksquare$$

5.3 GAUSSIAN AND WHITE PROCESSES

Gaussian processes play an important role in communication systems. The fundamental reason for their importance is that thermal noise in electronic devices, which is produced by the random movement of electrons due to thermal agitation, can be closely modeled by a Gaussian process.

In order to see the reason why thermal noise is a Gaussian process, consider a resistor. The free electrons in a resistor move as a result of thermal agitation. The movement of the electrons is quite random and can be in any direction; however, their velocity is a function of the ambient temperature. The higher the temperature, the higher the velocity of the electrons. The movement of these electrons generates a current with a random value. We can consider each electron in motion as a tiny current source, whose current is a random variable that can be positive or negative, depending on the direction of the movement of the electron. The total current generated by all electrons, which is the generated thermal noise, is the sum of the currents of all these current sources. We can assume that at least a majority of these sources behave independently and, therefore, the total current is the sum of a large number of independent and identically distributed random variables. Now by applying the central limit theorem, we conclude that this total current has a Gaussian distribution. This is the reason that thermal noise can be very well modeled by a Gaussian random process.

Gaussian processes provide rather good models for some information sources as well. Some interesting properties of the Gaussian processes, which will be discussed in this section, make these processes mathematically tractable and easy to use.

5.3.1 Gaussian Processes

In a Gaussian random process, we can look at different instances of time and the resulting random variables will be jointly Gaussian. We start our discussion with a formal definition of Gaussian processes.

Definition 5.3.1. A random process $X(t)$ is a *Gaussian process* if for all n and all $(t_1, t_2, \ldots, t_n)$, the random variables $\{X(t_i)\}_{i=1}^n$ have a jointly Gaussian density function. $\qquad \blacksquare$

From the preceding definition, we know that at any time instant t_0, the random variable $X(t_0)$ is Gaussian; at any two points t_1, t_2, random variables $(X(t_1), X(t_2))$ are distributed according to a two-dimensional jointly Gaussian random variable.

Example 5.3.1

Let $X(t)$ be a zero-mean stationary Gaussian random process with the power spectral density $S_X(t) = 5\Pi\left(\frac{f}{1000}\right)$. Determine the probability density function of the random variable $X(3)$.

Solution   Since $X(t)$ is a Gaussian random process, the probability density function of random variable $X(t)$ at any value of t is Gaussian. Therefore, $X(3) \sim \mathcal{N}(m, \sigma^2)$. Now we need to find m and σ^2. Since the process is zero mean, at any time instance t, we have $E[X(t)] = 0$; this means $m = E[X(3)] = 0$. To find the variance, we note that

$$\sigma^2 = \text{VAR}[X(3)]$$

$$= E[X^2(3)] - (E[X(3)])^2$$

$$= E[X(3)X(3)]$$

$$= R_X(0),$$

where, in the last step, we have used the fact that the process is stationary and hence $E[X(t_1)X(t_2)] = R_X(t_1 - t_2)$. But from Equation (5.2.12), we have

$$\sigma^2 = R_X(0)$$

$$= \int_{-\infty}^{\infty} S_X(f)\,df$$

$$= \int_{-500}^{500} 5\,df$$

$$= 5000;$$

therefore, $X(3) \sim \mathcal{N}(0, 5000)$, or the density function for $X(3)$ is

$$f(x) = \frac{1}{\sqrt{10000\pi}} e^{-\frac{x^2}{10000}}.$$ ∎

Just as we have defined jointly Gaussian random variables, we can also define *jointly Gaussian random processes.*

Definition 5.3.2.   The random processes $X(t)$ and $Y(t)$ are *jointly Gaussian* if for all n, m and all $(t_1, t_2, \ldots, t_n)$ and $(\tau_1, \tau_2, \ldots, \tau_m)$, the random vector $(X(t_1), X(t_2), \ldots, X(t_n), Y(\tau_1), Y(\tau_2), \ldots, Y(\tau_m))$ is distributed according to an $n + m$ dimensional jointly Gaussian distribution.   ∎

From this definition, it is obvious that if $X(t)$ and $Y(t)$ are jointly Gaussian, then each of them is individually Gaussian; but the converse is not always true. That is, two individually Gaussian random processes are not always jointly Gaussian.

Gaussian and jointly Gaussian random processes have some very important properties that are not shared by other families of random processes. Two of the most important properties are given here:

Property 1:   If the Gaussian process $X(t)$ is passed through an LTI system, then the output process $Y(t)$ will also be a Gaussian process. Moreover, $X(t)$ and $Y(t)$ will be jointly Gaussian processes.

This property follows directly from the fact that linear combinations of jointly Gaussian random variables are themselves jointly Gaussian.

This result is very important; it demonstrates one of the properties that makes Gaussian processes attractive. For a non-Gaussian process, knowledge of the statistical properties of the input process does not easily lead to the statistical properties of the output process. However, for a Gaussian process, we know that the output process of an LTI system will also be Gaussian.

Property 2:   For jointly Gaussian processes, uncorrelatedness and independence are equivalent.

This is also a straightforward consequence of the basic properties of Gaussian random variables, as outlined in our discussion of jointly Gaussian random variables.

Example 5.3.2

If the random process $X(t)$ in Example 5.3.1 is passed through a differentiator and the output process is denoted by $Y(t)$, i.e., $Y(t) = \frac{d}{dt}X(t)$, determine the probability density function $Y(3)$.

Solution   Since a differentiator is an LTI system, by Property 1, $Y(t)$ is a Gaussian process. This means that $Y(3)$ is a Gaussian random variable with mean m_Y and variance σ_Y^2. Also note that the impulse response of a differentiator is $h(t) = \delta'(t)$; hence, its transfer function is $H(f) = \mathcal{F}[h(t)] = j2\pi f$. In order to find m_Y, we use the result of Example 5.2.13, which shows $m_Y = 0$. To find σ_Y^2, we use a method similar to the one employed in Example 5.3.1. First we have to find $S_Y(f)$, which can be found using Equation (5.2.15) with $H(f) = j2\pi f$. This is very similar to Exercises 5.2.18 and 5.2.19. We have

$$\sigma_Y^2 = \int_{-\infty}^{\infty} S_Y(f)\,df$$

$$= \int_{-500}^{500} 5 \times 4\pi^2 f^2 \, df$$

$$= \left[\frac{20\pi^2}{3} f^3 \right]_{f=-500}^{500}$$

$$= \frac{20\pi^2}{3} \times 2 \times 500^3$$

$$\approx 1.64 \times 10^{10}.$$

Since $m_Y = 0$ and $\sigma_Y^2 = 1.64 \times 10^{10}$, we have $Y(3) \sim \mathcal{N}(0, 1.64 \times 10^{10})$.   ∎

5.3.2 White Processes

The term *white process* is used to denote processes in which all frequency components appear with equal power, i.e., the power spectral density is a constant for all frequencies. This parallels the notion of "white light," in which all colors exist.

Definition 5.3.3. A process $X(t)$ is called a *white process* if it has a flat spectral density, i.e., if $S_X(f)$ is a constant for all f.   ∎

In practice, the importance of white processes stems from the fact that thermal noise can be closely modeled as a white process over a wide range of frequencies. Also, a wide range of processes used to describe a variety of information sources can be modeled as the output of LTI systems driven by a white process. Figure 5.19 shows the power spectrum of a white process.

If we find the power content of a white process using $S_X(f) = C$, a constant, we will have

$$P_X = \int_{-\infty}^{\infty} S_X(f)\,df = \int_{-\infty}^{\infty} C\,df = \infty.$$

Obviously, no real physical process can have infinite power; therefore, a white process is not a meaningful physical process. However, quantum mechanical analysis of the thermal noise shows that it has a power spectral density given by

$$S_n(f) = \frac{\hbar f}{2(e^{\frac{\hbar f}{kT}} - 1)}, \tag{5.3.1}$$

in which $\hbar$ denotes *Planck's constant* (equal to 6.6×10^{-34} J × sec) and k is *Boltzmann's constant* (equal to 1.38×10^{-23} J/K). T denotes the temperature in degrees Kelvin. This power spectrum is shown in Figure 5.20.

The preceding spectrum achieves its maximum at $f = 0$, and the value of this maximum is $\frac{kT}{2}$. The spectrum goes to zero as f goes to infinity, but the rate of convergence to zero is very slow. For instance, at room temperature ($T = 300°K$) $S_n(f)$ drops to 90% of its maximum at about $f \approx 2 \times 10^{12}$ Hz, which is beyond the frequencies employed in conventional communication systems. Thus, we conclude

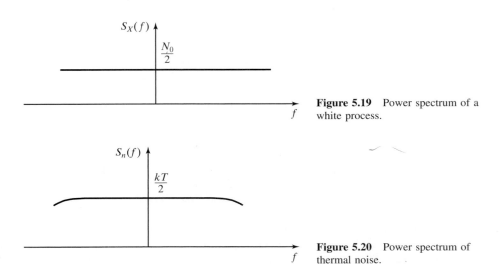

Figure 5.19 Power spectrum of a white process.

Figure 5.20 Power spectrum of thermal noise.

that thermal noise, though not precisely white, for all practical purposes can be modeled as a white process with a power spectrum equal to $\frac{kT}{2}$. The value kT is usually denoted by N_0; therefore, the power spectral density of thermal noise is usually given as $S_n(f) = \frac{N_0}{2}$. It is sometimes referred to as the *two-sided power spectral density*, emphasizing that this spectrum extends to both positive and negative frequencies. We will avoid this terminology throughout this text and simply use *power spectrum* or *power spectral density*.

Looking at the autocorrelation function for a white process, we see that

$$R_n(\tau) = \mathscr{F}^{-1}\left[\frac{N_0}{2}\right] = \frac{N_0}{2}\delta(\tau). \tag{5.3.2}$$

This shows that for all $\tau \neq 0$, we have $R_X(\tau) = 0$. Thus, if we sample a white process at two points t_1 and t_2 ($t_1 \neq t_2$), the resulting random variables will be uncorrelated. If the random process is white and also Gaussian, any pair of random variables $X(t_1)$, $X(t_2)$, where $t_1 \neq t_2$, will also be independent (recall that for jointly Gaussian random variables, uncorrelatedness and independence are equivalent.)

Properties of the Thermal Noise.   The thermal noise that we will use in subsequent chapters is assumed to have the following properties:

1. Thermal noise is a stationary process.
2. Thermal noise is a zero-mean process.
3. Thermal noise is a Gaussian process.
4. Thermal noise is a white process with a power spectral density $S_n(f) = \frac{kT}{2}$.

It is clear that the power spectral density of thermal noise increases with increasing the ambient temperature; therefore, keeping electric circuits cool makes their noise level low.

5.3.3 Filtered Noise Processes

In many cases, the white noise generated in one stage of the system is filtered by the next stage; therefore, in the following stage, we encounter filtered noise that is a bandpass process, i.e., its power spectral density is located away from the zero frequency and is mainly located around some frequency f_c, which is far from zero and larger than the bandwidth of the process.

In a bandpass process, sample functions are bandpass signals; like the bandpass signals discussed in Section 2.7, they can be expressed in terms of the in-phase and quadrature components. In this section, we study the main properties of bandpass noise processes and particularly study the main properties of the in-phase and quadrature processes.

Let us assume that the process $X(t)$ is the output of an ideal bandpass filter of bandwidth W which is located at frequencies around f_c. For example, one such filter can have a transfer function of the form

$$H_1(f) = \begin{cases} 1 & |f - f_c| \leq W \\ 0 & \text{otherwise} \end{cases}.$$ (5.3.3)

Another example of a such filter would be

$$H_2(f) = \begin{cases} 1 & f_c \leq |f| \leq f_c + W \\ 0 & \text{otherwise} \end{cases}.$$ (5.3.4)

Figure 5.21 shows plots of the transfer functions of these two filters.

Since thermal noise is white and Gaussian, the filtered thermal noise will be Gaussian but not white. The power spectral density of the filtered noise will be

$$S_X(f) = \frac{N_0}{2}|H(f)|^2 = \frac{N_0}{2}H(f),$$

where we have used the fact that for ideal filters $|H(f)|^2 = H(f)$. For the two filtered noise processes corresponding to $H_1(f)$ and $H_2(f)$, we have the following power spectral densities:

$$S_{X_1}(f) = \begin{cases} \frac{N_0}{2} & |f - f_c| \leq W \\ 0 & \text{otherwise} \end{cases}$$ (5.3.5)

and

$$S_{X_2}(f) = \begin{cases} \frac{N_0}{2} & f_c \leq |f| \leq f_c + W \\ 0 & \text{otherwise} \end{cases}.$$ (5.3.6)

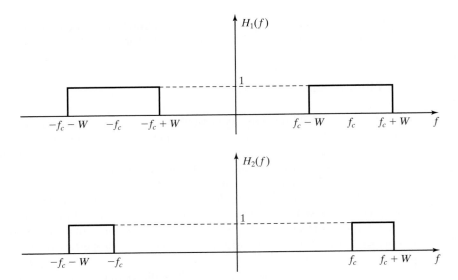

Figure 5.21   Filter transfer functions $H_1(f)$ and $H_2(f)$.

All bandpass filtered noise signals have an in-phase and quadrature component that are lowpass signals. This means that the bandpass random process $X(t)$ can be expressed in an equation that is similar to Equation (2.7.8), i.e.,

$$X(t) = X_c(t)\cos(2\pi f_c t) - X_s(t)\sin(2\pi f_c t), \tag{5.3.7}$$

where $X_c(t)$ and $X_s(t)$—the in-phase and quadrature components—are lowpass processes. Note that, as we did with Equation (2.7.17), we can also represent the filtered noise in terms of its envelope and phase as

$$X(t) = A(t)\cos(2\pi f_c t + \Theta(t)), \tag{5.3.8}$$

where $A(t)$ and $\Theta(t)$ are lowpass random processes.

Properties of the In-Phase and Quadrature Processes.   For filtered white Gaussian noise, the following properties for $X_c(t)$ and $X_s(t)$ can be proved (for proofs, see the references at the end of the chapter):

1. $X_c(t)$ and $X_s(t)$ are zero-mean, lowpass, jointly stationary, and jointly Gaussian random processes.
2. If the power in process $X(t)$ is P_X, then the power in each of the processes $X_c(t)$ and $X_s(t)$ is also P_X. In other words,

$$P_X = P_{X_c} = P_{X_s} = \int_{-\infty}^{\infty} S_X(f)\,df. \tag{5.3.9}$$

3. Processes $X_c(t)$ and $X_s(t)$ have a common power spectral density. This power spectral density is obtained by shifting the positive frequencies in $S_X(f)$ to the left by f_c, shifting the negative frequencies of $S_X(f)$ to the right by f_c, and adding the two shifted spectra. Therefore, if $H_1(f)$, given in Equation (5.3.3), is used, then

$$S_{X_{1c}}(f) = S_{X_{1s}}(f) = \begin{cases} N_0 & |f| \leq W \\ 0 & \text{otherwise} \end{cases}, \tag{5.3.10}$$

and if $H_2(f)$, given in Equation (5.3.4) is used, then

$$S_{X_{2c}}(f) = S_{X_{2s}}(f) = \begin{cases} \frac{N_0}{2} & |f| \leq W \\ 0 & \text{otherwise} \end{cases}. \tag{5.3.11}$$

The power spectral densities for these two cases are shown in Figure 5.22. It is obvious that in both cases, the power in the in-phase and quadrature components is equal to the power in the bandpass noise. When $H_1(f)$ is used, the power in the filtered noise from Equation (5.3.5) is $P_1 = \frac{N_0}{2} \times 4W = 2N_0W$, and when $H_2(f)$ is used, the power in the filtered noise from Equation (5.3.6) is $P_2 = \frac{N_0}{2} \times 2W = N_0W$. By integrating the power spectral density, we can see that the power in the in-phase and quadrature components in these two cases

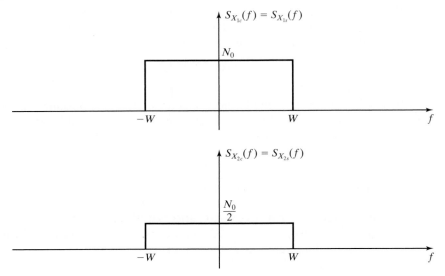

Figure 5.22   Power spectral densities of the in-phase and quadrature components of the filtered noise passed through $H(f)$ and $H_1(f)$.

are $2N_0W$ and N_0W, respectively, which are equal to the corresponding P_1 and P_2.

4. If $+f_c$ and $-f_c$ are the axis of symmetry of the positive and negative frequencies, respectively, in the spectrum of $H(f)$—as is the case for $H_1(f)$ in Equation (5.3.3) and is *not* the case for $H_2(f)$ in Equation (5.3.4)—then $X_c(t)$ and $X_s(t)$ will be independent processes.

Example 5.3.3

For the bandpass white noise at the output of filter $H_1(f)$, given in Equation (5.3.3), find the power spectral density of the process $Z(t) = aX_c(t) + bX_s(t)$.

Solution   Since f_c is the axis of symmetry of the noise power spectral density, the in-phase and quadrature components of the noise will be independent; therefore, we are dealing with the sum of two independent and zero-mean processes. This means that the conditions leading to Equation (5.2.23) are satisfied, and the power spectral density of $Z(t)$ is the sum of the power spectral densities of $aX_c(t)$ and $bX_s(t)$. But we already know that $S_{X_c}(f) = S_{X_s}(f) = S_X(f)$, therefore $S_Z(f) = a^2 S_X(f) + b^2 S_X(f) = (a^2 + b^2)S_X(f)$. Note that in the special case where $a = \cos\theta$ and $b = \sin\theta$, we have $S_Z(f) = S_X(f)$.   ∎

Noise Equivalent Bandwidth.   When a white Gaussian noise passes through a filter, the output process, although still Gaussian, will not be white anymore. The filter characteristic shapes the spectral properties of the output process, and we have

$$S_Y(f) = S_X(f)|H(f)|^2 = \frac{N_0}{2}|H(f)|^2.$$

Now if we want to find the power content of the output process, we have to integrate $S_Y(f)$. Thus,

$$P_Y = \int_{-\infty}^{\infty} S_Y(f)\, df = \frac{N_0}{2} \int_{-\infty}^{\infty} |H(f)|^2 \, df.$$

Therefore, to determine the output power, we have to evaluate the integral $\int_{-\infty}^{\infty} |H(f)|^2 \, df$. To do this calculation, we define B_{neq}, the *noise equivalent bandwidth* of a filter with the frequency response $H(f)$, as

$$B_{neq} = \frac{\int_{-\infty}^{\infty} |H(f)|^2 \, df}{2\, H_{max}^2}, \tag{5.3.12}$$

where H_{max} denotes the maximum of $|H(f)|$ in the passband of the filter. Figure 5.23 shows H_{max} and B_{neq} for a typical filter.

Using the preceding definition, we have

$$P_Y = \frac{N_0}{2} \int_{-\infty}^{\infty} |H(f)|^2 \, df$$

$$= \frac{N_0}{2} \times 2 B_{neq}\, H_{max}^2$$

$$= N_0\, B_{neq}\, H_{max}^2. \tag{5.3.13}$$

Therefore, when we have B_{neq}, finding the output noise power becomes a simple task. The noise equivalent bandwidth of filters and amplifiers is usually provided by the manufacturer.

Example 5.3.4

Find the noise equivalent bandwidth of a lowpass RC filter.

Solution   For this filter,

$$H(f) = \frac{1}{1 + j2\pi f RC},$$

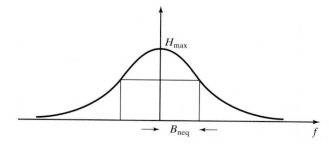

Figure 5.23   Noise equivalent bandwidth of a typical filter.

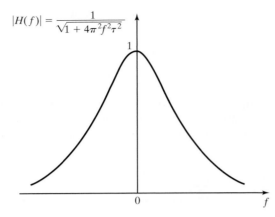

$|H(f)| = \dfrac{1}{\sqrt{1 + 4\pi^2 f^2 \tau^2}}$

Figure 5.24   Frequency response of a lowpass RC filter.

and is shown in Figure 5.24. Defining $\tau = RC$, we have

$$|H(f)| = \frac{1}{\sqrt{1 + 4\pi^2 f^2 \tau^2}},$$

and therefore $H_{\max} = 1$. We also have

$$
\begin{aligned}
\int_{-\infty}^{\infty} |H(f)|^2 \, df &= 2 \int_{0}^{\infty} \frac{1}{1 + 4\pi^2 f^2 \tau^2} \, df \\
&\overset{u=2\pi f\tau}{=} 2 \int_{0}^{\infty} \frac{1}{1 + u^2} \times \frac{du}{2\pi\tau} \\
&= \frac{1}{\pi\tau} \times \frac{\pi}{2} \\
&= \frac{1}{2\tau}.
\end{aligned}
$$

Hence,

$$B_{\text{neq}} = \frac{\frac{1}{2\tau}}{2 \times 1} = \frac{1}{4RC}. \qquad \blacksquare$$

5.4 FURTHER READING

The books by Leon Garcia (1994), Helstrom (1991), Davenport and Root (1987), Papoulis (1991), Nelson (1995), and Stark and Woods (1994) cover probability and random processes with an emphasis on electrical engineering applications. Gray and Davisson (1986) is particularly interesting, since it covers random processes for electrical engineers without compromising mathematical vigor. Advanced treatment of the material in this chapter, based on measure-theory concepts, can be found in the book by Wong and Hajek (1985). Details on properties of narrowband random processes can be found in Proakis and Salehi (2002).

PROBLEMS

5.1 A random experiment consists of drawing a ball from an urn that contains 4 red balls (numbered 1, 2, 3, 4) and three black balls numbered (1, 2, 3). Precisely state what outcomes are contained in the following events:

1. $E_1 =$ The number on the ball is even.
2. $E_2 =$ The color of the ball is red and its number is greater than 1.
3. $E_3 =$ The number on the ball is less than 3.
4. $E_4 = E_1 \cup E_3$.
5. $E_5 = E_1 \cup (E_2 \cap E_3)$.

5.2 If all balls in the preceding problem are equally likely to be drawn, find the probabilities of E_i, $1 \le i \le 5$.

5.3 In a certain city, three car brands A, B, and C have 20%, 30% and 50% of the market share, respectively. The probabilities that a car will need major repair during its first year of purchase are 5%, 10% and 15%, respectively.

1. What is the probability that a car in this city will need major repair during its first year of purchase?
2. If a car in this city needs major repair during its first year of purchase, what is the probability that it is made by manufacturer A?

5.4 Under what conditions can two disjoint events A and B be independent?

5.5 An information source produces 0 and 1 with probabilities 0.3 and 0.7, respectively. The output of the source is transmitted via a channel that has a probability of error (turning a 1 into a 0 or a 0 into a 1) equal to 0.2.

1. What is the probability that a 1 will be observed at the output?
2. What is the probability that a 1 was the output of the source if a 1 is observed at the output of the channel?

5.6 A coin is flipped three times, and the random variable X denotes the total number of heads that show up. The probability of landing on a head in one flip of this coin is denoted by p.

1. What values can the random variable X take?
2. What is the probability mass function of the random variable X?
3. Derive and plot the CDF of X.
4. What is the probability that X exceeds 1?

5.7 For coin A, the probability of landing on a head is equal to $\frac{1}{4}$ and the probability of landing on a tail is equal to $\frac{3}{4}$; coin B is a fair coin. Each coin is flipped four times. Let the random variable X denote the number of heads resulting from coin A, and Y denote the resulting number of heads from coin B.

1. What is the probability that $X = Y = 2$?
2. What is the probability that $X = Y$?
3. What is the probability that $X > Y$?
4. What is the probability that $X + Y \leq 5$?

5.8 A random variable X is defined by the CDF

$$F_X(x) = \begin{cases} 0, & x < 0 \\ \frac{1}{2}x, & 1 \leq x < 1 \\ K, & x \geq 1 \end{cases}.$$

1. Find the value of K.
2. Is this random variable discrete, continuous, or mixed?
3. What is the probability that $\frac{1}{2} < X \leq 1$?
4. What is the probability that $\frac{1}{2} < X < 1$?
5. What is the probability that X exceeds 2?

5.9 Random variable X is distributed according to $f_X(x) = \Lambda(x)$.

1. Determine $P(X > \frac{1}{2})$.
2. Determine $P(X > 0 | X < \frac{1}{2})$.
3. What is $f_X(x | X > \frac{1}{2})$?
4. What is $E(X | X > \frac{1}{2})$?

5.10 X is a Gaussian random variable with a mean 4 and a variance 9, i.e., $\mathcal{N}(4, 9)$. Determine the following probabilities:

1. $P(X > 7)$.
2. $P(0 < X < 9)$.

5.11 The noise voltage in an electric circuit can be modeled as a Gaussian random variable with a mean equal to zero and a variance equal to 10^{-8}.

1. What is the probability that the value of the noise exceeds 10^{-4}? What is the probability that it exceeds 4×10^{-4}? What is the probability that the noise value is between -2×10^{-4} and 10^{-4}?

2. Given that the value of the noise is positive, what is the probability that it exceeds 10^{-4}?

3. This noise passes through a half-wave rectifier with characteristics

$$g(x) = \begin{cases} x, & x > 0 \\ 0, & x \le 0. \end{cases}$$

Find the PDF of the rectified noise by first finding its CDF. Why can we not use the general expression in Equation (5.1.11)?

4. Find the expected value of the rectified noise in the previous part.

5. Now assume that the noise passes through a full-wave rectifier defined by $g(x) = |x|$. Find the density function of the rectified noise in this case. What is the expected value of the output noise in this case?

5.12 X is a $\mathcal{N}(0, \sigma^2)$ random variable. This random variable is passed through a system whose input–output relation is given by $y = g(x)$. Find the PDF or the PMF of the output random variable Y in each of the following cases:

1. Square law device, $g(x) = ax^2$.

2. Limiter,

$$g(x) = \begin{cases} -b, & x \le -b \\ b, & x \ge b \\ x, & |x| < b. \end{cases}$$

3. Hard limiter,

$$g(x) = \begin{cases} a, & x > 0 \\ 0, & x = 0 \\ b, & x < 0. \end{cases}$$

4. Quantizer, $g(x) = x_n$ for $a_n \le x < a_{n+1}$, $1 \le n \le N$, where x_n lies in the interval $[a_n, a_{n+1}]$, the sequence $\{a_1, a_2, \ldots, a_{N+1}\}$ satisfies the conditions $a_1 = -\infty$, $a_{N+1} = \infty$, and for $i > j$, we have $a_i > a_j$.

5.13 The random variable Φ is uniformly distributed on the interval $[-\frac{\pi}{2}, \frac{\pi}{2}]$. Find the probability density function of $X = \tan \Phi$. Find the mean and the variance of X.

5.14 Let Y be a positive-valued random variable, i.e., $f_Y(y) = 0$ for $y < 0$.

1. Let α be any positive constant, and show that $P(Y > \alpha) \le \frac{E[Y]}{\alpha}$ (Markov inequality).

2. Let X be any random variable with variance σ^2, and define $Y = (X - E[X])^2$ and $\alpha = \epsilon^2$ for some ϵ. Obviously, the conditions of the problem are satisfied for Y and α as chosen here. Derive the Chebychev inequality

$$P\left(|X - E[X]| > \epsilon\right) \leq \frac{\sigma^2}{\epsilon^2}.$$

5.15 Show that for a binomial random variable, the mean is given by np and the variance is given by $np(1 - p)$.

5.16 Show that for a Poisson random variable defined by the PMF $P(X = k) = \frac{\lambda^k}{k!}e^{-\lambda}$, where $k = 0, 1, 2, \ldots$ and $\lambda > 0$, we have $E[X] = \lambda$ and $\text{VAR}(X) = \lambda$.

5.17 Let X denote a Gaussian random variable with a mean equal to zero and a variance equal to σ^2. Show that

$$E[X^n] = \begin{cases} 0, & n = 2k + 1 \\ 1 \times 3 \times 5 \times \cdots \times (n - 1)\sigma^n, & n = 2k \end{cases}.$$

(Hint: Differentiate the identity $\int_0^\infty e^{-\frac{x^2}{2\sigma^2}}\, dx = \sqrt{2\pi\sigma^2}$, k times.)

5.18 Two random variables X and Y are distributed according to

$$f_{X,Y}(x, y) = \begin{cases} K(x + y), & 0 \leq x, y \leq 1 \\ 0, & \text{otherwise} \end{cases}.$$

1. Find K.

2. What is the probability that $X + Y > 1$?

3. Find $P(X > Y)$.

4. What is $P(X > Y | X + 2Y > 1)$?

5. Find $P(X = Y)$.

6. What is $P(X > 0.5 | X = Y)$?

7. Find $f_X(x)$ and $f_Y(y)$.

8. Find $f_X(x | X + 2Y > 1)$ and $E[X | X + 2Y > 1]$.

5.19 Let $X_1, X_2, \ldots, X_n$ denote independent and identically distributed random variables each with PDF $f_X(x)$.

1. If $Y = \min\{X_1, X_2, \ldots, X_n\}$, find the PDF of Y.

2. If $Z = \max\{X_1, X_2, \ldots, X_n\}$, find the PDF of Z.

5.20 Show that for the Rayleigh density function

$$f_X(x) = \begin{cases} \frac{x}{\sigma^2} e^{-\frac{x^2}{2\sigma^2}}, & x > 0 \\ 0, & \text{otherwise} \end{cases}$$

we have $E[X] = \sigma\sqrt{\frac{\pi}{2}}$ and $\text{VAR}(X) = \left(2 - \frac{\pi}{2}\right)\sigma^2$.

5.21 Let X and Y be independent random variables with

$$f_X(x) = \begin{cases} \alpha e^{-\alpha x}, & x > 0, \\ 0, & \text{otherwise} \end{cases}$$

and

$$f_Y(y) = \begin{cases} \beta e^{-\beta x}, & x > 0, \\ 0, & \text{otherwise} \end{cases}$$

where α and β are assumed to be positive constants. Find the PDF of $X + Y$ and treat the special case $\alpha = \beta$ separately.

5.22 Two random variables X and Y are distributed according to

$$f_{X,Y}(x, y) = \begin{cases} K e^{-x-y}, & x \geq y \geq 0 \\ 0, & \text{otherwise} \end{cases}.$$

1. Find the value of the constant K.
2. Find the marginal density functions of X and Y.
3. Are X and Y independent?
4. Find $f_{X|Y}(x|y)$.
5. Find $E[X|Y = y]$.
6. Find $\text{COV}(X, Y)$ and $\rho_{X,Y}$.

5.23 Let Θ be uniformly distributed on $[0, \pi]$, and let the random variables X and Y be defined by $X = \cos\Theta$ and $Y = \sin\Theta$. Show that X and Y are uncorrelated, but are not independent.

5.24 Let X and Y be two independent Gaussian random variables, each with mean zero and variance 1. Define the two events $E_1(r) = \{X > r \text{ and } Y > r\}$ and $E_2(r) = \{\sqrt{X^2 + Y^2} > 2r\}$, where r is any nonnegative constant.

1. Show that $E_1(r) \subseteq E_2(r)$; therefore, $P(E_1(r)) \leq P(E_2(r))$.
2. Show that $P(E_1(r)) = Q^2(r)$.

3. Use the relations for rectangular to polar transformation to find $P(E_2(r))$ and conclude with the bound

$$Q(r) \le \frac{1}{2} e^{-\frac{r^2}{2}}$$

on the Q function.

5.25 It can be shown that the Q function can be well approximated by

$$Q(x) \approx \frac{e^{-\frac{x^2}{2}}}{\sqrt{2\pi}} \left(b_1 t + b_2 t^2 + b_3 t^3 + b_4 t^4 + b_5 t^5\right),$$

where $t = \frac{1}{1+px}$ and

$$p = 0.2316419;$$
$$b_1 = 0.31981530;$$
$$b_2 = -0.356563782;$$
$$b_3 = 1.781477937;$$
$$b_4 = -1.821255978;$$
$$b_5 = 1.330274429.$$

Using this relation, write a computer program to compute the Q function at any given value of its argument. Compute $Q(x)$ for $x = 1, 1.5, 2, 2.5, 3, 3.5, 4, 4.5, 5$, and compare the results with those obtained from the table of the Q function.

5.26 Let X and Y be independent Gaussian random variables, each distributed according to $\mathcal{N}(0, \sigma^2)$. Define $Z = X + Y$ and $W = 2X - Y$. What can you say about the joint PDF of Z and W? What is the covariance of Z and W?

5.27 Let X and Y be two jointly Gaussian random variables with means m_X and m_Y, variances σ_X^2 and σ_Y^2, and a correlation coefficient $\rho_{X,Y}$. Show that $f_{X|Y}(x|y)$ is a Gaussian distribution with a mean $m_X + \rho \frac{\sigma_X}{\sigma_Y}(y - m_Y)$ and a variance $\sigma_X^2(1 - \rho_{X,Y}^2)$. What happens if $\rho = 0$? What happens if $\rho = \pm 1$?

5.28 Let X and Y be zero-mean jointly Gaussian random variables, each with variance σ^2. The correlation coefficient between X and Y is denoted by ρ. Random variables Z and W are defined by

$$\begin{cases} Z = X \cos\theta + Y \sin\theta \\ W = -X \sin\theta + Y \cos\theta \end{cases},$$

where θ is a constant angle.

1. Show that Z and W are jointly Gaussian random variables.

2. For what values of θ are the random variables Z and W independent?

5.29 Two random variables X and Y are distributed according to

$$f_{X,Y}(x, y) = \begin{cases} \frac{K}{\pi} e^{-\frac{x^2+y^2}{2}}, & xy \geq 0 \\ 0, & xy < 0 \end{cases}.$$

1. Find K.

2. Show that X and Y are each Gaussian random variables.

3. Show that X and Y are *not* jointly Gaussian.

4. Are X and Y independent?

5. Are X and Y uncorrelated?

6. Find $f_{X|Y}(x|y)$. Is this a Gaussian distribution?

5.30 Let X and Y be two independent Gaussian random variables with a common variance σ^2. The mean of X is m and Y is a zero-mean random variable. We define random variable V as $V = \sqrt{X^2 + Y^2}$. Show that

$$f_V(v) = \begin{cases} \frac{v}{\sigma^2} I_0\left(\frac{mv}{\sigma^2}\right) e^{-\frac{v^2+m^2}{2\sigma^2}}, & v > 0, \\ 0, & v \leq 0 \end{cases}$$

where

$$I_0(x) = \frac{1}{2\pi} \int_0^{2\pi} e^{x \cos u} \, du = \frac{1}{2\pi} \int_{-\pi}^{\pi} e^{x \cos u} \, du$$

is called the *modified Bessel function of the first kind and zero order*. The distribution of V is known as the *Rician distribution*. Show that, in the special case of $m = 0$, the Rician distribution simplifies to the Rayleigh distribution.

5.31 A coin has the probability of landing on a head equal to $\frac{1}{4}$ and is flipped 2000 times.

1. Using the law of large numbers, find a lower bound to the probability that the total number of heads lies between 480 and 520.

2. Using the central limit theorem, find the probability that the total number of heads lies between 480 and 520.

5.32 Find $m_X(t)$ for the random process $X(t)$ given in Example 5.2.3. Is it independent of t?

5.33 Let the random process $X(t)$ be defined by $X(t) = A + Bt$ where A and B are independent random variables, each uniformly distributed on $[-1, 1]$. Find $m_X(t)$ and $R_X(t_1, t_2)$.

5.34 Show that the process given in Example 5.2.3 is a stationary process.

5.35 Which one of the following functions can be the autocorrelation function of a random process and why?

    1. $f(\tau) = \sin(2\pi f_0 \tau)$.

    2. $f(\tau)$, as shown in Figure P-5.35.

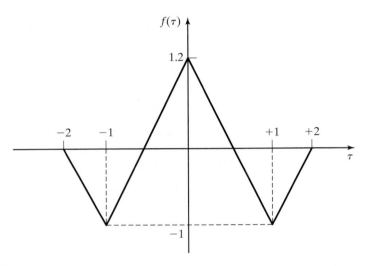

Figure P-5.35

5.36 Is the process of Example 5.2.5 stationary?

5.37 A random process $Z(t)$ takes the values 0 and 1. A transition from 0 to 1 or from 1 to 0 occurs randomly, and the probability of having n transitions in a time interval of duration τ, $(\tau > 0)$, is given by

$$p_N(n) = \frac{1}{1 + \alpha\tau}\left(\frac{\alpha\tau}{1 + \alpha\tau}\right)^n, \qquad n = 0, 1, 2, \ldots,$$

where $\alpha > 0$ is a constant. We further assume that at $t = 0$, $X(0)$ is equally likely to be 0 or 1.

    1. Find $m_Z(t)$.

    2. Find $R_Z(t + \tau, t)$. Is $Z(t)$ stationary?

    3. Determine the power spectral density of $Z(t)$.

5.38 The random process $X(t)$ is defined by

$$X(t) = X \cos 2\pi f_0 t + Y \sin 2\pi f_0 t,$$

where X and Y are two zero-mean independent Gaussian random variables each with the variance σ^2.

1. Find $m_X(t)$.
2. Find $R_X(t + \tau, t)$. Is $X(t)$ stationary?
3. Find the power spectral density of $X(t)$.
4. Answer Part 1 and 2 for the case where $\sigma_X^2 \neq \sigma_Y^2$.

5.39 Show that for jointly stationary processes $X(t)$ and $Y(t)$, we have $R_{XY}(\tau) = R_{YX}(-\tau)$. From this information, conclude that $S_{XY}(f) = S_{YX}^*(f)$.

5.40 A zero-mean white Gaussian noise process with the power spectral density of $\frac{N_0}{2}$ passes through an ideal lowpass filter with bandwidth B.

1. Find the autocorrelation of the output process $Y(t)$.
2. Assuming $\tau = \frac{1}{2B}$, find the joint probability density function of the random variables $Y(t)$ and $Y(t+\tau)$. Are these random variables independent?

5.41 Find the output autocorrelation function for a delay line with delay Δ when the input is a stationary process with the autocorrelation $R_X(\tau)$. Interpret the result.

5.42 We have proved that when the input to an LTI system is stationary, the output is also stationary. Is the converse of this theorem also true? That is, if we know that the output process is stationary, can we conclude that the input process is necessarily stationary?

5.43 Generalize the result of Example 5.2.20 when $X(t)$ is stationary.

1. Show that $X(t)$ and $\frac{d}{dt}X(t)$ are uncorrelated processes.
2. Show that the power spectrum of $Z(t) = X(t) + \frac{d}{dt}X(t)$ is the sum of the power spectra of $X(f)$ and $\frac{d}{dt}X(t)$.
3. Express the power spectrum of the sum in terms of the power spectrum of $X(t)$.

5.44 Assume $X(t)$ is a stationary process with the power spectral density $S_X(f)$. This process passes through the system shown in Figure P-5.44.

1. Is $Y(t)$ stationary? Why?
2. What is the power spectral density of $Y(t)$?

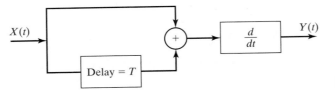

$X(t)$

$Y(t)$

$+$ $\dfrac{d}{dt}$

Delay $= T$

Figure P-5.44

3. What frequency components cannot be present in the output process and why?

5.45 The stationary random process $X(t)$ has a power spectral density denoted by $S_X(f)$.

    1. What is the power spectral density of $Y(t) = X(t) - X(t - T)$?

    2. What is the power spectral density of $Z(t) = X'(t) - X(t)$?

    3. What is the power spectral density of $W(t) = Y(t) + Z(t)$?

5.46 Show that for two jointly stationary processes $X(t)$ and $Y(t)$, we have

$$|R_{XY}(\tau)| \leq \sqrt{R_X(0)R_Y(0)} \leq \frac{1}{2}[R_X(0) + R_Y(0)].$$

5.47 The stationary process $X(t)$ is passed through an LTI system, and the output process is denoted by $Y(t)$. Find the output autocorrelation function and the cross-correlation function between the input and the output processes in each of the following cases:

    1. A delay system with delay Δ.

    2. A system with $h(t) = \frac{1}{t}$.

    3. A system with $h(t) = e^{-\alpha t}u(t)$, where $\alpha > 0$.

    4. A system described by the differential equation

$$\frac{d}{dt}Y(t) + Y(t) = \frac{d}{dt}X(t) - X(t).$$

    5. A finite-time average defined by the input–output relation

$$y(t) = \frac{1}{2T}\int_{t-T}^{t+T} x(\tau)\,d\tau,$$

where T is a constant.

5.48 Give an example of two processes $X(t)$ and $Y(t)$ for which $R_{XY}(t + \tau, t)$ is a function of τ, but $X(t)$ and $Y(t)$ are not stationary.

5.49 Find the power spectral density for each of the following processes:

1. $X(t) = A\cos(2\pi f_0 t + \Theta)$, where A is a constant and Θ is a random variable uniformly distributed on $[0, \frac{\pi}{4}]$.

2. $X(t) = X + Y$, where X and Y are independent, X is uniform on $[-1, 1]$, and Y is uniform on $[0, 1]$.

5.50 $X(t)$ is a stationary random process with the autocorrelation function $R_X(\tau) = e^{-\alpha|\tau|}$, where $\alpha > 0$. This process is applied to an LTI system with $h(t) = e^{-\beta t}u(t)$, where $\beta > 0$. Find the power spectral density of the output process $Y(t)$. Treat the cases $\alpha \neq \beta$ and $\alpha = \beta$ separately.

5.51 Let $Y(t) = X(t) + N(t)$, where $X(t)$ and $N(t)$ are signal and noise processes. It is assumed that $X(t)$ and $N(t)$ are jointly stationary with the autocorrelation functions $R_X(\tau)$ and $R_N(\tau)$ and cross-correlation function $R_{XN}(\tau)$. We want to separate the signal from the noise by passing $Y(t)$ through an LTI system with the impulse response $h(t)$ and the transfer function $H(f)$. The output process is denoted by $\hat{X}(t)$, which we want to be as close to $X(t)$ as possible.

1. Find the cross correlation between $\hat{X}(t)$ and $X(t)$ in terms of $h(\tau)$, $R_X(\tau)$, $R_N(\tau)$, and $R_{XN}(\tau)$.

2. Show that the LTI system that minimizes $E[X(t) - \hat{X}(t)]^2$ has a transfer function

$$H(f) = \frac{S_X(f) + S_{XN}(f)}{S_X(f) + S_N(f) + 2\text{Re}[S_{XN}(f)]}.$$

3. Now assume that $X(t)$ and $N(t)$ are independent and $N(t)$ is a zero-mean white Gaussian process with the power spectral density $\frac{N_0}{2}$. Find the optimal $H(f)$ under these conditions. What is the corresponding value of $E[X(t) - \hat{X}(t)]^2$ in this case?

4. In the special case of $S_N(f) = 1$, $S_X(f) = \frac{1}{1+f^2}$, and $S_{XN}(f) = 0$, find the optimal $H(f)$.

5.52 In this problem, we examine the estimation of a random process by observing another random process. Let $X(t)$ and $Z(t)$ be two jointly stationary random processes. We are interested in designing an LTI system with the impulse response $h(t)$, such that when $Z(t)$ is passed through it, the output process $\hat{X}(t)$ is as close to $X(t)$ as possible. In other words, we are interested in the best *linear* estimate of $X(t)$ based on the observation of $Y(t)$ in order to minimize $E[X(t) - \hat{X}(t)]^2$.

1. Let us assume we have two LTI systems with the impulse responses $h(t)$ and $g(t)$. $Z(t)$ is applied to both systems and the outputs are denoted

by $\hat{X}(t)$ and $\tilde{X}(t)$, respectively. The first filter is designed such that its output satisfies the condition

$$E\left[(X(t) - \hat{X}(t))Z(t - \tau)\right] = 0$$

for all values of τ and t, where as the second filter does not satisfy this property. Show that

$$E[X(t) - \tilde{X}(t)]^2 \geq E[X(t) - \hat{X}(t)]^2,$$

i.e., the necessary and sufficient condition for an optimal filter is that its output satisfies the *orthogonality condition*, as given by

$$E\left[(X(t) - \hat{X}(t))Z(t - \tau)\right] = 0,$$

which simply means that the estimation error $\epsilon(t) = X(t) - \hat{X}(t)$ must be orthogonal to the observable process $Z(t)$ at all times.

2. Show that the optimal $h(t)$ must satisfy

$$R_{XZ}(\tau) = R_Z(\tau) \star h(\tau).$$

3. Show that the optimal filter satisfies

$$H(f) = \frac{S_{XZ}(f)}{S_Z(f)}.$$

4. Derive an expression for $E[\epsilon^2(t)]$ when the optimal filter is employed.

5.53 What is the noise-equivalent bandwidth of an ideal bandpass filter with bandwidth W?

5.54 A zero-mean white Gaussian noise, $n_w(t)$, with power spectral density $\frac{N_0}{2}$ is passed through an ideal filter whose passband is from 3–11 kHz. The output process is denoted by $n(t)$.

1. If $f_0 = 7$ kHz, find $S_{n_c}(f)$, $S_{n_s}(f)$, and $R_{n_c n_s}(\tau)$, where $n_c(t)$ and $n_s(t)$ are the in-phase and quadrature components of $n(t)$.

2. Repeat Part 1 with $f_0 = 6$ kHz.

5.55 Let $p(t)$ be a bandpass signal with the in-phase and quadrature components $p_c(t)$ and $p_s(t)$; let $X(t) = \sum_{n=-\infty}^{\infty} A_n p(t - nT)$, where A_n's are independent random variables. Express $X_c(t)$ and $X_s(t)$ in terms of $p_c(t)$ and $p_s(t)$.

5.56 Let $X(t)$ be a bandpass process and let $V(t)$ denote its envelope. Show that for all choices of the center frequency f_0, $V(t)$ remains unchanged.

5.57 Let $n_w(t)$ be a zero-mean white Gaussian noise with the power spectral density $\frac{N_0}{2}$; let this noise be passed through an ideal bandpass filter with the bandwidth $2W$ centered at the frequency f_c. Denote the output process by $n(t)$.

1. Assuming $f_0 = f_c$, find the power content of the in-phase and quadrature components of $n(t)$.

2. Find the density function of $V(t)$, the envelope of $n(t)$.

3. Now assume $X(t) = A \cos 2\pi f_0 t + n(t)$, where A is a constant. What is the density function of the envelope of $X(t)$?

5.58 A noise process has a power spectral density given by

$$S_n(f) = \begin{cases} 10^{-8} \left(1 - \frac{|f|}{10^8}\right), & |f| < 10^8 \\ 0, & |f| > 10^8 \end{cases}.$$

This noise is passed through an ideal bandpass filter with a bandwidth of 2 MHz, centered at 50 MHz.

1. Find the power content of the output process.

2. Write the output process in terms of the in-phase and quadrature components, and find the power in each component. Assume $f_0 = 50$ MHz.

3. Find the power spectral density of the in-phase and quadrature components.

4. Now assume that the filter is not an ideal filter and is described by

$$|H(f)|^2 = \begin{cases} |f| - 49 & 49 \text{ MHz} < |f| < 51 \text{ MHz} \\ 0, & \text{otherwise} \end{cases}.$$

Repeat Parts 1, 2, and 3 with this assumption.

COMPUTER PROBLEMS

5.1 Generation of Random Variables

The objective of this problem is to generate a random variable X that has a linear probability density function as shown in Figure CP-5.1(a); that is,

$$f_X(x) = \begin{cases} \frac{x}{2} & 0 \le x \le 2 \\ 0 & \text{otherwise} \end{cases}.$$

The corresponding probability distribution function is

$$F_X(x) = P(X \le x) = \int_{-\infty}^{x} f_X(v)\, dv = \begin{cases} 0 & x < 0 \\ \frac{x^2}{4} & 0 \le x \le 2 \\ 1 & x > 2 \end{cases}.$$

$F_X(x)$ is shown in Figure CP-5.1(b). Note that $0 \le F_X(x) \le 1$.

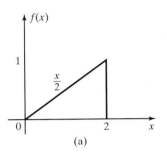
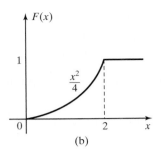

Figure CP-5.1   (a) A linear probability density function and (b) the corresponding probability distribution function.

To generate a sequence of samples of the random variable X, say $\{X_i\}$, we use MATLAB to generate uniformly distributed random variables $\{u_i\}$ in the range $(0, 1)$. Then, we set

$$F(x_i) = \frac{x_i^2}{4} = u_i$$

and solve for x_i. Thus, $x_i = 2\sqrt{u_i}$. Clearly, the range of x_i is $(0, 2)$.

1. Generate 10,000 samples of the random variable X by using the procedure previously described.

2. By subdividing the interval $(0, 2)$ into 20 equally spaced subintervals (bins), count the number of samples $\{x_i\}$ that fall into each bin and plot a histogram of the 10,000 randomly generated samples. Compare the histogram with the linear probability density function $f_X(x)$ and comment on how well the histogram matches $f_X(x)$.

5.2 Generation of Gaussian Random Variables

Additive noise encountered in communication systems is characterized by the Gaussian probability density function

$$f_X(x) = \frac{1}{\sqrt{2\pi}\,\sigma}e^{-\frac{x^2}{2\sigma^2}}, \quad -\infty < x < \infty,$$

as shown in Figure CP-5.2(a), where σ^2 is the variance of the random variable X. The probability distribution function, shown in Figure CP-5.2(b), is

$$F_X(x) = P(X \le x) = \int_{-\infty}^{x} f_X(v)\,dv, \quad -\infty < x < \infty.$$

The objective of this problem is to generate a sequence $\{x_i\}$ of Gaussian-distributed random variables on the computer. If there was a closed-form expression for $F_X(x)$, we could easily generate a sequence of uniformly distributed random variables $\{u_i\}$ and let $F_X(x_i) = u_i$. Then, we solve for x_i by performing the inverse mapping. However, in this case, $F_X(x)$ cannot be

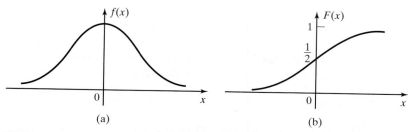

(a)   (b)

Figure CP-5.2 (a) The Gaussian probability density function and (b) the corresponding probability distribution function.

expressed in closed form. Instead, we circumvent this problem by first generating random variables that we related to Gaussian random variables. From probability theory, we know that a Rayleigh-distributed random variable R, which has a probability distribution function

$$F_R(r) = \begin{cases} 1 - e^{-\frac{r^2}{2\sigma^2}} & r \geq 0 \\ 0 & \text{otherwise} \end{cases}$$

is related to a pair of independent Gaussian random variables X and Y through the transformation

$$\begin{cases} X = R \cos \Theta \\ Y = R \sin \Theta \end{cases},$$

where Θ is a uniformly distributed random variable in the interval $(0, 2\pi)$ and is independent from R. The parameter σ^2 is the variance of X and Y. Therefore, we set $F_R(r_i) = u_i$, where $\{u_i\}$ is a sequence of uniformly distributed random variables on the interval $(0, 1)$. We can easily invert this equation to obtain

$$r_i = \sqrt{2\sigma^2 \ln\left(\frac{1}{1 - u_i}\right)}.$$

Next, we generate another sequence, say, $\{v_i\}$, of uniformly distributed random variables on the interval $(0, 1)$ and define

$$\theta_i = 2\pi v_i.$$

Finally, from $\{r_i\}$ and $\{\theta_i\}$, we generate the Gaussian random variables using the relations

$$\begin{cases} x_i = r_i \cos \theta_i \\ y_i = r_i \sin \theta_i \end{cases}.$$

1. By using the preceding procedure, generate 10,000 samples of a Gaussian random variable with a zero mean and a variance $\sigma^2 = 2$.

2. By subdividing the interval $(-10, 10)$ into 20 equally spaced subintervals (bins), count the number of samples of $\{x_i\}$ and $\{y_i\}$ that fall into each bin, and plot a histogram of the 10,000 randomly generated samples. Compare the histograms with the Gaussian probability density function $f_X(x)$, and comment on how well the histogram matches $f_X(x)$.

5.3 Autocorrelation Function and Power Spectrum

The objective of this problem is to compute the autocorrelation and the power spectrum of a sequence of random variables.

1. Generate a discrete-time sequence $\{x_n\}$ of $N = 1000$ statistically independent and identically distributed random variables, which are selected from a uniform distribution over the interval $(-\frac{1}{2}, \frac{1}{2})$. The estimate of the autocorrelation of the sequence $\{x_n\}$ is defined as

$$R_X(m) = \frac{1}{N-m} \sum_{n=1}^{N-m} x_n x_{n+m}, \quad m = 0, 1, \dots, M$$

$$= \frac{1}{N-|m|} \sum_{n=|m|}^{N} x_n x_{n+m}, \quad m = -1, -2, \dots, -M,$$

where $M = 100$. Compute $R_X(m)$ and plot it.

2. Determine and plot the power spectrum of the sequence $\{x_n\}$ by computing the discrete Fourier transform (DFT) of $R_X(m)$, which is defined as

$$S_X(f) = \sum_{m=-M}^{M} R_X(m) e^{-\frac{j 2\pi f m}{2M+1}}.$$

The FFT algorithm may be used to efficiently compute the DFT.

5.4 Filtered White Noise

A white random process $X(t)$ with the power spectrum $S_X(f) = 1$ for all f excites a linear filter with the impulse response

$$h(t) = \begin{cases} e^{-t} & t \geq 0 \\ 0 & \text{otherwise} \end{cases}.$$

1. Determine and plot the power spectrum $S_Y(f)$ of the filter output.

2. By using the inverse FFT algorithm on samples of $S_Y(f)$, compute and plot the autocorrelation function of the filter output $y(t)$. To be specific, use $N = 256$ frequency samples.

5.5 Generation of Lowpass Random Process

The objective of this problem is to generate samples of a lowpass random process by passing a white-noise sequence $\{x_n\}$ through a digital lowpass filter. The input to the filter is a sequence of statistically independent and identically distributed uniform random variables on the interval $(-\frac{1}{2}, \frac{1}{2})$. The digital lowpass filter has an impulse response

$$h(n) = \begin{cases} (0.9)^n & n \geq 0 \\ 0 & \text{otherwise} \end{cases}$$

and is characterized by the input–output recursive (difference) equation

$$y_n = 0.9 y_{n-1} + x_n, \quad n \geq 1, \, y_{-1} = 0.$$

1. Generate a sequence $\{x_n\}$ of 1000 samples and pass these through the filter to generate the sequence $\{y_n\}$.

2. Using the basic formula in Computer Problem CP-5.3, compute the autocorrelation functions $R_X(m)$ and $R_Y(m)$ for $|m| \leq 100$. Plot $R_X(m)$ and $R_Y(m)$ on separate graphs.

3. Compute and plot the power spectrum $S_X(f)$ and $S_Y(f)$ by evaluating the discrete Fourier transform of $R_X(m)$ and $R_Y(m)$.

5.6 Generation of Bandpass Random Process

A bandpass random process $X(t)$ can be represented as

$$X(t) = X_c(t) \cos 2\pi f_c t - X_s(t) \sin 2\pi f_c t,$$

where $X_c(t)$ and $X_s(t)$ are called the in-phase and quadrature components of $X(t)$. The random processes $X_c(t)$ and $X_s(t)$ are lowpass processes.

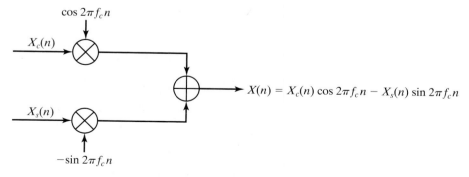

Figure CP-5.6 Generation of a bandpass random process.

1. Generate 1000 samples of a bandpass random process $X(t)$ by first generating 1000 samples of two statistically independent lowpass random processes $X_c(t)$ and $X_s(t)$, and then use these to modulate the quadrature carriers $\cos 2\pi f_c t$ and $\sin 2\pi f_c t$, as shown in Figure CP-5.6. The carrier frequency is $f_c = 1000/\pi$. The input sequence into the digital lowpass filter has statistically independent and identically distributed uniform random variables on the interval $(-\frac{1}{2}, \frac{1}{2})$.

2. Use the formula given in Computer Problem CP-5.3 to compute and plot the autocorrelations $R_{X_c}(m)$, $R_{X_s}(m)$, and $R_X(m)$ for $|m| \leq 100$.

3. Compute and plot the power spectra of $S_{X_c}(f)$, $S_{X_s}(f)$, and $S_X(f)$ by evaluating the discrete Fourier transform of $R_{X_c}(m)$, $R_{X_s}(m)$, and $R_X(m)$.

6

Effect of Noise on Analog Communication Systems

In Chapters 3 and 4, we studied the important characteristics of analog communication systems. These characteristics included time domain and frequency domain representations of the modulated signal, bandwidth requirements, power content of the modulated signal, and the modulator and demodulator implementation of various analog communication systems.

In this chapter, the effect of noise on various analog communication systems will be analyzed. As we will see, angle modulation systems and FM, in particular, can provide a high degree of noise immunity; therefore, they are desirable in cases of severe noise and/or low signal power. This noise immunity is obtained at the price of sacrificing channel bandwidth because, as we have seen in Chapter 4, the bandwidth requirements of angle modulation systems are considerably higher than the required bandwidth of amplitude modulation systems.

This chapter starts with a performance analysis of linear modulation systems in the presence of noise. Then, the effect of noise on angle modulation systems is discussed. Then, we will describe carrier-phase estimation using a phase-locked loop (PLL). Finally, we will analyze the effects of transmission loss and noise on analog communication systems in general.

6.1 EFFECT OF NOISE ON AMPLITUDE-MODULATION SYSTEMS

In this section, we determine the signal-to-noise ratio (SNR) of the output of the receiver that demodulates the amplitude-modulated signals. In evaluating the effect of noise on the various types of analog carrier-modulated signals, it is also interesting to compare the result with the effect of noise on an equivalent baseband communication system. We begin the evaluation of the effect of noise on a baseband system.

6.1.1 Effect of Noise on a Baseband System

Since baseband systems serve as a basis for comparison of various modulation systems, we begin with a noise analysis of a baseband system. In this case, there is no carrier demodulation to be performed. The receiver consists only of an ideal lowpass filter with the bandwidth W. The noise power at the output of the receiver, for a white noise input is therefore

$$P_{n_o} = \int_{-W}^{+W} \frac{N_0}{2} \, df$$
$$= N_0 W. \tag{6.1.1}$$

If we denote the received power by P_R, the baseband SNR is given by

$$\left(\frac{S}{N}\right)_b = \frac{P_R}{N_0 W}. \tag{6.1.2}$$

Example 6.1.1

Find the SNR in a baseband system with a bandwidth of 5 kHz and with $\frac{N_0}{2} = 10^{-14}$ W/Hz. The transmitter power is one kilowatt and the channel attenuation is 10^{-12}.

Solution   We have $P_R = 10^{-12} P_T = 10^{-12} \times 10^3 = 10^{-9}$ Watts. Therefore,

$$\left(\frac{S}{N}\right)_b = \frac{P_R}{N_0 W} = \frac{10^{-9}}{10^{-14} \times 5000} = 20.$$

This is equivalent to $10 \log_{10} 20 = 13$ dB.   ∎

6.1.2 Effect of Noise on DSB-SC AM

In DSB-SC AM, the transmitted signal is

$$u(t) = A_c m(t) \cos(2\pi f_c t); \tag{6.1.3}$$

therefore, the received signal at the output of the receiver noise-limiting filter is the sum of this signal and filtered noise. Recall from Section 5.3.3 that a filtered noise process can be expressed in terms if its in-phase and quadrature components, as given in Equation (5.3.7). Adding the filtered noise to the modulated signal, we can express the received signal as

$$r(t) = u(t) + n(t)$$
$$= A_c m(t) \cos(2\pi f_c t)$$
$$\quad + n_c(t) \cos(2\pi f_c t) - n_s(t) \sin(2\pi f_c t). \tag{6.1.4}$$

Suppose we demodulate the received signal by first multiplying $r(t)$ by a locally generated sinusoid $\cos(2\pi f_c t + \phi)$, where ϕ is the phase of the sinusoid, and then passing the product signal through an ideal lowpass filter having a bandwidth W. The multiplication of $r(t)$ with $\cos(2\pi f_c t + \phi)$ yields

$$
\begin{aligned}
r(t)\cos(2\pi f_c t + \phi) &= A_c m(t)\cos(2\pi f_c t)\cos(2\pi f_c t + \phi) \\
&\quad + n(t)\cos(2\pi f_c t + \phi) \\
&= \frac{1}{2}A_c m(t)\cos(\phi) + \frac{1}{2}A_c m(t)\cos(4\pi f_c t + \phi) \\
&\quad + \frac{1}{2}[n_c(t)\cos(\phi) + n_s(t)\sin(\phi)] \\
&\quad + \frac{1}{2}\Big[n_c(t)\cos(4\pi f_c t + \phi) \\
&\quad\quad - n_s(t)\sin(4\pi f_c t + \phi)\Big].
\end{aligned}
\tag{6.1.5}
$$

The lowpass filter rejects the double frequency components and passes only the lowpass components. Hence, its output is

$$
y(t) = \frac{1}{2}A_c m(t)\cos(\phi) + \frac{1}{2}[n_c(t)\cos(\phi) + n_s(t)\sin(\phi)].
\tag{6.1.6}
$$

As was discussed in Chapter 3, the effect of a phase difference between the received carrier and a locally generated carrier at the receiver is a drop equal to $\cos^2(\phi)$ in the received signal power. This can be avoided by employing a phase-locked loop, as described in Section 6.4. The effect of a phase-locked loop is to generate a sinusoidal carrier at the receiver with the same frequency and phase of the received carrier. If a phase-locked loop is employed, then $\phi = 0$ and the demodulator is called a *coherent* or *synchronous demodulator*. In our analysis in this section, we assume that we are employing a coherent demodulator. With this assumption, without loss of generality, we assume that $\phi = 0$; hence,

$$
y(t) = \frac{1}{2}[A_c m(t) + n_c(t)].
\tag{6.1.7}
$$

Therefore, at the receiver output, the message signal and the noise components are additive and we are able to define a meaningful SNR. The message signal power is given by

$$
P_o = \frac{A_c^2}{4}P_M,
\tag{6.1.8}
$$

where P_M is the power content of the message signal. The noise power is given by

$$
\begin{aligned}
P_{n_o} &= \frac{1}{4}P_{n_c} \\
&= \frac{1}{4}P_n,
\end{aligned}
\tag{6.1.9}
$$

where we have used the fact that the power contents of $n_c(t)$ and $n(t)$ are equal. This was discussed in Section 5.3.3, in Equation (5.3.9). The power content of $n(t)$ can be found by noting that it is the result of passing $n_w(t)$ through a filter with bandwidth B_c. Therefore, the power spectral density of $n(t)$ is given by

$$S_n(f) = \begin{cases} \frac{N_0}{2} & |f - f_c| < W \\ 0 & \text{otherwise} \end{cases}. \qquad (6.1.10)$$

The noise power is

$$\begin{aligned} P_n &= \int_{-\infty}^{\infty} S_n(f)\, df \\ &= \frac{N_0}{2} \times 4W \\ &= 2W N_0. \end{aligned} \qquad (6.1.11)$$

Now we can find the output SNR as

$$\begin{aligned} \left(\frac{S}{N}\right)_o &= \frac{P_o}{P_{n_o}} \\ &= \frac{\frac{A_c^2}{4} P_M}{\frac{1}{4} 2W N_0} \\ &= \frac{A_c^2 P_M}{2W N_0}. \end{aligned} \qquad (6.1.12)$$

In this case, the received signal power, as given by Equation (3.2.2), is $P_R = \frac{A_c^2 P_M}{2}$. Therefore, the output SNR in Equation (6.1.11) for DSB-SC AM may be expressed as

$$\left(\frac{S}{N}\right)_{o_{\text{DSB}}} = \frac{P_R}{N_0 W}, \qquad (6.1.13)$$

which is identical to $(S/N)_b$, which is given by Equation (6.1.2). Therefore, in DSB-SC AM, the output SNR is the same as the SNR for a baseband system. In other words, DSB-SC AM does not provide any SNR improvement over a simple baseband communication system.

6.1.3 Effect of Noise on SSB AM

In this case, the modulated signal, as given in Equation (3.2.8), is

$$u(t) = A_c m(t) \cos(2\pi f_c t) \pm A_c \hat{m}(t) \sin(2\pi f_c t). \qquad (6.1.14)$$

Therefore, the input to the demodulator is

$$r(t) = A_c m(t) \cos(2\pi f_c t) \pm A_c \hat{m}(t) \sin(2\pi f_c t) + n(t)$$
$$= (A_c m(t) + n_c(t)) \cos(2\pi f_c t) +$$
$$+ (\pm A_c \hat{m}(t) - n_s(t)) \sin(2\pi f_c t). \tag{6.1.15}$$

Here we assume that demodulation occurs with an ideal phase reference. Hence, the output of the lowpass filter is the in-phase component (with a coefficient of $\frac{1}{2}$) of the preceding signal. That is,

$$y(t) = \frac{A_c}{2} m(t) + \frac{1}{2} n_c(t). \tag{6.1.16}$$

We observe that, in this case again, the signal and the noise components are additive and a meaningful SNR at the receiver output can be defined. Parallel to our discussion of DSB, we have

$$P_o = \frac{A_c^2}{4} P_M \tag{6.1.17}$$

and

$$P_{n_o} = \frac{1}{4} P_{n_c} = \frac{1}{4} P_n, \tag{6.1.18}$$

where

$$P_n = \int_{-\infty}^{\infty} S_n(f) \, df = \frac{N_0}{2} \times 2W = W N_0. \tag{6.1.19}$$

Therefore,

$$\left(\frac{S}{N} \right)_o = \frac{P_o}{P_{n_o}} = \frac{A_c^2 P_M}{W N_0}. \tag{6.1.20}$$

But in this case,

$$P_R = P_U = A_c^2 P_M; \tag{6.1.21}$$

thus,

$$\left(\frac{S}{N} \right)_{oSSB} = \frac{P_R}{W N_0} = \left(\frac{S}{N} \right)_b. \tag{6.1.22}$$

Therefore, the signal-to-noise ratio in an SSB system is equivalent to that of a DSB system.

6.1.4 Effect of Noise on Conventional AM

In conventional DSB AM, the modulated signal was given in Equation (3.2.6) as

$$u(t) = A_c [1 + am(t)] \cos 2\pi f_c t. \tag{6.1.23}$$

Therefore, the received signal at the input to the demodulator is

$$r(t) = [A_c [1 + am_n(t)] + n_c(t)] \cos 2\pi f_c t - n_s(t) \sin 2\pi f_c t, \tag{6.1.24}$$

where a is the modulation index and $m_n(t)$ is normalized so that its minimum value is -1. If a synchronous demodulator is employed, the situation is basically similar to the DSB case, except that we have $1 + am_n(t)$ instead of $m(t)$. Therefore, after mixing and lowpass filtering, we have

$$y_1(t) = \frac{1}{2} [A_c [1 + am_n(t)] + n_c(t)]. \tag{6.1.25}$$

However, in this case, the desired signal is $m(t)$, not $1 + am_n(t)$. The DC component in the demodulated waveform is removed by a DC block and, hence, the lowpass filter output is

$$y(t) = \frac{1}{2} A_c a m_n(t) + \frac{n_c(t)}{2}. \tag{6.1.26}$$

In this case, the received signal power is given by

$$P_R = \frac{A_c^2}{2} \left[1 + a^2 P_{M_n} \right], \tag{6.1.27}$$

where we have assumed that the message process is zero mean. Now we can derive the output SNR as

$$
\begin{aligned}
\left(\frac{S}{N} \right)_{o_{AM}} &= \frac{\frac{1}{4} A_c^2 a^2 P_{M_n}}{\frac{1}{4} P_{n_c}} \\
&= \frac{A_c^2 a^2 P_{M_n}}{2 N_0 W} \\
&= \frac{a^2 P_{M_n}}{1 + a^2 P_{M_n}} \frac{\frac{A_c^2}{2} \left[1 + a^2 P_{M_n} \right]}{N_0 W} \\
&= \frac{a^2 P_{M_n}}{1 + a^2 P_{M_n}} \frac{P_R}{N_0 W} \\
&= \frac{a^2 P_{M_n}}{1 + a^2 P_{M_n}} \left(\frac{S}{N} \right)_b \\
&= \eta \left(\frac{S}{N} \right)_b, \tag{6.1.28}
\end{aligned}
$$

where we have used Equation (6.1.2) and η denotes the modulation efficiency.

We can see that, since $a^2 P_{M_n} < 1 + a^2 P_{M_n}$, the SNR in conventional AM is always smaller than the SNR in a baseband system. In practical applications, the modulation index a is in the range of 0.8–0.9. The power content of the normalized message process depends on the message source. For speech signals that usually have a large dynamic range, P_M is in the neighborhood of 0.1. This means that the overall loss in SNR, when compared to a baseband system, is a factor of 0.075 or equivalent to a loss of 11 dB. The reason for this loss is that a large part of the transmitter power is used to send the carrier component of the modulated signal and not the desired signal.

To analyze the envelope-detector performance in the presence of noise, we must use certain approximations. This is a result of the nonlinear structure of an envelope detector, which makes an exact analysis difficult. In this case, the demodulator detects the envelope of the received signal and the noise process. The input to the envelope detector is

$$r(t) = [A_c [1 + a m_n(t)] + n_c(t)] \cos 2\pi f_c t - n_s(t) \sin 2\pi f_c t; \qquad (6.1.29)$$

therefore, the envelope of $r(t)$ is given by

$$V_r(t) = \sqrt{[A_c [1 + a m_n(t)] + n_c(t)]^2 + n_s^2(t)}. \qquad (6.1.30)$$

Now we assume that the signal component in $r(t)$ is much stronger than the noise component. With this assumption, we have

$$P(n_c(t) \ll A_c [1 + a m_n(t)]) \approx 1; \qquad (6.1.31)$$

therefore, we have a high probability that

$$V_r(t) \approx A_c [1 + a m_n(t)] + n_c(t). \qquad (6.1.32)$$

After removing the DC component, we obtain

$$y(t) = A_c a m_n(t) + n_c(t), \qquad (6.1.33)$$

which is basically the same as $y(t)$ for the synchronous demodulation without the $\frac{1}{2}$ coefficient. This coefficient, of course, has no effect on the final SNR; therefore we conclude that, under the assumption of high SNR at the receiver input, the performance of synchronous and envelope demodulators is the same. However, if the preceding assumption is not true, we still have an additive signal and noise at the receiver output with synchronous demodulation, but the signal and noise become intermingled with envelope demodulation. To see this, let us assume that at the receiver input, the noise power[1] is much stronger than the signal power. This means that

$$
\begin{aligned}
V_r(t) &= \sqrt{\left[A_c\left[1 + am_n(t)\right] + n_c(t)\right]^2 + n_s^2(t)} \\
&= \sqrt{A_c^2(1 + am_n(t))^2 + n_c^2(t) + n_s^2(t) + 2A_c n_c(t)(1 + am_n(t))} \\
&\overset{a}{\approx} \sqrt{(n_c^2(t) + n_s^2(t))\left[1 + \frac{2A_c n_c(t)}{n_c^2(t) + n_s^2(t)}(1 + am_n(t))\right]} \\
&\overset{b}{\approx} V_n(t)\left[1 + \frac{A_c n_c(t)}{V_n^2(t)}(1 + am_n(t))\right] \\
&= V_n(t) + \frac{A_c n_c(t)}{V_n(t)}(1 + am_n(t)),
\end{aligned}
\tag{6.1.34}
$$

where (a) uses the fact that $A_c^2(1 + am_n(t))^2$ is small compared with the other components and (b) denotes $\sqrt{n_c^2(t) + n_s^2(t)}$ by $V_n(t)$, the envelope of the noise process; we have also used the approximation $\sqrt{1 + \epsilon} \approx 1 + \frac{\epsilon}{2}$, for small ϵ, where

$$
\epsilon = \frac{2A_c n_c(t)}{n_c^2(t) + n_s^2(t)}(1 + am_n(t)).
\tag{6.1.35}
$$

We observe that, at the demodulator output, the signal and the noise components are no longer additive. In fact, *the signal component is multiplied by noise* and is no longer distinguishable. In this case, no meaningful SNR can be defined. We say that this system *is operating below the threshold*. The subject of threshold and its effect on the performance of a communication system will be covered in more detail when we discuss the noise performance in angle modulation.

Example 6.1.2

We assume that the message is a wide-sense stationary random process $M(t)$ with the autocorrelation function

$$
R_M(\tau) = 16\,\text{sinc}^2(10{,}000\tau).
$$

We also know that all the realizations of the message process satisfy the condition $\max |m(t)| = 6$. We want to transmit this message to a destination via a channel with a 50 dB attenuation and additive white noise with the power spectral density $S_n(f) = \frac{N_0}{2} = 10^{-12}$ W/Hz. We also want to achieve an SNR at the modulator output of at least 50 dB. What is the required transmitter power and channel bandwidth if we employ the following modulation schemes?

1. DSB AM.
2. SSB AM.
3. Conventional AM with a modulation index equal to 0.8.

---

[1]By noise power at the receiver input, we mean the power of the noise within the bandwidth of the modulated signal, or equivalently, the noise power at the output of the noise-limiting filter.

Solution First, we determine the bandwidth of the message process. To do this, we obtain the power spectral density of the message process, namely,

$$S_M(f) = \mathscr{F}[R_M(\tau)] = \frac{16}{10,000} \Lambda \left(\frac{f}{10,000} \right),$$

which is nonzero for $-10,000 < f < 10,000$; therefore, $W = 10,000$ Hz. Now we can determine $\left(\frac{S}{N} \right)_b$ as a basis of comparison:

$$\left(\frac{S}{N} \right)_b = \frac{P_R}{N_0 W} = \frac{P_R}{2 \times 10^{-12} \times 10^4} = \frac{10^8 P_R}{2}.$$

Since the channel attenuation is 50 dB, it follows that

$$10 \log \frac{P_T}{P_R} = 50;$$

therefore,

$$P_R = 10^{-5} P_T.$$

Hence,

$$\left(\frac{S}{N} \right)_b = \frac{10^{-5} \times 10^8 P_T}{2} = \frac{10^3 P_T}{2}.$$

1. For DSB-SC AM, we have

$$\left(\frac{S}{N} \right)_o = \left(\frac{S}{N} \right)_b = \frac{10^3 P_T}{2} \sim 50 \text{ dB} = 10^5.$$

Therefore,

$$\frac{10^3 P_T}{2} = 10^5 \implies P_T = 200 \text{ Watt}$$

and

$$BW = 2W = 2 \times 10,000 = 20,000 \quad Hz \approx 20 \text{ kHz}.$$

2. For SSB AM,

$$\left(\frac{S}{N} \right)_o = \left(\frac{S}{N} \right)_b = \frac{10^3 P_T}{2} = 10^5 \implies P_T = 200 \text{ Watt}$$

and

$$BW = W = 10,000 \quad Hz = 10 \text{ kHz}.$$

3. For conventional AM, with $a = 0.8$,

$$\left(\frac{S}{N}\right)_o = \eta \left(\frac{S}{N}\right)_b = \eta \frac{10^3 P_T}{2},$$

where η is the modulation efficiency given by

$$\eta = \frac{a^2 P_{M_n}}{1 + a^2 P_{M_n}}.$$

First, we find P_{M_n}, the power content of the normalized message signal. Since $\max |m(t)| = 6$, we have

$$P_{M_n} = \frac{P_M}{(\max |m(t)|)^2} = \frac{P_M}{36}.$$

To determine P_M, we have

$$P_M = R_M(\tau)|_{\tau=0} = 16;$$

therefore,

$$P_{M_n} = \frac{16}{36} = \frac{4}{9}.$$

Hence,

$$\eta = \frac{0.8^2 \times \frac{4}{9}}{1 + 0.8^2 \times \frac{4}{9}} \approx 0.22.$$

Therefore,

$$\left(\frac{S}{N}\right)_o \approx 0.22 \frac{10^3 P_T}{2} = 0.11 \times 10^3 P_T = 10^5$$

or

$$P_T \approx 909 \text{ Watt.}$$

The bandwidth of conventional AM is equal to the bandwidth of DSB AM, i.e.,

$$\text{BW} = 2W = 20 \text{ kHz.} \qquad \blacksquare$$

6.2 EFFECT OF NOISE ON ANGLE MODULATION

In this section, we will study the performance of angle-modulated signals when contaminated by additive white Gaussian noise; we will also compare this performance with the performance of amplitude-modulated signals. Recall that in amplitude modulation, the message information is contained in the amplitude of the modulated signal; since noise is additive, the noise is directly added to the signal. However, in a frequency-modulated signal, the noise is added to the amplitude and the message information is contained in the frequency of the modulated signal. Therefore, the message is contaminated by the noise to the extent that the added noise changes the frequency of the modulated signal. The frequency of a signal can be described by

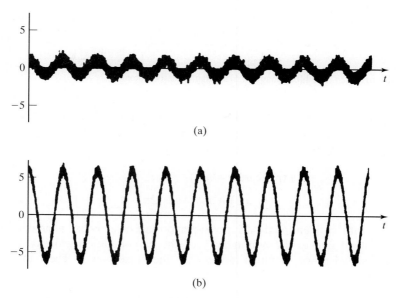

(a)

(b)

Figure 6.1   Effect of noise in frequency modulation.

its zero crossings. Therefore, the effect of additive noise on the demodulated FM signal can be described by the changes that it produces in the zero crossings of the modulated FM signal. Figure 6.1 shows the effect of additive noise on the zero crossings of two frequency-modulated signals, one with high power and the other with low power. From the previous discussion and also from Figure 6.1, it should be clear that the effect of noise in an FM system is different from that for an AM system. We also observe that the effect of noise in a low-power FM system is more severe than in a high-power FM system. In a low power signal, noise causes more changes in the zero crossings. The analysis that we present in this chapter verifies our intuition based on these observations.

The block diagram of the receiver for a general angle-modulated signal is shown in Figure 6.2. The angle-modulated signal is represented as[2]

$$u(t) = A_c \cos\left(2\pi f_c t + \phi(t)\right)$$

$$= \begin{cases} A_c \cos\left(2\pi f_c t + 2\pi k_f \int_{-\infty}^{t} m(\tau)\, d\tau\right) & \text{FM} \\ A_c \cos\left(2\pi f_c t + k_p m(t)\right) & \text{PM} \end{cases} . \qquad (6.2.1)$$

The additive white Gaussian noise $n_w(t)$ is added to $u(t)$, and the result is passed through a noise-limiting filter whose role is to remove the out-of-band noise. The bandwidth of this filter is equal to the bandwidth of the modulated signal;

[2]Throughout our noise analysis, when we refer to the modulated signal, we mean the signal as received by the receiver. Therefore, the signal power is the power in the received signal, not the transmitted power.

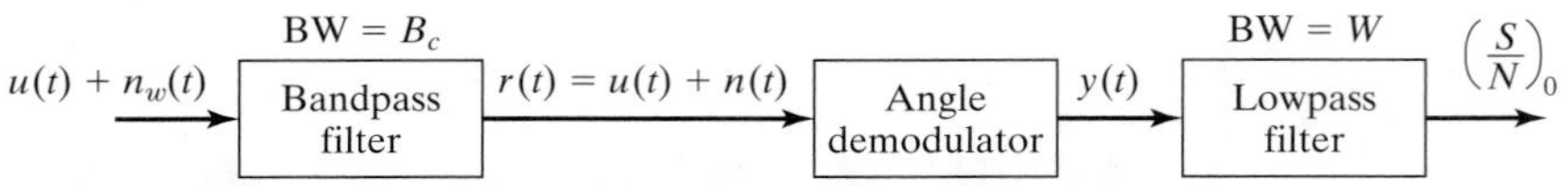

Figure 6.2   The block diagram of an angle demodulator.

therefore, it passes the modulated signal without distortion. However, it eliminates the out-of-band noise; hence, the noise output of the filter is a filtered noise process denoted by $n(t)$. The output of this filter is

$$r(t) = u(t) + n(t)$$
$$= u(t) + n_c(t)\cos(2\pi f_c t) - n_s(t)\sin(2\pi f_c t). \qquad (6.2.2)$$

As with conventional-AM noise-performance analysis, a precise analysis is quite involved due to the nonlinearity of the demodulation process. Let us assume that the signal power is much higher than the noise power. Then, the bandpass noise is represented as (see Equation 5.3.8)

$$n(t) = \sqrt{n_c^2(t) + n_s^2(t)}\, \cos\left(2\pi f_c t + \arctan\frac{n_s(t)}{n_c(t)}\right)$$
$$= V_n(t)\cos\left(2\pi f_c t + \Phi_n(t)\right), \qquad (6.2.3)$$

where $V_n(t)$ and $\Phi_n(t)$ represent the envelope and the phase of the bandpass noise process, respectively. The assumption that the signal is much larger than the noise means that

$$P(V_n(t) \ll A_c) \approx 1. \qquad (6.2.4)$$

Therefore, the phasor diagram of the signal and the noise are as shown in Figure 6.3. From this figure, it is obvious that we can write

$$r(t) \approx (A_c + V_n(t)\cos(\Phi_n(t) - \phi(t)))$$
$$\times \cos\left(2\pi f_c t + \phi(t)\right.$$
$$\left. + \arctan\frac{V_n(t)\sin(\Phi_n(t) - \phi(t))}{A_c + V_n(t)\cos(\Phi_n(t) - \phi(t))}\right)$$
$$\approx (A_c + V_n(t)\cos(\Phi_n(t) - \phi(t)))$$
$$\times \cos\left(2\pi f_c t + \phi(t) + \frac{V_n(t)}{A_c}\sin(\Phi_n(t) - \phi(t))\right). \qquad (6.2.5)$$

The demodulator processes this signal and, depending on whether it is a phase demodulator or a frequency demodulator, its output will be the phase or the instantaneous frequency of this signal (the instantaneous frequency is $\frac{1}{2\pi}$ times the derivative

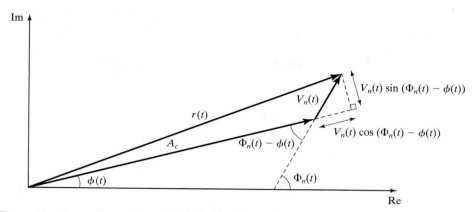

Figure 6.3   Phasor diagram of an angle-modulated signal when the signal is much stronger than the noise.

of the phase)[3]. Therefore, noting that

$$\phi(t) = \begin{cases} k_p m(t) & \text{PM} \\ 2\pi k_f \int_{-\infty}^{t} m(\tau)\, d\tau & \text{FM}, \end{cases} \qquad (6.2.6)$$

we see that the output of the demodulator is given by

$$y(t) = \begin{cases} \phi(t) + \frac{V_n(t)}{A_c} \sin\left(\Phi_n(t) - \phi(t)\right) & \text{PM} \\ \frac{1}{2\pi} \frac{d}{dt}\left(\phi(t) + \frac{V_n(t)}{A_c} \sin\left(\Phi_n(t) - \phi(t)\right)\right) & \text{FM} \end{cases}$$

$$= \begin{cases} k_p m(t) + \frac{V_n(t)}{A_c} \sin\left(\Phi_n(t) - \phi(t)\right) & \text{PM} \\ k_f m(t) + \frac{1}{2\pi} \frac{d}{dt} \frac{V_n(t)}{A_c} \sin\left(\Phi_n(t) - \phi(t)\right) & \text{FM} \end{cases}$$

$$= \begin{cases} k_p m(t) + Y_n(t) & \text{PM} \\ k_f m(t) + \frac{1}{2\pi} \frac{d}{dt} Y_n(t) & \text{FM} \end{cases}, \qquad (6.2.7)$$

where we have defined

$$Y_n(t) \overset{\text{def}}{=} \frac{V_n(t)}{A_c} \sin\left(\Phi_n(t) - \phi(t)\right). \qquad (6.2.8)$$

The first term in Equation (6.2.7) is the desired signal component, and the second term is the noise component. From this expression, we observe that the noise component is inversely proportional to the signal amplitude A_c. Hence, the higher the signal level, the lower the noise level. This is in agreement with the intuitive reasoning presented at the beginning of this section and based on Figure 6.1. Note that

---

[3]Of course, in the FM case, the demodulator output is the instantaneous frequency deviation of $v(t)$ from the carrier frequency f_c.

this is not the case with amplitude modulation. In AM systems, the noise component is independent of the signal component, and a scaling of the signal power does not affect the received noise power.

Let us study the properties of the noise component given by

$$Y_n(t) = \frac{V_n(t)}{A_c} \sin(\Phi_n(t) - \phi(t))$$

$$= \frac{1}{A_c} \Big[V_n(t) \sin \Phi_n(t) \cos \phi(t)$$

$$- V_n(t) \cos \Phi_n(t) \sin \phi(t) \Big]$$

$$= \frac{1}{A_c} [n_s(t) \cos \phi(t) - n_c(t) \sin \phi(t)]. \tag{6.2.9}$$

Here we provide an approximate and intuitive argument to determine the signal-to-noise ratio of the demodulated signal. A more rigorous derivation can be found in the references given at the end of this chapter.

Note that $\phi(t)$ is either proportional to the message signal or proportional to its integral. In both cases, it is a slowly varying signal compared to $n_c(t)$ and $n_s(t)$, which are the in-phase and quadrature components of the bandpass noise process at the receiver and have a much higher bandwidth than $\phi(t)$. In fact, the bandwidth of the filtered noise at the demodulator input is half of the bandwidth of the modulated signal, which is many times the bandwidth of the message signal. Therefore, we can say that when we compare variations in $n_c(t)$ and $n_s(t)$, we can assume that $\phi(t)$ is almost constant, i.e., $\phi(t) \approx \phi$. Therefore,

$$Y_n(t) = \frac{1}{A_c} [n_s(t) \cos \phi - n_c(t) \sin \phi]. \tag{6.2.10}$$

Now notice that in this case, f_c is the axis of symmetry of the bandpass noise process. Therefore, the conditions leading to the result of Exercise 5.3.3 are valid with $a = \frac{\cos \phi}{A_c}$ and $b = \frac{-\sin \phi}{A_c}$. By using the result of Exercise 5.3.3, we have

$$S_{Y_n}(f) = (a^2 + b^2)S_{n_c}(f) = \frac{S_{n_c}(f)}{A_c^2}, \tag{6.2.11}$$

where $S_{n_c}(f)$ is the power spectral density of the in-phase component of the filtered noise given in Equation (5.3.10). Note that the bandwidth of the filtered noise process extends from $f_c - \frac{B_c}{2}$ to $f_c + \frac{B_c}{2}$; hence, the spectrum of $n_c(t)$ extends from $-\frac{B_c}{2}$ to $\frac{B_c}{2}$. Therefore,

$$S_{X_{1c}}(f) = S_{X_{1s}}(f) = \begin{cases} N_0 & |f| \leq \frac{B_c}{2} \\ 0 & \text{otherwise} \end{cases}. \tag{6.2.12}$$

Substituting Equation (6.2.12) into Equation (6.2.10) results in

$$S_{Y_n}(f) = \begin{cases} \frac{N_0}{A_c^2} & |f| \leq \frac{B_c}{2} \\ 0 & \text{otherwise} \end{cases}. \tag{6.2.13}$$

This equation provides an expression for the power spectral density of the filtered noise at *the front end of the receiver*. After demodulation, another filtering is applied; this reduces the noise bandwidth to W, which is the bandwidth of the message signal. Note that in the case of FM modulation, as seen in Equation (6.2.7), the process $Y_n(t)$ is differentiated and scaled by $\frac{1}{2\pi}$. Using Equation (5.2.17) from Chapter 5, which gives the power spectral density of the derivative of a process, we conclude that the power spectral density of the process $\frac{1}{2\pi}\frac{d}{dt}Y_n(t)$ is given by (see Equation 5.2.17)

$$\frac{4\pi^2 f^2}{4\pi^2}S_{Y_n}(f) = f^2 S_{Y_n}(f) = \begin{cases} \frac{N_0}{A_c^2}f^2 & |f| \leq \frac{B_c}{2} \\ 0 & \text{otherwise} \end{cases}. \tag{6.2.14}$$

This means that in PM, the demodulated-noise power spectral density is given by Equation (6.2.13); in FM, it is given by Equation (6.2.14). In both cases, $\frac{B_c}{2}$ must be replaced by W to account for the additional postdemodulation filtering. In other words, for $|f| < W$, we have

$$S_{n_o}(f) = \begin{cases} \frac{N_0}{A_c^2} & \text{PM} \\ \frac{N_0}{A_c^2}f^2 & \text{FM} \end{cases}. \tag{6.2.15}$$

Figure 6.4 shows the power spectrum of the noise component at the output of the demodulator in the frequency interval $|f| < W$ for PM and FM.

It is interesting to note that PM has a flat noise spectrum and FM has a parabolic noise spectrum. Therefore, the effect of noise in FM for higher frequency components is much higher than the effect of noise on lower frequency components. The noise power at the output of the lowpass filter is the noise power in the frequency range $[W, +W]$. Therefore, it is given by

$$\begin{aligned} P_{n_o} &= \int_{-W}^{+W} S_{n_o}(f)\,df \\ &= \begin{cases} \int_{-W}^{+W} \frac{N_0}{A_c^2}\,df & \text{PM} \\ \int_{-W}^{+W} f^2\frac{N_0}{A_c^2}\,df & \text{FM} \end{cases} \\ &= \begin{cases} \frac{2WN_0}{A_c^2} & \text{PM} \\ \frac{2N_0 W^3}{3A_c^2} & \text{FM.} \end{cases} \end{aligned} \tag{6.2.16}$$

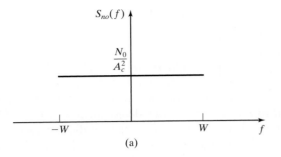

(a)

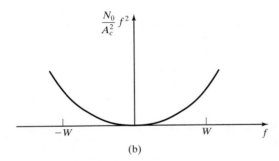

(b)

Figure 6.4 Noise power spectrum at demodulator output for $|f| < W$ in (a) PM and (b) FM.

Now we can use Equation (6.2.7) to determine the output signal-to-noise ratio in angle modulation. First, we have the output signal power

$$P_{s_o} = \begin{cases} k_p^2 P_M & \text{PM} \\ k_f^2 P_M & \text{FM} \end{cases}. \tag{6.2.17}$$

Then the signal-to-noise ratio, which is defined as

$$\left(\frac{S}{N}\right)_o \overset{\text{def}}{=} \frac{P_{s_o}}{P_{n_o}},$$

becomes

$$\left(\frac{S}{N}\right)_o = \begin{cases} \frac{k_p^2 A_c^2}{2} \frac{P_M}{N_0 W} & \text{PM} \\ \frac{3 k_f^2 A_c^2}{2 W^2} \frac{P_M}{N_0 W} & \text{FM} \end{cases}. \tag{6.2.18}$$

Noting that $\frac{A_c^2}{2}$ is the received signal power, denoted by P_R, and

$$\begin{cases} \beta_p = k_p \max |m(t)| & \text{PM} \\ \beta_f = \frac{k_f \max |m(t)|}{W} & \text{FM} \end{cases}, \tag{6.2.19}$$

we may express the output SNR as

$$
\left(\frac{S}{N}\right)_o = \begin{cases} P_R \left(\frac{\beta_p}{\max |m(t)|}\right)^2 \frac{P_M}{N_0 W} & \text{PM} \\[3mm] 3 P_R \left(\frac{\beta_f}{\max |m(t)|}\right)^2 \frac{P_M}{N_0 W} & \text{FM} \end{cases} . \tag{6.2.20}
$$

If we denote $\frac{P_R}{N_0 W}$ by $\left(\frac{S}{N}\right)_b$, the signal-to-noise ratio of a baseband system with the same received power, we obtain

$$
\left(\frac{S}{N}\right)_o = \begin{cases} \frac{P_M \beta_p^2}{(\max |m(t)|)^2} \left(\frac{S}{N}\right)_b & \text{PM} \\[3mm] 3 \frac{P_M \beta_f^2}{(\max |m(t)|)^2} \left(\frac{S}{N}\right)_b & \text{FM} \end{cases} . \tag{6.2.21}
$$

Note that in the preceding, expression, $\frac{P_M}{(\max |m(t)|)^2}$ is the average-to-peak-power-ratio of the message signal (or equivalently, the power content of the normalized message, P_{M_n}). Therefore,

$$
\left(\frac{S}{N}\right)_o = \begin{cases} \beta_p^2 P_{M_n} \left(\frac{S}{N}\right)_b & \text{PM} \\[3mm] 3\beta_f^2 P_{M_n} \left(\frac{S}{N}\right)_b & \text{FM} \end{cases} . \tag{6.2.22}
$$

Now using Carson's rule $B_c = 2(\beta + 1)W$, we can express the output SNR in terms of the bandwidth expansion factor, which is defined as the ratio of the channel bandwidth to the message bandwidth and is denoted by Ω:

$$
\Omega \overset{\text{def}}{=} \frac{B_c}{W} = 2(\beta + 1). \tag{6.2.23}
$$

From this relationship, we have $\beta = \frac{\Omega}{2} - 1$. Therefore,

$$
\left(\frac{S}{N}\right)_o = \begin{cases} P_M \left(\frac{\frac{\Omega}{2}-1}{\max |m(t)|}\right)^2 \left(\frac{S}{N}\right)_b & \text{PM} \\[3mm] 3 P_M \left(\frac{\frac{\Omega}{2}-1}{\max |m(t)|}\right)^2 \left(\frac{S}{N}\right)_b & \text{FM} \end{cases} . \tag{6.2.24}
$$

From Equation (6.2.20) and Equation (6.2.24), we can make several observations:

1. In both PM and FM, the output SNR is proportional to the square of the modulation index β. Therefore, increasing β increases the output SNR, even with low received power. This is in contrast to amplitude modulation, where such an increase in the received signal-to-noise ratio is not possible.

2. The increase in the received signal-to-noise ratio is obtained by increasing the bandwidth. Therefore, angle modulation provides a way to trade off bandwidth for transmitted power.

3. The relation between the output SNR and the bandwidth expansion factor Ω is a quadratic relation. This is far from optimal[4]. Information theoretical analysis of the performance of communication systems shows that the optimal relation between the output SNR and the bandwidth expansion ratio is an exponential relation.

4. Although we can increase the output signal-to-noise ratio by increasing β, having a large β means having a large B_c (by Carson's rule). Having a large B_c means having a large noise power at the input of the demodulator. This means that the approximation $P(V_n(t) \ll A_c) \approx 1$ will no longer apply and that the preceding analysis will not hold. In fact, if we increase β such that the preceding approximation does not hold, a phenomenon known as the *threshold effect* will occur and the signal will be lost in the noise. This means that although increasing the modulation index up to a certain value improves the performance of the system, this cannot continue indefinitely. After a certain point, increasing β will be harmful and deteriorates the performance of the system.

5. A comparison of the preceding result with the signal-to-noise ratio in amplitude modulation shows that, in both cases, increasing the transmitter power (and consequently the received power) will increase the output signal-to-noise ratio, but the mechanisms are totally different. In AM, any increase in the received power directly increases the signal power at the output of the demodulator. This is basically because the message is in the amplitude of the transmitted signal and an increase in the transmitted power directly affects the demodulated signal power. However, in angle modulation, the message is in the phase of the modulated signal and, consequently, increasing the transmitter power does not increase the demodulated message power. In angle modulation, the output signal-to-noise ratio is increased by a *decrease in the received noise power*, as seen from Equation (6.2.16) and Figure 6.1.

6. In FM, the effect of noise is higher at higher frequencies. This means that signal components at higher frequencies will suffer more from noise than signal components at lower frequencies. In some applications where FM is used to transmit SSB-FDM signals, those channels that are modulated on higher frequency carriers suffer from more noise. To compensate for this effect, such channels must have a higher signal level. The quadratic characteristics of the demodulated noise spectrum in FM is the basis of preemphasis and deemphasis filtering, which will be discussed later in this chapter.

Example 6.2.1

What is the required received power in an FM system with $\beta = 5$ if $W = 15$ kHz and $N_0 = 10^{-14}$ W/Hz? The power of the normalized message signal is assumed to be 0.1 Watt and the required SNR after demodulation is 60 dB.

---

[4]By optimal relation, we mean the maximum savings in transmitter power for a given expansion in bandwidth. An optimal system achieves the fundamental limits on communication, as predicted by information theory.

Solution   We use the relation

$$\left(\frac{S}{N}\right)_{\text{o}} = 3\beta^2 P_{M_n}\frac{P_R}{N_0 W},$$

with $\left(\frac{S}{N}\right)_{\text{o}} = 10^6$, $\beta = 5$, $P_{M_n} = 0.1$, $N_0 = 10^{-14}$, and $W = 15{,}000$, to obtain $P_R = 2 \times 10^{-5}$ or 20 microwatts.   ∎

6.2.1 Threshold Effect in Angle Modulation

The noise analysis of angle-demodulation schemes is based on the assumption that the signal-to-noise ratio at the demodulator input is high. With this crucial assumption, we observe that the signal and noise components at the demodulator output are additive and we are able to carry out the analysis. This assumption of high signal-to-noise ratio is a simplifying assumption that is usually made in the analysis of nonlinear modulation systems. Due to the nonlinear nature of the demodulation process, the additive signal and noise components at the input of the modulator do not result in additive signal and noise components at the output of the demodulator. In fact, this assumption is generally not correct. The signal and noise processes at the output of the demodulator are completely mixed in a single process by a complicated nonlinear functional. Only under the high signal-to-noise ratio assumption is this highly nonlinear functional approximated as an additive form. Particularly at low signal-to-noise ratios, signal and noise components are so intermingled that we cannot recognize the signal from the noise; therefore, no meaningful signal-to-noise ratio as a measure of performance can be defined.

In such cases, the signal is not distinguishable from the noise and a *mutilation* or *threshold effect* is present. There exists a specific signal-to-noise ratio at the input of the demodulator (known as the *threshold SNR*) below which signal mutilation occurs. The existence of the threshold effect places an upper limit on the tradeoff between bandwidth and power in an FM system. This limit is a practical limit in the value of the modulation index β_f. The analysis of the threshold effect and the derivation of the threshold index β_f is quite involved and beyond the scope of our analysis. The references cited at the end of this chapter can provide an analytic treatment of the subject. This text will only mention some results on the threshold effect in FM.

At threshold, the following approximate relation between $\frac{P_R}{N_0 W} = \left(\frac{S}{N}\right)_b$ and β_f holds in an FM system:

$$\left(\frac{S}{N}\right)_{b,\text{th}} = 20(\beta_f + 1). \tag{6.2.25}$$

From the this relation, given a received power P_R, we can calculate the maximum allowed β to make sure that the system works above threshold. Also, given a bandwidth allocation of B_c, we can find an appropriate β using Carson's rule $B_c = 2(\beta + 1)W$. Then, using the preceding threshold relation, we determine the required minimum received power to make the whole allocated bandwidth usable.

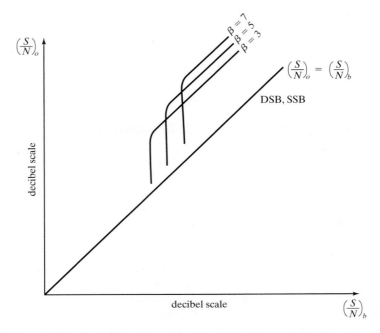

Figure 6.5   Output SNR of an FM system as a function of the baseband SNR for various values of β.

In general, there are two factors that limit the value of the modulation index β. The first is the limitations on channel bandwidth, which affect β through Carson's rule. The second is the limitation on the received power which limits the value of β to less than the value derived from Equation (6.2.25). Figure 6.5 shows plots of the output SNR in an FM system as a function of the baseband SNR. The output SNR values in these curves are in decibels, and different curves correspond to different values of β, as marked. The effect of threshold is apparent from the sudden drops in the output SNR. These plots are drawn for a sinusoidal message for which

$$\frac{P_M}{(\max |m(t)|)^2} = \frac{1}{2}. \tag{6.2.26}$$

In such a case,

$$\left(\frac{S}{N}\right)_o = \frac{3}{2}\beta^2 \left(\frac{S}{N}\right)_b. \tag{6.2.27}$$

As an example, for $\beta = 5$, this relation yields

$$\left.\left(\frac{S}{N}\right)_o\right|_{dB} = 15.7 + \left.\left(\frac{S}{N}\right)_b\right|_{dB} \tag{6.2.28}$$

and

$$\left(\frac{S}{N}\right)_{b,\text{th}} = 120 \sim 20.8 \quad \text{dB}. \tag{6.2.29}$$

For $\beta = 2$, we have

$$\left.\left(\frac{S}{N}\right)_o\right|_{\text{dB}} = 7.8 + \left.\left(\frac{S}{N}\right)_b\right|_{\text{dB}} \qquad (6.2.30)$$

$$\left(\frac{S}{N}\right)_{b,\text{th}} = 60 \approx 17.8 \qquad \text{dB}. \qquad (6.2.31)$$

It is apparent that if, for example, $\left(\frac{S}{N}\right)_b = 20$ dB, then regardless of the available bandwidth, we cannot use $\beta = 5$ for such a system because the system will operate below threshold. For this case, we can use $\beta = 2$. This yields an SNR equal to 27.8 dB at the output of the receiver. This is an improvement of 7.8 dB over a baseband system.

In general, if we want to employ the maximum available bandwidth, we must choose the largest possible β that guarantees that the system will operate above threshold. This is the value of β that satisfies

$$\left(\frac{S}{N}\right)_{b,\text{th}} = 20(\beta + 1). \qquad (6.2.32)$$

By substituting this value into Equation (6.2.22), we obtain

$$\left(\frac{S}{N}\right)_o = 60\beta^2(\beta + 1)P_{M_n}, \qquad (6.2.33)$$

which relates a desired output SNR to the highest possible β that achieves that SNR.

Example 6.2.2

Design an FM system that achieves an SNR at the receiver equal to 40 dB and requires the minimum amount of transmitter power. The bandwidth of the channel is 120 kHz; the message bandwidth is 10 kHz; the average-to-peak power ratio for the message, $P_{M_n} = \frac{P_M}{(\max |m(t)|)^2}$ is $\frac{1}{2}$; and the (one-sided) noise power spectral density is $N_0 = 10^{-8}$ W/Hz. What is the required transmitter power if the signal is attenuated by 40 dB in transmission through the channel?

Solution   First, we have to see whether the threshold or the bandwidth impose a more restrictive bound on the modulation index. By Carson's rule,

$$B_c = 2(\beta + 1)W$$

$$120,000 = 2(\beta + 1) \times 10,000,$$

from which we obtain $\beta = 5$. Using the relation

$$\left(\frac{S}{N}\right)_o = 60\beta^2(\beta + 1)P_{M_n} \qquad (6.2.34)$$

with $\left(\frac{S}{N}\right)_o = 10^4$, we obtain $\beta \approx 6.6$. Since the value of β given by the bandwidth constraint is less than the value of β given by the power constraint, we are limited in bandwidth (as opposed to being limited in power). Therefore, we choose $\beta = 5$, which, when substituted in the expansion for the output SNR,

$$\left(\frac{S}{N}\right)_o = \frac{3}{2}\beta^2\left(\frac{S}{N}\right)_b,$$

(6.2.35)

yields

$$\left(\frac{S}{N}\right)_b = \frac{800}{3} = 266.6 \approx 24.26 \text{ dB}.$$

(6.2.36)

Since $\left(\frac{S}{N}\right)_b = \frac{P_R}{N_0 W}$ with $W = 10000$ and $N_0 = 10^{-8}$, we obtain

$$P_R = \frac{8}{300} = 0.0266 \approx -15.74 \text{ dB}$$

(6.2.37)

and

$$P_T = -15.74 + 40 = 24.26 \quad \text{dB} \approx 266.66 \text{ Watts.}$$

(6.2.38)

Had there been no bandwidth constraint, we could have chosen $\beta = 6.6$, which would result in $\left(\frac{S}{N}\right)_b \approx 153$. In turn, we would have $P_R \approx 0.0153$ and $P_T \approx 153$ Watts. ∎

6.2.2 Preemphasis and Deemphasis Filtering

As observed in Figure 6.4, the noise power spectral density at the output of the demodulator in PM is flat within the message bandwidth; however, for FM, the noise power spectrum has a parabolic shape. This means that FM performs better in low-frequency components of the message signal and PM performs better in high-frequency components. Therefore, we want to design a system that performs frequency modulation for low-frequency components of the message signal, and works as a phase modulator for high-frequency components. Thus, we would have a better overall performance compared to each system alone. This is the idea behind preemphasis and deemphasis filtering techniques.

The objective in preemphasis and deemphasis filtering is to design a system which behaves like an ordinary frequency modulator–demodulator pair in the low-frequency band of the message signal, and like a phase modulator–demodulator pair in the high-frequency band of the message signal. Since a phase modulator is simply the cascade connection of a differentiator and a frequency modulator, we need a filter in cascade with the modulator that does not affect the signal at low frequencies and acts as a differentiator at high frequencies. A simple highpass filter is a very good approximation to such a system. Such a filter has a constant gain for low frequencies; at higher frequencies, it has frequency characteristics approximated by $K|f|$, which is the frequency characteristic of a differentiator. At the demodulator side, low frequencies have a simple FM demodulator and high frequency components have a phase demodulator, which is the cascade of a simple FM demodulator and an integrator. Therefore, the demodulator needs a filter that has a constant gain at low frequencies and behaves as an integrator at high frequencies. A good approximation

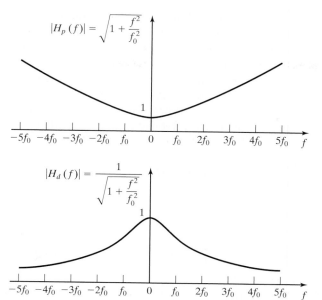

Figure 6.6 Preemphasis and deemphasis filter characteristics.

to such a filter is a simple lowpass filter. The modulator filter, which emphasizes high frequencies, is called the preemphasis filter; the demodulator filter, which is the inverse of the modulator filter, is called the deemphasis filter. Frequency responses of a preemphasis filter and a deemphasis filter are given in Figure 6.6.

Another way to look at preemphasis and deemphasis filtering is to note that due to the high level of noise at high-frequency components of the message in FM, we want to attenuate the high-frequency components of the demodulated signal. This results in a reduction in the noise level, but it causes the higher frequency components of the message signal to be also attenuated. To compensate for the attenuation of the higher components of the message signal, we can amplify these components at the transmitter before modulation. Therefore, we need a highpass filter at the transmitter and a lowpass filter at the receiver. The net effect of these filters should be a flat frequency response. Therefore, the receiver filter should be the inverse of the transmitter filter.

The characteristics of the preemphasis and deemphasis filters depend largely on the power spectral density of the message process. In commercial FM broadcasting of music and voice, first-order lowpass and highpass RC filters with a time constant of 75 μs are employed. In this case, the frequency response of the receiver (deemphasis) filter is given by

$$H_d(f) = \frac{1}{1 + j\frac{f}{f_0}}, \tag{6.2.39}$$

where $f_0 = \frac{1}{2\pi \times 75 \times 10^{-6}} \approx 2100$ Hz is the 3 dB frequency of the filter.

To analyze the effect of preemphasis and deemphasis filtering on the overall signal-to-noise ratio in FM broadcasting, we note that since the transmitter and the receiver filters cancel the effect of each other, the received power in the message signal remains unchanged, and we only have to consider the effect of filtering on the received noise. Of course, the only filter that has an effect on the received noise is the receiver filter, which shapes the power spectral density of the noise within the message bandwidth. The noise component before filtering has a parabolic power spectrum. Therefore, the noise component after the deemphasis filter has a power spectral density given by

$$S_{n_{PD}}(f) = S_{n_o}(f)|H_d(f)|^2$$
$$= \frac{N_0}{A_c^2} f^2 \frac{1}{1 + \frac{f^2}{f_0^2}}, \tag{6.2.40}$$

where we have used Equation (6.2.15). The noise power at the output of the demodulator can be obtained as

$$P_{n_{PD}} = \int_{-W}^{+W} S_{n_{PD}}(f)\, df$$
$$= \frac{N_0}{A_c^2} \int_{-W}^{+W} \frac{f^2}{1 + \frac{f^2}{f_0^2}}\, df$$
$$= \frac{2N_0 f_0^3}{A_c^2} \left[\frac{W}{f_0} - \arctan \frac{W}{f_0} \right]. \tag{6.2.41}$$

Since the demodulated message signal power in this case is equal to that of a simple FM system with no preemphasis and deemphasis filtering, the ratio of the output SNR's in these two cases is inversely proportional to the noise power ratios, i.e.,

$$\frac{\left(\frac{S}{N}\right)_{OPD}}{\left(\frac{S}{N}\right)_o} = \frac{P_{n_o}}{P_{n_{PD}}}$$

$$= \frac{\frac{2N_0 W^3}{3A_c^2}}{\frac{2N_0 f_0^3}{A_c^2}\left[\frac{W}{f_0} - \arctan \frac{W}{f_0}\right]}$$

$$= \frac{1}{3} \frac{\left(\frac{W}{f_0}\right)^3}{\frac{W}{f_0} - \arctan \frac{W}{f_0}}, \tag{6.2.42}$$

where we have used Equation (6.2.16). Hence, Equation (6.2.42) gives the improvement obtained by employing preemphasis and deemphasis filtering.

Example 6.2.3

In commercial FM broadcasting, $W = 15$ kHz, $f_0 = 2100$ Hz, and $\beta = 5$. Assuming that the average-to-peak power ratio of the message signal is 0.5, find the improvement in the output SNR of FM when we use preemphasis and deemphasis filtering rather than a baseband system.

Solution   From Equation (6.2.22), we have

$$\left(\frac{S}{N}\right)_o = 3 \times 5^2 \times 0.5 \times \left(\frac{S}{N}\right)_b$$

$$= 37.5 \left(\frac{S}{N}\right)_b$$

$$\approx 15.7 + \left(\frac{S}{N}\right)_b\bigg|_{dB} . \tag{6.2.43}$$

Therefore, FM with no preemphasis and deemphasis filtering performs 15.7 dB better than a baseband system. For FM with preemphasis and deemphasis filtering, we have

$$\left(\frac{S}{N}\right)_{oPD} = \frac{1}{3} \frac{\left(\frac{W}{f_0}\right)^3}{\frac{W}{f_0} - \arctan\frac{W}{f_0}} \left(\frac{S}{N}\right)_o$$

$$= \frac{1}{3} \frac{\left(\frac{15,000}{2100}\right)^3}{\frac{15,000}{2100} - \arctan\frac{15,000}{2100}} \left(\frac{S}{N}\right)_o$$

$$= 21.3 \left(\frac{S}{N}\right)_o$$

$$\approx 13.3 + \left(\frac{S}{N}\right)_o\bigg|_{dB}$$

$$\approx 13.3 + 15.7 + \left(\frac{S}{N}\right)_b\bigg|_{dB}$$

$$\approx 29 + \left(\frac{S}{N}\right)_b\bigg|_{dB} . \tag{6.2.44}$$

The overall improvement when using preemphasis and deemphasis filtering rather than a baseband system is, therefore, 29 dB.   ∎

6.3 COMPARISON OF ANALOG-MODULATION SYSTEMS

Now, we are at a point where we can present an overall comparison of different analog communication systems. The systems that we have studied include linear modulation systems (DSB-SC AM, conventional AM, SSB-SC AM, VSB) and nonlinear modulation systems (FM and PM).

The comparison of these systems can be done from various points of view. Here we present a comparison based on three important practical criteria:

1. The bandwidth efficiency of the system.
2. The power efficiency of the system as reflected in its performance in the presence of noise.
3. The ease of implementation of the system (transmitter and receiver).

Bandwidth Efficiency. The most bandwidth efficient analog communication system is the SSB-SC system with a transmission bandwidth equal to the signal bandwidth. This system is widely used in bandwidth critical applications, such as voice transmission over microwave and satellite links and some point-to-point communication systems in congested areas. Since SSB-SC cannot effectively transmit DC, it cannot be used for the transmission of signals that have a significant DC component, such as image signals. A good compromise is the VSB system, which has a bandwidth slightly larger than SSB and is capable of transmitting DC values. VSB is widely used in TV broadcasting and in some data communication systems. PM, and particularly FM, are the least favorable systems when bandwidth is the major concern, and their use is only justified by their high level of noise immunity.

Power Efficiency. A criterion for comparing the power efficiency of various systems is the comparison of their output signal-to-noise ratio at a given received signal power. We have already seen that angle-modulation schemes, and particularly FM, provide a high level of noise immunity and, therefore, power efficiency. FM is widely used on power-critical communication links, such as point-to-point communication systems and high-fidelity radio broadcasting. It is also used for transmission of voice (which has been already SSB/FDM multiplexed) on microwave line-of-sight and satellite links. Conventional AM and VSB+C are the least power-efficient systems and are not used when the transmitter power is a major concern. However, their use is justified by the simplicity of the receiver structure.

Ease of Implementation. The simplest receiver structure is the receiver for conventional AM, and the structure of the receiver for VSB+C system is only slightly more complicated. FM receivers are also easy to implement. These three systems are widely used for AM, TV, and high-fidelity FM broadcasting (including FM stereo). The power inefficiency of the AM transmitter is compensated by the extremely simple structure of literally hundreds of millions of receivers. DSB-SC and SSB-SC require synchronous demodulation and, therefore, their receiver structure is much more complicated. These systems are, therefore, never used for broadcasting purposes. Since the receiver structure of SSB-SC and DSB-SC have almost the same complexity and the transmitter of SSB-SC is only slightly more complicated than DSB-SC, DSB-SC is hardly used in analog signal transmission, due to its relative bandwidth inefficiency.

6.4 CARRIER-PHASE ESTIMATION WITH A PHASE-LOCKED LOOP (PLL)

In this section, we describe a method for generating a phase reference for synchronous demodulation of a DSB-SC AM signal. The received noise-corrupted signal at the input to the demodulator is given by

$$r(t) = u(t) + n(t)$$
$$= A_c m(t) \cos(2\pi f_c t + \phi) + n(t), \tag{6.4.1}$$

where $m(t)$ is the message signal, which is assumed to be a sample function of a zero-mean random process $M(t)$.

First, we note that the received signal $r(t)$ has a zero mean, since the message signal $m(t)$ is zero mean, i.e., $m(t)$ contains no *DC* component. Consequently, the average power at the output of a narrowband filter tuned to the carrier frequency f_c is zero. This fact implies that we cannot extract a carrier-signal component directly from $r(t)$.

If we square $r(t)$, the squared signal contains a spectral component at twice the carrier frequency. That is,

$$r^2(t) = A_c^2 m^2(t) \cos^2(2\pi f_c t + \phi) + \text{noise terms}$$
$$= \frac{1}{2} A_c^2 m^2(t) + \frac{1}{2} A_c^2 m^2(t) \cos(4\pi f_c t + 2\phi)$$
$$+ \text{noise terms.} \tag{6.4.2}$$

Since $E[M^2(t)] = R_M(0) > 0$, there is signal power at the frequency $2f_c$, which can be used to drive a phase-locked loop (PLL).

In order to isolate the desired double frequency component from the rest of the frequency components, the squared input signal is passed through a narrowband filter that is tuned to the frequency $2f_c$. The mean value of the output of such a filter is a sinusoid with frequency $2f_c$, phase 2ϕ, and amplitude $A_c^2 E[M^2(t)]H(2f_c)/2$, where $H(2f_c)$ is the gain (attenuation) of the filter at $f = 2f_c$. Thus, squaring the input signal produces a sinusoidal component at twice the carrier frequency, which can be used as the input to a PLL. The general configuration for the carrier-phase estimation system is illustrated in Figure 6.7.

The PLL consists of a multiplier, a loop filter, and a voltage-controlled oscillator (VCO), as shown in Figure 6.8. If the input to the PLL is the sinusoid $\cos(4\pi f_c t + 2\phi)$

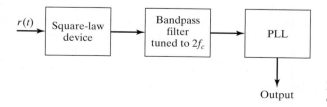

Output   **Figure 6.7**   System for
                      carrier-phase estimation.

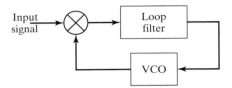

Figure 6.8   Basic elements of a PLL.

and the output of the VCO is $\sin(4\pi f_c t + 2\hat{\phi})$, where $\hat{\phi}$ represents the estimate of ϕ, the product of these two signals produces the signal

$$e(t) = \cos(4\pi f_c t + 2\phi)\sin(4\pi f_c t + 2\hat{\phi})$$

$$= \frac{1}{2}\sin 2(\hat{\phi} - \phi) + \frac{1}{2}\sin(8\pi f_c t + 2\hat{\phi} + 2\phi). \qquad (6.4.3)$$

Note that $e(t)$ contains a low-frequency term (DC) and a term at four times the carrier.

The loop filter is a lowpass filter that responds only to the low-frequency component $\sin(\hat{\phi} - \phi)$ and removes the component at $4f_c$. This filter is usually selected to have the relatively simple transfer function

$$G(s) = \frac{1 + \tau_2 s}{1 + \tau_1 s}, \qquad (6.4.4)$$

where the time constants τ_1 and τ_2 are design parameters ($\tau_1 \gg \tau_2$) that control the bandwidth of the loop. A higher-order filter that contains additional poles may be used, if necessary, to obtain a better loop response.

The output of the loop provides the control voltage for the VCO, whose implementation is described in Section 4.3 in the context of FM modulation. The VCO is basically a sinusoidal signal generator with an instantaneous phase given by

$$4\pi f_c t + 2\hat{\phi} = 4\pi f_c t + K_v \int_{-\infty}^{t} v(\tau)d\tau, \qquad (6.4.5)$$

where K_v is a gain constant in radians per volt-second. Hence, the carrier-phase estimate at the output of the VCO is

$$2\hat{\phi} = K_v \int_{-\infty}^{t} v(\tau)d\tau \qquad (6.4.6)$$

and the transfer function of the VCO is K_v/s.

Since the double-frequency term resulting from the multiplication of the input signal to the loop with the output of the VCO is removed by the loop filter, the PLL may be represented by the closed-loop system model shown in Figure 6.9.

The sine function of the phase difference $2(\hat{\phi} - \phi)$ makes the system nonlinear and, as a consequence, the analysis of the PLL performance in the presence of noise is somewhat involved, although it is mathematically tractable for simple loop filters.

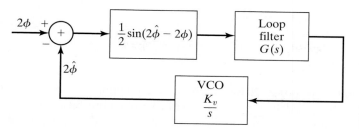

Figure 6.9 Model of a PLL.

In a steady-state operation when the loop is tracking the phase of the received carrier, the phase error $\hat{\phi} - \phi$ is small; hence,

$$\frac{1}{2} \sin 2(\hat{\phi} - \phi) \approx \hat{\phi} - \phi. \tag{6.4.7}$$

With this approximation, the PLL is represented by the *linear model* shown in Figure 6.10. This linear model has a closed-loop transfer function

$$H(s) = \frac{KG(s)/s}{1 + KG(s)/s}, \tag{6.4.8}$$

By substituting from Equation (6.4.4) for $G(s)$ into Equation (6.4.8), we obtain

$$H(s) = \frac{1 + \tau_2 s}{1 + \left(\tau_2 + \frac{1}{K}\right) s + \frac{\tau_1}{K} s^2}. \tag{6.4.9}$$

Hence, the closed-loop system function for the linearized PLL is second order when the loop filter has a single pole and a single zero. The parameter τ_2 determines the position of the zero in $H(s)$, while K, τ_1, and τ_2 control the position of the closed-loop system poles.

The denominator of $H(s)$ may be expressed in the standard form

$$D(s) = s^2 + 2\zeta \omega_n s + \omega_n^2, \tag{6.4.10}$$

where ζ is called the *loop-damping factor* and ω_n is the *natural frequency* of the loop. In terms of the loop parameters, $\omega_n = \sqrt{K/\tau_1}$ and $\zeta = \omega_n(\tau_2 + 1/K)/2$, the

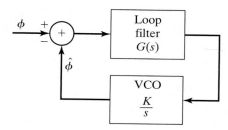

Figure 6.10 Linear model for a PLL.

closed-loop transfer function becomes

$$H(s) = \frac{(2\zeta\omega_n - \omega_n^2/K)s + \omega_n^2}{s^2 + 2\zeta\omega_n s + \omega_n^2}.$$ (6.4.11)

The magnitude response $20\log|H(j\omega)|$ as a function of the normalized frequency ω/ω_n is illustrated in Figure 6.11, with the damping factor as a parameter and $\tau_1 \gg 1$. Note that $\zeta = 1$ results in a critically damped loop response, $\zeta < 1$ produces an underdamped loop response, and $\zeta > 1$ yields an overdamped loop response.

The (one-sided) noise-equivalent bandwidth of the loop is (see Problem 6.6)

$$B_{\text{neq}} = \frac{\tau_2^2 \left(1/\tau_2^2 + K/\tau_1\right)}{4\left(\tau_2 + \frac{1}{K}\right)} = \frac{1 + (\tau_2\omega_n)^2}{8\zeta/\omega_n}.$$ (6.4.12)

In practice, the selection of the bandwidth of the PLL involves a trade-off between the speed of response and the noise in the phase estimate. On the one hand, we want to select the bandwidth of the loop to be sufficiently wide in order to track any time variations in the phase of the received carrier. On the other hand, a wideband PLL allows more noise to pass into the loop, which corrupts the phase estimate. Next, we assess the effects of noise in the quality of the phase estimate.

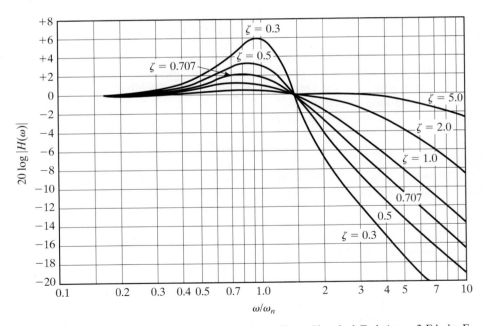

Figure 6.11 Frequency response of a second-order loop. (From *Phaselock Techniques*, 2 Ed., by F. M. Gardner; © 1979 by John Wiley & Sons. Reprinted with permission of the publisher.)

6.4.1 Effect of Additive Noise on Phase Estimation

In order to evaluate the effects of noise on the estimate of the carrier phase, let us assume that the PLL is tracking a sinusoid signal of the form

$$s(t) = A_c \cos[2\pi f_c t + \phi(t)], \tag{6.4.13}$$

which is corrupted by the additive narrowband noise

$$n(t) = n_c(t) \cos 2\pi f_c t - n_s(t) \sin 2\pi f_c t. \tag{6.4.14}$$

The in-phase and quadrature components of the noise are assumed to be statistically independent, stationary Gaussian noise processes with (two-sided) power spectral density $N_0/2$ W/Hz. By using simple trigonometric identities, the noise term in Equation (6.4.14) can be expressed as

$$n(t) = x_c(t) \cos[2\pi f_c t + \phi(t)] - x_s(t) \sin[2\pi f_c t + \phi(t)], \tag{6.4.15}$$

where

$$x_c(t) = n_c(t) \cos \phi(t) + n_s(t) \sin \phi(t)$$

and

$$x_s(t) = -n_c(t) \sin \phi(t) + n_s(t) \cos \phi(t). \tag{6.4.16}$$

We note that

$$x_c(t) + jx_s(t) = [n_c(t) + jn_s(t)]e^{-j\phi(t)}. \tag{6.4.17}$$

It is easy to verify that a phase shift does not change the first two moments of $n_c(t)$ and $n_s(t)$, so that the quadrature components $x_c(t)$ and $x_s(t)$ have exactly the same statistical characteristics as $n_c(t)$ and $n_s(t)$. (See Problem 5.28.)

Now, if $s(t) + n(t)$ is multiplied by the output of the VCO and the double frequency terms are neglected, the input to the loop filter is the noise-corrupted signal

$$e(t) = A_c \sin \Delta\phi + x_c(t) \sin \Delta\phi - x_s(t) \cos \Delta\phi, \tag{6.4.18}$$

where, by definition, $\Delta\phi = \hat{\phi} - \phi$ is the phase error. Thus, we have the equivalent model for the PLL with additive noise, as shown in Figure 6.12.

When the power $P_c = A_c^2/2$ of the incoming signal is much larger than the noise power, the phase estimate $\hat{\phi} \approx \phi$. Then, we may linearize the PLL; thus, we can easily determine the effect of the additive noise on the quality of the estimate $\hat{\phi}$. Under these conditions, the model for the linearized PLL with additive noise will appear as illustrated in Figure 6.13. Note that the gain parameter A_c may be normalized to unity, provided that the noise term is scaled by $1/A_c$. Thus, the noise term becomes

$$n_1(t) = \frac{x_c(t)}{A_c} \sin \Delta\phi - \frac{x_s(t)}{A_c} \cos \Delta\phi. \tag{6.4.19}$$

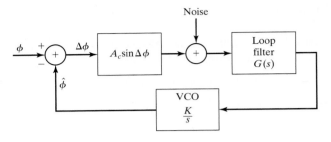

Figure 6.12 Equivalent model of PLL with additive noise.

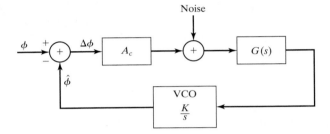

Figure 6.13 Linearized model of PLL with additive noise.

Since the noise $n_1(t)$ is additive at the input to the loop, the variance of the phase error $\Delta\phi$, which is also the variance of the VCO output phase, is

$$\sigma_{\hat{\phi}}^2 = \frac{N_0 B_{\text{neq}}}{A_c^2}, \tag{6.4.20}$$

where B_{neq} is the (one-sided) equivalent-noise bandwidth of the loop, given by Equation (6.4.12). Note that $A_c^2/2$ is the power of the input sinusoid and $\sigma_{\hat{\phi}}^2$ is simply the ratio of the total noise power within the bandwidth of the PLL divided by the input signal power. Hence,

$$\sigma_{\hat{\phi}}^2 = \frac{1}{\rho_L}, \tag{6.4.21}$$

where ρ_L is defined as the signal-to-noise ratio

$$\rho_L = \frac{A_c^2/2}{B_{\text{neq}} N_0/2}. \tag{6.4.22}$$

Thus, the variance of $\hat{\phi}$ is inversely proportional to the SNR.

The expression for the variance $\sigma_{\hat{\phi}}^2$ of the VCO phase error applies to the case where the SNR is sufficiently high so that the linear model for the PLL applies. An exact analysis based on the nonlinear PLL is mathematically tractable when $G(s) = 1$, which results in a first-order loop. In this case, the probability density function for the phase error can be derived [see Viterbi (1966)], and has the form

$$f(\Delta\phi) = \frac{\exp(\rho_L \cos \Delta\phi)}{2\pi I_0(\rho_L)}, \tag{6.4.23}$$

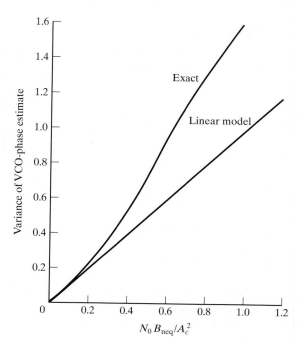

Figure 6.14 Comparison of VCO-phase variance for exact and approximate (linear model) first-order PLL. (From *Principles of Coherent Communication*, by A. J. Viterbi; © 1966 by McGraw-Hill. Reprinted with permission of the publisher.)

where ρ_L is the SNR defined in Equation (6.4.22), B_{neq} is the appropriate noise-equivalent bandwidth of the first-order loop, and $I_0(\cdot)$ is the modified Bessel function of order zero.

From the expression for $f(\Delta\phi)$, we may obtain the exact value of the variance $\sigma_{\hat{\phi}}^2$ for the phase error of a first-order PLL. This is plotted in Figure 6.14 as a function of $1/\rho_L$. Also shown for comparison is the result obtained with the linearized PLL model. Note that the variance for the linear model is close to the exact variance for $\rho_L > 3$. Hence, the linear model is adequate for practical purposes.

Approximate analysis of the statistical characteristics of the phase error for the nonlinear PLL have also been performed. Of particular importance is the transient behavior of the nonlinear PLL during initial acquisition. Another important problem is the behavior of the PLL at low SNR. It is known, for example, that when the SNR at the input to the PLL drops below a certain value, there is a rapid deterioration in the performance of the PLL. The loop begins to lose lock and an impulsive-type of noise, characterized as clicks, is generated; this degrades the performance of the loop. Results on these topics can be found in the texts by Viterbi (1966), Lindsey (1972), Lindsey and Simon (1973), and Gardner (1979), and in the survey papers by Gupta (1975) and Lindsey and Chie (1981).

Now that we have established the effect of noise on the performance of the PLL, let us return to the problem of carrier synchronization based on the system shown in Figure 6.15. The squaring of the received signal that produces the frequency

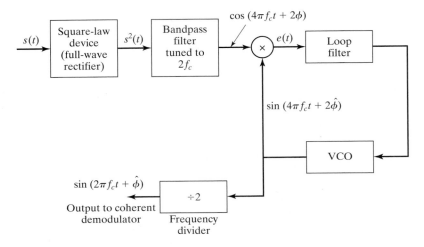

Figure 6.15   Carrier recovery using a square-law device.

component at $2f_c$ also enhances the noise power level at the input to the PLL; thus, it increases the variance of the phase error.

To elaborate on this point, let the input to the squarer be $u(t) + n(t)$. The output is

$$y(t) = u^2(t) + 2u(t)n(t) + n^2(t). \qquad (6.4.24)$$

The noise terms are $2u(t)n(t)$ and $n^2(t)$. By computing the autocorrelation and power spectral density of these two noise components, we can show that both components have spectral power in the frequency band centered at $2f_c$. Consequently, the bandpass filter with bandwidth B_{neq} centered at $2f_c$, which produces the desired sinusoidal signal component that drives the PLL, also passes noise due to these two noise terms.

Let us select the bandwidth of the loop to be significantly smaller than the bandwidth B_{bp} of the bandpass filter, so that the total noise spectrum at the input to the PLL may be approximated by a constant within the loop bandwidth. This approximation allows us to obtain a simple expression for the variance of the phase error as

$$\sigma_{\hat{\phi}}^2 = \frac{1}{\rho_L S_L}, \qquad (6.4.25)$$

where S_L is called the squaring loss and is given as

$$S_L = \frac{1}{1 + \frac{B_{\text{bp}}/2B_{\text{neq}}}{\rho_L}}. \qquad (6.4.26)$$

Since $S_L < 1$, we have an increase in the variance of the phase error; this is caused by the added noise power that results from the squaring operation. For example, when $\rho_L = B_{\text{bp}}/2B_{\text{neq}}$, the loss is 3 dB or, equivalently, the variance in the estimate increases by a factor of two.

Finally, we observe that the output of the VCO from the squaring loop must be frequency divided by a factor of two and phase shifted by 90° to generate the carrier signal for demodulating the received signal.

Costas Loop.   A second method for generating a properly phased carrier for a double sideband-suppressed carrier AM signal is illustrated by the block diagram in Figure 6.16.

The received signal

$$r(t) = A_c m(t) \cos(2\pi f_c t + \phi) + n(t)$$

is multiplied by $\cos(2\pi f_c t + \hat{\phi})$ and $\sin(2\pi f_c t + \hat{\phi})$, which are outputs from the VCO. The two products are

$$
\begin{aligned}
y_c(t) &= [A_c m(t) \cos(2\pi f_c t + \phi) \\
&\quad + n_c(t) \cos 2\pi f_c t - n_s(t) \sin 2\pi f_c t] \cos(2\pi f_c t + \hat{\phi}) \\
&= \frac{A_c}{2} m(t) \cos \Delta\phi + \frac{1}{2}[n_c(t) \cos \hat{\phi} + n_s(t) \sin \hat{\phi}] \\
&\quad + \text{double frequency terms}
\end{aligned}
\tag{6.4.27}
$$

$$
\begin{aligned}
y_s(t) &= [A_c m(t) \cos(2\pi f_c t + \phi) \\
&\quad + n_c(t) \cos 2\pi f_c t - n_s(t) \sin 2\pi f_c t] \sin(2\pi f_c t + \hat{\phi}) \\
&= \frac{A_c}{2} m(t) \sin \Delta\phi + \frac{1}{2}[n_c(t) \sin \hat{\phi} - n_s(t) \cos \hat{\phi}] \\
&\quad + \text{double frequency terms},
\end{aligned}
\tag{6.4.28}
$$

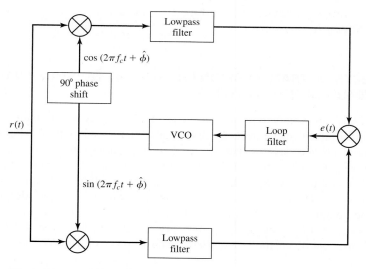

Figure 6.16   Block diagram of Costas loop.

where $\Delta\phi = \hat{\phi} - \phi$. The double frequency terms are eliminated by the lowpass filters following the multiplications.

An error signal is generated by multiplying the two outputs $y_c'(t)$ and $y_s'(t)$ of the lowpass filters. Thus,

$$
e(t) = y_c'(t) y_s'(t)
$$

$$
= \frac{A_c^2}{4} m^2(t) \sin 2\Delta\phi
$$

$$
+ \frac{A_c}{4} m(t)[n_c(t)\cos\hat{\phi} + n_s(t)\sin\hat{\phi}]\sin\Delta\phi
$$

$$
+ \frac{A_c}{4} m(t)[n_c(t)\sin\hat{\phi} - n_s(t)\cos\hat{\phi}]\cos\Delta\phi
$$

$$
+ \frac{1}{4}[n_c(t)\cos\hat{\phi} + n_s(t)\sin\hat{\phi}][n_c(t)\sin\hat{\phi} - n_s(t)\cos\hat{\phi}].
$$

This error signal is filtered by the loop filter whose output is the control voltage that drives the VCO.

We note that the error signal into the loop filter consists of the desired term $(A_c^2 m^2(t)/4) \sin 2\Delta\phi$, plus terms that involve signal × noise and noise × noise. These terms are similar to the two noise terms at the input of the PLL for the squaring method. In fact, if the loop filter in the Costas loop is identical to that used in the squaring loop, the two loops are equivalent. Under this condition, the probability density function of the phase error and the performance of the two loops are identical.

In conclusion, the squaring PLL and the Costas PLL are two practical methods for deriving a carrier-phase estimate for the synchronous demodulation of a DSB-SC AM signal.

6.5 EFFECTS OF TRANSMISSION LOSSES AND NOISE IN ANALOG COMMUNICATION SYSTEMS

In any communication system, there are usually two dominant factors that limit the performance of the system. One important factor is additive noise that is generated by electronic devices that are used to filter and amplify the communication signal. A second factor that affects the performance of a communication system is signal attenuation. Basically all physical channels, including wireline and radio channels, are lossy. Hence, the signal is attenuated (reduced in amplitude) as it travels through the channel. A simple mathematical model of the attenuation may be constructed, as shown in Figure 6.17, by multiplying the transmitted signal by the factor $\alpha < 1$. Consequently, if the transmitted signal is $s(t)$, the received signal is

$$
r(t) = \alpha s(t) + n(t). \tag{6.5.1}
$$

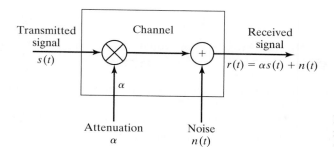

Figure 6.17 Mathematical model of channel with attenuation and additive noise.

Clearly, the effect of signal attenuation is to reduce the amplitude of the desired signal $s(t)$ and, thus, to render the communication signal more vulnerable to additive noise.

In many channels, such as wirelines and microwave line-of-sight channels, signal attenuation can be offset by using amplifiers to boost the level of the signal during transmission. However, an amplifier also introduces additive noise in the process of amplification and, thus, corrupts the signal. This additional noise must be taken into consideration in the design of the communication system.

In this section, we consider the effects of attenuation encountered in signal transmission through a channel and additive thermal noise generated in electronic amplifiers. We also demonstrate how these two factors influence the design of a communication system.

6.5.1 Characterization of Thermal Noise Sources

Any conductive two-terminal device is generally characterized as lossy and has some resistance, say R ohms. A resistor that is at a temperature T above absolute zero contains free electrons that exhibit random motion and, thus, result in a noise voltage across the terminals of the resistor. Such a noise voltage is called *thermal noise*.

In general, any physical resistor (or lossy device) may be modeled by a noise source in series with a noiseless resistor, as shown in Figure 6.18. The output $n(t)$ of the noise source is characterized as a sample function of a random process. Based on quantum mechanics, the power spectral density of thermal noise (see Section 5.3)

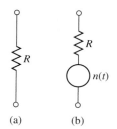

Figure 6.18 A physical resistor (a) is modeled as a noiseless resistor in series with a noise source (b).

is given as

$$S_R(f) = \frac{2R\hbar|f|}{\left(e^{\frac{\hbar|f|}{kT}} - 1\right)} \ \ (\text{volts})^2/\text{Hz}, \tag{6.5.2}$$

where h is Plank's constant, k is Boltzmann's constant, and T is the temperature of the resistor in degrees Kelvin, i.e., $T = 273 + C$, where C is in degrees Centigrade. As indicated in Section 5.3, at frequencies below 10^{12}Hz (which includes all conventional communication systems) and at room temperature,

$$e^{\frac{\hbar|f|}{kT}} \approx 1 + \frac{\hbar|f|}{kT}. \tag{6.5.3}$$

Consequently, the power spectral density is well approximated as

$$S_R(f) = 2RkT \ \ (\text{volts})^2/\text{Hz}. \tag{6.5.4}$$

When connected to a load resistance with value R_L, the noise voltage shown in Figure 6.19 delivers the maximum power when $R = R_L$. In such a case, the load is matched to the source and the maximum power delivered to the load is $E[N^2(t)]/4R_L$. Therefore, the power spectral density of the noise voltage across the load resistor is

$$S_n(f) = \frac{kT}{2} \ \ \text{W/Hz}. \tag{6.5.5}$$

As previously indicated in Section 5.3.2, kT is usually denoted by N_0. Hence, the power spectral density of thermal noise is generally expressed as

$$S_n(f) = \frac{N_0}{2} \ \ \text{W/Hz}. \tag{6.5.6}$$

For example, at room temperature $(T_0 = 290°$ K), $N_0 = 4 \times 10^{-21}$ W/Hz.

6.5.2 Effective Noise Temperature and Noise Figure

When we employ amplifiers in communication systems to boost the level of a signal, we are also amplifying the noise corrupting the signal. Since any amplifier has some finite passband, we may model an amplifier as a filter with the frequency response characteristic $H(f)$. Let us evaluate the effect of the amplifier on an input thermal noise source.

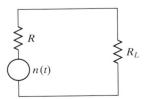

Figure 6.19   Noisy resistor connected to a load resistance R_L.

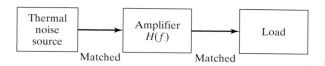

Figure 6.20   Thermal noise connected to amplifier and load.

Figure 6.20 illustrates a thermal noise source connected to a matched two-port network having the frequency response $H(f)$. The output of this network is connected to a matched load. First, we recall that the noise power at the output of the network is

$$P_{no} = \int_{-\infty}^{\infty} S_n(f)|H(f)|^2 df = \frac{N_0}{2} \int_{-\infty}^{\infty} |H(f)|^2 \, df. \qquad (6.5.7)$$

From Section 5.3.3, we recall that the noise equivalent bandwidth of the filter is defined as

$$B_{\text{neq}} = \frac{1}{2\mathscr{G}} \int_{-\infty}^{\infty} |H(f)|^2 \, df, \qquad (6.5.8)$$

where, by definition, $\mathscr{G} = |H(f)|^2_{\text{max}}$ is the *maximum available power gain* of the amplifier. Consequently, the output noise power from an ideal amplifier that introduces no additional noise may be expressed as

$$P_{no} = \mathscr{G} N_0 B_{\text{neq}}. \qquad (6.5.9)$$

Any practical amplifier introduces additional noise at its output due to internally generated noise. Hence, the noise power at its output may be expressed as

$$P_{no} = \mathscr{G} N_0 B_{\text{neq}} + P_{ni}$$
$$= \mathscr{G} k T B_{\text{neq}} + P_{ni}, \qquad (6.5.10)$$

where P_{ni} is the power of the amplifier output due to internally generated noise. Therefore,

$$P_{no} = \mathscr{G} k B_{\text{neq}} \left(T + \frac{P_{ni}}{\mathscr{G} k B_{\text{neq}}} \right). \qquad (6.5.11)$$

This leads us to define a quantity

$$T_e = \frac{P_{ni}}{\mathscr{G} k B_{\text{neq}}}, \qquad (6.5.12)$$

which we call the *effective noise temperature* of the two-port network (amplifier). Then

$$P_{no} = \mathscr{G} k B_{\text{neq}}(T + T_e). \qquad (6.5.13)$$

Thus, we interpret the output noise as originating from a thermal noise source at the temperature $T + T_e$.

A signal source at the input to the amplifier with power P_{si} will produce an output with power

$$P_{so} = \mathcal{G} P_{si}. \tag{6.5.14}$$

Hence, the output SNR from the two-port network is

$$\left(\frac{S}{N}\right)_0 = \frac{P_{so}}{P_{no}} = \frac{\mathcal{G} P_{si}}{\mathcal{G} kT B_{\text{neq}}(1 + T_e/T)}$$

$$= \frac{P_{si}}{N_0 B_{\text{neq}}(1 + T_e/T)}$$

$$= \frac{1}{1 + T_e/T} \left(\frac{S}{N}\right)_i, \tag{6.5.15}$$

where, by definition, $(S/N)_i$ is the input SNR to the two-port network. We observe that the SNR at the output of the amplifier is degraded (reduced) by the factor $(1 + T_e/T)$. Thus, T_e is a measure of the noisiness of the amplifier. An ideal amplifier is one for which $T_e = 0$.

When T is taken as room temperature T_0 (290°K), the factor $(1 + T_e/T_0)$ is called the *noise figure* of the amplifier. Specifically, the noise figure of a two-port network is defined as the ratio of the output noise power P_{no} to the output noise power of an ideal (noiseless) two-port network for which the thermal noise source is at room temperature ($T = 290°$K). Clearly, the ratio

$$F = \left(1 + \frac{T_e}{T_0}\right) \tag{6.5.16}$$

is the noise figure of the amplifier. Consequently, Equation (6.5.15) may be expressed as

$$\left(\frac{S}{N}\right)_0 = \frac{1}{F}\left(\frac{S}{N}\right)_i. \tag{6.5.17}$$

By taking the logarithm of both sides of Equation (6.5.17), we obtain

$$10\log\left(\frac{S}{N}\right)_0 = -10\log F + 10\log\left(\frac{S}{N}\right)_i. \tag{6.5.18}$$

Hence, $10\log F$ represents the loss in SNR due to the additional noise introduced by the amplifier. The noise figure for many low-noise amplifiers, such as traveling wave tubes, is below 3 dB. Conventional integrated circuit amplifiers have noise figures of 6 dB to 7 dB.

It is easy to show (see Problem 6.21) that the overall noise figure of a cascade of K amplifiers with gains $\mathcal{G}_k$ and corresponding noise figures F_k, $1 \leq k \leq K$ is

$$F = F_1 + \frac{F_2 - 1}{\mathcal{G}_1} + \frac{F_3 - 1}{\mathcal{G}_1 \mathcal{G}_2} + \cdots + \frac{F_K - 1}{\mathcal{G}_1 \mathcal{G}_2 \ldots \mathcal{G}_{K-1}}. \tag{6.5.19}$$

This expression is known as *Fries' formula*. We observe that the dominant term is F_1, which is the noise figure of the first amplifier stage. Therefore, the front end of a receiver should have a low noise figure and a high gain. In that case, the remaining terms in the sum will be negligible.

Example 6.5.1

Suppose an amplifier is designed with three identical stages, each of which has a gain of $\mathcal{G}_i = 5$ and a noise figure $F_i = 6$, $i = 1, 2, 3$. Determine the overall noise figure of the cascade of the three stages.

Solution   From Equation (6.5.19), we obtain

$$F = F_1 + \frac{F_2 - 1}{\mathcal{G}_1} + \frac{F_3 - 1}{\mathcal{G}_1 \mathcal{G}_2},$$

where $F_1 = F_2 = F_3 = 6$ and $\mathcal{G}_1 = \mathcal{G}_2 = 5$. Hence,

$$F_1 = 6 + 1 + 0.2 = 7.2,$$

or, equivalently, $F_{1\mathrm{dB}} = 8.57$ dB.   ∎

6.5.3 Transmission Losses

As we indicated previously, any physical channel attenuates the signal transmitted through it. The amount of signal attenuation generally depends on the physical medium, the frequency of operation, and the distance between the transmitter and the receiver. We define the loss $\mathcal{L}$ in signal transmission as the ratio of the input (transmitted) power to the output (received) power of the channel, i.e.,

$$\mathcal{L} = \frac{P_T}{P_R}, \tag{6.5.20}$$

or, in decibels,

$$\mathcal{L}_{\mathrm{dB}} \equiv 10 \log \mathcal{L} = 10 \log P_T - 10 \log P_R. \tag{6.5.21}$$

In wireline channels, the transmission loss is usually given in terms of decibels per unit length, e.g., dB/km. For example, the transmission loss in coaxial cable of 1 cm diameter is about 2 dB/km at a frequency of 1 Mhz. This loss generally increases with an increase in frequency.

Example 6.5.2

Determine the transmission loss for a 10 km and a 20 km coaxial cable if the loss per kilometer is 2 dB at the frequency operation.

Solution   The loss for the 10 km channel is $\mathcal{L}_{\mathrm{dB}} = 20$ dB. Hence, the output (received) power is $P_R = P_T/\mathcal{L} = 10^{-2} P_T$. For the 20 km channel, the loss is $\mathcal{L}_{\mathrm{dB}} = 40$ dB. Hence, $P_R = 10^{-4} P_T$. Note that doubling the cable length increases the attenuation by two orders of magnitude.   ∎

In line-of-sight radio systems, the transmission loss is given as

$$\mathscr{L} = \left(\frac{4\pi d}{\lambda}\right)^2, \tag{6.5.22}$$

where $\lambda = c/f$ is the wavelength of the transmitted signal, c is the speed of light $(3 \times 10^8 \text{m/s})$, f is the frequency of the transmitted signal, and d is the distance between the transmitter and the receiver in meters. In radio transmission, $\mathscr{L}$ is called the *free-space path loss*.

Example 6.5.3

Determine the free-space path loss for a signal transmitted at $f = 1$ MHz over distances of 10 km and 20 km.

Solution The loss given in Equation (6.5.22) for a signal at a wavelength $\lambda = 300$ m is

$$\mathscr{L}_{dB} = 20 \log_{10}(4\pi \times 10^4/300)$$
$$= 52.44 \text{ dB} \tag{6.5.23}$$

for the 10 km path and

$$\mathscr{L}_{dB} = 20 \log_{10}(8\pi \times 10^4/300)$$
$$= 58.44 \text{ dB} \tag{6.5.24}$$

for the 20 km path. It is interesting to note that doubling the distance in radio transmission increases the free-space path loss by 6 dB. ∎

Example 6.5.4

A signal is transmitted through a 10 km coaxial line channel, which exhibits a loss of 2 dB/km. The transmitted signal power is $P_{1dB} = -30$ dBW (-30 dBW means 30 dB below 1 watt or, simply, one milliwatt). Determine the received signal power and the power at the output of an amplifier that has a gain of $\mathscr{G}_{dB} = 15$ dB.

Solution The transmission loss for the 10 km channel is $\mathscr{L}_{dB} = 20$ dB. Hence, the received signal power is

$$P_{RdB} = P_{TdB} - \mathscr{L}_{dB} = -30 - 20 = -50 \text{ dBW} \tag{6.5.25}$$

The amplifier boosts the received signal power by 15 dB. Hence, the power at the output of the amplifier is

$$P_{0dB} = P_{RdB} + G_{dB}$$
$$= -50 + 15 = -35 \text{ dBW} \tag{6.5.26}$$

∎

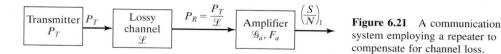

Figure 6.21 A communication system employing a repeater to compensate for channel loss.

6.5.4 Repeaters for Signal Transmission

Analog repeaters are basically amplifiers that are generally used in telephone wire-line channels and microwave line-of-sight radio channels to boost the signal level and, thus, to offset the effect of signal attenuation in transmission through the channel.

Figure 6.21 illustrates a system in which a repeater is used to amplify the signal that has been attenuated by the lossy transmission medium. Hence, the input signal power at the input to the repeater is

$$P_R = P_T/\mathcal{L}. \tag{6.5.27}$$

The output power from the repeater is

$$P_0 = \mathcal{G}_a \, P_R = \mathcal{G}_a \, P_T/\mathcal{L}. \tag{6.5.28}$$

We may select the amplifier gain $\mathcal{G}_a$ to offset the transmission loss. Hence, $\mathcal{G}_a = \mathcal{L}$ and $P_0 = P_T$.

Now, the SNR at the output of the repeater is

$$
\begin{aligned}
\left(\frac{S}{N}\right)_1 &= \frac{1}{F_a}\left(\frac{S}{N}\right)_i \\
&= \frac{1}{F_a}\left(\frac{P_R}{N_0 B_{\text{neq}}}\right) = \frac{1}{F_a}\left(\frac{P_T}{\mathcal{L} N_0 B_{\text{neq}}}\right) \\
&= \frac{1}{F_a \mathcal{L}}\left(\frac{P_T}{N_0 B_{\text{neq}}}\right).
\end{aligned}
\tag{6.5.29}
$$

Based on this result, we may view the lossy transmission medium followed by the amplifier as a cascade of two networks: one with a noise figure $\mathcal{L}$ and the other with a noise figure F_a. Then, for the cascade connection, the overall noise figure is

$$F = \mathcal{L} + \frac{F_a - 1}{\mathcal{G}_a}. \tag{6.5.30}$$

If we select $\mathcal{G}_a = 1/\mathcal{L}$, then

$$F = \mathcal{L} + \frac{F_a - 1}{1/\mathcal{L}} = \mathcal{L} F_a. \tag{6.5.31}$$

Hence, the cascade of the lossy transmission medium and the amplifier is equivalent to a single network with the noise figure $\mathcal{L} F_a$.

Now, suppose that we transmit the signal over K segments of the channel, where each segment has its own repeater, as shown in Figure 6.22. Then, if

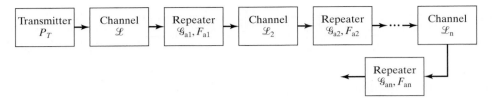

Figure 6.22   A communication system employing repeaters.

$F_i = \mathcal{L}_i F_{ai}$ is the noise figure of the ith section, the overall noise figure for the K sections is

$$F = \mathcal{L}_1 F_{a1} + \frac{\mathcal{L}_2 F_{a2} - 1}{\mathcal{G}_{a1}/\mathcal{L}_1} + \frac{\mathcal{L}_3 F_{a3} - 1}{(\mathcal{G}_{a1}/\mathcal{L}_1)(\mathcal{G}_{a2}/\mathcal{L}_2)} \cdots$$
$$+ \frac{\mathcal{L}_K F_{aK} - 1}{(\mathcal{G}_{a1}/\mathcal{L}_1)(\mathcal{G}_{a2}/\mathcal{L}_2)\ldots(\mathcal{G}_{aK}/\mathcal{L}_K)}. \tag{6.5.32}$$

Therefore, the signal-to-noise ratio at the output of the repeater (amplifier) at the receiver is

$$\left(\frac{S}{N}\right)_0 = \frac{1}{F}\left(\frac{S}{N}\right)_i$$
$$= \frac{1}{F}\left(\frac{P_T}{N_0 B_{neq}}\right). \tag{6.5.33}$$

In the important special case where the K segments are identical, i.e., $\mathcal{L}_i = \mathcal{L}$ for all i and $F_{ai} = F_a$ for all i, and where the amplifier gains are designed to offset the losses in each segment, i.e., $\mathcal{G}_{ai} = \mathcal{L}_i$ for all i, then the overall noise figure becomes

$$F = K\mathcal{L}F_a - (K - 1) \approx K\mathcal{L}F_a. \tag{6.5.34}$$

Hence,

$$\left(\frac{S}{N}\right)_0 \approx \frac{1}{K\mathcal{L}F_a}\left(\frac{P_T}{N_0 B_{neq}}\right). \tag{6.5.35}$$

Therefore, the overall noise figure for the cascade of the K identical segments is simply K times the noise figure of one segment.

Example 6.5.5

A signal with the bandwidth 4 kHz is to be transmitted a distance of 200 km over a wireline channel that has an attenuation of 2 dB/km. (a) Determine the transmitter power P_T required to achieve an SNR of $(S/N)_o = 30$ dB at the output of the receiver amplifier that has a noise figure $F_{adB} = 5$ dB. (b) Repeat the calculation when a repeater is inserted every 10 km in the wireline channel, where the repeater has a gain of 20 dB and a noise figure of $F_a = 5$ dB. Assume that the noise equivalent bandwidth of each repeater is $B_{neq} = 4$ kHz and that $N_0 = 4 \times 10^{-21}$ W/Hz.

Solution

1. The total loss in the 200 km wireline is 400 dB. From Equation (6.5.35), with $K = 1$, we have

$$10 \log(S/N) = -10 \log \mathscr{L} - 10 \log F_a - 10 \log(N_0 B_{neq}) + 10 \log P_T.$$

Hence,

$$P_{T\text{dB}} = (S/N)_{o\text{dB}} + F_{a\text{dB}} + (N_0 B_{neq})_{\text{dB}} + 10 \log \mathscr{L}$$
$$= 30 + 5 + 400 + (N_0 B_{neq})_{\text{dB}}.$$

But

$$(N_0 B_{neq})_{\text{dB}} = 10 \log(1.6 \times 10^{-17}) = -168 \text{ dBW},$$

where dBW denotes the power level relative to one Watt. Therefore,

$$P_{T\text{dB}} = 435 - 168 = 267 \text{ dBW}$$
$$P_T = 5 \times 10^{26} \text{ Watts},$$

which is an astronomical figure.

2. The use of a repeater every 10 km reduces the per segment loss to $\mathscr{L}_{\text{dB}} = 20$ dB. There are 20 repeaters and each repeater has a noise figure of 5 dB. Hence, Equation (6.5.35) yields

$$(S/N)_{0\text{dB}} = -10 \log K - 10 \log \mathscr{L} - 10 \log F_a - 10 \log(N_0 B_{neq}) + 10 \log P_T$$

and

$$30 = -13 - 20 - 5 + 168 + P_{T \text{ dB}}.$$

Therefore,

$$P_{T\text{dB}} = -100 \text{ dBW},$$

or equivalently,

$$P_T = 10^{-10} \text{ Watts} \quad (0.1 \text{ picowatts}). \qquad \blacksquare$$

The preceding example clearly illustrates the advantage of using analog repeaters in communication channels that span large distances. However, we also observed that analog repeaters add noise to the signal and, consequently, degrade the output SNR. It is clear from Equation (6.5.35) that the transmitted power P_T must be increased linearly with the number K of repeaters in order to maintain the same $(S/N)_o$ as K increases. Hence, for every factor of two increase in K, the transmitted power P_T must be increased by 3 dB.

6.6 FURTHER READING

Analysis of the effect of noise on analog communication systems can be found in many textbooks on communications, including Carlson (1986), Ziemer and Tranter (1990), Couch (1993), and Gibson (1993). The book by Sakrison (1968) provides a detailed analysis of FM in the presence of noise. Taub and Schilling (1986) provide in-depth treatment of the effect of threshold and various methods for threshold extension in FM.

PROBLEMS

6.1 The received signal $r(t) = s(t) + n(t)$ in a communication system is passed through an ideal LPF with bandwidth W and unity gain. The signal component $s(t)$ has a power spectral density

$$S_s(f) = \frac{P_0}{1 + (f/B)^2},$$

where B is the 3 dB bandwidth. The noise component $n(t)$ has a power spectral density $N_0/2$ for all frequencies. Determine and plot the SNR as a function of the ratio W/B. What is the filter bandwidth W that yields a maximum SNR?

6.2 The input to the system shown in Figure P-6.2 is the signal plus noise waveform

$$r(t) = A_c \cos 2\pi f_c t + n(t),$$

where $n(t)$ is a sample function of a white noise process with spectral density $N_0/2$.

1. Determine and sketch the frequency response of the RC filter.

2. Sketch the frequency response of the overall system.

3. Determine the SNR at the output of the ideal LPF assuming that $W > f_c$ where W denotes the bandwidth of the LPF. Sketch the SNR as a function of W for fixed values of R and C.

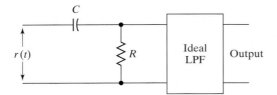

Figure P-6.2

6.3 A DSB amplitude-modulated signal with a power spectral density, as shown in Figure P-6.3(a), is corrupted with additive noise that has a power spectral density $N_0/2$ within the passband of the signal. The received signal plus noise is demodulated and the lowpass is filtered, as shown in Figure P-6.3(b). Determine the SNR at the output of the LPF.

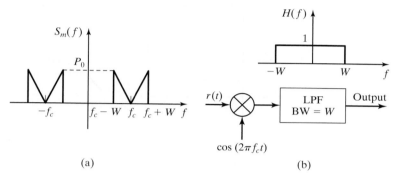

(a) (b)

Figure P-6.3

6.4 A certain communication channel is characterized by a 90 dB attenuation and additive white noise with the power spectral density of $\frac{N_0}{2} = 0.5 \times 10^{-14}$ W/Hz. The bandwidth of the message signal is 1.5 MHz, and its amplitude is uniformly distributed in the interval $[-1, 1]$. If we require that the SNR after demodulation be 30 dB, find the necessary transmitter power in each of the following cases:

   1. USSB modulation.

   2. Conventional AM with a modulation index of 0.5.

   3. DSB-SC modulation.

6.5 A sinusoidal message signal, whose frequency is less than 1000 Hz, modulates the carrier $c(t) = 10^{-3} \cos 2\pi f_c t$. The modulation scheme is conventional AM and the modulation index is 0.5. The channel noise is additive white noise with a power spectral density of $\frac{N_0}{2} = 10^{-12}$ W/Hz. At the receiver, the signal is processed, as shown in Figure P-6.5 (a). The frequency response of the bandpass noise-limiting filter is shown in Figure P-6.5 (b).

   1. Find the signal power and the noise power at the output of the noise-limiting filter.

   2. Find the output SNR.

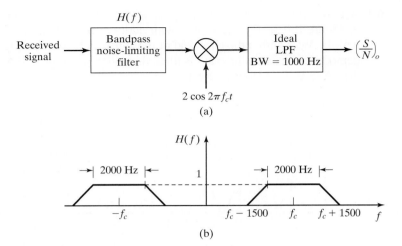

Figure P-6.5

6.6 Derive the expression for the (one-sided) noise equivalent bandwidth of the PLL given by Equation (6.4.12).

6.7 In an analog communication system, *demodulation gain* is defined as the ratio of the SNR at the output of the demodulator to the SNR at the output of the noise limiting filter at the receiver front end. Find expressions for the demodulation gain in each of the following cases:

    1. DSB.

    2. SSB.

    3. Conventional AM with a modulation index of a. What is the largest possible demodulation gain in this case?

    4. FM with a modulation index β_f.

    5. PM with a modulation index β_p.

6.8 In a broadcasting communication system, the transmitter power is 40 KW, the channel attenuation is 80 dB, and the noise power spectral density is 10^{-10} W/Hz. The message signal has a bandwidth of 10^4 Hz.

    1. Find the predetection signal-to-noise ratio (signal-to-noise ratio in $r(t) = \alpha u(t) + n(t)$).

    2. Find the output signal-to-noise ratio if the modulation is DSB.

    3. Find the output signal-to-noise ratio if the modulation is SSB.

    4. Find the output signal-to-noise ratio if the modulation is conventional AM with a modulation index of 0.85 and has a normalized message power of 0.2.

6.9 A communication channel has a bandwidth of 100 KHz. This channel is to be used for transmission of an analog source $m(t)$, where $|m(t)| < 1$, and its bandwidth is $W = 4$ KHz. The power content of the message signal is 0.1 Watts.

1. Find the ratio of the output SNR of an FM system that utilizes the whole bandwidth to the output SNR of a conventional AM system with a modulation index of $a = 0.85$. What is this ratio in dB?

2. Show that if an FM system and a PM system are employed and these systems have the same output signal-to-noise ratio, then

$$\frac{\text{BW}_{\text{PM}}}{\text{BW}_{\text{FM}}} = \frac{\sqrt{3}\beta_f + 1}{\beta_f + 1}.$$

6.10 The normalized message signal $m_n(t)$ has a bandwidth of 5000 Hz and power of 0.1 Watts, and the channel has a bandwidth of 100 KHz and attenuation of 80 dB. The noise is white with a power spectral density 0.5×10^{-12} W/Hz, and the transmitter power is 10 KW.

1. If AM with $a = 0.8$ is employed, what is $(\frac{S}{N})_o$?

2. If FM is employed, what is the highest possible $(\frac{S}{N})_o$?

6.11 A normalized message signal has a bandwidth of $W = 8$ KHz and a power of $P_{M_n} = \frac{1}{2}$. We must transmit this signal via a channel with an available bandwidth of 60 KHz and attenuation of 40 dB. The channel noise is additive and white with a power spectral density of $\frac{N_0}{2} = 10^{-12}$ W/Hz. A frequency modulation scheme, with no preemphasis/deemphasis filtering, has been proposed for this purpose.

1. If we want an SNR of at least 40 dB at the receiver output, what is the minimum required transmitter power and the corresponding modulation index?

2. If the minimum required SNR is increased to 60 dB, how would your answer change?

3. If in Part 2, we are allowed to employ preemphasis or deemphasis filters with a time constant of $\tau = 75\mu$sec, how would the answer change?

6.12 In transmission of telephone signals over line-of-sight microwave links, a combination of FDM-SSB and FM is often employed. A block diagram of such a system is shown in Figure P-6.12.

Each of the signals $m_i(t)$ is bandlimited to W Hz, and these signals are USSB modulated on carriers $c_i(t) = A_i \cos 2\pi f_{ci} t$, where $f_{ci} = (i-1)W$, $1 \leq i \leq K$, and $m(t)$ is the sum of all USSB-modulated signals. This signal FM modulates a carrier with frequency f_c with a modulation index of β.

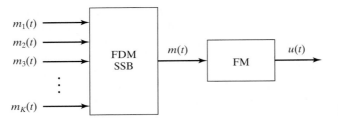

Figure P-6.12

1. Plot a typical spectrum of the USSB-modulated signal $m(t)$.

2. Determine the bandwidth of $m(t)$.

3. At the receiver side, the received signal $r(t) = u(t) + n_w(t)$ is first FM demodulated and then passed through a bank of USSB demodulators. Show that the noise power entering these demodulators depends on i.

4. Determine an expression for the ratio of the noise power entering the demodulator, whose carrier frequency is f_i to the noise power entering the demodulator with the carrier frequency f_j, $1 \leq i, j \leq K$.

5. How should the carrier amplitudes A_i be chosen to guarantee that, after USSB demodulation, the SNR for all channels is the same?

6.13 Suppose that the loop filter in Equation (6.4.4) for a PLL has the transfer function

$$G(s) = \frac{1}{s + \sqrt{2}}.$$

(1) Determine the closed-loop transfer function $H(s)$ and indicate if the loop is stable.

(2) Determine the damping factor and the natural frequency of the loop.

6.14 Consider the PLL for estimating the carrier phase of a signal in which the loop filter is specified as

$$G(s) = \frac{K}{1 + \tau_1 s}.$$

(1) Determine the closed-loop transfer function $H(s)$ and its gain at $f = 0$.

(2) For what range of value of τ_1 and K is the loop stable?

6.15 The loop filter $G(s)$ in a PLL is implemented by the circuit shown in Figure P-6.15. Determine the system function $G(s)$ and express the time constants τ_1 and τ_2, as shown in Equation (6.4.4), in terms of the circuit parameters.

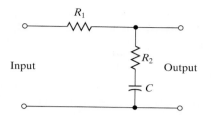

Figure P-6.15

6.16 The loop filter $G(s)$ in a PLL is implemented with the active filter shown in Figure P-6.16. Determine the system function $G(s)$ and express the time constants τ_1 and τ_2, as shown in Equation (6.4.4), in terms of the circuit parameters.

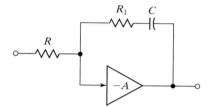

Figure P-6.16

6.17 A power meter that measures average power is connected to the output of a transmitter, as shown in Figure P-6.17. The meter reading is 20 Watts when it is connected to a 50 Ω load. Determine

    1. The voltage across the load resistance.

    2. The current through the load resistance.

    3. The power level in dBm units.

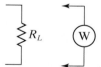

Figure P-6.17

6.18 A twisted-pair telephone wireline channel with characteristic impedance $Z_c = 300\ \Omega$ is terminated with a load of $Z_L = 300\ \Omega$. The telephone line is 200 km long and has a loss of 2 dB/km.

    1. If the average transmitted power $P_T = 10$ dBm, determine the received power P_R if the line contains no repeaters.

2. If repeaters with a gain of 20 dB are used to boost the signal on the channel, and if each repeater requires an input signal level of 10 dBm, determine the number of repeaters and their spacing. The noise figure of each repeater is 6 dB.

6.19 A radio antenna pointed in a direction of the sky has a noise temperature of $50°$ K. The antenna feeds the received signal to the pre-amplifier, which has a gain of 35 dB over a bandwidth of 10 MHz and a noise figure of 2 dB.

1. Determine the effective noise temperature at the input to the pre-amplifier.

2. Determine the noise power at the output of the pre-amplifier.

6.20 An amplifier has a noise equivalent bandwidth $B_{neq} = 25$ kHz and a maximum available power gain of $\mathcal{G} = 30$ dB. Its output noise power is $10^8 k T_0$, where T_0 denotes the ambient temperature. Determine the effective noise temperature and the noise figure.

6.21 Prove that the effective noise temperature of k two-port networks in cascade is

$$T_e = T_{e1} + \frac{T_{e2}}{\mathcal{G}_1} + \frac{T_{e3}}{\mathcal{G}_1 \mathcal{G}_2} + \cdots + \frac{T_{ek}}{\mathcal{G}_1 \mathcal{G}_2 \cdots \mathcal{G}_k}.$$

Using this relationship, prove Fries' formula, which is given in Equation (6.5.19).

COMPUTER PROBLEMS

6.1 Effect of Noise on DSB-SC AM

The transmitted message signal is given by

$$m(t) = \begin{cases} \text{sinc}(100t) & 0 \leq t \leq t_0 \\ 0 & \text{otherwise} \end{cases},$$

where $t_0 = 0.1$. The message signal modulates the carrier $c(t) = \cos 2\pi f_c t$, where $f_c = 250$ Hz, to produce the DSB-SC AM signal $u(t) = m(t)c(t)$.

1. By selecting the sampling interval $t_s = 0.0001$, generate 1000 samples of the message signal $m(t)$ and the modulated signal $u(t)$ at $t = n t_s$, $n = 0, 1, \ldots, 999$, and plot both signals.

2. Generate a sequence of 2000 zero-mean and unit-variance Gaussian random variables. Form the received signal sequence,

$$r(n t_s) = r(n) = u(n t_s) + \sigma \left[w_c(n t_s) \cos 2\pi f_c n t_s - w_s(n t_s) \sin 2\pi f_c n t_s \right]$$

$$= u(n) + \sigma \left[w_c(n) \cos 2\pi f_c n t_s - w_s(n) \sin 2\pi f_c n t_s \right],$$

where $w_c(t)$ and $w_s(t)$ represent the quadrature components of the additive Gaussian noise process and σ^2 is a scale factor that is proportional to the noise power. Generate and plot the received signal sequence $\{r(n)\}$ for the following values of σ: $\sigma = 0.1$, $\sigma = 1$, and $\sigma = 2$.

3. Demodulate the received signal sequence $\{r(n)\}$ by using the demodulator shown in Figure CP-6.1, and plot the received message signal $m_r(t)$ for each of the three values of σ. The lowpass filter is a linear phase FIR filter, which has 31 taps, a cutoff frequency (-3 dB) of 100 Hz, and a stopband attenuation of at least 30 dB. Comment on the effect of the additive noise on the demodulated signal $m_r(t)$ by comparing $m_r(t)$ with the transmitted message signal $m(t)$.

4. Determine the SNR at the receiver output for the three values of σ.

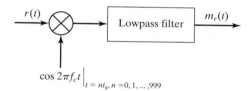

Figure CP-6.1 Demodulator for DSB-SC AM.

6.2 Effect of Noise on SSB-SC AM

Repeat Computer Problem 6.1 when the transmitted signal is an SSB-SC AM.

6.3 Effect of Noise on Conventional AM

A message signal is given by

$$m(t) = \begin{cases} \text{sinc}(100t) & 0 \le t \le t_0 \\ 0 & \text{otherwise} \end{cases},$$

where $t_0 = 0.1$. The message signal modulates the carrier $c(t) = \cos 2\pi f_c t$, where $f_c = 250$ Hz, to produce a conventional AM signal $u(t)$. The modulation index is $a = 0.8$.

1. By selecting the sampling interval $t_s = 0.0001$, generate 1000 samples of the message signal $m(t)$ and the modulated signal

$$u(t) = [1 + am(t)] \cos 2\pi f_c t \big|_{t=nt_s,\ n=0,1,\dots,999}.$$

Plot both signals.

2. Generate a sequence of 2000 zero-mean and unit-variance Gaussian random variables. Form the received signal sequence,

$$r(nt_s) = r(n) = u(nt_s) + \sigma \left[w_c(nt_s) \cos 2\pi f_c nt_s - w_s(nt_s) \sin 2\pi f_c nt_s \right]$$
$$= u(n) + \sigma \left[w_c(n) \cos 2\pi f_c nt_s - w_s(n) \sin 2\pi f_c nt_s \right],$$

where $w_c(t)$ and $w_s(t)$ represent the quadrature components of the additive Gaussian noise process and σ^2 is a scale factor that is proportional to the noise power. Generate and plot the received signal sequence $\{r(n)\}$ for the following values of σ: $\sigma = 0.1$, $\sigma = 1$, and $\sigma = 2$.

3. Demodulate the received signal sequence $\{r(n)\}$ by using an envelope detector that computes

$$e(t) = \sqrt{[1 + am(t) + w_c(t)]^2 + w_s^2(t)}\Big|_{t=nt_s, \ n=0,1,2,\dots,999}.$$

Plot $e(t)$ for each of the three values of σ. Compare $e(t)$ with the original message signal $m(t)$, and comment on the effect of the additive noise on the demodulated signal. Note that in the absence of noise, the message signal $m(t)$ can be obtained from the envelope signal $e(t)$ by subtracting the DC offset, which is equal to 1. In the presence of noise, what is the DC offset?

4. Determine the SNR at the receiver output for the three values of σ.

6.4 Effect of Noise on FM

The message signal

$$m(t) = \begin{cases} \text{sinc}(100t) & 0 \le t \le t_0 \\ 0 & \text{otherwise} \end{cases},$$

where $t_0 = 0.1$, frequency modulates the carrier $c(t) = \cos 2\pi f_c t$, in which $f_c = 250$ Hz. The frequency-deviation constant is $k_f = 100$. The frequency-modulated signal is

$$u(t) = \cos \left(2\pi f_c t + 2\pi k_f \int_{-\infty}^{t} m(\tau) \, d\tau \right).$$

1. By selecting the sampling interval $t_s = 0.0001$, generate 2000 samples of the message signal $m(t)$ and its integral. Plot them on separate graphs.

2. Generate and plot 2000 samples of the FM signal $u(t)$ in the time interval $|t| \le t_0$.

3. Use MATLAB's Fourier-transform routine to compute and plot the spectra of $m(t)$ and $u(t)$ on separate graphs.

4. Generate a sequence of 3998 zero-mean and unit-variance Gaussian random variables. Form the received signal sequence

$$r(nt_s) = r(n) = u(nt_s) + \sigma \left[w_c(nt_s) \cos 2\pi f_c nt_s - w_s(nt_s) \sin 2\pi f_c nt_s \right]$$

$$= u(n) + \sigma \left[w_c(n) \cos 2\pi f_c nt_s - w_s(n) \sin 2\pi f_c nt_s \right]$$

for $n = 0, \pm 1, \pm 2, \ldots, \pm 999$, where $w_c(t)$ and $w_s(t)$ represent the quadrature components of the additive Gaussian noise process and σ^2 is a scale factor that is proportional to the noise power. Generate and plot the received signal sequence $\{r(n)\}$ for $\sigma = 0.1$ and $\sigma = 1$.

5. Demodulate the received signal sequence $\{r(n)\}$ to obtain the received message signal $m_r(t)$. Compare the result with the original message signal. The FM signal can be demodulated by first finding the phase of $u(t)$, i.e., the integral of $m(t)$, which can be differentiated and divided by $2\pi k_f$ to yield $m(t)$. Use the MATLAB function unwrap.m to undo the effect of 2π phase foldings. Comment on how well the demodulated signal matches the original signal $m(t)$.

7

Analog-to-Digital Conversion

Communication systems are designed to transmit information. In any communication system, there exists an information source that produces the information; the purpose of the communication system is to transmit the output of the source to the destination. In radio broadcasting, for instance, the information source is either a speech source or a music source. In TV broadcasting, the information source is a video source that outputs a moving image. In FAX transmission, the information source produces a still image. In communication between computers, either binary data or ASCII characters are transmitted; therefore, the source can be modeled as a binary or ASCII source. In storage of binary data on a computer disk, the source is again a binary source.

In Chapters 3, 4, and 6, we studied the transmission of analog information using different types of analog modulation. In particular, we studied the trade-off between bandwidth and power in FM modulation. The rest of this book deals with transmission of digital data. Digital data transmission provides a higher level of noise immunity, more flexibility in the bandwidth–power trade-off, the possibility of applying cryptographic and antijamming techniques, and the ease of implementation using a large-scale integration of circuits. In order to employ the benefits of digital data transmission, we have to first convert analog information into digital form. Conversion of analog signals into digital data should be carried out with the goal of minimizing the signal distortion introduced in the conversion process.

In order to convert an analog signal to a digital signal, i.e., a stream of bits, three operations must be completed. First, the analog signal has to be sampled, so that we can obtain a *discrete-time continuous-valued signal* from the analog signal. This operation is called *sampling*. Then the sampled values, which can take an infinite number of values are *quantized*, i.e., rounded to a finite number of values. This is called the *quantization* process. After quantization, we have a discrete-time, discrete-amplitude signal. The third stage in analog-to-digital conversion is *encoding*. In encoding, a sequence of bits (ones and zeros) are assigned to different outputs of the quantizer. Since the possible outputs of the quantizer are finite, each sample of the signal can be represented by a finite number of bits. For instance, if the quantizer has $256 = 2^8$ possible levels, they can be represented by 8 bits.

7.1 SAMPLING OF SIGNALS AND SIGNAL RECONSTRUCTION FROM SAMPLES

The sampling theorem is one of the most important results in the analysis of signals; it has widespread applications in communications and signal processing. This theorem and its numerous applications clearly show how much can be gained by employing the frequency-domain methods and the insight provided by frequency-domain signal analysis. Many modern signal-processing techniques and the whole family of digital communication methods are based on the validity of this theorem and the insight provided by it. In fact, this theorem, together with results from signal quantization techniques, provide a bridge that connects the analog world to digital techniques.

7.1.1 The Sampling Theorem

The idea leading to the sampling theorem is very simple and quite intuitive. Let us assume that we have two signals, $x_1(t)$ and $x_2(t)$, as shown in Figure 7.1. The first signal $x_1(t)$ is a smooth signal, it varies very slowly; therefore, its main frequency content is at low frequencies. In contrast, $x_2(t)$ is a signal with rapid changes due to the presence of high-frequency components. We will approximate these signals with samples taken at regular intervals T_1 and T_2, respectively. To obtain an approximation of the original signal, we can use linear interpolation of the sampled values. It is obvious that the sampling interval for the signal $x_1(t)$ can be much larger than the sampling interval necessary to reconstruct signal $x_2(t)$ with comparable distortion.

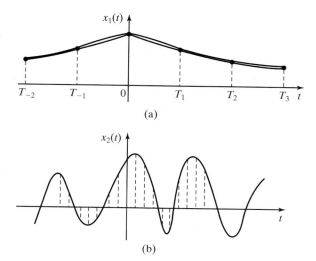

(a)

(b)

Figure 7.1   Sampling of signals.

This is a direct consequence of the smoothness of the signal $x_1(t)$ compared to $x_2(t)$. Therefore, the sampling interval for the signals of smaller bandwidth can be made larger, or the sampling frequency can be made smaller. The sampling theorem is a precise statement of this intuitive reasoning. It basically states two facts:

1. If the signal $x(t)$ is bandlimited to W, i.e., if $X(f) \equiv 0$ for $|f| \geq W$, then it is sufficient to sample at intervals $T_s = \frac{1}{2W}$.
2. If we are allowed to employ more sophisticated interpolating signals than linear interpolation, we are able to recover the exact original signal from the samples, as long as condition 1 is satisfied.

Obviously, the importance of the sampling theorem lies in the fact that it provides both a method to reconstruct the original signal from the sampled values, and it also gives a precise upper bound on the sampling interval (or equivalently, a lower bound on the sampling frequency) required for distortionless reconstruction.

Sampling Theorem.   Let the signal $x(t)$ have a bandwidth W, i.e., let $X(f) \equiv 0$ for $|f| \geq W$. Let $x(t)$ be sampled at multiples of some basic sampling interval T_s, where $T_s \leq \frac{1}{2W}$, to yield the sequence $\{x(nT_s)\}_{n=-\infty}^{+\infty}$. Then it is possible to reconstruct the original signal $x(t)$ from the sampled values by the reconstruction formula

$$x(t) = \sum_{n=-\infty}^{\infty} 2W'T_s x(nT_s) \mathrm{sinc}[2W'(t - nT_s)], \qquad (7.1.1)$$

where W' is any arbitrary number that satisfies the condition

$$W \leq W' \leq \frac{1}{T_s} - W.$$

In the special case where $T_s = \frac{1}{2W}$, we will have $W' = W = \frac{1}{2T_s}$ and the reconstruction relation simplifies to

$$x(t) = \sum_{n=-\infty}^{\infty} x(nT_s) \mathrm{sinc}\left(\frac{t}{T_s} - n\right).$$

Proof. Let $x_\delta(t)$ denote the result of sampling the original signal by impulses at nT_s time instants. Then

$$x_\delta(t) = \sum_{n=-\infty}^{\infty} x(nT_s)\delta(t - nT_s). \qquad (7.1.2)$$

We can write

$$x_\delta(t) = x(t) \sum_{n=-\infty}^{\infty} \delta(t - nT_s), \qquad (7.1.3)$$

where we have used the property that $x(t)\delta(t - nT_s) = x(nT_s)\delta(t - nT_s)$. Now if we find the Fourier transform of both sides of the preceding relation and apply the dual of the convolution theorem to the right-hand side, we obtain

$$X_\delta(f) = X(f) \star \mathscr{F}\left[\sum_{n=-\infty}^{\infty} \delta(t - nT_s)\right]. \tag{7.1.4}$$

Using Table 2.1 to find the Fourier transform of $\sum_{n=-\infty}^{\infty} \delta(t - nT_s)$, we obtain

$$\mathscr{F}\left[\sum_{n=-\infty}^{\infty} \delta(t - nT_s)\right] = \frac{1}{T_s}\sum_{n=-\infty}^{\infty} \delta\left(f - \frac{n}{T_s}\right). \tag{7.1.5}$$

Substituting Equation (7.1.5) into Equation (7.1.4), we obtain

$$X_\delta(f) = X(f) \star \frac{1}{T_s}\sum_{n=-\infty}^{\infty} \delta\left(f - \frac{n}{T_s}\right)$$

$$= \frac{1}{T_s}\sum_{n=-\infty}^{\infty} X\left(f - \frac{n}{T_s}\right), \tag{7.1.6}$$

where, in the last step, we have employed the convolution property of the impulse signal, which states $X(f) \star \delta\left(f - \frac{n}{T_s}\right) = X\left(f - \frac{n}{T_s}\right)$. This relation shows that $X_\delta(f)$, the Fourier transform of the impulse-sampled signal, is a replication of the Fourier transform of the original signal at a rate of $\frac{1}{T_s}$. Figure 7.2 shows a plot of $X_\delta(f)$. Now if $T_s > \frac{1}{2W}$, then the replicated spectrum of $x(t)$ overlaps and reconstruction of the original signal is not possible. This type of distortion, which results from undersampling, is known as *aliasing error* or *aliasing distortion*. However, if $T_s \leq \frac{1}{2W}$, no overlap occurs; and by employing an appropriate filter we can reconstruct the original signal. To get the original signal back, it is sufficient to filter the sampled signal through a lowpass filter with the frequency response characteristic

  1. $H(f) = T_s$ for $|f| < W$.
  2. $H(f) = 0$ for $|f| \geq \frac{1}{T_s} - W$.

For $W \leq |f| < \frac{1}{T_s} - W$, the filter can have any characteristics that makes its implementation easy. Of course, one obvious (though not practical) choice is an ideal lowpass filter with bandwidth W', where W' satisfies $W \leq W' < \frac{1}{T_s} - W$, i.e., by using a filter with a transfer function given by

$$H(f) = T_s \prod\left(\frac{f}{2W'}\right). \tag{7.1.7}$$

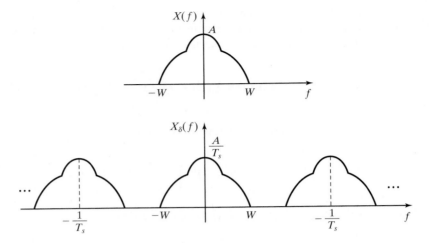

Figure 7.2 Frequency-domain representation of the sampled signal.

With this choice, we have

$$X(f) = X_\delta(f) T_s \prod \left(\frac{f}{2W'} \right). \tag{7.1.8}$$

Taking the inverse Fourier transform of both sides, we obtain

$$x(t) = x_\delta(t) \star 2W' T_s \text{sinc}(2W't)$$

$$= \left(\sum_{n=-\infty}^{\infty} x(nT_s) \delta(t - nT_s) \right) \star 2W' T_s \text{sinc}(2W't)$$

$$= \sum_{n=-\infty}^{\infty} 2W' T_s x(nT_s) \text{sinc} \left(2W'(t - nT_s) \right). \tag{7.1.9}$$

This relation shows that if we use sinc functions for interpolation of the sampled values, we can perfectly reconstruct the original signal. The sampling rate $f_s = \frac{1}{2W}$ is the minimum sampling rate at which no aliasing occurs. This sampling rate is known as the *Nyquist sampling rate*. If sampling is done at the Nyquist rate, then the only choice for the reconstruction filter is an ideal lowpass filter and $W' = W = \frac{1}{2T_s}$. In this case,

$$x(t) = \sum_{n=-\infty}^{\infty} x \left(\frac{n}{2W} \right) \text{sinc} \left(2Wt - n \right)$$

$$= \sum_{n=-\infty}^{\infty} x(nT_s) \text{sinc} \left(\frac{t}{T_s} - n \right). \tag{7.1.10}$$

In practical systems, sampling is done at a rate higher than the Nyquist rate. This allows for the reconstruction filter to be realizable and easier to build. In such cases, the distance between two adjacent replicated spectra in the frequency domain, i.e., $(\frac{1}{T_s} - W) - W = f_s - 2W$, is known as the *guard band*. Therefore, in systems with a guard band, we have $f_s = 2W + W_G$, where W is the bandwidth of the signal, W_G is the guard band, and f_s is the sampling frequency.

Note that there exists a strong similarity between our development of the sampling theorem and our previous development of the Fourier transform for periodic signals (or Fourier series). In the Fourier transform for periodic signals, we started with a (time) periodic signal and showed that its Fourier transform consists of a sequence of impulses. Therefore, to define the signal, it was enough to give the weights of these impulses (Fourier-series coefficients). In the sampling theorem, we started with an impulse-sampled signal, or a sequence of impulses in the time domain and showed that the Fourier transform is a periodic function in the frequency domain. Here again, the values of the samples are enough to define the signal completely. This similarity is a consequence of the duality between the time and frequency domains and the fact that both the Fourier-series expansion and the reconstruction from samples are orthogonal expansions, one in terms of the exponential signals and the other in terms of the sinc functions.

Example 7.1.1

In this development, we have assumed that samples are taken at multiples of T_s. What happens if we sample regularly with T_s as the sampling interval, but the first sample is taken at some $0 < t_0 < T_s$?

Solution   We define a new signal $y(t) = x(t + t_0)$. Then $y(t)$ is bandlimited with $Y(f) = e^{j2\pi f t_0} X(f)$ and the samples of $y(t)$ at $\{kT_s\}_{k=-\infty}^{\infty}$ are equal to the samples of $x(t)$ at $\{t_0 + kT_s\}_{k=-\infty}^{\infty}$. Applying the sampling theorem to the reconstruction of $y(t)$, we have

$$y(t) = \sum_{k=-\infty}^{\infty} y(kT_s)\operatorname{sinc}(2W(t - kT_s))$$

$$= \sum_{k=-\infty}^{\infty} x(t_0 + kT_s)\operatorname{sinc}(2W(t - kT_s));$$

hence,

$$x(t + t_0) = \sum_{k=-\infty}^{\infty} x(t_0 + kT_s)\operatorname{sinc}(2W(t - kT_s)).$$

Substituting $t = -t_0$, we obtain the following important interpolation relation:

$$x(0) = \sum_{k=-\infty}^{\infty} x(t_0 + kT_s)\operatorname{sinc}(2W(t_0 + kT_s)). \tag{7.1.11}$$

∎

Example 7.1.2

A bandlimited signal has a bandwidth equal to 3400 Hz. What sampling rate should be used to guarantee a guard band of 1200 Hz?

Solution We have

$$f_s = 2W + W_G;$$

therefore,

$$f_s = 2 \times 3400 + 1200 = 8000. \qquad \blacksquare$$

After sampling, the continuous-time signal is transformed to a discrete-time signal. In other words, the time-axis has been quantized. After this step, we have samples taken at discrete times, but the amplitude of these samples is still continuous. The next step in analog-to-digital conversion is the quantization of the signal amplitudes. This step results in a signal that is quantized in both time and amplitude.

7.2 QUANTIZATION

After sampling, we have a discrete-time signal, i.e., a signal with values at integer multiples of T_s. The amplitudes of these signals are still continuous, however. Transmission of real numbers requires an infinite number of bits, since generally the base 2 representation of real numbers has infinite length. After sampling, we will use quantization, in which the amplitude becomes discrete as well. As a result, after the quantization step, we will deal with a discrete-time, finite-amplitude signal, in which each sample is represented by a finite number of bits. In this section, we will study different quantization methods. We begin with scalar quantization, in which samples are quantized individually; then, we will explore vector quantization, in which blocks of samples are quantized at a time.

7.2.1 Scalar Quantization

In scalar quantization, each sample is quantized into one of a finite number of levels, which is then encoded into a binary representation. The quantization process is a rounding process; each sampled signal point is rounded to the "nearest" value from a finite set of possible quantization levels. In scalar quantization, the set of real numbers $\mathbb{R}$ is partitioned into N disjoint subsets denoted by $\mathcal{R}_k$, $1 \le k \le N$ (each called a *quantization region*). Corresponding to each subset $\mathcal{R}_k$, a *representation point* (or *quantization level*) $\hat{x}_k$ is chosen, which usually belongs to $\mathcal{R}_k$. If the sampled signal at time i, x_i belongs to $\mathcal{R}_k$, then it is represented by $\hat{x}_k$, which is the *quantized version* of x. Then, $\hat{x}_k$ is represented by a binary sequence and transmitted. This latter step is called *encoding*. Since there are N possibilities for the quantized levels, $\log_2 N$ bits are enough to encode these levels into binary sequences.[1] Therefore, the

---

[1] N is generally chosen to be a power of 2. If it is not, then the number of required bits would be $\lceil \log_2 N \rceil$, where $\lceil x \rceil$ denotes the smallest integer greater than or equal to x. To make the development easier, we always assume that N is a power of 2.

number of bits required to transmit each source output is $R = \log_2 N$ bits. The price that we have paid for representing (rounding) every sample that falls in the region $\mathcal{R}_k$ by a single point $\hat{x}_k$ is the introduction of distortion.

   Figure 7.3 shows an example of an 8-level quantization scheme. In this scheme, the eight regions are defined as $\mathcal{R}_1 = (-\infty, a_1]$, $\mathcal{R}_2 = (a_1, a_2]$, ..., $\mathcal{R}_8 = (a_7, +\infty)$. The representation point (or quantized value) in each region is denoted by $\hat{x}_i$ and is shown in the figure. The quantization function Q is defined by

$$Q(x) = \hat{x}_i \quad \text{for all } x \in \mathcal{R}_i. \tag{7.2.1}$$

This function is also shown in the figure.

   Depending on the measure of distortion employed, we can define the average distortion resulting from quantization. A popular measure of distortion, used widely in practice, is the *squared error distortion* defined as $(x - \hat{x})^2$. In this expression x is the sampled signal value and $\hat{x}$ is the quantized value, i.e., $\hat{x} = Q(x)$. If we are using the squared error distortion measure, then

$$d(x, \hat{x}) = (x - Q(x))^2 = \tilde{x}^2,$$

where $\tilde{x} = x - \hat{x} = x - Q(x)$. Since X is a random variable, so are $\hat{X}$ and $\tilde{X}$; therefore, the average (mean squared error) distortion is given by

$$D = E[d(X, \hat{X})] = E[(X - Q(X))^2].$$

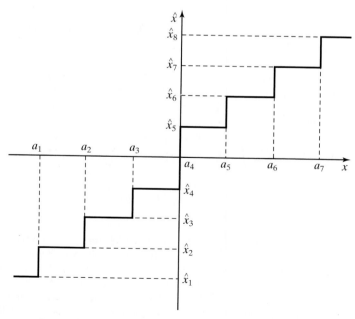

Figure 7.3   Example of an 8-level quantization scheme.

Example 7.2.1

The source $X(t)$ is a stationary Gaussian source with mean zero and power spectral density

$$S_x(f) = \begin{cases} 2 & |f| < 100 \text{ Hz} \\ 0 & \text{otherwise} \end{cases}.$$

The source is sampled at the Nyquist rate and each sample is quantized using the 8-level quantizer which is shown in Figure 7.3 This figure has $a_1 = -60$, $a_2 = -40$, $a_3 = -20$, $a_4 = 0$, $a_5 = 20$, $a_6 = 40$, $a_7 = 60$, and $\hat{x}_1 = -70$, $\hat{x}_2 = -50$, $\hat{x}_3 = -30$, $\hat{x}_4 = -10$, $\hat{x}_5 = 10$, $\hat{x}_6 = 30$, $\hat{x}_7 = 50$, and $\hat{x}_8 = 70$. What is the resulting distortion and rate?

Solution   The sampling frequency is $f_s = 200$ Hz. Each sample is a zero-mean Gaussian random variable with variance

$$\sigma^2 = E(X_i^2) = R_X(\tau)|_{\tau=0} = \int_{-\infty}^{\infty} S_X(f) df = \int_{-100}^{100} 2 df = 400,$$

where we have used Equations (5.2.10) and (5.2.13). Since each sample is quantized into 8 levels, $\log_2 8 = 3$ bits are sufficient to represent (encode) the sample; therefore, the resulting rate is

$$R = 3 f_s = 600 \text{ bits/sec.}$$

To find the distortion, we have to evaluate $E(X - \hat{X})^2$ for each sample. Thus, we have

$$D = E(X - \hat{X})^2 = \int_{-\infty}^{\infty} (x - Q(x))^2 f_X(x) dx,$$

where $f_X(x)$ denotes the probability density function of the random variable X. From here, we have

$$D = \sum_{i=1}^{8} \int_{\mathcal{R}_i} (x - Q(x))^2 f_X(x) dx,$$

or equivalently,

$$D = \int_{-\infty}^{a_1} (x - \hat{x}_1)^2 f_X(x) dx + \sum_{i=2}^{7} \int_{a_{i-1}}^{a_i} (x - \hat{x}_i)^2 f_X(x) dx$$

$$+ \int_{a_7}^{\infty} (x - \hat{x}_8)^2 f_X(x) dx, \tag{7.2.2}$$

where $f_X(x)$ is $\frac{1}{\sqrt{2\pi 400}} e^{-\frac{x^2}{800}}$. Substituting $\{a_i\}_{i=1}^{7}$ and $\{\hat{x}_i\}_{i=1}^{8}$ in this integral and evaluating the result with the Q-function table, we obtain $D \approx 33.38$. Note that if we were to use zero bits per source output, then the best strategy would be to set the reconstructed signal equal to zero. In this case, we would have a distortion of $D = E(X - 0)^2 = \sigma^2 = 400$. This quantization scheme and transmission of 3 bits per source output has enabled us to reduce the distortion to 33.38, which is a factor of 11.98 reduction or 10.78 dB. ∎

   In the preceding example, we have chosen $E(X - Q(X))^2$, which is the mean squared distortion, or *quantization noise* as the measure of performance. A more meaningful measure of performance is a normalized version of the quantization noise, and it is normalized with respect to the power of the original signal.

   Definition 7.2.1.   If the random variable X is quantized to $Q(X)$, the *signal-to-quantization noise ratio* (SQNR) is defined by

$$\text{SQNR} = \frac{E[X^2]}{E[X - Q(X)]^2}. \tag{7.2.3}$$

   ∎

When dealing with signals, the quantization noise power is

$$P_{\tilde{X}} = \lim_{T \to \infty} \frac{1}{T} \int_{-\frac{T}{2}}^{\frac{T}{2}} E[X(t) - Q(X(t))]^2 \, dt \tag{7.2.4}$$

and the signal power is

$$P_X = \lim_{T \to \infty} \frac{1}{T} \int_{-\frac{T}{2}}^{\frac{T}{2}} E[X^2(t)] dt. \tag{7.2.5}$$

Hence, the signal-to-quantization noise ratio is

$$\text{SQNR} = \frac{P_X}{P_{\tilde{X}}}. \tag{7.2.6}$$

If $X(t)$ is stationary, then this relation simplifies to Equation (7.2.3), where X is the random variable representing $X(t)$ at any point.

Example 7.2.2
   Determine the SQNR for the quantization scheme given in Example 7.2.1.

   Solution   From Example 7.2.1, we have $P_X = 400$ and $P_{\tilde{X}} = D = 33.38$. Therefore,

$$\text{SQNR} = \frac{P_X}{P_{\tilde{X}}} = \frac{400}{33.38} = 11.98 \approx 10.78 \quad \text{dB.} \qquad \blacksquare$$

   Uniform Quantization.   Uniform quantizers are the simplest examples of scalar quantizers. In a uniform quantizer, the entire real line is partitioned into N regions. All regions except $\mathcal{R}_1$ and $\mathcal{R}_N$ are of equal length, which is denoted by Δ. This means that for all $1 \leq i \leq N - 2$, we have $a_{i+1} - a_i = \Delta$. It is further assumed that the quantization levels are at a distance of $\frac{\Delta}{2}$ from the boundaries $a_1, a_2, \ldots, a_{N-1}$. Figure 7.3 is an example of an 8-level uniform quantizer. In a uniform quantizer, the mean squared error distortion is given by

$$
D = \int_{-\infty}^{a_1} (x - (a_1 - \Delta/2))^2 f_X(x)dx
$$

$$
+ \sum_{i=1}^{N-2} \int_{a_1+(i-1)\Delta}^{a_1+i\Delta} (x - (a_1 + i\Delta - \Delta/2))^2 f_X(x)dx
$$

$$
+ \int_{a_1+(N-2)\Delta}^{\infty} (x - (a_1 + (N-2)\Delta + \Delta/2))^2 f_X(x)dx. \qquad (7.2.7)
$$

Thus, D is a function of two design parameters, namely, a_1 and Δ. In order to design the optimal uniform quantizer, we have to differentiate D with respect to these variables and find the values that minimize D.

Minimization of distortion is generally a tedious task and is done mainly by numerical techniques. Table 7.1 gives the optimal quantization level spacing for a zero-mean unit-variance Gaussian random variable. The last column in the table gives the entropy after quantization which is discussed in Chapter 12.

Nonuniform Quantization. If we relax the condition that the quantization regions (except for the first and the last one) be of equal length, then we are minimizing the distortion with less constraints; therefore, the resulting quantizer will perform better than a uniform quantizer with the same number of levels. Let us assume that we are interested in designing the optimal mean squared error quantizer with N levels of quantization with no other constraint on the regions. The average distortion will be given by

$$
D = \int_{-\infty}^{a_1} (x - \hat{x}_1)^2 f_X(x)dx + \sum_{i=1}^{N-2} \int_{a_i}^{a_{i+1}} (x - \hat{x}_{i+1})^2 f_X(x)dx
$$

$$
+ \int_{a_{N-1}}^{\infty} (x - \hat{x}_N)^2 f_X(x)dx. \qquad (7.2.8)
$$

There exists a total of $2N - 1$ variables in this expression ($a_1, a_2, \ldots, a_{N-1}$ and $\hat{x}_1, \hat{x}_2, \ldots, \hat{x}_N$) and the minimization of D is to be done with respect to these variables. Differentiating with respect to a_i yields

$$
\frac{\partial}{\partial a_i} D = f_X(a_i)[(a_i - \hat{x}_i)^2 - (a_i - \hat{x}_{i+1})^2] = 0, \qquad (7.2.9)
$$

which results in

$$
a_i = \frac{1}{2}(\hat{x}_i + \hat{x}_{i+1}). \qquad (7.2.10)
$$

This result simply means that, in an optimal quantizer, *the boundaries of the quantization regions are the midpoints of the quantized values.* Because quantization is done on a minimum distance basis, each x value is quantized to the nearest $\{\hat{x}_i\}_{i=1}^N$.

TABLE 7.1 OPTIMAL UNIFORM QUANTIZER FOR A GAUSSIAN SOURCE

Number Output Levels N	Output-level Spacing Δ	Mean-squared Error D	Entropy $H(\hat{x})$
1	—	1.000	0.0
2	1.596	0.3634	1.000
3	1.224	0.1902	1.536
4	0.9957	0.1188	1.904
5	0.8430	0.08218	2.183
6	0.7334	0.06065	2.409
7	0.6508	0.04686	2.598
8	0.5860	0.03744	2.761
9	0.5338	0.03069	2.904
10	0.4908	0.02568	3.032
11	0.4546	0.02185	3.148
12	0.4238	0.01885	3.253
13	0.3972	0.01645	3.350
14	0.3739	0.01450	3.440
15	0.3534	0.01289	3.524
16	0.3352	0.01154	3.602
17	0.3189	0.01040	3.676
18	0.3042	0.009430	3.746
19	0.2909	0.008594	3.811
20	0.2788	0.007869	3.874
21	0.2678	0.007235	3.933
22	0.2576	0.006678	3.990
23	0.2482	0.006185	4.045
24	0.2396	0.005747	4.097
25	0.2315	0.005355	4.146
26	0.2240	0.005004	4.194
27	0.2171	0.004687	4.241
28	0.2105	0.004401	4.285
29	0.2044	0.004141	4.328
30	0.1987	0.003905	4.370
31	0.1932	0.003688	4.410
32	0.1881	0.003490	4.449
33	0.1833	0.003308	4.487
34	0.1787	0.003141	4.524
35	0.1744	0.002986	4.560
36	0.1703	0.002843	4.594

From Max (1960); © IEEE.

To determine the quantized values $\hat{x}_i$, we differentiate D with respect to $\hat{x}_i$ and define $a_0 = -\infty$ and $a_N = +\infty$. Thus, we obtain

$$\frac{\partial D}{\partial \hat{x}_i} = \int_{a_{i-1}}^{a_i} 2(x - \hat{x}_i) f_X(x) dx = 0, \qquad (7.2.11)$$

which results in

$$\hat{x}_i = \frac{\int_{a_{i-1}}^{a_i} x f_X(x)dx}{\int_{a_{i-1}}^{a_i} f_X(x)dx}. \tag{7.2.12}$$

Equation (7.2.12) shows that in an optimal quantizer, *the quantized value (or representation point) for a region should be chosen to be the centroid of that region.* Equations (7.2.10) and (7.2.12) give the necessary conditions for a scalar quantizer to be optimal; they are known as the Lloyd–Max conditions. The criteria for optimal quantization (the Lloyd–Max conditions) can then be summarized as follows:

1. The boundaries of the quantization regions are the midpoints of the corresponding quantized values (nearest neighbor law).
2. The quantized values are the centroids of the quantization regions.

Although these rules are very simple, they do not result in analytical solutions to the optimal quantizer design. The usual method of designing the optimal quantizer is to start with a set of quantization regions and then, using the second criterion, to find the quantized values. Then, we design new quantization regions for the new quantized values, and alternate between the two steps until the distortion does not change much from one step to the next. Based on this method, we can design the optimal quantizer for various source statistics. Table 7.2 shows the optimal nonuniform quantizers for various values of N for a zero-mean unit-variance Gaussian source. If, instead of this source, a general Gaussian source with mean m and variance σ^2 is used, then the values of a_i and $\hat{x}_i$ read from Table 7.2 are replaced with $m + \sigma a_i$ and $m + \sigma \hat{x}_i$, respectively, and the value of the distortion D will be replaced by $\sigma^2 D$.

Example 7.2.3

How would the results of Example 7.2.1 change if, instead of the uniform quantizer shown in Figure 7.3, we used an optimal nonuniform quantizer with the same number of levels?

Solution   We can find the quantization regions and the quantized values from Table 7.2 with $N = 8$, and then use the fact that our source is an $N(0, 400)$ source, i.e., $m = 0$ and $\sigma = 20$. Therefore, all a_i and $\hat{x}_i$ values read from the table should be multiplied by $\sigma = 20$ and the distortion has to be multiplied by 400. This gives us the values $a_1 = -a_7 = -34.96$, $a_2 = -a_6 = -21$, $a_3 = -a_5 = -10.012$, $a_4 = 0$ and $\hat{x}_1 = -\hat{x}_8 = -43.04$, $\hat{x}_2 = -\hat{x}_7 = -26.88$, $\hat{x}_3 = -\hat{x}_6 = -15.12$, $\hat{x}_4 = -\hat{x}_5 = -4.902$ and a distortion of $D = 13.816$. The SQNR is

$$\text{SQNR} = \frac{400}{13.816} = 28.95 \approx 14.62 \text{ dB},$$

which is 3.84 dB better than the SQNR of the uniform quantizer.   ∎

TABLE 7.2 OPTIMAL NON-UNIFORM QUANTIZER FOR A GAUSSIAN SOURCE

N	$\pm a_i$	$\pm \hat{x}_i$	D	$H(\hat{X})$
1	—	0	1	0
2	0	0.7980	0.3634	1
3	0.6120	0, 1.224	0.1902	1.536
4	0, 0.9816	0.4528, 1.510	0.1175	1.911
5	0.3823, 1.244	0, 0.7646, 1.724	0.07994	2.203
6	0, 0.6589, 1.447	0.3177, 1.000, 1.894	0.05798	2.443
7	0.2803, 0.8744, 1.611	0, 0.5606, 1.188, 2.033	0.04400	2.647
8	0, 0.5006, 1.050, 1.748	0.2451, 0.7560, 1.344, 2.152	0.03454	2.825
9	0.2218, 0.6812, 1.198, 1.866	0, 0.4436, 0.9188, 1.476, 2.255	0.02785	2.983
10	0, 0.4047, 0.8339, 1.325, 1.968	0.1996, 0.6099, 1.058, 1.591, 2.345	0.02293	3.125
11	0.1837, 0.5599, 0.9656, 1.436, 2.059	0, 0.3675, 0.7524, 1.179, 1.693, 2.426	0.01922	3.253
12	0, 0.3401, 0.6943, 1.081, 1.534, 2.141	0.1684, 0.5119, 0.8768, 1.286, 1.783, 2.499	0.01634	3.372
13	0.1569, 0.4760, 0.8126, 1.184, 1.623, 2.215	0, 0.3138, 0.6383, 0.9870, 1.381, 1.865, 2.565	0.01406	3.481
14	0, 0.2935, 0.5959, 0.9181, 1.277, 1.703, 2.282	0.1457, 0.4413, 0.7505, 1.086, 1.468, 1.939, 2.625	0.01223	3.582
15	0.1369, 0.4143, 0.7030, 1.013, 1.361, 1.776, 2.344	0, 0.2739, 0.5548, 0.8512, 1.175, 1.546, 2.007, 2.681	0.01073	3.677
16	0, 0.2582, 0.5224, 0.7996, 1.099, 1.437, 1.844, 2.401	0.1284, 0.3881, 0.6568, 0.9424, 1.256, 1.618, 2.069, 2.733	0.009497	3.765
17	0.1215, 0.3670, 0.6201, 0.8875, 1.178, 1.508, 1.906, 2.454	0, 0.2430, 0.4909, 0.7493, 1.026, 1.331, 1.685, 2.127, 2.781	0.008463	3.849
18	0, 0.2306, 0.4653, 0.7091, 0.9680, 1.251, 1.573, 1.964, 2.504	0.1148, 0.3464, 0.5843, 0.8339, 1.102, 1.400, 1.746, 2.181, 2.826	0.007589	3.928
19	0.1092, 0.3294, 0.5551, 0.7908, 1.042, 1.318, 1.634, 2.018, 2.55	0, 0.2184, 0.4404, 0.6698, 0.9117, 1.173, 1.464, 1.803, 2.232, 2.869	0.006844	4.002
20	0, 0.2083, 0.4197, 0.6375, 0.8661, 1.111, 1.381, 1.690, 2.068, 2.594	0.1038, 0.3128, 0.5265, 0.7486, 0.9837, 1.239, 1.524, 1.857, 2.279, 2.908	0.006203	4.074
21	0.09918, 0.2989, 0.5027, 0.7137, 0.9361, 1.175, 1.440, 1.743, 2.116, 2.635	0, 0.1984, 0.3994, 0.6059, 0.8215, 1.051, 1.300, 1.579, 1.908, 2.324, 2.946	0.005648	4.141
22	0, 0.1900, 0.3822, 0.5794, 0.7844, 1.001, 1.235, 1.495, 1.793, 2.160, 2.674	0.09469, 0.2852, 0.4793, 0.6795, 0.8893, 1.113, 1.357, 1.632, 1.955, 2.366, 2.982	0.005165	4.206

(continued overleaf)

TABLE 7.2   (*CONTINUED*)

N	$\pm a_i$	$\pm \hat{x}_i$	D	$H(\hat{X})$
23	0.09085, 0.2736, 0.4594, 0.6507, 0.8504, 1.062, 1.291, 1.546, 1.841, 2.203, 2.711	0, 0.1817, 0.3654, 0.5534, 0.7481, 0.9527, 1.172, 1.411, 1.682, 2.000, 2.406, 3.016	0.004741	4.268
24	0, 0.1746, 0.3510, 0.5312, 0.7173, 0.9122, 1.119, 1.344, 1.595, 1.885, 2.243, 2.746	0.08708, 0.2621, 0.4399, 0.6224, 0.8122, 1.012, 1.227, 1.462, 1.728, 2.042, 2.444, 3.048	0.004367	4.327
25	0.08381, 0.2522, 0.4231, 0.5982, 0.7797, 0.9702, 1.173, 1.394, 1.641, 1.927, 2.281, 2.779	0, 0.1676, 0.3368, 0.5093, 0.6870, 0.8723, 1.068, 1.279, 1.510, 1.772, 2.083, 2.480, 3.079	0.004036	4.384
26	0, 0.1616, 0.3245, 0.4905, 0.6610, 0.8383, 1.025, 1.224, 1.442, 1.685, 1.968, 2.318, 2.811	0.08060, 0.2425, 0.4066, 0.5743, 0.7477, 0.9289, 1.121, 1.328, 1.556, 1.814, 2.121, 2.514, 3.109	0.003741	4.439
27	0.07779, 0.2340, 0.3921, 0.5587, 0.7202, 0.8936, 1.077, 1.273, 1.487, 1.727, 2.006, 2.352, 2.842	0, 0.1556, 0.3124, 0.4719, 0.6354, 0.8049, 0.9824, 1.171, 1.374, 1.599, 1.854, 2.158, 2.547, 3.137	0.003477	4.491
28	0, 0.1503, 0.3018, 0.04556, 0.6132, 0.7760, 0.9460, 1.126, 1.319, 1.529, 1.766, 2.042, 2.385, 2.871	0.07502, 0.2256, 0.3780, 0.5333, 0.6930, 0.8589, 1.033, 1.118, 1.419, 1.640, 1.892, 2.193, 2.578, 3.164	0.003240	4.542

From Max (1960); © IEEE.

7.2.2 Vector Quantization

In scalar quantization, each output of the discrete-time source (which is usually the result of sampling of a continuous-time source) is quantized separately and then encoded. For example, if we are using a 4-level scalar quantizer and encoding each level into two bits, we are using two bits per each source output. This quantization scheme is shown in Figure 7.4.

Now if we consider two samples of the source at each time, and we interpret these two samples as a point in a plane, the quantizer partitions the entire plane into 16 quantization regions, as show in Figure 7.5. We can see that the regions in the two-dimensional space are all of rectangular shape. If we allow 16 regions of any shape in the two-dimensional space, we are capable of obtaining better results. This means that we are quantizing two source outputs at a time by using 16 regions, which is equivalent to four bits per two source outputs or two bits per each source output. Therefore, the number of bits per source output for quantizing two samples at a time is equal to the number of bits per source output obtained in the scalar case. Because we are relaxing the requirement of having rectangular regions, the performance may improve. Now, if we take three samples at a time and quantize the

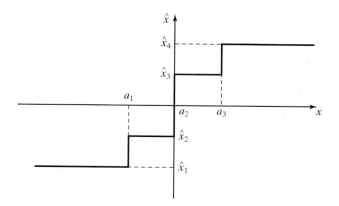

Figure 7.4   A 4-level scalar quantizer.

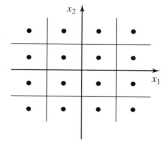

Figure 7.5   A scalar 4-level quantization applied to two samples.

entire three-dimensional space into 64 regions, we will have even less distortion with the same number of bits per source output. The idea of vector quantization is to take blocks of source outputs of length n, and design the quantizer in the n-dimensional Euclidean space, rather than doing the quantization based on single samples in a one-dimensional space.

Let us assume that the quantization regions in the n-dimensional space are denoted by $\mathcal{R}_i$, $1 \leq i \leq K$. These K regions partition the n-dimensional space. Each block of source output of length n is denoted by the n-dimensional vector $\mathbf{x} \in R^n$; if $\mathbf{x} \in \mathcal{R}_i$, it is quantized to $Q(\mathbf{x}) = \hat{\mathbf{x}}_i$. Figure 7.6 shows this quantization scheme for $n = 2$. Now, since there are a total of K quantized values, $\log K$ bits are enough to represent these values. This means that we require $\log K$ bits per n source outputs, or the rate of the source code is

$$R = \frac{\log K}{n} \quad \text{bits/source output.} \tag{7.2.13}$$

The optimal vector quantizer of dimension n and number of levels K chooses the regions $\mathcal{R}_i$'s and the quantized values $\hat{\mathbf{x}}_i$'s such that the resulting distortion is minimized. Applying the same procedure that we used for the case of scalar quantization, we obtain the following criteria for an optimal vector quantizer design:

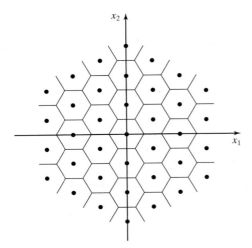

Figure 7.6   Vector quantization in two dimensions.

1. Region $\mathcal{R}_i$ is the set of all points in the n-dimensional space that are closer to $\hat{\mathbf{x}}_i$ than any other $\hat{\mathbf{x}}_j$, for all $j \neq i$; that is,

$$\mathcal{R}_i = \{\mathbf{x} \in R^n : \|\mathbf{x} - \hat{\mathbf{x}}_i\| < \|\mathbf{x} - \hat{\mathbf{x}}_j\|, \ \forall j \neq i\};$$

2. $\hat{\mathbf{x}}_i$ is the centroid of the region $\mathcal{R}_i$; that is,

$$\hat{\mathbf{x}}_i = \frac{1}{P(\mathbf{X} \in \mathcal{R}_i)} \int \cdots \int_{\mathcal{R}_i} \mathbf{x} f_{\mathbf{X}}(\mathbf{x}) d\mathbf{x}.$$

A practical approach to designing optimal vector quantizers is on the basis of the same approach employed in designing optimal scalar quantizers. Starting from a given set of quantization regions, we derive the optimal quantized vectors for these regions by using criterion 2. Then we repartition the space using the first criterion, and go back and forth until the changes in distortion are negligible.

Vector quantization has found widespread applications in speech and image coding; numerous algorithms for reducing its computational complexity have been proposed.

7.3 ENCODING

In the encoding process, a sequence of bits are assigned to different quantization values. Since there are a total of $N = 2^\nu$ quantization levels, ν bits are sufficient for the encoding process. In this way, we have ν bits corresponding to each sample; since the sampling rate is f_s samples/second, we will have a bit rate of $R = \nu f_s$ bits/second.

The assignment of bits to quantization levels can be done in a variety of ways. In scalar quantization, a natural way of encoding is to assign the values of 0 to

TABLE 7.3   NBC AND GRAY CODES FOR A 16-LEVEL QUANTIZATION

Quantization level	Level order	NBC code	Gray code
$\hat{x}_1$	0	0000	0000
$\hat{x}_2$	1	0001	0010
$\hat{x}_3$	2	0010	0011
$\hat{x}_4$	3	0011	0001
$\hat{x}_5$	4	0100	0101
$\hat{x}_6$	5	0101	0100
$\hat{x}_7$	6	0110	0110
$\hat{x}_8$	7	0111	0111
$\hat{x}_9$	8	1000	1111
$\hat{x}_{10}$	9	1001	1110
$\hat{x}_{11}$	10	1010	1100
$\hat{x}_{12}$	11	1011	1101
$\hat{x}_{13}$	12	1100	1001
$\hat{x}_{14}$	13	1101	1000
$\hat{x}_{15}$	14	1110	1010
$\hat{x}_{16}$	15	1111	1011

$N - 1$ to different quantization levels starting from the lowest level to the highest level in order of increasing level value. Then we can assign the binary expansion of the numbers 0 to $N - 1$ to these levels. Thus, v zeros are assigned to the lowest quantization level, $\underbrace{0 \ldots 0}_{v-1} 1$ to the second lowest quantization level, $\underbrace{0 \ldots 0}_{v-2} 10$ to the next level, $\ldots$ and $\underbrace{1 \ldots 1}_{v}$ to the highest quantization level. This type of encoding is called *natural binary coding* or NBC for short. Another approach to coding is to encode the quantized levels in a way that adjacent levels differ only in one bit. This type of coding is called Gray Coding.

Table 7.3 gives an example of NBC and Gray coding for a quantizer with $N = 16$ levels.

7.4 WAVEFORM CODING

Waveform coding schemes are designed to reproduce the waveform output of the source at the destination with as little distortion as possible. In these techniques, no attention is paid to the mechanism that produces the waveform; all attempts are directed at reproducing the source output at the destination with high fidelity. The structure of the source plays no role in the design of waveform coders and only properties of the waveform affect the design. Thus, waveform coders are robust and can be used with a variety of sources as long as the waveforms produced by the sources have certain similarities. In this section, we study some basic waveform coding methods that are widely applied to a variety of sources.

7.4.1 Pulse Code Modulation (PCM)

Pulse code modulation is the simplest and oldest waveform coding scheme. A pulse code modulator consists of three basic sections: a sampler, a quantizer and an encoder. A functional block diagram of a PCM system is shown in Figure 7.7. In PCM, we make the following assumptions:

1. The waveform (signal) is bandlimited with a maximum frequency of W. Therefore, it can be fully reconstructed from samples taken at a rate of $f_s = 2W$ or higher.
2. The signal is of finite amplitude. In other words, there exists a maximum amplitude x_{max} such that for all t, we have $|x(t)| \leq x_{max}$.
3. The quantization is done with a large number of quantization levels N, which is a power of 2 ($N = 2^v$).

The waveform entering the sampler is a bandlimited waveform with the bandwidth W. Usually, there exists a filter with bandwidth W prior to the sampler to prevent any components beyond W from entering the sampler. This filter is called the presampling filter. The sampling is done at a rate higher than the Nyquist rate; this allows for some guardband. The sampled values then enter a scalar quantizer. The quantizer is either a uniform quantizer, which results in a uniform PCM system, or a nonuniform quantizer. The choice of the quantizer is based on the characteristics of the source output. The output of the quantizer is then encoded into a binary sequence of length v, where $N = 2^v$ is the number of quantization levels.

Uniform PCM.   In uniform PCM, we assume that the quantizer is a uniform quantizer. Since the range of the input samples is $[-x_{max}, +x_{max}]$ and the number of quantization levels is N, the length of each quantization region is given by

$$\Delta = \frac{2x_{max}}{N} = \frac{x_{max}}{2^{v-1}}. \tag{7.4.1}$$

$$x(t) \longrightarrow \boxed{\text{Sampler}} \xrightarrow{\{x_n\}} \boxed{\text{Quantizer}} \xrightarrow{\{\hat{x}_n\}} \boxed{\text{Encoder}} \xrightarrow{\dots 0\,1\,1\,0\dots}$$

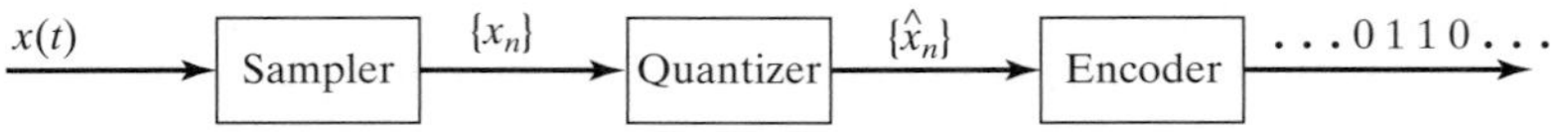

Figure 7.7   Block diagram of a PCM system.

The quantized values in uniform PCM are chosen to be the midpoints of the quantization regions; therefore, the error $\tilde{x} = x - Q(x)$ is a random variable taking values in the interval $(-\frac{\Delta}{2}, +\frac{\Delta}{2}]$. In ordinary PCM applications, the number of levels (N) is usually high and the range of variations of the input signal (amplitude variations x_{max}) is small. This means that the length of each quantization region (Δ) is small. Under these assumptions, in each quantization region, the error $\tilde{X} = X - Q(X)$ can be approximated by a uniformly distributed random variable on $(-\frac{\Delta}{2}, +\frac{\Delta}{2}]$. In other words,

$$f(\tilde{x}) = \begin{cases} \frac{1}{\Delta} & -\frac{\Delta}{2} \leq \tilde{x} \leq \frac{\Delta}{2} \\ 0 & \text{otherwise} \end{cases}.$$

The distortion introduced by quantization (quantization noise) is therefore

$$E[\tilde{X}^2] = \int_{-\frac{\Delta}{2}}^{+\frac{\Delta}{2}} \frac{1}{\Delta} \tilde{x}^2 d\tilde{x} = \frac{\Delta^2}{12} = \frac{x_{max}^2}{3N^2} = \frac{x_{max}^2}{3 \times 4^\nu}, \qquad (7.4.2)$$

where ν is the number of bits/source sample and we have employed Equation (7.4.1). The signal-to-quantization noise ratio then becomes

$$\text{SQNR} = \frac{P_X}{\tilde{X}^2} = \frac{3 \times N^2 P_X}{x_{max}^2} = \frac{3 \times 4^\nu P_X}{x_{max}^2}, \qquad (7.4.3)$$

where P_X is the power in each sample. In the case where $X(t)$ is a wide-sense stationary process, P_X can be found using any of the following relations:

$$P_X = R_X(\tau)|_{\tau=0}$$

$$= \int_{-\infty}^{\infty} S_X(f)\, df$$

$$= \int_{-\infty}^{\infty} x^2 f_X(x)\, dx.$$

Note that since x_{max} is the maximum possible value for X, we always have $P_X = E[X^2] \leq x_{max}^2$. This means that $\frac{P_X}{x_{max}^2} < 1$ (usually $\frac{P_X}{x_{max}^2} \ll 1$); hence, $3N^2 = 3 \times 4^\nu$ is an upperbound to the SQNR in uniform PCM. This also means that SQNR in uniform PCM deteriorates as the dynamic range of the source increases because an increase in the dynamic range of the source results in a decrease in $\frac{P_X}{x_{max}^2}$.

Expressing SQNR in decibels, we obtain

$$\text{SQNR}\Big|_{\text{dB}} \approx 10 \log_{10} \frac{P_X}{x_{max}^2} + 6\nu + 4.8. \qquad (7.4.4)$$

We can see that each extra bit (increase in ν by one) increases the SQNR by 6 dB. This is a very useful strategy for estimating how many extra bits are required to achieve a desired SQNR.

Example 7.4.1

What is the resulting SQNR for a signal uniformly distributed on $[-1, 1]$, when uniform PCM with 256 levels is employed?

Solution We have $P_X = \int_{-1}^{1} \frac{1}{2} x^2 dx = \frac{1}{3}$. Therefore, using $x_{\max} = 1$ and $v = \log 256 = 8$, we have

$$\text{SQNR} = 3 \times 4^v \times P_X = 4^v = 65536 \approx 48.16 \text{ dB}.$$ ■

The issue of bandwidth requirements of pulse transmission systems, of which PCM is an example, is dealt with in detail in Chapter 9. In this chapter, we briefly discuss some results concerning the bandwidth requirements of a PCM system. If a signal has a bandwidth of W, then the minimum number of samples for perfect reconstruction of the signal is given by the sampling theorem, and it is equal to $2W$ samples/sec. If some guardband is required, then the number of samples per second is f_s, which is more than $2W$. For each sample, v bits are used; therefore, a total of vf_s bits/sec are required for transmission of the PCM signal. In the case of sampling at the Nyquist rate, this is equal to $2vW$ bits/sec. The minimum bandwidth requirement for binary transmission of R bits/sec (or, more precisely, R pulses/sec) is $\frac{R}{2}$. (See Chapter 9.)[2] Therefore, the minimum bandwidth requirement of a PCM system is

$$\text{BW}_{\text{req}} = \frac{vf_s}{2}, \tag{7.4.5}$$

which, in the case of sampling at the Nyquist rate, gives the absolute minimum bandwidth requirement as

$$\text{BW}_{\text{req}} = vW. \tag{7.4.6}$$

This means that a PCM system expands the bandwidth of the original signal by a factor of at least v.

Nonuniform PCM.
As long as the statistics of the input signal are close to the uniform distribution, uniform PCM works fine. However, in coding of certain signals such as speech, the input distribution is far from uniformly distributed. For a speech waveform, in particular, there exists a higher probability for smaller amplitudes and a lower probability for larger amplitudes. Therefore, it makes sense to design a quantizer with more quantization regions at lower amplitudes and fewer quantization regions at larger amplitudes. The resulting quantizer will be a nonuniform quantizer that has quantization regions of various sizes.

The usual method for performing nonuniform quantization[3] is to first pass the samples through a nonlinear element that compresses the large amplitudes (reduces the dynamic range of the signal) and then performs a uniform quantization on

---

[2]A more practical bandwidth requirement is $\frac{R}{\alpha}$, where $1 < \alpha < 2$.

[3]Sometimes, the term *nonlinear quantization* is used. This is misleading, because all quantization schemes, uniform or nonuniform, are nonlinear.

Figure 7.8   Block diagram of a nonuniform PCM system.

the output. At the receiving end, the inverse (expansion) of this nonlinear operation is applied to obtain the sampled value. This technique is called *companding* (*comp*ressing–exp*anding*). A block diagram of this system is shown in Figure 7.8.

There are two types of companders that are widely used for speech coding. The μ-law compander, used in the United States and Canada, employs the logarithmic function at the transmitting side, where $|x| \le 1$:

$$g(x) = \frac{\log(1 + \mu|x|)}{\log(1 + \mu)}\, \text{sgn}(x). \tag{7.4.7}$$

The parameter μ controls the amount of compression and expansion. The standard PCM system in the United States and Canada employs a compressor with $\mu = 255$ followed by a uniform quantizer with 8 bits/sample. Use of a compander in this system improves the performance of the system by about 24 dB. Figure 7.9 illustrates the μ-law compander characteristics for $\mu = 0, 5$, and 255.

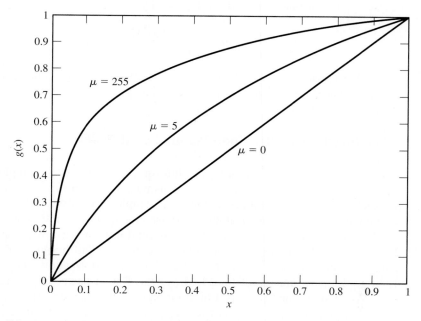

Figure 7.9   A graph of μ-law compander characteristics.

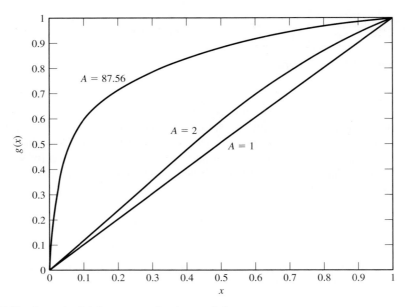

Figure 7.10   A graph of A-law compander characteristics.

The second widely used logarithmic compressor is the A-law compander. The characteristics of this compander are given by

$$g(x) = \frac{1 + \log A|x|}{1 + \log A} \, \text{sgn}(x), \tag{7.4.8}$$

where A is chosen to be 87.56. The performance of this compander is comparable to the performance of the μ-law compander. Figure 7.10 illustrates the characteristics of this compander for $A = 1, 2$, and 87.56.

7.4.2 Differential Pulse Code Modulation (DPCM)

In a PCM system, after sampling the information signal, each sample is quantized independently using a scalar quantizer. This means that previous sample values have no effect on the quantization of the new samples. However, when a bandlimited random process is sampled at the Nyquist rate or faster, the sampled values are usually correlated random variables. The exception is the case when the spectrum of the process is flat within its bandwidth. This means that the previous samples give some information about the next sample; thus, this information can be employed to improve the performance of the PCM system. For instance, if the previous sample values were small, and there is a high probability that the next sample value will be small as well, then it is not necessary to quantize a wide range of values to achieve a good performance.

In the simplest form of differential pulse code modulation (DPCM), the difference between two adjacent samples is quantized. Because two adjacent samples are highly correlated, their difference has small variations; therefore, to achieve a certain level of performance, fewer levels (and therefore fewer bits) are required to quantize it. This means that DPCM can achieve performance levels at lower bit rates than PCM.

Figure 7.11 shows a block diagram of this simple DPCM scheme. As seen in the figure, the input to the quantizer is not simply $X_n - X_{n-1}$ but rather $X_n - \hat{Y}'_{n-1}$. We will see that $\hat{Y}'_{n-1}$ is closely related to X_{n-1}, and this choice has an advantage because the accumulation of quantization noise is prevented. The input to the quantizer Y_n is quantized by a scalar quantizer (uniform or nonuniform) to produce $\hat{Y}_n$. Using the relations

$$Y_n = X_n - \hat{Y}'_{n-1} \tag{7.4.9}$$

and

$$\hat{Y}'_n = \hat{Y}_n + \hat{Y}'_{n-1}, \tag{7.4.10}$$

we obtain the quantization error between the input and the output of the quantizer as

$$\hat{Y}_n - Y_n = \hat{Y}_n - (X_n - \hat{Y}'_{n-1})$$
$$= \hat{Y}_n - X_n + \hat{Y}'_{n-1}$$
$$= \hat{Y}'_n - X_n. \tag{7.4.11}$$

At the receiving end, we have

$$\hat{X}_n = \hat{Y}_n + \hat{X}_{n-1}. \tag{7.4.12}$$

Comparing Equation (7.4.10) and Equation (7.4.12) we see that $\hat{Y}'_n$ and $\hat{X}_n$ satisfy the same difference equation with the same excitation function ($\hat{Y}_n$). Therefore, if the initial conditions of $\hat{Y}'_n$ and $\hat{X}_n$ are chosen to be the same, they will be equal.

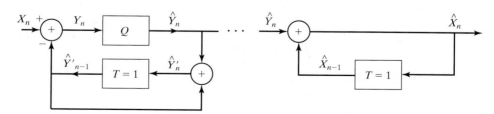

Figure 7.11 A simple DPCM encoder and decoder.

For instance, if we let $\hat{Y}'_{-1} = \hat{X}_{-1} = 0$, then all n will have $\hat{Y}'_n = \hat{X}_n$. Substituting this in Equation (7.4.11), we obtain

$$\hat{Y}_n - Y_n = \hat{X}_n - X_n. \tag{7.4.13}$$

This shows that the quantization error between X_n and its reproduction $\hat{X}_n$ is the same as the quantization error between the input and the output of the quantizer. However, the range of variations of Y_n is usually much smaller than that of X_n; therefore, Y_n can be quantized with fewer bits.

7.4.3 Delta Modulation (DM)

Delta modulation is a simplified version of the DPCM system shown in Figure 7.11. In delta modulation, the quantizer is a one-bit (two-level) quantizer with magnitudes $\pm\Delta$. A block diagram of a DM system is shown in Figure 7.12. The same analysis that was applied to the simple DPCM system is valid here.

In delta modulation only one bit per sample is employed, so the quantization noise will be high unless the dynamic range of Y_n is very low. This, in turn, means that X_n and X_{n-1} must have a very high correlation coefficient. To have a high correlation between X_n and X_{n-1}, we have to sample at rates much higher than the Nyquist rate. Therefore, in DM, the sampling rate is usually much higher than the Nyquist rate, but since the number of bits per sample is only one, the total number of bits per second required to transmit a waveform is lower than that of a PCM system.

A major advantage of delta modulation is the very simple structure of the system. At the receiving end, we have the following relation for the reconstruction of $\hat{X}_n$:

$$\hat{X}_n - \hat{X}_{n-1} = \hat{Y}_n. \tag{7.4.14}$$

Solving this equation for $\hat{X}_n$, and assuming zero initial conditions, we obtain

$$\hat{X}_n = \sum_{i=0}^{n} \hat{Y}_i. \tag{7.4.15}$$

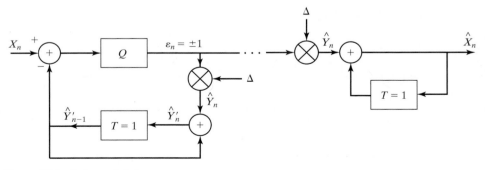

Figure 7.12   Delta modulation.

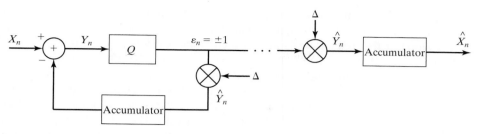

Figure 7.13   Delta modulation with integrators.

This means that to obtain $\hat{X}_n$, we only have to accumulate the values of $\hat{Y}_n$. If the sampled values are represented by impulses, the accumulator will be a simple integrator. This simplifies the block diagram of a DM system, as shown in Figure 7.13.

The step size Δ is a very important parameter in designing a delta modulator system. Large values of Δ cause the modulator to follow rapid changes in the input signal; but at the same time, they cause excessive quantization noise when the input changes slowly. This case is shown in Figure 7.14. For large Δ, when the input varies slowly, a large quantization noise occurs; this is known as *granular noise*. The case of a too small Δ is shown in Figure 7.15. In this case, we have a problem with rapid changes in the input. When the input changes rapidly (high-input slope), it takes a

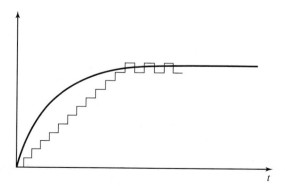

Figure 7.14   Large Δ and granular noise.

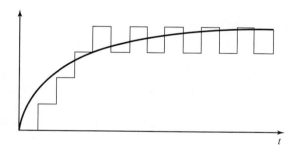

Figure 7.15   Small Δ and slope overload distortion.

rather long time for the output to follow the input, and an excessive quantization noise is caused in this period. This type of distortion, which is caused by the high slope of the input waveform, is called *slope overload distortion*.

Adaptive Delta Modulation. We have seen that a step size that is too large causes granular noise, and a step size too small results in slope overload distortion. This means that a good choice for Δ is a "medium" value; but in some cases, the performance of the best medium value (i.e., the one minimizing the mean squared distortion) is not satisfactory. An approach that works well in these cases is to change the step size according to changes in the input. If the input tends to change rapidly, the step size must be large so that the output can follow the input quickly and no slope overload distortion results. When the input is more or less flat (slowly varying), the step size changed to a small value to prevent granular noise. Such changes in the step size are shown in Figure 7.16.

To adaptively change the step size, we have to design a mechanism for recognizing large and small input slopes. If the slope of the input is small, the output of the quantizer $\hat{Y}$ alternates between Δ and $-\Delta$, as shown in Figure 7.16. This is the case where granular noise is the main source of noise, and we have to decrease the step size. However, in the case of slope overload, the output cannot follow the input rapidly and the output of the quantizer will be a succession of $+\Delta$'s or $-\Delta$'s. We can see that the sign of two successive $\hat{Y}_n$'s is a good criterion for changing the step size. If the two successive outputs have the same sign, the step size should be increased; if they are of opposite signs, it should be decreased.

A particularly simple rule to change the step size is given by

$$\Delta_n = \Delta_{n-1} K^{\epsilon_n \times \epsilon_{n-1}}, \tag{7.4.16}$$

where ϵ_n is the output of the quantizer before being scaled by the step size and K is some constant larger than one. It has been verified that in the 20–60 kilobits/sec range, with a choice of $K = 1.5$, the performance of adaptive delta modulation systems is 5–10 dB better than the performance of delta modulation when applied to speech sources.

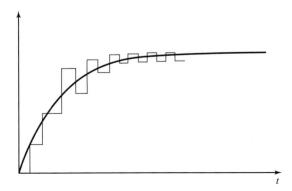

Figure 7.16 Performance of adaptive delta modulation.

7.5 ANALYSIS–SYNTHESIS TECHNIQUES

In contrast to waveform coding, analysis–synthesis techniques are methods that are based on a model for the mechanism that produces the waveform. The parameters of the model that are extracted from the source waveform are quantized, encoded, and transmitted to the receiving end. At the receiving end, based on the received information, the same model is synthesized and used to generate an output similar to the original waveform. These systems are mainly used for speech coding. In this section, we will briefly treat a system known as *linear predictive coding* (LPC).

Speech is produced as a result of excitation of the vocal tract by the vocal cords. This mechanism can be modeled as a time-varying filter (the vocal tract) excited by a signal generator. The vocal tract is a combination of the throat, the mouth, the tongue, the lips, and the nose. They change shape during generation of speech; therefore, the vocal tract is modeled as a time-varying system. The properties of the excitation signal highly depend on the type of speech sounds; they can be either *voiced* or *unvoiced*. For voiced speech, the excitation can be modeled as a periodic sequence of impulses at a frequency f_0, the value of which depends on the speaker. The reciprocal $\frac{1}{f_0}$ is called the *pitch period*. For unvoiced speech, the excitation is well modeled as a white noise. This model is shown in Figure 7.17. The vocal tract filter is usually modeled as an all-pole filter described by the difference equation

$$X_n = \sum_{i=1}^{p} a_i X_{n-i} + G w_n, \tag{7.5.1}$$

where w_n denotes the input sequence (white noise or impulses), G is a gain parameter, $\{a_i\}$ are the filter coefficients, and p is the number of poles of the filter. The process w_n, which represents the part of X_n that is not contained in the previous p samples, is called the *innovation process*.

Speech signals are known to be stationary for short periods of time, such as 20–30 msec. This characteristic behavior follows from the observation that the vocal tract cannot change instantaneously. Hence, over 20–30 msec intervals, the all-pole filter coefficients may be assumed to be fixed. At the encoder, we observe a 20–30 msec record of speech from which we estimate the model parameters $\{a_i\}$, the

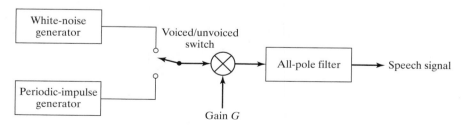

Figure 7.17   Model for speech generation mechanism.

type of excitation signal (white noise or impulse), the pitch period $\frac{1}{f_0}$ if the speech is voiced, and the gain parameter G.

To elaborate on this process, suppose that the speech signal is filtered to 3 kHz and sampled at a rate of 8000 samples/sec. The signal samples are subdivided into blocks of 160 samples, corresponding to 20 msec intervals. Let $\{x_n, 1 \leq n \leq 160\}$ be the sequence of samples for a block. The encoder must measure the model parameters to be transmitted to the receiver.

Linear prediction is used to determine the filter coefficients at the encoder. A linear predictor of order p is an all-zero digital filter with input $\{x_n\}$ and output

$$\hat{x}_n = \sum_{k=1}^{p} a_k x_{n-k} \qquad \text{for } 1 \leq n \leq N, \tag{7.5.2}$$

where we have assumed that $x_n = 0$ is outside the interval of interest. Figure 7.18 illustrates the functional block diagram for the prediction process. The difference between the actual speech sample x_n and the predicted value $\hat{x}_n$ constitutes the prediction error e_n, i.e.,

$$e_n = x_n - \hat{x}_n$$

$$= x_n - \sum_{k=1}^{p} a_k x_{n-k}. \tag{7.5.3}$$

In order to extract as much information as possible from the previous values of X_n, we choose the coefficients $\{a_i\}$ so that the average of the squared error terms, i.e.,

$$\mathcal{E}_p = \frac{1}{N} \sum_{n=1}^{N} e_n^2$$

$$= \frac{1}{N} \sum_{n=1}^{N} \left(x_n - \sum_{k=1}^{p} a_k x_{n-k} \right)^2, \tag{7.5.4}$$

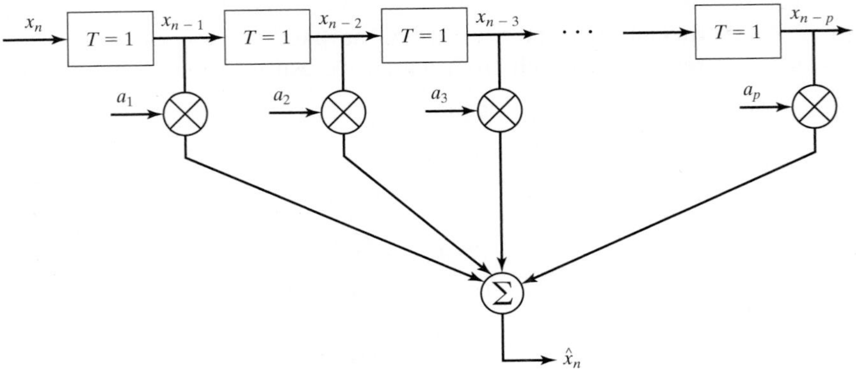

Figure 7.18   Functional block diagram of linear prediction.

is minimized. Differentiating $\mathscr{E}_p$ with respect to each of the prediction filter coefficients $\{a_i\}$ and setting the derivative to zero, we obtain a set of linear equations for the filter coefficients, i.e.,

$$\frac{1}{N} \sum_{n=1}^{N} x_n x_{n-i} = \frac{1}{N} \sum_{n=1}^{N} \sum_{k=1}^{p} a_k x_{n-i} x_{n-k} \qquad \text{for } 1 \leq i \leq p. \tag{7.5.5}$$

Since we have assumed that outside the stationary interval, $1 \leq n \leq N$, we have $x_n = 0$, we can write the preceding relation as

$$\frac{1}{N} \sum_{n=-\infty}^{\infty} x_n x_{n-i} = \frac{1}{N} \sum_{n=-\infty}^{\infty} \sum_{k=1}^{p} a_k x_{n-i} x_{n-k}$$

$$= \sum_{k=1}^{p} a_k \left[\frac{1}{N} \sum_{n=-\infty}^{\infty} x_{n-i} x_{n-k} \right]. \tag{7.5.6}$$

Now if we define

$$\hat{R}_i = \frac{1}{N} \sum_{n=-\infty}^{\infty} x_n x_{n-i}, \tag{7.5.7}$$

we can write Equation (7.5.6) as

$$\hat{R}_i = \sum_{k=1}^{p} a_k \hat{R}_{i-k} \qquad \text{for } 1 \leq i \leq p. \tag{7.5.8}$$

These equations are called the Yule–Walker equations. We can express them in the matrix form

$$r = \hat{R}a, \tag{7.5.9}$$

where a is the vector of the linear predictor coefficients, $\hat{R}$ is a $p \times p$ matrix whose $(i, j)^{\text{th}}$ element is $\hat{R}_{i-j}$, and r is a vector whose components are $\hat{R}_i$'s. It can be easily verified from the definition of $\hat{R}_i$ that

$$\hat{R}_i = \hat{R}_{-i}; \tag{7.5.10}$$

therefore, the matrix $\hat{R}$ is a symmetric matrix. Also, it is obvious that all elements of $\hat{R}$ that are on a parallel line to the diagonal elements are equal. Such a matrix is called a *Toeplitz matrix*, and efficient recursive algorithms exist for finding its inverse and, thus, solving Equation (7.5.9) for the vector of predictor coefficients. One such algorithm is the well-known Levinson–Durbin algorithm. Refer to the references at the end of this chapter for the details of this algorithm.

   For the optimal choice of the predictor coefficients, the squared error term can be shown to be

$$\mathscr{E}_p^{\min} = \hat{R}_0 - \sum_{k=1}^{p} \hat{R}_k. \tag{7.5.11}$$

According to the speech production model,

$$\mathscr{E}_p^{\min} = \frac{1}{N} \sum_{n=1}^{N} \left[x_n - \sum_{k=1}^{p} a_k x_{n-k} \right]^2$$

$$= G^2 \frac{1}{N} \sum_{n=1}^{N} w_n^2. \tag{7.5.12}$$

If we normalize the excitation sequence $\{w_n\}$ such that $\frac{1}{N} \sum_{n=1}^{N} w_n^2 = 1$, we obtain the value of the gain parameter as

$$G = \sqrt{\mathscr{E}_p^{\min}}. \tag{7.5.13}$$

The estimation of the type of excitation (impulsive or noise), as well as the estimate of the pitch period $\frac{1}{f_0}$ (when the excitation consists of impulses), may be accomplished by various algorithms. One simple approach is to transform the speech data into the frequency domain and look for sharp peaks in the signal spectrum. If the spectrum exhibits peaks at some fundamental frequency f_0, the excitation is taken to be a periodic impulse train with the period $\frac{1}{f_0}$. If the spectrum of the speech samples exhibit no sharp peaks, the excitation is taken as white noise.

The prediction filter coefficients, gain, voiced–unvoiced information, and pitch $\frac{1}{f_0}$ are quantized and transmitted to the receiver for each block of sampled speech. The speech signal is synthesized from these parameters using the system model shown in Figure 7.17. Typically, the voiced–unvoiced information requires one bit, the pitch frequency is represented by 6 bits, the gain parameter can be represented by 5 bits using logarithmic companding, and the prediction coefficients require 8–10 bits/coefficient. Based on linear predictive coding, speech can be compressed to bit rates as low as 2400 bits/sec. We could alternatively use vector quantization when quantizing the LPC parameters. This would further reduce the bit rate. In contrast, PCM applied to speech has a bit rate of 64,000 bits/sec.

LPC is widely used in speech coding to reduce the bandwidth. By vector quantizing the LPC parameters, good quality speech can be achieved at bit rates of about 4800 bits/sec. One version of LPC with vector quantization has been adopted as a standard for speech compression in mobile (cellular) telephone systems. Efficient speech coding is a very active area for research, and we expect to see further reduction in the bit rate of commercial speech encoders over the coming years.

7.6 DIGITAL AUDIO TRANSMISSION AND DIGITAL AUDIO RECORDING

Audio signals constitute a large part of our daily communications. Today, thousands of radio stations broadcast audio signals in analog form. The quality of voice signal broadcasting is generally acceptable, as long as the voice signal is intelligible. On the other hand, the quality of music signals that are broadcast via AM radio is relatively low fidelity because the bandwidth of the transmitted signal is restricted through regulation (by the Federal Communication Commission in the United States). The FM radio broadcast of analog signals provides higher fidelity by using a significantly larger channel bandwidth for signal transmission. Commercial radio broadcasting of audio signals in digital form has already begun with the advent of satellite radio systems.

In the transmission of audio signals on telephone channels, the conversion from analog-to-digital transmission, which has been taking place over the past several decades, is now nearly complete. We will describe some of the current developments in the digital encoding of audio signals for telephone transmission.

The entertainment industry has experienced the most dramatic changes and benefits in the conversion of analog audio signals to digital form. The development of the compact disc (CD) player and the digital audio tape recorder have rendered the previous analog recording systems technically obsolete. We shall use the CD player as a case study of the sophisticated source encoding/decoding and channel encoding/decoding methods that have been developed over the past few years for digital audio systems.

7.6.1 Digital Audio in Telephone Transmission Systems

Nearly all of the transmission of speech signals over telephone channels is currently digital. The encoding of speech signals for transmission over telephone channels has been a topic of intense research for over 50 years. A wide variety of methods for speech source encoding have been developed over the years; many of these methods are in use today.

The general configuration for a speech signal encoder is shown in Figure 7.19. Because the frequency content of speech signals is limited to below 3200 Hz, the speech signal is first passed through an antialiasing lowpass filter and then sampled. To ensure that aliasing is negligible, a sampling rate of 8000 Hz or higher is typically selected. The analog samples are then quantized and represented in digital form for transmission over telephone channels.

PCM and DPCM are widely used waveform encoding methods for digital speech transmission. Logarithmic $\mu = 255$ compression, given by Equation (7.4.7),

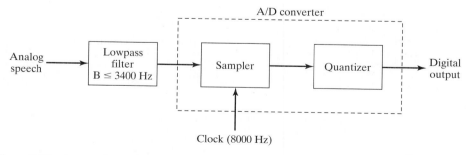

Figure 7.19   A general speech encoder.

is generally used for achieving nonuniform quantization. The typical bit rate for PCM is 64,000 bps, while the typical bit rate for DPCM is 32,000 bps.

PCM and DPCM encoding and decoding are generally performed in a telephone central office, where telephone lines from subscribers in a common geographical area are connected to the telephone transmission system. The PCM or DPCM encoded speech signals are transmitted from one telephone central office to another in digital form over so-called *trunk lines*, which are capable of carrying the digitized speech signals of many subscribers. The method for simultaneous transmission of several signals over a common communication channel is called *multiplexing*. In the case of PCM and DPCM transmission, the signals from different subscribers are multiplexed in time; hence, we have the term *time-division multiplexing* (TDM). In TDM, a given time interval T_f is selected as a frame. Each frame is subdivided into N subintervals of duration T_f/N, where N corresponds to the number of users who will use the common communication channel. Then, each subscriber who wishes to use the channel for transmission is assigned a subinterval within each frame. In PCM, each user transmits one 8-bit sample in each subinterval.

In digital speech transmission over telephone lines via PCM, there is a standard TDM hierarchy that has been established for accommodating multiple subscribers. In the first level of the TDM hierarchy, 24 digital subscriber signals are time-division multiplexed into a single high-speed data stream of 1.544 Mbps (24 × 64 kbps plus a few additional bits for control purposes). The resulting combined TDM signal is usually called a DS-1 channel. In the second level of TDM, four DS-1 channels are multiplexed into a DS-2 channel, each having the bit rate of 6.312 Mbps. In a third level of hierarchy, seven DS-2 channels are combined via TDM to produce a DS-3 channel, which has a bit rate of 44.736 Mbps. Beyond DS-3, there are two more levels of TDM hierarchy. Figure 7.20 illustrates the TDM hierarchy for the North American telephone system.

In mobile cellular radio systems (see Section 4.5 for a description) for transmission of speech signals, the available bit rate per user is small and cannot support the high bit rates required by waveform encoding methods, such as PCM and DPCM. For this application, the analysis–synthesis method based on linear predictive coding

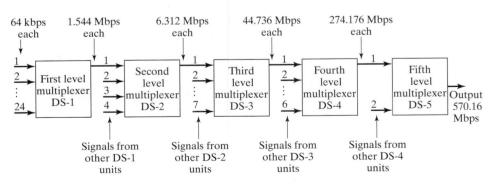

Figure 7.20   The TDM hierarchy for the North American telephone system.

(LPC), as described in Section 7.5, is used to estimate the set of model parameters from short segments of the speech signal. The speech model parameters are then transmitted over the channel using vector quantization. Thus, a bit rate in the range of 4800–9600 bps is achieved with LPC.

In mobile cellular communication systems, the base station in each cell serves as the interface to the terrestrial telephone system. LPC speech compression is only required for the radio transmission between the mobile subscriber and the base station in any cell. At the base station interface, the LPC-encoded speech is converted to PCM or DPCM for transmission over the terrestrial telephone system at a bit rate of 64,000 bps or 32,000 bps, respectively. Hence, we note that a speech signal transmitted from a mobile subscriber to a fixed subscriber will undergo two different types of encoding; however a speech signal communication between two mobiles serviced by different base stations, connected via the terrestrial telephone system, will undergo four encoding and decoding operations.

7.6.2 Digital Audio Recording

Audio recording became a reality with the invention of the phonograph during the second half of the nineteenth century. The phonograph had a lifetime of approximately 100 years, before it was supplanted by the compact disc that was introduced in 1982. During the 100-year period, we witnessed the introduction of a wide variety of records, the most popular of which proved to be the long-playing (LP) record that was introduced in 1948. LP records provide relatively high-quality analog audio recording.

In spite of their wide acceptance and popularity, analog audio recordings have a number of limitations, including a limited dynamic range (typically about 70 dB) and a relatively low signal-to-noise ratio (typically about 60 dB). By comparison, the dynamic range of orchestral music is in the range of 100–120 dB. This means that if we record the music in analog form, at low music levels, noise will be audible and, if we wish to prevent this noise, saturation will occur at high music levels.

Digital audio recording and playback allows us to improve the fidelity of recorded music by increasing the dynamic range and the signal-to-noise ratio. Furthermore, digital recordings are generally more durable and do not deteriorate with playing time, as do analog recordings. Next, we will describe a compact disc (CD) system as an example of a commercially successful digital audio system that was introduced in 1982. Table 7.4 provides a comparison of some important specifications of an LP record and a CD system. The advantages of the latter are clearly evident.

From a systems point of view, the CD system embodies most of the elements of a modern digital communication system. These include analog-to-digital (A/D) and digital-to-analog (D/A) conversion, interpolation, modulation/demodulation, and channel coding/decoding. A general block diagram of the elements of a CD digital audio system are illustrated in Figure 7.21. Now, we will describe the main features of the source encoder and decoder.

The two audio signals from the left (L) and right (R) microphones in a recording studio or a concert hall are sampled and digitized by passing them through an A/D

TABLE 7.4   COMPARISON OF AN LP RECORD AND A CD SYSTEM

Specification/Feature	LP record	CD system
Frequency response	30 Hz–20 kHZ ±3 dB	20 Hz–20 kHz +0.5/−1 dB
Dynamic range	70 dB 1 kHz	>90 dB
Signal-to-noise ratio	60 dB	>90 dB
Harmonic distortion	1%–2%	0.005%
Durability	High-frequency response degrades with playing	Permanent
Stylus life	500–600 hours	5000 hours

converter. Recall that the frequency band of audible sound is limited to approximately 20 kHz. Therefore, the corresponding Nyquist sampling rate is 40 kHz. To allow for some frequency guardband and to prevent aliasing, the sampling rate in a CD system has been selected to be 44.1 kHz. This frequency is compatible with video recording equipment that is commonly used for the digital recording of audio signals on magnetic tape.

The samples of both the L and R signals are quantized using uniform PCM with 16 bits/sample. According to the formula for SQNR given by Equation (7.4.4), 16-bit uniform quantization results in an SQNR of over 90 dB. In addition, the total harmonic distortion achieved is 0.005%. The PCM bytes from the digital recorder are encoded to provide protection against channel errors in the readback process and passed to the modulator.

At the modulator, digital control and display information is added, including a table of contents of the disc. This information allows for programmability of the CD player.

Using a laser, the digital signal from the modulator is optically recorded on the surface of a glass disc that is coated with photoresist. This results in a master disc, which is used to produce CD's by a series of processes that ultimately convert the information into tiny pits on the plastic disc. The disc is coated with a reflective aluminum coating and then with a protective lacquer.

In the CD player, a laser is used to optically scan a track on the disc at a constant velocity of 1.25 m/sec and, thus, to read the digitally recorded signal. After the L and R signal are demodulated and passed through the channel decoder, the digital audio signal is converted back to an analog audio signal by means of a D/A converter.

The conversion of L and R digital audio signals into the D/A converter has a precision of 16 bits. In principle, the digital-to-analog conversion of the two 16-bit signals at the 44.1 kHz sampling rate is relatively simple. However, the practical

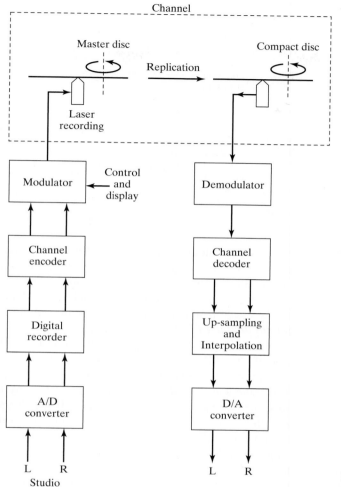

Figure 7.21   CD Digital Audio System.

implementation of a 16-bit D/A converter is very expensive. On the other hand, inexpensive D/A converters with 12-bit (or less) precision are readily available. The problem is to devise a method for D/A conversion that employs low precision and, hence, results in a low-cost D/A converter, while maintaining the 16-bit precision of the digital audio signal.

The practical solution to this problem is to expand the bandwidth of the digital audio signal by oversampling through interpolation and digital filtering prior to analog conversion. The basic approach is shown in the block diagram given in Figure 7.22. The 16-bit L and R digital audio signals are up-sampled by some multiple U by inserting $U - 1$ zeros between successive 16-bit signal samples. This process effectively increases the sampling rate to $U \times 44.1$ kHz. The high rate L and

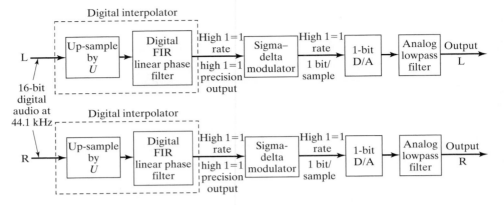

Figure 7.22   Oversampling and digital filtering.

R signals are then filtered by a finite-duration impulse response (FIR) digital filter, which produces a high-rate, high-precision output. The combination of up-sampling and filtering is a practical method for realizing a *digital interpolator*. The FIR filter is designed to have a linear phase and a bandwidth of approximately 20 kHz. It serves the purpose of eliminating the spectral images created by the up-sampling process and is sometimes called an *antiimaging filter*.

If we observe the high sample rate and high precision of the L and R digital audio signals of the output of the FIR digital filter, we will find that successive samples are nearly the same; they differ only in the low-order bits. Consequently, it is possible to represent successive samples of the digital audio signals by their differences and, thus, to reduce the dynamic range of the signals. If the oversampling factor U is sufficiently large, delta modulation may be employed to reduce the quantized output to a precision of 1 bit/sample. Thus, the D/A converter is considerably simplified. An oversampling factor $U = 256$ is normally chosen in practice. This raises the sampling rate to 11.2896 MHz.

Recall that the general configuration for the conventional delta modulation system is as shown in Figure 7.23. Suppose we move the integrator from the decoder to the input of the delta modulator (DM). This has two effects. First, it preemphasizes the low frequencies in the input signal; thus, it increases the correlation of the signal into the DM. Secondly, it simplifies the DM decoder because the differentiator (the inverse system) required at the decoder is canceled by the integrator. Hence, the decoder is reduced to a simple lowpass filter. Furthermore, the two integrators at the encoder can be replaced by a single integrator placed before the quantizer. The resulting system, shown in Figure 7.24, is called a *sigma–delta modulator* (SDM). Figure 7.25 illustrates an SDM that employs a single digital integrator (first-order SDM) with a system function

$$H(z) = \frac{z^{-1}}{1 - z^{-1}}.$$

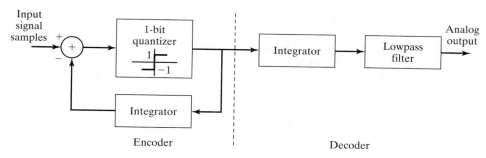

Figure 7.23   Delta modulation.

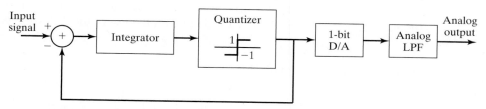

Figure 7.24   A sigma–delta modulator.

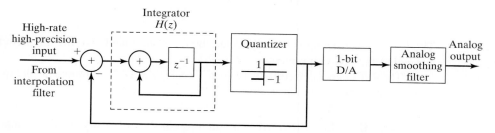

Figure 7.25   A first-order sigma–delta modulator.

Thus, the SDM simplifies the D/A conversion process by requiring only a 1-bit D/A followed by a conventional analog filter (a Butterworth filter, for example) for providing antialiasing protection and signal smoothing. The output analog filters have a passband of approximately 20 kHz; thus, they eliminate any noise above the desired signal band. In modern CD players, the interpolator, the SDM, the 1-bit D/A converter, and the lowpass smoothing filter are generally implemented on a single integrated chip.

7.7 THE JPEG IMAGE-CODING STANDARD

The JPEG standard, adopted by the Joint Photographic Experts Group, is a widely used standard for lossy compression of still images. Although several standards for image compression exist, JPEG is by far the most widely accepted. The JPEG standard achieves very good-to-excellent image quality and is applicable to both color and gray-scale images. The standard is also rather easy to implement and can be implemented in software with acceptable computational complexity.

JPEG belongs to the class of transform-coding techniques, i.e., coding techniques that do not compress the signal (in this case, an image signal) directly, but compress the transform of it. The most widely used transform technique in image coding is DCT (Discrete Cosine Transform). The major benefits of DCT are its high degree of energy compaction properties and the availability of a fast algorithm for computation of the transform. The energy compaction property of the DCT results in transform coefficients, in which only a few of them have significant values, so that nearly all of the energy is contained in those particular components.

The discrete cosine transform of an $N \times N$ picture with luminance function $x(m, n)$, $0 \leq m, n \leq N - 1$ can be obtained with the use of the following equations:

$$X(0, 0) = \frac{1}{N} \sum_{k=0}^{N-1} \sum_{l=0}^{N-1} x(k, l) \tag{7.7.1}$$

$$X(u, v) = \frac{2}{N} \sum_{k=0}^{N-1} \sum_{l=0}^{N-1} x(k, l) \cos\left[\frac{(2k+1)u\pi}{2N}\right] \cos\left[\frac{(2l+1)v\pi}{2N}\right] \quad u, v \neq 0. \tag{7.7.2}$$

The $X(0, 0)$ coefficient is usually called the DC component, and the other coefficients are called the AC components.

The JPEG encoder consists of three blocks: the DCT component, the quantizer, and the encoder, as shown in Figure 7.26.

The DCT Component.   A picture consists of many pixels arranged in an $m \times n$ array. The first step in DCT transformation of the image is to divide the picture array into 8×8 subarrays. This size of the subarrays has been chosen as a compromise of complexity and quality. In some other standards, 4×4 or 16×16 subarrays are chosen. If the number of rows or columns (m or n) is not a multiple of 8, then the last row (or column) is replicated to make it a multiple of 8. The replications are removed at the decoder.

After generating the subarrays, the DCT of each subarray is computed. This process generates 64 DCT coefficients for each subarray starting from the DC component $X(0, 0)$ and going up to $X(7, 7)$. The process is shown in Figure 7.27.

The Quantizer.   Due to the energy compaction property of the DCT, only low-frequency components of the DCT coefficients have significant values.

Since the DC component carries most of the energy and since there exists a strong correlation between the DC component of a subarray and the DC component of the preceding subarray, a uniform differential quantization scheme is employed for quantization of DC components. The AC components are quantized using a

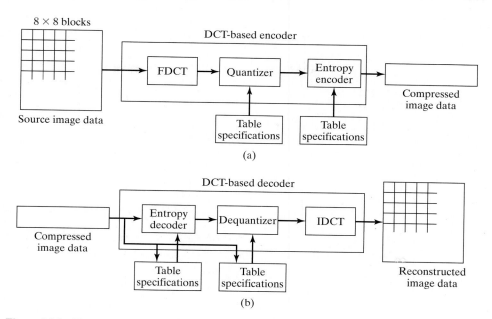

Figure 7.26   The block diagram of a JPEG encoder (a) and decoder (b).

TABLE 7.5   QUANTIZATION TABLE FOR JPEG.

16	11	10	16	24	40	51	61
12	12	14	19	26	58	60	55
14	13	16	24	40	57	69	56
14	17	22	29	51	87	80	62
18	22	37	56	68	109	103	77
24	35	55	64	81	104	113	92
49	64	78	87	103	121	120	101
72	92	95	98	112	100	103	99

uniform quantization scheme. Although all components are quantized using a uniform scheme, different uniform quantization schemes use different step sizes. All quantizers, however, have the same number of quantization regions, namely 256.

A 64-element quantization table determines the step size for uniform quantization of each DCT component. These step sizes are obtained using psychovisual experiments. The result of the quantization step is an 8 × 8 array with nonzero elements only in the top left corner and many zero elements in other locations. A sample quantization table illustrating the quantization steps for different coefficients is shown in Table 7.5.

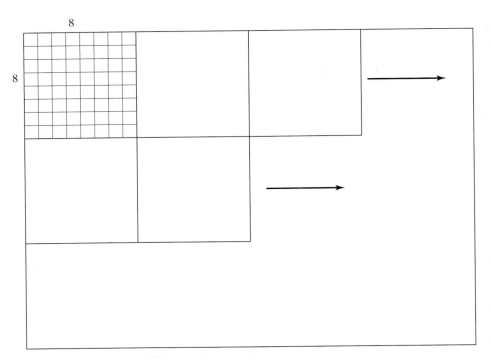

Figure 7.27   The DCT transformation in JPEG.

After the quantization process, the quantized DCT coefficients are arranged in a vector by zig-zag sampling, as shown in Figure 7.28. Using this type of sampling, we obtain a vector $\mathbf{X}$ of length 64 with nonzero values only in the first few components.

The Encoding.   The quantization step provides lossy compression of the image using the method described previously. After this step, entropy coding (as will be discussed in Chapter 12) is employed to provide lossless compression of the quantized values. One of the entropy-coding methods specified in the JPEG standard is Huffman coding; this will be discussed in Section 12.3.1. In this case, Huffman codes are based on tables specifying codewords for different amplitudes. Since the quantized subarrays contain a large number of zeros, some form of runlength coding is used to compress these zeros. Refer to the references at the end of this chapter for further details.

Compression and Picture Quality in JPEG.   Depending on the rate, JPEG can achieve high-compression ratios with moderate-to-excellent image quality for both gray-scale and color images. At rates of 0.2–0.5 bits/pixel, moderate-to-good quality pictures can be obtained that are sufficient for some applications. Increasing the rate to 0.5–0.75 bits/pixel results in good-to-very good quality images that are sufficient for many applications. At 0.75–1.5 bits/pixel, excellent quality images are

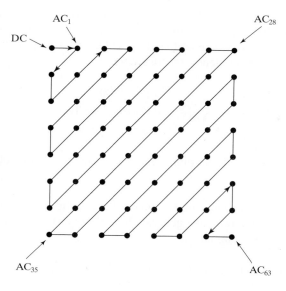

Figure 7.28 Zig-zag sampling of the DCT coefficients.

obtained that are sufficient for most applications. Finally, at rates of 1.5–2 bits/pixel, the resulting image is practically indistinguishable from the original. These rates are sufficient for the most demanding applications.

7.8 FURTHER READING

Jayant and Noll (1984) and Gersho and Gray (1992) examine various quantization- and waveform-coding techniques in detail. Gersho and Gray (1992) include detailed treatment of vector quantization. Analysis–synthesis techniques and linear-predictive coding are treated in books on speech coding, specifically Markel and Gray (1976); Rabiner and Schafer (1979); and Deller, Proakis, and Hansen (2000). The JPEG standard is described in detail in the book by Gibson, et. al. (1998).

PROBLEMS

7.1 Assume $x(t)$ has a bandwidth of 40 kHz.

1. What is the minimum sampling rate for this signal?

2. What is the minimum sampling rate if a guardband of 10 kHz is required?

3. What is the maximum sampling interval for the signal $x_1(t) = x(t)\cos(80,000\pi t)$?

7.2 For a lowpass signal with a bandwidth of 6000 Hz, what is the minimum sampling frequency for perfect reconstruction of the signal? What is the minimum required sampling frequency if a guardband of 2000 Hz is required? If the reconstruction filter has the frequency response

$$H(f) = \begin{cases} K & |f| < 7000 \\ K - K\frac{|f|-7000}{3000} & 7000 < |f| < 10000 , \\ 0 & \text{otherwise} \end{cases}$$

what is the minimum required sampling frequency and the value of K for perfect reconstruction?

7.3 Let the signal $x(t) = A\,\text{sinc}(1000t)$ be sampled with a sampling frequency of 2000 samples/sec. Determine the most general class of reconstruction filters for perfect reconstruction of this signal.

7.4 The lowpass signal $x(t)$ with a bandwidth of W is sampled with a sampling interval of T_s, and the signal

$$x_p(t) = \sum_{n=-\infty}^{\infty} x(nT_s)p(t - nT_s)$$

is reconstructed from the samples, where $p(t)$ is an arbitrary-shaped pulse (not necessarily time limited to the interval $[0, T_s]$).

1. Find the Fourier transform of $x_p(t)$.

2. Find the conditions for perfect reconstruction of $x(t)$ from $x_p(t)$.

3. Determine the required reconstruction filter.

7.5 The lowpass signal $x(t)$ with a bandwidth of W is sampled at the Nyquist rate, and the signal

$$x_1(t) = \sum_{n=-\infty}^{\infty} (-1)^n x(nT_s)\delta(t - nT_s)$$

is generated.

1. Find the Fourier transform of $x_1(t)$.

2. Can $x(t)$ be reconstructed from $x_1(t)$ by using an LTI system? Why?

3. Can $x(t)$ be reconstructed from $x_1(t)$ by using a linear time-varying system? How?

7.6 A lowpass signal $x(t)$ with bandwidth W is sampled with a sampling interval T_s, and the sampled values are denoted by $x(nT_s)$. A new signal $x_1(t)$ is generated by linear interpolation of the sampled values, i.e.,

$$x_1(t) = x(nT_s) + \frac{t - nT_s}{T_s}(x((n+1)T_s) - x(nT_s)) \qquad nT_s \leq t \leq (n+1)T_s.$$

1. Find the power spectrum of $x_1(t)$.

2. Under what conditions can the original signal be reconstructed from the sampled signal and what is the required reconstruction filter?

7.7 A lowpass signal $x(t)$ with bandwidth of 50 Hz is sampled at the Nyquist rate and the resulting sampled values are

$$x(nT_s) = \begin{cases} -1 & -4 \leq n < 0 \\ 1 & 0 < n \leq 4 \\ 0 & \text{otherwise} \end{cases}.$$

1. Find $x(.005)$.

2. Is this signal-power type or energy type? Find its power or energy content.

7.8 Let W be arbitrary and $x(t)$ be a lowpass signal with a bandwidth W.

1. Show that the set of signals $\{\phi_n(t)\}_{n=-\infty}^{\infty}$, where $\phi_n = \text{sinc}(2Wt - n)$ represents an orthogonal signal set. How should these signals be weighted to generate an orthonormal set?

2. Conclude that the reconstruction from the samples relation

$$x(t) = \sum_{n=-\infty}^{\infty} x(nT_s)\text{sinc}(2Wt - n)$$

is an orthogonal expansion relation.

3. From Part 2, show that for all n,

$$\int_{-\infty}^{\infty} x(t)\text{sinc}(2Wt - n)\,dt = Kx(nT_s),$$

and find K.

7.9 Let $X(t)$ denote a zero-mean, wide-sense stationary Gaussian process with $P_X = 10$.

1. Using Table 7.1, design a 16-level optimal uniform quantizer for this source.

2. What is the resulting distortion if the quantizer in Part 1 is employed?

3. What is the amount of improvement in SQNR (in decibels) that results from doubling the number of quantization levels from 8 to 16?

7.10 Using Table 7.1, design an optimal quantizer for the source given in Example 7.2.1. Compare the distortion of this quantizer to the distortion obtained there.

7.11 Solve Problem 7.9 by using Table 7.2 instead of Table 7.1 to design an optimal nonuniform quantizer for the Gaussian source.

7.12 Consider the encoding of the two random variables X and Y, which are uniformly distributed on the region between the two squares, as shown in Figure P-7.12.

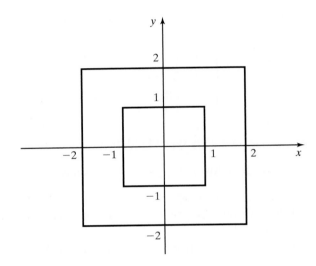

Figure P-7.12

   1. Find $f_X(x)$ and $f_Y(y)$.

   2. Assume each of the random variables X and Y are quantized using four level uniform quantizers. What is the resulting distortion? What is the resulting number of bits per (X, Y) pair?

   3. Now assume that instead of scalar quantizers for X and Y, we employ a vector quantizer to achieve the same level of distortion as in Part 2. What is the resulting number of bits/source output pair (X, Y)?

7.13 Two random variables X and Y are uniformly distributed on the square shown in Figure P-7.13.

   1. Find $f_X(x)$ and $f_Y(y)$.

   2. Assume that each of the random variables X and Y are quantized using four level uniform quantizers. What is the resulting distortion? What is the resulting number of bits per (X, Y) pair?

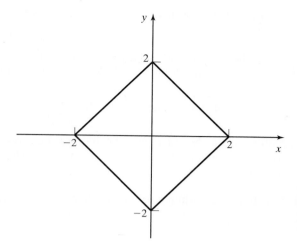

Figure P-7.13

3. Now assume that instead of scalar quantizers for X and Y, we employ a vector quantizer with the same number of bits/source output pair (X, Y) as in Part 2. What is the resulting distortion for this vector quantizer?

7.14 Solve Example 7.4.1 when the samples are uniformly distributed on $[-2, 2]$.

7.15 A stationary random process has an autocorrelation function given by $R_X(\tau) = \frac{A^2}{2} e^{-|\tau|} \cos 2\pi f_0 \tau$; we know that the random process never exceeds 6 in magnitude. Assume that $A = 6$.

    1. How many quantization levels are required to guarantee an SQNR of at least 60 dB?

    2. Assuming that the signal is quantized to satisfy the condition of Part 1 and assuming the approximate bandwidth of the signal is W, what is the minimum required bandwidth for the transmission of a binary PCM signal based on this quantization scheme?

7.16 A signal can be modeled as a lowpass stationary process $X(t)$, whose probability density function at any time t_0 is shown in Figure P-7.16.

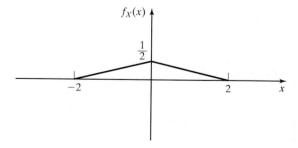

Figure P-7.16

The bandwidth of this process is 5 kHz, and we desire to transmit it using a PCM system.

1. If sampling is done at the Nyquist rate and a uniform quantizer with 32 levels is employed, what is the resulting SQNR? What is the resulting bit rate?

2. If the available bandwidth of the channel is 40 kHz, what is the highest achievable SQNR?

3. If, instead of sampling at the Nyquist rate, we require a guardband of at least 2 kHz and the bandwidth of the channel is 40 kHz, what is the highest achievable SQNR?

7.17 A stationary source is distributed according a triangular PDF, $f_X(x) = \frac{1}{2}\Lambda\left(\frac{x}{2}\right)$. This source is quantized using the 4-level uniform quantizer

$$Q(x) = \begin{cases} 1.5, & 1 < x \leq 2 \\ 0.5, & 0 < x \leq 1 \\ -0.5, & -1 < x \leq 0 \\ -1.5, & -2 \leq x \leq -1 \end{cases}.$$

1. Find the entropy of the quantized source.

2. Determine the PDF of the random variable representing the quantization error, i.e., $\tilde{X} = X - Q(X)$.

7.18 The random process $X(t)$ is defined by $X(t) = Y\cos(2\pi f_0 t + \Theta)$, where Y and Θ are two independent random variables, Y uniform on $[-3, 3]$ and Θ uniform on $[0, 2\pi]$.

1. Find the autocorrelation function of $X(t)$ and its power spectral density.

2. If $X(t)$ is to be transmitted to maintain an SQNR of at least 40 dB using a uniform PCM system, what is the required number of bits/sample and the least bandwidth requirement (in terms of f_0)?

3. If the SQNR is to be increased by 24 dB, how many more bits/sample must be introduced, and what is the new minimum bandwidth requirement in this case?

7.19 A zero-mean stationary information source $X(t)$ has a power spectral density given by

$$S_X(f) = \begin{cases} \frac{1}{\pi}\frac{1}{1+f^2} & |f| \leq 200 \text{ Hz} \\ 0 & \text{otherwise} \end{cases}.$$

The amplitude of this source is limited to 10 in magnitude. This source is sampled at the Nyquist rate and the samples are coded using an 8 bit/sample uniform PCM system.

1. Determine the resulting SQNR in decibels.

2. If we want to increase the SQNR by at least 20 dB, how should the required number of quantization levels change?

3. In Part 2, what is the minimum required bandwidth for the transmission of the PCM signal?

(Hint: $\frac{d}{dx} \arctan x = \frac{1}{1+x^2}$, and for $x > 20$, you can use the approximation $\arctan x \approx \pi/2$.)

7.20 Signal $X(t)$ has a bandwidth of 12,000 Hz, and its amplitude at any time is a random variable whose PDF is shown in Figure P-7.20. We want to transmit this signal using a uniform PCM system.

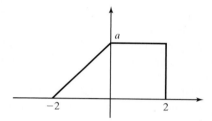

Figure P-7.20

1. Show that $a = \frac{1}{3}$.

2. Determine the power in $X(t)$.

3. What is the SQNR in decibels if a PCM system with 32 levels is employed?

4. What is the minimum required transmission bandwidth in Part 3?

5. If we need to increase the SQNR by at least 20 dB, by how much should the transmission bandwidth increase?

7.21 The power spectral density of a zero-mean WSS random process $X(t)$ is given by

$$S_X(f) = \begin{cases} \frac{f+5000}{5000} & -5000 \le f \le 0 \\ \frac{-f+5000}{5000} & 0 < f \le 5000 \\ 0 & \text{otherwise} \end{cases},$$

and the maximum amplitude of this process is 600.

1. What is the power content of this process?

2. If this process is sampled at rate f_s to guarantee a guardband of 2000 Hz, what is f_s?

3. If we use a PCM system with 256 quantization levels on this process (sampled at the rate you found in Part 2), what is the resulting SQNR (in decibels)?

4. In Part 3, what is the resulting bit rate?

5. If the output of the PCM system is to be transmitted using a binary system, what is the required minimum transmission bandwidth?

6. If we need to increase the SQNR by at least 25 dB, what is the required number of quantization levels, the resulting SQNR, and the required transmission bandwidth?

7.22 In our analysis of PCM systems, we always assumed that the transmitted bits were received with no errors. However, practical channels cause errors. Let us assume that the output of a PCM system is transmitted via a channel whose error probability is denoted by P_2. We further assume that P_2 is small enough such that, in transmission of the v bits resulting from encoding of each quantized sample, either no error occurs or, at most, one error occurs. This means that the probability of each transmitted bit being in error is P_2, and the probability of no error in transmission of v bits is roughly $1 - vP_2$. We also assume that for the binary representation of each quantized value, natural binary coding (NBC) is employed, i.e., the lowest quantized level is mapped into a sequence of zeros and the largest level is mapped into a sequence of all ones, and all the other levels are mapped according to their relative value.

1. Show that, if an error occurs in the least significant bit, its effect on the quantized value is equivalent to Δ, the spacing between the levels; if an error occurs in the next bit, its effect on the quantized value is 2Δ; ..., if an error occurs in the most significant bit, its effect on the quantized value is $2^{v-1}\Delta$.

2. From Part 1, show that the mean squared error resulting from channel errors is given by

$$D_{\text{channel}} = P_2 \Delta^2 \frac{4^v - 1}{3},$$

where $\Delta = \frac{2x_{\max}}{N} = \frac{x_{\max}}{2^{v-1}}$ is the spacing between adjacent levels.

3. From Part 2, conclude that the total distortion, which is the sum of the quantization distortion and the transmission distortion due to channel errors, can be expressed by

$$D_{\text{total}} = \frac{x_{\max}^2}{3 \times N^2} \left(1 + 4P_2(N^2 - 1)\right) = \frac{x_{\max}^2}{3 \times 4^v} \left(1 + 4P_2(4^v - 1)\right).$$

4. Finally, show that the SNR, defined as the ratio of the signal power to the total noise power, is given by

$$\text{SNR} = \frac{3N^2 \overline{\check{X}^2}}{1 + 4P_2(N^2 - 1)} = \frac{3 \times 4^v \overline{\check{X}^2}}{1 + 4P_2(4^v - 1)},$$

where

$$\check{X} = \frac{X}{x_{\max}}.$$

7.23 In a CD player, the sampling rate is 44.1 kHz, and the samples are quantized using a 16 bit/sample quantizer. Determine the resulting number of bits for a piece of music with a duration of 50 minutes.

COMPUTER PROBLEMS

7.1 Determining the Centroids

Use MATLAB to determine the centroids of the quantization regions for a zero-mean unit-variance Gaussian distribution, where the boundaries of the quantization regions are given by $(-5, -4, -2, 0, 1, 3, 5)$. The Gaussian distribution is given in the m-file `normal.m`. Although the support of the Gaussian distribution is $(-\infty, \infty)$, for practical purposes, it is sufficient to use a range that is many times the standard deviation of the distribution. For example, $(m - 10\sigma, m + 10\sigma)$, where m is the mean and σ is the standard deviation (σ^2 is the variance) of the Gaussian random variable.

7.2 Uniform Quantizer Distortion

The objective of this problem is to use MATLAB to determine the mean squared error for a uniform quantizer with 12 quantization levels, each of length 1, designed for a zero-mean Gaussian source with $\sigma^2 = 4$. The quantization regions are symmetric with respect to the mean of the distribution.

1. Specify the boundaries of the quantization regions.
2. Specify the 12 quantization regions.
3. Determine the 12 quantization values corresponding to the quantization regions and the resulting mean-squared distortion.

7.3 Design of Lloyd–Max Quantizer

The objective of this problem is to use MATLAB to design a 10-level Lloyd–Max (nonuniform) quantizer for a zero-mean, unit-variance Gaussian source.

1. Determine the quantization boundaries and the quantization levels.
2. Determine the mean squared distortion.

7.4 Uniform PCM

The objective of this exercise is to investigate the error in the quantization of a sinusoidal signal using uniform PCM.

1. Quantize the sinusoidal signal

$$s(t) = \sin t, \quad 0 \le t \le 10$$

once to 8 levels and once to 16 levels for a sampling interval $T_s = 0.1$. Plot the original signal and the two quantized versions on the same graph, and observe the results.

2. Compute the SQNR for the 8 level and 16 level quantizers.

7.5 Quantization Error in Uniform PCM

The objective of this problem is to evaluate the quantization error in quantizing a Gaussian source using uniform PCM.

1. Generate 500 zero-mean, unit-variance Gaussian random variables and quantize them by using a uniform 64-level PCM quantizer and encoder. Plot the 500-point sequence generated.

2. Determine the SQNR for the 64-level quantizer.

3. Determine the first five values of the sequence, the corresponding quantized values, and the corresponding PCM codewords.

4. Plot the quantization error, defined as the difference between the input value and the quantized value, for the 500-point sequence.

7.6 Nonuniform PCM

The objective of this problem is to evaluate the quantization error in quantizing the output of a Gaussian source with nonuniform PCM.

1. Generate 500 zero-mean, unit-variance Gaussian random variables and quantize them using a 16-, 64-, and 128-level quantizer and a $\mu = 255$ nonlinearity. Plot the input–output quantizer characteristic for each case and the corresponding error sequence.

2. Determine the SQNR for each quantizer.

8

Digital Modulation in an Additive White Gaussian Noise Baseband Channel

In Chapter 7, we described methods for converting the output of a signal source into a sequence of binary digits. In this chapter, we consider the transmission of the digital information sequence over communication channels that are characterized as *additive white Gaussian noise* (AWGN) channels. The AWGN channel is one of the simplest mathematical models for various physical communication channels, including wirelines and some radio channels. Such channels are basically analog channels, which means that the digital information sequence to be transmitted must be mapped into analog signal waveforms.

Our treatment focuses on the characterization and the design of analog signal waveforms that carry digital information and their performance on an AWGN channel. In this chapter, we consider *baseband channels*, i.e., channels having frequency passbands that usually include zero frequency ($f = 0$). When the digital information is transmitted through a baseband channel, there is no need to use a carrier frequency for the transmission of the digitally modulated signals. On the other hand, there are many communication channels (including telephone channels, radio channels, and satellite channels) that have frequency passbands that are far removed from $f = 0$. These types of channels are called *bandpass channels*. In such channels, the information-bearing signal is impressed on a sinusoidal carrier, which shifts the frequency content of the information-bearing signal to the appropriate frequency band that is passed by the channel. Thus, the signal is transmitted by carrier modulation. Digital transmission through bandpass channels is treated in Chapter 10.

Throughout this chapter, we study only one-shot communications, i.e., transmission of only one signal corresponding to a single message, with no transmission following this single transmission. We study sequential transmission of digital data and intersymbol interference in Chapter 9.

We begin by developing a geometric representation of these types of signals, which is useful in assessing their performance characteristics. Then, we describe several different types of analog signal waveforms for transmitting digital information, and we give their geometric representation. The optimum demodulation and detection of these signals is then described, and their performance on the AWGN channel is evaluated in terms of the probability of error. In Chapter 10, we compare the various baseband and carrier modulation methods on the basis of their performance characteristics, their bandwidth requirements, and their implementation complexity.

8.1 GEOMETRIC REPRESENTATION OF SIGNAL WAVEFORMS

In this section, we develop a geometric representation of signal waveforms as points in a signal space. Such a representation provides a compact characterization of signal sets for transmitting digital information over a channel, and it simplifies the analysis of their performance. Using vector representation, waveform communication channels are represented by vector channels. This reduces the complexity of analysis considerably.

In a digital communication system, the modulator input is typically a sequence of binary information digits. The modulator may map each information bit to be transmitted into one of two possible distinct signal waveforms, say $s_1(t)$ or $s_2(t)$. Thus, a zero is represented by the transmitted signal waveform $s_1(t)$, and a one is represented by the transmitted signal waveform $s_2(t)$. This type of digital modulation is called *binary modulation*. Alternatively, the modulator may transmit k bits ($k > 1$) at a time by employing $M = 2^k$ distinct signal waveforms, say $s_m(t)$, $1 \leq m \leq M$. This type of digital modulation is called *M*-ary (nonbinary) modulation. In this section, we develop a vector representation of such digital signal waveforms.

Suppose we have a set of M signal waveforms $s_m(t)$, $1 \leq m \leq M$, which are to be used for transmitting information over a communication channel. From the set of M waveforms, we first construct a set of $N \leq M$ orthonormal waveforms, where N is the dimension of the signal space. For this purpose, we use the Gram–Schmidt orthogonalization procedure.

Gram–Schmidt Orthogonalization Procedure. We begin with the first waveform $s_1(t)$, which is assumed to have energy $\mathscr{E}_1$. The first waveform of the orthonormal set is constructed simply as

$$\psi_1(t) = \frac{s_1(t)}{\sqrt{\mathscr{E}_1}}. \tag{8.1.1}$$

Thus, $\psi_1(t)$ is simply $s_1(t)$ normalized to unit energy.

The second waveform is constructed from $s_2(t)$ by first computing the projection of $s_2(t)$ onto $\psi_1(t)$, which is

$$c_{21} = \int_{-\infty}^{\infty} s_2(t)\psi_1(t)dt. \tag{8.1.2}$$

Then, $c_{21}\psi_1(t)$ is subtracted from $s_2(t)$ to yield

$$d_2(t) = s_2(t) - c_{21}\psi_1(t). \tag{8.1.3}$$

Now, $d_2(t)$ is orthogonal to $\psi_1(t)$, but it does not possess unit energy. If $\mathscr{E}_2$ denotes the energy in $d_2(t)$, then the energy-normalized waveform that is orthogonal to $\psi_1(t)$ is

$$\psi_2(t) = \frac{d_2(t)}{\sqrt{\mathscr{E}_2}};\tag{8.1.4}$$

$$\mathscr{E}_2 = \int_{-\infty}^{\infty} d_2^2(t)\,dt.\tag{8.1.5}$$

In general, the orthogonalization of the kth function leads to

$$\psi_k(t) = \frac{d_k(t)}{\sqrt{\mathscr{E}_k}},\tag{8.1.6}$$

where

$$d_k(t) = s_k(t) - \sum_{i=1}^{k-1} c_{ki}\psi_i(t),\tag{8.1.7}$$

$$\mathscr{E}_k = \int_{-\infty}^{\infty} d_k^2(t)\,dt,\tag{8.1.8}$$

and

$$c_{ki} = \int_{-\infty}^{\infty} s_k(t)\psi_i(t)dt, \qquad i = 1, 2, \ldots k-1.\tag{8.1.9}$$

Thus, the orthogonalization process is continued until all the M signal waveforms $\{s_m(t)\}$ have been exhausted and $N \leq M$ orthonormal waveforms have been constructed. If at any step $d_k(t) = 0$, then there will be no new $\psi(t)$; hence, no new dimension is introduced. The N orthonormal waveforms $\{\psi_n(t)\}$ form an *orthonormal basis* in the N-dimensional signal space. The dimensionality N of the signal space will be equal to M if all the M signal waveforms are linearly independent, i.e., if none of the signal waveforms is a linear combination of the other signal waveforms.

Example 8.1.1

Let us apply the Gram–Schmidt procedure to the set of four waveforms illustrated in Figure 8.1(a). The waveform $s_1(t)$ has energy $\mathscr{E}_1 = 2$, so that $\psi_1(t) = s_1(t)/\sqrt{2}$. Next, we observe that $c_{21} = 0$, so that $\psi_1(t)$ and $s_2(t)$ are orthogonal. Therefore, $\psi_2(t) = s_2(t)/\sqrt{\mathscr{E}_2} = s_2(t)/\sqrt{2}$. To obtain $\psi_3(t)$, we compute c_{31} and c_{32}, which are $c_{31} = 0$ and $c_{32} = -\sqrt{2}$. Hence,

$$d_3(t) = s_3(t) + \sqrt{2}\psi_2(t).$$

Since $d_3(t)$ has unit energy, it follows that $\psi_3(t) = d_3(t)$. Finally, we find that $c_{41} = \sqrt{2}, c_{42} = 0, c_{43} = 1$. Hence,

$$d_4(t) = s_4(t) - \sqrt{2}\psi_1(t) - \psi_3(t) = 0.$$

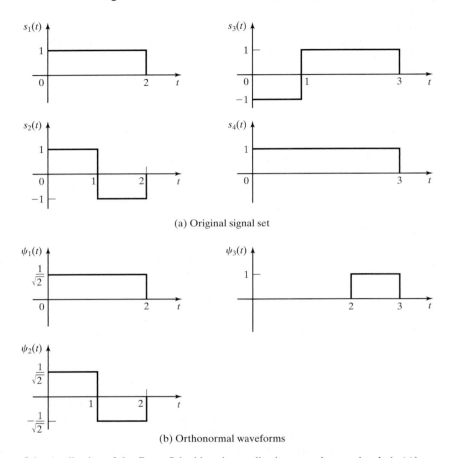

(a) Original signal set

(b) Orthonormal waveforms

Figure 8.1 Application of the Gram–Schmidt orthogonalization procedure to signals $\{s_i(t)\}$.

Thus, $s_4(t)$ is a linear combination of $\psi_1(t)$ and $\psi_3(t)$; consequently, the dimensionality of the signal set is $N = 3$. The functions $\psi_1(t)$, $\psi_2(t)$, and $\psi_3(t)$ are shown in Figure 8.1(b).   ∎

Once we have constructed the set of orthogonal waveforms $\{\psi_n(t)\}$, we can express the M signals $\{s_m(t)\}$ as exact linear combinations of the $\{\psi_n(t)\}$. Hence, we may write

$$s_m(t) = \sum_{n=1}^{N} s_{mn}\psi_n(t), \qquad m = 1, 2, \ldots M, \tag{8.1.10}$$

where the weighting coefficients in this linear combination are given as

$$s_{mn} = \int_{-\infty}^{\infty} s_m(t)\psi_n(t)\,dt. \tag{8.1.11}$$

Since the basis functions $\{\psi_n(t)\}$ are orthonormal, the energy of each signal waveform is related to the weighting coefficients as follows:

$$\mathscr{E}_m = \int_{-\infty}^{\infty} s_m^2(t)\, dt = \sum_{n=1}^{N} s_{mn}^2.$$

On the basis of expression in Equation (8.1.10), each signal waveform may be represented by the vector

$$s_m = (s_{m1}, s_{m2} \ldots, s_{mN}), \tag{8.1.12}$$

or, equivalently, as a point in N-dimensional signal space with coordinates $\{s_{mi}, i = 1, 2, \ldots, N\}$. The energy of the mth signal waveform is simply the square of the length of the vector or, equivalently, the square of the Euclidean distance from the origin to the point in the N-dimensional space. We can also show that the inner product of two signals is equal to the inner product of their vector representations, i.e.,

$$\int_{-\infty}^{\infty} s_m(t)s_n(t)\, dt = s_m \cdot s_n. \tag{8.1.13}$$

Thus, any N-dimensional signal can be represented geometrically as a point in the signal space spanned by the N orthonormal functions $\{\psi_n(t)\}$.

Example 8.1.2

Let us determine the vector representations of the four signals shown in Figure 8.1(a) by using the orthonormal set of functions in Figure 8.1(b). Since the dimensionality of the signal space is $N = 3$, each signal is described by three components, which are obtained by projecting each of the four signal waveforms on the three orthonormal basis functions $\psi_1(t), \psi_2(t), \psi_3(t)$. Thus, we obtain $s_1 = (\sqrt{2}, 0, 0), s_2 = (0, \sqrt{2}, 0), s_3 = (0 - \sqrt{2}, 1), s_4 = (\sqrt{2}, 0, 1)$. These signal vectors are shown in Figure 8.2. ∎

Finally, we should observe that the set of basis functions $\{\psi_n(t)\}$ obtained by the Gram–Schmidt procedure is not unique. For example, another set of basis functions that span the three-dimensional space is shown in Figure 8.3. For this basis, the signal vectors are $s_1 = (1, 1, 0), s_2 = (1, -1, 0), s_3 = (-1, 1, 1,), $ and $s_4 = (1, 1, 1)$. We should note that the change in the basis functions does not change the dimensionality of the space N, the lengths (energies) of the signal vectors, or the inner product of any two vectors. A change in the basis is essentially a rotation of the signal points around the origin.

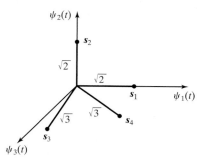

Figure 8.2 Signal vectors corresponding to the signals $s_i(t), i = 1, 2, 3, 4$.

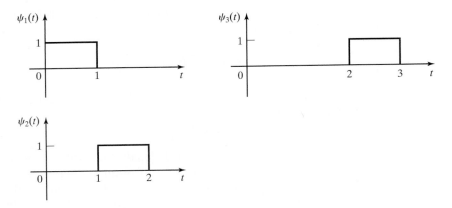

Figure 8.3 Alternate set of basis functions.

Although the Gram–Schmidt orthogonalization procedure is guaranteed to generate an orthonormal basis for representation of the signal set, in many cases, it is easier to use a method based on inspection to generate the orthonormal basis. We explore this method in the problems at the end of this chapter.

8.2 BINARY PULSE MODULATION

In this section, we consider two different binary pulse modulation methods: binary pulse amplitude modulation (PAM) and binary pulse position modulation (PPM). We assume that the information to be transmitted is a binary sequence that consists of zeros and ones, and occurs at the bit rate R_b bits/sec (bps).

8.2.1 Binary Pulse Amplitude Modulation

Binary PAM is the simplest digital modulation method. In binary PAM, the information bit 1 may be represented by a pulse of amplitude A, and the information bit 0 may be represented by a pulse of amplitude $-A$, as shown in Figure 8.4. Since one signal pulse is the negative of the other, this type of signaling is also called *binary antipodal signaling*. Pulses are transmitted at a bit rate $R_b = 1/T_b$ bits/sec, where T_b is called the *bit interval*. Although the pulses are shown as rectangular,

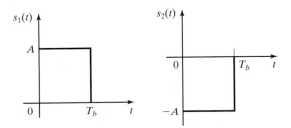

Figure 8.4 Binary PAM signals.

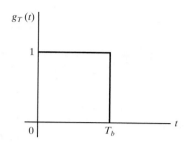

Figure 8.5   A rectangular pulse of unit amplitude and duration T_b.

in practical systems the rise time and decay time are nonzero and the pulses are generally smoother. The pulse shape determines the spectral characteristics of the transmitted signal, as described in Chapter 9.

The binary PAM signal waveforms are expressed as

$$s_m(t) = A_m g_T(t), \;\; 0 \le t \le T_b \;, \;\; m = 1, 2, \tag{8.2.1}$$

where A_m takes one of two possible values (A for $m = 1$, and $-A$ for $m = 2$), and $g_T(t)$ is a rectangular pulse of unit amplitude, as shown in Figure 8.5. The signal energy in each of the two waveforms is

$$
\begin{aligned}
\mathcal{E}_m &= \int_0^{T_b} s_m^2(t)\, dt \;, \; m = 1, 2 \\
&= A^2 \int_0^{T_b} g_T^2(t)\, dt \\
&= A^2 T_b.
\end{aligned}
\tag{8.2.2}
$$

Hence, the two signal waveforms have equal energy, i.e., $\mathcal{E}_m = A^2 T_b$, for $m = 1, 2$. Each signal waveform carries one bit of information. Therefore, we define the signal energy per bit of information as $\mathcal{E}_b$. Thus, we have $A = \sqrt{\mathcal{E}_b / T_b}$.

The two signal waveforms in binary PAM have a very simple geometric representation. The signal waveforms are expressed as

$$s_m(t) = s_m \psi(t), \;\; m = 1, 2, \tag{8.2.3}$$

where $\psi(t)$ is the unit energy rectangular pulse shown in Figure 8.6, and $s_1 = \sqrt{\mathcal{E}_b}$, $s_2 = -\sqrt{\mathcal{E}_b}$. We note that the binary PAM signal waveforms can be uniquely

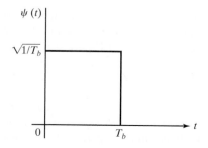

Figure 8.6   Unit energy basis function for binary PAM.

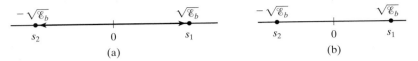

Figure 8.7 Geometric representation of binary PAM.

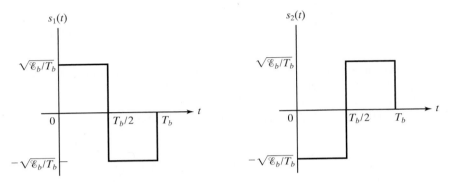

Figure 8.8 Binary antipodal signals in Example 8.2.1.

represented geometrically in one dimension (on the real line) as two vectors, and each has the amplitude $\sqrt{\mathcal{E}_b}$, as shown in Figure 8.7(a). For simplicity, we usually omit drawing the vectors from the origin, and we simply display the two endpoints at $\sqrt{\mathcal{E}_b}$ and $-\sqrt{\mathcal{E}_b}$, as shown in Figure 8.7(b).

Example 8.2.1

Consider the two antipodal signal waveforms shown in Figure 8.8. Show that these signals have exactly the same geometric representation as the two rectangular pulses in Figure 8.4.

Solution   Note that the two signal pulses have energy $\mathcal{E}_b$. The waveforms of these signals may also be represented as

$$s_m(t) = s_m \psi(t), \quad m = 1, 2,$$

where $\psi(t)$ is the unit energy waveform shown in Figure 8.9, and $s_1 = \sqrt{\mathcal{E}_b}, s_2 = -\sqrt{\mathcal{E}_b}$. Therefore, the two antipodal signal waveforms in Figure 8.8 have exactly the same geometric signal representation as those shown in Figure 8.4. ∎

From this discussion, we conclude that any pair of antipodal signal waveforms can be represented geometrically as two vectors (two signal points) on the real line, where one vector is the negative of the other, as shown in Figure 8.7.

8.2.2 Binary Pulse Position Modulation

Another type of pulse modulation that may be used to transmit a binary information sequence is binary PPM. In binary PPM, we employ two signal waveforms, $s_1(t)$ and $s_2(t)$, which are shown in Figure 8.10. It is interesting to note that these two

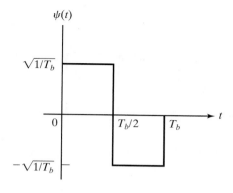

Figure 8.9   Unit energy basis function for the antipodal signals in Figure 8.8.

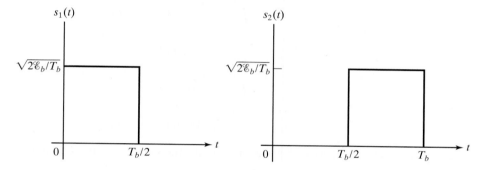

Figure 8.10   Signal pulses in binary PPM (orthogonal signals).

signal waveforms are orthogonal, i.e.,

$$\int_0^{T_b} s_1(t)s_2(t)\,dt = 0. \tag{8.2.4}$$

As we may observe by inspection, the two signal waveforms have identical energies, i.e.,

$$\mathscr{E}_b = \int_0^{T_b} s_1^2(t)dt = \int_0^{T_b} s_2^2(t)\,dt. \tag{8.2.5}$$

In order to represent these two waveforms geometrically as vectors, we need two nonoverlapping (orthogonal) basis functions, and each must be normalized to unit energy. These two waveforms, $\psi_1(t)$ and $\psi_2(t)$, are shown in Figure 8.11. Consequently, the signal waveforms $s_1(t)$ and $s_2(t)$ may be expressed as

$$\begin{aligned} s_1(t) &= s_{11}\psi_1(t) + s_{12}\psi_2(t) \\ s_2(t) &= s_{21}\psi_1(t) + s_{22}\psi_2(t), \end{aligned} \tag{8.2.6}$$

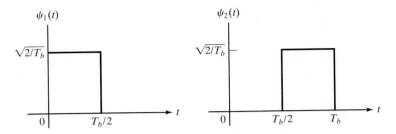

Figure 8.11   Two orthogonal basis functions for binary PPM signals.

where we can easily observe that

$$s_{11} = \int_0^{T_b} s_1(t)\psi_1(t)dt = \sqrt{\mathcal{E}_b};$$

$$s_{12} = \int_0^{T_b} s_1(t)\psi_2(t)dt = 0;$$

$$s_{21} = \int_0^{T_b} s_2(t)\psi_1(t)dt = 0; \qquad (8.2.7)$$

$$s_{22} = \int_0^{T_b} s_2(t)\psi_2(t)dt = \sqrt{\mathcal{E}_b}.$$

In this case, the two signal waveforms are represented as two-dimensional vectors s_1 and s_2 when

$$s_1 = (s_{11}, 0) = \left(\sqrt{\mathcal{E}_b}, 0\right) \qquad (8.2.8)$$

and

$$s_2 = (0, s_{22}) = \left(0, \sqrt{\mathcal{E}_b}\right), \qquad (8.2.9)$$

as shown in Figure 8.12(a) or as shown in Figure 8.12(b) as two signal points in two-dimensional space. We observe that the two signal vectors are perpendicular; hence, they are orthogonal, i.e., their dot product is equal to zero.

Example 8.2.2

Consider the two orthogonal signal waveforms shown in Figure 8.13. Show that these two signal waveforms have a similar geometric representation as the two PPM pulses shown in Figure 8.10.

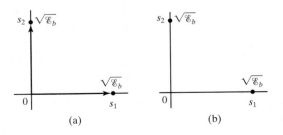

(a)   (b)

Figure 8.12   Geometric representation of binary orthogonal (PPM) signal waveforms.

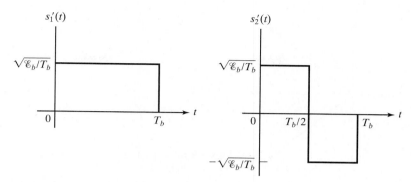

Figure 8.13   Two orthogonal signal waveforms.

Solution   By inspection, the two signal waveforms satisfy the orthogonality condition given by Equation (8.2.4), and they have energy $\mathcal{E}_b$. By using the orthonormal basis waveforms $\psi_1(t)$ and $\psi_2(t)$ in Figure 8.11, the signal waveforms $s_1'(t)$ and s_2' are expressed as

$$s_1'(t) = s_{11}'\psi_1(t) + s_{12}'\psi_2(t)$$

and

$$s_2'(t) = s_{12}'\psi_1(t) + s_{22}'\psi_2(t), \tag{8.2.10}$$

where

$$s_{11}' = \int_0^{T_b} s_1'(t)\psi_1(t)dt = \sqrt{\mathcal{E}_b/2};$$

$$s_{12}' = \int_0^{T_b} s_1'(t)\psi_2(t)dt = \sqrt{\mathcal{E}_b/2};$$

$$\tag{8.2.11}$$

$$s_{11}' = \int_0^{T_b} s_2'(t)\psi_1(t)dt = \sqrt{\mathcal{E}_b/2};$$

$$s_{22}' = \int_0^{T_b} s_2'(t)\psi_2(t)dt = -\sqrt{\mathcal{E}_b/2}.$$

The vectors $s_1' = \left(\sqrt{\mathcal{E}_b/2}, \sqrt{\mathcal{E}_b/2}\right)$ and $s_2' = \left(\sqrt{\mathcal{E}_b/2}, -\sqrt{\mathcal{E}_b/2}\right)$ are shown in Figure 8.14. We observe that s_1' and s_2' are perpendicular (orthogonal vectors) and are simply a phase-rotated version of the orthogonal vectors shown in Figure 8.12(a).   ∎

8.3 OPTIMUM RECEIVER FOR BINARY MODULATED SIGNALS IN ADDITIVE WHITE GAUSSIAN NOISE

In this section, we describe the signal processing operations performed at the receiver to recover the transmitted information. We begin by describing the channel that corrupts the transmitted signal by the addition of noise.

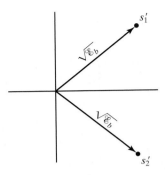

Figure 8.14   Signal vectors for the binary orthogonal waveforms shown in Figure 8.13.

Additive White Gaussian Noise Channel.

The communication channel is assumed to corrupt the transmitted signal by the addition of white Gaussian noise, as shown in Figure 8.15. Thus, the received signal in a signal interval of duration T_b may be expressed as

$$r(t) = s_m(t) + n(t), \qquad m = 1, 2, \tag{8.3.1}$$

where $n(t)$ denotes the sample function of the additive white Gaussian noise (AWGN) process with the power spectral density $S_n(f) = N_0/2$ W/Hz. Based on the observation of $r(t)$ over the signal interval, we wish to design a receiver that is optimum in the sense that it minimizes the probability of making an error. To be specific, we focus on the processing of the received signal $r(t)$ in the interval $0 \leq t \leq T_b$.

It is convenient to subdivide the receiver into two parts, the signal demodulator and the detector, as shown in Figure 8.16. In Section 8.4.6, we will show that such a subdivision does not affect the optimality of the overall system. The signal demodulator's function is to convert the received signal waveform $r(t)$ into a vector y, whose dimension is equal to the dimension of the transmitted signal waveforms. The detector's function is to decide which of the two possible signal waveforms was transmitted; this decision is based on observation of the vector y.

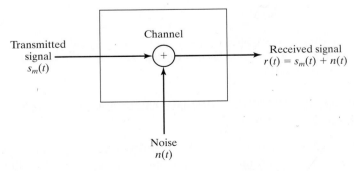

Figure 8.15   Model for the received signal passed through an AWGN channel.

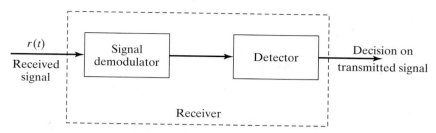

Figure 8.16 Receiver for digitally modulated signals.

Two realizations of the signal demodulator are described in Section 8.3.1 and Section 8.3.2. The first is based on the use of signal correlators. The second is based on the use of matched filters. The optimum detector that follows the signal demodulator is designed to minimize the probability of error.

8.3.1 Correlation-Type Demodulator

In this section, we describe the processing of the received signal by a correlation-type demodulator for binary antipodal signals (binary PAM) and binary orthogonal signals (binary PPM). We begin with binary antipodal signals.

Binary Antipodal Signals.   Let us consider the binary antipodal signals generally represented as

$$s_m(t) = s_m \psi(t), \quad m = 1, 2, \tag{8.3.2}$$

where $\psi(t)$ is the unit energy rectangular pulse shown in Figure 8.6 and $s_1 = \sqrt{\mathcal{E}_b}$, $s_2 = -\sqrt{\mathcal{E}_b}$. Therefore, the received signal is

$$r(t) = s_m \psi(t) + n(t), \quad 0 \le t \le T_b, \quad m = 1, 2. \tag{8.3.3}$$

In a correlation-type demodulator, the received signal $r(t)$ is multiplied by the signal waveform $\psi(t)$ and the product is integrated over the interval $0 \le t \le T_b$, as illustrated in Figure 8.17. We say that $r(t)$ is cross-correlated with $\psi(t)$.

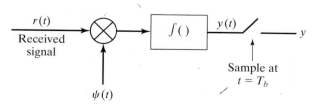

Figure 8.17 Cross-correlator for binary antipodal signals.

Mathematically, this cross-correlation operation produces the output

$$
\begin{aligned}
y(t) &= \int_0^t r(\tau)\psi(\tau)\,d\tau \\
&= \int_0^t [s_m\psi(\tau) + n(\tau)]\psi(\tau)\,d\tau \\
&= s_m \int_0^t \psi^2(\tau)\,d\tau + \int_0^t n(\tau)\psi(\tau)\,d\tau.
\end{aligned}
\tag{8.3.4}
$$

We sample the output of the correlator at $t = T_b$. Thus, we obtain

$$
y(T_b) = s_m + n,
\tag{8.3.5}
$$

where n is the additive noise term defined as

$$
n = \int_0^{T_b} \psi(\tau)n(\tau)\,d\tau.
\tag{8.3.6}
$$

Since $n(t)$ is a sample function of a white Gaussian noise process, the noise term n is a Gaussian random variable with zero mean and with variance

$$
\begin{aligned}
\sigma_n^2 = E(n^2) &= \int_0^{T_b}\int_0^{T_b} E[n(t)n(\tau)]\psi(t)\psi(\tau)\,dt\,d\tau \\
&= \int_0^{T_b}\int_0^{T_b} \frac{N_0}{2}\delta(t - \tau)\psi(t)\psi(\tau)\,dt\,d\tau \\
&= \frac{N_0}{2}\int_0^{T_b} \psi^2(t)\,dt = \frac{N_0}{2},
\end{aligned}
\tag{8.3.7}
$$

where we have used Equation (5.3.2) for the autocorrelation function of white noise. Therefore, for a given signal transmission (given s_m), the output of the correlator ($y = y(T_b)$) is a Gaussian random variable with mean s_m and variance $N_0/2$, i.e.,

$$
f(y|s_m) = \frac{1}{\sqrt{\pi N_o}}e^{-(y-s_m)^2/N_0}, \qquad m = 1, 2.
\tag{8.3.8}
$$

These two conditional probability density functions are illustrated in Figure 8.18. This correlator output is fed to the detector, which decides whether the transmitted bit is a zero or a one, as described in Section 8.3.3.

Example 8.3.1

Sketch the noise-free output of the correlator for the rectangular pulse $\psi(t)$, as shown in Figure 8.6, when $s_1(t)$ and $s_2(t)$ are transmitted.

Solution   With $n(t) = 0$, the signal waveform at the output of the correlator is

$$
y(t) = \int_0^t s_m\psi^2(\tau)\,d\tau = s_m \int_0^t \psi^2(\tau)\,d\tau.
$$

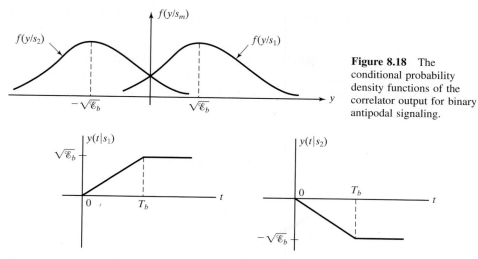

Figure 8.18 The conditional probability density functions of the correlator output for binary antipodal signaling.

Figure 8.19   Noise-free cross-correlator outputs when $s_1(t)$ and $s_2(t)$ are transmitted.

The graphs of $y(t)$ for $s_1 = \sqrt{\mathcal{E}_b}$ and $s_2 = -\sqrt{\mathcal{E}_b}$ are shown in Figure 8.19. Since the pulse $\psi(t)$ is constant over the integration interval $0 \le t \le T_b$, the correlator is just a simple integrator. We observe that the maximum signal at the output of the correlator occurs at $t = T_b$. We also observe that the correlator must be reset to zero at the end of each bit interval T_b, so that it can be used in the demodulation of the received signal in the next signal interval. Such an integrator is called an *integrate-and-dump filter*. ∎

Example 8.3.2

If the binary antipodal signals illustrated in Figure 8.8 are used for the transmission of information, demonstrate that the output of the correlation-type demodulator is exactly the same as that given for the previously discussed rectangular pulse signals.

Solution   In this case, $\psi(t)$ is the signal waveform shown in Figure 8.9, and the received signal is

$$r(t) = s_m \psi(t) + n(t).$$

The cross-correlator output is

$$y(t) = \int_0^t r(\tau)\psi(\tau)\,d\tau$$

$$= s_m \int_0^t \psi^2(\tau)\,d\tau + \int_0^t n(\tau)\psi(\tau)\,d\tau;$$

at $t = T_b$, we have

$$y(T_b) = s_m + n.$$

The signal component s_m in the cross-correlator output is the same as that obtained for the rectangular pulse, and the noise term n has exactly the same mean (zero) and variance $\sigma_n^2 = N_0/2$. ∎

Binary Orthogonal Signals.   Let us consider the two orthogonal signals given by Equation (8.2.6) and illustrated in Figure 8.10, where $\psi_1(t)$ and $\psi_2(t)$ are the orthogonal basis functions shown in Figure 8.11, and $s_1 = \left(\sqrt{\mathcal{E}_b}, 0\right)$ and $s_2 = \left(0, \sqrt{\mathcal{E}_b}\right)$ are the signal vectors. These two signals are used to transmit a binary information sequence, which consists of zeros and ones and occurs at a rate $R_b = 1/T_b$ bits/sec. In the presence of AWGN, the received signal has the form

$$r(t) = s_m(t) + n(t), \quad 0 \le t \le T_b, \quad m = 1, 2. \tag{8.3.9}$$

Since the two possible transmitted signals are two-dimensional, the received signal $r(t)$ is cross-correlated with each of the two basis signal waveforms $\psi_1(t)$ and $\psi_2(t)$, as shown in Figure 8.20. The correlator output waveforms are

$$y_m(t) = \int_0^t r(\tau)\psi_m(\tau)\,d\tau, \quad m = 1, 2, \tag{8.3.10}$$

which, when sampled at $t = T_b$, result in the outputs

$$y_m = y_m(T_b) = \int_0^{T_b} r(\tau)\psi_m(\tau)d\tau, \quad m = 1, 2. \tag{8.3.11}$$

Now, suppose the transmitted signal is $s_1(t) = s_{11}\psi_1(t)$, so that $r(t) = s_{11}\psi_1(t) + n(t)$. The output of the first correlator is

$$\begin{aligned}
y_1 &= \int_0^{Tb} [s_{11}\psi_1(\tau) + n(\tau)]\psi_1(\tau)\,d\tau \\
&= s_{11} + n_1 = \sqrt{\mathcal{E}_b} + n_1,
\end{aligned} \tag{8.3.12}$$

where $s_{11} = \sqrt{\mathcal{E}_b}$ is the signal component and n_1 is the noise component, defined as

$$n_1 = \int_0^{T_b} n(\tau)\psi_1(\tau)\,d\tau. \tag{8.3.13}$$

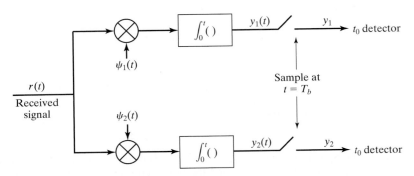

Figure 8.20   Correlation-type demodulator for binary orthogonal signals.

The output of the second correlator is

$$y_2 = \int_0^{T_b} [s_{11}\psi_1(\tau) + n(\tau)]\psi_2(\tau)\, d\tau$$

$$= s_{11} \int_0^{T_b} \psi_1(\tau)\psi_2(\tau)\, d\tau + \int_0^{T_b} n(\tau)\psi_2(\tau)\, d\tau \tag{8.3.14}$$

$$= \int_0^{T_b} n(\tau)\psi_2(\tau)\, d\tau = n_2.$$

The output of the second correlator only includes the noise component n_2, because $\psi_1(t)$ and $\psi_2(t)$ are orthogonal. Consequently, the received signal vector is

$$\begin{aligned} \mathbf{y} &= (y_1, y_2) \\ &= \left(\sqrt{\mathscr{E}_b} + n_1, n_2\right). \end{aligned} \tag{8.3.15}$$

It is easy to verify that when the signal $s_2(t) = s_{22}\psi_2(t)$ is transmitted, the outputs of the two correlators are $y_1 = n_1$ and $y_2 = s_{22} + n_2 = \sqrt{\mathscr{E}_b} + n_2$. Hence, the received signal vector is

$$\begin{aligned} \mathbf{y} &= (y_1, y_2) \\ &= \left(n_1, \sqrt{\mathscr{E}_b} + n_2\right). \end{aligned} \tag{8.3.16}$$

The vector $\mathbf{y}$ at the output of the cross-correlators is fed to the detector, which decides whether the received signal vector corresponds to the transmission of a one or a zero.

The statistical characteristics of the observed signal vector $\mathbf{y}$ are easily determined. Since $n(t)$ is a sample function of a white Gaussian noise process, the noise terms n_1 and n_2 are zero-mean Gaussian random variables with variance $\sigma_n^2 = N_0/2$. Furthermore, the correlation between n_1 and n_2 is

$$E(n_1 n_2) = \int_0^{T_b} \int_0^{T_b} E[n(t)n(\tau)]\psi_1(t)\psi_2(\tau)\, dt\, d\tau$$

$$= \int_0^{T_b} \int_0^{T_b} \frac{N_0}{2}\delta(t - \tau)\psi_1(t)\psi_2(\tau)\, dt\, d\tau \tag{8.3.17}$$

$$= \frac{N_0}{2} \int_0^{T_b} \psi_1(t)\psi_2(t)\, dt = 0.$$

Therefore, n_1 and n_2 are uncorrelated; since they are Gaussian, n_1 and n_2 are statistically independent. Consequently, when the transmitted signal is $s_1(t)$, the conditional joint probability density function of the correlator output components (y_1, y_2) is

$$f(y_1, y_2 | s_1) = \left(\frac{1}{\sqrt{\pi N_0}}\right)^2 e^{-\left(y_1 - \sqrt{\mathscr{E}_b}\right)^2 / N_0} e^{-y_2^2 / N_0}. \tag{8.3.18}$$

When $s_2(t)$ is transmitted, the conditional joint probability density function of the correlator output components (y_1, y_2) is

$$f(y_1, y_2 | s_2) = \left(\frac{1}{\sqrt{\pi N_0}}\right)^2 e^{-y_1^2/N_0} e^{-\left(y_2 - \sqrt{\mathcal{E}_b}\right)^2/N_0}. \tag{8.3.19}$$

Since the noise components n_1 and n_2 are statistically independent, we observe that the joint probability density functions of (y_1, y_2), which are given by Equation (8.3.18) and Equation (8.3.19), factor into a product of marginal probability density functions, i.e.,

$$f(y_1, y_2 | s_m) = f(y_1 | s_m) f(y_2 | s_m), \quad m = 1, 2. \tag{8.3.20}$$

For example, Figure 8.21 shows the probability density functions $f(y_1 | s_1)$ and $f(y_2 | s_1)$ when $s_1(t)$ is transmitted.

Example 8.3.3

Assuming that the transmitted signal waveform is $s_2(t) = \sqrt{\mathcal{E}_b}\psi_2(t)$, in the absence of additive noise, sketch the output waveforms of the two correlators shown in Figure 8.20. The other signal waveform is $s_1(t) = \sqrt{\mathcal{E}_b}\psi_1(t)$, where $\psi_1(t)$ and $\psi_2(t)$ are the basis functions shown in Figure 8.11.

Solution When $s_2(t)$ is transmitted in the absence of noise, the outputs of the two correlators shown in Figure 8.20 are

$$y_1(t) = \int_0^t s_2(\tau)\psi_1(\tau)\, d\tau = \sqrt{\mathcal{E}_b} \int_0^t \psi_2(\tau)\psi_1(\tau)\, d\tau$$

$$y_2(t) = \int_0^t s_2(\tau)\psi_2(\tau)\, d\tau = \sqrt{\mathcal{E}_b} \int_0^t \psi_2^2(\tau)\, d\tau.$$

The graphs of the correlator outputs are illustrated in Figure 8.22. Note that the noise-free output of the first correlator is zero for $0 \le t \le T_b$ because $\psi_1(t)$ and $\psi_2(t)$ are nonoverlapping orthogonal waveforms. ∎

Example 8.3.4

If the binary orthogonal signals illustrated in Figure 8.13 are used for the transmission of binary information, determine the outputs of the two correlators when the signal waveform $s_1'(t)$ is transmitted.

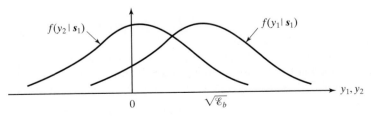

Figure 8.21 The conditional probability density functions of the outputs (y_1, y_2) from the cross-correlators of two orthogonal signals.

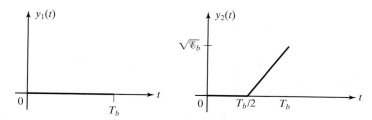

Figure 8.22   Correlator output signal waveforms when $s_2(t)$ is the transmitted signal.

Solution   The two orthogonal signals are expressed as a linear combination of the two nonoverlapping basis functions $\psi_1(t)$ and $\psi_2(t)$, as shown in Figure 8.11, as

$$s_1'(t) = s_{11}\psi_1(t) + s_{12}\psi_2(t) = \sqrt{\mathcal{E}_b/2}\,\psi_1(t) + \sqrt{\mathcal{E}_b/2}\,\psi_2(t)$$

and

$$s_2'(t) = s_{21}\psi_1(t) + s_{22}\psi_2(t) = \sqrt{\mathcal{E}_b/2}\,\psi_1(t) - \sqrt{\mathcal{E}_b/2}\,\psi_2(t).$$

When $s_1'(t)$ is transmitted, the received signal is $r(t) = s_1'(t) + n(t)$. The two correlator outputs are

$$y_m = \int_0^{T_b} r(\tau)\psi_m(\tau)\,d\tau, \qquad m = 1, 2.$$

By substituting for $s_1'(t)$ in the preceding integral and using the orthogonality property of $\psi_1(t)$ and $\psi_2(t)$, we obtain

$$y_1 = \sqrt{\mathcal{E}_b/2} + n_1$$

and

$$y_2 = \sqrt{\mathcal{E}_b/2} + n_2,$$

where the noise components are defined as

$$n_m = \int_0^{T_b} n(\tau)\psi_m(\tau)\,d\tau, \qquad m = 1, 2.$$

As demonstrated, n_1 and n_2 are zero-mean uncorrelated Gaussian random variables, each having a variance $\sigma_n^2 = N_0/2$. Therefore, the joint probability density function of y_1 and y_2, conditioned on $s_1'(t)$ being transmitted, is

$$f\left(y_1, y_2 | s_1'\right) = \left(\frac{1}{\sqrt{\pi N_0}}\right)^2 e^{-\left(y_1 - \sqrt{\mathcal{E}_b/2}\right)^2 / N_0}\, e^{-\left(y_2 - \sqrt{\mathcal{E}_b/2}\right)^2 / N_0}. \qquad \blacksquare$$

8.3.2 Matched-Filter-Type Demodulator

Instead of using cross-correlation to perform the demodulation, we may use a filter-type demodulator, as described in the next paragraph. As in the preceding section, we describe this demodulator for binary antipodal and binary orthogonal signals.

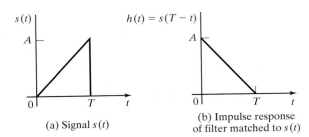

(a) Signal $s(t)$

(b) Impulse response
of filter matched to $s(t)$

Figure 8.23 Signal $s(t)$ and filter
matched to $s(t)$.

Binary Antipodal Signals.   As we have observed, in the case of binary antipodal signals, the received signal is

$$r(t) = s_m \psi(t) + n(t), \ \ 0 \le t \le T_b, \ \ m = 1, 2, \tag{8.3.21}$$

where $\psi(t)$ is a unit energy rectangular pulse. Suppose we pass the received signal $r(t)$ through a linear, time-invariant filter with impulse response

$$h(t) = \psi(T_b - t), \ \ 0 \le t \le T_b. \tag{8.3.22}$$

The filter output is

$$y(t) = \int_0^t r(\tau) h(t - \tau) \, d\tau. \tag{8.3.23}$$

If we sample the output of the filter at $t = T_b$, we obtain

$$y(T_b) = \int_0^{T_b} r(\tau) h(T_b - \tau) \, d\tau.$$

But $h(T_b - \tau) = \psi(\tau)$. Therefore,

$$
\begin{aligned}
y(T_b) &= \int_0^{T_b} [s_m \psi(\tau) + n(\tau)] \psi(\tau) \, d\tau \\
&= s_m \int_0^{T_b} \psi^2(\tau) \, d\tau + \int_0^{T_b} n(\tau) \psi(\tau) \, d\tau \\
&= s_m + n.
\end{aligned}
\tag{8.3.24}
$$

Hence, the output of the filter at $t = T_b$ is exactly the same as the output obtained with a cross-correlator.

A filter whose impulse response $h(t) = s(T - t)$, where $s(t)$ is assumed to be confined to the time interval $0 \le t \le T$, is called the *matched filter* to the signal $s(t)$. An example of a signal and its matched filter are shown in Figure 8.23. The response of $h(t) = s(T - t)$ to the signal $s(t)$ is

$$y(t) = \int_0^t s(\tau) s(T - t + \tau) \, d\tau, \tag{8.3.25}$$

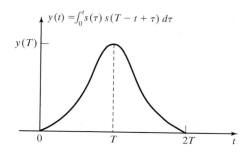

$$y(t) = \int_0^t s(\tau)\, s(T - t + \tau)\, d\tau$$

Figure 8.24   The matched filter output is the autocorrelation function of $s(t)$.

which is basically the time-autocorrelation function of the signal $s(t)$, as shown in Equation (2.3.37). Figure 8.24 illustrates $y(t)$ for the triangular signal pulse shown in Figure 8.23. We note that the autocorrelation function $y(t)$ is an even function of t, which attains a peak at $t = T$. The peak value $y(T)$ is equal to the energy of the signal $s(t)$.

Binary Orthogonal Signals.   As we have observed in the previous section, orthogonal signals are two-dimensional signals; hence, two correlators are needed to perform the demodulation. In place of the correlators, two linear time-invariant filters may be employed. To be specific, consider the received signal

$$r(t) = s_m(t) + n(t), \quad 0 \le t \le T_b, \quad m = 1, 2, \tag{8.3.26}$$

where $s_m(t)$ and $m = 1, 2$ are the two orthogonal waveforms given by Equation (8.2.6) and illustrated in Figure 8.10, and $\psi_m(t)$ and $m = 1, 2$ are the two orthogonal basis functions shown in Figure 8.11. The impulse responses of the two filters matched to $\psi_1(t)$ and $\psi_2(t)$ are defined as

$$h_1(t) = \psi_1(T_b - t), \quad 0 \le t \le T_b$$

and

$$h_2(t) = \psi_2(T_b - t), \quad 0 \le t \le T_b. \tag{8.3.27}$$

When the received signal $r(t)$ is passed through the two filters, their outputs are

$$y_m(t) = \int_0^t r(\tau) h_m(t - \tau)\, d\tau, \quad m = 1, 2. \tag{8.3.28}$$

If we sample the outputs of these filters at $t = T_b$, we obtain

$$y_m = y_m(T_b) = \int_0^{T_b} r(\tau) h_m(T_b - \tau)\, d\tau$$

$$= \int_0^{T_b} r(\tau) \psi_m(\tau)\, d\tau, \quad m = 1, 2. \tag{8.3.29}$$

By comparing the outputs of the two matched filters at $t = T_b$ with the outputs obtained from the cross-correlators given by Equation (8.3.11), we observe that the outputs are identical. Therefore, the correlation-type demodulator and the matched filter-type demodulator yield identical outputs at $t = T_b$.

Properties of the Matched Filter.   A matched filter has some interesting properties. Let us prove the most important property, which may be stated as follows: If a signal $s(t)$ is corrupted by AWGN, the filter with the impulse response matched to $s(t)$ maximizes the output signal-to-noise ratio (SNR).

To prove this property, let us assume that the received signal $r(t)$ consists of the signal $s(t)$ and AWGN $n(t)$, which has zero mean and a power-spectral density $S_n(f) = N_0/2$ W/Hz. Suppose the signal $r(t)$ is passed through a filter with the impulse response $h(t), 0 \leq t \leq T$, and its output is sampled at time $t = T$. The filter response to the signal and noise components is

$$y(t) = \int_0^t r(\tau)h(t - \tau)\,d\tau$$

$$= \int_0^t s(\tau)h(t - \tau)\,d\tau + \int_0^t n(\tau)h(t - \tau)\,d\tau.$$

$$(8.3.30)$$

At the sampling instant $t = T$, the signal and noise components are

$$y(T) = \int_0^T s(\tau)h(T - \tau)\,d\tau + \int_0^T n(\tau)h(T - \tau)\,d\tau$$

$$= y_s(T) + y_n(T),$$

$$(8.3.31)$$

where $y_s(T)$ represents the signal component and $y_n(T)$ represents the noise component. The problem is to select the filter impulse response that maximizes the output SNR, defined as

$$\left(\frac{S}{N}\right)_0 = \frac{y_s^2(T)}{E\left[y_n^2(T)\right]}.$$

$$(8.3.32)$$

The denominator in Equation (8.3.32) is simply the variance of the noise term at the output of the filter. Let us evaluate $E\left[y_n^2(T)\right]$. We have

$$E\left[y_n^2(T)\right] = \int_0^T \int_0^T E[n(\tau)n(t)]h(T - \tau)h(T - t)\,dt\,d\tau$$

$$= \frac{N_0}{2} \int_0^T \int_0^T \delta(t - \tau)h(T - \tau)h(T - t)\,dt\,d\tau$$

$$= \frac{N_0}{2} \int_0^T h^2(T - t)\,dt.$$

$$(8.3.33)$$

Note that the variance depends on the power spectral density of the noise and the energy in the impulse response $h(t)$.

By substituting for $y_s(T)$ and $E\left[y_n^2(T)\right]$ in Equation (8.3.32), we obtain the expression for the output SNR as

$$\left(\frac{S}{N}\right)_0 = \frac{\left[\int_0^T s(\tau)h(T-\tau)d\tau\right]^2}{\frac{N_0}{2}\int_0^T h^2(T-t)\,dt} = \frac{\left[\int_0^T h(\tau)s(T-\tau)d\tau\right]^2}{\frac{N_0}{2}\int_0^T h^2(T-t)\,dt}. \tag{8.3.34}$$

Since the denominator of the SNR depends on the energy in $h(t)$, the maximum output SNR over $h(t)$ is obtained by maximizing the numerator of $(S/N)_0$ subject to the constraint that the denominator is held constant. The maximization of the numerator is most easily performed by using the Cauchy–Schwartz inequality, which generally states that if $g_1(t)$ and $g_2(t)$ are finite energy signals, then

$$\left[\int_{-\infty}^{\infty} g_1(t)g_2(t)\,dt\right]^2 \le \int_{-\infty}^{\infty} g_1^2(t)\,dt \int_{-\infty}^{\infty} g_2^2(t)\,dt, \tag{8.3.35}$$

where equality holds when $g_1(t) = Cg_2(t)$ for any arbitrary constant C. If we set $g_1(t) = h(t)$ and $g_2(t) = s(T-t)$, it is clear that the $(S/N)_0$ is maximized when $h(t) = Cs(T-t)$, i.e., $h(t)$ is matched to the signal $s(t)$. The scale factor C^2 drops out of the expression for $(S/N)_0$, since it appears in both the numerator and the denominator.

The output (maximum) SNR obtained with the matched filter is

$$\left(\frac{S}{N}\right)_0 = \frac{2}{N_0}\int_0^T s^2(t)\,dt$$

$$= \frac{2\mathscr{E}_s}{N_0}, \tag{8.3.36}$$

where $\mathscr{E}_s$ is the energy of the signal $s(t)$. Note that the output SNR from the matched filter depends on the energy of the waveform $s(t)$, but not on the detailed characteristics of $s(t)$. This is another interesting property of the matched filter.

Frequency Domain Interpretation of the Matched Filter.   The matched filter has an interesting frequency domain interpretation. Since $h(t) = s(T-t)$, the Fourier transform of this relationship is

$$H(f) = \int_0^T s(T-t)\,e^{-j2\pi ft}\,dt$$

$$= \left[\int_0^T s(\tau)\,e^{j2\pi f\tau}d\tau\right]e^{-j2\pi fT} \tag{8.3.37}$$

$$= S^*(f)e^{-j2\pi fT}.$$

We observe that the matched filter has a frequency response that is the complex conjugate of the transmitted signal spectrum multiplied by the phase factor $e^{-j2\pi fT}$, which represents the sampling delay of T. In other words, $|H(f)| = |S(f)|$, so that the magnitude response of the matched filter is identical to the transmitted signal spectrum. On the other hand, the phase of $H(f)$ is the negative of the phase of $S(f)$, shifted by a linear function of T.

Now, if the signal $s(t)$, with spectrum $S(f)$, is passed through the matched filter, the filter output has a spectrum $Y(f) = |S(f)|^2 e^{-j2\pi fT}$. Hence, the output waveform is

$$
\begin{aligned}
y_s(t) &= \int_{-\infty}^{\infty} Y(f)\, e^{j2\pi ft}\, df \\
&= \int_{-\infty}^{\infty} |S(f)|^2\, e^{-j2\pi fT}\, e^{j2\pi ft}\, df.
\end{aligned}
\tag{8.3.38}
$$

By sampling the output of the matched filter at $t = T$, we obtain

$$
y_s(T) = \int_{-\infty}^{\infty} |S(f)|^2 df = \int_0^T s^2(t)\, dt = \mathscr{E}_s,
\tag{8.3.39}
$$

where the last step follows from Parseval's relation.

The noise of the output of the matched filter has a power spectral density

$$
S_0(f) = |H(f)|^2 N_0/2.
\tag{8.3.40}
$$

Hence, the total noise power at the output of the matched filter is

$$
\begin{aligned}
P_n &= \int_{-\infty}^{\infty} S_0(f)\, df \\
&= \int_{-\infty}^{\infty} \frac{N_0}{2} |H(f)|^2\, df = \frac{N_0}{2} \int_{-\infty}^{\infty} |S(f)|^2\, df = \frac{\mathscr{E}_s N_0}{2}.
\end{aligned}
\tag{8.3.41}
$$

The output SNR is simply the ratio of the signal power

$$
P_s = y_s^2(T)
\tag{8.3.42}
$$

to the noise power P_n. Hence,

$$
\left(\frac{S}{N}\right)_0 = \frac{P_s}{P_n} = \frac{\mathscr{E}_s^2}{\mathscr{E}_s N_0/2} = \frac{2\mathscr{E}_s}{N_0},
\tag{8.3.43}
$$

which agrees with the result given by Equation (8.3.36).

Example 8.3.5

Consider the binary orthogonal PPM signals, which are shown in Figure 8.10, for transmitting information over an AWGN channel. The noise is assumed to have zero mean and power spectral density $N_0/2$. Determine the impulse response of the matched filter demodulators, and the output waveforms of the matched filter demodulators when the transmitted signal is $s_1(t)$.

Solution   The binary PPM signals have dimension $N = 2$. Hence, two basis functions are needed to represent the signals. From Figure 8.11, we choose $\psi_1(t)$ and $\psi_2(t)$ as

$$\psi_1(t) = \begin{cases} \sqrt{\dfrac{2}{T_b}}, & 0 \leq t \leq \dfrac{T_b}{2} \\ 0, & \text{otherwise} \end{cases} ;$$

$$\psi_2(t) = \begin{cases} \sqrt{\dfrac{2}{T_b}}, & \dfrac{T_b}{2} \leq t \leq T_b \\ 0, & \text{otherwise} \end{cases} .$$

These waveforms are illustrated in Figure 8.25(a). The impulse responses of the two matched filters are

$$h_1(t) = \psi_1(T_b - t) = \begin{cases} \sqrt{\dfrac{2}{T_b}}, & \dfrac{T_b}{2} \leq t \leq T_b \\ 0, & \text{otherwise} \end{cases}$$

and

$$h_2(t) = \psi_2(T_b - t) = \begin{cases} \sqrt{\dfrac{2}{T_b}}, & 0 \leq t \leq T_b/2 \\ 0, & \text{otherwise} \end{cases}$$

and are illustrated in Figure 8.25(b).

If $s_1(t)$ is transmitted, the (noise-free) responses of the two matched filters are as shown in Figure 8.25(c). Since $y_1(t)$ and $y_2(t)$ are sampled at $t = T_b$, we observe that $y_{1s}(T_b) = \sqrt{\mathcal{E}_b}$ and $y_{2s}(T_b) = 0$. Hence, the received vector formed from the two matched filter outputs at the sampling instant $t = T_b$ is

$$\mathbf{y} = (y_1, y_2) = \left(\sqrt{\mathcal{E}_b} + n_1, n_2 \right), \tag{8.3.44}$$

where $n_1 = y_{1n}(T_b)$ and $n_2 = y_{2n}(T_b)$ are the noise components at the outputs of the matched filters, given by

$$y_{kn}(T_b) = \int_0^{T_b} n(t)\psi_k(t)\,dt, \qquad k = 1, 2. \tag{8.3.45}$$

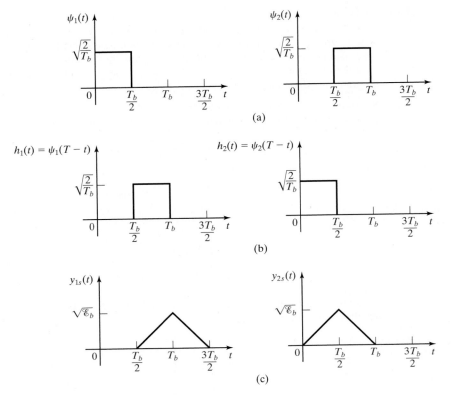

Figure 8.25 Basis functions and matched filter responses for Example 8.3.5.

Clearly, $E[n_k] = E[y_{kn}(T_b)] = 0$. Their variance is

$$\sigma_n^2 = E\left[y_{kn}^2(T_b)\right] = \int_0^{T_b} \int_0^{T_b} E[n(t)n(\tau)]\psi_k(t)\psi_k(\tau)\, dt\, d\tau$$

$$= \frac{N_0}{2} \int_0^{T_b} \int_0^{T_b} \delta(t - \tau)\psi_k(\tau)\psi_k(t)\, dt\, d\tau \tag{8.3.46}$$

$$= \frac{N_0}{2} \int_0^{T_b} \psi_k^2(t)\, dt = \frac{N_0}{2}.$$

Observe that for the first matched filter,

$$\left(\frac{S}{N}\right)_0 = \frac{\left(\sqrt{\mathcal{E}_b}\right)^2}{N_0/2} = \frac{2\mathcal{E}_b}{N_0},$$

which agrees with our previous result, since $\mathcal{E}_b$ is the signal energy per information bit. Also, note that the outputs of the two matched filters corresponding to the transmitted signal $s_2(t)$ are $(y_1, y_2) = \left(n_1, \sqrt{\mathcal{E}_b} + n_2\right)$. ∎

8.3.3 The Performance of the Optimum Detector for Binary Signals

In this section, we describe the optimum decision rule, which is employed by the detector to make decisions based on the output from the demodulator. For this development, we assume that the signals received in successive signal intervals are statistically independent, so the detector only needs to consider its input in a given bit interval when making a decision on the transmitted signal in that bit interval.

Binary Antipodal Signals. As we have observed, the output of the demodulator in any signal (bit) interval is

$$y = s_m + n, \qquad m = 1, 2, \qquad (8.3.47)$$

where $s_m = \pm\sqrt{\mathcal{E}_b}$ and n is a zero-mean Gaussian random variable with variance $\sigma_n^2 = N_0/2$. The conditional probability density functions $p(y|s_m)$, $m = 1, 2$, are given by Equation (8.3.8) and illustrated in Figure 8.18.

Since the input to the detector is a scalar, it is apparent that the detector compares y with a threshold α, determines whether $y > \alpha$ and declares that the signal $s_1(t)$ was transmitted. Otherwise, it declares that $s_2(t)$ was transmitted. The optimality of this scheme will be shown in Section 8.4.6. Later in this chapter, we demonstrate that this decision rule maximizes the probability of making a correct decision, or equivalently, that it minimizes the probability of error.

For the binary antipodal signals, the average probability of error as a function of the threshold α is

$$P_2(\alpha) = P(s_1) \int_{-\infty}^{\alpha} f(y|s_1)dy + P(s_2) \int_{\alpha}^{\infty} f(y|s_2)\,dy, \qquad (8.3.48)$$

where $P(s_1)$ and $P(s_2)$ are the *a priori* probabilities of the two possible transmitted signals. Let us determine the value of the threshold α, say α^*, that minimizes the average probability of error. Differentiating $P_2(\alpha)$ with respect to α and setting the derivative to zero, we obtain

$$P(s_1)f(\alpha|s_1) - P(s_2)f(\alpha|s_2) = 0,$$

or equivalently,

$$\frac{f(\alpha|s_1)}{f(\alpha|s_2)} = \frac{P(s_2)}{P(s_1)}. \qquad (8.3.49)$$

Substituting the conditional probability density functions given by Equation (8.3.8) into Equation (8.3.49) with $s_1 = \sqrt{\mathcal{E}_b}$ and $s_2 = -\sqrt{\mathcal{E}_b}$, we have

$$e^{-\left(\alpha-\sqrt{\mathcal{E}_b}\right)^2/N_0}\, e^{\left(\alpha+\sqrt{\mathcal{E}_b}\right)^2/N_0} = \frac{P(s_2)}{P(s_1)}, \qquad (8.3.50)$$

or equivalently,

$$e^{4\alpha\sqrt{\mathcal{E}_b}/N_0} = \frac{P(s_2)}{P(s_1)}.$$

Clearly, the optimum value of the threshold is

$$\alpha^* = \frac{N_0}{4\sqrt{\mathcal{E}_b}} \ln \frac{P(s_2)}{P(s_1)}. \tag{8.3.51}$$

We observe that if $P(s_1) > P(s_2)$, then $\alpha^* < 0$, and if $P(s_2) > P(s_1)$, then $\alpha^* > 0$. In practice, the two possible signals are usually equally probable, i.e., the a priori probabilities $P(s_1) = P(s_2) = 1/2$. Here, Equation (8.3.51) yields the threshold $\alpha^* = 0$. For this case, the average probability of error is

$$P_2 = \frac{1}{2} \int_{-\infty}^0 f(y|s_1)dy + \frac{1}{2} \int_0^\infty f(y|s_2)dy$$

$$= \int_{-\infty}^0 f(y|s_1)dy = \frac{1}{\sqrt{\pi N_0}} \int_{-\infty}^0 e^{-\left(y - \sqrt{\mathcal{E}_b}\right)^2/N_0} dy. \tag{8.3.52}$$

By making a simple change in variable, specifically $x = \left(y - \sqrt{\mathcal{E}_b}\right)/\sqrt{N_0/2}$, the integral may be expressed as

$$P_2 = \frac{1}{\sqrt{2\pi}} \int_{-\infty}^{-\sqrt{2\mathcal{E}_b/N_0}} e^{-x^2/2} dx$$

$$= Q\left(\sqrt{\frac{2\mathcal{E}_b}{N_0}}\right), \tag{8.3.53}$$

where $Q(x)$ is the area under the tail of the normal (Gaussian) probability density function, defined as

$$Q(x) = \frac{1}{\sqrt{2\pi}} \int_{-\infty}^{-x} e^{-u^2/2} du = \frac{1}{\sqrt{2\pi}} \int_x^\infty e^{-u^2/2} du. \tag{8.3.54}$$

For more properties of $Q(x)$, see the discussion following Equation (5.1.7). We observe that the probability of error depends only on the signal-to-noise ratio $2\mathcal{E}_b/N_0$ and not on the detailed characteristics of the pulse waveforms. Also, Equations (5.1.8–5.1.10) clearly show that the error probability tends to zero exponentially as SNR increases.

Binary Orthogonal Signals.   In the case of binary orthogonal signals, the output of the demodulator is the two-dimensional vector $y = (y_1, y_2)$, where y_1 and y_2 are the outputs of the two cross-correlators or the two matched filters. Recall that if the signal $s_1(t)$ is transmitted, the demodulator outputs are

$$y_1 = \sqrt{\mathcal{E}_b} + n_1$$

and

$$y_2 = n_2,$$

where the noise components n_1 and n_2 are statistically independent, zero-mean Gaussian variables with variance $\sigma_n^2 = N_0/2$. For the important case in which the two orthogonal signals are equally probable, i.e., $P(s_1) = P(s_2) = 1/2$, the detector that minimizes the average probability of error simply compares y_1 with y_2. If $y_1 > y_2$, the detector declares that $s_1(t)$ was transmitted. Otherwise, it declares that $s_2(t)$ was transmitted. The optimality of this scheme follows from the discussion in Section 8.4.6. Based on this decision rule, assuming that $s_1(t)$ was transmitted, the probability of error is simply the probability that $y_1 - y_2 < 0$. Since y_1 and y_2 are Gaussian with equal variance $\sigma_n^2 = N_0/2$ and statistically independent, the difference

$$z = y_1 - y_2$$
$$= \sqrt{\mathcal{E}_b} + n_1 - n_2 \tag{8.3.55}$$

is also a Gaussian random variable with mean $\sqrt{\mathcal{E}_b}$ and variance $\sigma_z^2 = N_0$. Consequently, the probability density function of z is

$$f(z) = \frac{1}{\sqrt{2\pi N_0}} e^{-\left(z - \sqrt{\mathcal{E}_b}\right)^2 / 2N_0}, \tag{8.3.56}$$

and the average probability of error is

$$P_2 = P(z < 0) = \int_{-\infty}^{0} f(z)\, dz$$

$$= \frac{1}{\sqrt{2\pi}} \int_{-\infty}^{-\sqrt{\mathcal{E}_b/N_0}} e^{-x^2/2}\, dx = Q\left(\sqrt{\frac{\mathcal{E}_b}{N_0}}\right). \tag{8.3.57}$$

When we compare the average probability of error of binary antipodal signals given by Equation (8.3.53) to that of binary orthogonal signals, we observe that, for the same error probability P_2, the binary antipodal signals require a factor of two (3 dB) less signal energy than orthogonal signals. The graphs of P_2 for these two signal types are shown in Figure 8.26.

8.4 *M*-ARY PULSE MODULATION

In Sections 8.2 and 8.3, we described the transmission of one bit at a time by employing two signal waveforms that are either antipodal or orthogonal. In this section, we consider the simultaneous transmission of multiple bits by using more than two signal waveforms. In this case, the binary information sequence is subdivided into blocks of k bits, called symbols, and each block (or symbol) is represented by one of $M = 2^k$ signal waveforms, each of duration T. This type of modulation is called *M*-ary modulation. Here we define the signaling (symbol) rate, R_s, as the number of signals (or symbols) transmitted per second. Clearly,

$$R_s = \frac{1}{T},$$

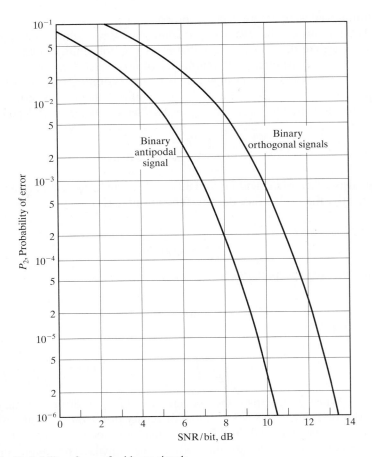

Figure 8.26   Probability of error for binary signals.

and since each signal carries $k = \log_2 M$ bits of information, the bit rate is given by

$$R_b = k R_s = \frac{k}{T}.$$

The bit interval is

$$T_b = \frac{1}{R_b} = \frac{T}{k};$$

it is shown in Figure 8.27.

The M signal waveforms may be one-dimensional or multi-dimensional. The one-dimensional M-ary signals are a generalization of the binary PAM (antipodal) signals. The multi-dimensional signals are a generalization of the binary PPM (orthogonal) signals. We begin with the description of M-ary PAM.

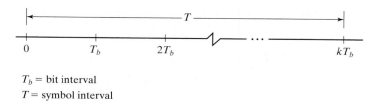

T_b = bit interval
T = symbol interval

Figure 8.27   Relationship between the symbol interval and the bit interval.

8.4.1 *M*-ary Pulse Amplitude Modulation

The generalization of binary PAM to *M*-ary PAM is relatively straightforward. The *k*-bit symbols are used to select $M = 2^k$ signal amplitudes. Hence, in general, the *M*-ary PAM signal waveforms may be expressed as

$$\begin{aligned} s_m(t) &= A_m g_T(t), \ \ 0 \le t \le T, \ \ m = 1, 2, \dots, M \\ &= s_m \psi(t), \ \ 0 \le t \le T, \ \ m = 1, 2, \dots, M, \end{aligned} \tag{8.4.1}$$

where the pulse $g_T(t)$ and the basis function $\psi(t)$ are shown in Figures 8.28(a) and 8.28(b), respectively. We observe that all *M* signal waveforms have the same pulse shape. Hence, they are one-dimensional signals. We also note that $s_m = A_m \sqrt{T}$. An important feature of these PAM signals is that they have different energies. That is,

$$\mathscr{E}_m = \int_0^T s_m^2(t)dt = s_m^2 \int_0^T \psi^2(t)dt = s_m^2 = A_m^2 T. \tag{8.4.2}$$

Assuming that all *k*-bit symbols are equally probable, the average energy of the transmitted signals is

$$\mathscr{E}_{av} = \frac{1}{M} \sum_{m=1}^{M} \mathscr{E}_m = \frac{T}{M} \sum_{m=1}^{M} A_m^2. \tag{8.4.3}$$

In order to minimize the average transmitted energy and to avoid transmitting signals with a DC component, we want to select the *M* signal amplitudes to be symmetric about the origin and equally spaced. That is,

$$A_m = (2m - 1 - M)A, \ \ m = 1, 2, \dots, M, \tag{8.4.4}$$

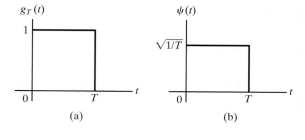

Figure 8.28   Rectangular pulse $g_T(t)$ and basis function $\psi(t)$ for *M*-ary PAM.

where A is an arbitrary scale factor. The corresponding average energy, assuming that all k-bit symbols are equally probable, is

$$\mathcal{E}_{av} = \frac{A^2 T}{M} \sum_{m=1}^{M} (2m - 1 - M)^2 \tag{8.4.5}$$

$$= A^2 T (M^2 - 1)/3.$$

For the signal amplitudes given in Equation (8.4.4), the corresponding signal constellation points in a geometric representation of the M-ary PAM signals are given as

$$\begin{aligned} s_m &= A_m \sqrt{T} \\ &= A\sqrt{T}(2m - 1 - M), \quad m = 1, 2, .., M. \end{aligned} \tag{8.4.6}$$

It is convenient to define the distance parameter d as $d = A\sqrt{T}$, so that

$$s_m = (2m - 1 - M)d, \quad m = 1, 2, \ldots, M. \tag{8.4.7}$$

The signal constellation point diagram is shown in Figure 8.29. Note that the distance between two adjacent signal points is $2d$.

Example 8.4.1

Sketch the signal waveforms for $M = 4$ PAM, which are described by Equation (8.4.1), and determine the average transmitted signal energy.

Solution   The four signal waveforms are shown in Figure 8.30. The average energy, based on equally probable signals according to Equation (8.4.5) is

$$\mathcal{E}_{av} = 5A^2 T = 5d^2,$$

where $d^2 = A^2 T$ by definition.   ■

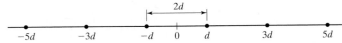

Figure 8.29   Signal point constellation for M-ary PAM.

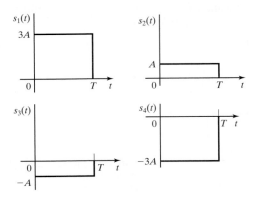

Figure 8.30 $M = 4$ PAM signal waveforms.

8.4.2 *M*-ary Orthogonal Signals

M-ary orthogonal signal waveforms at baseband can be constructed in a variety of ways. For example, Figure 8.31 illustrates two sets of $M = 4$ orthogonal signal waveforms. We observe that the signal waveforms $s_i(t)$ and $i = 1, 2, 3, 4$ in Figure 8.31(a) do not overlap over the time interval $(0, T)$, and they represent pulse position modulation (PPM) for $M = 4$. The signal waveforms $s_i'(t)$ and $i = 1, 2, 3, 4$ in Figure 8.31(b) are overlapping in time, but still satisfy the orthogonality condition, namely,

$$\int_0^T s_i'(t)s_j'(t)dt = 0, \quad i \neq j. \tag{8.4.8}$$

The number of dimensions required to represent a set of M orthogonal waveforms is $N = M$. Hence, a set of M orthogonal signal waveforms can be represented geometrically by M orthogonal vectors in M-dimensional space. To be specific, let us consider M-ary PPM signal waveforms. Such a set of M baseband PPM signals are expressed mathematically as

$$s_m(t) = \sqrt{\mathscr{E}_s}\psi_m(t), \quad m = 1, 2, .., M, \tag{8.4.9}$$

where $\psi_m(t)$ and $m = 1, 2, .., M$ are a set of M orthogonal basis waveforms. These waveform are defined as

$$\psi_m(t) = g_T\left(t - \frac{(m-1)T}{M}\right), \quad \frac{(m-1)T}{M} \leq t \leq \frac{mT}{M}, \tag{8.4.10}$$

in which $g_T(t)$ is a unit energy pulse, which is nonzero over the time interval $0 \leq t \leq T/M$ and the basis functions $\psi_m(t)$ and $m = 1, 2, \ldots, M$ are simply time-shifted replicas of $g_T(t)$, as illustrated in Figure 8.32. Each signal waveform $s_m(t)$ has energy

$$\int_0^T s_m^2(t)dt = \mathscr{E}_s \int_0^T \psi_m^2(t)dt = \mathscr{E}_s, \text{ all } m.$$

$\mathscr{E}_s$ denotes the energy of each of the signal waveforms representing k-bit symbols. Consequently, M-ary PPM signal waveforms are represented geometrically by the following M-dimensional vectors:

$$\begin{aligned}
s_1 &= \left(\sqrt{\mathscr{E}_s}, 0, 0, \ldots, 0\right); \\
s_2 &= \left(0, \sqrt{\mathscr{E}_s}, 0, \ldots, 0\right); \\
&\vdots \\
s_M &= \left(0, 0, 0, \ldots, \sqrt{\mathscr{E}_s}\right).
\end{aligned} \tag{8.4.11}$$

Clearly, these vectors are orthogonal, i.e., $s_i \cdot s_j = 0$ when $i \neq j$. It is also interesting to note that the M signal vectors are mutually equidistant, i.e.,

$$d_{mn} = \sqrt{\| s_m - s_n \|^2} = \sqrt{2\mathscr{E}_s}, \quad \text{for all } m \neq n. \tag{8.4.12}$$

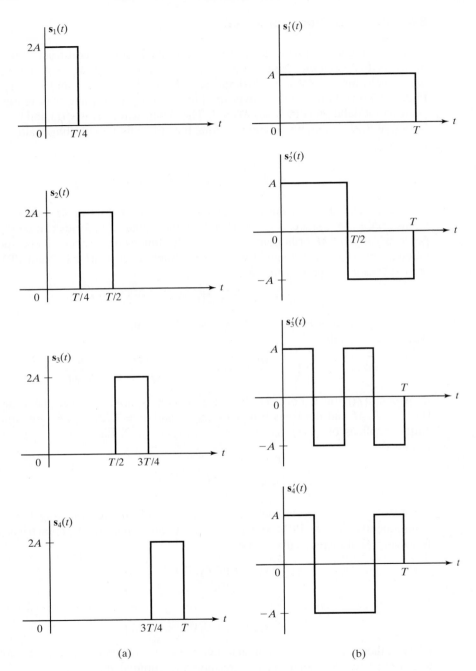

(a)   (b)

Figure 8.31 Two sets of $M = 4$ orthogonal signal waveforms.

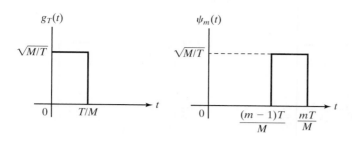

Figure 8.32   Rectangular pulse $g_T(t)$ and basis function $\psi_m(t)$ for *M*-ary PPM signal waveforms.

Example 8.4.2

Determine the vectors in a geometric representation of the $M = 4$ signal waveforms $s_i'(t)$ and $i = 1, 2, 3, 4$, that are shown in Figure 8.31(b). Use the basis waveforms $\psi_m(t)$ that are shown in Figure 8.32

Solution   We note that the four orthogonal waveforms have equal energy, given by

$$\int_0^T \left[s_i'(t) \right]^2 dt = \frac{\mathcal{E}_s}{T} \int_0^T dt = \mathcal{E}_s.$$

By computing the projection of each signal waveform on the four basis waveforms $\psi_m(t)$, i.e.,

$$\int_0^T s_i'(t)\psi_m(t)dt, \quad m = 1, 2, 3, 4,$$

we obtain the vector s_i'. Thus, we obtain

$$
\begin{aligned}
s_1' &= \left(\sqrt{\mathcal{E}_s/4}, \; \sqrt{\mathcal{E}_s/4}, \; \sqrt{\mathcal{E}_s/4}, \; \sqrt{\mathcal{E}_s/4} \right); \\
s_2' &= \left(\sqrt{\mathcal{E}_s/4}, \; \sqrt{\mathcal{E}_s/4}, \; -\sqrt{\mathcal{E}_s/4}, \; -\sqrt{\mathcal{E}_s/4} \right); \\
s_3' &= \left(\sqrt{\mathcal{E}_s/4}, \; -\sqrt{\mathcal{E}_s/4}, \; \sqrt{\mathcal{E}_s/4}, \; -\sqrt{\mathcal{E}_s/4} \right); \\
s_4' &= \left(\sqrt{\mathcal{E}_s/4}, \; -\sqrt{\mathcal{E}_s/4}, \; -\sqrt{\mathcal{E}_s/4}, \; \sqrt{\mathcal{E}_s/4} \right).
\end{aligned}
$$

We observe that these four signal vectors are orthogonal, i.e., $s_i' \cdot s_j' = 0$ for $i \neq j$.   ∎

8.4.3 Biorthogonal Signals

In general, a set of M biorthogonal signals are constructed from a set of $M/2$ orthogonal signals $[s_i(t), \; i = 1, 2, \ldots, M/2]$ and their negatives $[-s_i(t), i = 1, 2, \ldots, M/2]$. As we shall observe in the next chapter, the channel bandwidth required to transmit the information sequence via biorthogonal signals is just one-half of that required to transmit M orthogonal signals. For this reason, biorthogonal signals are preferred over orthogonal signals in some applications.

To develop the geometric representation of M-ary biorthogonal signals, let us consider PPM signals. We begin with $M/2$ PPM signals of the form

$$s_m(t) = \sqrt{\mathcal{E}_s}\,\psi_m(t), \quad m = 1, 2, \ldots, M/2, \quad 0 \leq t \leq T, \tag{8.4.13}$$

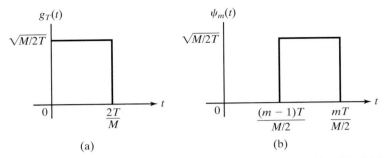

Figure 8.33 Rectangular pulse $g_T(t)$ and basis function $\psi_m(t)$ for M-ary biorthogonal PPM signal waveforms.

where $\psi_m(t), m = 1, 2, \ldots$, and $M/2$ are a set of $M/2$ basis waveforms as illustrated in Figure 8.33 and defined mathematically as

$$\psi_m(t) = g_T\left(t - \frac{(m-1)T}{M/2}\right), \quad \frac{(m-1)T}{M/2} \leq t \leq \frac{mT}{M/2}, \qquad (8.4.14)$$

where $g_T(t)$ is also shown in Figure 8.33. The negatives of these signal waveforms are

$$s_{\frac{M}{2}+m}(t) = -s_m(t) = -\sqrt{\mathcal{E}_s}\psi_m(t), \quad m = 1, 2, .., M/2. \qquad (8.4.15)$$

We observe that the number of dimensions needed to represent the $M/2$ orthogonal signals is $N = M/2$. In the case of PPM signals, the signal vectors of $M/2$ dimensions are

$$\begin{aligned} s_1 &= \left(\sqrt{\mathcal{E}_s}, 0, 0, \ldots, 0\right); \\ s_2 &= \left(0, \sqrt{\mathcal{E}_s}, 0, \ldots, 0\right); \\ &\vdots \\ s_{M/2} &= \left(0, 0, 0, \ldots, \sqrt{\mathcal{E}_s}\right); \end{aligned} \qquad (8.4.16)$$

the other $M/2$ signal vectors corresponding to the signal waveforms that are the negatives of the $M/2$ PPM signal waveforms are

$$\begin{aligned} s_{\frac{M}{2}+1} &= \left(-\sqrt{\mathcal{E}_s}, 0, 0, \ldots, 0\right); \\ s_{\frac{M}{2}+2} &= \left(0, -\sqrt{\mathcal{E}_s}, 0, \ldots, 0\right); \\ &\vdots \\ s_M &= \left(0, 0, 0, \ldots, -\sqrt{\mathcal{E}_s}\right). \end{aligned} \qquad (8.4.17)$$

Example 8.4.3

Determine the vector representation of the four biorthogonal signal waveforms shown in Figure 8.34(a).

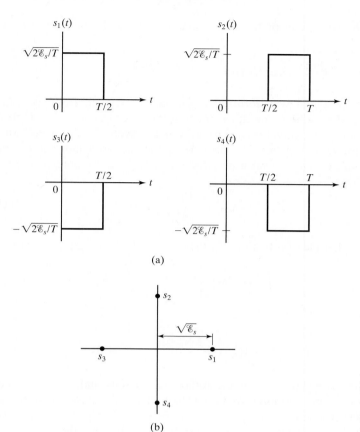

Figure 8.34 $M = 4$ biorthogonal signal waveforms and signal point constellation for Example 8.4.3.

Solution   The signal waveform $s_1(t)$ and $s_2(t)$ are orthogonal and are represented by the vectors

$$\boldsymbol{s}_1 = \left(\sqrt{\mathscr{E}_s}, 0\right)$$

and

$$\boldsymbol{s}_2 = \left(0, \sqrt{\mathscr{E}_s}\right).$$

The signal waveforms $s_3(t) = -s_1(t)$ and $s_4(t) = -s_2(t)$. Therefore, the geometric representation of $s_3(t)$ and $s_4(t)$ is

$$\boldsymbol{s}_3 = \left(-\sqrt{\mathscr{E}_s}, 0\right)$$

and

$$\boldsymbol{s}_4 = \left(0, -\sqrt{\mathscr{E}_s}\right).$$

The signal points corresponding to these vectors are shown in Figure 8.34(b).   ∎

In a general set of M biorthogonal signals, the distance between any pair of signal vectors is either $d_{mn} = \sqrt{2\mathscr{E}_s}$ when the two signal vectors are orthogonal or $d_{mn} = 2\sqrt{\mathscr{E}_s}$ when one of the two signal vectors is the negative of the other (antipodal).

8.4.4 Simplex Signal Waveforms

Another set of M signal waveforms that can be constructed from M orthogonal signals is obtained by subtracting the average of the M orthogonal signals from each of the orthogonal signal waveforms. The M signal waveforms that result are called *simplex signal waveforms*. Thus, if we have M orthogonal baseband signal waveforms $\{s_m(t)\}$, the simplex signal waveforms, denoted as $\{s'_m(t)\}$, are constructed as

$$s'_m(t) = s_m(t) - \frac{1}{M} \sum_{k=1}^{M} s_k(t). \tag{8.4.18}$$

Then, it follows that (see Problem 8.26) the energy of these signals $s'_m(t)$ is

$$\mathscr{E}'_s = \int_0^T \left[s'_m(t) \right]^2 dt = \left(1 - \frac{1}{M} \right) \mathscr{E}_s \tag{8.4.19}$$

and

$$\int_0^T s'_m(t)s'_n(t)\, dt = -\frac{\mathscr{E}_s}{M-1}, \quad m \neq n, \tag{8.4.20}$$

where $\mathscr{E}_s$ is the energy of each of the orthogonal signals and $\mathscr{E}'_s$ is the energy of each of the signals in the simplex signal set. Note that the waveforms in the simplex set have smaller energy than the waveforms in the orthogonal signal set. Second, we note that simplex signal waveforms are not orthogonal. Instead, they have a negative correlation, which is equal for all pairs of signal waveforms. We surmise that among all the possible M-ary signal waveforms of equal energy $\mathscr{E}_s$, the simplex signal set results in the smallest probability of error when used to transmit information on an additive white Gaussian noise channel.

The geometric representation of a set of M simplex signals is obtained by subtracting the mean-signal vector from a set of M orthogonal vectors. Thus, we have

$$s'_m = s_m - \frac{1}{M} \sum_{k=1}^{M} s_k, \quad m = 1, 2, \ldots, M. \tag{8.4.21}$$

The effect of subtracting the mean signal

$$\bar{s} = \frac{1}{M} \sum_{k=1}^{M} s_k \tag{8.4.22}$$

from each orthogonal vector is to translate the origin of the M orthogonal signals to the point $\bar{s}$ and to minimize the energy in the signal set $\{s'_m\}$.

If the energy per signal for the orthogonal signals is $\mathscr{E}_s = \| s_m \|^2$, then the energy for the simplex signals is

$$\mathscr{E}'_s = \| s'_m \|^2 = \| s_m - \bar{s} \|^2$$

$$= \left(1 - \frac{1}{M}\right) \mathscr{E}_s. \tag{8.4.23}$$

The distance between any two signal points is not changed by the translation of the origin, i.e., the distance between signal points remains at $d = \sqrt{2\mathscr{E}_s}$. Finally, as indicated, the M simplex signals are correlated. The cross-correlation coefficient (normalized cross correlation) between the mth and nth signals is

$$\gamma_{mn} = \frac{s'_m \cdot s'_m}{\| s'_m \| \| s'_n \|}$$

$$= \frac{-1/M}{(1 - 1/M)} = \frac{-1}{M - 1}. \tag{8.4.24}$$

Hence, all the signals have the same pair-wise correlation. Figure 8.35 illustrates a set of $M = 4$ simplex signals.

8.4.5 Binary-Coded Signal Waveforms

Signal waveforms for transmitting digital information may also be constructed from a set of M binary code words of the form

$$c_m = (c_{m1}, c_{m2}, \ldots, c_{mN}), \quad m = 1, 2, \ldots, M, \tag{8.4.25}$$

where $c_{mj} = 0$ or 1 for all m and j. In this form, N is called the block length, or dimension, of the code words. Given M code words, we can construct M signal waveforms by mapping a code bit $c_{mj} = 1$ into a pulse $g_T(t)$ of duration T/N and a code bit $c_{mj} = 0$ into the negative pulse $-g_T(t)$.

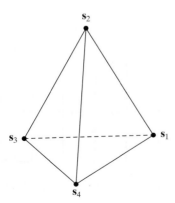

Figure 8.35 Signal constellation for the $M = 4$ simplex signals.

Example 8.4.4

Given the code words

$$
\begin{aligned}
\mathbf{c}_1 &= [1\ 1\ 1\ 1\ 0] \\
\mathbf{c}_2 &= [1\ 1\ 0\ 0\ 1] \\
\mathbf{c}_3 &= [1\ 0\ 1\ 0\ 1] \\
\mathbf{c}_4 &= [0\ 1\ 0\ 1\ 0]
\end{aligned}
$$

construct a set of $M - 4$ signal waveforms, as previously described, using a rectangular pulse $g_T(t)$.

Solution   As indicated, a code bit 1 is mapped into the rectangular pulse $g_T(t)$ of duration $T/5$, and a code bit 0 is mapped into the rectangular pulse $-g_T(t)$. Thus, we construct the four waveforms shown in Figure 8.36 that correspond to the four code words.   ■

Let us consider the geometric representation of a set of M signal waveforms generated from a set of M binary words of the form

$$\mathbf{c}_m = (c_{m1}, c_{m2}, \dots, c_{mN}), \quad m = 1, 2, \dots, M, \tag{8.4.26}$$

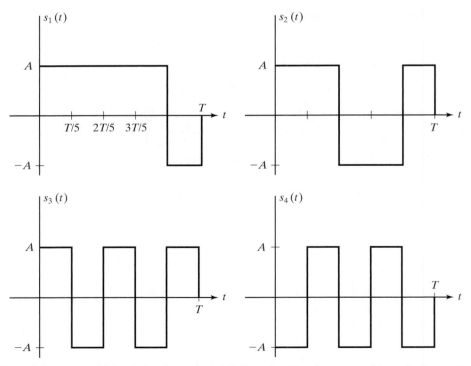

Figure 8.36   A set of $M = 4$ signal waveforms of dimension $N = 5$ constructed from the four code words in Example 8.4.4.

where $c_{mj} = 0$ or 1 for all m and j. The M signal waveforms are of dimension N and are represented geometrically in vector form as

$$s_m = (s_{m1}, s_{m2}, \ldots, s_{mN}), \quad m = 1, 2, .., M, \tag{8.4.27}$$

where $s_{mj} = \pm\sqrt{\mathcal{E}_s/N}$ for all m and j. Thus, the energy of each signal waveform is $\mathcal{E}_s$.

In general, there are 2^N possible signals that can be constructed from the 2^N possible binary code words. The M code words are a subset of the 2^N possible binary code words. We also observe that the 2^N possible signal points correspond to the vertices of an N-dimensional hypercube with its center at the origin. Figure 8.37 illustrates the signal points in $N = 2$ and $N = 3$ dimensions.

The M signals constructed in this manner have equal energy $\mathcal{E}_s$. We determine the cross-correlation coefficient between any pair of signals based on the method used to select the M signals from the 2^N possible signals. This topic is treated in Chapter 13. We can show that any adjacent signal points have a cross-correlation coefficient of (see Problem 8.25)

$$\gamma = \frac{N - 2}{N} \tag{8.4.28}$$

and a corresponding Euclidean distance

$$d = 2\sqrt{\mathcal{E}_s/N}. \tag{8.4.29}$$

The distance between adjacent signal points affects the performance of the detector; this will be discussed later.

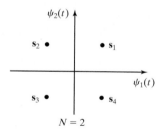

$N = 2$

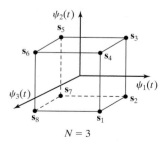

$N = 3$

Figure 8.37   Signal points for signals generated from binary codes.

8.4.6 The Optimum Receiver for *M*-ary Signals in AWGN

In Section 8.3, we described the optimum receiver for binary antipodal and binary orthogonal signals in AWGN. In this section, we derive the optimum receiver for *M*-ary signals corrupted by AWGN.

We assume that the digital communication system transmits digital information by using any of the *M*-ary signal waveforms described in the preceding sections. As described previously, the input sequence to the modulator is subdivided into k-bit blocks, called symbols, and each of the $M = 2^k$ symbols is associated with a corresponding baseband signal waveform from the set $\{s_m(t), m = 1, 2, .., M\}$. Each signal waveform is transmitted within the symbol (signaling) interval or time slot of duration T. To be specific, we consider the transmission of information over the interval $0 \le t \le T$.

The channel is assumed to corrupt the signal by the addition of white Gaussian noise, as shown in Figure 8.38. Thus, the received signal in the interval $0 \le t \le T$ may be expressed as

$$r(t) = s_m(t) + n(t), \quad 0 \le t \le T, \tag{8.4.30}$$

where $n(t)$ denotes the sample function of the additive white Gaussian noise (AWGN) process with the power spectral density $S_n(f) = \frac{N_0}{2}$ W/Hz. Based on the observation of $r(t)$ over the signal interval, we wish to design a receiver that is optimum in the sense that it minimizes the probability of making an error.

The Signal Demodulator.   As in the case of binary signals, it is convenient to subdivide the receiver into two parts: the signal demodulator and the detector. The function of the signal demodulator is to convert the received waveform $r(t)$ into an N-dimensional vector $\mathbf{y} = (y_1, y_2, \ldots, y_N)$, where N is the dimension of the transmitted signal waveforms. The function of the detector is to decide which of the M possible signal waveforms was transmitted based on observation of the vector $\mathbf{y}$.

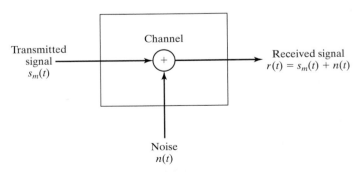

Figure 8.38   Model for received signal passed through an AWGN channel.

We have shown that the *M*-ary signal waveforms, each of which is *N*-dimensional, may be represented in general as

$$s_m(t) = \sum_{k=1}^{N} s_{mk} \psi_k(t), \quad 0 \le t \le T, \tag{8.4.31}$$
$$m = 1, 2 \dots, M,$$

where $\{s_{mk}\}$ are the coordinates of the signal vector

$$\boldsymbol{s}_m = (s_{m1}, s_{m2}, .., s_{mN}), \quad m = 1, 2, \dots, M, \tag{8.4.32}$$

and $\psi_k(t)$ and $k = 1, 2, \dots, N$ are *N* orthonormal basis waveforms that span the *N*-dimensional signal space. Consequently, every one of the *M* possible transmitted signals of the set $\{s_m(t), 1 \le m \le M\}$ can be represented as a weighted linear combination of the *N* basis waveforms $\{\psi_k(t)\}$.

In the case of the noise waveform $n(t)$ in the received signal $r(t)$, the functions $\{\psi_k(t)\}$ do not span the noise space. However, we will show that the noise terms that fall outside the signal space are irrelevant to the detection of the signal.

As a generalization of the demodulator for binary signals, we employ either a correlation-type demodulator that consists of *N* cross correlators in parallel, as shown in Figure 8.39, or a matched-filter-type demodulator that consists of *N* matched filters in parallel, as shown in Figure 8.40. To be specific, suppose the received signal $r(t)$

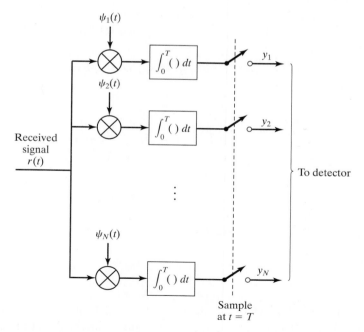

Figure 8.39 Correlation-type demodulator.

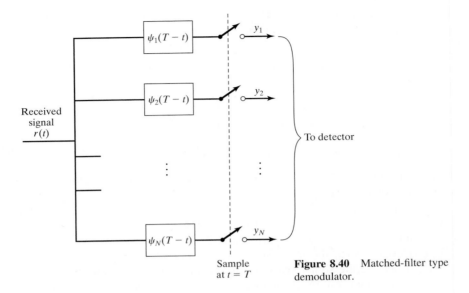

Figure 8.40   Matched-filter type demodulator.

is passed through a parallel bank of N cross correlators. The correlator outputs at the end of the signal interval are

$$\int_0^T r(t)\psi_k(t)dt \; = \int_0^T [s_m(t) + n(t)]\psi_k(t)\, dt \tag{8.4.33}$$

$$y_k \; = s_{mk} + n_k, \quad k = 1, 2, \ldots, N,$$

where

$$s_{mk} = \int_0^T s_m(t)\psi_k(t)\, dt, \quad k = 1, 2, \ldots, N \tag{8.4.34}$$

$$n_k = \int_0^T n(t)\psi_k(t)\, dt, \quad k = 1, 2, \ldots, N.$$

Equation (8.4.33) is equivalent to

$$\mathbf{y} = \mathbf{s}_m + \mathbf{n}, \tag{8.4.35}$$

where the signal is now represented by the vector $\mathbf{s}_m$ with components s_{mk} and $k = 1, 2, \ldots, N$. Their values depend on which of the M signals was transmitted. The components of $\mathbf{n}$, i.e., $\{n_k\}$, are random variables that arise from the presence of the additive noise.

In fact, we can express the received signal $r(t)$ in the interval $0 \leq t \leq T$ as

$$
\begin{aligned}
r(t) &= \sum_{k=1}^{N} s_{mk} \psi_k(t) + \sum_{k=1}^{N} n_k \psi_k(t) + n'(t) \\
&= \sum_{k=1}^{N} y_k \psi_k(t) + n'(t).
\end{aligned}
\tag{8.4.36}
$$

The term $n'(t)$, defined as

$$
n'(t) = n(t) - \sum_{k=1}^{N} n_k \psi_k(t),
\tag{8.4.37}
$$

is a zero-mean Gaussian noise process that represents the difference between the original noise process $n(t)$ and the part that corresponds to the projection of $n(t)$ onto the basis functions $\{\psi_k(t)\}$. We will show that $n'(t)$ is irrelevant when we decide which signal was transmitted. Consequently, the decision may be based entirely on the correlator output signal and noise components $y_k = s_{mk} + n_k, k = 1, 2, \ldots, N$.

Since the signals $\{s_m(t)\}$ are deterministic, the signal components $\{s_{mk}\}$ are deterministic. The noise components $\{n_k\}$ are Gaussian. Their mean values are

$$
E[n_k] = \int_0^T E[n(t)] \psi_k(t) \, dt = 0
\tag{8.4.38}
$$

for all k. Their covariances are

$$
\begin{aligned}
E[n_k n_m] &= \int_0^T \int_0^T E[n(t) n(\tau)] \psi_k(t) \psi_m(t) \, dt \, d\tau \\
&= \int_0^T \int_0^T \frac{N_0}{2} \delta(t - t) \psi_k(t) \psi_m(\tau) \, dt \, d\tau \\
&= \frac{N_0}{2} \int_0^T \psi_k(t) \psi_m(t) \, dt \\
&= \frac{N_0}{2} \delta_{mk},
\end{aligned}
\tag{8.4.39}
$$

where $\delta_{mk} = 1$ when $m = k$ and will otherwise be zero. Therefore, the N noise components $\{n_k\}$ are zero-mean uncorrelated Gaussian random variables with a common variance $\sigma_n^2 = N_0/2$, and it follows that

$$
f(\boldsymbol{n}) = \prod_{i=1}^{N} f(n_i) = \frac{1}{(\pi N_0)^{N/2}} e^{-\sum_{i=1}^{N} \frac{n_i^2}{N_0}}.
\tag{8.4.40}
$$

From the previous development, the correlator outputs $\{y_k\}$ conditioned on the mth signal being transmitted are Gaussian random variables with mean

$$E[y_k] = E[s_{mk} + n_k] = s_{mk} \tag{8.4.41}$$

and equal variance

$$\sigma_y^2 = \sigma_n^2 = N_0/2. \tag{8.4.42}$$

Since the noise components $\{n_k\}$ are uncorrelated Gaussian random variables, they are also statistically independent. As a consequence, the correlator outputs $\{y_k\}$ conditioned on the mth transmitted signal are statistically independent Gaussian variables. Hence, the conditional probability density functions (PDF's) of the random variables $(y_1, y_2, \ldots, y_N) = \boldsymbol{y}$ are simply

$$f(\boldsymbol{y}|s_m) = \prod_{k=1}^{N} f(y_k|s_{mk}), \quad m = 1, 2, \ldots, M, \tag{8.4.43}$$

where

$$f(y_k|s_{mk}) = \frac{1}{\sqrt{\pi N_0}} e^{-(y_k - s_{mk})^2/N_0}, \quad k = 1, 2, \ldots, N. \tag{8.4.44}$$

Substituting Equation (8.4.44) into Equation (8.4.43), we obtain the joint conditional PDF's as

$$f(\boldsymbol{y}|s_m) = \frac{1}{(\pi N_0)^{N/2}} \exp\left[-\sum_{k=1}^{N}(y_k - s_{mk})^2/N_0\right] \tag{8.4.45}$$

$$= \frac{1}{(\pi N_0)^{N/2}} \exp[-\parallel \boldsymbol{y} - \boldsymbol{s}_m \parallel^2 /N_0], \quad m = 1, 2, \ldots, M. \tag{8.4.46}$$

As a final point, we wish to show that the correlator outputs $(y_1, y_2, \ldots, y_N)$ are *sufficient statistics* for reaching a decision on which of the M signals was transmitted. In other words, we wish to show that no additional relevant information can be extracted from the remaining noise process $n'(t)$. Indeed, $n'(t)$ is uncorrelated with the N correlator outputs $\{y_k\}$, i.e.,

$$E[n'(t)y_k] = E[n'(t)]s_{mk} + E[n'(t)n_k]$$

$$= E[n'(t)n_k]$$

$$= E\left[\left(n(t) - \sum_{j=1}^{N} n_j \psi_j(t)\right) n_k\right]$$

$$= \int_0^T E[n(t)n(\tau)]\psi_k(\tau)\,dt - \sum_{j=1}^N E(n_j n_k)\psi_j(t)$$

$$= \frac{N_0}{2}\psi_k(t) - \frac{N_0}{2}\psi_k(t) = 0. \tag{8.4.47}$$

This means $n'(t)$ and r_k are uncorrelated because $n'(t)$ is zero-mean. Since $n'(t)$ and $\{y_k\}$ are Gaussian and uncorrelated, they are also statistically independent. Consequently, $n'(t)$ does not contain any information that is relevant to the decision as to which signal waveform was transmitted. All the relevant information is contained in the correlator outputs $\{y_k\}$. Hence, $n'(t)$ may be ignored.

Example 8.4.5

Consider an M-ary PAM signal set given by Equation (8.4.1), where the signal points $\{s_m\}$ are defined by Equation (8.4.7) and the basis waveform $\psi(t)$ is shown in Figure 8.28. The additive noise is a zero-mean white Gaussian noise process with the spectral density $N_0/2$. Determine the PDF of the received signal at the output of the demodulator and sketch the PDFs for the case $M = 4$.

Solution   The received signal is expressed as

$$r(t) = s_m \psi(t) + n(t)$$

and the output of the demodulator is

$$y(T) = \int_0^T r(t)\psi(t)\,dt = \int_0^T [s_m \psi(t) + n(t)]\psi(t)\,dt$$

$$= s_m + n,$$

where n is a zero-mean Gaussian random variable with variance $\sigma_n^2 = N_0/2$. Therefore, the PDF of $y \equiv y(T)$ is

$$f(y|s_m) = \frac{1}{\sqrt{\pi N_0}} e^{-(y-s_m)^2/N_0}, \quad m = 1, 2, \ldots, M$$

and $s_m = (2m - 1 - M)d$. The PDF's for $M = 4$ PAM are shown in Figure 8.41. ∎

Example 8.4.6

Consider the $M = 4$ orthogonal PPM signal waveforms shown in Figure 8.31(a), where the signal constellation points $\{s_m\}$ are given by Equation (8.4.11), and the basis waveforms $\{\psi_m(t)\}$ are shown in Figure 8.32. The additive noise is a zero-mean white Gaussian noise process with a spectral density $N_0/2$. Determine the PDF of the received

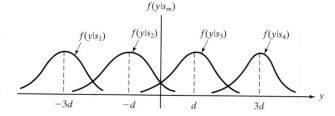

Figure 8.41   PDF's for $M = 4$ received PAM signals in additive white Gaussian noise.

signal vector y at the output of the demodulator, assuming that the signal $s_1(t)$ was transmitted and sketch the PDF's of each of the components of the vector y.

Solution   The signal vector corresponding to the transmitted $s_1(t)$ is

$$s_1 = \left(\sqrt{\mathcal{E}_s}, \ 0, \ 0, \ 0 \right)$$

and the received signal vector is, according to Equation (8.4.35),

$$\begin{aligned} y &= s_1 + n \\ &= \left(\sqrt{\mathcal{E}_s} + n_1, n_2, n_3, n_4 \right), \end{aligned}$$

where the noise components n_1, n_2, n_3, n_4 are mutually statistically independent, zero-mean Gaussian random variables with identical variance $\sigma_n^2 = N_0/2$. Therefore, the joint PDF of the vector components y_1, y_2, y_3, y_4 is, according to Equation (8.4.45) with $N = 4$,

$$f(y|s_1) = f(y_1, y_2, y_3, y_4|s_1) = \frac{1}{(\pi N_0)^2} e^{-\left[\left(y_1 - \sqrt{\mathcal{E}_s} \right)^2 + y_2^2 + y_3^2 + y_4^2 \right]/N_0}.$$

The PDF's of each of the components of the vector y are shown in Figure 8.42. We note that the PDF's of y_2, y_3, y_4 are identical.   ■

The Optimum Detector.   As we have observed from this development, when a signal is transmitted over an AWGN channel, either a correlation-type demodulator or a matched-filter-type demodulator, produces the vector $y = (y_1, y_2, \ldots, y_N)$, which contains all the relevant information in the received signal waveform. The received vector y is the sum of two vectors. The first vector is s_m, the signal vector that is equivalent to the transmitted signal, and the second vector is n, the noise vector. The vector s_m is a point in the signal constellation, and the vector n is an N-dimensional random vector with statistically independent and identically distributed components, where each component is a Gaussian random variable with mean 0 and variance $N_0/2$. Since the components of the noise are independent and have the same mean and variance, the distribution of the noise vector n in the N-dimensional space has spherical symmetry. When s_m is transmitted, the received vector y, which represents the transmitted signal s_m plus the spherically symmetric noise n, can be represented by a spherical cloud centered at s_m. The density of this cloud is higher at the center and becomes less as we depart from s_m, i.e., these points become less likely to be received. The variance of the noise $N_0/2$ determines the density of the noise cloud around the center signal s_m. For low $N_0/2$, the cloud is closely centered

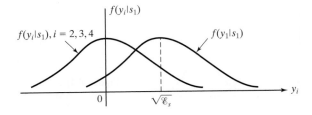

Figure 8.42   The PDF's of the received signal components of the vector y in Example 8.4.6.

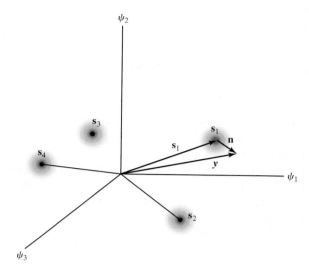

Figure 8.43 Signal constellation, noise cloud, and received vector for $N = 3$ and $M = 4$. It is assumed that s_i is transmitted.

around s_m and its density (representing the probability) reduces sharply as the distance from the center is increased. For high $N_0/2$, the cloud is spread and larger distances have a higher probability compared with the low $N_0/2$ case. The signal constellation, the noise cloud, and the received vector are shown in Figure 8.43 for the case of $N = 3$ and $M = 4$.

We wish to design a signal detector that makes a decision on the transmitted signal in each signal interval based on the observation of the vector y in each interval, such that the probability of a correct decision is maximized. With this goal in mind, we consider a decision rule based on the computation of the *posterior probabilities* defined as

$$P \text{ (signal } s_m \text{ was transmitted } |y) \qquad m = 1, 2, \ldots, M, \qquad (8.4.48)$$

which we abbreviate as $P(s_m|y)$. The decision criterion is based on selecting the signal corresponding to the maximum of the set of posterior probabilities $\{P(s_m|y)\}$. At the end of this section, we show that this criterion maximizes the probability of a correct decision; hence, it minimizes the probability of error. Intuitively, this decision is the best possible decision that minimizes the error probability. It is clear that in the absence of any received information y, the best decision is to choose the signal s_m that has the highest prior probability $P(s_m)$. After receiving the information y, the prior probabilities $P(s_m)$ are replaced with the *posterior* (conditional) probabilities $P(s_m|y)$, and the receiver chooses the s_m that maximizes $P(s_m|y)$. This decision criterion is called the *maximum a posteriori probability* (MAP) criterion.

Using Bayes's rule, we may express the posterior probabilities as

$$P(s_m|y) = \frac{f(y|s_m)P(s_m)}{f(y)}, \qquad (8.4.49)$$

where $f(y|s_m)$ is the conditional PDF of the observed vector given s_m, and $P(s_m)$ is the *a priori probability* of the mth signal being transmitted. The denominator of Equation (8.4.49) may be expressed as

$$f(y) = \sum_{m=1}^{M} f(y|s_m) P(s_m).$$

(8.4.50)

From Equations (8.4.49) and (8.4.50), we observe that the computation of the posterior probabilities $P(s_m|y)$ requires knowledge of the *a priori probabilities* $P(s_m)$ and the conditional PDF's $f(y|s_m)$ for $m = 1, 2, \ldots, M$.

Some simplification occurs in the MAP criterion when the M signals are equally probable a priori, i.e., $P(s_m) = 1/M$ for all M. Furthermore, we note that the denominator in Equation (8.4.49) is independent of which signal is transmitted. Consequently, the decision rule based on finding the signal that maximizes $P(s_m|y)$ is equivalent to finding the signal that maximizes $f(y|s_m)$.

The conditional PDF $f(y|s_m)$, or any monotonic function of it, is usually called the *likelihood function*. The decision criterion based on the maximum of $f(y|s_m)$ over the M signals is called *the maximum-likelihood (ML) criterion*. We observe that a detector based on the MAP criterion and one that is based on the ML criterion make the same decisions, as long as the apriori probabilities $P(s_m)$ are all equal; in other words, the signals $\{s_m\}$ are equiprobable.

In the case of an AWGN channel, the likelihood function $f(y|s_m)$ is given by Equation (8.4.45). To simplify the computations, we may work with the natural logarithm of $f(y|s_m)$, which is a monotonic function. Thus,

$$\ln f(y|s_m) = \frac{-N}{2} \ln(\pi N_0) - \frac{1}{N_0} \sum_{k=1}^{N} (y_k - s_{mk})^2.$$

(8.4.51)

The maximum of $\ln f(y|s_m)$ over s_m is equivalent to finding the signals s_m that minimize the Euclidean distance

$$D(y, s_m) = \sum_{k=1}^{N} (y_k - s_{mk})^2.$$

(8.4.52)

We call $D(y, s_m)$ $m = 1, 2, \ldots, M$, the *distance metrics*. Hence, for the AWGN channel, the decision rule based on the ML criterion reduces to finding the signal s_m that is closest in distance to the received signal vector y. We will refer to this decision rule as *minimum distance detection*.

Another interpretation of the optimum decision rule based on the ML criterion is obtained by expanding the distance metrics in Equation (8.4.52) as

$$
\begin{aligned}
D(y, s_m) &= \sum_{n=1}^{N} y_n^2 - 2 \sum_{n=1}^{N} y_n s_{mn} + \sum_{n=1}^{N} s_{mn}^2 \\
&= \| y \|^2 - 2y \cdot s_m + \| s_m \|^2, \qquad m = 1, 2, \ldots, M.
\end{aligned}
$$

(8.4.53)

The term $\| y \|^2$ is common to all decision metrics; hence, it may be ignored in the computations of the metrics. The result is a set of modified distance metrics

$$D'(y, s_m) = -2y \cdot s_m + \| s_m \|^2 . \tag{8.4.54}$$

Note that selecting the signal s_m that minimizes $D'(y, s_m)$ is equivalent to selecting the signal that maximizes the metric $C(y, s_m) = -D'(y, s_m)$, i.e.,

$$C(y, s_m) = 2y \cdot s_m - \| s_m \|^2 . \tag{8.4.55}$$

The term $y \cdot s_m$ represents the projection of the received signal vector onto each of the M possible transmitted signal vectors. The value of each of these projections is a measure of the correlation between the received vector and the mth signal. For this reason, we call $C(y, s_m), m = 1, 2, \ldots , M$, the *correlation metrics* for deciding which of the M signals was transmitted. Finally, the terms $\| s_m \|^2 = \mathscr{E}_m, m = 1, 2, \ldots M$, may be viewed as bias terms that serve as compensation for signal sets that have unequal energies, such as PAM. If all signals have the same energy, $\| s_m \|^2$ may also be ignored in the computation of the correlation metrics $C(y, s_m)$ and the distance metrics $D'(y, s_m)$.

In summary, we have demonstrated that the optimum ML detector computes a set of M distances $D(y, s_m)$ or $D'(y, s_m)$ and selects the signal corresponding to the smallest (distance) metric. Equivalently, the optimum ML detector computes a set of M correlation metrics $C(y, s_m)$ and selects the signal corresponding to the largest correlation metric.

This development for the optimum detector treated the important case in which all signals are equally probable. In this case, the MAP criterion is equivalent to the ML criterion. However, when the signals are not equally probable, the optimum MAP detector bases its decision on the probabilities $P(s_m|y), m = 1, 2, \ldots , M$ given by Equation (8.4.49), or equivalently, on the *posterior probability metrics*,

$$\text{PM}(y, s_m) = f(y|s_m)P(s_m). \tag{8.4.56}$$

The following example illustrates this computation for binary PAM signals.

Example 8.4.7

Consider the case of binary PAM signals in which the two possible signal points are $s_1 = -s_2 = \sqrt{\mathscr{E}_b}$, where $\mathscr{E}_b$ is the energy per bit. The prior probabilities are $P(s_1)$ and $P(s_2)$. Determine the metrics for the optimum MAP detector when the transmitted signal is corrupted with AWGN.

Solution The received signal vector (which is one dimensional) for binary PAM is

$$y = \pm\sqrt{\mathscr{E}_b} + n,$$

where $n = y_n(T)$ is a zero-mean Gaussian random variable with a variance $\sigma_n^2 = N_0/2$. Consequently, the conditional PDF's $f(y|s_m)$ for the two signals are

$$f(y|s_1) = \frac{1}{\sqrt{2\pi\sigma_n^2}} e^{-\left(r-\sqrt{\mathcal{E}_b}\right)^2/2\sigma_n^2}$$

and

$$f(y|s_2) = \frac{1}{\sqrt{2\pi\sigma_n^2}} e^{-\left(r+\sqrt{\mathcal{E}_b}\right)^2/2\sigma_n^2}.$$

Then the metrics PM (y, s_1) and PM (y, s_2) defined by Equation (8.4.56) are

$$\text{PM } (y, s_1) = f(y|s_1)P(s_1)$$

and

$$\text{PM } (y, s_2) = f(y|s_2)P(s_2).$$

If PM $(y, s_1) > $ PM (y, s_2), we select s_1 as the transmitted signal; otherwise, we select s_2. This decision rule may be expressed as

$$\frac{\text{PM } (y, s_1)}{\text{PM } (y, s_2)} \underset{s_2}{\overset{s_1}{\gtrless}} 1. \tag{8.4.57}$$

But

$$\frac{\text{PM } (y, s_1)}{\text{PM } (y, s_2)} = \frac{P(s_1)}{P(s_2)} \exp\left\{\left[\left(y + \sqrt{\mathcal{E}_b}\right)^2 - \left(y - \sqrt{\mathcal{E}_b}\right)^2\right]/2\sigma_n^2\right\},$$

so Equation (8.4.57) may be expressed as

$$\frac{\left(y + \sqrt{\mathcal{E}_b}\right)^2 - \left(y - \sqrt{\mathcal{E}_b}\right)^2}{2\sigma_n^2} \underset{s_2}{\overset{s_1}{\gtrless}} \ln\frac{P(s_2)}{P(s_1)},$$

or equivalently,

$$y \underset{s_2}{\overset{s_1}{\gtrless}} \frac{N_0}{4\sqrt{\mathcal{E}_b}}\ln\frac{P(s_2)}{P(s_1)}. \tag{8.4.58}$$

This is the final form for the optimum detector. Its input y from the demodulator is compared with the threshold $\left(N_0/4\sqrt{\mathcal{E}_b}\right)\ln\left(P(s_2)/P(s_1)\right)$. We note that this is exactly the same detection rule that was obtained in Section 8.3.3, Equation (8.3.51), for binary antipodal signals (binary PAM).

It is interesting to note that in the case of unequal prior probabilities, it is necessary to know not only the values of the prior probabilities but also the value of the power spectral density N_0 and the signal energy $\mathcal{E}_b$ in order to compute the threshold. When the two signals are equally probable, the threshold is zero. Consequently, the detector does not need to know the values of these parameters. ■

We conclude this section with the proof that the decision rule based on the maximum-likelihood criterion minimizes the probability of error when the M signals are equally probable a priori. Let us denote by R_m the region in the N-dimensional space for which we decide that signal $s_m(t)$ was transmitted when the vector $y = (y_1, y_2, \ldots y_N)$ is received. The probability of a decision error given that $s_m(t)$ was transmitted is

$$P(e|s_m) = \int_{R_m^c} f(y|s_m)dy, \tag{8.4.59}$$

where R_m^c is the complement of R_m. The average probability of error is

$$P_M = \sum_{m=1}^{M} \frac{1}{M} P(e|s_m)$$

$$= \sum_{m=1}^{M} \frac{1}{M} \int_{R_m^c} f(y|s_m)dy \qquad (8.4.60)$$

$$= \sum_{m=1}^{M} \frac{1}{M} \left[1 - \int_{R_m} f(y|s_m)dy \right].$$

We note that P_M is minimized by selecting the signal s_m if $f(y|s_m)$ is larger than $f(y|s_k)$ for all $m \neq k$.

Similarly for the MAP criterion, when the M signals are not equally probable, the average probability of error is

$$P_M = 1 - \sum_{m=1}^{M} \int_{R_m} P(s_m|y) f(y)dy.$$

P_M is a minimum when the points that are to be included in each particular region R_m are those for which $P(s_m|y)$ exceeds all other posterior probabilities.

8.5 PROBABILITY OF ERROR FOR M-ARY PULSE MODULATION

In this section, we evaluate the performance of the detector in terms of the probability of error when the additive noise is white and Gaussian. We begin with pulse amplitude modulation.

8.5.1 Probability of Error for M-ary Pulse Amplitude Modulation

Recall that binary PAM signals are antipodal signals; hence, the probability of error of the optimum detector for equally probable binary PAM signals, previously derived in Section 8.3.3, is

$$P_2 = Q\left(\sqrt{\frac{2\mathscr{E}_b}{N_0}} \right), \qquad (8.5.1)$$

where $Q(x)$ is defined in Equation (8.3.54), $\mathscr{E}_b$ is the signal energy per bit, and $N_0/2$, is the power spectral density of the AWGN.

We make two interesting observations about the form of P_2. First, we note that the probability of error depends only on the ratio $\mathscr{E}_b/N_0$ and not on any other detailed characteristics of the signals and the noise. Second, we note that $2\mathscr{E}_b/N_0$

is also the output SNR from the matched filter (and correlation-type) demodulator. The ratio $\mathcal{E}_b/N_0$ is usually called the *signal-to-noise ratio* (SNR) or SNR/bit.

We also observe that the probability of error may be expressed in terms of the distance between the two signals s_1 and s_2. From Figure 8.7, we observe that the two signals are separated by the distance $d_{12} = 2\sqrt{\mathcal{E}_b}$. Substituting $\mathcal{E}_b = d_{12}^2/4$ in Equation (8.5.1), we obtain

$$P_2 = Q\left(\sqrt{\frac{d_{12}^2}{2N_0}}\right). \qquad (8.5.2)$$

This expression illustrates the dependence of the error probability on the distance between the two signal points. Equation (8.5.2) can be used for computing the error probability *of any binary communication system with two equiprobable messages corrupted by AWGN.*

In the case of M-ary PAM, the input to the detector is

$$y = s_m + n, \qquad (8.5.3)$$

where s_m denotes the mth transmitted amplitude level, defined previously in Equation (8.4.7), and n is a Gaussian random variable with zero mean and variance $\sigma_n^2 = N_0/2$. The average probability of error for equally probable amplitude levels can be determined from the decision rule that finds the maximum of the correlation metrics:

$$\begin{aligned} C(y, s_m) &= 2ys_m - s_m^2 \\ &= 2(y - s_m/2)s_m. \end{aligned} \qquad (8.5.4)$$

Equivalently, the optimum detector may compare the input y with a set of $M - 1$ thresholds, which are placed at the midpoints of successive amplitude levels, as shown in Figure 8.44. Thus, a decision is made in favor of the amplitude level that is closest to y.

The placing of the thresholds as shown in Figure 8.44 helps evaluate the probability of error. On the basis that all amplitude levels are equally likely a priori, the average probability of a symbol error is simply the probability that the noise variable n exceeds in magnitude one-half of the distance between levels. However, when either one of the two outside levels $\pm(M - 1)$ is transmitted, an error can

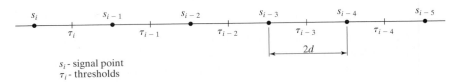

s_i - signal point
τ_i - thresholds

Figure 8.44   Placement of thresholds at midpoints of successive amplitude levels.

occur in one direction only. Thus, we have

$$P_M = \frac{M-1}{M} P(|y - s_m| > d)$$

$$= \frac{M-1}{M} \frac{2}{\sqrt{\pi N_0}} \int_d^\infty e^{-x^2/N_0}\, dx$$

$$\text{(8.5.5)}$$

$$= \frac{M-1}{M} \frac{2}{\sqrt{2\pi}} \int_{\sqrt{2d^2/N_0}}^\infty e^{-x^2/2}\, dx$$

$$= \frac{2(M-1)}{M} Q\left(\sqrt{2d^2/N_0}\right),$$

where $2d$ is the distance between adjacent signal points.

The distance parameter d is easily related to the average transmitted signal energy. Recall that $d = A\sqrt{T}$ and $\mathscr{E}_{av}$ is given by Equation (8.4.5) as $\mathscr{E}_{av} = A^2 T (M^2 - 1)/3$. Hence,

$$\mathscr{E}_{av} = d^2(M^2 - 1)/3, \qquad \text{(8.5.6)}$$

and the average probability of error is expressed as

$$P_M = \frac{2(M-1)}{M} Q\left(\sqrt{\frac{6\mathscr{E}_{av}}{(M^2 - 1)N_0}}\right). \qquad \text{(8.5.7)}$$

Since the average transmitted signal energy $\mathscr{E}_{av} = T P_{av}$, where P_{av} is the average transmitted power, P_M may also be expressed as a function of P_{av}.

In plotting the probability of a symbol error for M-ary signals such as M-ary PAM, it is customary to use the average SNR/bit as the basic parameter. Since each signal carries $k = \log_2 M$ bits of information, the average energy per bit $\mathscr{E}_{bav}$ is given by $\mathscr{E}_{bav} = \mathscr{E}_{av}/k$ and $k = \log_2 M$. Equation (8.5.7) may be expressed as

$$P_M = \frac{2(M-1)}{M} Q\left(\sqrt{\frac{6(\log_2 M)\mathscr{E}_{bav}}{(M^2 - 1)N_0}}\right), \qquad \text{(8.5.8)}$$

where $\mathscr{E}_{bav}/N_0$ is the average SNR/bit. Figure 8.45 illustrates the probability of a symbol error as a function of $10\log_{10}\mathscr{E}_{bav}/N_0$ with M as a parameter. Note that the case $M = 2$ corresponds to the error probability for binary antipodal signals. We also observe that for a fixed error probability, say $P_M = 10^{-5}$, the SNR/bit increases by over 4 dB for every factor of two increase in M. For large M, the additional SNR/bit required to increase M by a factor of two approaches 6 dB. Thus, each time we double the number of amplitude levels M, we can transmit one additional information bit; the cost measured in terms of additional transmitter power is 6 dB/bit for large M, i.e., the transmitter power must be increased by a factor of 4 ($\log_{10} 4 = 6$ dB) for each additional transmitted bit in order to achieve a specified value of P_M.

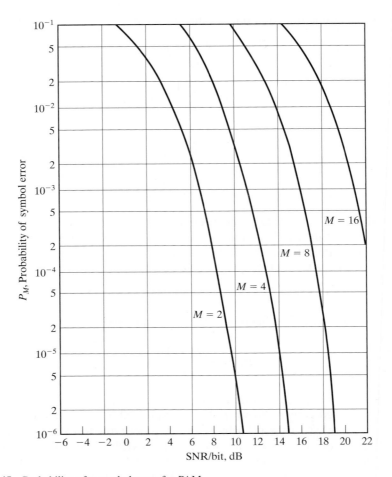

Figure 8.45   Probability of a symbol error for PAM.

Example 8.5.1

Using Figure 8.45, determine (approximately) the SNR/bit required to achieve a symbol error probability of $P_M = 10^{-6}$ for $M = 2, 4$, and 8.

Solution   From observation of Figure 8.45, we know that the required SNR/bit is (approximately) as follows:

$$10.5\,\text{dB} \quad \text{for } M = 2 \quad (1 \text{ bit/symbol});$$
$$14.8\,\text{dB} \quad \text{for } M = 4 \quad (2 \text{ bits/symbol});$$
$$19.2\,\text{dB} \quad \text{for } M = 8 \quad (3 \text{ bits/symbol}).$$

Note that for these small values of M, each additional bit by which we increase (double) the number of amplitudes requires an increase in the transmitted signal power (or energy) by a little over 4 dB in order to maintain the same symbol error probability. For large values of M, the argument of the function $Q(x)$ in Equation (8.5.8) is the dominant term in the expression for the error probability. Since $M = 2^k$, where k is the number of bits/symbol, increasing k by 1 bit to $k + 1$ requires that the energy/bit must be increased by a factor of 4 (6 dB) in order to have the same value of the argument in $Q(x)$.

■

8.5.2 Probability of Error for *M*-ary Orthogonal Signals

In deriving the probability of error for the general class of M-ary orthogonal signals, it is convenient to use PPM signals, which have the simple vector representation given in Equation (8.4.11), where $\mathscr{E}_s$ is the energy for each of the signal waveforms.

For equal energy orthogonal signals, the optimum detector selects the signal resulting in the largest cross correlation between the received vector $\boldsymbol{y}$ and each of the M possible transmitted signal vectors $\{\boldsymbol{s}_m\}$, i.e.,

$$C(\boldsymbol{y}, \boldsymbol{s}_m) = \boldsymbol{y} \cdot \boldsymbol{s}_m = \sum_{k=1}^{M} y_k s_{mk}, \qquad m = 1, 2, \ldots, M. \tag{8.5.9}$$

To evaluate the probability of error, let us suppose that the signal $\boldsymbol{s}_1$ is transmitted. Then the vector at the input to the detector is

$$\boldsymbol{y} = \left(\sqrt{\mathscr{E}_s} + n_1, n_2, n_3, \ldots, n_M \right), \tag{8.5.10}$$

where $n_1, n_2, \ldots, n_M$ are zero-mean, mutually statistically independent Gaussian random variables with equal variance $\sigma_n^2 = N_0/2$. Substituting Equation (8.5.10) into Equation (8.5.9), we obtain

$$\begin{aligned}
C(\boldsymbol{y}, \boldsymbol{s}_1) &= \sqrt{\mathscr{E}_s} \left(\sqrt{\mathscr{E}_s} + n_1 \right); \\
C(\boldsymbol{y}, \boldsymbol{s}_2) &= \sqrt{\mathscr{E}_s} n_2; \\
&\vdots \\
C(\boldsymbol{y}, \boldsymbol{s}_M) &= \sqrt{\mathscr{E}_s} n_M.
\end{aligned} \tag{8.5.11}$$

Note that the scale factor $\sqrt{\mathscr{E}_s}$ may be eliminated from the correlator outputs by dividing each output by $\sqrt{\mathscr{E}_s}$. Then, with this normalization, the PDF of the first correlator output $(y_1 = \sqrt{\mathscr{E}_s} + n_1)$ is

$$f(y_1) = \frac{1}{\sqrt{\pi N_0}} e^{-\left(y_1 - \sqrt{\mathscr{E}_s}\right)^2 / N_0}, \tag{8.5.12}$$

and the PDF's of the other $M - 1$ correlator outputs are

$$f(y_m) = \frac{1}{\sqrt{\pi N_0}} e^{-y_m^2 / N_0}, \qquad m = 2, 3, \ldots, M. \tag{8.5.13}$$

It is mathematically convenient to first derive the probability that the detector makes a correct decision. This is the probability that y_1 is larger than each of the other $M - 1$ correlator outputs $n_2, n_3, \ldots, n_M$. This probability may be expressed as

$$P_c = \int_{-\infty}^{\infty} P(n_2 < y_1, n_3 < y_1, \ldots, n_M < y_1 | y_1) f_{y_1}(y_1) dy_1, \qquad (8.5.14)$$

where $P(n_2 < y_1, n_3 < y_1, \ldots, n_M < y_1 | y_1)$ denotes the joint probability that $n_2, n_3, \ldots, n_M$ are all less than y_1, conditioned on any given y_1. Then this joint probability is averaged over all y_1. Since the $\{y_m\}$ are statistically independent, the joint probability factors into a product of $M - 1$ marginal probabilities of the form

$$P(n_m < y_1 | y_1) = \int_{-\infty}^{y_1} f(y_m) dy_m, \quad m = 2, 3, \ldots, M$$

$$= \frac{1}{2\pi} \int_{-\infty}^{\sqrt{2y_1^2/N_0}} e^{-y_m^2/2} dy_m \qquad (8.5.15)$$

$$= 1 - Q\left(\sqrt{\frac{2y_1^2}{N_0}}\right).$$

These probabilities are identical for $m = 2, 3, \ldots, M$; hence, the joint probability under consideration is simply the result in Equation (8.5.15) raised to the $(M - 1)$ power. Thus, the probability of a correct decision is

$$P_c = \int_{-\infty}^{\infty} \left[1 - Q\left(\sqrt{\frac{2y_1^2}{N_0}}\right)\right]^{M-1} f(y_1) dy_1 \qquad (8.5.16)$$

and the probability of a k-bit symbol error is

$$P_M = 1 - P_c. \qquad (8.5.17)$$

Therefore,

$$P_M = \frac{1}{\sqrt{2\pi}} \int_{-\infty}^{\infty} \left\{1 - [1 - Q(x)]^{M-1}\right\} e^{-\left(x - \sqrt{2\mathscr{E}_s/N_0}\right)^2/2} dx. \qquad (8.5.18)$$

The same expression for the probability of error is obtained when any one of the other $M - 1$ signals is transmitted. Since all the M signals are equally likely, the expression for P_M given in Equation (8.5.18) is the average probability of a symbol error. This expression can be evaluated numerically.

To make a fair comparison of communication systems, we want to have the probability of error expressed in terms of the SNR/bit, $\mathscr{E}_b/N_0$, instead of the SNR/symbol, $\mathscr{E}_s/N_0$. This is important because, depending on the size of constellation, different systems carry a different number of bits/signal. With $M = 2^k$, each symbol conveys k bits of information; hence, $\mathscr{E}_s = k\mathscr{E}_b$. Thus, Equation (8.5.18) may be expressed in terms of $\mathscr{E}_b/N_0$ by substituting for $\mathscr{E}_s$.

Sometimes, we also want to convert the probability of a symbol error into an equivalent probability of a binary digit error. For equiprobable orthogonal signals, all symbol errors are equiprobable and occur with probability

$$\frac{P_M}{M-1} = \frac{P_M}{2^k - 1}. \qquad (8.5.19)$$

Furthermore, there are $\binom{k}{n}$ ways in which n bits out of k may be in error. Hence, the average number of bit errors per k-bit symbol is

$$\sum_{n=1}^{k} n \binom{k}{n} \frac{P_M}{2^k - 1} = k \frac{2^{k-1}}{2^k - 1} P_M, \qquad (8.5.20)$$

and the average bit-error probability is simply the result in Equation (8.5.20) divided by k, the number of bits/symbol. Thus,

$$P_b = \frac{2^{k-1}}{2^k - 1} P_M \approx \frac{P_M}{2}, \quad k \gg 1. \qquad (8.5.21)$$

The graphs of the probability of a binary digit error as a function of the SNR/bit, $\mathscr{E}_b/N_0$, are shown in Figure 8.46 for $M = 2, 4, 8, 16, 32, 64$. This figure illustrates that by increasing the number M of waveforms, we can reduce the SNR/bit required to achieve a given probability of a bit error. For example, to achieve a $P_b = 10^{-5}$, the required SNR/bit is a little more than 12 dB for $M = 2$, but if M is increased to 64 signal waveforms ($k = 6$ bits/symbol), the required SNR/bit is approximately 6 dB. Thus, a savings of over 6 dB (a factor of four reduction) is realized in the transmitter power (or energy) required to achieve a $P_b = 10^{-5}$ by increasing M from $M = 2$ to $M = 64$. Note that this behavior is in complete contrast with M-ary PAM, where increasing M increases the error probability.

What is the minimum required $\mathscr{E}_b/N_0$ to achieve an arbitrarily small probability of error as $M \to \infty$? To answer this question, we first derive an important bound on the error probability of a general M-ary communication system, called the *union bound*. Then, we apply this bound to an M-ary orthogonal signaling system.

8.5.3 A Union Bound on the Probability of Error

In the preceding sections, we have introduced a number of binary and M-ary communication systems and their corresponding optimal detection schemes. We have also derived expressions for the error probability of these systems in terms of their SNR/bit ($\mathscr{E}_b/N_0$). Most of the error probability expressions are derived for equiprobable messages, where the optimal decision rule is the maximum-likelihood (ML) rule.

We have seen that for binary equiprobable signaling over an AWGN channel, regardless of the signaling type, the error probability can be expressed as

$$P_2 = Q\left(\frac{d}{\sqrt{2N_0}}\right), \qquad (8.5.22)$$

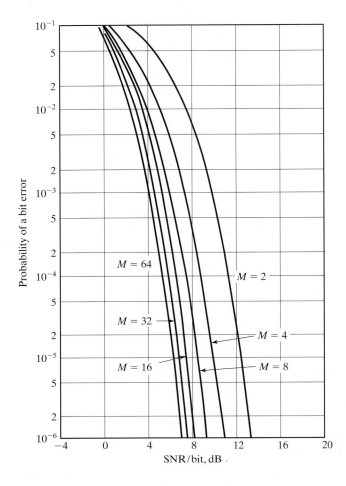

Figure 8.46 Probability of a bit error for optimum detection of orthogonal signals.

where d is the (Euclidean) distance between the two signal points in the constellation and is related to the signal waveforms via

$$d^2 = \int_{-\infty}^{\infty} (s_1(t) - s_2(t))^2 \, dt.$$

A natural question here is whether a similar simple expression exists for the error probability of the general equiprobable M-ary signaling. The answer to this question is negative. We have already seen that, even in rather simple signaling schemes such as orthogonal signaling, the expression for the error probability cannot be given in closed form. However, there is a simple upper bound to the error probability of general equiprobable M-ary signaling systems. This upper bound is known as the *union bound*; we will derive this upper bound next.

Let us assume that $s_m(t)$ is transmitted in an *M*-ary equiprobable signaling scheme. The error probability is the probability that the receiver detects a signal other than $s_m(t)$. Let E_i denote the event that message i is detected at the receiver. Then

$$P_m = P(\text{error}|\ s_m(t) \text{ sent}) = P\left(\bigcup_{\substack{i=1 \\ i\neq m}}^{M} E_i \,\middle|\, s_m(t) \text{ sent}\right) \leq \sum_{\substack{i=1 \\ i\neq m}}^{M} P\left(E_i|\ s_m(t) \text{ sent}\right).$$

(8.5.23)

A necessary (but not sufficient) condition for message $s_i(t)$ to be detected at the receiver when signal $s_m(t)$ is sent is that $\mathbf{y}$ be closer to s_i than to s_m, i.e.,

$$D(\mathbf{y}, s_i) < D(\mathbf{y}, s_m).$$

Thus, we conclude that

$$P\left(E_i|\ s_m(t) \text{ sent}\right) \leq P\left(D(\mathbf{y}, s_i) < D(\mathbf{y}, s_m)\right). \qquad (8.5.24)$$

But $P\left(D(\mathbf{y}, s_i) < D(\mathbf{y}, s_m)\right)$ is the probability of error in a binary equiprobable signaling system, which, by Equation (8.5.22), is given by

$$P\left(D(\mathbf{y}, s_i) < D(\mathbf{y}, s_m)\right) = Q\left(\frac{d_{mi}}{\sqrt{2N_0}}\right). \qquad (8.5.25)$$

From Equations (8.5.24) and (8.5.25), we conclude that

$$P\left(E_i|\ m \text{ sent}\right) \leq Q\left(\frac{d_{mi}}{\sqrt{2N_0}}\right). \qquad (8.5.26)$$

Substituting Equation (8.5.26) into Equation (8.5.23) yields

$$P_m \leq \sum_{\substack{i=1 \\ i\neq m}}^{M} Q\left(\frac{d_{mi}}{\sqrt{2N_0}}\right). \qquad (8.5.27)$$

We define the *minimum distance* of a signal set as the minimum of the distances between any two points in the corresponding signal constellation. In other words,

$$d_{\min} = \min_{\substack{1\leq m, m'\leq M \\ m'\neq m}} d_{mm'}. \qquad (8.5.28)$$

Then, for any $1 \leq i, m \leq M$, we have $d_{mi} \geq d_{\min}$, and since the Q-function is a decreasing function [see the discussion following Equation (5.1.7)], we have

$$Q\left(\frac{d_{mi}}{\sqrt{2N_0}}\right) \leq Q\left(\frac{d_{\min}}{\sqrt{2N_0}}\right). \qquad (8.5.29)$$

Using the inequality alternative of Equation (8.5.29) in Equation (8.5.27) yields

$$P_m \leq \sum_{\substack{i=1 \\ i \neq m}}^{M} Q\left(\frac{d_{\min}}{\sqrt{2N_0}}\right) = (M-1)Q\left(\frac{d_{\min}}{\sqrt{2N_0}}\right). \tag{8.5.30}$$

It is clear that this bound is independent of m. Therefore, it is valid independently of which signal was transmitted. Hence, we conclude that

$$P_M = \frac{1}{M}\sum_{m=1}^{M} P_m \leq (M-1)Q\left(\frac{d_{\min}}{\sqrt{2N_0}}\right) \leq \frac{M-1}{2}e^{-\frac{d_{\min}^2}{4N_0}}, \tag{8.5.31}$$

where the last step uses the inequality $Q(x) \leq 1/2e^{-x^2/2}$ from Equation (5.1.8). Equation (8.5.31) is known as the *union bound* for the error probability.

As can be seen from Equation (8.5.31), there are two versions of the union bound. The first version is in terms of the Q-function, and the second version is in terms of the exponential function. The union bound provides a very useful bound on the error probability, particularly at high signal-to-noise ratios. At lower signal-to-noise ratios, the bound becomes loose and useless. To analyze the performance of a system at low signal-to-noise ratios, more powerful bounding techniques are needed.

The union bound also signifies the role of the minimum distance of a signal set on its performance, particularly at large signal-to-noise ratios. A good signal set should provide the maximum possible value of $d_{\min}$. In other words, to design a good signal set, the points in the corresponding constellation should be maximally apart.

A Union Bound on the Error Probability of M-ary Orthogonal Signaling.

Let us investigate the effects of increasing M on the probability of error for orthogonal signals. To simplify the mathematical development, we use the union bound to derive an upper bound on the probability of a symbol error. This is much easier than the exact form given in Equation (8.5.18).

As we discussed earlier [see Equation (8.4.12)], M-ary orthogonal signals are equidistant with

$$d_{mm'}^2 = \|s_m - s_{m'}\|^2 = 2\mathscr{E}_s; \tag{8.5.32}$$

therefore,

$$d_{\min} = \sqrt{2\mathscr{E}_s}. \tag{8.5.33}$$

Using this value of $d_{\min}$ in Equation (8.5.31), we obtain the union bound on the error probability of an M-ary orthogonal signaling system as

$$P_M \leq \frac{M-1}{2}e^{-\frac{\mathscr{E}_s}{2N_0}} \leq Me^{-\frac{\mathscr{E}_s}{2N_0}}. \tag{8.5.34}$$

Thus, using $M = 2^k$ and $\mathscr{E}_s = k\mathscr{E}_b$, we have

$$P_M \leq 2^k e^{-k\mathscr{E}_b/2N_0}$$

$$= e^{-k(\mathscr{E}_b/N_0 - 2\ln 2)/2}. \tag{8.5.35}$$

As $k \to \infty$ or equivalently, as $M \to \infty$, the probability of error approaches zero exponentially, provided that $\mathscr{E}_b/N_0$ is greater than $2\ln 2$, i.e.,

$$\frac{\mathscr{E}_b}{N_0} > 2\ln 2 = 1.39 \approx 1.42\,\text{dB}. \tag{8.5.36}$$

The simple upper bound on the probability of error given by Equation (8.5.35) implies that as long as SNR ≥ 1.42 dB, we can achieve an arbitrarily low P_M. However, this union bound is not a very tight upper bound at low SNR values. In fact, by more elaborate bounding techniques, it can be shown that $P_M \to 0$ as $k \to \infty$, provided that

$$\frac{\mathscr{E}_b}{N_0} > \ln 2 = 0.693 \approx -1.6\,\text{dB}. \tag{8.5.37}$$

Hence, -1.6 dB is the minimum required SNR/bit to achieve an arbitrarily small probability of error in the limit as $k \to \infty$ ($M \to \infty$). This minimum SNR/bit (-1.6 dB) is called the *Shannon limit* for an additive white Gaussian noise channel. More discussion on the Shannon limit and how the value of -1.6 dB is derived is given in Section 12.6 in Equation (12.6.7) and the discussion leading to it.

8.5.4 Probability of Error for M-ary Biorthogonal Signals

As previously indicated in Section 8.4.3, a set of $M = 2^k$ biorthogonal signals are constructed from $M/2$ orthogonal signals by including the negatives of the orthogonal signals. Thus, we achieve a reduction in the complexity of the demodulator for the biorthogonal signals relative to that for the orthogonal signals. This occurs because the former is implemented with $M/2$ cross correlators or matched filters, whereas the latter requires M matched filters or cross correlators. Biorthogonal signals are also more bandwidth efficient than orthogonal signals.

   The vector representation of biorthogonal signals is given by Equations (8.4.16) and (8.4.17).

   To evaluate the probability of error for the optimum detector, let us assume that the signal $s_1(t)$ corresponding to the vector $\mathbf{s}_1 = (\mathscr{E}_s, 0, 0, \ldots, 0)$ was transmitted. Then, the received signal vector is

$$\mathbf{y} = \left(\sqrt{\mathscr{E}_s} + n_1, n_2, \ldots, n_{M/2}\right), \tag{8.5.38}$$

where the $\{n_m\}$ are zero-mean, mutually statistically independent, and identically distributed Gaussian random variables with the variance $\sigma_n^2 = N_0/2$. The optimum detector decides in favor of the signal corresponding to the largest in magnitude of the cross correlators, or

$$C(\mathbf{y}, \mathbf{s}_m) = \mathbf{y} \cdot \mathbf{s}_m = \sum_{k=1}^{M/2} y_k s_{mk}, \qquad m = 1, 2, \ldots, M/2, \qquad (8.5.39)$$

while the sign of this largest term is used to decide whether $s_m(t)$ or $-s_m(t)$ was transmitted. According to this decision rule, the probability of a correct decision is equal to the probability that $y_1 = \sqrt{\mathcal{E}_s} + n_1 > 0$ and y_1 exceeds $|y_m| = |n_m|$ for $m = 2, 3, \ldots, M/2$. But

$$P(|n_m| < y_1 | y_1 > 0) = \frac{1}{\sqrt{\pi N_0}} \int_{-y_1}^{y_1} e^{-x^2/N_0}\, dx$$

$$= \frac{1}{\sqrt{2\pi}} \int_{-y_1/\sqrt{N_0/2}}^{y_1/\sqrt{N_0/2}} e^{-x^2/2}\, dx, \qquad (8.5.40)$$

so the probability of a correct decision is

$$P_c = \int_0^\infty \left[\frac{1}{\sqrt{2\pi}} \int_{-y_1/\sqrt{N_0/2}}^{y_1/\sqrt{N_0/2}} e^{-x^2/2}\, dx \right]^{\frac{M}{2}-1} f(y_1)\, dy_1.$$

Upon substitution for $f(y_1)$, we obtain

$$P_c = \int_{-\sqrt{2\mathcal{E}_s/N_0}}^\infty \left[\frac{1}{\sqrt{2\pi}} \int_{-\left(v+\sqrt{2\mathcal{E}_s/N_0}\right)}^{v+\sqrt{2\mathcal{E}_s/N_0}} e^{-x^2/2}\, dx \right]^{\frac{M}{2}-1} \frac{1}{\sqrt{2\pi}} e^{-v^2/2}\, dv,$$

(8.5.41)

where we have used the PDF of y_1 given in Equation (8.5.12). Finally, the probability of a symbol error $P_M = 1 - P_c$.

P_c and, hence, P_M may be evaluated numerically for different values of M from Equation (8.5.41). The graph shown in Figure 8.47 illustrates P_M as function of $\mathcal{E}_b/N_0$, where $\mathcal{E}_s = k\mathcal{E}_b$ for $M = 2, 4, 8, 16$, and 32. We observe that this graph is similar to that for orthogonal signals (see Figure 8.46). However, in this case, the probability of error for $M = 4$ is greater than that for $M = 2$. This is because we have plotted the symbol error probability P_M in Figure 8.47. If we plot the equivalent bit error probability, we would find that the graphs for $M = 2$ and $M = 4$ coincide. As in the case of orthogonal signals, as $M \to \infty$ (or $k \to \infty$), the minimum required $\mathcal{E}_b/N_0$ to achieve an arbitrarily small probability of error is -1.6 dB, the Shannon limit.

8.5.5 Probability of Error for *M*-ary Simplex Signals

Next, we consider the probability of error for M simplex signals. Recall from Section 8.4.4 that simplex signals are a set of M equally correlated signals with a mutual cross-correlation coefficient $\gamma_{mn} = -1/(M-1)$. These signals have the

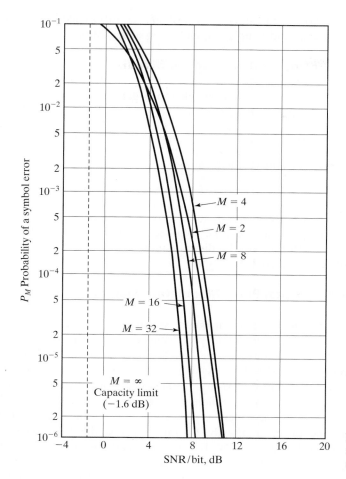

Figure 8.47 Probability of symbol error for biorthogonal signals.

same minimum separation of $\sqrt{2\mathscr{E}_s}$ between adjacent signal points in *M*-dimensional space as orthogonal signals. They achieve this mutual separation with a transmitted energy of $\mathscr{E}_s(M-1)/M$, which is less than that required for orthogonal signals by a factor of $(M-1)/M$. Consequently, the probability of error for simplex signals is identical to the probability of error for orthogonal signals, but this performance is achieved with a savings of

$$10\log(1-\gamma_{mn}) = 10\log\frac{M}{M-1}\ \text{dB} \tag{8.5.42}$$

in SNR. Notice that for $M=2$, i.e., binary modulation, the simplex signals become antipodal; consequently, the transmitted signal energy (or power) required to achieve the same performance as binary orthogonal signals is 3 dB less. Therefore, the result in Equation (8.5.42) is consistent with our previous comparison of binary orthogonal signals and antipodal signals. For large values of M, the difference in performance between orthogonal signals and simplex signals approaches zero.

8.5.6 Probability of Error for Binary-Coded Signals

Recall that a set of M signal waveforms that are generated from a set of N-dimensional binary codewords of the form

$$\mathbf{c}_m = (c_{m1}, c_{m2}, \ldots, c_{nN}), \quad m = 1, 2, \ldots, M \tag{8.5.43}$$

are represented geometrically in vector form as

$$\mathbf{s}_m = (s_{m1}, s_{m2}, \ldots, s_{mN}) , \quad m = 1, 2, \ldots, M, \tag{8.5.44}$$

where $s_{mj} = \pm\sqrt{\mathcal{E}_s/N}$ for all m and j, and $\mathcal{E}_s$ is the energy per waveform. The error probability of a coded system can be obtained using the union bound given in Equation (8.5.31). The performance of a digital communication system that employs binary-coded signals is considered in Chapter 13.

8.5.7 Comparison of Digital Pulse Modulation Methods

In comparing the probability of error for binary signals, we observed that antipodal signals provide a 3 dB improvement in performance over orthogonal signals. In the case of M-ary orthogonal, biorthogonal, and simplex signals, we demonstrated that as M is increased, the SNR/bit ($\mathcal{E}_b/N_0$) required to achieve a specified bit error probability decreases. Hence, M-ary orthogonal, biorthogonal, and simplex signals become more power or energy efficient as M increases than binary antipodal signals. As M approaches infinity, there is a limiting value of the SNR/bit required to achieve a decreasing bit error probability; this is called the Shannon limit. By using a union bound on the error probability, we obtained the limit on the SNR/bit as $\mathcal{E}_b/N_0 > 1.39 \approx (1.42$ dB). We also indicated that a tighter bound, which is derived in Chapter 12, yields the Shannon limit as $\mathcal{E}_b/N_0 > 0.69 \approx (-1.6$ dB).

We should emphasize that the reduction in the SNR/bit as M increases is achieved at the cost of increasing the dimensionality of the transmitted signals, i.e., increasing the length of the signal vectors $\mathbf{s}_m$. For orthogonal signals, the number of dimensions is $N = M$. In Chapter 10, we will observe that increasing the dimensionality of the signals increases the channel bandwidth required to transmit the signals.

In contrast to orthogonal, biorthogonal, and simplex signals, whose dimensionality increases as M increases, we observe that in M-ary PAM, the dimensionality of the signal waveforms remains the same and is independent of the value of M. Thus, increasing the value of M in PAM does not require an increase in the channel bandwidth to transmit the signal waveforms. However, the SNR/bit required to achieve a specified error probability increases as M is increased.

A more thorough comparison of these digital modulation methods is given in Chapter 10, which includes carrier modulation methods.

8.6 SYMBOL SYNCHRONIZATION

In a communication system, the output of the receiving filter $y(t)$ must be sampled periodically at the symbol rate and at the precise sampling time instants $t_m = mT + \tau_0$, where T is the symbol interval and τ_0 is a nominal time delay that depends on the propagation time of the signal from the transmitter to the receiver. To perform this periodic sampling, we require a clock signal at the receiver. The process of extracting such a clock signal at the receiver is usually called *symbol synchronization* or *timing recovery*.

Timing recovery is one of the most critical functions performed at the receiver of a synchronous digital communication system. We should note that the receiver must know not only the frequency $(1/T)$ at which the outputs of the matched filters or correlators are sampled, but also where to take the samples within each symbol interval. The choice of sampling instant within the symbol interval of duration T is called the *timing phase*.

The best timing corresponds to the time instant within the symbol interval where the signal output of the receiver filter is a maximum. In a practical communication system, the receiver clock must be continuously adjusted in frequency $(1/T)$ and in timing phase τ_0 to compensate for frequency drifts between the oscillators used in the transmitter and receiver clocks; thus, it will optimize the sampling time instants of the matched filter or correlator outputs.

Symbol synchronization can be accomplished in one of several ways. In some communication systems, the transmitter and receiver clocks are synchronized to a master clock, which provides a very precise timing signal. In this case, the receiver must estimate and compensate for the relative time delay between the transmitted and received signals. Such may be the case for radio communication systems where precise clock signals are transmitted from a master radio station.

Another method for achieving symbol synchronization is for the transmitter to simultaneously transmit the clock frequency $1/T$ or a multiple of $1/T$ along with the information signal. The receiver may simply employ a narrowband filter tuned to the transmitted clock frequency; thus, it can extract the clock signal for sampling. This approach has the advantage of being simple to implement. There are several disadvantages, however. One is that the transmitter must allocate some of its available power to the transmission of the clock signal. Another is that some small fraction of the available channel bandwidth must be allocated for the transmission of the clock signal. In spite of these disadvantages, this method is frequently used in telephone transmission systems that employ large bandwidths to transmit the signals of many users. In such a case, the transmission of a clock signal is shared in the demodulation of the signals among the many users. Through this shared use of the clock signal, the penalty in transmitter power and in bandwidth allocation is reduced proportionally by the number of users.

A clock signal can also be extracted from the received data signal. There are a number of different methods that can be used at the receiver to achieve self-synchronization. Next, we will consider four approaches to the problem of achieving symbol synchronization from the received signal.

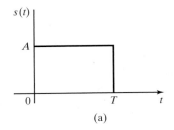

(a)

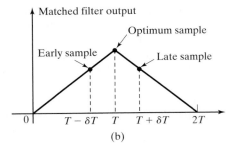

(b)

Figure 8.48 (a) Rectangular signal pulse and (b) its matched filter output.

8.6.1 Early–Late Gate Synchronizers

One method for generating a symbol timing signal at the receiver exploits the symmetry properties of the signal at the output of the matched filter or correlator. To describe this method, let us consider the rectangular pulse $s(t)$, $0 \le t \le T$, shown in Figure 8.48(a). The output of the filter matched to $s(t)$ attains its maximum value at time $t = T$, as shown in Figure 8.48(b). Thus, the output of the matched filter is the time autocorrelation function of the pulse $s(t)$. Of course, this statement holds for any arbitrary pulse shape, so the approach that we describe generally applies to any signal pulse. Clearly, the proper time to sample the output of the matched filter for a maximum output is at $t = T$, i.e., at the peak of the correlation function.

In the presence of noise, the identification of the peak value of the signal is generally difficult. Instead of sampling the signal at the peak, suppose we sample early (at $t = T - \delta T$) and late (at $t = T + \delta T$). The absolute values of the early samples $|y[m(T - \delta T)]|$ and the late samples $|y[m(T + \delta T)]|$ will be smaller (on the average in the presence of noise) than the samples of the peak value $|y(mT)|$. Since the autocorrelation function is even with respect to the optimum sampling time $t = T$, the absolute values of the correlation function at $t = T - \delta T$ and $t = T + \delta T$ are equal. Under this condition, the proper sampling time is the midpoint between $t = T - \delta T$ and $t = T + \delta T$. This condition forms the basis for the early–late gate symbol synchronizer.

Figure 8.49 illustrates the block diagram of an early–late gate synchronizer. In this figure, correlators are used in place of the equivalent matched filters. The two correlators integrate over the symbol interval T, but one correlator starts integrating δT early relative to the estimated optimum sampling time, and the other integrator

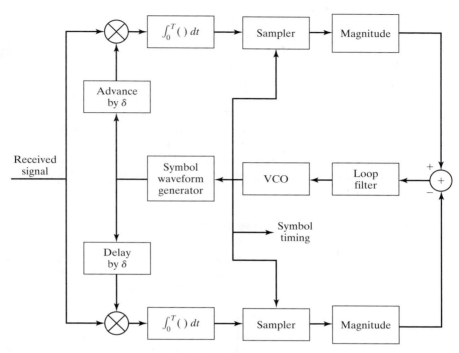

Figure 8.49   Block diagram of early–late gate synchronizer.

starts integrating δT late relative to the estimated optimum sampling time. An error signal is formed by taking the difference between the absolute values of the two correlator outputs. To smooth the noise corrupting the signal samples, the error signal is passed through a lowpass filter. If the timing is off relative to the optimum sampling time, the average error signal at the output of the lowpass filter is nonzero, and the clock signal is either retarded or advanced, depending on the sign of the error. Thus, the smoothed error signal is used to drive a voltage-controlled oscillator (VCO), whose output is the desired clock signal that is used for sampling. The output of the VCO is also used as a clock signal for a symbol waveform generator, which puts out the same basic pulse waveform as the transmitting filter. This pulse waveform is advanced and delayed, and then it is fed to the two correlators, as shown in Figure 8.49. Note that if the signal pulses are rectangular, there is no need for a signal pulse generator within the tracking loop.

We observe that the early–late gate synchronizer is basically a closed-loop control system whose bandwidth is relatively narrow compared to the symbol rate $1/T$. The bandwidth of the loop determines the quality of the timing estimate. A narrowband loop provides more averaging over the additive noise; thus, it improves the quality of the estimated sampling instants, provided that the channel propagation delay is constant and the clock oscillator at the transmitter is not drifting with time (drifting very slowly with time). On the other hand, if the channel propagation delay is changing with time or the transmitter clock is also drifting with time, then the bandwidth of the loop must be increased to provide for faster tracking of time variations in symbol timing. This increases the noise in the loop and degrades the quality of the timing estimate.

In the tracking mode, the two correlators are affected by adjacent symbols. However, if the sequence of information symbols has zero mean, as is the case for PAM and some other signal modulations, the contribution to the output of the correlators from adjacent symbols averages out to zero in the lowpass filter.

An equivalent realization of the early–late gate synchronizer that is somewhat easier to implement is shown in Figure 8.50. In this case, the clock from the VCO is advanced and delayed by δT, and these clock signals are used to sample the outputs of the two correlators.

8.6.2 Minimum Mean-Square-Error Method

Another approach to the problem of timing recovery from the received signal is based on the minimization of the mean-square-error (MSE) between the samples at the output of the receiver filter and the desired symbols. We assume that the baseband

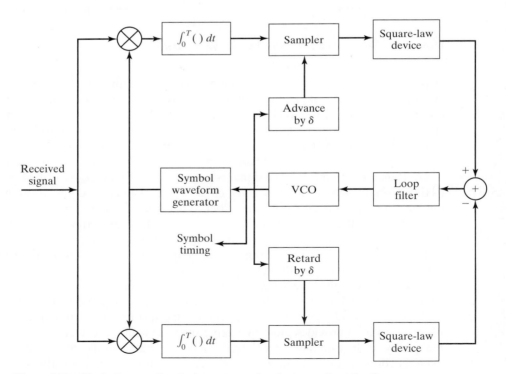

Figure 8.50   Block diagram of early–late gate synchronizer—an alternative form.

signal at the transmitter is of the form

$$v(t) = \sum_{n=-\infty}^{\infty} a_n g_T(t - nT),$$ (8.6.1)

where $\{a_n\}$ is the data sequence and T is the symbol interval. To be specific, let us assume that $v(t)$ is a PAM baseband signal and the data sequence $\{a_n\}$ is a zero-mean, stationary sequence with statistically independent and identically distributed elements. Therefore, the signal $v(t)$ has zero mean, i.e., $E[v(t)] = 0$. Furthermore, the autocorrelation function of $v(t)$ is periodic in T; hence, $v(t)$ is a periodically stationary process.

The received signal at the output of the matched filter at the receiver may be expressed as

$$y(t) = \sum_{n=-\infty}^{\infty} a_n x(t - nT - \tau_0) + w(t),$$ (8.6.2)

where $x(t) = g_T(t) * g_R(t)$, the asterisk denotes convolution, $g_R(t)$ is the impulse response of the receiver filter, $w(t)$ represents the noise at the output of the receiver filter, and τ_0 ($\tau_0 < T$) represents the timing phase.

The MSE between the output of the receiver filter and the desired symbol at the mth symbol interval is defined as

$$\text{MSE} = E\{[y_m(\tau_0) - a_m]^2\},$$ (8.6.3)

where

$$y_m(\tau_0) = \sum_{n=-\infty}^{\infty} a_n x(mT - nT - \tau_0) + w(mT).$$ (8.6.4)

Since the desired symbol a_m is not known a priori at the receiver, we may use the output of the detector, denoted as $\hat{a}_m$, for the mth symbol. Thus, we substitute $\hat{a}_m$ for a_m in the MSE expression. Hence, the MSE is redefined as

$$\text{MSE} = E\{[y_m(\tau_0) - \hat{a}_m]^2\}.$$ (8.6.5)

The minimum of MSE with respect to the timing phase τ_0 is found by differentiating Equation (8.6.5) with respect to τ_0. Thus, we obtain the necessary condition

$$\sum_m [y_m(\tau_0) - \hat{a}_m] \frac{dy_m(\tau_0)}{d\tau_0} = 0.$$ (8.6.6)

An interpretation of the necessary condition in Equation (8.6.6) is that the optimum sampling time corresponds to the condition that the error signal $[y_m(\tau_0) - \hat{a}_m]$ is uncorrelated to the derivative $dy_m(\tau_0)/d\tau_0$. Since the detector output is used in the formation of the error signal $[y_m(\tau_0) - \hat{a}_m]$, this timing phase-estimation method is said to be *decision directed*.

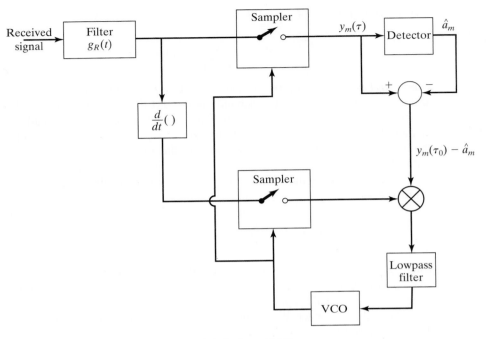

Figure 8.51   Timing recovery based on minimization of MSE.

Figure 8.51 illustrates an implementation of the system that is based on the condition given in Equation (8.6.6). Note that the summation operation is implemented as a lowpass filter, which averages a number of symbols. The averaging time is roughly equal to the reciprocal of the bandwidth of the filter. The filter output drives the voltage-controlled oscillator (VCO), which provides the best MSE estimate of the timing phase τ_0.

8.6.3 Maximum-Likelihood Method

In the ML method, the optimum symbol timing is obtained by maximizing the *likelihood function*, defined as

$$\Lambda(\tau_0) = \sum_m a_m y_m(\tau_0), \qquad (8.6.7)$$

where $y_m(\tau_0)$ is the sampled output of the receiving filter given by Equation (8.6.4). From a mathematical viewpoint, the likelihood function can be shown to be proportional to the probability of the received signal (vector) conditioned on a known transmitted signal. Physically, $\Lambda(\tau_0)$ is simply the output of the matched filter or correlator at the receiver averaged over a number of symbols.

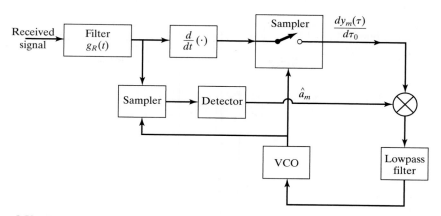

Figure 8.52 Decision-directed ML timing recovery method for baseband PAM.

A necessary condition for τ_0 to be the ML estimate is that

$$\frac{d\Lambda(\tau_0)}{d\tau_0} = \sum_m a_m \frac{dy_m(\tau_0)}{d\tau_0} = 0. \tag{8.6.8}$$

This result suggests the implementation of the tracking loop shown in Figure 8.52. We observe that the product of the detector output $\hat{a}_m$ with $dy_m(\tau_0)/d\tau_0$ is averaged by a lowpass filter that drives the VCO. Since the detector output is used in the estimation method, the estimate $\hat{\tau}_0$ is decision directed.

As an alternative to using the output symbols from the detector, we may use a *nondecision-directed method* that does not require knowledge of the information symbols. This method is based on averaging over the statistics of the symbols. For example, we may square the output of the receiving filter and maximize the function

$$\Lambda_2(\tau_0) = \sum_m y_m^2(\tau_0) \tag{8.6.9}$$

with respect to τ_0. Thus, we obtain

$$\frac{d\Lambda_2(\tau_0)}{d\tau_0} = 2 \sum_m y_m(\tau_0) \frac{dy_m(\tau_0)}{d\tau_0} = 0. \tag{8.6.10}$$

The condition for the optimum τ_0 given by Equation (8.6.10) may be satisfied by the implementation shown in Figure 8.53. In this case, there is no need to know the data sequence $\{a_m\}$. Hence, the method is nondecision directed.

8.6.4 Spectral-Line Method

Since the signal component at the output of the receiver filter is periodic with period T, we can recover a clock signal with frequency $1/T$ by filtering out a signal component at $f = 1/T$. We observe, however, that $E[y(t)] = 0$ because $E(a_n) = 0$.

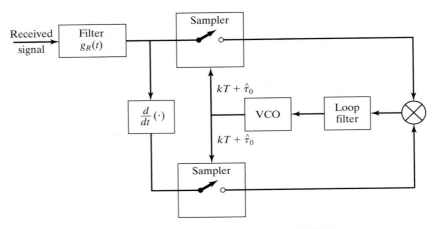

Figure 8.53   Nondecision-directed estimation of timing for baseband PAM.

Therefore, $y(t)$ cannot be used directly to generate a frequency component at $f = 1/T$. On the other hand, we may perform a nonlinear operation on $y(t)$ to generate power at $f = 1/T$ and its harmonics.

Let us consider a square-law nonlinearity. If we square the signal $y(t)$ given by Equation (8.6.2) and take the expected value with respect to the data sequence $\{a_n\}$, we obtain

$$E[y^2(t)] = E\left[\sum_n \sum_m a_n a_m x(t - mT - \tau_0)x(t - nT - \tau_0)\right] + \text{noise component}$$

$$= \sigma_a^2 \sum_{n=-\infty}^{\infty} x^2(t - nT - \tau_0) + \text{noise component}, \tag{8.6.11}$$

where $\sigma_a^2 = E\left[a_n^2\right]$. Since $E\left[y^2(t)\right] > 0$, we may use $y^2(t)$ to generate the desired frequency component.

Let us use Poisson's sum formula on the signal component (see Problem 2.56) to express Equation (8.6.11) in the form of a Fourier series. Hence,

$$\sigma_a^2 \sum_n x^2(t - nT - \tau_0) = \frac{\sigma_a^2}{T} \sum_n c_m\, e^{j2\pi m(t-\tau_0)/T}, \tag{8.6.12}$$

where

$$c_m = \int_{-\infty}^{\infty} X(f)X\left(\frac{m}{T} - f\right) df. \tag{8.6.13}$$

By design, we assume that the transmitted signal spectrum is confined to frequencies below $1/T$. Hence, $X(f) = 0$ for $|f| > 1/T$; consequently, there are only 3 nonzero terms ($m = 0, \pm 1$) in Equation (8.6.12). Therefore, the square of the signal component contains a DC component and a component at the frequency $1/T$.

This development suggests that we square the signal $y(t)$ at the output of the receiving filter and filter $y^2(t)$ with a narrowband filter $B(f)$ tuned to the symbol rate $1/T$. If we set the filter response $B(1/T) = 1$, then

$$\frac{\sigma_a^2}{T}\text{Re}\left[c_1\, e^{j2\pi(t-\tau_0)/T}\right] = \frac{\sigma_a^2}{T}c_1 \cos\frac{2\pi}{T}(t-\tau_0), \qquad (8.6.14)$$

so that the timing signal is a sinusoid with a phase of $-2\pi\tau_0/T$, assuming that $X(f)$ is real. We may use alternate zero crossings of the timing signal as an indication of the correct sampling times. However, the alternate zero crossings of the signal given by Equation (8.6.14) occur at

$$\frac{2\pi}{T}(t-\tau_0) = (4k+1)\frac{\pi}{2}, \qquad (8.6.15)$$

or equivalently, at

$$t = kT + \tau_0 + \frac{T}{4}, \qquad (8.6.16)$$

which is offset in time by $T/4$ relative to the desired zero crossings. In a practical system, the timing offset can be easily compensated either by relatively simple clock circuitry or by designing the bandpass filter $B(f)$ to have a $\pi/2$ phase shift at $f = 1/T$. Figure 8.54 illustrates this method for generating a timing signal at the receiver.

The additive noise that corrupts the signal will generally cause fluctuations in the zero crossings of the desired signal. The effect of the fluctuations will depend on the amplitude c_1 of the mean timing sinusoidal signal given by Equation (8.6.14). We note that the signal amplitude c_1 is proportional to the slope of the timing signal in the vicinity of the zero crossing, as shown in Figure 8.55. Therefore, when the amplitude c_1 is larger, the slope will be larger; consequently, the timing errors due to the noise will be smaller. From Equation (8.6.13), we observe that c_1 depends on the amount of spectral overlap of $X(f)$ and $X(1/T - f)$. Thus, c_1 depends on the amount by which the bandwidth of $X(f)$ exceeds the (Nyquist) bandwidth $1/2T$. In other words, c_1 depends on the excess bandwidth of the signal, which is defined

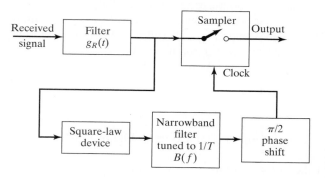

Figure 8.54 Symbol timing based on the spectral-line method.

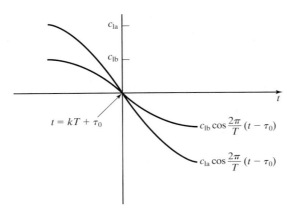

Figure 8.55 Illustration of the slope of the sinusoid at the zero crossing as a function of the amplitude.

as the band of frequencies of the signal $X(f)$ beyond $f = 1/2T$. If the excess bandwidth is zero, i.e., $X(f) = 0$ for $|f| > 1/2T$, then $c_1 = 0$; thus, this method fails to provide a timing signal. If the excess bandwidth is large, such as $\alpha/2T$ where $\alpha = 1/2$ or 1, the timing signal amplitude will be sufficiently large to yield relatively accurate symbol timing estimates.

8.7 FURTHER READING

The geometrical representation of digital signals as vectors was first used by Kotelnikov (1947) and by Shannon (1948a, 1948b) (in his classic papers). This approach was popularized by Wozencraft and Jacobs (1965). Today, this approach to signal analysis and design is widely used. Treatments similar to those given in the text may be found in most books on digital communications.

The matched filter was introduced by North (1943), who showed that it maximized the SNR. Analysis of various binary and M-ary modulation signals in AWGN were performed in the two decades following Shannon's work. Treatments similar to those given in this chapter may be found in most books on digital communications.

A number of books and tutorial papers have been published on the topic of time synchronization. Books that cover both carrier-phase recovery and time synchronization have been written by Stiffler (1971), Lindsey (1972), Lindsey and Simon (1973), Meyr and Ascheid (1990), and Mengali and D'Andrea (1997). The tutorial paper by Franks (1980) presents a very readable introduction to this topic.

PROBLEMS

8.1 Consider the three waveforms $\psi_n(t)$ shown in Figure P-8.1.

    1. Show that these waveforms are orthonormal.

2. Express the waveform $x(t)$ as a weighted linear combination of $\psi_n(t)$, $n = 1, 2, 3$ if

$$x(t) = \begin{cases} -1, & 0 \le t \le 1 \\ 1, & 1 \le t \le 3 \\ -1, & 3 \le t \le 4 \end{cases}$$

and determine the weighting coefficients.

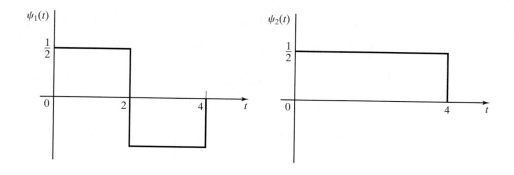

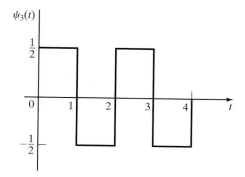

Figure P-8.1

8.2 Use the orthonormal waveforms in Figure P-8.1 to approximate the function

$$x(t) = \sin(\pi t/4)$$

over the interval $0 \le t \le 4$ by the linear combination

$$\widehat{x}(t) = \sum_{n=1}^{3} c_n \psi_n(t).$$

1. Determine the expansion coefficients $\{c_n\}$ that minimize the mean square approximation error

$$E = \int_0^4 \left[x(t) - \hat{x}(t)\right]^2 dt.$$

2. Determine the residual mean square error $E_{\min}$.

8.3 Consider the four waveforms shown in Figure P-8.3.

1. Determine the dimensionality of the waveforms and a set of basis functions.
2. Use the basis functions to represent the four waveforms by vectors s_1, s_2, s_3, s_4.
3. Determine the minimum distance between any pair of vectors.

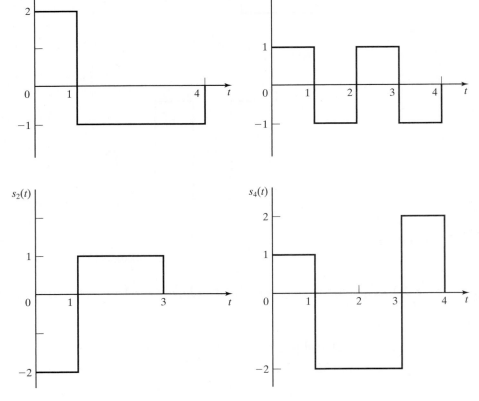

Figure P-8.3

8.4 Determine a set of orthonormal functions for the four signals shown in Figure P-8.4.

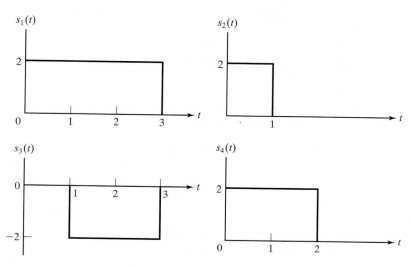

Figure P-8.4

8.5 The received signal in a binary communication system that employs antipodal signals is

$$r(t) = s(t) + n(t),$$

where $s(t)$ is shown in Figure P-8.5 and $n(t)$ is AWGN with power spectral density $N_0/2$ W/Hz.

1. Sketch the impulse responses of the filter matched to $s(t)$.
2. Sketch the output of the matched filter to the input $s(t)$.
3. Determine the variance at the noise of the output of the matched filter at $t = 3$.
4. Determine the probability of error as a function of A and N_0.

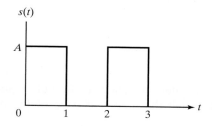

Figure P-8.5

8.6 A matched filter has the frequency response

$$H(f) = \frac{1 - e^{-j2\pi f T}}{j2\pi f}.$$

1. Determine the impulse response $h(t)$ corresponding to $H(f)$.
2. Determine the signal waveform to which the filter characteristic is matched.

8.7 Prove that when a sinc pulse $g_T(t)$ is passed through its matched filter, the output is the same sinc pulse.

8.8 The demodulation of the binary antipodal signals

$$s_1(t) = -s_2(t) = \begin{cases} \sqrt{\frac{\mathcal{E}_b}{T}}, & 0 \leq t \leq T \\ 0, & \text{otherwise} \end{cases}$$

can be accomplished by use of a single integrator, as shown in Figure P-8.8, which is sampled periodically at $t = kT$, $k = 0, \pm 1, \pm 2, \ldots$. The additive noise is zero-mean Gaussian with power spectral density of $\frac{N_0}{2}$ W/Hz.

1. Determine the output SNR of the demodulator at $t = T$.
2. If the ideal integrator is replaced by the RC filter shown in Figure P-8.8, determine the output SNR as a function of the time constant RC.
3. Determine the value of RC that maximizes the output SNR.

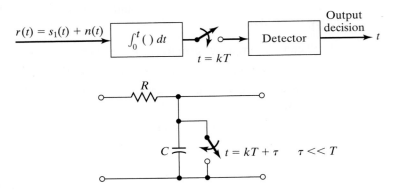

Figure P-8.8

8.9 Sketch the impulse responses of the filters matched to the pulses shown in Figure P-8.9. Also determine and sketch the output of each of the matched filters.

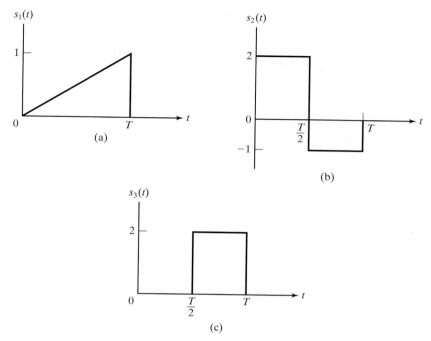

Figure P-8.9

8.10 A binary PAM communication system employs rectangular pulses of duration T_b and amplitudes $\pm A$ to transmit digital information at a rate $R = 10^5$ bits/sec. If the power spectral density of the additive Gaussian noise is $N_0/2$, where $N_0 = 10^{-2}$ W/Hz, determine the value of A that is required to achieve an error probability of $P_2 = 10^{-6}$.

8.11 In a binary PAM communication system for which the two signals occur with unequal probabilities (p and $1 - p$), the optimum detector compares the output of the correlator, which takes the values

$$y = \pm\sqrt{\mathcal{E}_b} + n$$

with a threshold α, which is given by Equation (8.3.51). The resulting error probability of the detector is given by Equation (8.3.48).

    1. Determine the value of the threshold $\alpha = \alpha^*$ for $p = 0.3$, $\mathcal{E}_b = 1$ and $N_0 = 0.1$.

    2. Determine the average probability of error when $\alpha = \alpha^*$.

8.12 Suppose that two signal waveforms $s_1(t)$ and $s_2(t)$ are orthogonal over the interval $(0, T)$. A sample function $n(t)$ of a zero-mean, white noise process is cross-correlated with $s_1(t)$ and $s_2(t)$ to yield

$$n_1 = \int_0^T s_1(t)\, n(t)\, dt$$

and

$$n_2 = \int_0^T s_2(t)\, n(t)\, dt.$$

Prove that $E(n_1 n_2) = 0$.

8.13 A binary PAM communication system is used to transmit data over an AWGN channel. The prior probabilities for the bits are $P(a_m = 1) = 1/3$ and $P(a_m = -1) = 2/3$.

    1. Determine the optimum threshold at the detector.

    2. Determine the average probability of error.

    3. Evaluate the average probability of error when $\mathscr{E}_b = 1$ and $N_0 = 0.1$.

8.14 In a binary antipodal signaling scheme, the signals are given by

$$s_1(t) = -s_2(t) = \begin{cases} \frac{2At}{T} & 0 \leq t \leq \frac{T}{2} \\[2mm] 2A\left(1 - \frac{t}{T}\right) & \frac{T}{2} \leq t \leq T. \\[2mm] 0 & \text{otherwise} \end{cases}$$

The channel is AWGN and $\mathcal{S}_n(f) = \frac{N_0}{2}$. The two signals have prior probabilities p and $1 - p$.

    1. Determine the structure of the optimal receiver.

    2. Determine an expression for the error probability.

    3. Plot the error probability as a function of p for $0 \leq p \leq 1$.

8.15 In an additive white Gaussian noise channel with noise power spectral density of $\frac{N_0}{2}$, two equiprobable messages are transmitted by

$$s_1(t) = \begin{cases} \frac{At}{T}, & 0 \leq t \leq T \\ 0, & \text{otherwise} \end{cases}.$$

$$s_2(t) = \begin{cases} A\left(1 - \frac{t}{T}\right), & 0 \leq t \leq T \\ 0, & \text{otherwise} \end{cases}$$

    1. Determine the structure of the optimal receiver.

    2. Determine the probability of error.

8.16 Suppose that binary PAM is used for transmitting information over an AWGN with power spectral density of $N_0/2 = 10^{-10}$ W/Hz. The transmitted signal energy is $\mathcal{E}_b = A^2 T/2$, where T is the bit interval and A is the signal amplitude. Determine the signal amplitude required to achieve an error probability of 10^{-6} if the data rate is (a)10 kbps, (b)100 kbps, and (c) 1 Mbps.

8.17 A Manchester encoder maps an information 1 into 10 and a 0 into 01. The signal waveforms corresponding to the Manchester code are shown in Figure P-8.17. Determine the probability of error if the two signals are equally probable.

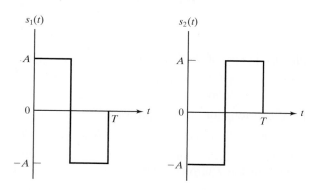

Figure P-8.17

8.18 The signal waveform

$$s(t) = \begin{cases} e^{-t} & 0 \le t \le T \\ 0 & \text{otherwise} \end{cases}$$

is passed through its matched filter, $h(t) = s(T - t)$. Determine the output of the matched filter.

8.19 The data rate in a binary PAM communication system with AWGN is 2 Mbps. If the desired average error probability is 10^{-6}, determine the SNR/bit, $\mathcal{E}_b/N_0$, and the power-to-noise ratio P_{av}/N_0.

8.20 A binary digital communication system employs the signals

$$s_0(t) = 0, \quad 0 \le t \le T$$

and

$$s_1(t) = A, \quad 0 \le t \le T$$

for transmitting the information. This is called *on–off signaling*. The demodulator cross-correlates the received signal $r(t)$ with $s(t)$ and samples the output of the correlator at $t = T$.

    1. Determine the optimum detector for an AWGN channel and the optimum threshold, assuming that the signals are equally probable.

2. Determine the probability of error as a function of the SNR. How does on–off signaling compare with antipodal signaling?

8.21 Three messages, m_1, m_2, and m_3, are to be transmitted over an AWGN channel with noise power spectral density $\frac{N_0}{2}$. The messages are

$$s_1(t) = \begin{cases} 1 & 0 \leq t \leq T \\ 0 & \text{otherwise} \end{cases}$$

and

$$s_2(t) = -s_3(t) = \begin{cases} 1 & 0 \leq t \leq \frac{T}{2} \\ -1 & \frac{T}{2} \leq t \leq T \\ 0 & \text{otherwise} \end{cases}.$$

1. What is the dimensionality of the signal space?
2. Find an appropriate basis for the signal space. (Hint: You can find the basis without using the Gram–Schmidt procedure.)
3. Sketch the signal constellation for this problem.
4. Derive and sketch the optimal decision regions R_1, R_2, and R_3.
5. Which of the three messages is more vulnerable to errors and why? In other words, which of the $P(\text{error}|m_i$ transmitted), $i = 1, 2, 3$ is larger?
6. Using the union bound, find an upper bound on the error probability of this signaling scheme.

8.22 A three-level PAM system is used to transmit the output of a memoryless ternary source whose rate is 2000 symbols/sec. The signal constellation is shown in Figure P-8.22. Determine the input to the detector, the optimum threshold that minimizes the average probability of error, and the average probability of error.

    $-A$ 0 A **Figure P-8.22**

8.23 Consider a signal detector with an input

$$r = \pm A + n,$$

where $+A$ and $-A$ occur with equal probability and the noise variable n is characterized by the (Laplacian) PDF shown in Figure P-8.23.

1. Determine the probability of error as a function of the parameters A and σ.
2. Determine the "SNR" required to achieve an error probability of 10^{-5}. How does the SNR compare with the result for a Gaussian PDF?

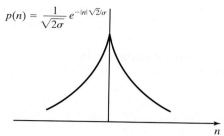

$$p(n) = \frac{1}{\sqrt{2}\sigma} e^{-|n|\sqrt{2}/\sigma}$$

n **Figure P-8.23**

8.24 Determine the average energy of a set of M PAM signals of the form

$$s_m(t) = s_m \psi(t), \quad m = 1, 2, \ldots, M$$
$$0 \le t \le T,$$

where

$$s_m = (2m - 1 - M)d, \quad m = 1, 2, \ldots, M.$$

The signals are equally probable with amplitudes that are symmetric about zero and are uniformly spaced with distance d between adjacent amplitudes, as shown in Figure 8.29.

8.25 Show that the correlation coefficient of two adjacent signal points corresponding to the vertices of an N-dimensional hypercube with its center at the origin is given by

$$\gamma = \frac{N - 2}{N},$$

and their Euclidean distance is

$$d = 2\sqrt{\mathscr{E}_s/N}.$$

8.26 Consider a set of M orthogonal signal waveforms $s_m(t), 1 \le m \le M$, and $0 \le t \le T$, all of which have the same energy $\mathscr{E}$. Define a new set of M waveforms as

$$s'_m(t) = s_m(t) - \frac{1}{M} \sum_{k=1}^{M} s_k(t), \quad 1 \le m \le M$$
$$0 \le t \le T.$$

Show that the M signal waveform $\left\{ s'_m(t) \right\}$ have equal energy, given by

$$\mathscr{E}' = (M - 1)\mathscr{E}/M,$$

and are equally correlated, with correlation coefficient

$$\gamma_{mn} = \frac{1}{\mathscr{E}'} \int_0^T s'_m(t) s'_n(t)\, dt = -\frac{1}{M - 1}.$$

8.27 In this chapter, we showed that an optimal demodulator can be realized as a correlation-type demodulator and a matched filter-type demodulator, where in both cases, $\psi_j(t)$ and $1 \le j \le N$ were used for correlating $r(t)$ or designing the matched filters. Show that an optimal demodulator for a general M-ary communication system can also be designed based on correlating $r(t)$ with $s_i(t)$ and $1 \le i \le M$ or designing filters that are matched to $s_i(t)$'s and $1 \le i \le M$. Precisely describe the structure of such demodulators by giving their block diagram and all relevant design parameters, and compare their complexity with the complexity of the demodulators obtained in the text.

8.28 Consider a biorthogonal signal set with $M = 8$ signal points. Determine a union bound for the probability of a symbol error as a function of $\mathcal{E}_b/N_0$. The signal points are equally likely a priori.

8.29 Consider an M-ary digital communication system where $M = 2^N$, and N is the dimension of the signal space. Suppose that the M signal vectors lie on the vertices of a hypercube that is centered at the origin, as illustrated in Figure 8.37. Determine the average probability of a symbol error as a function of $\mathcal{E}_s/N_0$, where $\mathcal{E}_s$ is the energy per symbol, $N_0/2$ is the power spectral density of the AWGN, and all signal points are equally probable.

8.30 Consider the signal waveform

$$s(t) = \sum_{k=1}^{n} c_i p(t - nT_c),$$

where $p(t)$ is a rectangular pulse of unit amplitude and duration T_c. The $\{c_i\}$ may be viewed as a code vector $c = [c_1, c_2, \dots, c_n]$, where the elements $c_i = \pm 1$. Show that the filter matched to the waveform $s(t)$ may be realized as a cascade of a filter matched to $p(t)$ followed by a discrete-time filter matched to the vector c. Determine the value of the output of the matched filter at the sampling instant $t = nT_c$.

8.31 A speech signal is sampled at a rate of 8 kHz, logarithmically compressed and encoded into a PCM format using 8 bits/sample. The PCM data is transmitted through an AWGN baseband channel via M-level PAM. Determine the symbol rate required for transmission when (a)$M = 4$,(b) $M = 8$, and (c) $M = 16$.

8.32 Two equiprobable messages are transmitted via an additive white Gaussian noise channel with a noise power spectral density of $\frac{N_0}{2} = 1$. The messages are transmitted by the signals

$$s_1(t) = \begin{cases} 1 & 0 \le t \le 1 \\ 0 & \text{otherwise} \end{cases}$$

and $s_2(t) = s_1(t - 1)$.

We intended to implement the receiver using a correlation type structure, but due to imperfections in the design of the correlators, the structure shown in Figure P-8.32 has been implemented. The imperfection appears in the integrator in the upper branch where we have $\int_0^{1.5}$ instead of $\int_0^1$. The decision box, therefore, observes r_1 and r_2; based on this observation, it has to decide which message was transmitted. What decision rule should be adopted by the decision box for an optimal decision?

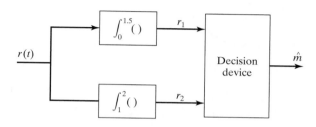

Figure P-8.32

8.33 A Hadamard matrix is defined as a matrix whose elements are ± 1 and row vectors are pairwise orthogonal. In the case when n is a power of 2, an $n \times n$ Hadamard matrix is constructed by means of the recursion

$$H_2 = \begin{bmatrix} 1 & 1 \\ 1 & -1 \end{bmatrix} \qquad H_{2n} = \begin{bmatrix} H_n & H_n \\ H_n & -H_n \end{bmatrix}.$$

1. Let c_i denote the ith row of an $n \times n$ Hadamard matrix as previously defined. Show that the waveforms constructed as

$$s_i(t) = \sum_{k=1}^{n} c_{ik} p(t - kT_c), \quad i = 1, 2, \ldots, n$$

are orthogonal, where $p(t)$ is an arbitrary pulse confined to the time interval $0 \le t \le T_c$.

2. Show that the matched filters (or cross correlators) for the n waveforms $\{s_i(t)\}$ can be realized by a single filter (or correlator) matched to the pulse $p(t)$ followed by a set of n cross correlators using the code words $\{c_i\}$.

8.34 The discrete sequence

$$r_k = \sqrt{\mathscr{E}_c} c_k + n_k, \quad k = 1, 2, \ldots, n$$

represents the output sequence of samples from a demodulator, where $c_k = \pm 1$ are elements of one of two possible codewords, $c_1 = [1, 1, \ldots, 1]$ and $c_2 = [1, 1, \ldots, 1, -1, \ldots, -1]$. The codeword c_2 has w elements, which are $+1$, and $n - w$ elements, which are -1, where w is some positive integer. The noise sequence $\{n_k\}$ is white Gaussian with variance σ^2.

1. What is the optimum maximum likelihood detector for the two possible transmitted signals?

2. Determine the probability error as a function of the parameters $(\sigma^2, \mathcal{E}_b, w)$.

3. What is the value of w that minimizes the error probability?

8.35 Show that the early–late gate synchronizer illustrated in Figure 8.49 is a close approximation to the timing recovery system illustrated in Figure P-8.35.

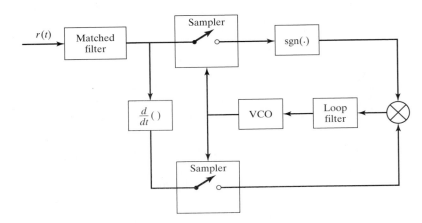

Figure P-8.35

COMPUTER PROBLEMS

8.1 Simulation of Binary Communication Systems Using Orthogonal Signals

The purpose of this problem is to estimate and plot the probability of error as a function of the SNR for a binary communication system that employs binary orthogonal signals that are transmitted over an additive white Gaussian noise channel. The model of the binary communication system employing orthogonal signals is shown in Figure CP-8.1. As shown, we simulate the generation of a binary sequence of 0's and 1's that occur with equal probability and are mutually statistically independent. To accomplish this task, we use a random number generator that generates uniform random numbers in the range $(0, 1)$. If a number generated is in the range $(0, 0.5)$, the binary source output is a 0. Otherwise, it is a 1. If a zero is generated, then $y_0 = \sqrt{\mathcal{E}_b} + n_0$ and $y_1 = n_1$, where y_0 and y_1 represent the outputs of the two matched filters or the two correlators for the binary orthogonal signals. If a 1 is generated, then $y_0 = n_0$ and $y_1 = \sqrt{\mathcal{E}_b} + n_1$ are generated.

The additive noise components n_0 and n_1 are generated by means of two Gaussian noise generators. Their means are zero and their variances are $\sigma^2 = N_0/2$.

For convenience, we may normalize the signal energy $\mathcal{E}_b$ to unity ($\mathcal{E}_b = 1$) and vary σ^2. Note that the SNR, which is defined as $\mathcal{E}_b/N_0$, is then equal to $1/2\sigma^2$. The matched filter or correlator outputs y_0 and y_1 are fed to the detector, which decides whether a 0 or a 1 has been received. The detector output is compared with the binary transmitted data sequence and an error counter is used to count the number of bit errors.

Perform the simulation for the transmission of 10,000 bits at several different values of SNR, which cover the range of SNR's $0 \leq 10 \log \mathcal{E}_b/N_0 < 10$ dB. Plot the error probability as a function of the SNR. Compare the estimated error probability with the theoretical error probability given by the formula

$$P_2 = Q\left(\sqrt{\frac{\mathcal{E}_b}{N_0}}\right).$$

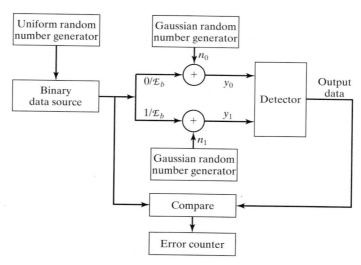

Figure CP-8.1 Simulation model for binary orthogonal signals.

8.2 Simulation of the Performance of Binary Antipodal Signals

The purpose of this problem is to estimate and plot the probability of error when binary antipodal signals are transmitted over an additive white Gaussian noise channel. The model of the binary communication system employing antipodal signals is shown in Figure CP-8.2. As shown, we simulate the generation of the random variable y, which is the input to the detector. A uniform random number generator is used to generate the binary information sequence of 0's and 1's from the data source. The sequence of 0's and 1's is mapped into a sequence of $\pm\mathcal{E}_b$, where $\mathcal{E}_b$ represents the signal energy per bit. A Gaussian noise generator is used to generate a sequence of zero-mean Gaussian numbers with variance σ^2. For equally probable 0's and 1's, the detector compares the random variable y with the threshold zero. If $y > 0$, the decision is made that the transmitted bit is a zero. If $y < 0$, the decision is made that the transmitted bit is a 1. The output of the detector is compared with the transmitted sequence of information bits, and the bit errors are counted.

Perform the simulation for the transmission of 10,000 bits at several different values of SNR, which covers the range of SNR's $0 < 10 \log_{10} \mathscr{E}_b/N_0 \leq 7$. Plot the error probability as a function of the SNR. Compare the estimated error probability with the theoretical error probability given by the formula

$$P_2 = Q\left(\sqrt{\frac{2\mathscr{E}_b}{N_0}}\right).$$

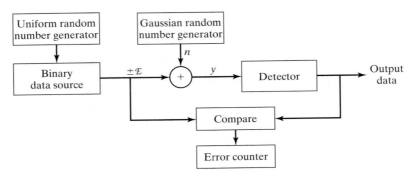

Figure CP-8.2   Model of the binary communication system employing antipodal signals.

8.3 Simulation of Binary Communication System Using On–Off Signals

The model for the system to be simulated is similar to that shown in Figure CP-8.2, except that one of the signals is zero. Thus, we generate a sequence of random variables $\{y_i\}$, where either $y_i = n_i$ when no signal is transmitted or $y_i = \sqrt{\mathscr{E}_b} + n_i$ when the signal is transmitted, and n_i is a zero-mean Gaussian random variable with variance $\sigma^2 = N_0/2$. Hence, the two possible PDF's or y_i are shown in Figure CP-8.3. The detector compares each of the random variables $\{y_i\}$ to the optimum threshold $\sqrt{\mathscr{E}_b}/2$ and makes the decision that a 0 was transmitted when $y_i < \sqrt{\mathscr{E}_b}/2$ and that a 1 was transmitted when $y_i > \sqrt{\mathscr{E}_b}/2$.

Perform the simulation for the transmission of 10,000 bits at several different values of SNR that covers the range of SNR's $0 \leq \log_{10} \mathscr{E}_b/N_0 \leq 13 \ dB$. Plot the error probability as a function of the SNR. Compare the estimated error

probability obtained from the simulation with the theoretical error probability
given by the formula

$$P_2 = Q\left(\sqrt{\frac{\mathscr{E}_b}{2N_0}}\right).$$

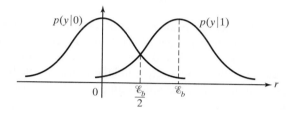

Figure CP-8.3 The probability density
functions for the received signal at the
output of the correlator for on-off signals.

8.4 Effect of Noise in a Binary Communication System

The effect of noise on the performance of a binary communication system can
be observed from the received signal plus noise at the input to the detector.
For example, consider the binary communication system that employs binary
orthogonal signals, for which the input to the detector consists of a pair of
random variables (y_0, y_1), where either

$$(y_0, y_1) = \left(\sqrt{\mathscr{E}_b} + n_0, n_1\right)$$

or

$$(y_0, y_1) = \left(n_0, \sqrt{\mathscr{E}_b} + n_1\right).$$

The noise variables n_0 and n_1 are zero-mean, independent Gaussian random
variables with equal variance σ^2.

Perform the simulation as described in CP-8.1 to generate 100 samples of the
received sequence (y_0, y_1) for each value of $\sigma = 0.1, \sigma = 0.3$, and $\sigma = 0.5$,
where $\mathscr{E}_b$ is normalized to unity. Plot these 100 samples for each σ as separate
two-dimensional plots with coordinates (y_0, y_1). Comment on the results.

8.5 Simulation of 4-Amplitude PAM System

The purpose of this problem is to estimate and plot the probability of error as a
function of the SNR for a digital communication system that employs four-level
PAM. The receiver is assumed to consist of a signal correlator or a matched
filter followed by a detector. The model for the system to be simulated is shown
in Figure CP-8.5. As shown, we simulate the generation of the random variable
y, which is the output of the signal correlator and the input to the detector.

We begin by generating a sequence of quaternary symbols that are mapped into corresponding amplitude levels $\{A_m\}$. To accomplish this task, we use a random number generator that generates a uniform random number in the range $(0, 1)$. This range is subdivided into four equal intervals $(0, 0.25)$, $(0.25, 0.5)$, $(0.5, 0.75)$ and $(0.75, 1)$, where the subintervals correspond to the symbols (pairs of information bits) 00, 01, 11, and 10, respectively. Thus, the output of the uniform random number generator is mapped into the corresponding signal amplitude levels $(-3d, -d, d,$ and $3d)$, respectively.

The additive noise component having mean zero and variance σ^2 is generated by means of a Gaussian random number generator. For convenience, we may normalize the distance parameter $d = 1$ and vary σ^2. The detector observes $y = A_m + n$ and computes the distance between y and the four possible transmitted signal amplitudes. Its output $\hat{A}_m$ is the signal amplitude level corresponding to the smallest distance. $\hat{A}_m$ is compared with the actual transmitted signal amplitude, and an error counter is used to count the errors made by the detector.

Perform a simulation for the transmission of 10,000 symbols at different values of the average bit SNR, which is defined as

$$\frac{\mathscr{E}_{\text{avb}}}{N_0} = \frac{5}{4}\left(\frac{d^2}{\sigma^2}\right),$$

and plot the estimated probability of a symbol error as a function of $\mathscr{E}_{\text{avb}}/N_0$. Also plot the theoretical expression for the probability of error and compare the simulated performance to the theoretical performance.

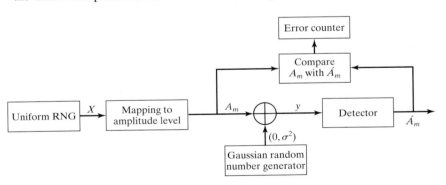

Figure CP-8.5   Block diagram of four-level PAM for Monte Carlo simulation.

8.6 Simulation of 16-Amplitude PAM System

Modify the simulation described in CP-8.5 to transmit 16-level PAM. In this case, the 16-ary symbols may be generated directly by subdividing the interval $(0, 1)$ into 16 equal-width subintervals and mapping the 16-ary symbols into the 16-ary signal amplitudes. Perform the simulation and plot the estimated error probability. Also, plot the theoretical probability of a symbol error given by the expression in the text and compare the estimated values to the theoretical values.

8.7 Simulation of a Communication System Employing $M = 4$ Orthogonal Signals

The purpose of this problem is to estimate and plot the probability of error as a function of the SNR for a digital communication system that employs $M = 4$ orthogonal signals and transmits them over an additive white Gaussian noise channel. The model of the system to be simulated is shown in Figure CP-8.7.

As shown, we simulate the generation of the random variables y_0, y_1, y_2, y_3, which constitute the input to the detector. First, we may generate a binary sequence of 0's and 1's that occur with equal probability and are mutually statistically independent. The binary sequence is grouped into pairs of bits, which are mapped into the corresponding signal components. As an alternative to generating the individual bits, we may generate pairs of bits, as described in CP 8.5. In either case, we have the mapping of the four symbols into the signal points.

$$00 \rightarrow s_0 = \left(\sqrt{\mathscr{E}_s}, 0, 0, 0\right);$$

$$01 \rightarrow s_1 = \left(0, \sqrt{\mathscr{E}_s}, 0, 0\right);$$

$$10 \rightarrow s_2 = \left(0, 0, \sqrt{\mathscr{E}_s}, 0\right);$$

$$11 \rightarrow s_3 = \left(0, 0, 0, \sqrt{\mathscr{E}_s}\right).$$

The additive noise components n_0, n_1, n_2, n_3 are generated by means of four Gaussian noise generators, each having zero mean and variance $\sigma^2 = N_0/2$. For convenience, we may normalize $\mathscr{E}_s = 1$ and vary σ^2. Since $\mathscr{E}_s = 2\mathscr{E}_b$, it follows that $\mathscr{E}_b = 1/2$. The detector input consists of the received signal vector $y = s_i + n, \quad i = 0, 1, 2, 3$. The detector computes the correlation metrics $y \cdot s_j, \quad j = 0, 1, 2, 3$, and detects the signal having the largest correlation value. An error counter is used to count the number of bit errors.

Perform the simulation for the transmission of 10,000 symbols (20,000 bits) at several different values of SNR/bit that covers the range $0 \leq 10 \log \mathscr{E}_b/N_0 \leq 8$ dB. Plot the estimated probability of a bit error and compare the estimated values with the theoretical expression for the probability of a bit error. Comment on the results.

8.8 Simulation of a Communication System Employing $M = 8$ Orthogonal Signals

Modify the simulation described in CP-8.7 to transmit $M = 8$ orthogonal signals. Perform the simulation and plot the results.

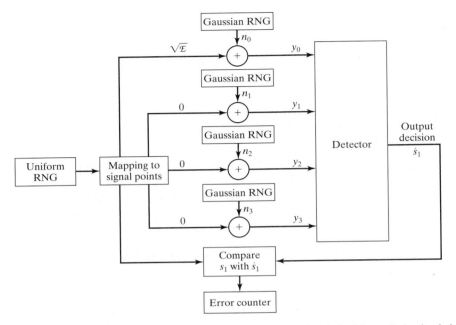

Figure CP-8.7   Block diagram of system with $M = 4$ orthogonal signals for Monte Carlo simulation.

8.9 Simulation of a Communication System Employing $M = 4$ Biorthogonal Signals

The purpose of this problem is to estimate and plot the probability of error as a function of the SNR for a digital communication system that employs $M = 4$ biorthogonal signals and transmits them over an additive white Gaussian noise channel. The model for the system to be simulated is shown in Figure CP-8.9.

As shown, we simulate the generation of the random variables y_0 and y_1, which constitute the input to the detector. We begin by generating a binary sequence of 0's and 1's that occur with equal probability and are mutually statistically independent, as described in CP 8.1. The binary sequence is grouped into pairs of bits, which are mapped into the corresponding signals as follows:

$$00 \rightarrow s_0 = \left(\sqrt{\mathcal{E}_s}, 0 \right);$$

$$01 \rightarrow s_1 = \left(0, \sqrt{\mathcal{E}_s} \right);$$

$$10 \rightarrow s_2 = \left(0, -\sqrt{\mathcal{E}_s} \right);$$

$$11 \rightarrow s_3 = \left(-\sqrt{\mathcal{E}_s}, 0 \right).$$

Since $s_2 = -s_1$ and $s_3 = -s_0$, the demodulation requires two correlators or two matched filters, whose outputs are y_0 and y_1. The additive noise components n_0 and n_1 are generated by means of two Gaussian noise generators, each having mean zero and variance $\sigma^2 = N_0/2$. For convenience, we may normalize the symbol energy to $\mathcal{E}_s = 1$ and vary the noise variance σ^2. Since $\mathcal{E}_s = 2\mathcal{E}_b$, it follows that $\mathcal{E}_b = 1/2$. The detector output is compared with the transmitted sequence of bits and an error counter is used to count the number of bit errors and the number of symbol errors.

Perform the simulation for the transmission of 10,000 symbols (20,000 bits) at several different values of SNR/bit that covers the range $0 \le 10 \log_{10} \mathcal{E}_b/N_0 \le 8 \, dB$. Plot the estimated probability of a symbol error and compare the estimated values with the theoretical expression for the probability of a symbol error. Comment on the results.

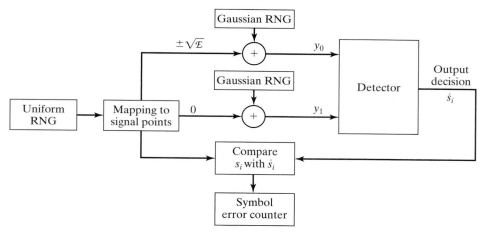

Figure CP-8.9 Block diagram of the system with $M = 4$ biorthogonal signals for Monte Carlo simulation.

8.10 Early–Late Gate Synchronizer for PAM

The objective of this problem is to simulate the operation of an early–late gate synchronizer for a binary PAM system. The basic pulse used in PAM has a raised-cosine spectrum with a roll-off factor of 0.4. The system transmission rate $1/T = 4800$ bits/s. Consequently, the pulse is given as

$$x(t) = \text{sinc}(4800t) \frac{\cos 1920\pi t}{1 - 1.4746 \times 10^7 t^2}.$$

This signal pulse extends from $-\infty$ to $+\infty$.

Plot $x(t)$ and verify that, for all practical purposes, it is sufficient to consider only the interval $|t| \le 0.6 \times 10^{-3}$, which is roughly $[-3T, 3T]$. Truncate the pulse to this interval and compute the autocorrelation function. Plot the autocorrelation, and determine its length in samples and the position of its maximum, i.e., the optimum sampling time. Select the sampling rate to be 40 samples per bit interval T.

Simulate the early–late gate synchronizer when the incorrect sampling is to the right or to the left of the peak by 10 samples and verify that the synchronizer finds the correct sampling time, i.e., the maximum of the pulse.

9

Digital Transmission through Bandlimited AWGN Channels

In the preceding chapter, we considered digital communication over an AWGN channel and evaluated the probability of error performance of the optimum receiver for several different types of baseband digital modulation techniques. In this chapter, we treat digital communication over a channel that is modeled as a linear filter with a bandwidth limitation. The bandlimited channels most frequently encountered in practice are telephone channels, microwave line-of-sight (LOS) radio channels, satellite channels, and underwater acoustic channels.

In general, a linear filter channel imposes more stringent requirements on the design of modulation signals. Specifically, the transmitted signals must be designed to satisfy the bandwidth constraint imposed by the channel. The bandwidth constraint generally precludes the use of rectangular pulses at the output of the modulator. Instead, the transmitted signals must be shaped to restrict their bandwidth to that available on the channel. The design of bandlimited signals is one of the topics treated in this chapter.

We will see that a linear filter channel distorts the transmitted signal. The channel distortion results in intersymbol interference at the output of the demodulator and leads to an increase in the probability of error at the detector. Devices or methods for correcting or undoing the channel distortion, called *channel equalizers*, are then described.

9.1 DIGITAL TRANSMISSION THROUGH BANDLIMITED CHANNELS

A bandlimited channel such as a telephone wireline is characterized as a linear filter with impulse response $c(t)$ and frequency response $C(f)$, where

$$C(f) = \int_{-\infty}^{\infty} c(t)e^{-j2\pi ft}\,dt. \tag{9.1.1}$$

475

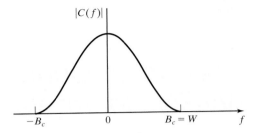

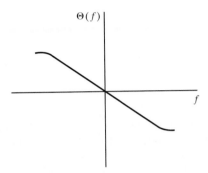

Figure 9.1 Magnitude and phase responses of bandlimited channel.

If the channel is a baseband channel that is bandlimited to B_c Hz, then $C(f) = 0$ for $|f| > B_c$. Any frequency components at the input to the channel that are higher than B_c Hz will not be passed by the channel. For this reason, we consider the design of signals for transmission through the channel that are bandlimited to $W = B_c$ Hz, as shown in Figure 9.1. Henceforth, W will denote the bandwidth limitation of the signal and the channel.

Now, suppose that the input to a bandlimited channel is a signal waveform $g_T(t)$, where the subscript T denotes that the signal waveform is the output of the transmitter. Then, the response of the channel is the convolution of $g_T(t)$ with $c(t)$, i.e.,

$$h(t) = \int_{-\infty}^{\infty} c(\tau) g_T(t - \tau) d\tau = c(t) * g_T(t), \qquad (9.1.2)$$

or, when expressed in the frequency domain, we have

$$H(f) = C(f)G_T(f), \qquad (9.1.3)$$

where $G_T(f)$ is the spectrum (Fourier transform) of the signal $g_T(t)$ and $H(f)$ is the spectrum of $h(t)$. Thus, the channel alters or distorts the transmitted signal $g_T(t)$.

Let us assume that the signal at the output of the channel is corrupted by AWGN. Then, the signal at the input to the demodulator is of the form $h(t) + n(t)$, where $n(t)$ denotes the AWGN. The linear filter channel model is shown in Figure 9.2.

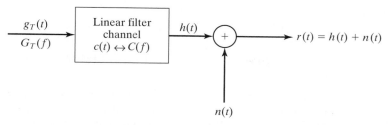

Figure 9.2   Linear filter model for a bandlimited channel.

From the preceding chapter, we recall that in the presence of AWGN, a demod-ulator that employs a filter that is matched to the signal $h(t)$ maximizes the SNR at its output. Therefore, let us pass the received signal $r(t) = h(t) + n(t)$ through a filter that has a frequency response

$$G_R(f) = H^*(f)e^{-j2\pi f t_0}, \tag{9.1.4}$$

where t_0 is some nominal time delay at which we sample the filter output. The subscript R denotes that the matched filter is at the receiver.

The signal component at the output of the matched filter at the sampling instant $t = t_0$ is

$$y_s(t_0) = \int_{-\infty}^{\infty} |H(f)|^2 df = \mathscr{E}_h, \tag{9.1.5}$$

which is the energy in the channel output waveform $h(t)$. The noise component at the output of the matched filter has a zero mean and a power-spectral density

$$S_n(f) = \frac{N_0}{2}|H(f)|^2. \tag{9.1.6}$$

Hence, the noise power at the output of the matched filter has a variance

$$\sigma_n^2 = \int_{-\infty}^{\infty} S_n(f)df = \frac{N_0}{2}\int_{-\infty}^{\infty} |H(f)|^2 df = \frac{N_0 \mathscr{E}_h}{2}. \tag{9.1.7}$$

Then the SNR at the output of the matched filter is

$$\left(\frac{S}{N}\right)_0 = \frac{\mathscr{E}_h^2}{N_0 \mathscr{E}_h/2} = \frac{2\mathscr{E}_h}{N_0}. \tag{9.1.8}$$

This is the result for the SNR at the output of the matched filter that was obtained in Chapter 8, except the received signal energy $\mathscr{E}_h$ has replaced the transmitted signal energy $\mathscr{E}_s$. Compared to the previous result, the major difference in this development is that the filter impulse response is matched to the received signal $h(t)$ instead of the transmitted signal. Note that the implementation of the matched filter at the receiver requires that $h(t)$ or, equivalently, the channel impulse response $c(t)$ must be known to the receiver.

Example 9.1.1

The signal pulse

$$g_T(t) = \frac{1}{2}\left[1 + \cos\frac{2\pi}{T}\left(t - \frac{T}{2}\right)\right], \quad 0 \le t \le T$$

is transmitted through a baseband channel with a frequency-response characteristic as shown in Figure 9.3(a). The signal pulse is illustrated in Figure 9.3(b). The channel output is corrupted by AWGN with the power spectral density $N_0/2$. Determine the matched filter to the received signal and the output SNR.

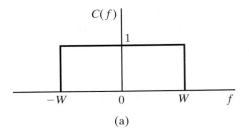

(a)

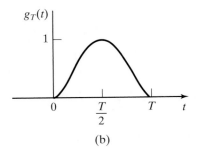

(b)

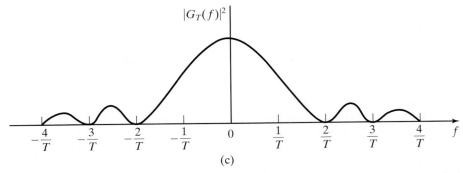

(c)

Figure 9.3 The signal pulse in (b) is transmitted through the ideal bandlimited channel shown in (a). The spectrum of $g_T(t)$ is shown in (c).

Solution   This problem is most easily solved in the frequency domain. First, the spectrum of the transmitted signal pulse is

$$G_T(f) = \frac{T}{2} \frac{\sin \pi f T}{\pi f T (1 - f^2 T^2)} e^{-j\pi f T}$$

$$= \frac{T}{2} \frac{\text{sinc } f T}{(1 - f^2 T^2)} e^{-j\pi f T}.$$

The spectrum $|G_T(f)|^2$ is shown in Figure 9.3(c). Hence,

$$H(f) = C(f) G_T(f)$$

$$= \begin{cases} G_T(f), & |f| \leq W \\ 0, & \text{otherwise} \end{cases}.$$

Then, the signal component at the output of the filter matched to $H(f)$ is

$$\mathcal{E}_h = \int_{-W}^{W} |G_T(f)|^2 \, df$$

$$= \frac{1}{(2\pi)^2} \int_{-W}^{W} \frac{(\sin \pi f T)^2}{f^2 (1 - f^2 T^2)^2} \, df$$

$$= \frac{T}{(2\pi)^2} \int_{-WT}^{WT} \frac{\sin^2 \pi \alpha}{\alpha^2 (1 - \alpha^2)^2} \, d\alpha.$$

The variance of the noise component is

$$\sigma_n^2 = \frac{N_0}{2} \int_{-W}^{W} |G_T(f)|^2 \, df = \frac{N_0 \mathcal{E}_h}{2}.$$

Hence, the output SNR is

$$\left(\frac{S}{N}\right)_0 = \frac{2 \mathcal{E}_h}{N_0}.$$

In this example, we observe that the signal at the input to the channel is not bandlimited. Hence, only a part of the transmitted signal energy is received, i.e., only the signal energy that falls within the passband $|f| \leq W$ of the channel. The amount of signal energy at the output of the matched filter depends on the value of the channel bandwidth W when the signal pulse duration is fixed (see Problem 9.1). The maximum value of $\mathcal{E}_h$, obtained as $W \to \infty$, is

$$\max \mathcal{E}_h = \int_{-\infty}^{\infty} |G_T(f)|^2 = \int_0^T g_T^2(t) \, dt = \mathcal{E}_g,$$

where $\mathcal{E}_g$ is the energy of the signal pulse $g_T(t)$.   ∎

In this development, we considered the transmission and reception of only a single isolated signal waveform $g_T(t)$ through a bandlimited channel with the impulse response $c(t)$. We observed that the performance of the system is determined by $\mathcal{E}_h$, the energy in the received signal $h(t)$. To maximize the received SNR, we must make sure that the spectrum of the transmitted signal waveform $g_T(t)$ is limited to the bandwidth of the channel. The impact of the channel bandwidth limitation is felt when we consider the transmission of a sequence of signal waveforms. This problem is treated in the next section.

9.1.1 Digital PAM Transmission through Bandlimited Baseband Channels

Let us consider the baseband PAM communication system illustrated by the functional block diagram in Figure 9.4. The system consists of a transmitting filter having an impulse response $g_T(t)$, the linear filter channel with AWGN, a receiving filter with an impulse response $g_R(t)$, a sampler that periodically samples the output of the receiving filter, and a symbol detector. The sampler requires the extraction of a timing signal from the received signal as described in Section 8.6. This timing signal serves as a clock that specifies the appropriate time instants for sampling the output of the receiving filter.

Let us consider digital communication by means of M-ary PAM. Hence, the input binary data sequence is subdivided into k-bit symbols, and each symbol is mapped into a corresponding amplitude level that amplitude modulates the output of the transmitting filter. The baseband signal at the output of the transmitting filter (the input to the channel) may be expressed as

$$v(t) = \sum_{n=-\infty}^{\infty} a_n g_T(t - nT), \qquad (9.1.9)$$

where $T = k/R_b$ is the symbol interval ($1/T = R_b/k$ is the symbol rate), R_b is the bit rate, and $\{a_n\}$ is a sequence of amplitude levels corresponding to the sequence of k-bit blocks of information bits.

The channel output, which is the received signal at the demodulator, may be expressed as

$$r(t) = \sum_{n=-\infty}^{\infty} a_n h(t - nT) + n(t), \qquad (9.1.10)$$

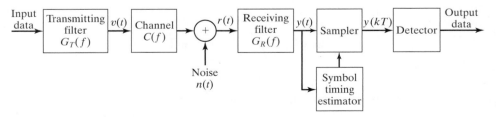

Figure 9.4   Block diagram of a digital PAM system.

where $h(t)$ is the impulse response of the cascade of the transmitting filter and the channel. Thus, $h(t) = c(t) * g_T(t)$, $c(t)$ is the impulse response of the channel, and $n(t)$ represents the AWGN.

The received signal is passed through a linear receiving filter with the impulse response $g_R(t)$ and frequency response $G_R(f)$. If $g_R(t)$ is matched to $h(t)$, then its output SNR is a maximum at the proper sampling instant. The output of the receiving filter may be expressed as

$$y(t) = \sum_{n=-\infty}^{\infty} a_n x(t - nT) + w(t), \tag{9.1.11}$$

where $x(t) = h(t) * g_R(t) = g_T(t) * c(t) * g_R(t)$ and $w(t) = n(t) * g_R(t)$ denotes the additive noise at the output of the receiving filter.

To recover the information symbols $\{a_n\}$, the output of the receiving filter is sampled periodically, every T seconds. Thus, the sampler produces

$$y(mT) = \sum_{n=-\infty}^{\infty} a_n x(mT - nT) + w(mT), \tag{9.1.12}$$

or equivalently,

$$y_m = \sum_{n=-\infty}^{\infty} a_n x_{m-n} + w_m$$

$$= x_0 a_m + \sum_{n \neq m} a_n x_{m-n} + w_m, \tag{9.1.13}$$

where $x_m = x(mT)$, $w_m = w(mT)$, and $m = 0, \pm 1, \pm 2, \ldots$.

The first term on the right-hand side (RHS) of Equation (9.1.13) is the desired symbol a_m, scaled by the gain parameter x_0. When the receiving filter is matched to the received signal $h(t)$, the scale factor is

$$x_0 = \int_{-\infty}^{\infty} h^2(t) dt = \int_{-\infty}^{\infty} |H(f)|^2 df$$

$$= \int_{-W}^{W} |G_T(f)|^2 |C(f)|^2 df = \mathscr{E}_h, \tag{9.1.14}$$

as indicated by the development of Equations (9.1.4) and (9.1.5). The second term on the RHS of Equation (9.1.13) represents the effect of the other symbols at the sampling instant $t = mT$, called the *intersymbol interference* (ISI). In general, ISI causes a degradation in the performance of the digital communication system. Finally, the third term, w_m, which represents the additive noise, is a zero-mean Gaussian random variable with variance $\sigma_w^2 = N_0 \mathscr{E}_h / 2$, previously given by Equation (9.1.7).

By appropriately designing the transmitting and receiving filters, we can satisfy the condition $x_n = 0$ for $n \neq 0$, so that the ISI term vanishes. In this case, the only term that can cause errors in the received digital sequence is the additive noise. The design of transmitting and receiving filters is considered in the next section.

9.2 SIGNAL DESIGN FOR BANDLIMITED CHANNELS

In this section, we consider the problem of designing a bandlimited transmitting filter. First, the design will be done under the condition that there is no channel distortion. Later, we consider the problem of filter design when the channel distorts the transmitted signal. Since $H(f) = C(f)G_T(f)$, the condition for distortion-free transmission is that the frequency response characteristic $C(f)$ of the channel must have a constant magnitude and a linear phase over the bandwidth of the transmitted signal, i.e.,

$$C(f) = \begin{cases} C_0 e^{-j2\pi f t_0}, & |f| \le W \\ 0, & |f| > W \end{cases}, \tag{9.2.1}$$

where W is the available channel bandwidth, t_0 represents an arbitrary finite delay, which we set to zero for convenience, and C_0 is a constant gain factor, which we set to unity for convenience. Thus, under the condition that the channel is distortion-free and the bandwidth of $g_T(t)$ is limited to W, we have $H(f) = G_T(f)$. Consequently, the matched filter at the receiver has a frequency response $G_R(f) = G_T^*(f)$, and its output at the periodic sampling times $t = mT$ has the form

$$y(mT) = x(0)a_m + \sum_{n \ne m} a_n x(mT - nT) + w(mT), \tag{9.2.2}$$

or more simply,

$$y_m = x_0 a_m + \sum_{n \ne m} a_n x_{m-n} + w_m, \tag{9.2.3}$$

where $x(t) = g_T(t) * g_R(t)$ and $w(t)$ is the output response of the matched filter to the input AWGN process $n(t)$.

The middle term on the RHS of Equation (9.2.3) represents the ISI. The amount of ISI and noise that is present in the received signal can be viewed on an oscilloscope. Specifically, we may display the received signal on the vertical input with the horizontal sweep rate set at $1/T$. The resulting oscilloscope display is called an *eye pattern* because of its resemblance to the human eye. Examples of two eye patterns, one for binary PAM and the other for quaternary ($M = 4$) PAM, are illustrated in Figure 9.5(a).

The effect of ISI is to cause the eye to close, thereby reducing the margin for additive noise to cause errors. Figure 9.5(b) illustrates the effect of ISI in reducing the opening of the eye. Note that ISI distorts the position of the zero crossings and causes a reduction in the eye opening. As a consequence, the system is more sensitive to a synchronization error and exhibits a smaller margin against additive noise.

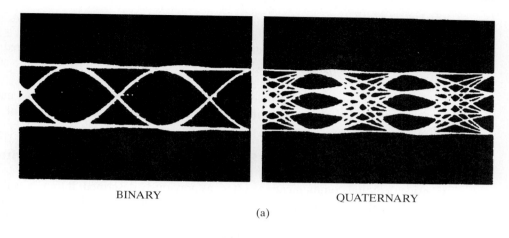

BINARY   QUATERNARY

(a)

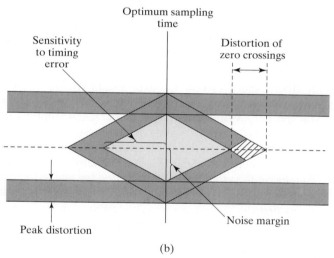

(b)

Figure 9.5 Eye patterns. (a) Examples of eye patterns for binary and quaternary PAM and (b) the effect of ISI on eye opening.

Example 9.2.1

Consider a binary PAM system that transmits data at a rate of $1/T$ bits/sec through an ideal channel of bandwidth W. The sampled output from the matched filter at the receiver is

$$y_m = a_m + 0.2a_{m-1} - 0.3a_{m-2} + w_m,$$

where $a_m = \pm 1$, with equal probability. Determine the peak value of the ISI and the noise margin, as defined in Figure 9.5(b).

Solution If we compare the matched filter output y_m with that given by Equation (9.2.3), it is apparent that $x_0 = 1$, $x_1 = 0.2$, $x_2 = -0.3$, and $x_m = 0$, otherwise. The peak value of the ISI occurs when $a_{m-1} = -a_{m-2}$, so that the ISI term will take the peak value of $+0.5$. Since $x_0 = 1$ and $a_m = \pm 1$, the ISI causes a 50% reduction in the eye opening at the sampling times $t = mT$, $m = 0, \pm 1, \pm 2, \ldots$. Hence, the noise margin is reduced by 50% to a value of 0.5. Thus, compared to the case in which there is no ISI, a noise component that is 50% smaller will cause an error at the detector. $\blacksquare$

Next, we consider the problem of signal design under two conditions, namely, that there is no ISI at the sampling instants and that a controlled amount of ISI is allowed.

9.2.1 Design of Bandlimited Signals for Zero ISI — The Nyquist Criterion

Consider a digital communication system as previously described. It transmits through an ideal bandlimited channel, when the bandwidth of $g_T(t)$ is less than or equal to W. Thus, the bandwidth of the channel, which is the Fourier transform of the signal at the output of the receiving filter, is given as

$$
\begin{aligned}
X(f) &= G_T(f)C(f)G_R(f) \\
&= G_T(f)G_R(f)C_0\, e^{-j2\pi f t_0} \\
&= G_T(f)G_R(f),
\end{aligned} \tag{9.2.4}
$$

where $G_T(f)$ and $G_R(f)$ denote the frequency responses of the transmitter and receiver filters and $C(f) = C_0 \exp(-j2\pi f t_0)$, $|f| \leq W$ denotes the frequency response of the channel. For convenience, we set $C_0 = 1$ and $t_0 = 0$. We have also seen that the output of the receiving filter, when sampled periodically at $t = mT$, $m = \ldots, -2 - 1, 0, 1, 2 \ldots$ yields the expression given by Equation (9.2.3). In this equation, the first term on the right-hand side of the equation is the desired symbol, the second term constitutes the ISI, and the third term is the additive noise.

To remove the effect of ISI, it is necessary and sufficient that $x(mT - nT) = 0$ for $n \neq m$ and $x(0) \neq 0$, where without loss of generality, we can assume $x(0) = 1$. This means that the overall communication system has to be designed such that

$$
x(nT) = \begin{cases} 1, & n = 0 \\ 0, & n \neq 0 \end{cases}. \tag{9.2.5}
$$

In this section, we derive the necessary and sufficient condition for $X(f)$ so $x(t)$ can satisfy the preceding relation. This condition is known as the *Nyquist pulse-shaping criterion* or *Nyquist condition for zero ISI*.

Nyquist Condition for Zero ISI.   A necessary and sufficient condition for $x(t)$ to satisfy

$$
x(nT) = \begin{cases} 1, & n = 0 \\ 0, & n \neq 0 \end{cases} \tag{9.2.6}
$$

is that its Fourier transform $X(f)$ must satisfy

$$
\sum_{m=-\infty}^{\infty} X\left(f + \frac{m}{T}\right) = T. \tag{9.2.7}
$$

Proof.   In general, $x(t)$ is the inverse Fourier transform of $X(f)$. Hence,

$$x(t) = \int_{-\infty}^{\infty} X(f)e^{j2\pi ft}\,df. \tag{9.2.8}$$

At the sampling instants $t = nT$, this relation becomes

$$x(nT) = \int_{-\infty}^{\infty} X(f)e^{j2\pi fnT}\,df. \tag{9.2.9}$$

Let us break the integral in Equation (9.2.9) into integrals covering the finite range of $1/T$. Thus, we obtain

$$x(nT) = \sum_{m=-\infty}^{\infty} \int_{(2m-1)/2T}^{(2m+1)/2T} X(f)e^{j2\pi fnT}\,df$$

$$= \sum_{m=-\infty}^{\infty} \int_{-1/2T}^{1/2T} X\left(f + \frac{m}{T}\right)e^{j2\pi fnT}\,df$$

$$= \int_{-1/2T}^{1/2T} \left[\sum_{m=-\infty}^{\infty} X\left(f + \frac{m}{T}\right)\right]e^{j2\pi fnT}\,df$$

$$= \int_{-1/2T}^{1/2T} Z(f)e^{j2\pi fnT}\,df, \tag{9.2.10}$$

where we have defined $Z(f)$ by

$$Z(f) = \sum_{m=-\infty}^{\infty} X\left(f + \frac{m}{T}\right). \tag{9.2.11}$$

Obviously, $Z(f)$ is a periodic function with period $\frac{1}{T}$; therefore, it can be expanded in terms of its Fourier series coefficients $\{z_n\}$ as

$$Z(f) = \sum_{n=-\infty}^{\infty} z_n e^{j2\pi nfT}, \tag{9.2.12}$$

where

$$z_n = T \int_{-\frac{1}{2T}}^{\frac{1}{2T}} Z(f)e^{-j2\pi nfT}\,df. \tag{9.2.13}$$

Comparing Equations (9.2.13) and (9.2.10), we obtain

$$z_n = Tx(-nT). \tag{9.2.14}$$

Therefore, the necessary and sufficient conditions for Equation (9.2.6) to be satisfied is that

$$z_n = \begin{cases} T & n = 0 \\ 0, & n \neq 0 \end{cases}, \tag{9.2.15}$$

which, when substituted into Equation (9.2.12), yields

$$Z(f) = T, \tag{9.2.16}$$

or equivalently,

$$\sum_{m=-\infty}^{\infty} X\left(f + \frac{m}{T}\right) = T. \tag{9.2.17}$$

This concludes the proof for the condition that $X(f)$ must satisfy to obtain zero ISI.

Now, suppose that the channel has a bandwidth of W. Then, $C(f) = 0$ for $|f| > W$; consequently, $X(f) = 0$ for $|f| > W$. We distinguish three cases:

1. In this case, $T < \frac{1}{2W}$, or equivalently, $\frac{1}{T} > 2W$. Since $Z(f) = \sum_{n=-\infty}^{+\infty} X\left(f + \frac{n}{T}\right)$ consists of nonoverlapping replicas of $X(f)$, which are separated by $\frac{1}{T}$ as shown in Figure 9.6, there is no choice for $X(f)$ to ensure $Z(f) = T$ in this case. Thus, there is no way that we can design a system with no ISI.

2. In this case, $T = \frac{1}{2W}$, or equivalently, $\frac{1}{T} = 2W$ (the Nyquist rate). The replications of $X(f)$, separated by $\frac{1}{T}$, are about to overlap, as shown in Figure 9.7.

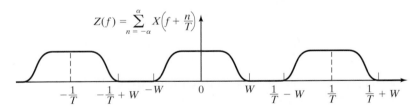

Figure 9.6 Plot of $Z(f)$ for the case $T < \frac{1}{2W}$.

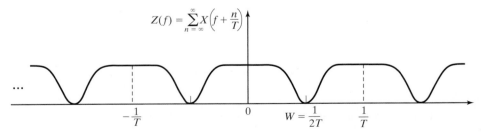

Figure 9.7 Plot of $Z(f)$ for the case $T = \frac{1}{2W}$.

It is clear that there exists only one $X(f)$ that results in $Z(f) = T$, namely,

$$X(f) = \begin{cases} T & |f| < W, \\ 0, & \text{otherwise} \end{cases} \qquad (9.2.18)$$

or $X(f) = T\Pi\left(\frac{f}{2W}\right)$, which results in

$$x(t) = \text{sinc}\left(\frac{t}{T}\right). \qquad (9.2.19)$$

This means that the smallest value of T for which transmission with zero ISI is possible is $T = \frac{1}{2W}$; for this value, $x(t)$ has to be a sinc function. The difficulty with this choice of $x(t)$ is that it is concausal and therefore nonrealizable. To make it realizable, usually a delayed version of it, i.e., $\text{sinc}\left(\frac{t-t_0}{T}\right)$, is used and t_0 is chosen such that for $t < 0$, we have $\text{sinc}\left(\frac{t-t_0}{T}\right) \approx 0$. Of course with this choice of $x(t)$, the sampling time must also be shifted to $mT + t_0$. A second difficulty with this pulse shape is that its rate of convergence to zero is slow. The tails of $x(t)$ decay as $1/t$; consequently, a small mistiming error in sampling the output of the matched filter at the demodulator results in an infinite series of ISI components. Such a series is not absolutely summable because of the $1/t$ rate of decay of the pulse; hence, the sum of the resulting ISI does not converge.

3. In this case, for $T > \frac{1}{2W}$, $Z(f)$ consists of overlapping replications of $X(f)$ separated by $\frac{1}{T}$, as shown in Figure 9.8. In this case, there exists an infinite number of choices for $X(f)$, such that $Z(f) \equiv T$.

For the $T > \frac{1}{2W}$ case, a particular pulse spectrum that has desirable spectral properties and has been widely used in practice is the raised cosine spectrum. The raised cosine frequency characteristic (see Problem 9.5) is given as

$$X_{\text{rc}}(f) = \begin{cases} T, & 0 \le |f| \le (1-\alpha)/2T \\ \frac{T}{2}\left[1 + \cos\left(\frac{\pi T}{\alpha}\left(|f| - \frac{1-\alpha}{2T}\right)\right)\right], & \frac{1-\alpha}{2T} \le |f| \le \frac{1+\alpha}{2T} \\ 0, & |f| > \dfrac{1+\alpha}{2T} \end{cases} , \qquad (9.2.20)$$

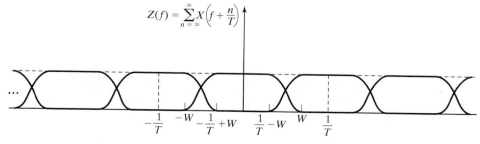

$$Z(f) = \sum_{n=\infty}^{\infty} X\left(f + \frac{n}{T}\right)$$

$-\frac{1}{T}$ $-W$ $-\frac{1}{T}+W$ $\frac{1}{T}-W$ W $\frac{1}{T}$

Figure 9.8   Plot of $Z(f)$ for the case $T > \frac{1}{2W}$.

where α is called the *roll-off factor*, which takes values in the range $0 \le \alpha \le 1$. The bandwidth occupied by the signal beyond the Nyquist frequency $\frac{1}{2T}$ is called the excess bandwidth and is usually expressed as a percentage of the Nyquist frequency. For example, when $\alpha = \frac{1}{2}$, the excess bandwidth is 50%; when $\alpha = 1$, the excess bandwidth is 100%. The pulse $x(t)$ having the raised cosine spectrum is

$$x(t) = \frac{\sin \pi t/T}{\pi t/T} \frac{\cos(\pi \alpha t/T)}{1 - 4\alpha^2 t^2/T^2}$$

$$= \mathrm{sinc}(t/T) \frac{\cos(\pi \alpha t/T)}{1 - 4\alpha^2 t^2/T^2}. \tag{9.2.21}$$

Note that $x(t)$ is normalized so that $x(0) = 1$. Figure 9.9 illustrates the raised cosine spectral characteristics and the corresponding pulses for $\alpha = 0, 1/2, 1$. We note that for $\alpha = 0$, the pulse reduces to $x(t) = \mathrm{sinc}(t/T)$ and the symbol rate is $1/T = 2W$. When $\alpha = 1$, the symbol rate is $1/T = W$. In general, the tails of $x(t)$ decay as $1/t^3$ for $\alpha > 0$. Consequently, a mistiming error in sampling leads to a series of intersymbol interference components that converges to a finite value.

Due to the smooth characteristics of the raised cosine spectrum, it is possible to design practical filters for the transmitter and the receiver that approximate the overall desired frequency response. In the special case where the channel is ideal

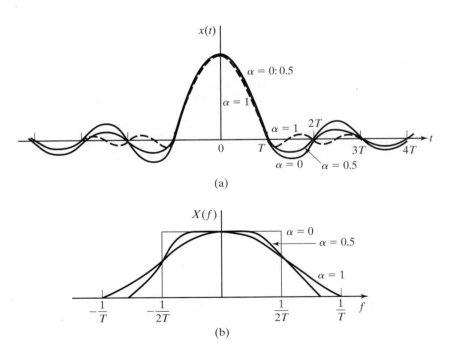

(a)

(b)

Figure 9.9 Pulses having a raised cosine spectrum.

with $C(f) = \Pi\left(\frac{f}{2W}\right)$, we have

$$X_{\text{rc}}(f) = G_T(f)G_R(f). \tag{9.2.22}$$

In this case, if the receiver filter is matched to the transmitter filter, we have $X_{\text{rc}}(f) = G_T(f)G_R(f) = |G_T(f)|^2$. Ideally,

$$G_T(f) = \sqrt{|X_{\text{rc}}(f)|}\, e^{-j2\pi f t_0} \tag{9.2.23}$$

and $G_R(f) = G_T^*(f)$, where t_0 is some nominal delay that is required to assure physical realizability of the filter. Thus, the overall raised cosine spectral characteristic is split evenly between the transmitting filter and the receiving filter. We should also note that an additional delay is necessary to ensure the physical realizability of the receiving filter.

Example 9.2.2

An ideal channel has the frequency response characteristic shown in Figure 9.10. Determine the frequency response characteristics $G_T(f)$ and $G_R(f)$ of the transmit and receiver filters, such that $G_T(f)G_R(f) = X_{\text{rc}}(f)$, where $X_{\text{rc}}(f)$ is the raised cosine spectral characteristic given by Equation (9.2.20), and the desired roll-off factor is selected to be $\alpha = 1/2$. Also, determine the symbol rate $1/T$, and compare it with the Nyquist rate.

Solution   Since the passband of the channel is limited to $|f| < 1200$ Hz and $\alpha = 1/2$, we have

$$\frac{1+\alpha}{2T} = \frac{3/2}{2T} = 1200.$$

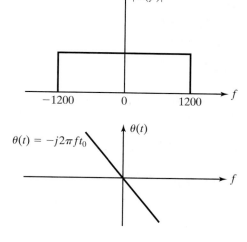

Figure 9.10   Frequency response of ideal channel in Example 9.2.2.

Hence, the symbol rate $1/T = 1600$ symbols/sec. In contrast, the Nyquist rate is 2400 symbols/sec. The frequency response $X_{rc}(f)$ is given as (with $T = 1/1600$),

$$X_{rc}(f) = \begin{cases} T, & 0 \le |f| \le 400 \\ \frac{T}{2}\left[1 + \cos\left(\frac{\pi}{800}\left(|f| - 400\right)\right)\right], & 400 \le |f| \le 1200 \\ 0, & |f| \ge 1200 \end{cases}.$$

Then,

$$|G_T(f)| = |G_R(f)| = \sqrt{X_{rc}(f)}.$$

The phase characteristics of $G_T(f)$ and $G_R(f)$ can be selected to be linear, with $\theta_R(f) = -\theta_T(f)$. ∎

9.2.2 Design of Bandlimited Signals with Controlled ISI — Partial Response Signals

As we have observed from our discussion of signal design for zero ISI, it is necessary to reduce the symbol rate $1/T$ below the Nyquist rate of $2W$ symbols/sec in order to realize practical transmitting and receiving filters. On the other hand, suppose we choose to relax the condition of zero ISI and, thus, achieve a symbol transmission rate of $2W$ symbols/sec. By allowing for a controlled amount of ISI, we can achieve this symbol rate.

We have already seen that the condition of zero ISI is $x(nT) = 0$ for $n \ne 0$. However, suppose that we design the bandlimited signal to have controlled ISI at one time instant. This means that we allow one additional nonzero value in the samples $\{x(nT)\}$. The ISI that we introduce is deterministic or "controlled"; hence, it can be taken into account at the receiver. We will discuss this case next.

One special case that leads to (approximately) physically realizable transmitting and receiving filters is specified by the samples[1]

$$x(nT) = \begin{cases} 1, & n = 0, 1 \\ 0, & \text{otherwise} \end{cases}. \tag{9.2.24}$$

Now, using Equation (9.2.14), we obtain

$$z_n = \begin{cases} T & n = 0, -1 \\ 0, & \text{otherwise} \end{cases}, \tag{9.2.25}$$

which, when substituted into Equation (9.2.12), yields

$$Z(f) = T + T\, e^{-j2\pi fT}. \tag{9.2.26}$$

---

[1]It is convenient to deal with samples of $x(t)$ that are normalized to unity for $n = 0, 1$.

As in the preceding section, it is impossible to satisfy this equation for $T < \frac{1}{2W}$. However, for $T = \frac{1}{2W}$, we obtain

$$X(f) = \begin{cases} \frac{1}{2W}\left[1 + e^{-j\pi f/W}\right], & |f| < W \\ 0, & \text{otherwise} \end{cases}$$

$$= \begin{cases} \frac{1}{W}e^{-j\pi f/2W}\cos\left(\frac{\pi f}{2W}\right), & |f| < W \\ 0, & \text{otherwise} \end{cases} . \qquad (9.2.27)$$

Therefore, $x(t)$ is given by

$$x(t) = \text{sinc}\ (2Wt) + \text{sinc}(2Wt - 1). \qquad (9.2.28)$$

This pulse is called a duobinary signal pulse. It is illustrated, along with its magnitude spectrum, in Figure 9.11. We note that the spectrum decays to zero smoothly, which means that physically realizable filters can be designed to approximate this spectrum very closely. Thus, a symbol rate of $2W$ is achieved.

Another special case that leads to (approximately) physically realizable transmitting and receiving filters is specified by the samples

$$x(nT) = \begin{cases} 1, & n = -1 \\ -1, & n = 1 \\ 0, & \text{otherwise} \end{cases} . \qquad (9.2.29)$$

The corresponding pulse $x(t)$ is given as

$$x(t) = \text{sinc}\ (t + T)/T - \text{sinc}\ (t - T)/T, \qquad (9.2.30)$$

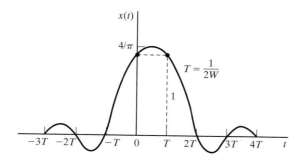

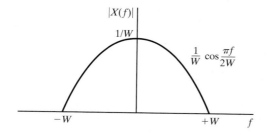

Figure 9.11 Time domain and frequency domain characteristics of a duobinary signal.

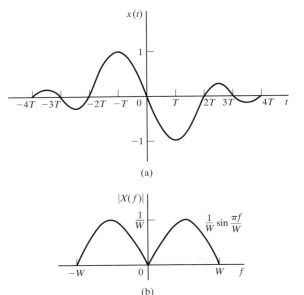

Figure 9.12 Time domain and frequency domain characteristics of a modified duobinary signal.

and its spectrum is

$$X(f) = \begin{cases} \frac{1}{2W}\left[e^{j\pi f/W} - e^{-j\pi f/W}\right] = \frac{j}{W}\sin\frac{\pi f}{W}, & |f| \le W \\ 0, & |f| > W \end{cases}. \qquad (9.2.31)$$

This pulse and its magnitude spectrum are illustrated in Figure 9.12. It is called a *modified duobinary signal pulse*. It is interesting to note that the spectrum of this signal has a zero at $f = 0$, making it suitable for transmission over a channel that does not pass DC.

We can obtain other interesting and physically realizable filter characteristics by selecting different values for the samples $\{x(nT)\}$ and by selecting more than two nonzero samples. However, as we select more nonzero samples, the problem of unraveling the controlled ISI becomes more cumbersome and impractical.

In general, the class of bandlimited signals pulses that have the form

$$x(t) = \sum_{n=-\infty}^{\infty} x\left(\frac{n}{2W}\right) \frac{\sin 2\pi W(t - n/2W)}{2\pi W(t - n/2W)}$$

and their corresponding spectra

$$X(f) = \begin{cases} \frac{1}{2W}\sum_{n=-\infty}^{\infty} x\left(\frac{n}{2W}\right)e^{-jn\pi f/W}, & |f| \le W \\ 0, & |f| > W \end{cases}$$

are called *partial response signals* when controlled ISI is purposely introduced by selecting two or more nonzero samples from the set $\{x(n/2W)\}$. The resulting signal pulses allow us to transmit information symbols at the Nyquist rate of $2W$ symbols/sec. The detection of the received symbols in the presence of controlled ISI is described in Section 9.3.2.

9.3 PROBABILITY OF ERROR FOR DETECTION OF DIGITAL PAM

In this section, we evaluate the performance of the receiver for demodulating and detecting an M-ary PAM signal in the presence of additive white Gaussian noise at its input. First, we consider the case in which the transmitter and receiver filters $G_T(f)$ and $G_R(f)$ are designed for zero ISI. Then, we consider the case in which $G_T(f)$ and $G_R(f)$ are designed such that $x(t) = g_T(t) * g_R(t)$ is either a duobinary signal or a modified duobinary signal.

9.3.1 Probability of Error for Detection of Digital PAM with Zero ISI

In the absence of ISI, the received signal sample at the output of the receiving matched filter has the form

$$y_m = x_0 a_m + w_m,\tag{9.3.1}$$

where, from Equation (9.1.5) with $H(f) = G_T(f)$,

$$x_0 = \int_{-W}^{W} |G_T(f)|^2\, df = \mathscr{E}_g\tag{9.3.2}$$

and w_m is the additive Gaussian noise which has zero mean and has variance (Equation (9.1.7))

$$\sigma_w^2 = \mathscr{E}_g N_0/2.\tag{9.3.3}$$

In general, a_m takes one of M possible equally spaced amplitude values with equal probability. Given a particular amplitude level, the problem is to determine the probability of error.

In the absence of ISI, the problem of evaluating the probability of error for digital PAM in a bandlimited, additive white Gaussian noise channel is identical to the evaluation of the error probability for M-ary PAM, as given in Section 8.5.1. The final result that is obtained from the derivation is

$$P_M = \frac{2(M-1)}{M} Q\left(\sqrt{\frac{2\mathscr{E}_g}{N_0}}\right).\tag{9.3.4}$$

But $\mathscr{E}_g = 3\mathscr{E}_{av}/(M^2 - 1)$, $\mathscr{E}_{av} = k\mathscr{E}_{bav}$ is the average energy/symbol and $\mathscr{E}_{bav}$ is the average energy/bit. Hence,

$$P_M = \frac{2(M-1)}{M} Q\left(\sqrt{\frac{6(\log_2 M)\mathscr{E}_{bav}}{(M^2-1)N_0}}\right). \tag{9.3.5}$$

This is exactly the form for the probability of error of M-ary PAM derived in Section 8.5.1. In the treatment of PAM in this chapter, we imposed the additional constraint that the transmitted signal is bandlimited to the bandwidth allocated for the channel. Consequently, the transmitted signal pulses were designed to be bandlimited and to have zero ISI.

In contrast, no bandwidth constraint was imposed on the PAM signals considered in Section 8.5.1. Nevertheless, the receivers (demodulators and detectors) in both cases are optimum (matched filters) for the corresponding transmitted signals. Consequently, no loss in error-rate performance results from the bandwidth constraint when the signal pulse is designed for zero ISI and the channel does not distort the transmitted signal.

9.3.2 Symbol-by-Symbol Detection of Data with Controlled ISI

In this section, we describe a symbol-by-symbol method for detecting the information symbols at the demodulator when the received signal contains controlled ISI. This symbol detection method is relatively easy to implement. A second method, based on the maximum-likelihood criterion for detecting a sequence of symbols, is described in Section 9.3.4. This second method minimizes the probability of error but is a little more complex to implement. In particular, we consider the detection of the duobinary and the modified duobinary partial response signals. In both cases, we assume that the desired spectral characteristic $X(f)$ for the partial response signal is split evenly between the transmitting and receiving filters, i.e., $|G_T(f)| = |G_R(f)| = |X(f)|^{1/2}$.

For the duobinary signal pulse, $x(nT) = 1$, for $n = 0, 1$ and zero otherwise. Hence, the samples at the output of the receiving filter have the form

$$y_m = b_m + w_m = a_m + a_{m-1} + w_m, \tag{9.3.6}$$

where $\{a_m\}$ is the transmitted sequence of amplitudes and $\{w_m\}$ is the sequence of additive Gaussian noise samples. Let us ignore the noise for the moment and consider the binary case where $a_m = \pm 1$ with equal probability. Then, b_m takes on one of three possible values, namely, $b_m = -2, 0, 2$, with corresponding probabilities $1/4, 1/2, 1/4$. If a_{m-1} is the detected symbol from the signaling interval beginning at $(m-1)$, its effect on b_m, the received signal in the mth signaling interval, can be eliminated by subtraction, thus allowing a_m to be detected. This process can be repeated sequentially for every received symbol.

The major problem with this procedure is that errors arising from the additive noise tend to propagate. For example, if the detector makes an error in detecting a_{m-1}, its effect on b_m is not eliminated; in fact, it is reinforced by the incorrect subtraction. Consequently, the detection of a_m is also likely to be in error.

Error propagation can be avoided by *precoding* the data at the transmitter instead of eliminating the controlled ISI by subtraction at the receiver. The precoding is performed on the binary data sequence prior to modulation. From the data sequence $\{d_n\}$ of 1's and 0's that is to be transmitted, a new sequence $\{p_n\}$ called the precoded sequence is generated. For the duobinary signal, the precoded sequence is defined as

$$p_m = d_m \ominus p_{m-1}, \qquad m = 1, 2, \ldots, \tag{9.3.7}$$

where the symbol $\ominus$ denotes modulo-2 subtraction.[2] Then, we set $a_m = -1$ if $p_m = 0$, and $a_m = 1$ if $p_m = 1$, i.e., $a_m = 2p_m - 1$.

The noise-free samples at the output of the receiving filter are given as

$$
\begin{aligned}
b_m &= a_m + a_{m-1} \\
&= (2p_m - 1) + (2p_{m-1} - 1) \\
&= 2(p_m + p_{m-1} - 1).
\end{aligned}
\tag{9.3.8}
$$

Consequently,

$$p_m + p_{m-1} = \frac{b_m}{2} + 1. \tag{9.3.9}$$

Since $d_m = p_m \oplus p_{m-1}$, it follows that the data sequence d_m is obtained from b_m by using the relation

$$d_m = \frac{b_m}{2} + 1 \ (\text{mod } 2). \tag{9.3.10}$$

Consequently, if $b_m = \pm 2, d_m = 0$ and if $b_m = 0, d_m = 1$. The effect of precoding is clear from Equation (9.3.10). The received level at the mth transmission b_m is directly related d_m, the data at the same transmission time. Therefore, an error in reception of b_m only affects the corresponding data d_m, and no error propagation occurs.

Example 9.3.1

For the binary data sequence $\{d_n\}$ given as

1 1 1 0 1 0 0 1 0 0 0 1 1 0 1,

determine the precoded sequence $\{p_n\}$, the transmitted sequence $\{a_n\}$, the received sequence $\{b_n\}$, and the decoded sequence $\{d_n\}$.

Solution By using the Equations (9.3.7), (9.3.8), and (9.3.10), we obtain the desired sequences, which are given in Table 9.1. ∎

In the preceding derivation, we neglected the effect of the additive noise on the detection method. In the presence of additive noise, the sampled outputs from the receiving filter are given by Equation (9.3.6). In this case, $y_m = b_m + w_m$ is

---

[2]Although this is identical to modulo-2 addition, it is convenient to view the precoding operation for duobinary in terms of modulo-2 subtraction. In the M-ary case, modulo-M addition and subtraction are clearly different.

TABLE 9.1 BINARY SIGNALING WITH DUOBINARY PULSES

Data sequence d_n	1	1	1	0	1	0	0	1	0	0	0	1	1	0	1
Precoded sequence p_n	0	1	0	1	1	0	0	0	1	1	1	0	1	1	0
Transmitted sequence a_n	-1	1	-1	1	1	-1	-1	-1	1	1	1	1	-1	1	1	-1
Received sequence b_n		0	0	0	2	0	-2	-2	0	2	2	2	0	0	2	0
Decoded sequence d_n		1	1	1	0	1	0	0	1	0	0	0	1	1	0	1

compared with the two thresholds set at $+1$ and -1. The data sequence $\{d_n\}$ is obtained according to the detection rule

$$d_m = \begin{cases} 1, & \text{if } -1 < y_m < 1 \\ 0, & \text{if } |y_m| \geq 1 \end{cases}. \tag{9.3.11}$$

The extension from binary PAM to multilevel PAM signaling using the duobinary pulses is straightforward. In this case, the M-level amplitude sequence $\{a_m\}$ results in a (noise-free) sequence

$$b_m = a_m + a_{m-1}, \qquad m = 1, 2, \ldots, \tag{9.3.12}$$

which has $2M - 1$ possible equally spaced levels. The amplitude levels are determined from the relation

$$a_m = 2p_m - (M - 1), \tag{9.3.13}$$

where $\{p_m\}$ is the precoded sequence that is obtained from an M-level data sequence $\{d_m\}$ according to the relation

$$p_m = d_m \ominus p_{m-1} (\text{mod } M), \tag{9.3.14}$$

where the possible values of the sequence $\{d_m\}$ are $0, 1, 2, \ldots, M - 1$.

In the absence of noise, the samples at the output of the receiving filter may be expressed as

$$\begin{aligned} b_m &= a_m + a_{m-1} \\ &= 2[p_m + p_{m-1} - (M - 1)]. \end{aligned} \tag{9.3.15}$$

Hence,

$$p_m + p_{m-1} = \frac{b_m}{2} + (M - 1). \tag{9.3.16}$$

Since $d_m = p_m + p_{m-1}$, it follows that

$$d_m = \frac{b_m}{2} + (M - 1) \ (\text{mod } M). \tag{9.3.17}$$

Here again, we see that error propagation has been prevented by using precoding.

TABLE 9.2 FOUR-LEVEL TRANSMISSION WITH DUOBINARY PULSES

Data sequence d_n		0	0	1	3	1	2	0	3	3	2	0	1	0
Precoded sequence p_n	0	0	0	1	2	3	3	1	2	1	1	3	2	2
Transmitted sequence a_n	−3	−3	−3	−1	1	3	3	−1	1	−1	−1	3	1	1
Received sequence b_n		−6	−6	−4	0	4	6	2	0	0	−2	2	4	2
Decoded sequence d_n		0	0	1	3	1	2	0	3	3	2	0	1	0

Example 9.3.2

Consider the four-level data sequence $\{d_n\}$

$$0\,0\,1\,3\,1\,2\,0\,3\,3\,2\,0\,1\,0,$$

which was obtained by mapping two bits into four-level symbols, i.e., $00 \to 0$, $01 \to 1$, $10 \to 2$, and $11 \to 3$. Determine the precoded sequence $\{p_n\}$, the transmitted sequence $\{a_n\}$, the received sequence $\{b_n\}$, and the decoded sequence $\{d_n\}$.

Solution By using Equation (9.3.12) through (9.3.17), we obtain the desired sequences, which are given in Table 9.2. ∎

In the presence of noise, the received signal-plus-noise is quantized to the nearest possible signal level and the preceding rule is used on the quantized values to recover the data sequence.

In the case of the modified duobinary pulse, the controlled ISI is specified by the values $x(n/2W) = -1$ for $n = 1$, $x(n/2W) = 1$ for $n = -1$, and zero otherwise. Consequently, the noise-free sampled output from the receiving filter is given as

$$b_m = a_m - a_{m-2}, \tag{9.3.18}$$

where the M-level sequence $\{a_n\}$ is obtained by mapping a precoded sequence according to the relation Equation (9.3.13) and

$$p_m = d_m \oplus p_{m-2} \ (\text{mod } M). \tag{9.3.19}$$

From these relations, it is easy to show that the detection rule for receiving the data sequence $\{d_m\}$ from $\{b_m\}$ in the absence of noise is

$$d_m = \frac{b_m}{2} \ (\text{mod } M). \tag{9.3.20}$$

As demonstrated, the precoding of the data at the transmitter makes it possible to detect the received data on a symbol-by-symbol basis without having to look back at previously detected symbols. Thus, error propagation is avoided.

The probability of error of the symbol-by-symbol detector previously described is determined in the following section.

9.3.3 Probability of Error for Symbol-by-Symbol Detection of Partial Response Signals

In this section, we determine the probability of error for the symbol-by-symbol detection of digital M-ary PAM signaling using duobinary and modified duobinary pulses. The channel is assumed to be an ideal bandlimited channel with additive white Gaussian noise. The model for the communication system is shown in Figure 9.13.

Symbol-by-Symbol Detector. At the transmitter, the M-level data sequence $\{d_n\}$ is precoded as previously described. The precoder output is mapped into one of M possible amplitude levels. Then the transmitting filter with frequency response $G_T(f)$ has an output

$$v(t) = \sum_{n=-\infty}^{\infty} a_n g_T(t - nT). \tag{9.3.21}$$

The partial-response function $X(f)$ is divided equally between the transmitting and receiving filters. Hence, the receiving filter is matched to the transmitted pulse, and the cascade of the two filters results in the frequency characteristic

$$|G_T(f)G_R(f)| = |X(f)|. \tag{9.3.22}$$

The matched filter output is sampled at $t = nT = n/2W$ and the samples are fed to the decoder. For the duobinary signal, the output of the matched filter at the sampling instant may be expressed as

$$\begin{aligned} y_m &= a_m + a_{m-1} + w_m \\ &= b_m + w_m, \end{aligned} \tag{9.3.23}$$

where w_m is the additive noise component. Similarly, the output of the matched filter for the modified duobinary signal is

$$\begin{aligned} y_m &= a_m - a_{m-2} + w_m \\ &= b_m + w_m. \end{aligned} \tag{9.3.24}$$

For binary transmission, let $a_m = \pm d$, where $2d$ is the distance between signal levels. Then, the corresponding values of b_m are $(2d, 0, -2d)$. For M-ary PAM signal transmission, where $a_m = \pm d, \pm 3d, \ldots, \pm(M - 1)d$, the received signal levels

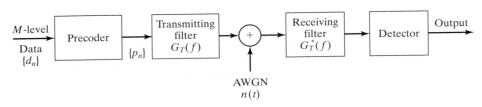

Figure 9.13 Block diagram of the modulator and demodulator for partial response signals.

are $b_m = 0, \pm 2d, \pm 4d, \ldots, \pm 2(M-1)d$. Hence, the number of received levels is $2M - 1$. The input transmitted symbols $\{a_m\}$ are assumed to be equally probable. Then, for duobinary and modified duobinary signals, it is easily demonstrated that, in the absence of noise, the received output levels have a (triangular) probability mass function of the form

$$P(b = 2md) = \frac{M - |m|}{M^2}, \qquad m = 0, \pm 1, \pm 2, \ldots \pm (M-1), \qquad (9.3.25)$$

where b denotes the noise-free received level and $2d$ is the distance between any two adjacent received signal levels.

The channel corrupts the signal transmitted through it by the addition of white Gaussian noise with zero-mean and a power-spectral density $N_0/2$. We assume that a symbol error is committed whenever the magnitude of the additive noise exceeds the distance d. This assumption neglects the rare event that a large noise component with magnitude exceeding d may result in a received signal level that yields a correct symbol decision. The noise component w_m is zero-mean Gaussian with variance

$$\sigma_w^2 = \frac{N_0}{2} \int_{-W}^{W} |G_R(f)|^2 \, df$$

$$= \frac{N_0}{2} \int_{-W}^{W} |X(f)| df = 2N_0/\pi, \qquad (9.3.26)$$

where we have used Equations (9.2.27) and (9.2.31) to compute the integral. Equation (9.3.26) applies to both duobinary and modified duobinary signals. Hence, an upper bound on the symbol probability of error is

$$P_M < \sum_{m=-(M-2)}^{M-2} P(|y - 2md| > d | b = 2md) P(b = 2md)$$

$$+ 2P(y + 2(M-1)d > d | b = -2(M-1)d) P(b = -2(M-1)d)$$

$$= P(|y| > d | b = 0) \left[2 \sum_{m=0}^{M-1} P(b = 2md) - P(b = 0) - P(b = -2(M-1)d) \right]$$

$$= \left(1 - \frac{1}{M^2} \right) P(|y| > d | b = 0). \qquad (9.3.27)$$

But

$$P(|y| > d | b = 0) = \frac{2}{\sqrt{2\pi}\sigma_w} \int_{d}^{\infty} e^{-x^2/2\sigma_w^2} \, dx$$

$$= 2Q \left(\sqrt{\frac{\pi d^2}{2N_0}} \right). \qquad (9.3.28)$$

Therefore, the average probability of a symbol error is upper-bounded as

$$P_M < 2\left(1 - \frac{1}{M^2}\right) Q\left(\sqrt{\frac{\pi d^2}{2N_0}}\right). \tag{9.3.29}$$

The scale factor d in Equation (9.3.29) can be eliminated by expressing d in terms of the average power transmitted into the channel. For the M-ary PAM signal in which the transmitted levels are equally probable, the average power at the output of the transmitting filter is

$$\begin{aligned}
P_{av} &= \frac{E\left(a_m^2\right)}{T} \int_{-W}^{W} |G_T(f)|^2 \, df \\
&= \frac{E\left(a_m^2\right)}{T} \int_{-W}^{W} |X(f)| \, df \\
&= \frac{4}{\pi T} E\left(a_m^2\right),
\end{aligned} \tag{9.3.30}$$

where $E\left(a_m^2\right)$ is the mean square value of the M signal levels, which is

$$E\left(a_m^2\right) = \frac{d^2(M^2 - 1)}{3}. \tag{9.3.31}$$

Therefore,

$$d^2 = \frac{3\pi P_{av} T}{4(M^2 - 1)}. \tag{9.3.32}$$

By substituting the value of d^2 from Equation (9.3.32) into Equation (9.3.29), we obtain the upper bound for the symbol error probability as

$$P_M < 2\left(1 - \frac{1}{M^2}\right) Q\left(\sqrt{\left(\frac{\pi}{4}\right)^2 \frac{6}{M^2 - 1} \frac{\mathscr{E}_{av}}{N_0}}\right), \tag{9.3.33}$$

where $\mathscr{E}_{av} = P_{av} T$ is the average energy/transmitted symbol, which can also be expressed in terms of the average bit energy as $\mathscr{E}_{av} = k\mathscr{E}_{bav} = (\log_2 M)\mathscr{E}_{bav}$.

The expression in Equation (9.3.33) for the probability of error of M-ary PAM holds for both a duobinary and a modified duobinary partial response signal. If we compare this result with the error probability of M-ary PAM with zero ISI, which can be obtained by using a signal pulse with a raised cosine spectrum, we note that the performance of partial response duobinary or modified duobinary has a loss of $(\pi/4)^2$ or 2.1 dB. This loss in SNR is due to the fact that the detector for the partial response signals makes decisions on a symbol-by-symbol basis; thus, it ignores the inherent memory contained in the received signal at the input to the detector.

9.3.4 Maximum-Likelihood Sequence Detection of Partial Response Signals

When the received signal sequence has no memory, the symbol-by-symbol detector that was described in Sections 8.3.3 and 8.4.6 is optimum in the sense of minimizing the probability of a symbol error. On the other hand, when the received symbol sequence has memory, i.e., the received symbols in successive symbol–time intervals are statistically interdependent, the optimum detector bases its decisions on the observation of a sequence of received symbols over successive symbol–time intervals.

The sequence of received symbols resulting from the transmission of partial response signal waveforms is characterized as a sequence having memory between successive symbols. To observe the memory in the received sequence, let us look at the noise-free received sequence for binary transmission given in Table 9.1. The sequence $\{b_m\}$ is 0, 0, 0, 2, 0, -2, -2, 0, 2, 2.... We note that it is not possible to have a transition from -2 to $+2$ or from $+2$ to -2 in one symbol interval. For example, if the signal level at the input to the detector is -2, the next signal level can be either -2 or 0. Similarly, if the signal level at a given sampling instant is 2, the signal level in the next time instant can be either 2 or 0. In other words, it is not possible to encounter a transition from -2 to 2 or vice versa between two successive received samples from the matched filter. However, a symbol-by-symbol detector does not exploit this constraint or inherent memory in the received sequence.

The memory that is inherent in the received sequence $\{y_m\}$ resulting from transmission of a partial response signal waveform is conveniently represented by a trellis diagram. For example, the trellis for the duobinary partial response signal for binary data transmission is illustrated in Figure 9.14. For binary modulation, this trellis contains two states; each state correspond to the two possible input values of a_m, i.e., $a_m = \pm 1$. Each branch in the trellis is labeled by two numbers. The first number on the left is the new data bit, i.e., $a_{m+1} = \pm 1$. This number determines the transition to the new state. The number on the right is the received signal level.

The duobinary signal has a memory of length $L = 1$. Hence, for binary modulation, the trellis has $S_t = 2$ states. For M-ary PAM modulation, the number of trellis states is M.

The optimum sequence detector that exploits the memory inherent in the received sequence of symbols may be based on the maximum a posteriori probability (MAP) criterion (see Section 8.4.6). For example, consider the transmission of a sequence of N symbols $\{a_m, m = 1, 2, \ldots, N\}$. If each symbol can take one

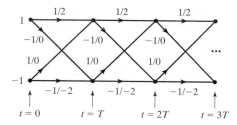

Figure 9.14 Trellis for duobinary partial response signals.

of M possible values, then there are M^N possible transmitted sequences, denoted as $a^{(k)}$, $1 \le k \le M^N$. The received sequence is $\{y_m, 1 \le m \le N\}$, and it is denoted as y. A detector that bases its decision on the MAP criterion computes a posteriori probabilities

$$P(a^{(k)} \text{ was transmitted } | y), \ 1 \le k \le M^N \qquad (9.3.34)$$

and selects the particular sequence that yields the highest probability. But the probabilities in Equation (9.3.34) can be expressed as

$$P\left(a^{(k)} \text{ was transmitted } | y\right) = \frac{f\left(y|a^{(k)}\right) P\left(a^{(k)}\right)}{f(y)}. \qquad (9.3.35)$$

When all the symbol sequences $a^{(k)}$ are equally likely to be transmitted, which is usually the case in practice, selecting the sequence that maximizes the a posteriori probability is equivalent to finding the sequence $a^{(k)}$ that maximizes the conditional probability density function $f\left(y|a^{(k)}\right)$ over all possible M^N sequences. Thus, the optimum detection criterion is the maximum-likelihood (ML) criterion. In the case of additive Gaussian noise, $f\left(y|a^{(k)}\right)$ is a multi-dimensional Gaussian probability density function.

As previously described, the optimum ML sequence detector computes M^N probabilities, which includes one probability for each of the M^N possible transmitted sequences. When M and N are large, the computational complexity of the ML detector becomes prohibitive. However, the computational complexity of the detector can be reduced significantly. A computationally efficient algorithm for performing ML sequence detection, invented by Andrew Viterbi during the late 1960s, allows us to reduce the computational burden by eliminating sequences as new data is received. In effect, the Viterbi algorithm (VA) is a sequential trellis search algorithm, which is described in more detail in Chapter 13 (Section 13.3.2) as a decoding algorithm for convolutional codes.

In the case of the trellis shown in Figure 9.14 for the duobinary partial response signal, we observe that there are two states, labeled as $+1$ and -1, as well as nodes having two incoming signal paths and two outgoing signal paths. At each node of the trellis, the VA computes two probabilities corresponding to each of the two incoming signal paths. One of the two paths is selected as the more probable (based on the values of the corresponding probabilities) and the other path is discarded. The surviving path at each node is then extended to two new paths, which includes one path for each of the two possible input symbols, and the search process continuous. Thus, the VA reduces the computational complexity of the ML detector.

We can show that by replacing the symbol-by-symbol detector with the ML sequence detector for the duobinary and modified duobinary signals, the 2.1 dB loss in performance is completely eliminated. Therefore, with the use of the Viterbi algorithm, which performs ML sequence detection, the partial response signals achieve the same performance as conventional PAM, which employs a raised cosine signal pulse to avoid ISI.

9.4 SYSTEM DESIGN IN THE PRESENCE OF CHANNEL DISTORTION

In Section 9.2.1, we described a signal design criterion that results in zero ISI at the output of the receiving filter. Recall that a signal pulse $x(t)$ will satisfy the condition of zero ISI at the sampling instants $t = nT, n = \pm 1, \pm 2, \ldots$, if its spectrum $X(f)$ satisfies the condition given by Equation (9.2.7). From this condition, we concluded that, for ISI-free transmission over a channel, the transmitter–receiver filters and the channel transfer function must satisfy

$$G_T(f)C(f)G_R(f) = X_{\mathrm{rc}}(f), \tag{9.4.1}$$

where $X_{\mathrm{rc}}(f)$ denotes the Fourier transform of an appropriate raised cosine pulse, whose parameters depend on the channel bandwidth W and the transmission interval T. In this section, we are concerned with the design of a digital communication system that suppresses ISI in a channel with distortion. We first present a brief coverage of various types of channel distortion, and then we consider the design of transmitter and receiver filters.

We distinguish two types of distortion. *Amplitude distortion* results when the amplitude characteristic $|C(f)|$ is not constant for $|f| \leq W$, as illustrated in Figure 9.15(a). The second type of distortion, called *phase distortion*, results when the phase characteristic $\theta_c(f)$ is nonlinear in frequency, as illustrated in Figure 9.15(b).

Another view of phase distortion is obtained by considering the derivative of $\theta_c(f)$. Thus, we define the envelope delay characteristic as we did in Problem 2.66:

$$\tau(f) = \frac{-1}{2\pi} \frac{d\theta_c(f)}{df}. \tag{9.4.2}$$

When $\theta_c(f)$ is linear in f, the envelope delay is constant for all frequencies. In this case, all frequencies in the transmitted signal pass through the channel with the same fixed-time delay. In such a case, there is no phase distortion. However, when $\theta_c(f)$ is nonlinear, the envelope delay $\tau(f)$ varies with frequency and the various

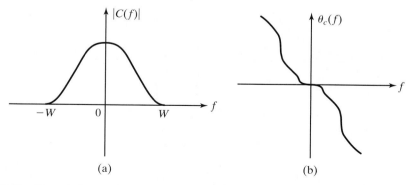

(a)   (b)

Figure 9.15   Channel characteristics illustrating (a) amplitude distortion and (b) phase distortion.

frequency components in the input signal undergo different delays in passing through the channel. In such a case, we say that the transmitted signal has suffered from *delay distortion*.

Both amplitude and delay distortion cause intersymbol interference in the received signal. For example, let us assume that we have designed a pulse with a raised cosine spectrum that has zero ISI at the sampling instants. An example of such a pulse is illustrated in Figure 9.16(a). When the pulse is passed through a channel filter with a constant amplitude $|C(f)| = 1$ for $|f| < W$ and a quadratic phase characteristic (linear envelope delay), the received pulse at the output of the channel is as shown in Figure 9.16(b). Note that the periodic zero crossings have been shifted by the delay distortion, so that the resulting pulse suffers from ISI. Consequently, a sequence of successive pulses would be smeared into one another, and the peaks of the pulses would no longer be distinguishable due to the ISI.

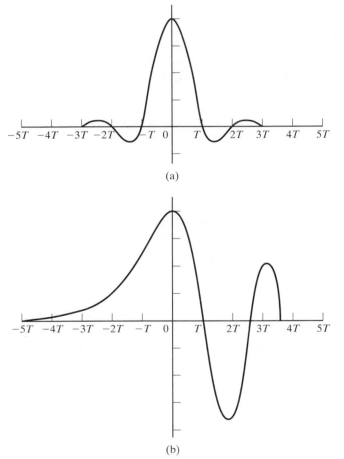

(a)

(b)

Figure 9.16 Effect of channel distortion in (a) channel input and (b) channel output.

In the next section, we consider two design problems. First, we consider the design of transmitting and receiving filters in the presence of channel distortion when the channel characteristics are known. Second, we consider the design of special filters, called channel equalizers, that automatically and adaptively correct for the channel distortion when the channel characteristics, i.e., $|C(f)|$ and $\theta_c(f)$, are unknown.

9.4.1 Design of Transmitting and Receiving Filters for a Known Channel

In this section, we assume that the channel frequency response characteristic $C(f)$ is known; thus, we consider the problem of designing a transmitting filter and a receiving filter that maximize the SNR at the output of the receiving filter and result in zero ISI. Figure 9.17 illustrates the overall system under consideration.

For the signal component, we must satisfy the condition

$$G_T(f)C(f)G_R(f) = X_{\text{rc}}(f)e^{-j2\pi f t_0}, \qquad |f| \le W, \qquad (9.4.3)$$

where $X_{\text{rc}}(f)$ is the desired raised cosine spectrum that yields zero ISI at the sampling instants and t_0 is a time delay, which is necessary to ensure the physical realizability of the transmitter and receiver filters.

The noise at the output of the receiving filter may be expressed as

$$w(t) = \int_{-\infty}^{\infty} n(t - \tau)g_R(\tau)\,d\tau, \qquad (9.4.4)$$

where $n(t)$ is the input to the filter. The noise $n(t)$ is assumed to be zero-mean Gaussian. Hence, $w(t)$ is zero-mean Gaussian, with a power-spectral density

$$S_w(f) = S_n(f)|G_R(f)|^2, \qquad (9.4.5)$$

where $S_n(f)$ is the spectral density of the noise process $n(t)$.

For simplicity, we consider binary PAM transmission. Then, the sampled output of the matched filter is

$$y_m = x_0 a_m + w_m = a_m + w_m, \qquad (9.4.6)$$

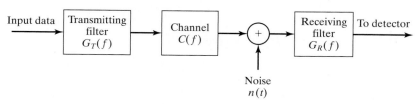

Figure 9.17   System configuration for the design of $G_T(f)$ and $G_R(f)$.

where x_0 is normalized to unity, $a_m = \pm d$, and w_m represents the noise term that is zero-mean Gaussian with variance

$$\sigma_w^2 = \int_{-\infty}^{\infty} S_n(f) |G_R(f)|^2 \, df. \tag{9.4.7}$$

Consequently, the probability of error is

$$P_2 = \frac{1}{\sqrt{2\pi}} \int_{d/\sigma_w}^{\infty} e^{-y^2/2} \, dy = Q\left(\sqrt{\frac{d^2}{\sigma_w^2}} \right). \tag{9.4.8}$$

Now, suppose that we select the filter at the transmitter to have the frequency response

$$G_T(f) = \frac{\sqrt{X_{\text{rc}}(f)}}{C(f)} e^{-j2\pi f t_0}, \tag{9.4.9}$$

where t_0 is a suitable delay to ensure causality. Then, the cascade of the transmit filter and the channel results in the frequency response

$$G_T(f) C(f) = \sqrt{X_{\text{rc}}(f)} \, e^{-j2\pi f t_0}. \tag{9.4.10}$$

In the presence of additive white Gaussian noise, the filter at the receiver is designed to be matched to the received signal pulse. Hence, its frequency response is

$$G_R(f) = \sqrt{X_{\text{rc}}(f)} \, e^{-j2\pi f t_r}, \tag{9.4.11}$$

where t_r is an appropriate delay.

Let us compute the SNR d^2/σ_w^2 for these filter characteristics. The noise variance is

$$\sigma_w^2 = \frac{N_0}{2} \int_{-\infty}^{\infty} |G_R(f)|^2 df = \frac{N_0}{2} \int_{-W}^{W} X_{\text{rc}}(f) = \frac{N_0}{2}. \tag{9.4.12}$$

The average power transmitted is

$$P_{\text{av}} = \frac{E\left(a_m^2\right)}{T} \int_{-\infty}^{\infty} g_T^2(t) \, dt = \frac{d^2}{T} \int_{-W}^{W} \frac{X_{\text{rc}}(f)}{|C(f)|^2} df; \tag{9.4.13}$$

hence,

$$d^2 = P_{\text{av}} T \left[\int_{-W}^{W} \frac{X_{\text{rc}}(f)}{|C(f)|^2} df \right]^{-1}. \tag{9.4.14}$$

Therefore, the SNR d^2/σ_w^2 is given as

$$\frac{d^2}{\sigma_w^2} = \frac{2 P_{\text{av}} T}{N_0} \left[\int_{-W}^{W} \frac{X_{\text{rc}}(f)}{|C(f)|^2} df \right]^{-1}. \tag{9.4.15}$$

We note that the term

$$10 \, \log_{10} \int_{-W}^{W} \frac{X_{\text{rc}}(f)}{|C(f)|^2} df, \tag{9.4.16}$$

with $|C(f)| \le 1$ for $|f| \le W$, represents the loss in performance in dB of the communication system due to channel distortion. When the channel is ideal, $|C(f)| = 1$ for $|f| \le W$; hence, there is no performance loss. We also note that this loss is entirely due to amplitude distortion in the channel because the phase distortion has been totally compensated by the transmit filter.

Example 9.4.1

Determine the magnitude of the transmitting and receiving filter characteristics for a binary communication system that transmits data at a rate of 4800 bits/sec over a channel with frequency response

$$|C(f)| = \frac{1}{\sqrt{1 + \left(\frac{f}{W}\right)^2}}, \qquad |f| \le W,$$

where $W = 4800$ Hz. The additive noise is zero-mean white Gaussian with a spectral density $N_0/2$.

Solution   Since $W = 1/T = 4800$, we use a signal pulse with a raised cosine spectrum and $\alpha = 1$. Thus,

$$X_{\text{rc}}(f) = \frac{T}{2} \left[1 + \cos(\pi T |f|)\right]$$

$$= T \cos^2 \frac{\pi |f|}{9600}.$$

Then,

$$|G_T(f)| = \sqrt{T \left[1 + \left(\frac{f}{W}\right)^2\right]} \cos \frac{\pi |f|}{9600}, \qquad |f| \le 4800 \text{ Hz}$$

$$|G_R(f)| = \sqrt{T} \cos \frac{\pi |f|}{9600}, \qquad |f| \le 4800 \text{ Hz}$$

and

$$|G_T(f)| = |G_R(f)| = 0 \text{ for } |f| > 4800 \text{ Hz}. \qquad \blacksquare$$

9.4.2 Channel Equalization

In the preceding section, we described the design of transmitting and receiving filters for digital PAM transmission when the frequency response characteristics of the channel are known. Our objective was to design these filters for zero ISI at the sampling instants. This design methodology is appropriate when the channel is precisely known and its characteristics do not change with time.

In practice, we often encounter channels whose frequency response characteristics are either unknown or change with time. For example, in data transmission over the dial-up telephone network, the communication channel will be different every time we dial a number because the channel route will be different. Once a connection is made, however, the channel will be time-invariant for a relatively long period of time. This is an example of a channel whose characteristics are unknown a priori. Examples of time-varying channels are radio channels, such as ionospheric propagation channels. These channels are characterized by time-varying frequency response characteristics. These types of channels demonstrate where the optimization of the transmitting and receiving filters, as described in Section 9.4.1, is not possible.

Under these circumstances, we may design the transmitting filter to have a square-root raised cosine frequency response, i.e.,

$$G_T(f) = \begin{cases} \sqrt{X_{rc}(f)}e^{-j2\pi f t_0}, & |f| \le W \\ 0, & |f| > W \end{cases},$$

and the receiving filter, with frequency response $G_R(f)$, to be matched to $G_T(f)$. Therefore,

$$|G_T(f)||G_R(f)| = X_{rc}(f). \tag{9.4.17}$$

Then, due to channel distortion, the output of the receiving filter is

$$y(t) = \sum_{n=-\infty}^{\infty} a_n x(t - nT) + w(t), \tag{9.4.18}$$

where $x(t) = g_T(t) * c(t) * g_R(t)$. The filter output may be sampled periodically to produce the sequence

$$y_m = \sum_{n=-\infty}^{\infty} a_n x_{m-n} + w_m$$

$$= x_0 a_m + \sum_{\substack{n=-\infty \\ n \ne m}}^{+\infty} a_n x_{m-n} + w_m, \tag{9.4.19}$$

where $x_n = x(nT), n = 0, \pm 1, \pm 2, \ldots$. The middle term on the right-hand side of Equation (9.4.19) represents the ISI.

In any practical system, it is reasonable to assume that the ISI affects a finite number of symbols. Hence, we may assume that $x_n = 0$ for $n < -L_1$ and $n > L_2$, where L_1 and L_2 are finite, positive integers Consequently, the ISI observed at the output of the receiving filter may be viewed as being generated by passing the data sequence $\{a_m\}$ through an FIR filter with coefficients $\{x_n, -L_1 \le n \le L_2\}$, as

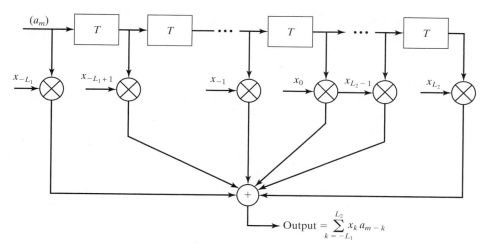

Figure 9.18   Equivalent discrete-time channel filter.

shown in Figure 9.18. This filter is called the *equivalent discrete-time channel filter*. Since its input is the discrete information sequence (binary or M-ary), the output of the discrete-time channel filter may be characterized as the output of a finite-state machine with $L = L_1 + L_2$ states, corrupted by additive Gaussian noise. Hence, the noise-free output of the filter is described by a trellis having M^L states.

Maximum-Likelihood Sequence Detection.   The optimum detector for the information sequence $\{a_m\}$, which is based on the observation of the received sequence $\{y_m\}$ and given by Equation (9.4.19), is an ML sequence detector. The detector is akin to the ML sequence detector described in the context of detecting partial response signals that have controlled ISI. The Viterbi algorithm described in Section 13.3.2 provides a method for searching through the trellis for the ML signal path. To accomplish this search, the equivalent channel filter coefficients $\{x_n\}$ must be known or measured by some method. At each stage of the trellis search, there are M^L surviving sequences with M^L corresponding Euclidean distance path metrics.

Due to the exponential increase in the computational complexity of the Viterbi algorithm with the span (length L) of the ISI, this type of detection is practical only when M and L are small. For example, in mobile cellular telephone systems that employ digital transmission of speech signals, M is usually selected to be small, i.e., $M = 2$ or 4, and $2 \leq L \leq 5$. In this case, the ML sequence detector may be implemented with reasonable complexity. However, when M and L are large, the ML sequence detector becomes impractical. In such a case, other more practical but suboptimum methods are used to detect the information sequence $\{a_m\}$ in the presence of ISI. Nevertheless, the performance of the ML sequence detector for a channel with ISI serves as a benchmark for comparing its performance with that of suboptimum methods. Two suboptimum methods are described next.

Linear Equalizers.   For channels whose frequency response characteristics are unknown and time variant, we may employ a linear filter with adjustable parameters, which are updated on a periodic basis to compensate for the channel distortion. Such a filter, having parameters that are adjusted periodically, is called an *adaptive equalizer.*

First, we consider the design characteristics for a linear equalizer from a frequency domain viewpoint. Figure 9.19 shows a block diagram of a system that employs a linear filter as a channel equalizer.

The demodulator consists of a receiving filter with the frequency response $G_R(f)$ in cascade with a channel equalizing filter that has a frequency response $G_E(f)$. Since $G_R(f)$ is matched to $G_T(f)$ and they are designed so that their product satisfies Equation (9.4.17), $|G_E(f)|$ must compensate for the channel distortion. Hence, the equalizer frequency response must equal the inverse of the channel response, i.e.,

$$G_E(f) = \frac{1}{C(f)} = \frac{1}{|C(f)|}e^{-j\theta_c(f)}, \quad |f| \leq W, \tag{9.4.20}$$

where $|G_E(f)| = 1/|C(f)|$ and the equalizer phase characteristic $\theta_E(f) = -\theta_c(f)$. In this case, the equalizer is said to be the inverse channel filter to the channel response.

We note that the inverse channel filter completely eliminates ISI caused by the channel. Since it forces the ISI to be zero at the sampling times $t = nT$, the equalizer is called a *zero-forcing equalizer.* Hence, the input to the detector is of the form

$$y_m = a_m + w_m,$$

where w_m is the noise component, which is zero-mean Gaussian with a variance

$$\sigma_w^2 = \int_{-\infty}^{\infty} S_n(f)|G_R(f)|^2|G_E(f)|^2 \, df$$

$$= \int_{-W}^{W} \frac{S_n(f)|X_{\text{rc}}(f)|}{|C(f)|^2} \, df, \tag{9.4.21}$$

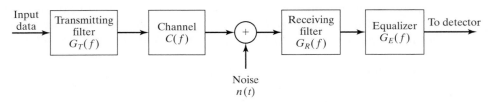

Figure 9.19   Block diagram of a system with equalizer.

in which $S_n(f)$ is the power spectral density of the noise. When the noise is white, $S_n(f) = N_0/2$ and the variance becomes

$$\sigma_w^2 = \frac{N_0}{2} \int_{-W}^{W} \frac{|X_{\text{rc}}(f)|}{|C(f)|^2} \, df. \tag{9.4.22}$$

In general, the noise variance at the output of the zero-forcing equalizer is higher than the noise variance at the output of the optimum receiving filter $|G_R(f)|$, which is given by Equation (9.4.12) for the case in which the channel is known.

Example 9.4.2

The channel given in Example 9.4.1 is equalized by a zero-forcing equalizer. Assuming that the transmitting and receiving filters satisfy Equation (9.4.17), determine the value of the noise variance at the sampling instants and the probability of error.

Solution   When the noise is white, the variance of the noise at the output of the zero-forcing equalizer (input to the detector) is given by Equation (9.4.22). Hence,

$$\sigma_w^2 = \frac{N_0}{2} \int_{-W}^{W} \frac{|X_{\text{rc}}(f)|}{|C(f)|^2} \, df$$

$$= \frac{T N_0}{2} \int_{-W}^{W} \left[1 + (fW)^2\right] \cos^2 \frac{\pi |f|}{2W} \, df$$

$$= N_0 \int_{0}^{1} (1 + x^2) \cos^2 \frac{\pi x}{2} \, dx$$

$$= \left(\frac{2}{3} - \frac{1}{\pi^2}\right) N_0.$$

The average transmitted power is

$$P_{\text{av}} = \frac{(M^2 - 1)d^2}{3T} \int_{-W}^{W} |G_T(f)|^2 \, df$$

$$= \frac{(M^2 - 1)d^2}{3T} \int_{-W}^{W} |X_{\text{rc}}(f) \, df$$

$$= \frac{(M^2 - 1)d^2}{3T}.$$

The general expression for the probability of error is given as

$$P_M = \frac{2(M - 1)}{M} Q \left(\sqrt{\frac{3 P_{\text{av}} T}{(M^2 - 1)(2/3 - 1/\pi^2) N_0}} \right).$$

If the channel were ideal, the argument of the Q-function would be $\frac{6 P_{\text{av}} T}{(M^2 - 1) N_0}$. Hence, the loss in performance due to the nonideal channel is given by the factor $2\left(2/3 - \frac{1}{\pi^2}\right) = 1.133$ or 0.54 dB.   ∎

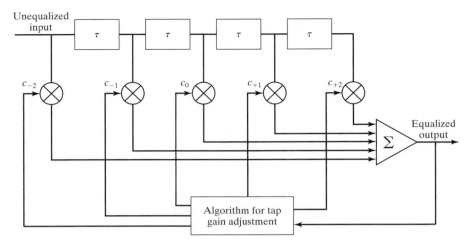

Figure 9.20   Linear transversal filter.

Let us now consider the design of a linear equalizer from a time-domain viewpoint. We noted previously that in real channels, the ISI is limited to a finite number of samples, L samples. As a consequence, in practice for example, the channel equalizer is approximated by a finite duration impulse response (FIR) filter, or transversal filter, with adjustable tap coefficients $\{c_n\}$ as illustrated in Figure 9.20. The time delay τ between adjacent taps may be selected as large as T, the symbol interval, in which case the FIR equalizer is called a *symbol-spaced equalizer*. In this case, the input to the equalizer is the sampled sequence given by Equation (9.4.19). However, we note that when $1/T < 2W$, frequencies in the received signal that are above the folding frequency $1/T$ are aliased into frequencies below $1/T$. In this case, the equalizer compensates for the aliased channel-distorted signal.

On the other hand, when the time delay τ between adjacent taps is selected such that $1/\tau \geq 2W > 1/T$, no aliasing occurs; hence, the inverse channel equalizer compensates for the true channel distortion. Since $\tau < T$, the channel equalizer is said to have *fractionally spaced taps*, and it is called a *fractionally spaced equalizer*. In practice, τ is often selected as $\tau = T/2$. Notice that, in this case, the sampling rate at the output of the filter $G_R(f)$ is $\frac{2}{T}$.

The impulse response of the FIR equalizer is

$$g_E(t) = \sum_{n=-N}^{N} c_n \delta(t - n\tau), \tag{9.4.23}$$

and the corresponding frequency response is

$$G_E(f) = \sum_{n=-N}^{N} c_n e^{-j2\pi f n\tau}, \tag{9.4.24}$$

where $\{c_n\}$ are the $(2N + 1)$ equalizer coefficients and N is chosen sufficiently large so that the equalizer spans the length of the ISI, i.e., $2N + 1 \geq L$. Since

$X(f) = G_T(f)C(f)G_R(f)$ and $x(t)$ is the signal pulse corresponding to $X(f)$, the equalized output signal pulse is

$$q(t) = \sum_{n=-N}^{N} c_n x(t - n\tau). \tag{9.4.25}$$

The zero-forcing condition can now be applied to the samples of $q(t)$ taken at times $t = mT$. These samples are

$$q(mT) = \sum_{n=-N}^{N} c_n x(mT - n\tau), \qquad m = 0, \pm 1, \dots, \pm N. \tag{9.4.26}$$

Since there are $2N + 1$ equalizer coefficients, we can control only $2N + 1$ sampled values of $q(t)$. Specifically, we may force the conditions

$$q(mT) = \sum_{n=-N}^{N} c_n x(mT - n\tau) = \begin{cases} 1, & m = 0 \\ 0, & m = \pm 1, +2, \dots, \pm N \end{cases}, \tag{9.4.27}$$

which may be expressed in matrix form as $Xc = q$, where X is a $(2N+1) \times (2N+1)$ matrix with elements $\{x(mT - nt)\}$, c is the $(2N+1)$ coefficient vector, and q is the $(2N+1)$ column vector with one nonzero element. Thus, we obtain a set of $2N + 1$ linear equations for the coefficients of the zero-forcing equalizer.

We should emphasize that the FIR zero-forcing equalizer does not completely eliminate ISI because it has a finite length. However, as N is increased, the residual ISI can be reduced; in a limit as $N \to \infty$, the ISI is completely eliminated.

Example 9.4.3

Consider a channel distorted pulse $x(t)$, at the input to the equalizer, given by the expression

$$x(t) = \frac{1}{1 + \left(\frac{2t}{T}\right)^2},$$

where $1/T$ is the symbol rate. The pulse is sampled at the rate $2/T$ and equalized by a zero-forcing equalizer. Determine the coefficients of a five-tap zero-forcing equalizer.

Solution   According to Equation (9.4.27), the zero-forcing equalizer must satisfy the equations

$$q(mT) = \sum_{n=-2}^{2} c_n x(mT - nT/2) = \begin{cases} 1, & m = 0 \\ 0, & m = \pm 1, \pm 2 \end{cases}.$$

The matrix X with elements $x(mT - nT/2)$ is given as

$$X = \begin{bmatrix} \frac{1}{5} & \frac{1}{10} & \frac{1}{17} & \frac{1}{26} & \frac{1}{37} \\ 1 & \frac{1}{2} & \frac{1}{5} & \frac{1}{10} & \frac{1}{17} \\ \frac{1}{5} & \frac{1}{2} & 1 & \frac{1}{2} & \frac{1}{5} \\ \frac{1}{17} & \frac{1}{10} & \frac{1}{5} & \frac{1}{2} & 1 \\ \frac{1}{37} & \frac{1}{26} & \frac{1}{17} & \frac{1}{10} & \frac{1}{5} \end{bmatrix}.$$

The coefficient vector c and the vector q are given as

$$c = \begin{bmatrix} c_{-2} \\ c_{-1} \\ c_0 \\ c_1 \\ c_2 \end{bmatrix} \qquad q = \begin{bmatrix} 0 \\ 0 \\ 1 \\ 0 \\ 0 \end{bmatrix}.$$

Then, the linear equations $Xc = q$ can be solved by inverting the matrix X. Thus, we obtain

$$c_{\text{opt}} = X^{-1}q = \begin{bmatrix} -2.2 \\ 4.9 \\ -3 \\ 4.9 \\ -2.2 \end{bmatrix}.$$

■

One drawback to the zero-forcing equalizer is that it ignores the presence of additive noise. As a consequence, its use may result in significant noise enhancement. This is easily seen by noting that in a frequency range where $C(f)$ is small, the channel equalizer $G_E(f) = 1/C(f)$ compensates by placing a large gain in that frequency range. Consequently, the noise in that frequency range is greatly enhanced. An alternative is to relax the zero ISI condition and select the channel equalizer characteristic such that the combined power in the residual ISI and the additive noise at the output of the equalizer is minimized. A channel equalizer that is optimized based on the minimum mean-square-error (MMSE) criterion accomplishes the desired goal.

To elaborate, let us consider the noise-corrupted output of the FIR equalizer, which is

$$z(t) = \sum_{n=-N}^{N} c_n y(t - n\tau), \qquad (9.4.28)$$

where $y(t)$ is the input to the equalizer, which is given by Equation (9.4.18). The output is sampled at times $t = mT$. Thus, we obtain

$$z(mT) = \sum_{n=-N}^{N} c_n y(mT - n\tau). \qquad (9.4.29)$$

The desired response sample at the output of the equalizer at $t = mT$ is the transmitted symbol a_m. The error is defined as the difference between a_m and $z(mT)$. Then, the mean-square-error between the actual output sample $z(mT)$ and the desired values a_m is

$$\text{MSE} = E[z(mT) - a_m]^2$$

$$= E\left[\sum_{n=-N}^{N} c_n y(mT - n\tau) - a_m \right]^2$$

$$= \sum_{n=-N}^{N} \sum_{k=-N}^{N} c_n c_k R_Y(n - k) - 2 \sum_{k=-N}^{N} c_k R_{AY}(k) + E\left(a_m^2\right), \qquad (9.4.30)$$

where the correlations are defined as

$$R_Y(n - k) = E\left[y(mT - n\tau)y(mT - k\tau)\right]$$

and

$$R_{AY}(k) = E[y(mT - k\tau)a_m] \tag{9.4.31}$$

and the expectation is taken with respect to the random information sequence $\{a_m\}$ and the additive noise.

The MMSE solution is obtained by differentiating Equation (9.4.30) with respect to the equalizer coefficients $\{c_n\}$. Thus, we obtain the necessary conditions for the MMSE as

$$\sum_{n=-N}^{N} c_n R_Y(n - k) = R_{AY}(k), \qquad k = 0, \pm 1, 2, \ldots, \pm N. \tag{9.4.32}$$

There are $(2N + 1)$ linear equations for the equalizer coefficients. In contrast to the zero-forcing solution previously described, these equations depend on the statistical properties (the autocorrelation) of the noise as well as the ISI through the autocorrelation $R_Y(n)$.

In practice, we would not normally know the autocorrelation $R_Y(n)$ and the cross correlation $R_{AY}(n)$. However, these correlation sequences can be estimated by transmitting a test signal over the channel and using the time average estimates

$$\hat{R}_Y(n) = \frac{1}{K} \sum_{k=1}^{K} y(kT - n\tau)y(kT)$$

and

$$\hat{R}_{AY}(n) = \frac{1}{K} \sum_{k=1}^{K} y(kT - n\tau)a_k \tag{9.4.33}$$

in place of the ensemble averages to solve for the equalizer coefficients given by Equation (9.4.32).

Adaptive Equalizers.   We have shown that the tap coefficients of a linear equalizer can be determined by solving a set of linear equations. In the zero-forcing optimization criterion, the linear equations are given by Equation (9.4.27). On the other hand, if the optimization criterion is based on minimizing the MSE, the optimum equalizer coefficients are determined by solving the set of linear equations given by Equation (9.4.32).

In both cases, we may express the set of linear equations in the general matrix form

$$Bc = d, \tag{9.4.34}$$

where B is a $(2N + 1)$ x $(2N + 1)$ matrix, c is a column vector representing the $2N + 1$ equalizer coefficients, and d is a $(2N + 1)$-dimensional column vector. The solution of Equation (9.4.34) yields

$$c_{\text{opt}} = B^{-1}d. \tag{9.4.35}$$

In practical implementations of equalizers, the solution of Equation (9.4.34) for the optimum coefficient vector is usually obtained by an iterative procedure that avoids the explicit computation of the inverse of the matrix B. The simplest iterative procedure is the method of steepest descent, in which one begins by choosing arbitrarily the coefficient vector c, for example c_0. This initial choice of coefficients corresponds to a point on the criterion function that is being optimized. For example, in the case of the MSE criterion, the initial guess c_0 corresponds to a point on the quadratic MSE surface in the $(2N+1)$-dimensional space of coefficients. The gradient vector, defined as g_0, which is the derivative of the MSE with respect to the $2N + 1$ filter coefficients, is then computed at this point on the criterion surface and each tap coefficient is changed in the direction opposite to its corresponding gradient component. The change in the jth tap coefficient is proportional to the size of the jth gradient component.

For example, the gradient vector, denoted as g_k, for the MSE criterion, found by taking the derivatives of the MSE with respect to each of the $2N+1$ coefficients, is

$$g_k = Bc_k - d, \qquad k = 0, 1, 2, \ldots . \tag{9.4.36}$$

Then the coefficient vector c_k is updated according to the relation

$$c_{k+1} = c_k - \Delta g_k, \tag{9.4.37}$$

where Δ is the *step-size parameter* for the iterative procedure. To ensure convergence of the iterative procedure, Δ is chosen to be a small positive number. In such a case, the gradient vector g_k converges toward zero, i.e., $g_k \to 0$ as $k \to \infty$, and the coefficient vector $c_k \to c_{opt}$, as illustrated in Figure 9.21 for two-dimensional optimization. In general, convergence of the equalizer tap coefficients to c_{opt} cannot be attained in a finite number of iterations with the steepest-descent method. However, the optimum solution c_{opt} can be approached as closely as desired in a few hundred iterations. In digital communication systems that employ channel equalizers, each iteration corresponds to a time interval for sending one symbol; hence, a few hundred iterations to achieve convergence to c_{opt} corresponds to a fraction of a second.

Adaptive channel equalization is required for channels whose characteristics change with time. In such a case, the ISI varies with time. The channel equalizer must track such time variations in the channel response and adapt its coefficients to reduce the ISI. In the context of the preceding discussion, the optimum coefficient vector c_{opt} varies with time due to time variations in the matrix B and, for the case of the MSE criterion, time variations in the vector d. Under these conditions, the iterative method previously described can be modified to use estimates of the gradient components. Thus, the algorithm for adjusting the equalizer tap coefficients may be expressed as

$$\hat{c}_{k+1} = \hat{c}_k - \Delta \hat{g}_k, \tag{9.4.38}$$

where $\hat{g}_k$ denotes an estimate of the gradient vector g_k and $\hat{c}_k$ denotes the estimate of the tap coefficient vector.

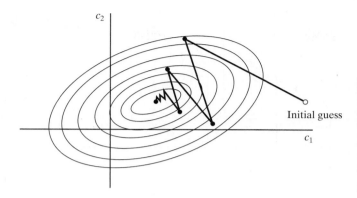

Figure 9.21   Example of convergence characteristics of a gradient algorithm. (From *Introduction to Adaptive Arrays,* by R.A. Monzigo and T.W. Miller; ©1980 by John Wiley & Sons. Reprinted with permission of the publisher.)

In the case of the MSE criterion, the gradient vector $\mathbf{g}_k$ given by Equation (9.4.36) may also be expressed as (see Problem 9.25)

$$\mathbf{g}_k = -E(e_k \mathbf{y}_k).$$

An estimate of the gradient vector at the kth iteration is computed as

$$\hat{\mathbf{g}}_k = -e_k \mathbf{y}_k, \tag{9.4.39}$$

where e_k denotes the difference between the desired output from the equalizer at the kth time instant and the actual output $z(kT)$ and $\mathbf{y}_k$ denotes the column vector of $2N+1$ received signal values contained in the equalizer at time instant k. The *error signal* is expressed as

$$e_k = a_k - z_k, \tag{9.4.40}$$

where $z_k = z(kT)$ is the equalizer output given by Equation (9.4.29) and a_k is the desired symbol. Hence, by substituting Equation (9.4.39) into Equation (9.4.38), we obtain the adaptive algorithm for optimizing the tap coefficients (based on the MSE criterion) as

$$\hat{\mathbf{c}}_{k+1} = \hat{\mathbf{c}}_k + \Delta e_k \mathbf{y}_k. \tag{9.4.41}$$

Since an estimate of the gradient vector is used in Equation (9.4.41), the algorithm is called a *stochastic gradient algorithm*. It is also known as the *LMS algorithm*.

A block diagram of an adaptive equalizer that adapts its tap coefficients according to Equation (9.4.41) is illustrated in Figure 9.22. Note that the difference between the desired output a_k and the actual output z_k from the equalizer is used to form the error signal e_k. This error is scaled by the step-size parameter Δ, and the scaled error signal Δe_k multiplies the received signal values $\{y(kT - n\tau)\}$ at the $2N+1$ taps. The products $\Delta e_k y(kT - n\tau)$ at the $(2N+1)$ taps are then added to the previous values of the tap coefficients to obtain the updated tap coefficients, according to Equation (9.4.41). This computation is repeated for each received symbol. Thus, the equalizer coefficients are updated at the symbol rate.

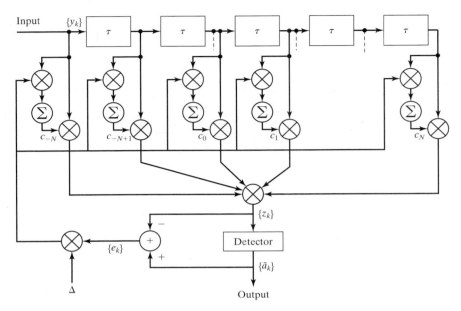

Figure 9.22   Linear adaptive equalizer based on the MSE criterion.

Initially, the adaptive equalizer is trained by the transmission of a known pseudo-random sequence $\{a_m\}$ over the channel. At the demodulator, the equalizer employs the known sequence to adjust its coefficients. Upon initial adjustment, the adaptive equalizer switches from a *training mode* to a *decision-directed mode*, in which case the decisions at the output of the detector are sufficiently reliable so that the error signal is formed by computing the difference between the detector output and the equalizer output, i.e.,

$$e_k = \tilde{a}_k - z_k, \tag{9.4.42}$$

where $\tilde{a}_k$ is the output of the detector. In general, decision errors at the output of the detector occur infrequently; consequently, such errors have little effect on the performance of the tracking algorithm given by Equation (9.4.41).

A rule of thumb for selecting the step-size parameter to ensure convergence and good tracking capabilities in slowly varying channels is

$$\Delta = \frac{1}{5(2N+1)P_R}, \tag{9.4.43}$$

where P_R denotes the received signal-plus-noise power, which can be estimated from the received signal.

The convergence characteristics of the stochastic gradient algorithm in Equation (9.4.41) is illustrated in Figure 9.23. These graphs were obtained from a computer simulation of an 11-tap adaptive equalizer operating on a channel with a

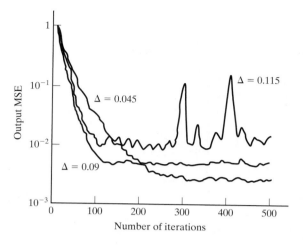

Figure 9.23 Initial convergence characteristics of the LMS algorithm with different step sizes. (From *Digital Signal Processing* by J. G. Proakis and D. G. Manolakis; ©1988, Macmillan. Reprinted with permission of the publisher.)

rather modest amount of ISI. The input signal-plus-noise power P_R was normalized to unity. The rule of thumb given in Equation (9.4.43) for selecting the step size gives $\Delta = 0.018$. The effect of making Δ too large is illustrated by the large jumps in MSE, as shown for $\Delta = 0.115$. As Δ is decreased, the convergence is slowed somewhat, but a lower MSE is achieved; this indicates that the estimated coefficients are closer to c_{opt}.

Although we have described the operation of an adaptive equalizer that is optimized on the basis of the MSE criterion, the operation of an adaptive equalizer based on the zero-forcing method is very similar. The major difference lies in the method for estimating the gradient vectors g_k at each iteration. A block diagram of an adaptive zero-forcing equalizer is shown in Figure 9.24.

Decision-Feedback Equalizer.   The linear filter equalizers previously described are very effective on channels, such as wireline telephone channels, where the ISI is not severe. The severity of the ISI is directly related to the spectral characteristics, and is not necessarily related to the time span of the ISI. For example, consider the ISI resulting from two channels illustrated in Figure 9.25. The time span for the ISI in channel A is 5 symbol intervals on each side of the desired signal component, which has a value of 0.72. On the other hand, the time span for the ISI in channel B is one symbol interval on each side of the desired signal component, which has a value of 0.815. The energy of the total response is normalized to unity for both channels.

In spite of the shorter ISI span, channel B results in more severe ISI. This is evidenced in the frequency response characteristics of these channels, which are shown in Figure 9.26. We observe that channel B has a spectral null (the frequency response $C(f) = 0$ for some frequencies in the band $|f| \leq W$) at $f = 1/2T$, whereas this does not occur in the case of channel A. Consequently, a linear equalizer will introduce a large gain in its frequency response to compensate for the channel null. Thus, the noise in channel B will be enhanced much more than in channel A. This

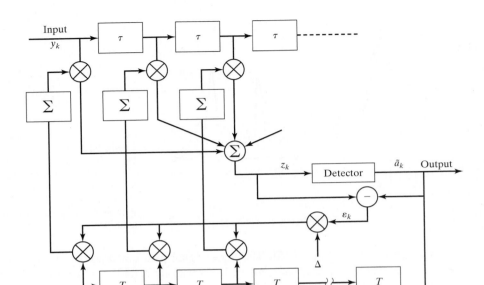

Figure 9.24 An adaptive zero-forcing equalizer.

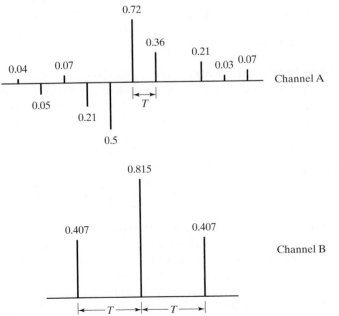

Figure 9.25 Two channels with ISI.

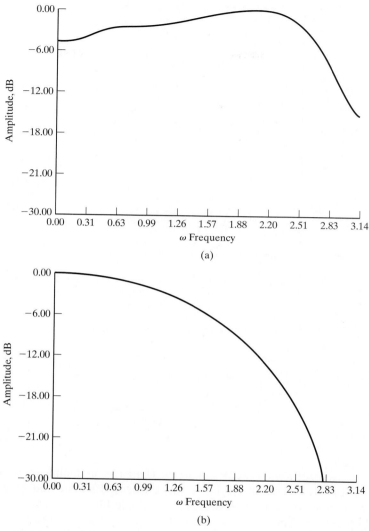

Figure 9.26 Amplitude spectra for (a) channel A shown in Figure 9.25(a) and (b) channel B shown in Figure 9.25(b).

implies that the performance of the linear equalizer for channel B will be significantly worse than that for channel A. This fact is borne out by the computer simulation results for the performance of the linear equalizer for the two channels, as shown in Figure 9.27. Hence, the basic limitation of a linear equalizer is that it performs poorly on channels having spectral nulls. Such channels are often encountered in radio communications, such as ionospheric transmission at frequencies below 30 MHz, and mobile radio channels, such as those used for cellular radio communications.

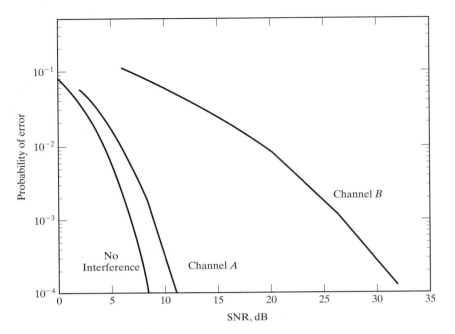

Figure 9.27 Error-rate performance of linear MSE equalizer.

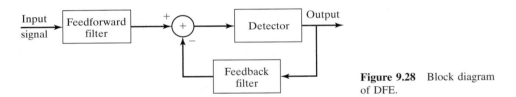

Figure 9.28 Block diagram of DFE.

A *decision-feedback equalizer* (DFE) is a nonlinear equalizer that employs previous decisions to eliminate the ISI caused by previously detected symbols on the current symbol to be detected. A simple block diagram for a DFE is shown in Figure 9.28. The DFE consists of two filters. The first filter is called a *feedforward filter*, and it is generally a fractionally spaced FIR filter with adjustable tap coefficients. This filter is identical in form to the linear equalizer previously described. Its input is the received filtered signal $y(t)$. The second filter is a *feedback filter*. It is implemented as an FIR filter with symbol-spaced taps having adjustable coefficients. Its input is the set of previously detected symbols. The output of the feedback filter is subtracted from the output of the feedforward filter to form the input to the detector. Thus, we have

$$z_m = \sum_{n=0}^{N_1} c_n y(mT - n\tau) - \sum_{n=1}^{N_2} b_n \tilde{a}_{m-n}, \qquad (9.4.44)$$

where $\{c_n\}$ and $\{b_n\}$ are the adjustable coefficients of the feedforward and feedback filters, respectively, $\tilde{a}_{m-n}$, $n = 1, 2, \ldots, N_2$ are the previously detected symbols, $N_1 + 1$ is the length of the feedforward filter, and N_2 is the length of the feedback filter. Based on the input z_m, the detector determines which of the possible transmitted symbols is closest in distance to the input signal z_m. Thus, it makes its decision and outputs $\tilde{a}_m$. What makes the DFE nonlinear is the nonlinear characteristic of the detector, which provides the input to the feedback filter.

The tap coefficients of the feedforward and feedback filters are selected to optimize some desired performance measure. For mathematical simplicity, the MSE criterion is usually applied and a stochastic gradient algorithm is commonly used to implement an adaptive DFE. Figure 9.29 illustrates the block diagram of an adaptive DFE whose tap coefficients are adjusted by means of the LMS stochastic gradient algorithm. Figure 9.30 illustrates the probability of error performance of the DFE, obtained by computer simulation, for binary PAM transmission over channel B. The gain in performance relative to that of a linear equalizer is clearly evident.

We should mention that decision errors from the detector that are fed to the feedback filter have a small effect on the performance of the DFE. In general, there is a small loss in performance of 1 to 2 dB at error rates below 10^{-2}; this is due to decision errors, but these decision errors in the feedback filters are not catastrophic. The effect of decision errors in the feedback filter for channel B is also shown in Figure 9.30.

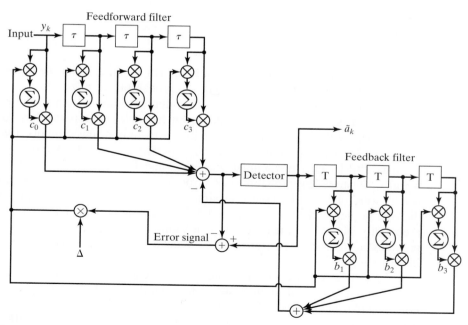

Figure 9.29   Adaptive DFE.

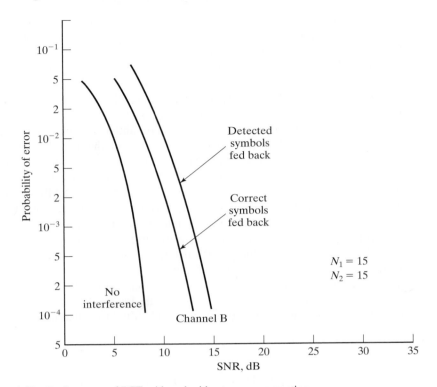

Figure 9.30 Performance of DFE with and without error propagation.

Although the DFE outperforms a linear equalizer, it is not the optimum equalizer from the viewpoint of minimizing the probability of error. As indicated previously, the optimum detector in a digital communication system in the presence of ISI is an ML symbol sequence detector. Such a detector is particularly appropriate for channels with severe ISI when the ISI spans only a few symbols. For example, Figure 9.31 illustrates the error probability performance of the Viterbi algorithm for a binary PAM signal transmitted through channel B. For purposes of comparison, we also illustrate the probability of error for a decision feedback equalizer. Both results were obtained by computer simulation. We observe that the performance of the ML sequence detector is about 4.5 dB better than that of the DFE at an error probability of 10^{-4}. Hence, this is one example where the ML sequence detector provides a significant performance gain on a channel that has a relatively short ISI span.

In conclusion, we mention that adaptive equalizers are widely used in high-speed digital communication systems for radio channels and telephone channels. For example, high-speed telephone line modems (with a bit rate above 2400 bps) generally include an adaptive equalizer that is implemented as an FIR filter with coefficients that are adjusted based on the MMSE criterion. Depending on the data

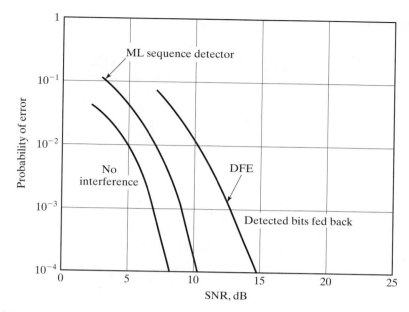

Figure 9.31   Performance of Viterbi detector and DFE for channel B.

speed, the equalizer typically spans between 20 and 70 symbols. The LMS algorithm given by Equation (9.4.41) is usually employed for the adjustment of the equalizer coefficients adaptively.

9.5 FURTHER READING

The pioneering work on signal design for bandwidth-constrained channels was done by Nyquist (1928). The use of binary partial response signals was originally proposed in the paper by Lender (1963) and was later generalized by Kretzmer (1966). The problem of optimum transmitter and receiver filter design was investigated by Gerst and Diamond (1961), Tufts (1965), Smith (1965), and Berger and Tufts (1967).

Adaptive equalization for digital communication was introduced by Lucky (1965, 1966). Widrow (1966) devised the LMS algorithm for adaptively adjusting the equalizer coefficients.

The Viterbi algorithm was devised by Viterbi (1967) for the purpose of decoding convolutional codes, which are described in Chapter 13. Its use as the ML sequence detector for partial response signals and, more generally, for symbols corrupted by intersymbol interference, was proposed and analyzed by Forney (1972) and Omura (1971). A comprehensive treatment of adaptive equalization algorithms is given in the book by Proakis (2001).

PROBLEMS

9.1 In Example 9.1.1, the ideal channel of bandwidth W limits the transmitted signal energy that passes through the channel. The received signal energy as a function of the channel bandwidth is

$$\mathcal{E}_h(W) = \frac{T}{(2\pi)^2} \int_{-WT}^{WT} \frac{\sin^2 \pi \alpha}{\alpha^2 (1-\alpha^2)^2} \, d\alpha,$$

where $\alpha = fT$.

1. Evaluate (numerically) $\mathcal{E}_h(W)$ for $W = \frac{1}{2T}, \frac{1}{T}, \frac{1.5}{T}, \frac{2}{T}, \frac{2.5}{T}, \frac{3}{T}$, and plot $\frac{\mathcal{E}_h(W)}{T}$ as a function of W.

2. Determine the value of $\mathcal{E}_h(W)$ in the limit as $W \to \infty$. For the computation, you may use the time-domain relation

$$\lim_{W \to \infty} \mathcal{E}_h(W) = \int_{-\infty}^{+\infty} g_T^2(t) \, dt.$$

9.2 In a binary PAM system, the input to the detector is

$$y_m = a_m + n_m + i_m,$$

where $a_m = \pm 1$ is the desired signal, n_m is a zero-mean Gaussian random variable with variance σ_n^2, and i_m represents the ISI due to channel distortion. The ISI term is a random variable which takes the values $-\frac{1}{2}, 0, \frac{1}{2}$ with probabilities $\frac{1}{4}, \frac{1}{2}, \frac{1}{4}$, respectively. Determine the average probability of error as a function of σ_n^2.

9.3 In a binary PAM system, the clock that specifies the sampling of the correlator output is offset from the optimum sampling time by 10%.

1. If the signal pulse used is rectangular, determine the loss in SNR due to the mistiming.

2. Determine the amount of intersymbol interference introduced by the mistiming, and determine its effect on performance.

9.4 The frequency response characteristic of a lowpass channel can be approximated by

$$H(f) = \begin{cases} 1 + \alpha \cos 2\pi f t_0, & |\alpha| < 1, \ |f| \le W \\ 0, & \text{otherwise} \end{cases},$$

where W is the channel bandwidth. An input signal $s(t)$ whose spectrum is bandlimited to W/Hz, is passed through the channel.

1. Show that the channel output is

$$y(t) = s(t) + \frac{\alpha}{2}\left[s(t - t_0) + s(t + t_0)\right].$$

Thus, the channel produces a pair of echoes.

2. Suppose the received signal $y(t)$ is passed through a filter matched to $s(t)$. Determine the output of the matched filter at $t = kT$, $k = 0, \pm1, \pm2, \ldots$, where T is the symbol duration.

3. What is the ISI pattern resulting from the channel if $t_0 = T$?

9.5 Show that a pulse having the raised cosine spectrum given by Equation (9.2.20) satisfies the Nyquist criterion given by Equation (9.2.7) for any value of the roll-off factor α.

9.6 Show that for any value of α, the raised cosine spectrum given by Equation (9.2.20) satisfies

$$\int_{-\infty}^{+\infty} X_{\text{rc}}(f)\,df = 1.$$

(Hint: Use the fact that $X_{\text{rc}}(f)$ satisfies the Nyquist criterion given by Equation (9.2.7).)

9.7 Equation (9.2.7) gives the necessary and sufficient condition for the spectrum $X(f)$ of the pulse $x(t)$ that yields zero ISI. Prove that, for any pulse that is bandlimited to $|f| < 1/T$, the zero ISI condition is satisfied if $Re\left[X(f)\right]$ for $f > 0$ consists of a rectangular function plus an arbitrary odd function about $f = 1/2T$, and $Im\left[X(f)\right]$ is any arbitrary even function about $f = 1/2T$.

9.8 A channel has a passband characteristic in the frequency range $|f| \le 1400$ Hz.

1. Select a symbol rate and a PAM signal constellation size to achieve a 9600 bps signal transmission.

2. If a square-root raised cosine pulse is used for the transmitter pulse $g_T(t)$, select the roll-off factor. Assume that the channel has an ideal frequency response characteristic.

9.9 Design an M-ary PAM system that transmits digital information over an ideal channel with bandwidth $W = 2400$ Hz. The bit rate is 14,400 bits/sec. Specify the number of transmitted points, the number of received signal points using a duobinary signal pulse, and the required $\mathcal{E}_b$ to achieve an error probability of 10^{-6}. The additive noise is zero-mean Gaussian with a power spectral density 10^{-4} W/Hz.

9.10 When the additive noise at the input to the modulator is colored, the filter matched to the signal no longer maximizes the output SNR. In such a case, we may consider the use of a prefilter that "whitens" the colored noise. The prefilter is followed by a filter matched to the prefiltered signal. Towards this end, consider the configuration shown in Figure P-9.10.

1. Determine the frequency response characteristic of the prefilter that whitens the noise.

2. Determine the frequency response characteristic of the filter matched to $\tilde{s}(t)$.

3. Consider the prefilter and the matched filter as a single "generalized matched filter." What is the frequency response characteristic of this filter?

4. Determine the SNR at the input to the detector.

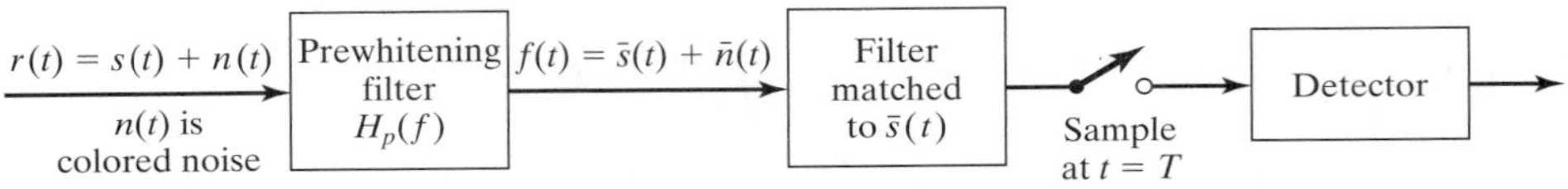

$r(t) = s(t) + n(t)$ Prewhitening filter $H_p(f)$ $f(t) = \bar{s}(t) + \bar{n}(t)$ Filter matched to $\bar{s}(t)$ Sample at $t = T$ Detector

$n(t)$ is colored noise

Figure P-9.10

9.11 Consider the transmission of data via PAM over a channel that has a bandwidth of 1500 Hz. Show how the symbol rate varies as a function of the excess bandwidth. In particular, determine the symbol rate for excess bandwidths of 25%, 33%, 50%, 67%, 75%, and 100%.

9.12 The binary sequence 10010110010 is the input to the precoder whose output is used to modulate a duobinary transmitting filter. Construct a table as in Table 9.2; show the precoded sequence, the transmitted amplitude levels, the received signal levels, and the decoded sequence.

9.13 Repeat the Problem 9.12 for a modified duobinary signal pulse.

9.14 A precoder for a partial response signal fails to work if the desired partial response at $n = 0$ is zero modulo M. For example, consider the desired response for $M = 2$:

$$x(nT) = \begin{cases} 2, & n = 0 \\ 1, & n = 1 \\ -1, & n = 2 \\ 0, & \text{otherwise} \end{cases}.$$

Show why this response cannot be precoded.

9.15 A baseband digital communication system employs the signals shown in Figure P-9.15(a) for the transmission of two equiprobable messages. It is assumed that the communication problem studied here is a "one shot" communication problem, that is, the messages are transmitted just once and no transmission takes place afterwards. The channel has no attenuation and the noise is AWG with power spectral density $\frac{N_0}{2}$.

(1) Find an appropriate orthonormal basis for the representation of the signals.

(2) In a block diagram, give the precise specifications of the optimal receiver using matched filters. Label the block diagram carefully.

(3) Find the error probability of the optimal receiver.

(4) Show that the optimal receiver can be implemented by using just *one* filter. (See the block diagram shown in Figure P-9.15(b).) What are the characteristics of the matched filter and the sampler and decision device?

(5) Now assume the channel is not ideal, but has an impulse response of $c(t) = \delta(t) + \frac{1}{2}\delta(t - \frac{T}{2})$. Using the same matched filter as the previous Part, design an optimal receiver.

(6) Assume that the channel impulse response is $c(t) = \delta(t) + a\delta(t - \frac{T}{2})$, where a is a random variable uniformly distributed on $[0, 1]$. Using the same matched filter, design the optimal receiver.

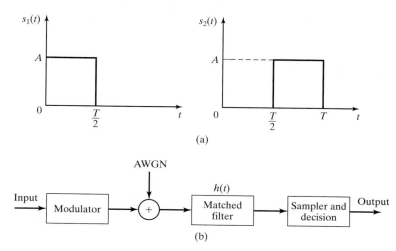

(a)

(b)

Figure P-9.15

9.16 Sketch and label the trellis for a duobinary signal waveform used in conjunction with the precoding given by Equation (9.3.7). Repeat this for the modified duobinary signal waveform with the precoder given by Equation (9.3.19). Comment on any similarities and differences.

9.17 Consider the use of a (square-root) raised cosine signal pulse with a roll-off factor of unity for the transmission of binary PAM over an ideal bandlimited channel that passes the pulse without distortion. Thus, the transmitted signal is

$$v(t) = \sum_{k=-\infty}^{\infty} a_k g_T(t - kT_b),$$

where the signal interval $T_b = T/2$ and the symbol rate doubles for no ISI.

1. Determine the ISI values at the output of a matched filter demodulator.

2. Sketch the trellis for the maximum likelihood sequence detector. Label the states.

9.18 A binary antipodal signal is transmitted over a nonideal bandlimited channel, which introduces ISI over two adjacent symbols. For an isolated transmitted signal pulse $s(t)$, the (noise-free) output of the demodulator is $\sqrt{\mathcal{E}_b}$ at $t = T$ and $\sqrt{\mathcal{E}_b}/4$ at $t = 2T$ at zero for $t = kT, k > 2$, where $\mathcal{E}_b$ is the signal energy and T is the signaling interval.

1. Determine the average probability of error assuming that the two signals are equally probable and the additive noise is white and Gaussian.

2. By plotting the error probability obtained in Part 1 and the error probability for the case of no ISI, determine the relative difference in SNR of the error probability of 10^{-6}.

9.19 Determine the frequency response characteristics for the RC circuit shown in Figure P-9.19. Also determine the expression for the envelope delay.

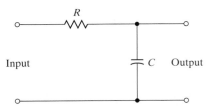

Figure P-9.19

9.20 Consider the RC lowpass filter shown in Figure P-9.19 where $\tau = RC = 10^{-6}$.

1. Determine and sketch the envelope (group) delay of the filter as a function of frequency.

2. Suppose that the input to the filter is a lowpass signal of bandwidth $\Delta f = 1k$ Hz. Determine the effect of the RC filter on this signal.

9.21 A microwave radio channel has a frequency response

$$C(f) = 1 + 0.3 \cos 2\pi f T.$$

Determine the frequency response characteristic for the optimum transmitting and receiving filters that yield zero ISI at a rate of $1/T$ symbols/sec and have an excess bandwidth of 50%. Assume that the additive noise spectrum is flat.

9.22 An $M = 4$ PAM modulation is used for transmitting at a bit rate of 9600 bits/sec on a channel having a frequency response

$$C(f) = \frac{1}{1 + j \frac{f}{2400}},$$

where $|f| \leq 2400$, and $C(f) = 0$, otherwise. The additive noise is zero-mean, white Gaussian with power spectral density $\frac{N_0}{2}$ W/Hz. Determine the (magnitude) frequency response characteristic of the optimum transmitting and receiving filters.

9.23 Binary PAM is used to transmit information over an unequalized linear filter channel. When $a = 1$ is transmitted, the noise-free output of the demodulator is

$$x_m = \begin{cases} 0.3, & m = 1 \\ 0.9, & m = 0 \\ 0.3, & m = -1 \\ 0, & \text{otherwise} \end{cases}.$$

1. Design a three-tap zero forcing linear equalizer so that the output is

$$q_m = \begin{cases} 1, & m = 0 \\ 0, & m = \pm 1 \end{cases}.$$

2. Determine q_m for $m = \pm 2, \pm 3$ by convolving the impulse response of the equalizer with the channel response.

9.24 The transmission of a signal pulse with a raised cosine spectrum through a channel results in the following (noise-free) sampled output from the demodulator:

$$x_k = \begin{cases} -0.5, & k = -2 \\ 0.1, & k = -1 \\ 1, & k = 0 \\ -0.2, & k = 1 \\ 0.05, & k = 2 \\ 0, & \text{otherwise} \end{cases}.$$

1. Determine the tap coefficients of a three-tap linear equalizer based on the zero-forcing criterion.

2. For the coefficients determined in Part 1, determine the output of the equalizer for the case of the isolated pulse. Thus, determine the residual ISI and its span in time.

9.25 Show that the gradient vector in the minimization of the MSE may be expressed as

$$g_k = -E[e_k y_k],$$

where the error $e_k = a_k - z_k$ and the estimate of g_k, i.e.,

$$\hat{g}_k = -e_k y_k,$$

satisfies the condition that $E[\hat{g}_k] = g_k$.

9.26 A nonideal bandlimited channel introduces ISI over three successive symbols. The (noise-free) response of the matched filter demodulator sampled at the sampling time $t = kT$ is

$$\int_{-\infty}^{\infty} s(t)s(t-kT)dt = \begin{cases} \mathcal{E}_b, & k = 0 \\ 0.9\mathcal{E}_b, & k = \pm 1 \\ 0.1\mathcal{E}_b, & k = \pm 2 \\ 0, & \text{otherwise} \end{cases}.$$

Determine the tap coefficients of a three-tap linear equalizer that equalizes the channel (received signal) response to an equivalent partial response (duobinary) signal

$$y_k = \begin{cases} \mathcal{E}_b, & k = 0, 1 \\ 0, & \text{otherwise} \end{cases}.$$

9.27 Determine the tap weight coefficients of a three-tap zero-forcing equalizer if the ISI spans three symbols and is characterized by the values $x(0) = 1$, $x(-1) = 0.3$, and $x(1) = 0.2$. Also, determine the residual ISI at the output of the equalizer for the optimum tap coefficients.

9.28 In line-of-sight microwave radio transmission, the signal arrives at the receiver via two propagation paths: the direct path and a delayed path that occurs due to signal reflection from the surrounding terrain. Suppose that the received signal has the form

$$r(t) = s(t) + \alpha s(t-T) + n(t),$$

where $s(t)$ is the transmitted signal, α is the attenuation ($\alpha < 1$) of the secondary path, and $n(t)$ is AWGN.

1. Determine the output of the demodulator at $t = T$ and $t = 2T$ that employs a filter matched to $s(t)$.

2. Determine the probability of error for a symbol-by-symbol detector if the transmitted signal is binary antipodal and the detector ignores the ISI.

3. What is the error-rate performance of a simple (one-tap) DFE that estimates α and removes the ISI? Sketch the detector structure that employs a DFE.

9.29 Repeat Problem 9.27 and use the MMSE criterion for optimizing the tap coefficients. Assume that the noise power spectrum is 0.1 W/Hz.

COMPUTER PROBLEMS

9.1 Linear Filter Model of a Communication Channel

As indicated in this chapter, a bandlimited communication channel can be modeled as a linear filter whose frequency response characteristics match the frequency response characteristics of the channel. Therefore, we may design digital finite-duration impulse response (FIR) or infinite-duration impulse response (IIR) filters that approximate the frequency response characteristics of analog communication channels.

Suppose that we wish to model an ideal channel that has an amplitude response $A(f) = 1$ for $|f| \leq 2000$ Hz and $A(f) = 0$ for $|f| > 2000$ Hz, and a constant delay (linear phase) for $|f| \leq 2000$ Hz. The sampling rate for the digital filter is selected as $F_s = 10,000$ Hz. Since the desired phase response is linear, only an FIR filter could satisfy this condition. However, it is not possible to achieve a zero response in the stopband. Instead, we select the stopband response to be -40 dB and the stopband frequency to be 2500 Hz. In addition, we allow for a small amount, 0.5 dB, of ripple in the passband.

Design an FIR filter with these characteristics. Plot the frequency and phase response of the filter in the frequency band $0 \leq f \leq 5000$ Hz. Note that 5000 Hz is the folding frequency.

9.2 Simulation of a Time-Variant Multipath Channel

A two-path (multipath) radio channel can be modeled in the time-domain as illustrated in Figure CP-9.2. It has a time-variant impulse response, which may be expressed as

$$c(t; \tau) = b_1(t)\delta(\tau) + b_2(t)\delta(t - \tau_d),$$

where $b_1(t)$ and $b_2(t)$ represent the time-varying propagation behavior of the channel and τ_d is the delay between the two multipath components. The time-varying gains $b_1(t)$ and $b_2(t)$ of the two paths are typically characterized as random processes. Suppose we model $b_1(t)$ and $b_2(t)$ as Gaussian random processes that are generated by passing white Gaussian noise through lowpass filters. In discrete time, we can use relatively simple digital IIR filters excited by white Gaussian noise (WGN) sequences. For example, a simple lowpass IIR digital filter having two identical poles is described by the z-transform

$$H(z) = \frac{(1-p)^2}{(1-pz^{-1})^2}$$

or the corresponding difference equation

$$b_n = 2pb_{n-1} - p^2 b_{n-2} + (1-p)^2 w_n,$$

where $\{w_n\}$ is the input WGN sequence, $\{b_n\}$ is the output sequence, and $0 < p < 1$ is the pole position. The position of the pole controls the bandwidth of the filter; hence, the rate of variation of $\{b_n\}$. When p is close to unity, the filter bandwidth is narrow, whereas when p is close to zero, the bandwidth is wide. Hence, when p is close to unity, the filter output changes more slowly than the case when the pole is close to the origin in the z-plane.

Generate and plot the output sequences $\{b_{1n}\}$, $\{b_{2n}\}$ of the two filters when $p = 0.5$ and $p = 0.99$ for 1000 samples. Let

$$c_n = b_{1,n} + b_{2,n-d}$$

be the output channel sequence, where $d = 5$ samples of delay. Plot the output sequence $\{c_n\}$ for 1000 samples when $p = 0.5$ and $p = 0.99$.

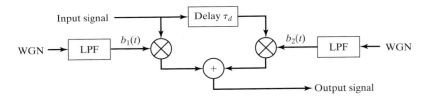

Figure CP-9.2   Two-path radio channel model.

9.3 Effect of Intersymbol Interference

The purpose of this problem is to view the effect of intersymbol interference (ISI) on the received signal sequence $\{y_n\}$ for two channels that are characterized by the discrete time responses $\{x_n\}$ as follows:

Channel 1

$$x_n = \begin{cases} 1, & n = 0 \\ -0.25, & n = \pm 1 \\ 0.1, & n = \pm 2 \\ 0, & \text{otherwise} \end{cases}$$

and

Channel 2

$$x_n = \begin{cases} 1, & n = 0 \\ 0.5, & n = \pm 1 \\ -0.2, & n = \pm 2 \\ 0, & \text{otherwise} \end{cases} .$$

Note that in these channels, the ISI is limited to two symbols on either side of the desired transmitted signal. Hence, the cascade of the transmitter and receiver filters and the channel at the sampling instants are represented by the equivalent discrete-time FIR channel filter shown in Figure CP-9.3.

Suppose that the transmitted signal sequence is binary (± 1), and let the received sequence at the output of the equivalent discrete-time FIR channel filter be denoted as $\{y_k\}$. Therefore, the output sequence $\{y_k\}$ can be expressed as

$$y_k = a_k + x_1 a_{k-1} + x_{-1} a_{k+1} + x_2 a_{k-2} + x_{-2} a_{k+2}, \quad k = 1, 2, \ldots,$$

where $\{a_k = \pm 1\}$ is the input data sequence. Assuming that $a_k = 1$, compute and plot the values of $\{y_k\}$ for the 16 possible data sequences $\{a_{k-1}, a_{k+1}, a_{k-2}, a_{k+2}\}$ for channel 1 and channel 2. Use separate plots for the two channels. Repeat the experiment when $a_k = -1$. Which channel characteristic results in detector errors due to ISI even in the absence of noise?

Repeat the preceding experiments when the channel output sequence is corrupted by additive zero-mean white Gaussian noise with the variance $\sigma^2 = 0.1$. Compare these results with the noiseless output and comment on the effect of the additive noise on the two channel output sequence.

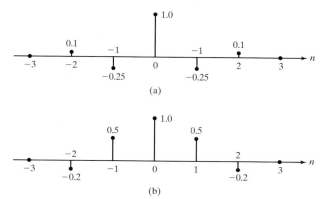

Figure CP-9.3 FIR channel models with ISI: (a) channel 1; (b) channel 2.

9.4 Design of Optimum Transmitter and Receiver Filters

The objective of this problem is to design a digital implementation of the transmitter and receiver filters, $G_T(f)$ and $G_R(f)$, such that

$$|G_T(f)||G_R(f)| = X_{\text{rc}}(f)$$

and $G_R(f)$ is the matched filter to $G_T(f)$.

The simplest way to design and implement the transmitter and receiver filters in digital form is to employ FIR filters with linear phase (symmetric impulse response). If $X_{\mathrm{rc}}(f)$ is the desired raised cosine frequency response with a specified roll-off parameter, the magnitude response of the transmitter and receiver filters is

$$|G_T(f)| = |G_R(f)| = \sqrt{X_{\mathrm{rc}}(f)}.$$

The frequency response of the filters is related to the impulse response by the Fourier transform relation

$$G_T(f) = \sum_{n=-(N-1)/2}^{(N-1)/2} g_T(n) \, e^{-j2\pi f n T_s}, \tag{A}$$

where T_s is the sampling interval and N is the length of the filter. Note that N is odd. Since $G_T(f)$ is bandlimited, we may select the sampling frequency F_s to be at least $2/T$. In particular, let

$$F_s = \frac{1}{T_s} = \frac{4}{T},$$

or equivalently, $T_s = T/4$. Hence, the folding frequency is $F_s/2 = 2/T$. Since $|G_T(f)| = \sqrt{X_{\mathrm{rc}}(f)}$, we may sample $X_{\mathrm{rc}}(f)$ at equally spaced points in frequency, with the frequency separation $\Delta f = F_s/N$. Thus, we have

$$\sqrt{X_{\mathrm{rc}}(m\Delta f)} = \sqrt{X_{\mathrm{rc}}\left(\frac{m F_s}{N}\right)} = \sum_{n=-(N-1)/2}^{(N-1)/2} g_T(n) \, e^{-j2\pi mn/N}. \tag{B}$$

The inverse transform relation is used to obtain the impulse response, i.e.,

$$g_T(n) = \sum_{m=-(N-1)/2}^{(N-1)/2} \sqrt{X_{\mathrm{rc}}\left(\frac{4m}{NT}\right)} e^{j2\pi mn/N}, \quad n = 0, \pm 1, \ldots \pm \left(\frac{N-1}{2}\right). \tag{C}$$

Since $g_T(n)$ is symmetric, the impulse response of the desired linear phase transmitter filter is obtained by delaying $g_T(n)$ by $(N-1)/2$ samples, i.e., we have $g_T\left(n - \frac{N-1}{2}\right)$, $n = 0, 1, \ldots, N-1$.

1. Determine and plot the impulse response $g_T(n)$ of the transmitter filter of length $N = 31$ when $X_{\mathrm{rc}}(f)$ has a roll-off factor $\alpha = 1/4$, and $1/T = 1800$ Hz.

2. Determine and plot the frequency response $|G_T(f)|$ for $0 \leq f \leq 0.5$ from Equation (A). Does $G_T(f) = 0$ for $|f| \geq (1+\alpha)/T$? Explain why it does or does not.

3. Plot $|G_T(f)|^2$ and $X_{\mathrm{rc}}(f)$ on the same graph and compare the two graphs. Explain the difference.

4. Repeat Steps 1, 2, and 3 for $\alpha = 1/4$, $1/T = 1800$ Hz, and $N = 41$. Compare and comment in the results.

9.5 Design of a Duobinary Signal Filter

The objective of this problem is to design a digital implementation of the transmitter and receiver filters $G_T(f)$ and $G_R(f)$ such that their product is equal to the spectrum of a duobinary pulse and $G_R(f)$ is the matched filter to $G_T(f)$.

To satisfy the frequency domain specification, we have

$$|G_T(f)||G_R(f)| = \begin{cases} \dfrac{1}{W}\cos\left(\dfrac{\pi f}{2W}\right), & |f| \leq W \\ \\ 0, & |f| > W \end{cases}.$$

Therefore,

$$|G_T(f)| = \begin{cases} \sqrt{\dfrac{1}{W}\cos\left(\dfrac{\pi f}{2W}\right)}, & |f| \leq W \\ \\ 0, & |f| > W \end{cases}.$$

By following the same approach as in CP-9.4, we obtain the impulse responses for a linear phase FIR implementation of the transmitter and receiver filters. Therefore, with $W = 1/2T$ and $F_s = 4/T$, we have

$$g_T(n) = \sum_{n=-(N-1)/2}^{(N-1)/2} |G_T\left(\dfrac{4m}{NT}\right)| e^{j2\pi mn/N}, \quad n = 0, \pm 1, \ldots, \pm\left(\dfrac{N-1}{2}\right)$$

and $g_R(n) = g_T(n)$.

1. Determine and plot $g_T\left(n - \frac{N-1}{2}\right)$ for $W = 1800$ and $N = 31$.
2. By using Equation (A) of Problem CP-9.4, determine and plot $|G_T(f)|$.
3. Plot $|G_T(f)|^2$ and the ideal duobinary (cosine) spectrum on the same graph and compare the two spectra. Explain the difference.

9.6 Precoding for Duobinary Signals

Write a MATLAB program that takes a binary data sequence $\{d_k\}$, precodes it for a duobinary pulse transmission system to produce the sequence $\{p_k\}$, and maps the precoded sequence into the transmitted amplitude levels $\{a_k\}$. Then, from the transmitted sequence $\{a_k\}$, from the received noise-free sequences $\{b_k\}$, and from $\{b_k\}$, recover the original data sequence $\{d_k\}$. Verify the operation of your program using the data sequence $\{1\ 0\ 0\ 1\ 0\ 1\ 1\ 1\ 0\ 1\ 1\ 0\}$.

9.7 Simulation of a Binary Communication System Using Duobinary Signal

The objective of this problem is to simulate a binary PAM communication system that employs a duobinary signal pulse, where the precoding and amplitude conversion that yields the sequence $\{a_n\}$ are performed as prescribed in the text. Hence, the input to the detector is the sequence

$$y_k = b_k + n_k$$
$$= a_k + a_{k-1} + n_k, \quad k + 1, 2,$$

where the sequence $\{n_k\}$ is zero-mean, Gaussian, and uncorrelated. The variance of n_k is σ^2.

Perform the simulation for 10,000 bits and measure the bit-error probability for $\sigma^2 = 0.1, \sigma^2 = 0.5$, and $\sigma^2 = 1$. Plot the theoretical error probability for binary PAM with no ISI, and compare the simulation results with this ideal performance. You should observe some degradation in the performance of the duobinary system. What are the approximate values of the degradation for $\sigma^2 = 0.1, \sigma^2 = 0.5$, and $\sigma^2 = 1$?

9.8 Zero-Forcing Equalizer

The objective of this problem is to design a zero-forcing equalizer for the channel-distorted pulse $x(t)$ at the input to the equalizer given by the expression

$$x(t) = \frac{1}{1 + \left(\frac{2t}{T}\right)^2},$$

where $1/T$ is the symbol rate. The pulse is sampled at the rate $2/T$, which is the input rate to the equalizer.

As described in the text, the zero-forcing equalizer must satisfy the equations

$$q(mT) = \sum_{n=-K}^{K} c_n x\left(mT - \frac{nT}{2}\right) = \begin{cases} 1, & m = 0 \\ 0, & m = \pm 1, \pm 2, \ldots, \pm K \end{cases},$$

where $2K + 1$ is the number of taps in the equalizer and $\{c_n\}$ are the equalizer coefficients.

Write a program that computes the equalizer coefficients for any value of K. Compute the equalizer coefficients for $K = 2, 4$, and 6. Plot the input sequence to the equalizer for the pulse $x(t)$, and plot the equalizer output samples for each $K = 2, 4$, and 6. Compare how well the equalizer performs for each value of K by comparing the residual ISI at the output of the equalizer.

9.9 MSE Equalizer

The objective of this problem is to design an equalizer, based on the mean-square-error criterion, for the channel distorted pulse described by the expression

$$x(t) = \frac{1}{1 + (2t/T)^2},$$

where $1/T$ is the symbol rate. The pulse is sampled at the rate $2/T$, which is the input rate to the equalizer. The information symbols have zero mean, unit variance, and are uncorrelated, i.e.,

$$E(a_n) = 0;$$

$$E(|a_n|^2) = 1;$$

$$E(a_n a_m) = 0, \quad n \neq m.$$

The additive noise samples at the input to the equalizer are zero-mean uncorrelated Gaussian with variance $\sigma^2 = 0.01$ and $\sigma^2 = 0.1$.

Write a program that computes the coefficients of the equalizer having $2K + 1$ taps. Compute the equalizer coefficients for $K = 2$, 4, and 6. Plot the sequence to the equalizer for the pulse $x(t)$, and plot the equalizer output samples for each $K = 2$, 4, and 6. Compare how well the equalizer performs for each value of K by computing the residual ISI at the output of the equalizer.

9.10 Adaptive Equalizer

The objective of this problem is to implement an adaptive equalizer based on the LMS algorithm. The channel is modeled as an FIR filter with symbol-spaced

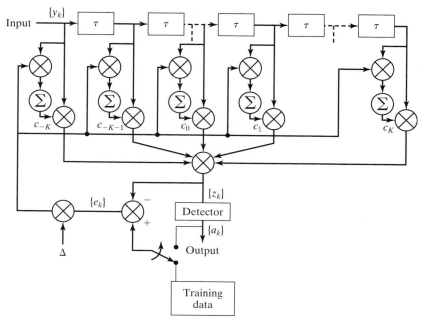

Figure CP-9.10 Linear adaptive equalizer based on the MSE criterion.

values that are given as follows:

$$x = [0.05, -0.063, 0.088, -0.126, -0.25, 0.9047, 0.25, 0.126, 0.038, 0.088].$$

The MSE equalizer is also an FIR filter with symbol-spaced tap coefficients. Training symbols are transmitted initially to train the equalizer. In the data mode, the equalizer employs the output of the detector in forming the error signal used in the LMS algorithm. The block diagram of the system is shown in Figure CP-9.10.

Write a program that performs the simulation of the system in Figure CP-9.10. Use 1000 training (binary) symbols and 10,000 binary data symbols for the FIR channel model previously given. Use $\sigma^2 = 0.01$, $\sigma^2 = 0.1$, and $\sigma^2 = 1$ for the variance of the additive, zero-mean, Gaussian noise sequence. Compare the measured error rate with that of an ideal channel with no ISI.

10

Transmission of Digital Information via Carrier Modulation

In Chapters 8 and 9, our treatment of the transmission of digital information was focused on *baseband channels*, i.e., channels having passbands that include zero frequency ($f = 0$). The design of signal waveforms for the transmission of digital information over such channels does not require the use of a sinusoidal carrier signal. In this chapter, we focus on the design and transmission of signals that are suitable for *bandpass channels*, i.e., channels that have frequency passbands far removed from $f = 0$. In such channels, which include wireline channels, radio channels, and satellite channels, the information-bearing signal is impressed on a sinusoidal carrier, which shifts the frequency content of the information-bearing signal to the appropriate frequency band that is passed by the channel. Thus, the signal is transmitted by carrier modulation.

The types of carrier modulation treated in this chapter are pulse-amplitude modulation, phase modulation (phase-shift keying), quadrature-amplitude modulation (QAM), and frequency modulation (frequency-shift keying).

As in the case of baseband signals, the geometric representation of carrier-modulated digital signals is useful in describing the processing of the received signals and assessing their performance in the presence of additive Gaussian noise. We begin with amplitude-modulated signals.

10.1 AMPLITUDE-MODULATED DIGITAL SIGNALS

To transmit the digital signal waveforms through a bandpass channel by amplitude modulation, the baseband signal waveforms $s_m(t), m = 1, 2, \ldots, M$ are multiplied by a sinusoidal carrier of the form $\cos 2\pi f_c t$, as shown in Figure 10.1. In this figure, f_c is the carrier frequency and corresponds to the center frequency in the passband

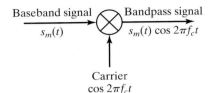

Figure 10.1   Amplitude modulation of the sinusoidal carrier by the baseband signal.

of the channel. Thus, the transmitted signal waveforms may be expressed as

$$u_m(t) = s_m(t) \cos 2\pi f_c t, \qquad m = 1, 2, \ldots, M. \qquad (10.1.1)$$

As previously described in Section 3.2, amplitude modulation of the carrier $\cos 2\pi f_c t$ by the baseband signal waveforms $\{s_m(t)\}$ shifts the spectrum of the baseband signal by an amount f_c; thus it places the signal into the passband of the channel. Recall that the Fourier transform of the carrier is $\frac{1}{2}[\delta(f - f_c) + \delta(f + f_c)]$. Because multiplication of two signals in the time domain corresponds to the convolution of their spectra in the frequency domain, the spectrum of the amplitude-modulated signal given by Equation (10.1.1) is

$$U_m(f) = \frac{1}{2}[S_m(f - f_c) + S_m(f + f_c)]. \qquad (10.1.2)$$

Thus, the spectrum of the baseband signal $s_m(t)$ is shifted in frequency by an amount equal to the carrier frequency f_c. The result is a DSB-SC AM signal, as illustrated in Figure 10.2. The upper sideband of the carrier-modulated signal contains the frequency content of $u_m(t)$ for $|f| > f_c$, i.e., for $f_c < |f| \le f_c + W$.

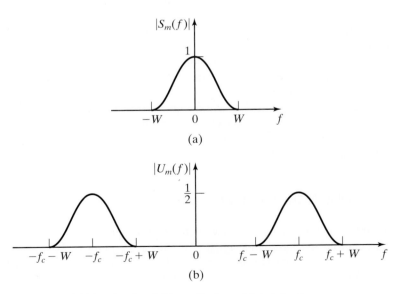

Figure 10.2   Spectra of (a) baseband and (b) amplitude-modulated signals.

The lower sideband of $u_m(t)$ includes the frequency content for $|f| < f_c$, i.e., for $f_c - W \leq |f| < f_c$. Hence, the DSB-SC amplitude-modulated signal occupies a channel bandwidth of $2W$, which is twice the bandwidth required to transmit the baseband signal.

The energy of the bandpass signal waveforms $u_m(t), m = 1, 2, \ldots, M$, which are given by Equation (10.1.1), is defined as

$$\mathcal{E}_m = \int_{-\infty}^{\infty} u_m^2(t)dt = \int_{-\infty}^{\infty} s_m^2(t) \cos^2 2\pi f_c t \, dt$$

$$= \frac{1}{2} \int_{\infty}^{\infty} s_m^2(t)dt + \frac{1}{2} \int_{-\infty}^{\infty} s_m^2(t) \cos 4\pi f_c t \, dt. \tag{10.1.3}$$

We note that when $f_c \gg W$, the term

$$\frac{1}{2} \int_{-\infty}^{\infty} s_m^2(t) \cos 4\pi f_c t \, dt \tag{10.1.4}$$

involves the integration of the product of a slowly varying function, i.e., $s_m^2(t)$, with a rapidly varying sinusoidal term, i.e., $\cos 4\pi f_c t$. With an argument similar to the argument following Equation (3.2.2), we conclude that the integral in Equation (10.1.4) over a single cycle of $\cos 4\pi f_c t$ is zero (see Figure 3.6); hence, the integral over an arbitrary number of cycles is also zero. Consequently, the energy of the carrier-modulated signal is

$$\mathcal{E}_m = \frac{1}{2} \int_{-\infty}^{\infty} s_m^2(t) \, dt. \tag{10.1.5}$$

Thus, we have shown that the energy in the bandpass signal is one-half of the energy in the baseband signal. The scale factor of $1/2$ is due to the carrier component $\cos 2\pi f_c t$, which has an average power of $1/2$.

We should emphasize that any of the baseband pulse signals described in Chapter 8, including binary antipodal, orthogonal, and M-ary signals, e.g., orthogonal (PPM), biorthogonal, and simplex signals, can be used to amplitude modulate a carrier; thus, they can be transformed to bandpass signals that are appropriate for digital information transmission on a bandpass channel.

Geometric Representation of Bandpass Signals.   The geometric representation of amplitude-modulated signals remains the same as that of the equivalent baseband signals. For example, suppose that a set of M-dimensional orthogonal baseband signal waveforms is represented as

$$s_m(t) = \sqrt{\mathcal{E}_s} \psi_m(t), \qquad m = 1, 2, \ldots, M, \tag{10.1.6}$$

where $\mathscr{E}_s$ is the energy of each waveform and $\psi_m(t)$ is one of M baseband (orthogonal) basis pulse waveforms, each of which has unit energy. If this set of M signal waveforms amplitude modulates a carrier, the modulated signal is

$$u_m(t) = s_m(t)\cos 2\pi f_c t$$
$$= \sqrt{\mathscr{E}_s}\,\psi_m(t)\cos 2\pi f_c t, \quad m = 1, 2, \ldots, M. \tag{10.1.7}$$

It is easy to show that if the original signals are orthogonal, then the modulated signals will also be orthogonal. Let us define a new set of bandpass (carrier modulated) basis functions as

$$\psi_{cm}(t) = \sqrt{2}\,\psi_m(t)\cos 2\pi f_c t, \quad m = 1, 2, \ldots, M. \tag{10.1.8}$$

These new functions also have unit energy. That is,

$$\int_{-\infty}^{\infty} \psi_{cm}^2(t)\,dt = 2\int_{-\infty}^{\infty} \psi_m^2(t)\cos^2 2\pi f_c t\,dt$$
$$= \int_{-\infty}^{\infty} \psi_m^2(t)\,dt + \int_{-\infty}^{\infty} \psi_m^2(t)\cos 4\pi f_c t\,dt$$
$$= 1,$$

where the second term on the right-hand side integrates to zero as in Equation (10.1.4). Consequently, the bandpass waveforms given by Equation (10.1.7) are represented by the carrier-modulated basis functions as

$$u_m(t) = \sqrt{\mathscr{E}_s/2}\,\psi_{cm}(t), \quad m = 1, 2, \ldots, M. \tag{10.1.9}$$

Therefore, if s_m is the vector corresponding to the geometric representation of a baseband signal, the vector representation of the carrier amplitude modulated signal is $s_{cm} = s_m/\sqrt{2}$. Hence, the only change in the vector representation of the carrier-modulated signal waveforms is the scale factor of $\sqrt{2}$ that relates the elements of the vector of the baseband signal to the elements of the vector of the carrier-modulated signal.

Example 10.1.1

A carrier amplitude modulated PAM signal may be expressed as

$$u_m(t) = s_m(t)\cos 2\pi f_c t$$
$$= s_m\psi(t)\cos 2\pi f_c t, \ 0 \le t \le T,$$

where $s_m = (2m - 1 - M)d$, $m = 1, 2, \ldots, M$, and $d = A\sqrt{T}$, as given by Equation (8.4.7), and $\psi(t)$ is a unit energy pulse defined over the interval $0 \le t \le T$. Determine the basis function for the carrier-modulated PAM signal and the corresponding signal points in a vector representation.

Solution   The carrier amplitude PAM signal may be expressed as

$$u_m(t) = \frac{s_m}{\sqrt{2}} \psi_c(t) = s_{cm} \psi_c(t) \ , \quad m = 1, 2, \ldots, M,$$

where

$$\psi_c(t) = \sqrt{2} \psi(t) \cos 2\pi f_c t$$

and

$$s_{cm} = s_m/\sqrt{2} = (2m - 1, M)d/\sqrt{2}, \ m = 1, 2, \ldots, M. \qquad \blacksquare$$

Bandlimited Channels.   Usually, a bandpass channel has a limited bandwidth, which defines the range of frequencies that are passed by the channel. Outside of this frequency range, any frequencies contained in the transmitted signal are severely attenuated, resulting in signal distortion. Just as in the case of baseband signal transmission, we can shape the spectrum of the baseband signals prior to amplitude modulation. This is shown in Figure 10.3 (for PAM signals), where the signal-shaping filter is designed according to the Nyquist criterion for zero intersymbol interference. Thus, the frequency content of the transmitted signal is confined to the range of frequencies passed by the channel.

10.1.1 Demodulation and Detection of Amplitude-Modulated Signals

The demodulation of a bandpass signal may be accomplished by means of correlation or matched filtering, as described for baseband channels. However, as we will observe from the development that follows, the presence of the carrier introduces an additional complication in the demodulation of the signal.

The transmitted signal in a signaling interval may be expressed as

$$u_m(t) = s_m(t) \cos 2\pi f_c t, \ m = 1, 2, \ldots, M, \qquad (10.1.10)$$

and the received signal has the form

$$r(t) = u_m(t) + n(t), \qquad (10.1.11)$$

where

$$n(t) = n_c(t) \cos 2\pi f_c t - n_s(t) \sin 2\pi f_c t \qquad (10.1.12)$$

is a bandpass noise process, as discussed in Section 5.3.3.

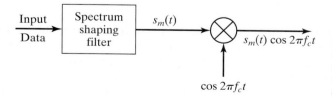

Figure 10.3   Shaping the spectrum of the amplitude-modulated signal for transmission over a bandlimited channel.

In particular, consider the transmission of a PAM signal. In this case, the baseband signal is

$$s_m(t) = s_m \psi(t) \tag{10.1.13}$$

and the received signal is

$$r(t) = s_m \psi(t) \cos 2\pi f_c t + n(t). \tag{10.1.14}$$

By cross correlating the received signal $r(t)$ with the basis function $\psi_c(t) = \sqrt{2}\psi(t)$ $\cos 2\pi f_c t$, we obtain, at the sampling instant $t = T$,

$$
\begin{aligned}
y(T) &= \int_0^T r(t)\psi_c(t)\, dt \\
&= \sqrt{2}s_m \int_0^T \psi^2(t) \cos^2 2\pi f_c t\, dt + \sqrt{2}\int_0^T n(t)\psi(t) \cos 2\pi f_c t\, dt \\
&= \frac{s_m}{\sqrt{2}} + n = s_{cm} + n, \tag{10.1.15}
\end{aligned}
$$

where n represents the additive noise component at the output of the correlator. An identical result is obtained if a matched filter replaces the correlator to demodulate the received signal.

Carrier-Phase Recovery.   In the preceding development, we assumed that the function $\psi(t)$ is perfectly synchronized with the signal component of $r(t)$ in both time and carrier phase, as shown in Figure 10.4 for PAM. In practice, however, these ideal conditions do not hold. First, the propagation delay encountered in transmitting a signal through the channel results in a carrier-phase offset in the received signal. Second, the oscillator that generates the carrier signal $\cos 2\pi f_c t$ at the receiver is not generally phase-locked to the oscillator used at the transmitter. Practical oscillators usually drift in frequency and phase. Consequently, the demodulation of the bandpass PAM signal, as illustrated in Figure 10.4, is ideal, but not practical. In a practical system, it is necessary to generate a phase-coherent carrier at the receiver to perform the demodulation of the received signal.

In general, the received signal has a carrier phase offset ϕ. To estimate ϕ from the received signal, we must observe $r(t)$ over many signal intervals. When observed over many signal intervals, the message signal is zero mean due to the randomness in the signal amplitude values $\{s_m\}$; thus, the transmitted DSB-SC AM signal has zero average power at the carrier frequency f_c. Consequently, it is not possible to estimate the carrier phase directly from $r(t)$. However, if we square $r(t)$, we generate a frequency component at $f = 2f_c$ that has nonzero average power. This component

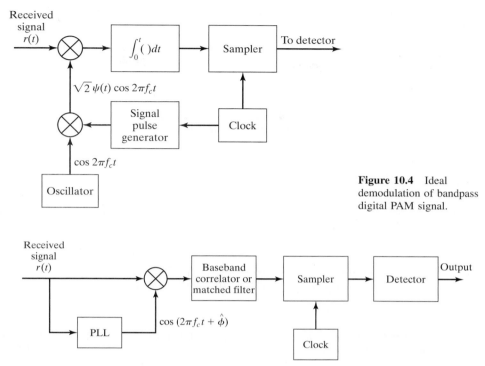

Figure 10.4   Ideal demodulation of bandpass digital PAM signal.

Figure 10.5   Demodulation of carrier-amplitude modulated signal by using a PLL to acquire the carrier phase.

can be filtered out by a narrowband filter tuned to $2f_c$, which can be used to drive a PLL, as described in Section 6.4. A functional block diagram of the receiver that employs a PLL for estimating the carrier phase is shown in Figure 10.5.

The Costas loop, also described in Section 6.4, is an alternative method for estimating the carrier phase from the received signal $r(t)$. The PLL and the Costas loop yield phase estimates that are comparable in quality in the presence of additive channel noise.

As an alternative to performing the correlation or matched filtering at the baseband as shown in Figure 10.5, we may perform cross correlation or matched filtering either at the bandpass or at some convenient intermediate frequency. In particular, a bandpass correlator may be used to multiply the received signal $r(t)$ by the amplitude-modulated carrier $\sqrt{2}\psi(t)\cos(2\pi f_c t + \hat{\phi})$, where $\cos(2\pi f_c t + \hat{\phi})$ is the output of the PLL. The product signal is integrated over the signaling interval T, the output of the integrator is sampled at $t = T$, and the sample is passed to the detector. If we use a matched filter instead of a correlator, the filter impulse response is $\psi(T-t)\cos[2\pi f_c(t-T)+\hat{\phi}]$. The functional block diagrams for these demodulators are shown in Figure 10.6.

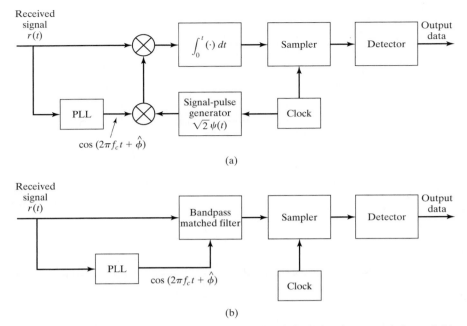

Figure 10.6   Bandpass demodulation of a digital PAM signal via (a) bandpass correlation and (b) bandpass matched filtering.

Optimum Detector.   In the case of a perfect (noise-free) carrier-phase estimate, $\phi = \hat{\phi}$, the input to the detector is the signal-plus-noise term given by Equation (10.1.15). As in the case of baseband PAM, for equiprobable messages, the optimum detector bases its decision on the distance metrics

$$D(y, s_m) = (y - s_{cm})^2, \quad m = 1, 2, \ldots, M \tag{10.1.16}$$

or, equivalently, on the correlation metrics

$$C(y, s_m) = 2y s_{cm} - s_{cm}^2. \tag{10.1.17}$$

The optimum detector for amplitude-modulated binary orthogonal and M-ary orthogonal, biorthogonal, and simplex signals is basically the same as that for the equivalent baseband signals. Consequently, the probability of error of the optimum detector for binary and M-ary amplitude-modulated signals is exactly the same as that of the equivalent baseband signals described in Chapter 8, as shown in Equations (8.5.7) and (8.5.8).

10.2 PHASE-MODULATED DIGITAL SIGNALS

In the preceding section, we observed that bandpass signal waveforms, which are appropriate for transmission on bandpass channels, were generated by taking a set of baseband signals and impressing them on the amplitude of a carrier. In this section, we generate bandpass signal waveforms by digitally modulating the phase of the carrier.

Let us begin by considering a set of M two-dimensional baseband signal waveforms, say, $s_m(t)$, $m = 1, 2, \ldots, M$. If these waveforms are used to amplitude modulate a carrier, we obtain the following set of M bandpass signal waveforms:

$$u_m(t) = s_m(t) \cos 2\pi f_c t, \quad m = 1, 2, \ldots, M, \quad 0 \le t \le T. \tag{10.2.1}$$

Now, consider the special case in which the M two-dimensional bandpass signal waveforms are constrained to have the same energy, i.e.,

$$\mathcal{E}_m = \int_0^T u_m^2(t)\, dt = \int_0^T s_m^2(t) \cos^2 2\pi f_c t\, dt$$

$$= \frac{1}{2} \int_0^T s_m^2(t)\, dt + \frac{1}{2} \int_0^T s_m^2(t) \cos 4\pi f_c t\, dt. \tag{10.2.2}$$

As indicated previously, the integral of the double-frequency component in Equation (10.2.2) averages to zero when $f_c \gg W$, where W is the bandwidth of the baseband signal $s_m(t)$. Hence,

$$\mathcal{E}_m = \frac{1}{2} \int_0^T s_m^2(t)\, dt = \mathcal{E}_s \text{ for all } m, \tag{10.2.3}$$

where $\mathcal{E}_s$ denotes the energy/signal or symbol.

Since all the two-dimensional signal waveforms have the same energy, the corresponding signal points in the vector representation of the signal waveforms fall on a circle of radius $\mathcal{E}_s$. For example, in the case of the four biorthogonal waveforms, the signal points appear as shown in Figure 10.7 or any phase-rotated version of these signal points.

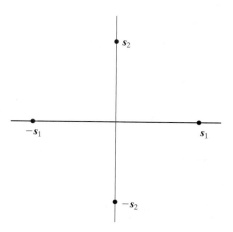

Figure 10.7 Signal points for $M = 4$ biorthogonal signal waveforms.

From this geometric representation for $M = 4$, we observe that the signal points are equivalent to a single signal whose phase is shifted by multiples of $\pi/2$. In other words, a bandpass signal of the form $s(t)\cos(2\pi f_c t + \pi m/2)$, $m = 0, 1, 2, 3$, has the same geometric representation as an $M = 4$ general biorthogonal signal set. Therefore, a simple way to generate a set of M two-dimensional bandpass signals having equal energy is to impress the information on the phase of the carrier. Thus, we have a carrier-phase modulated signal.

The general representation of a set of M carrier-phase modulated signal waveforms is

$$u_m(t) = g_T(t)\cos\left(2\pi f_c t + \frac{2\pi m}{M}\right), \quad m = 0, 1, \ldots, M - 1, \quad 0 \le t \le T,$$

$$(10.2.4)$$

where $g_T(t)$ is a baseband pulse shape, which determines the spectral characteristics of the transmitted signal. When $g_T(t)$ is a rectangular pulse, defined as

$$g_T(t) = \sqrt{\frac{2\mathcal{E}_s}{T}}, \quad 0 \le t \le T, \qquad (10.2.5)$$

the corresponding transmitted signal waveforms

$$u_m(t) = \sqrt{\frac{2\mathcal{E}_s}{T}}\cos\left(2\pi f_c t + \frac{2\pi m}{M}\right), \quad m = 0, 1, \ldots, M - 1, \quad 0 \le t \le T,$$

$$(10.2.6)$$

have a constant envelope (the pulse shape $g_T(t)$ is a rectangular pulse) and the carrier phase changes abruptly at the beginning of each signal interval. This type of digital-phase modulation is called *phase-shift keying* (PSK). Figure 10.8 illustrates a four-phase ($M = 4$) PSK signal waveform, usually called a quadrature PSK (QPSK) signal.

By viewing the angle of the cosine function in Equation (10.2.6) as the sum of two angles, we may express the waveforms in Equation (10.2.6) as

$$u_m(t) = g_T(t)A_{mc}\cos 2\pi f_c t - g_T(t)A_{ms}\sin 2\pi f_c t, \qquad (10.2.7)$$

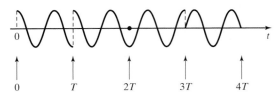

Figure 10.8   Example of a four-phase PSK signal.

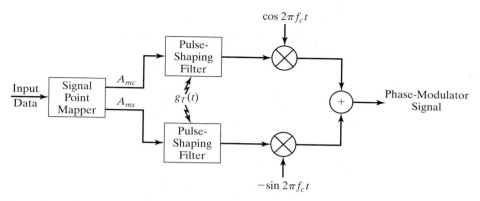

Figure 10.9   Block diagram of a digital-phase modulator.

where

$$A_{mc} = \cos 2\pi m/M, \quad m = 0, 1, \ldots, M - 1$$
$$A_{ms} = \sin 2\pi m/M, \quad m = 0, 1, \ldots, M - 1. \quad (10.2.8)$$

Therefore, the modulator for the phase-modulated signal may be implemented as shown in Figure 10.9, employing two quadrature carrier signals, where each quadrature carrier is amplitude modulated by the information-bearing signal. The pulse-shaping filters are designed to limit the spectrum of the transmitted signal to the allocated channel bandwidth.

It follows from Equation (10.2.7) that digital phase-modulated signals can be represented geometrically as two-dimensional vectors with components $\sqrt{\mathscr{E}_s} \cos 2\pi m/M$ and $\sqrt{\mathscr{E}_s} \sin 2\pi m/M$, i.e.,

$$s_m = \left(\sqrt{\mathscr{E}_s} \cos 2\pi m/M, \quad \sqrt{\mathscr{E}_s} \sin 2\pi m/M \right). \quad (10.2.9)$$

Note that the orthogonal basis functions are

$$\psi_1(t) = \sqrt{\frac{1}{\mathscr{E}_s}} g_T(t) \cos 2\pi f_c t,$$

and

$$\psi_2(t) = -\sqrt{\frac{1}{\mathscr{E}_s}} g_T(t) \sin 2\pi f_c t,$$

where the energy of the pulse $g_T(t)$ is normalized to $2\mathscr{E}_s$. Signal-point constellations for $M = 2, 4$, and 8 are illustrated in Figure 10.10. We observe that binary-phase modulation is identical to binary PAM.

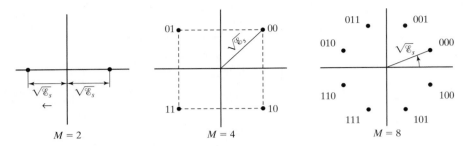

Figure 10.10   PSK signal constellations.

The mapping or assignment of k information bits into the $M = 2^k$ possible phases may be done in a number of ways. The preferred assignment is to use *Gray encoding*, in which adjacent phases differ by one binary digit, as illustrated in Figure 10.10. Because the most likely errors caused by noise involve the erroneous selection of an adjacent phase to the transmitted phase, only a single bit error occurs in the k-bit sequence with Gray encoding.

The Euclidean distance between any two signal points in the constellation is

$$d_{mn} = \sqrt{\| s_m - s_n \|^2}$$

$$= \sqrt{2\mathscr{E}_s \left(1 - \cos \frac{2\pi(m-n)}{M} \right)}, \qquad (10.2.10)$$

and the minimum Euclidean distance (the distance between two adjacent signal points) is simply

$$d_{\min} = \sqrt{2\mathscr{E}_s \left(1 - \cos \frac{2\pi}{M} \right)} = 2\sqrt{\mathscr{E}_s} \sin \frac{\pi}{M}. \qquad (10.2.11)$$

The minimum Euclidean distance $d_{\min}$ plays an important role in determining the error-rate performance of the receiver that demodulates and detects the information in the presence of additive Gaussian noise, as we have previously observed in Section 8.5.3.

Example 10.2.1

Consider the $M = 2, 4$, and 8 PSK signal constellations shown in Figure 10.10; all have the same transmitted signal energy $\mathscr{E}_s$. Determine the minimum distance $d_{\min}$ between adjacent signal points. For $M = 8$, determine by how many dB the transmitted signal energy $\mathscr{E}_s$ must be increased to achieve the same $d_{\min}$ as $M = 4$.

Solution   For $M = 2$, $d_{\min} = 2\sqrt{\mathscr{E}_s}$; for $M = 4$, $d_{\min} = \sqrt{2\mathscr{E}_s}$; and for $M = 8$,

$$d_{\min} = 2\sqrt{\mathscr{E}_s} \sin \pi/8.$$

$$= \sqrt{0.586\mathscr{E}_s}$$

For $M = 8$, the energy $\mathcal{E}_s$ must be increased by the factor of $2/0.586 = 3.413$ or 5.33 dB, in order to achieve the same minimum distance as $M = 4$.

For large values of M, $\sin \pi/M \approx \pi/M$, so

$$d_{\min} \approx \frac{2\pi}{M} \sqrt{\mathcal{E}_s}, \quad M \gg 2.$$

Consequently, when the number M of signal points is doubled, which allows us to transmit one additional information bit per symbol, the signal energy $\mathcal{E}_s$ must be increased by a factor of 4, or 6 dB, in order to maintain the same minimum distance between adjacent signal points.

∎

10.2.1 Demodulation and Detection of Phase-Modulated Signals

The received bandpass signal from an AWGN channel in a signaling interval $0 \leq t \leq T$ may be expressed as

$$\begin{aligned} r(t) &= u_m(t) + n(t) \\ &= [A_{mc}g_T(t) + n_c(t)] \cos 2\pi f_c t - [A_{ms}g_T(t) + n_s(t)] \sin 2\pi f_c t \\ & \quad m = 0, 1, 2, \dots, M-1, \end{aligned} \qquad (10.2.12)$$

where $n(t)$ is the additive bandpass Gaussian noise represented in terms of its quadrature components $n_c(t)$ and $n_s(t)$ as

$$n(t) = n_c(t) \cos 2\pi f_c t - n_s(t) \sin 2\pi f_c t$$

and A_{mc} and A_{ms} are the information-bearing signal components that are related to the transmitted carrier phase by Equation (10.2.8).

The received signal may be correlated with

$$\psi_1(t) = \sqrt{\frac{1}{\mathcal{E}_s}} g_T(t) \cos 2\pi f_c t$$

and

$$\psi_2(t) = -\sqrt{\frac{1}{\mathcal{E}_s}} g_T(t) \sin 2\pi f_c t.$$

The outputs of the two correlators yield the two noise-corrupted signal components, which may be expressed as

$$\begin{aligned} \mathbf{y} &= \mathbf{s}_m + \mathbf{n} \\ &= \left(\sqrt{\mathcal{E}_s} \cos 2\pi m/M + n_c, \quad \sqrt{\mathcal{E}_s} \sin 2\pi m/M + n_s \right), \end{aligned} \qquad (10.2.13)$$

where, by definition,

$$n_c = \frac{1}{\sqrt{4\mathcal{E}_s}} \int_0^T g_T(t) n_c(t)\, dt$$

$$(10.2.14)$$

$$n_s = \frac{1}{\sqrt{4\mathcal{E}_s}} \int_0^T n_s(t) g_T(t)\, dt.$$

Because the quadrature noise components $n_c(t)$ and $n_s(t)$ are zero mean and uncorrelated (see Property 4 of filtered noise on page 252), it follows that $E[n_c] = E[n_s] = 0$ and $E[n_c n_s] = 0$.

The variance of the noise components is

$$E\left[n_c^2\right] = E\left[n_s^2\right] = \frac{1}{4\mathcal{E}_s} \int_0^T \int_0^T g_T(t) g_T(\tau) E[n_c(t) n_c(\tau)]\, dt\, d\tau$$

$$= \frac{N_0}{4\mathcal{E}_s} \int_0^T g_T^2(t)\, dt$$

$$(10.2.15)$$

$$= N_0/2.$$

The optimum detector projects the received signal vector onto each of the M possible transmitted signal vectors $\{s_m\}$ and selects the vector corresponding to the largest projection; thus, we compute the correlation metrics

$$C(\mathbf{y}, \mathbf{s}_m) = \mathbf{y} \cdot \mathbf{s}_m, \quad m = 0, 1, \ldots, M-1$$

$$(10.2.16)$$

and select the signal vector that results in the largest correlation.

Because all signals have equal energy, an equivalent detector metric for digital phase modulation is to compute the phase of the received signal vector $\mathbf{y} = (y_1, y_2)$, which is

$$\Theta = \tan^{-1} \frac{y_2}{y_1},$$

$$(10.2.17)$$

and select the signal from the set $\{s_m\}$ whose phase is closest to Θ. In the next section, we evaluate the probability of error based on the phase metric given by Equation (10.2.17).

Carrier-Phase Estimation.   As previously indicated, in any carrier-modulation system, the oscillators employed at the transmitter and the receiver are not generally phase-locked. As a consequence, the received signal will be of the form

$$r(t) = A_{mc} g_T(t) \cos(2\pi f_c t + \phi) - A_{ms} g_T(t) \sin(2\pi f_c t + \phi) + n(t), \quad (10.2.18)$$

where ϕ is the carrier-phase offset. This phase offset must be estimated at the receiver, and the phase estimate must be used in the demodulation of the received signal. Hence, the received signal must be correlated with the two orthogonal basis functions

$$\psi_1(t) = \sqrt{\frac{1}{\mathcal{E}_s}} g_T(t) \cos(2\pi f_c t + \hat{\phi})$$

and   (10.2.19)

$$\psi_2(t) = -\sqrt{\frac{1}{\mathcal{E}_s}} g_T(t) \sin(2\pi f_c t + \hat{\phi}),$$

where $\hat{\phi}$ is the estimate of the carrier phase obtained by a PLL, as shown in Figure 10.11. When $g_T(t)$ is a rectangular pulse, the signal pulse generator may be eliminated.

When the digital information is transmitted via the M-phase modulation of a carrier, a PLL may be used to estimate the carrier-phase offset. For $M = 2$, the squaring PLL and the Costas loop described in Section 6.4 are directly applicable.

For $M > 2$, the received signal may first be raised to the Mth power, as shown in Figure 10.12. Thus, if the received signal $r(t)$ has the form

$$r(t) = s_m(t) + n(t)$$

$$= g_T(t) \cos\left(2\pi f_c t + \phi + \frac{2\pi m}{M}\right) + n(t),$$ (10.2.20)

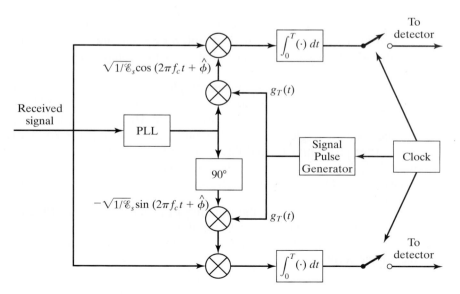

Figure 10.11   Demodulator for an M-ary phase modulated signal.

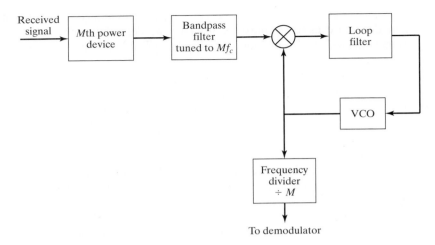

Figure 10.12   Carrier-phase estimation for an M-ary PSK signal.

and we pass $r(t)$ through an Mth power device, the output signal will contain harmonics of the carrier f_c. The harmonic that we wish to select is $\cos(2\pi M f_c t + M\phi)$ for driving the PLL. We note that

$$M\left(\frac{2\pi m}{M}\right) = 2\pi m = 0 (\mathrm{mod}\, 2\pi), \quad m = 1, 2, \ldots, M. \tag{10.2.21}$$

Thus, the information is removed from the Mth harmonic. The bandpass filter tuned to the frequency $M f_c$ produces the desired frequency component $\cos(2\pi M f_c t + M\phi)$, which is driving the PLL. The VCO output is $\sin(2\pi M f_c t + M\hat{\phi})$, so this output is divided in frequency by M to yield $\sin(2\pi f_c t + \hat{\phi})$ and is phase-shifted by $\pi/2$ to yield $\cos(2\pi f_c t + \hat{\phi})$. The two quadrature-carrier components are then passed to the demodulator.

We should note that the quadrature-phase carrier components generated as previously described, contain phase ambiguities of multiples of $2\pi/M$ that result from multiplying the carrier phase ϕ by M. Because $M\phi (\mathrm{mod}\, 2\pi)$ is less than 2π, dividing the resulting angle by M yields a phase estimate of $|\hat{\phi}| < 2\pi/M$. However, the true carrier phase may exceed this estimate by multiples of $2\pi/M$, i.e., by $2\pi k/M$, for $k = 1, 2, \ldots, M - 1$. Such phase ambiguities can be overcome by differentially encoding the data at the transmitter and differentially decoding at the detector, as described next.

Just as in the case of the squaring PLL, the Mth power PLL operates in the presence of noise that has been enhanced by the Mth power-law device. The variance of the phase error in the PLL resulting from the additive noise may be expressed in the simple form

$$\sigma_{\hat{\phi}}^2 = \frac{1}{S_{ML}\rho_L}, \tag{10.2.22}$$

where ρ_L is the loop SNR and S_{ML} is the *M-phase power loss*. S_{ML} has been evaluated by Lindsey and Simon (1973) for $M = 4$ and $M = 8$.

   Another method for extracting a carrier-phase estimate $\hat{\phi}$ from the received signal for M-ary phase modulation is the decision-feedback PLL (DFPLL), which is shown in Figure 10.13. The received signal is demodulated by using the two quadrature phase-locked carriers given by Equation (10.2.19) to yield $\boldsymbol{y} = (y_1, y_2)$ at the sampling instants. The phase estimate $\hat{\theta} = \tan^{-1} y_2/y_1$ is computed at the detector and quantized to the nearest of the M possible transmitted phases, which we denote as $\hat{\theta}_m$. The two outputs of the quadrature multipliers are delayed by one

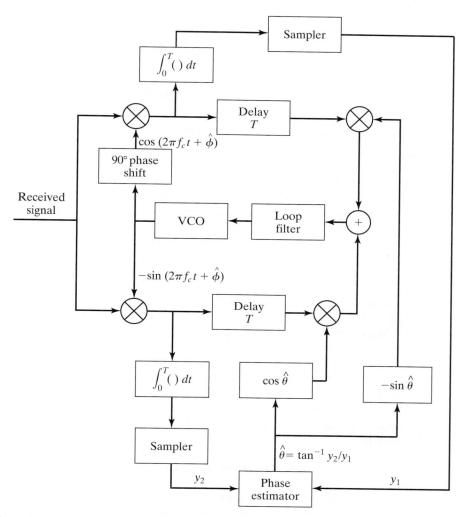

Figure 10.13   Carrier recovery for an *M*-ary PSK using a DFPLL.

symbol interval T and multiplied by $\cos\hat{\theta}_m$ and $-\sin\hat{\theta}_m$. Thus, we obtain

$$-r(t)\cos(2\pi f_c t + \hat{\phi})\sin\theta_m = -\frac{1}{2}[g_T(t)\cos\theta_m + n_c(t)]\sin\theta_m\cos(\phi - \hat{\phi})$$

$$+ \frac{1}{2}[g_T(t)\sin\theta_m + n_s(t)]\sin\theta_m\sin(\phi - \hat{\phi})$$

$$+ \text{double-frequency terms,}$$

where we assume that $\hat{\theta}_m = \theta_m$, and

$$-r(t)\sin(2\pi f_c t + \hat{\phi})\cos\theta_m = \frac{1}{2}[g_T(t)\cos\theta_m + n_c(t)]\cos\theta_m\sin(\phi - \hat{\phi})$$

$$+ \frac{1}{2}[g_T(t)\sin\theta_m + n_s(t)]\cos\theta_m\cos(\phi - \hat{\phi})$$

$$+ \text{double-frequency terms.}$$

These two signals are added together to generate the error signal

$$e(t) = \tfrac{1}{2}g_T(t)\sin(\phi - \hat{\phi}) - \frac{1}{2}n_c(t)\sin(\phi - \hat{\phi} - \theta_m)$$

$$- \frac{1}{2}n_c(t)\cos(\phi - \hat{\phi} - \theta_m) \qquad (10.2.23)$$

$$+ \text{double frequency terms.}$$

This error signal is the input to the loop filter that provides the control signal for the VCO.

We observe that the two quadrature noise components in Equation (10.2.23) appear as additive terms and no term involves a product of two noise components, as in the output of the Mth power-law device. Consequently, there is no power loss resulting from nonlinear operations on the received signal in the DFPLL. The M-phase decision-feedback tracking loop also has phase ambiguities of $2\pi k/M$, necessitating the need for differentially encoding the information sequence prior to transmission and differentially decoding the received sequence at the detector to recover the information.

10.2.2 Differential-Phase Modulation and Demodulation

The performance of ideal, coherent phase modulation and demodulation is closely attained in communication systems that transmit a carrier signal along with the information signal. The carrier-signal component, usually referred to as a pilot signal, may be filtered from the received signal and used to perform phase-coherent demodulation. However, when no separate carrier signal is transmitted, the receiver must estimate the carrier phase from the received signal. As indicated in Section 10.2.1, the phase at the output of a PLL has ambiguities of multiples of $2\pi/M$, necessitating the need to differentially encode the data prior to modulation. This differential encoding allows us to decode the received data at the detector in the presence of the phase ambiguities.

In differential encoding, the information is conveyed by phase shifts between any two successive signal intervals. For example, in binary-phase modulation, the information bit 1 may be transmitted by shifting the phase of the carrier by $180°$ relative to the previous carrier phase, while the information bit 0 is transmitted by a zero-phase shift relative to the phase in the preceding signaling interval. In four-phase modulation, the relative phase shifts between successive intervals are $0°, 90°, 180°$, and $270°$, corresponding to the information bits $00, 01, 11$, and 10, respectively. The generalization of differential encoding for $M > 4$ is straightforward. The phase-modulated signals resulting from this encoding process are called *differentially encoded*. The encoding is performed by a relatively simple logic circuit preceding the modulator.

Demodulation and detection of the differentially encoded phase-modulated signal may be performed as described in the preceding section using the output of a PLL to perform the demodulation. The received signal phase $\Theta = \tan^{-1} y_2/y_1$ at the detector is mapped into one of the M possible transmitted signal phases $\{\theta\}$ that is closest to Θ. Following the detector, there is a relatively simple phase comparator that compares the phases of the detected signal over two consecutive intervals to extract the transmitted information. Thus, phase ambiguities of $2\pi/M$ are rendered irrelevant.

A differentially encoded phase-modulated signal also allows a type of demodulation that does not require the estimation of the carrier phase. Instead, the phase of the received signal in any given signaling interval is compared to the phase of the received signal from the preceding signaling interval. To elaborate, suppose that we demodulate the differentially encoded signal by multiplying $r(t)$ by $\cos 2\pi f_c t$ and $\sin 2\pi f_c t$ and integrating the two products over the interval T. At the kth signaling interval, the demodulator output is given by the complex number

$$y_k = \sqrt{\mathcal{E}_s}\, e^{j(\theta_k - \phi)} + n_k, \tag{10.2.24}$$

where θ_k is the phase angle of the transmitted signal at the kth signaling interval, ϕ is the carrier phase, and $n_k = n_{kc} + jn_{ks}$ is the noise vector. Similarly, the received signal vector at the output of the demodulator in the preceding signaling interval is the complex number

$$y_{k-1} = \sqrt{\mathcal{E}_s}\, e^{j(\theta_{k-1} - \phi)} + n_{k-1}. \tag{10.2.25}$$

The decision variable for the phase detector is the phase difference between these two complex numbers. Equivalently, we can project y_k onto y_{k-1} and use the phase

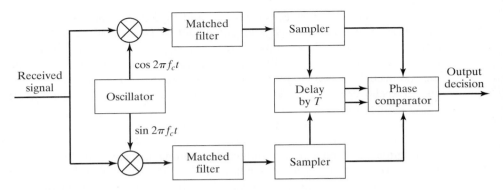

Figure 10.14   Block diagram of a DPSK demodulator.

of the resulting complex number, i.e.,

$$y_k y_{k-1}^* = \mathcal{E}_s\, e^{j(\theta_k - \theta_{k-1})} + \sqrt{\mathcal{E}_s}\, e^{j(\theta_k - \phi)} n_{k-1}^* + \sqrt{\mathcal{E}_s} e^{-j(\theta_{k-1} - \phi)}\, n_k + n_k n_{k-1}^*,$$

$$(10.2.26)$$

which, in the absence of noise, yields the phase difference $\theta_k - \theta_{k-1}$. Thus, the mean value of $y_k y_{k-1}^*$ is independent of the carrier phase. Differentially encoded PSK signaling that is demodulated and detected as just described is called *differential PSK* (DPSK).

The demodulation and detection of DPSK using matched filters is illustrated in Figure 10.14. If the pulse $g_T(t)$ is rectangular, the matched filters may be replaced by integrators, which are also called *integrate-and-dump filters*.

10.2.3 Probability of Error for Phase-Coherent PSK Modulation

In this section, we shall evaluate the probability of error for M-ary phase modulation in AWGN with the optimum demodulator and detector. The optimum detector based on the phase metric given by Equation (10.2.17) will be used in the computation. We assume that a perfect estimate of the received carrier phase is available. Consequently, the performance that we derive is for ideal phase-coherent modulation.

Consider the case in which the transmitted signal phase is $\theta = 0$, corresponding to the signal $u_0(t)$. Hence, the transmitted signal vector is

$$s_0 = \left(\sqrt{\mathcal{E}_s}, 0 \right) \tag{10.2.27}$$

and the received signal vector has the components

$$\begin{aligned} y_1 &= \sqrt{\mathcal{E}_s} + n_c; \\ y_2 &= n_s. \end{aligned} \tag{10.2.28}$$

Because n_c and n_s are jointly Gaussian random variables, it follows that y_1 and y_2 are jointly Gaussian random variables with $E[y_1] = \sqrt{\mathcal{E}_s}$, $E[y_2] = 0$, and $\sigma_{y_1}^2 = \sigma_{y_2}^2 = N_0/2 = \sigma_y^2$. Consequently,

$$f_{\mathbf{y}}(y_1, y_2) = \frac{1}{2\pi\sigma_y^2} e^{-\left[\left(y_1 - \sqrt{\mathcal{E}_s}\right)^2 + y_2^2\right]/2\sigma_y^2}. \tag{10.2.29}$$

The detector metric is the phase $\Theta = \tan^{-1} y_2/y_1$. The PDF of Θ is obtained by a change in variables from (y_1, y_2) to

$$V = \sqrt{y_1^2 + y_2^2}$$

and

$$\Theta = \tan^{-1} \frac{y_2}{y_1}. \tag{10.2.30}$$

This change in variables yields the joint PDF

$$f_{V,\Theta}(v, \theta) = \frac{v}{2\pi\sigma_y^2} e^{-(v^2 + \mathcal{E}_s - 2\sqrt{\mathcal{E}_s}\,v\cos\theta)/2\sigma_y^2}. \tag{10.2.31}$$

Integration of $f_{V,\Theta}(v, \theta)$ over the range of v yields $f_{\Theta}(\theta)$, i.e.,

$$f_{\Theta}(\theta) = \int_0^{\infty} f_{V,\Theta}(v, \theta)\, dv \tag{10.2.32}$$

$$= \frac{1}{2\pi} e^{-\rho_s \sin^2\theta} \int_0^{\infty} v e^{-(v - \sqrt{2\rho_s}\cos\theta)^2/2}\, dv, \tag{10.2.33}$$

where, for convenience, we have defined the symbol SNR as $\rho_s = \mathcal{E}_s/N_0$. Figure 10.15 illustrates $f_{\Theta}(\theta)$ for several values of the SNR parameter ρ_s when the transmitted phase is zero. Note that $f_{\Theta}(\theta)$ becomes narrower and more peaked about $\theta = 0$ as the SNR ρ_s increases.

When $u_0(t)$ is transmitted, a decision error is made if the noise causes the phase to fall outside the range $[-\pi/M, \pi/M]$. Hence, the probability of a symbol error is

$$P_M = 1 - \int_{-\pi/M}^{\pi/M} f_{\Theta}(\theta)\, d\theta. \tag{10.2.34}$$

In general, the integral of $f_{\Theta}(\theta)$ does not reduce to a simple form and must be evaluated numerically, except for $M = 2$ and $M = 4$.

For binary-phase modulation, the two signals $u_0(t)$ and $u_1(t)$, are antipodal; hence, the error probability is

$$P_2 = Q\left(\sqrt{\frac{2\mathcal{E}_b}{N_0}}\right). \tag{10.2.35}$$

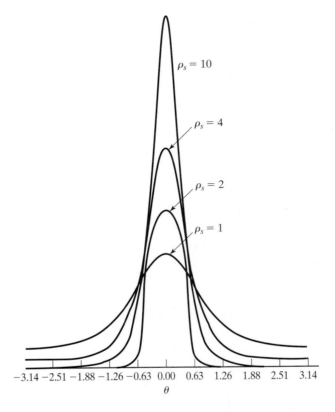

$\rho_s = 10$

$\rho_s = 4$

$\rho_s = 2$

$\rho_s = 1$

−3.14 −2.51 −1.88 −1.26 −0.63 0.00   0.63   1.26   1.88   2.51   3.14

θ

Figure 10.15   Probability density function $f_\Theta(\theta)$ for $\rho_s = 1, 2, 4, 10$.

When $M = 4$, we basically have two binary-phase modulation signals in phase quadrature. With a perfect estimate of the carrier phase, there is no crosstalk or interference between the signals on the two quadrature carriers; hence, the bit error probability is identical to that in Equation (10.2.35). On the other hand, the symbol error probability for $M = 4$ is determined by noting that

$$P_c = (1 - P_2)^2 = \left[1 - Q\left(\sqrt{\frac{2\mathscr{E}_b}{N_0}} \right) \right]^2, \tag{10.2.36}$$

where P_c is the probability of a correct decision for the 2-bit symbol. The result in Equation (10.2.36) follows from the statistical independence of the noise on the quadrature carriers. Therefore, the symbol error probability for $M = 4$ is

$$P_4 = 1 - P_c \tag{10.2.37}$$

$$P_4 = 2Q\left(\sqrt{\frac{2\mathscr{E}_b}{N_0}} \right) \left[1 - \frac{1}{2} Q\left(\sqrt{\frac{2\mathscr{E}_b}{N_0}} \right) \right]. \tag{10.2.38}$$

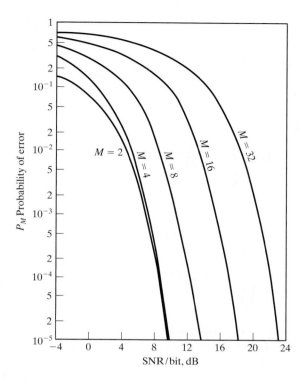

Figure 10.16 Probability of a symbol error for PSK signals.

If the signal-to-noise ratio is not too low, then $\frac{1}{2}Q\left(\sqrt{2\mathscr{E}_b/N_0}\right) \ll 1$, and we have

$$P_4 \approx 2Q\left(\sqrt{\frac{2\mathscr{E}_b}{N_0}}\right). \tag{10.2.39}$$

For $M > 4$, the symbol error probability P_M is obtained by numerically integrating Equation (10.2.34). Figure 10.16 illustrates this error probability as a function of the SNR/bit for $M = 2, 4, 8, 16,$ and 32. The graphs clearly illustrate the penalty in the SNR/bit as M increases beyond $M = 4$. For example, at $P_M = 10^{-5}$, the difference between $M = 4$ and $M = 8$ is approximately 4 dB, and the difference between $M = 8$ and $M = 16$ is approximately 5 dB. For large values of M, doubling the number of phases requires an additional 6 dB/bit to achieve the same performance.

To approximate the error probability for large values of M and large SNR, we must first approximate $f_\Theta(\theta)$. For $\mathscr{E}_s/N_0 \gg 1$ and $|\Theta| \leq \pi/2$, $f_\Theta(\theta)$ is approximated as

$$f_\Theta(\theta) \approx \sqrt{\frac{\rho_s}{\pi}} \cos\theta\, e^{-\rho_s \sin^2\theta}. \tag{10.2.40}$$

By substituting for $f_\Theta(\theta)$ in Equation (10.2.34) and performing the change in variable from θ to $u = \sqrt{\rho_s} \sin\theta$, we find that

$$P_M \approx 1 - \int_{-\pi/M}^{\pi/M} \sqrt{\frac{\rho_s}{\pi}} \cos\theta \, e^{-\rho_s \sin^2\theta} \, d\theta$$

$$\approx \frac{2}{\sqrt{2\pi}} \int_{\sqrt{2\rho_s} \sin\pi/M}^{\infty} e^{-u^2/2} \, du$$

$$= 2Q\left(\sqrt{2\rho_s} \sin\frac{\pi}{M}\right)$$

$$= 2Q\left(\sqrt{2k\rho_b} \sin\frac{\pi}{M}\right) \tag{10.2.41}$$

$$= 2Q\left(\sqrt{2k \sin^2\left(\frac{\pi}{M}\right) \frac{\mathcal{E}_b}{N_0}}\right) \tag{10.2.42}$$

$$\approx 2Q\left(\sqrt{\frac{2\pi^2 \log_2 M}{M^2} \frac{\mathcal{E}_b}{N_0}}\right), \tag{10.2.43}$$

where $k = \log_2 M$ and $\rho_s = k\rho_b$. Note that the last approximation holds for large M, where $\sin\frac{\pi}{M} \approx \frac{\pi}{M}$. We note that the approximations to the error probability given in Equations (10.2.41) and (10.2.44) are good for all values of M. For example, when $M = 2$ and $M = 4$, we have $P_2 = P_4 = 2Q\left(\sqrt{2\rho_b}\right)$, which compares favorably (a factor of 2 difference) with the exact error probability given by Equation (10.2.35). Also, from Equation (10.2.43), it is clear that due to the presence of M^2 in the denominator, doubling M deteriorates the performance by a factor of 4 (6 dB). This is similar to the performance of baseband and carrier-modulated PAM signals.

Example 10.2.2

By using the binary event error probability given in Equation (8.5.2) and the Euclidean distance between two adjacent signal points in a PSK signal constellation, which is given by Equation (10.2.11), determine an approximation to the symbol error probability. Compare the result with that given by Equation (10.2.41).

Solution   First, we determine the error probability in selecting a particular signal point other than the transmitted signal point when the signal is corrupted by AWGN. From Equation (8.5.2), we have

$$P_2 = Q\left(\sqrt{\frac{d_{12}^2}{2N_0}}\right),$$

where d_{12}^2 is the square of the Euclidean distance between the transmitted signal point and the particular erroneously selected signal point. In the case of a PSK signal constellation, the error probability is dominated by the erroneous selection of either one of the two signal points adjacent to the transmitted signal point. Consequently, an approximation to the symbol error probability is

$$P_M \approx 2Q \left(\sqrt{\frac{d_{\min}^2}{2N_0}} \right),$$

where $d_{\min}$ is given by Equation (10.2.11). By substituting for $d_{\min}$ in the approximation to P_M, we obtain

$$P_M \approx 2Q \left(\sqrt{2\rho_s} \sin \frac{\pi}{M} \right),$$

which is identical to the expression given in Equation (10.2.41).   ■

The equivalent bit error probability for M-ary phase modulation is rather tedious to derive due to its dependence on the mapping of k-bit symbols into the corresponding signal phases. When the Gray code is used in the mapping, two k-bit symbols corresponding to adjacent signal phases differ in only a single bit. Because the most probable errors due to noise result in the erroneous selection of an adjacent phase to the true phase, most k-bit symbol errors contain only a single bit error. Hence, the equivalent bit-error probability for M-ary phase modulation is well approximated as

$$P_b \approx \frac{1}{k} P_M. \tag{10.2.44}$$

The performance analysis just given applies to phase-coherent demodulation with conventional (absolute) phase mapping of the information into signal phases. As indicated in Section 10.2.2, when phase ambiguities result in the estimation of the carrier phase, the information symbols are differentially encoded at the transmitter and differentially decoded at the receiver. Coherent demodulation of differentially encoded phase-modulated signals results in a higher probability of error than the error probability derived for absolute-phase encoding. With differentially encoded signals, an error in the detected phase (due to noise) will frequently result in decoding errors over two consecutive signaling intervals. This is especially the case for error probabilities below 10^{-1}. Therefore, the probability of error for differentially encoded M-ary phase-modulated signals is approximately twice the probability of error for M-ary phase modulation with absolute-phase encoding. However, a factor-of-2 increase in the error probability translates into a relatively small loss in SNR, as depicted in Figure 10.16.

10.2.4 Probability of Error for DPSK

Now consider the evaluation of the error probability performance of a DPSK demodulator and detector. The derivation of the exact value of the probability of error for M-ary DPSK is extremely difficult, except for $M = 2$. The major difficulty is encountered in the determination of the PDF for the phase of the random variable

$y_k y_{k-1}^*$ given by Equation (10.2.26). However, an approximation to the performance of DPSK is easily obtained, as we now demonstrate.

Without loss of generality, suppose the phase difference $\theta_k - \theta_{k-1} = 0$. Furthermore, the exponential factors $e^{-j(\theta_{k-1}-\phi)}$ and $e^{j(\theta_k-\phi)}$ in Equation (10.2.26) can be absorbed into the Gaussian noise components n_{k-1} and n_k, as shown in Problem 5.28, without changing their statistical properties. Therefore, $y_k y_{k-1}^*$ in Equation (10.2.26) can be expressed as

$$y_k y_{k-1}^* = \mathcal{E}_s + \sqrt{\mathcal{E}_s}(n_k + n_{k-1}^*) + n_k n_{k-1}^*. \tag{10.2.45}$$

The complication in determining the PDF of the phase is the term $n_k n_{k-1}^*$. However, at SNR's of practical interest, the term $n_k n_{k-1}^*$ is small relative to the dominant noise term $\sqrt{\mathcal{E}_s}(n_k + n_{k-1}^*)$. If we neglect the term $n_k n_{k-1}^*$ and we also normalize $y_k y_{k-1}^*$ by dividing through by $\sqrt{\mathcal{E}_s}$, the new set of decision metrics becomes

$$\begin{aligned} y_1 &= \sqrt{\mathcal{E}_s} + \mathrm{Re}(n_k + n_{k-1}^*) \\ y_2 &= \mathrm{Im}(n_k + n_{k-1}^*). \end{aligned} \tag{10.2.46}$$

The variables y_1 and y_2 are uncorrelated Gaussian random variables with identical variances $\sigma_n^2 = N_0$. The phase is

$$\Theta = \tan^{-1} \frac{y_2}{y_1}. \tag{10.2.47}$$

At this stage, we have a problem that is identical to the one we previously solved for phase-coherent demodulation and detection. The only cdifference is that the noise variance is now twice as large as in the case of PSK. Thus, we conclude that the performance of DPSK is 3 dB worse than that for PSK. This result is relatively good for $M \geq 4$, but it is pessimistic for $M = 2$ in the sense that the loss in binary DPSK, relative to binary PSK, is less than 3 dB at large SNR.

In binary DPSK, the two possible transmitted phase differences are zero and π radians. As a consequence, only the real part of $y_k y_{k-1}^*$ is needed for recovering the information. We express the real part as

$$\mathrm{Re}(y_k y_{k-1}^*) = \frac{1}{2}(y_k y_{k-1}^* + y_k^* y_{k-1}). \tag{10.2.48}$$

Because the phase difference between the two successive signaling intervals is zero, an error is made if $\mathrm{Re}(y_k y_{k-1}^*)$ is less than zero. A rigorous analysis of the error probability of DPSK based on computing the probability of $y_k y_{k-1}^* + y_k^* y_{k-1} < 0$ results in the following expression:

$$P_2 = \frac{1}{2}e^{-\rho_b}. \tag{10.2.49}$$

Here, $\rho_b = \mathcal{E}_b/N_0$ is the SNR/bit.

The graph of Equation (10.2.49) is shown in Figure 10.17. Also shown in this figure is the probability of error for binary PSK. We observe that, at error probabilities

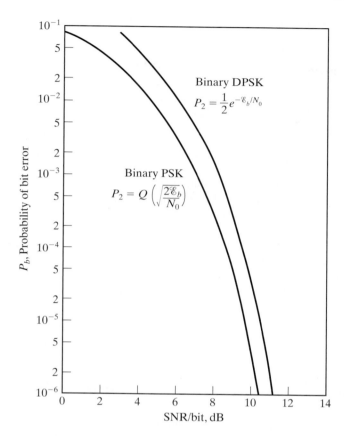

Figure 10.17 Probability of error for binary PSK and DPSK.

below 10^{-4}, the difference in SNR between binary PSK and binary DPSK is less than 1 dB. Since DPSK does not require the estimation of the carrier phase, the need for a phased-locked loop is eliminated, and the implementation of the demodulator is simplified. Given the relatively small difference in performance between binary DPSK and binary PSK, binary DPSK is often preferable in practice.

10.3 QUADRATURE AMPLITUDE-MODULATED DIGITAL SIGNALS

In our discussion of carrier-phase modulation, we observed that the bandpass signal waveforms may be represented as given by Equation (10.2.7), in which the signal waveforms are viewed as two orthogonal carrier signals, $\cos 2\pi f_c t$ and $\sin 2\pi f_c t$, which are amplitude modulated by the information bits. However, the carrier-phase modulation signal waveforms are constrained to have equal energy $\mathcal{E}_s$, which implies that the signal points in the geometric representation of the signal waveforms lie on a circle of radius $\mathcal{E}_s$. If we remove the constant energy restriction, we can construct two-dimensional signal waveforms whose vector representation is not constrained to fall on a circle.

The easiest way to construct such signals is to impress separate information bits on each of the quadrature carriers, $\cos 2\pi f_c t$ and $\sin 2\pi f_c t$. This type of digital modulation is called quadrature amplitude modulation (QAM). We may view this method of information transmission as a form of quadrature-carrier multiplexing, previously described in Section 3.4.2.

The transmitted signal waveforms have the form

$$u_m(t) = A_{mc} g_T(t) \cos 2\pi f_c t + A_{ms} g_T(t) \sin 2\pi f_c t, \quad m = 1, 2, \dots, M, \quad (10.3.1)$$

where $\{A_{mc}\}$ and $\{A_{ms}\}$ are the sets of amplitude levels that are obtained by mapping k-bit sequences into signal amplitudes. For example, Figure 10.18 illustrates a 16-QAM signal constellation that is obtained by amplitude modulating each quadrature carrier by $M = 4$ PAM. In general, rectangular signal constellations result when two quadrature carriers are each modulated by PAM. Figure 10.19 illustrates the functional block diagram of a modulator for QAM implementation.

More generally, QAM may be viewed as a form of combined digital-amplitude and digital-phase modulation. Thus, the transmitted QAM signal waveforms may be expressed as

$$u_{mn}(t) = A_m g_T(t) \cos(2\pi f_c t + \theta_n), \quad \begin{aligned} m &= 1, 2, \dots, M_1 \\ n &= 1, 2, \dots, M_2. \end{aligned} \quad (10.3.2)$$

If $M_1 = 2^{k_1}$ and $M_2 = 2^{k_2}$, the combined amplitude- and phase-modulation method results in the simultaneous transmission of $k_1 + k_2 = \log_2 M_1 M_2$ binary digits occurring at a symbol rate $R_b/(k_1 + k_2)$.

It is clear that the geometric signal representation of the signals given by Equations (10.3.1) and (10.3.2) is in terms of two-dimensional signal vectors of the form

$$s_m = \left(\sqrt{\mathscr{E}_s} A_{mc}, \sqrt{\mathscr{E}_s} A_{ms} \right), \quad m = 1, 2, \dots, M, \quad (10.3.3)$$

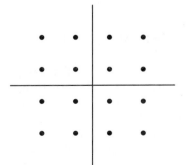

Figure 10.18 $M = 16$-QAM signal constellation.

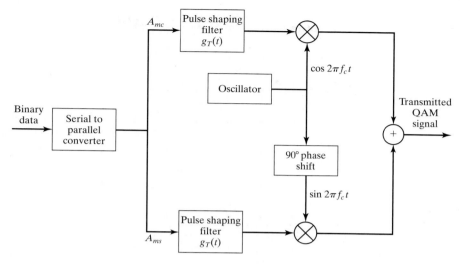

Figure 10.19   Functional block diagram of a modulator for QAM.

and the orthogonal basis functions are

$$\psi_1(t) = \sqrt{\frac{1}{\mathscr{E}_s}} g_T(t) \cos 2\pi f_c t$$

$$\psi_2(t) = \sqrt{\frac{1}{\mathscr{E}_s}} g_T(t) \sin 2\pi f_c t.$$

It should be noted that $M = 4$ rectangular QAM and $M = 4$ PSK are identical signal constellations.

Examples of signal-space constellations for QAM are shown in Figure 10.20.

The average transmitted energy for those signal constellations is simply the sum of the average energies in each of the quadrature carriers. For the signal constellations, as shown in Figure 10.20, the average energy per symbol is given as

$$\mathscr{E}_{av} = \frac{1}{M} \sum_{i=1}^{M} \| s_i \|^2. \tag{10.3.4}$$

The distance between any pair of signals points is

$$d_{mn} = \sqrt{\| s_m - s_n \|^2}. \tag{10.3.5}$$

Example 10.3.1

Consider the $M = 8$ rectangular signal constellation and the $M = 8$ combined PAM–PSK signal constellation, as shown in Figure 10.21(a) and (b). Assuming that adjacent signal points in each of the two signal constellations are separated by a distance

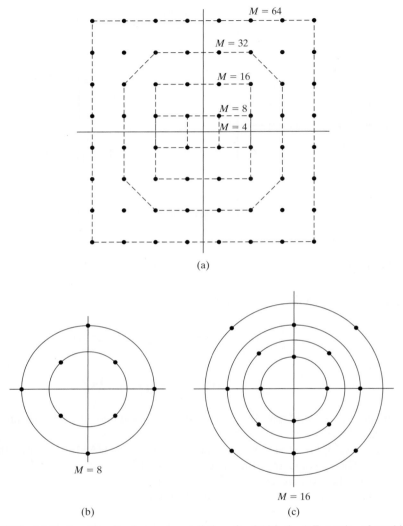

Figure 10.20   (a) Rectangular signal-space constellations for QAM. (b, c) Examples of combined PAM–PSK signal-space constellations.

$d = 2$, determine the average energy per symbol for each constellation. Which signal constellation is preferable? Why?

Solution   For the $M = 8$ signal constellation shown in Figure 10.21(a), the coordinates of the signal points are $(\pm 1, \pm 1)$ and $(\pm 3, \pm 1)$. Therefore, the average energy per symbol is

$$\mathscr{E}_{av} = \frac{1}{8}[4(2) + 4(10)] = 6.$$

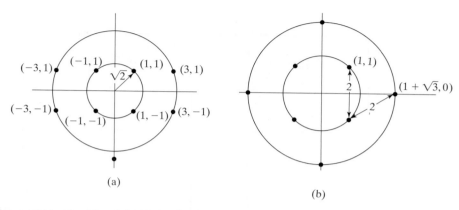

Figure 10.21   Two $M = 8$ QAM signal point constellations.

For the $M = 8$ signal constellation shown in Figure 10.21(b), the coordinates of the signal points are $(\pm 1, \pm 1)$ and $(1 + \sqrt{3}, 0)$, $(-1 - \sqrt{3}, 0)$, $(0, 1 + \sqrt{3})$, and $(0, -1 - \sqrt{3})$. Therefore, the average energy per symbol is

$$\mathscr{E}_{av} = \frac{1}{8}[4(2) + 4(7.464)] = 4.73.$$

This signal constellation is preferable because it requires 1 dB less energy per symbol to achieve the same performance as that of the rectangular signal constellation.

∎

10.3.1 Demodulation and Detection of Quadrature Amplitude-Modulated Signals

Let us assume that a carrier-phase offset is introduced in the transmission of the signal through the channel. In addition, the received signal is corrupted by additive Gaussian noise. Hence, $r(t)$ may be expressed as

$$r(t) = A_{mc}g_T(t)\cos(2\pi f_c t + \phi) + A_{ms}g_T(t)\sin(2\pi f_c t + \phi) + n(t). \qquad (10.3.6)$$

Suppose that an estimate $\hat{\phi}$ of the carrier phase is available at the demodulator. Then, the received signal may be correlated with the two basis functions

$$\psi_1(t) = \sqrt{\frac{1}{\mathscr{E}_s}}g_T(t)\cos(2\pi f_c t + \hat{\phi})$$

and

$$\psi_2(t) = \sqrt{\frac{1}{\mathscr{E}_s}}g_T(t)\sin(2\pi f_c t + \hat{\phi}), \qquad (10.3.7)$$

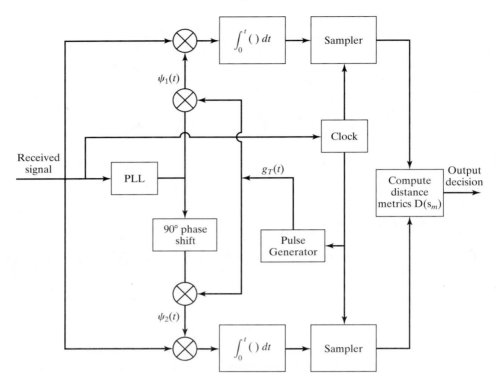

Figure 10.22   Demodulation and detection of QAM signals.

as illustrated in Figure 10.22, and the outputs of the correlators are sampled and passed to the detector.

The input to the detector consists of the two sampled components:

$$
\begin{aligned}
y_1 &= A_{mc}\sqrt{\mathcal{E}_s}\cos(\phi - \hat{\phi}) - A_{ms}\sqrt{\mathcal{E}_s}\sin(\phi - \hat{\phi}) + n_c\cos\hat{\phi} + n_s\sin\hat{\phi} \\
y_2 &= A_{mc}\sqrt{\mathcal{E}_s}\sin(\phi - \hat{\phi}) + A_{ms}\sqrt{\mathcal{E}_s}\cos(\phi - \hat{\phi}) + n_s\cos\hat{\phi} - n_c\sin\hat{\phi}.
\end{aligned}
\tag{10.3.8}
$$

We observe that the effect of an imperfect phase estimate is twofold. First, the desired signal components in y_1 and y_2 are reduced in amplitude by the factor $\cos(\phi - \hat{\phi})$. In turn, this reduces the SNR by the factor $\cos^2(\phi - \hat{\phi})$. Second, there is a leakage of the quadrature signal components into the desired signal. This signal leakage, which is scaled by $\sin(\phi - \hat{\phi})$, causes a significant performance degradation unless $\phi - \hat{\phi}$ is very small. This point serves to emphasize the importance of having an accurate carrier-phase estimate in order to demodulate the QAM signal.

The optimum detector computes the distance metrics

$$
D(\mathbf{y}, \mathbf{s}_m) = \parallel \mathbf{y} - \mathbf{s}_m \parallel^2, \quad m = 1, 2, \ldots, M
\tag{10.3.9}
$$

and selects the signal corresponding to the smallest value of $D(\mathbf{y}, \mathbf{s}_m)$. If a correlation metric is used in place of a distance metric, it is important to recognize that correlation metrics must employ bias correction because the QAM signals are not equal energy signals.

Carrier-Phase Estimation.   As we have previously indicated, the demodulation of a QAM signal requires a carrier that is phase-locked to the phase of the received carrier signal. Carrier-phase estimation for QAM can be accomplished in a number of different ways depending on the signal-point constellation and the phase relationships of the various signal points.

For example, consider the eight-point QAM signal constellation shown in Figure 10.20(b). The signal points in this constellation have one of two possible amplitude values and eight possible phases. The phases are spaced 45° apart. This phase symmetry allows us to use a PLL driven by the output of an 8th power-law device that generates carrier components at $8f_c$, where f_c is the carrier frequency. Thus, the method illustrated in Figure 10.12 may be generally used for any QAM signal constellation that contains signal points with phases that are multiples of some phase angle θ, where $L\theta = 360°$ for some integer L.

Another method for extracting a carrier-phase estimate $\hat{\phi}$ from the received M-ary QAM signal is the DFPLL previously described in Section 10.2.1. The basic idea in the DFPLL is to estimate the phase of the QAM signal in each signal interval and remove the phase modulation from the carrier. The DFPLL may be used with any QAM signal, regardless of the phase relationships among the signal points. To be specific, we express the received QAM signal as

$$r(t) = A_m g_T(t) \cos(2\pi f_c t + \theta_n + \phi) + n(t), \qquad (10.3.10)$$

where θ_n is the phase of the signal point and ϕ is the carrier phase. This signal is demodulated by cross correlating $r(t)$ with $\psi_1(t)$ and $\psi_2(t)$, which are given by Equation (10.3.7). The sampled values at the output of the correlators are

$$y_1 = A_m \sqrt{\mathcal{E}_s} \cos(\theta_n + \phi - \hat{\phi}) + n_c \cos(\theta_n + \phi - \hat{\phi}) - n_s \sin(\theta_n + \phi - \hat{\phi})$$
$$y_2 = A_m \sqrt{\mathcal{E}_s} \sin(\theta_n + \phi - \hat{\phi}) + n_c \sin(\theta_n + \phi - \hat{\phi}) - n_s \cos(\theta_n + \phi - \hat{\phi}).$$
$$(10.3.11)$$

Now suppose that the detector, which is based on y_1 and y_2, has made the correct decision on the transmitted signal point. Then we multiply y_1 by $-\sin\theta_n$ and y_2 by $\cos\theta_n$. Thus, we obtain

$$
\begin{aligned}
-y_1 \sin\theta_n &= -A_m\sqrt{\mathcal{E}_s}\cos(\theta_n + \phi - \hat{\phi})\sin\theta_n + \text{ noise component} \\
&= A_m\sqrt{\mathcal{E}_s}[-\sin\theta_n\cos\theta_n\cos(\phi - \hat{\phi}) + \sin^2\theta_n\sin(\phi - \hat{\phi}) \\
&\quad + \text{ noise component} \\
y_2\cos\theta_n &= A_m\sqrt{\mathcal{E}_s}\sin(\theta_n + \phi - \hat{\phi})\cos\theta_n + \text{ noise component} \\
&= A_m\sqrt{\mathcal{E}_s}[\sin\theta_n\cos\theta_n\cos(\phi - \hat{\phi}) + \cos^2\theta_n\sin(\phi - \hat{\phi})] \\
&\quad + \text{ noise component}.
\end{aligned}
\qquad (10.3.12)
$$

Adding these two terms, we obtain an error signal $e(T)$, given as

$$e(T) = y_2 \cos \theta_n - y_1 \sin \theta_n$$
$$= A_m \sqrt{\mathcal{E}_s} \sin(\phi - \hat{\phi}) + \text{noise components.} \tag{10.3.13}$$

This error signal is now passed to the loop filter that drives the VCO. Thus, only the phase θ_n of the QAM signal is used to obtain an estimate of the carrier phase. Consequently, the general block diagram for the DFPLL, which is given in Figure 10.13, also applies to carrier-phase estimation for an M-ary QAM signal.

As in the case of digitally phase-modulated signals, this method for carrier-phase recovery results in phase ambiguities. This problem is solved generally by differential encoding of the data sequence at the input to the modulator.

10.3.2 Probability of Error for QAM

To determine the probability of error for QAM, we must specify the signal-point constellation. We begin with QAM signal sets that have $M = 4$ points. Figure 10.23 illustrates two four-point signal sets. The first is a four-phase modulated signal and the second is a QAM signal with two amplitude levels, labeled $\sqrt{\mathcal{E}_1}$ and $\sqrt{\mathcal{E}_2}$, and four phases. Because the probability of error is dominated by the minimum distance between pairs of signal points, we impose the condition that $d_{\min} = 2\sqrt{\mathcal{E}_s}$ for both signal constellations; we also evaluate the average energy, based on the premise that all signal points are equally probable. For the four-phase signal, we have

$$\mathcal{E}_{\text{av}} = \frac{1}{4}(4)2\mathcal{E}_s = 2\mathcal{E}_s. \tag{10.3.14}$$

For the two-amplitude, four-phase QAM, we place the points on circles of radii $\sqrt{\mathcal{E}_1}$ and $\sqrt{\mathcal{E}_2}$. Thus, with the constraint that $d_{\min} = 2\sqrt{\mathcal{E}}$, the average energy is

$$\mathcal{E}_{\text{av}} = \frac{1}{4}[2(3)\mathcal{E}_s + 2\mathcal{E}_s] = 2\mathcal{E}_s, \tag{10.3.15}$$

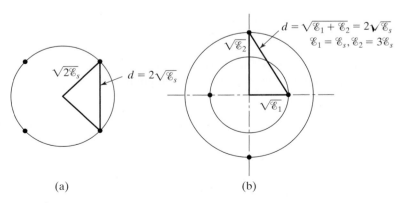

(a)   (b)

Figure 10.23   Two four-point signal constellations.

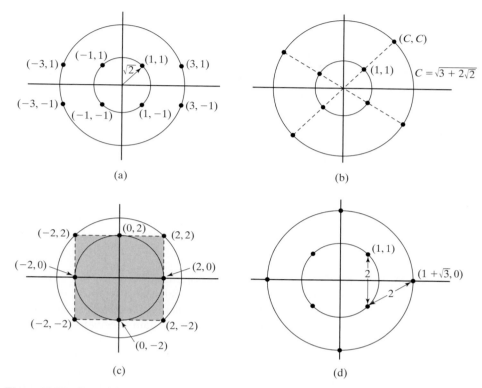

Figure 10.24 Four eight-point QAM signal constellations.

which is the same average energy as the $M = 4$-phase signal constellation. Hence, for all practical purposes, the error-rate performance of the two signal sets is the same. In other words, the two-amplitude QAM signal set has no advantage over $M = 4$-phase modulation.

Next, consider $M = 8$ QAM. In this case, there are many possible signal constellations. The four signal constellations shown in Figure 10.24 consist of two amplitudes and have a minimum distance between signal points of $2\sqrt{\mathscr{E}_s}$. The coordinates (A_{mc}, A_{ms}) for each signal point, normalized by $\sqrt{\mathscr{E}_s}$, are given in the figure. Assuming that the signal points are equally probable, the average transmitted signal energy is

$$\mathscr{E}_{av} = \frac{\mathscr{E}_s}{M} \sum_{m=1}^{M} \left(a_{mc}^2 + a_{ms}^2 \right), \qquad (10.3.16)$$

where (a_{mc}, a_{ms}) are the normalized coordinates of the signal points.

In Figure 10.24, the two signal sets (a) and (c) contain signal points that fall on a rectangular grid and have $\mathscr{E}_{av} = 6\mathscr{E}_s$. The signal set (b) requires an average

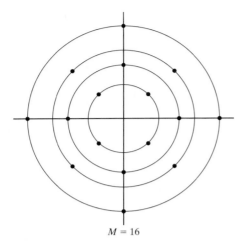

$M = 16$

Figure 10.25   Circular 16-point QAM signal constellation.

transmitted signal $\mathscr{E}_{av} = 6.82\mathscr{E}_s$, and the fourth requires $\mathscr{E}_{av} = 4.73\mathscr{E}_s$. Therefore, to achieve the same probability of error, the signal set (d) requires approximately 1 dB less energy than (a) and (c) and 1.6 dB less energy than (b). This signal constellation is known to be the best eight-point QAM constellation because it requires the least energy for a given minimum distance between signal points.

For $M \geq 16$, there are many more possibilities for selecting the QAM signal points in the two-dimensional space. For example, we may choose a circular multiamplitude constellation for $M = 16$, as shown in Figure 10.25. In this case, the signal points at a given amplitude level are phase-rotated by $\pi/4$ relative to the signal points at adjacent amplitude levels. This 16-QAM constellation is a generalization of the optimum 8-QAM constellation. However, the circular 16-QAM constellation is not the best 16-QAM signal constellation for the AWGN channel.

Rectangular QAM signal constellations have the distinct advantage of being easily generated as two PAM signals impressed on phase-quadrature carriers. In addition, they are easily demodulated as previously described. Although they are not the best M-ary QAM signal constellations for $M \geq 16$, the average transmitted energy required to achieve a given minimum distance is only slightly greater than the average energy required for the best M-ary QAM signal constellation. For these reasons, rectangular M-ary QAM signals are most frequently used in practice.

For rectangular signal constellations in which $M = 2^k$ where k is even, the QAM signal constellation is equivalent to two PAM signals on quadrature carriers, each having $\sqrt{M} = 2^{k/2}$ signal points. Because the signals in the phase-quadrature components are perfectly separated by coherent detection when $\phi = \hat{\phi}$, the probability of error for QAM is easily determined from the probability of error for PAM. Specifically, the probability of a correct decision for the M-ary QAM system is

$$P_c = \left(1 - P_{\sqrt{M}}\right)^2, \qquad (10.3.17)$$

where $P_{\sqrt{M}}$ is the probability of error of a $\sqrt{M}$-ary with one-half the average energy in each quadrature signal of the equivalent QAM system. By appropriately modifying the probability of error for M-ary PAM, we obtain

$$P_{\sqrt{M}} = 2 \left(1 - \frac{1}{\sqrt{M}} \right) Q \left(\sqrt{\frac{3}{M-1} \frac{\mathscr{E}_{av}}{N_0}} \right), \tag{10.3.18}$$

where $\mathscr{E}_{av}/N_0$ is the average SNR/symbol. Therefore, the probability of a symbol error for the M-ary QAM is

$$P_M = 1 - \left(1 - P_{\sqrt{M}} \right)^2. \tag{10.3.19}$$

We note that this result is exact for $M = 2^k$ when k is even. On the other hand, when k is odd, there is no equivalent M-ary PAM system. This is no problem, however, because it is rather easy to determine the error rate for a rectangular signal set. If we employ the optimum detector that bases its decisions on the optimum distance metrics given by Equation (10.3.9), it is relatively straightforward to show that the symbol error probability is tightly upper-bounded as

$$P_M \leq 1 - \left[1 - 2Q \left(\sqrt{\frac{3\mathscr{E}_{av}}{(M-1)N_0}} \right) \right]^2$$

$$\leq 4Q \left(\sqrt{\frac{3k\mathscr{E}_{bav}}{(M-1)N_0}} \right) \tag{10.3.20}$$

for any $k \geq 1$, where $\mathscr{E}_{bav}/N_0$ is the average SNR/bit. The probability of a symbol error is plotted in Figure 10.26 as a function of the average SNR/bit.

It is interesting to compare the performance of QAM with that of phase modulation for any given signal size M, because both types of signals are two-dimensional. Recall that for M-ary phase modulation, the probability of a symbol error is approximated as

$$P_M \approx 2Q \left(\sqrt{2\rho_s} \sin \frac{\pi}{M} \right), \tag{10.3.21}$$

where ρ_s is the SNR/symbol. For M-ary QAM, we may use the expression in Equation (10.3.20). Because the error probability is dominated by the argument of the Q-function, we may simply compare the arguments of Q for the two signal formats. Thus, the ratio of these two arguments is

$$\mathscr{R}_M = \frac{3/(M-1)}{2\sin^2 \pi/M}. \tag{10.3.22}$$

For example, when $M = 4$, we have $\mathscr{R}_M = 1$. Hence, 4-PSK and 4-QAM yield comparable performance for the same SNR/symbol. On the other hand, when $M > 4$, we find that $\mathscr{R}_M > 1$; thus, M-ary QAM yields better performance than M-ary PSK. Table 10.1 illustrates the SNR advantage of QAM over PSK for several values

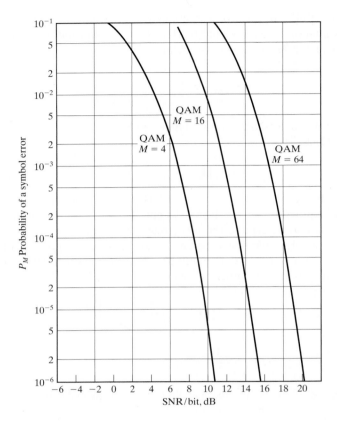

Figure 10.26 Probability of a symbol error for QAM.

TABLE 10.1 SNR
ADVANTAGE OF M-ARY QAM
OVER M-ARY PSK.

M	$10 \log_{10} \mathscr{R}_M$
8	1.65
16	4.20
32	7.02
64	9.95

of M. For example, we observe that 32-QAM has a 7-dB SNR advantage over
32-PSK.

Example 10.3.2

From the expression for the symbol error probability of QAM, which is given by
Equation (10.3.20), determine the increase in the average energy per bit $\mathscr{E}_{bav}$ required
to maintain the same performance (error probability) if the number of bits per symbol
is increased from k to $k + 1$ where k is large.

Solution Since $M = 2^k$, increasing k by one additional bit means that the number of signal points must be increased from M to $2M$. For M large, we observe that increasing M to $2M$ in the argument of the Q-function in Equation (10.3.20) requires that we increase the average energy per bit from $\mathscr{E}_{bav}$ to $2\mathscr{E}_{bav}$; this allows the error probability to remain the same. Therefore, in QAM, increasing the number of bits per symbol by one bit causes an additional 3 dB increase in the transmitted signal energy to maintain the same error probability. This is a factor of two smaller in energy than PSK for large M. Therefore, QAM is more energy efficient than PSK for $M > 4$, as indicated in Table 10.1. ∎

10.4 FREQUENCY-MODULATED DIGITAL SIGNALS

The simplest form of frequency modulation for digital transmission is binary frequency-shift keying (FSK). In binary FSK, we employ two different frequencies, such as f_0 and $f_1 = f_0 + \Delta f$, to transmit a binary information sequence. The choice of frequency separation $\Delta F = f_1 - f_0$ is considered shortly. Thus, the two signal waveforms may be expressed as

$$u_0(t) = \sqrt{\frac{2\mathscr{E}_b}{T_b}} \cos 2\pi f_0 t, \quad 0 \le t \le T_b$$

$$u_1(t) = \sqrt{\frac{2\mathscr{E}_b}{T_b}} \cos 2\pi f_1 t, \quad 0 \le t \le T_b, \tag{10.4.1}$$

where $\mathscr{E}_b$ is the signal energy/bit and T_b is the duration of the bit interval.

More generally, M-ary FSK may be used to transmit a block of $k = \log_2 M$ bits/signal waveform. In this case, the M signal waveforms may be expressed as

$$u_m(t) = \sqrt{\frac{2\mathscr{E}_s}{T}} \cos(2\pi f_c t + 2\pi m \Delta f t), \quad m = 0, 1, \ldots, M-1, \quad 0 \le t \le T, \tag{10.4.2}$$

where $\mathscr{E}_s = k\mathscr{E}_b$ is the energy per symbol, $T = kT_b$ is the symbol interval, and Δf is the frequency separation between successive frequencies, i.e., $\Delta f = f_m - f_{m-1}$, where $f_m = f_c + m \Delta f$.

Note that the M FSK waveforms have equal energy $\mathscr{E}_s$. The frequency separation Δf determines the degree to which we can discriminate among the M possible transmitted signals. As a measure of the similarity (or dissimilarity) between a pair of signal waveforms, we define the correlation coefficients

$$\gamma_{mn} = \frac{1}{\mathscr{E}_s} \int_0^T u_m(t)u_n(t) \, dt. \tag{10.4.3}$$

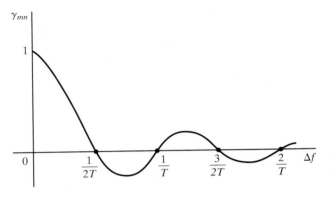

Figure 10.27 Cross-correlation coefficient as a function of frequency separation for FSK signals.

Substituting for $u_m(t)$ and $u_n(t)$ in Equation (10.4.3), we obtain

$$\gamma_{mn} = \frac{1}{\mathcal{E}_s} \int_0^T \frac{2\mathcal{E}_s}{T} \cos(2\pi f_c t + 2\pi m \Delta f t) \cos(2\pi f_c t + 2\pi n \Delta f t)\, dt$$

$$= \frac{1}{T} \int_0^T \cos 2\pi (m-n) \Delta f t\, dt + \frac{1}{T} \int_0^T \cos[4\pi f_c t + 2\pi (m+n) \Delta f t]\, dt$$

$$= \frac{\sin 2\pi (m-n) \Delta f T}{2\pi (m-n) \Delta f T},$$

(10.4.4)

where the second integral vanishes when $f_c \gg 1/T$. A plot of γ_{mn} as a function of the frequency separation Δf is given in Figure 10.27. We observe that the signal waveforms are orthogonal when Δf is a multiple of $1/2T$. Hence, the minimum frequency separation between successive frequencies for orthogonality is $1/2T$. We also note that the minimum value of the correlation coefficient is $\gamma_{mn} = -0.217$, which occurs at the frequency separation $\Delta f = 0.715/T$.

M-ary orthogonal FSK waveforms have a geometric representation as M M-dimensional orthogonal vectors, given as

$$s_0 = \left(\sqrt{\mathcal{E}_s}, 0, 0, \ldots, 0 \right)$$

$$s_1 = \left(0, \sqrt{\mathcal{E}_s}, 0, \ldots, 0 \right)$$

$$\vdots$$

(10.4.5)

$$s_{M-1} = \left(0, 0, \ldots, 0, \sqrt{\mathcal{E}_s} \right),$$

where the basis functions are $\psi_m(t) = \sqrt{2/T} \cos 2\pi (f_c + m \Delta f) t$. The distance between the pairs of signal vectors is $d = \sqrt{2\mathcal{E}_s}$ for all m, n, which is also the minimum distance among the M signals.

10.4.1 Demodulation and Detection of Frequency-Modulated Signals

Assume that the FSK signals are transmitted through an additive white Gaussian noise channel. Furthermore, we assume that each signal is phase-shifted in the transmission through the channel. Consequently, the filtered received signal at the input to the demodulator may be expressed as

$$r(t) = \sqrt{\frac{2\mathcal{E}_s}{T}} \cos(2\pi f_c t + 2\pi m \Delta f t + \phi_m) + n(t), \qquad (10.4.6)$$

where ϕ_m denotes the phase shift of the mth signal and $n(t)$ represents the additive bandpass noise, which may be expressed as

$$n(t) = n_c(t) \cos 2\pi f_c t - n_s(t) \sin 2\pi f_c t. \qquad (10.4.7)$$

The demodulation and detection of the M FSK signals may be accomplished by one of two methods. One approach is to estimate the M carrier-phase shifts $\{\phi_m\}$ and perform *phase-coherent demodulation and detection*. As an alternative method, the carrier phases may be ignored in the demodulation and detection of the FSK signals. The latter method is called *noncoherent demodulation and detection*.

In phase-coherent demodulation, the received signal $r(t)$ is correlated with each of the M possible received signals $\cos(2\pi f_c t + 2\pi m \Delta f t + \hat{\phi}_m)$, $m = 0, 1, \ldots, M-1$, where $\{\hat{\phi}_m\}$ are the carrier phase estimates. A block diagram illustrating this type of demodulation is shown in Figure 10.28. It is interesting to note that when $\hat{\phi}_m \neq \phi_m, m = 0, 1, \ldots, M - 1$ (imperfect phase estimates), the frequency separation required for signal orthogonality at the demodulator is $\Delta f = 1/T$ (see Problem 10.17), which is twice the minimum separation for orthogonality when $\phi = \hat{\phi}$.

The requirement for estimating M carrier phases makes coherent demodulation of FSK signals extremely complex and impractical, especially when the number of signals is large. Therefore, we shall not consider the coherent detection of FSK signals.

Instead, we now consider a method for demodulation and detection that does not require knowledge of the carrier phases. The demodulation may be accomplished as shown in Figure 10.29. In this case, there are two correlators per signal waveform, or a total of $2M$ correlators, in general. The received signal is correlated with the basis functions (quadrature carriers)

$$\sqrt{\frac{2}{T}} \cos(2\pi f_c t + 2\pi m \Delta f t)$$

and

$$\sqrt{\frac{2}{T}} \sin(2\pi f_c t + 2\pi m \Delta f t) \text{ for } m = 0, 1, \ldots, M - 1.$$

The $2M$ outputs of the correlators are sampled at the end of the signal interval and the $2M$ samples are passed to the detector. Thus, if the mth signal is transmitted,

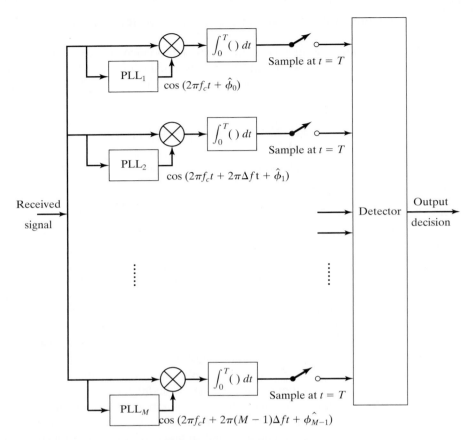

Figure 10.28   Phase coherent demodulation of *M*-ary FSK signals.

the $2M$ samples at the detector may be expressed as

$$y_{kc} = \sqrt{\mathscr{E}_s} \left[\frac{\sin 2\pi (k-m)\Delta f T}{2\pi (k-m)\Delta f T} \cos \phi_m - \frac{\cos 2\pi (k-m)\Delta f T - 1}{2\pi (k-m)\Delta f T} \sin \phi_m \right] + n_{kc}$$

$$y_{ks} = \sqrt{\mathscr{E}_s} \left[\frac{\cos 2\pi (k-m)\Delta f T - 1}{2\pi (k-m)\Delta f T} \cos \phi_m + \frac{\sin 2\pi (k-m)\Delta f T}{2\pi (k-m)\Delta f T} \sin \phi_m \right] + n_{ks},$$

$$(10.4.8)$$

where n_{kc} and n_{ks} denote the Gaussian noise components in the sampled outputs. We observe that when $k = m$, the sampled values to the detector are

$$y_{mc} = \sqrt{\mathscr{E}_s} \cos \phi_m + n_{mc}$$
$$y_{ms} = \sqrt{\mathscr{E}_s} \sin \phi_m + n_{ms}.$$

$$(10.4.9)$$

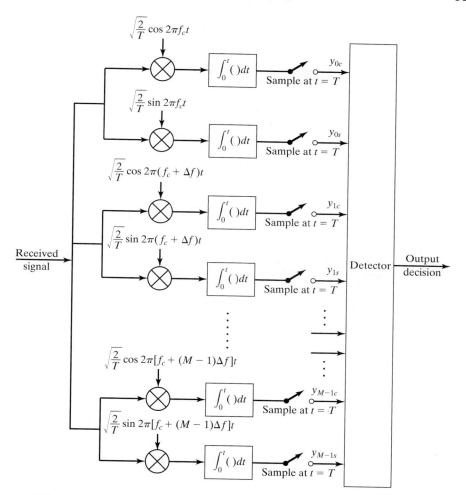

Figure 10.29   Demodulation of M-ary FSK signals for noncoherent detection.

Furthermore, we observe that when $k \neq m$, the signal components in the samples y_{kc} and y_{ks} will vanish, independent of the values of the phase shift ϕ_k, provided that the frequency separation between successive frequencies is $\Delta f = 1/T$. In such a case, the other $2(M-1)$ correlator outputs consist of noise only, i.e.,

$$y_{kc} = n_{kc}, \quad y_{ks} = n_{ks}, \quad k \neq m. \tag{10.4.10}$$

In the development that follows, we assume that $\Delta f = 1/T$, so that the signals are orthogonal.

It is easily shown (see Problem 10.18) that the $2M$ noise samples $\{n_{kc}\}$ and $\{n_{ks}\}$ are zero-mean, mutually uncorrelated Gaussian random variables with an equal variance $\sigma^2 = N_0/2$. Consequently, the joint PDF for y_{mc} and y_{ms} conditioned on ϕ_m is

$$f_{Y_m}(y_{mc}, y_{ms}|\phi_m) = \frac{1}{2\pi\sigma^2} e^{-[(y_{mc}-\sqrt{\mathscr{E}_s}\cos\phi_m)^2+(y_{ms}-\sqrt{\mathscr{E}_s}\sin\phi_m)^2]/2\sigma^2}, \quad (10.4.11)$$

and for $m \neq k$, we have

$$f_{Y_k}(y_{kc}, y_{ks}) = \frac{1}{2\pi\sigma^2} e^{-\left(y_{kc}^2+y_{ks}^2\right)/2\sigma^2}. \quad (10.4.12)$$

Given the $2M$ observed random variables $\{y_{kc}, y_{ks}, k = 0, 1, \ldots, M-1)$, the optimum detector selects the signal that corresponds to the maximum of the posterior probabilities

$$P[s_m \text{ was transmitted } |y] \equiv P(s_m|y), \quad m = 0, 1, \ldots, M-1, \quad (10.4.13)$$

where y is the $2M$ dimensional vector with elements $\{y_{kc}, y_{ks}, k = 0, 1, \ldots, M-1\}$. We derive the form for the optimum noncoherent detector for the case of binary FSK. The generalization to M-ary FSK is straightforward.

Optimum Detector for Binary FSK.   In binary orthogonal FSK, the two posterior probabilities are

$$P(s_0|y) = \frac{f_Y(y|s_0)P(s_0)}{f_Y(y)}$$

$$(10.4.14)$$

$$P(s_1|y) = \frac{f_Y(y|s_1)P(s_1)}{f_Y(y)};$$

hence, the optimum detection rule may be expressed as

$$P(s_0|y) \underset{s_1}{\overset{s_0}{\gtrless}} P(s_1|y) \quad (10.4.15)$$

or, equivalently,

$$\frac{f_Y(y|s_0)P(s_0)}{f_Y(y)} \underset{s_1}{\overset{s_0}{\gtrless}} \frac{f_Y(y|s_1)P(s_1)}{f_Y(y)}, \quad (10.4.16)$$

where y is the four-dimensional vector $y = (y_{0c}, y_{0s}, y_{1c}, y_{1s})$. The relation in Equation (10.4.16) simplifies to the detection rule

$$\frac{f_Y(y|s_0)}{f_Y(y|s_1)} \underset{s_1}{\overset{s_0}{\gtrless}} \frac{P(s_1)}{P(s_0)}. \quad (10.4.17)$$

The ratio of PDF's in the left-hand side of Equation (10.4.17) is the likelihood ratio, which we denote as

$$\Lambda(y) = \frac{f_Y(y|s_0)}{f_Y(y|s_1)}. \quad (10.4.18)$$

The right-hand side of Equation (10.4.17) is the ratio of the two prior probabilities, which takes the value of unity when the two signals are equally probable.

The PDF's $f_Y(y|s_0)$ and $f_Y(y|s_1)$ in the likelihood ratio may be expressed as

$$f_Y(y|s_0) = f_{Y_1}(y_{1c}, y_{1s}) \int_0^{2\pi} f_{Y_0}(y_{0c}, y_{0s}|\phi_0) f_\Phi(\phi_0) d\phi_0$$

$$f_Y(y|s_1) = f_{Y_0}(y_{0c}, y_{0s}) \int_0^{2\pi} f_{Y_1}(y_{1c}, y_{1s}|\phi_1) f_\Phi(\phi_1) d\phi_1, \tag{10.4.19}$$

where $f_{Y_m}(y_{mc}, y_{ms}|\phi_m)$ and $f_{Y_k}(y_{kc}, y_{ks})$, $m \neq k$ are given by Equations (10.4.11) and (10.4.12), respectively. Thus, the carrier phases ϕ_0 and ϕ_1 are eliminated by simply averaging $f_{Y_m}(y_{mc}, y_{ms}|\phi_m)$.

The uniform PDF for ϕ_m represents the most ignorance regarding the phases of the carriers. This is called the *least favorable* PDF for ϕ_m. When $f_{\Phi_m}(\phi_m) = 1/2\pi$, $0 \leq \phi_m \leq 2\pi$, is substituted into the integrals given in Equation (10.4.19), we obtain

$$\frac{1}{2\pi} \int_0^{2\pi} f_{Y_m}(y_{mc}, y_{ms}|\phi_m) d\phi_m$$

$$= \frac{1}{2\pi\sigma^2} e^{-(y_{mc}^2 + y_{ms}^2 + \mathscr{E}_s)/2\sigma^2} \frac{1}{2\pi} \int_0^{2\pi} e^{\sqrt{\mathscr{E}_s}(y_{mc}\cos\phi_m + y_{ms}\sin\phi_m)/\sigma^2} d\phi_m. \tag{10.4.20}$$

But

$$\frac{1}{2\pi} \int_0^{2\pi} e^{\sqrt{\mathscr{E}_s}(y_{mc}\cos\phi_m + y_{ms}\sin\phi_m)/\sigma^2} d\phi_m = I_0 \left(\frac{\sqrt{\mathscr{E}_s(y_{mc}^2 + y_{ms}^2)}}{\sigma^2} \right), \tag{10.4.21}$$

where $I_0(x)$ is the modified Bessel function of order zero. This function is a monotonically increasing function of its argument, as illustrated in Figure 10.30. $I_0(x)$ has the power series expansion

$$I_0(x) = \sum_{k=0}^{\infty} \frac{x^{2k}}{2^{2k}(k!)^2}. \tag{10.4.22}$$

From Equations (10.4.17)–(10.4.19), we obtain the likelihood ratio $\Lambda(\mathbf{r})$ in the form

$$\Lambda(y) = \frac{I_0\left(\sqrt{\mathscr{E}_s(y_{0c}^2 + y_{0s}^2)}/\sigma^2\right)}{I_0\left(\sqrt{\mathscr{E}_s(y_{1c}^2 + y_{1s}^2)}/\sigma^2\right)} \underset{s_1}{\overset{s_0}{\gtrless}} \frac{P(s_1)}{P(s_0)}. \tag{10.4.23}$$

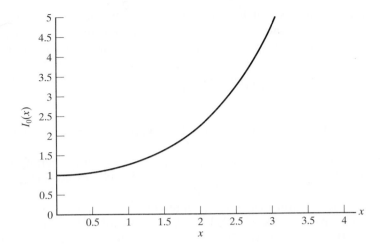

Figure 10.30   Graph of $I_0(x)$.

Thus, the optimum detector computes the two envelopes

$$y_0 = \sqrt{y_{0c}^2 + y_{0s}^2}$$

and

$$y_1 = \sqrt{y_{1c}^2 + y_{1s}^2}$$

and the corresponding values of the Bessel function

$$I_0\left(\sqrt{\mathscr{E}_s y_0^2 / \sigma^2}\right)$$

and

$$I_0\left(\sqrt{\mathscr{E}_s y_1^2 / \sigma^2}\right)$$

to form the likelihood ratio. We observe that this computation requires knowledge of the noise variance σ^2 and the signal energy $\mathscr{E}_s$. The likelihood ratio is then compared with the threshold $P(s_1)/P(s_0)$ to determine which signal was transmitted.

A significant simplification in the implementation of the optimum detector occurs when the two signals are equally probable. In such a case, the threshold becomes unity and, due to the monotonicity of the Bessel function, the optimum detector rule simplifies to

$$\sqrt{y_{0c}^2 + y_{0s}^2} \underset{s_1}{\overset{s_0}{\gtrless}} \sqrt{y_{1c}^2 + y_{1s}^2}. \tag{10.4.24}$$

Thus, the optimum detector bases its decision on the two envelopes

$$y_0 = \sqrt{y_{0c}^2 + y_{0s}^2}$$

and

$$y_1 = \sqrt{y_{1c}^2 + y_{1s}^2};$$

hence, it is called an *envelope detector*.

The computation of the envelopes of the received signal samples at the demodulator's output renders the carrier signal phases $\{\phi_m\}$ irrelevant in the decision as to which signal was transmitted. Equivalently, the decision may be based on the computation of the squared envelopes y_0^2 and y_1^2; in this case, the detector is called a *square-law detector*. Figure 10.31 shows the block diagram of the demodulator and the square-law detector.

The generalization of the optimum demodulator and detector to M-ary orthogonal FSK signals is straightforward. As illustrated in Figure 10.29, the output of the optimum demodulator at the sampling instant consists of the $2M$ vector components $(y_{0c}, y_{0s}, y_{1c}, y_{1s}, \ldots, y_{M-1c}, y_{M-1s})$. Then the optimum noncoherent detector

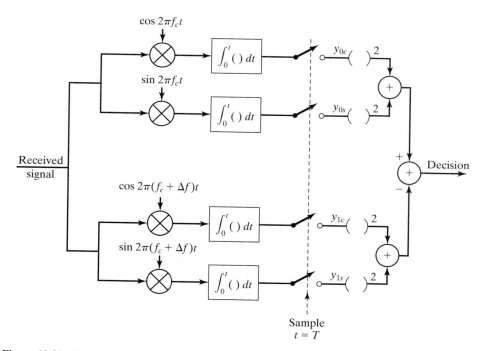

Figure 10.31   Demodulation and square-law detection of binary FSK signals.

computes the M envelopes as

$$y_m = \sqrt{y_{mc}^2 + y_{ms}^2}, \quad m = 0, 1, \ldots, M - 1. \tag{10.4.25}$$

Thus, the unknown carrier phases of the received signals are rendered irrelevant to the decision as to which signal was transmitted. When all the M signals are equally likely to be transmitted, the optimum detector selects the signal corresponding to the largest envelope (or squared envelope). In the case of nonequally probable transmitted signals, the optimum detector must compute the M posterior probabilities in Equation (10.4.13) and then select the signal corresponding to the largest posterior probability.

10.4.2 Probability of Error for Noncoherent Detection of FSK

Consider M-ary orthogonal FSK signals that are detected noncoherently. We assume that the M signals are equally probable a priori and that $u_0(t)$ was transmitted in the interval $0 \leq 1 \leq T$.

The M-decision metrics at the detector are the M envelopes

$$y_m = \sqrt{y_{mc}^2 + y_{ms}^2} \quad m = 0, 1, \ldots, M - 1, \tag{10.4.26}$$

where

$$\begin{aligned} y_{0c} &= \sqrt{\mathscr{E}_s} \cos \phi_0 + n_{0c} \\ y_{0s} &= \sqrt{\mathscr{E}_s} \sin \phi_0 + n_{0s} \end{aligned} \tag{10.4.27}$$

and

$$\begin{aligned} y_{mc} &= n_{mc}, \quad m = 1, 2, \ldots, M - 1 \\ y_{ms} &= n_{ms}, \quad m = 1, 2, \ldots, M - 1. \end{aligned} \tag{10.4.28}$$

The additive noise components $\{n_{mc}\}$ and $\{n_{ms}\}$ are mutually statistically independent zero-mean Gaussian variables with equal variance $\sigma^2 = N_0/2$. Thus, the PDF's of the random variables at the input to the detector are

$$f_{Y_0}(y_{0c}, y_{0s}) = \frac{1}{2\pi\sigma^2} e^{-\left(y_{0c}^2 + y_{0s}^2 + \mathscr{E}_s\right)/2\sigma^2} I_0 \left(\frac{\sqrt{\mathscr{E}_s \left(y_{0c}^2 + y_{0s}^2\right)}}{\sigma^2} \right) \tag{10.4.29}$$

$$f_{Y_m}(y_{mc}, y_{ms}) = \frac{1}{2\pi\sigma^2} e^{-\left(y_{mc}^2 + y_{ms}^2\right)/2\sigma^2}, \quad m = 1, 2, \ldots, M - 1. \tag{10.4.30}$$

Let us make a change in variables in the joint PDF's given by Equations (10.4.29) and (10.4.30). We define the normalized variables

$$R_m = \frac{\sqrt{Y_{mc}^2 + Y_{ms}^2}}{\sigma}$$

$$\Theta_m = \tan^{-1} \frac{Y_{ms}}{Y_{mc}}. \tag{10.4.31}$$

Clearly, $Y_{mc} = \sigma R_m \cos \Theta_m$ and $Y_{ms} = \sigma R_m \sin \Theta_m$. The Jacobian of this transformation is

$$|J| = \begin{vmatrix} \sigma \cos \Theta_m & \sigma \sin \Theta_m \\ -\sigma R_m \sin \Theta_m & \sigma R_m \cos \Theta_m \end{vmatrix} = \sigma^2 R_m. \tag{10.4.32}$$

Consequently,

$$f_{R_0 \Theta_0}(r_0, \theta_0) = \frac{r_0}{2\pi} e^{-(r_0^2 + 2\mathcal{E}_s/N_0)/2} I_0 \left(\sqrt{\frac{2\mathcal{E}_s}{N_0}} r_0 \right) \tag{10.4.33}$$

$$f_{R_m \Theta_m}(r_m, \theta_m) = \frac{r_m}{2\pi} e^{-r_m^2/2}, \quad m = 1, 2, \ldots, M - 1. \tag{10.4.34}$$

Finally, by averaging $f_{R_m \Theta_m}(r_m, \theta_m)$ over Θ_m, the factor of 2π is (see Example 5.1.11 and Problem 5.30) eliminated from Equations (10.4.33) and (10.4.34). Thus, we find that R_0 has a Rice probability distribution and R_m, $m = 1, 2, \ldots, M - 1$ are each Rayleigh distributed.

The probability of a correct decision is simply the probability that $R_0 > R_1$, $R_0 > R_2, \ldots$, and $R_0 > R_{M-1}$. Hence,

$$P_c = P(R_1 < R_0, R_2 < R_0, \ldots, R_{M-1} < R_0)$$

$$= \int_0^\infty P(R_1 < R_0, R_2 < R_0, \ldots, R_{M-1} < R_0 | R_0 = x) f_{R_0}(x) \, dx. \tag{10.4.35}$$

Because the random variables R_m, $m = 1, 2, \ldots, M - 1$ are statistically i.i.d., the joint probability in Equation (10.4.35) conditioned on R_0 factors into a product of $M - 1$ identical terms. Thus,

$$P_c = \int_0^\infty [P(R_1 < R_0 | R_0 = x)]^{M-1} f_{R_0}(x) \, dx, \tag{10.4.36}$$

where

$$P(R_1 < R_0 | R_0 = x) = \int_0^x f_{R_1}(r_1) \, dr_1$$

$$= 1 - e^{-x^2/2}. \tag{10.4.37}$$

The $(M - 1)$st power of Equation (10.4.37) may be expressed as

$$\left[1 - e^{-x^2/2} \right]^{M-1} = \sum_{n=0}^{M-1} (-1)^n \binom{M-1}{n} e^{-nx^2/2}. \tag{10.4.38}$$

Substitution of this result in Equation (10.4.36) and integration over x yields the probability of a correct decision as

$$P_c = \sum_{n=0}^{M-1} (-1)^n \binom{M-1}{n} \frac{1}{n+1} e^{-n\rho_s/(n+1)}, \tag{10.4.39}$$

where $\rho_s = \mathscr{E}_s/N_0$ is the SNR/symbol. Then, the probability of a symbol error, which is $P_M = 1 - P_c$, becomes

$$P_M = \sum_{n=1}^{M-1} (-1)^{n+1} \binom{M-1}{n} \frac{1}{n+1} e^{-nk\rho_b/(n+1)}, \qquad (10.4.40)$$

where $\rho_b = \mathscr{E}_b/N_0$ is the SNR/bit.

For binary FSK ($M = 2$), Equation (10.4.40) reduces to the simple form

$$P_2 = \frac{1}{2} e^{-\rho_b/2}. \qquad (10.4.41)$$

For $M > 2$, we may compute the probability of a bit error by using the relationship

$$P_b = \frac{2^{k-1}}{2^k - 1} P_M, \qquad (10.4.42)$$

which was established in Section 8.5.2. Figure 10.32 shows the bit-error probability as function of the SNR/bit ρ_b for $M = 2, 4, 8, 16$, and 32. Just as in the case of

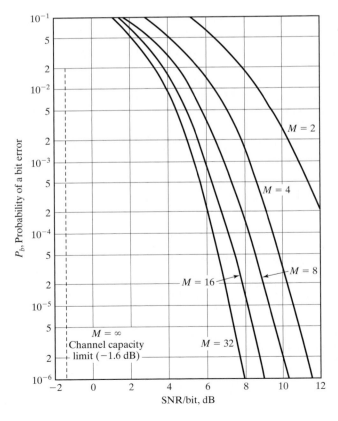

Figure 10.32 Probability of a bit error for noncoherent detection of orthogonal FSK signals.

coherent detection of M-ary orthogonal signals (see Section 8.5.2), we observe that for any given bit-error probability, the SNR/bit decreases as M increases. It will be shown in Chapter 1 that, in the limit as $M \rightarrow \infty$ (or $k = \log_2 M \rightarrow \infty$), the probability of a bit-error P_b can be made arbitrarily small provided that the SNR/bit is greater than the Shannon limit of -1.6 dB.

Example 10.4.1

> Compare the error probability of binary FSK, which is given by Equation (10.4.41) with that for binary DPSK, which is given by Equation (10.2.49).
>
> **Solution**   The error probability for binary DPSK is

$$P_2 = \frac{1}{2} e^{-\rho_b}.$$

> Comparing this expression with that for binary FSK, we observe that binary FSK requires twice the transmitted signal energy to achieve the same performance as binary DPSK.   ∎

10.4.3 Continuous-Phase FSK (CPFSK)

Ordinary FSK signals may be generated by having $M = 2^k$ separate oscillators tuned to the desired frequencies $f_c + m\Delta f \equiv f_m$ and selecting one of the M frequencies according to the particular k-bit symbol that is to be transmitted in a signal interval. However, abruptly switching from one oscillator output to another in successive signaling intervals results in relatively large spectral sidelobes outside of the main spectral band of the signal, which decay slowly with frequency separation. Consequently, this method wastes bandwidth.

To avoid the use of signals with large spectral sidelobes, we may use the information-bearing signal to frequency modulate a single carrier whose phase is changed in a continuous manner. The resulting frequency-modulated signal is phase continuous; hence, it is called *continuous-phase* FSK (CPFSK).

In order to represent a CPFSK signal, we begin with a PAM signal

$$v(t) = \sum_n a_n g_T(t - nT), \tag{10.4.43}$$

where the amplitudes are obtained by mapping k-bit blocks of binary digits from the information sequence into the amplitude levels $\pm 1, \pm 3, \ldots, \pm(M - 1)$, and $g_T(t)$ is a rectangular pulse of amplitude $1/2T$ and duration T. The signal $v(t)$ is used to frequency modulate the carrier. Consequently, the frequency-modulated carrier is

$$u(t) = \sqrt{\frac{2\mathscr{E}_s}{T}} \cos \left[2\pi f_c t + 4\pi T f_d \int_{-\infty}^{t} v(\tau) d\tau + \phi_0 \right], \tag{10.4.44}$$

where f_d is the *peak-frequency derivation* and ϕ_0 is an arbitrary initial phase of the carrier. Note that the instantaneous frequency of the carrier is $f_c + 2T f_d v(t)$.

We observe that, although $v(t)$ contains discontinuities, the integral of $v(t)$ is continuous. We may denote the phase of the carrier as

$$\theta(t; \boldsymbol{a}) = 4\pi T f_d \int_{-\infty}^{t} v(\tau)d\tau, \qquad (10.4.45)$$

where $\boldsymbol{a}$ denotes the sequence of signal amplitudes. Since $\theta(t; \boldsymbol{a})$ is a continuous function of t, we have a continuous-phase signal.

The phase of the carrier in the interval $nT \le t \le (n+1)T$ is determined by the integral in Equation (10.4.45). Thus,

$$\theta(t; \boldsymbol{a}) = 2\pi f_d T \sum_{k=-\infty}^{n-1} a_k + 2\pi (t - nT) f_d a_n \qquad (10.4.46)$$

$$= \theta_n + 2\pi h a_n q(t - nT),$$

where h, θ_n, and $q(t)$ are defined as

$$h = 2 f_d T \qquad (10.4.47)$$

$$\theta_n = \pi h \sum_{k=-\infty}^{n-1} a_k \qquad (10.4.48)$$

$$q(t) = \begin{cases} 0, & t < 0 \\ t/2T, & 0 \le t \le T. \\ \frac{1}{2}, & t > T \end{cases} \qquad (10.4.49)$$

The parameter h is called the *modulation index*. We observe that θ_n represents the phase accumulation (memory) from all symbols up to time $(n-1)T$. The signal $q(t)$ is simply the integral of the rectangular pulse, as illustrated in Figure 10.33.

It is instructive to sketch the set of all phase trajectories $\theta(t; \boldsymbol{a})$ generated by all possible values of the information sequence $\{a_n\}$. For example, with binary symbols, $a_n = \pm 1$, the set of phase trajectories beginning at time $t = 0$ is shown in Figure 10.34. For comparison, the phase trajectories for quaternary CPFSK ($a_n = \pm 1, \pm 3$) are illustrated in Figure 10.35. These phase diagrams are called *phase trees*.

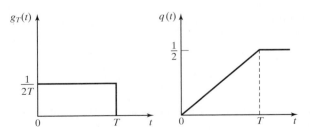

Figure 10.33   The signal plot $g_T(t)$ and its integral $q(t)$.

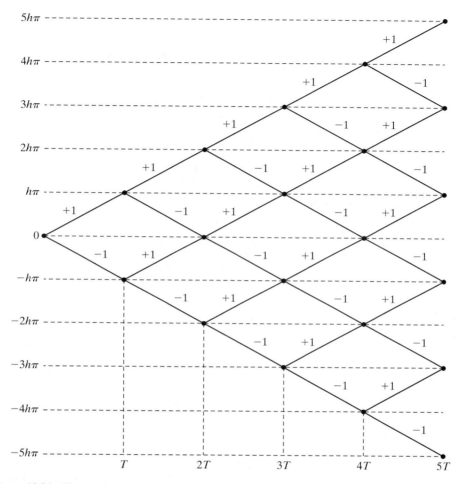

Figure 10.34 Phase trajectory for binary CPFSK.

We observe that the phase trees are piecewise linear because the pulse $g_T(t)$ is rectangular. Smoother phase trajectories and phase trees may be obtained by using pulses that do not contain discontinuities.

The phase trees shown in these figures grow with time. However, the phase of the carrier is unique only in the range $\theta = 0$ to $\theta = 2\pi$ or, equivalently, from $\theta = -\pi$ to $\theta = \pi$. When the phase trajectories are plotted modulo 2π, e.g., in the range $(-\pi, \pi)$, the phase tree collapses into a structure called a *phase trellis*.

Simpler representations for the phase trajectories can be obtained by displaying only the terminal values of the signal phase at the time instants $t = nT$. In this case, we restrict the modulation index h to be rational. In particular, let us assume that $h = m/p$, where m and p are relatively prime integers. Then, at the time instants

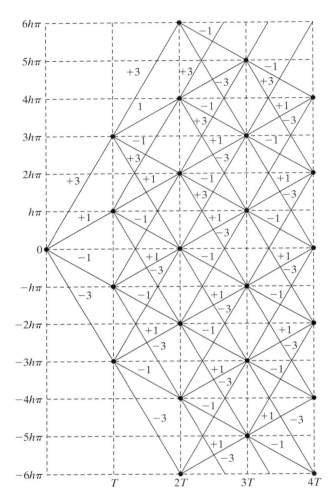

Figure 10.35 Phase trajectory for quaternary CPFSK.

$t = nT$, the terminal phase states for m even are

$$\Theta_s = \left\{0, \; \frac{\pi m}{p}, \; \frac{2\pi m}{p}, \ldots, \; \frac{(p-1)\pi m}{p}\right\} \tag{10.4.50}$$

and for m odd are

$$\Theta_s = \left\{0, \; \frac{\pi m}{p}, \; \frac{2\pi m}{p}, \ldots, \; \frac{(2p-1)\pi m}{p}\right\}. \tag{10.4.51}$$

Hence, there are p terminal phase states when m is even and $2p$ terminal phase states when m is odd. For example, binary CPFSK with $h = 1/2$ has four terminal phase states, namely, $0, \pi/2, \pi, 3\pi/2$. The *state trellis* for this signal is illustrated in Figure 10.36. In the state trellis, the phase transitions from one state to another are

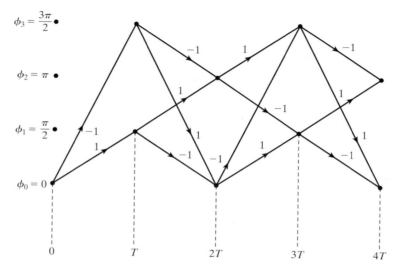

Figure 10.36   State trellis for binary CPFSK with $h = 1/2$.

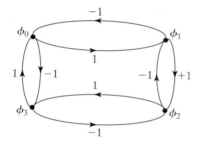

Figure 10.37   State diagram for binary CPFSK with $h = 1/2$.

not true phase trajectories. They represent phase transitions to the terminal states at the time instants $t = nT$.

An alternative representation to the state trellis is the *state diagram*, which also illustrates the state transitions at the time instants $t = nT$. This is an even more compact representation of the CPFSK signal. Only the possible terminal phase states and their transitions are displayed in the state diagram. Time does not appear explicitly as a variable. For example, the state diagram for the CPFSK signal with $h = 1/2$ is shown in Figure 10.37.

We should emphasize that a CPFSK signal cannot be represented by discrete points in signal space as in the case of PAM, PSK, and QAM, because the phase of the carrier is time variant. Instead, the constant amplitude CPFSK signal may be represented in two-dimensional space by a circle, where points on the circle represent the combined amplitude and phase trajectory of the carrier as a function of time. For example, Figure 10.38 illustrates the signal-space diagrams for binary CPFSK with $h = 1/2$ and $h = 1/4$. The dots at $\theta = 0, \pi/2, \pi, 3\pi/2$ and

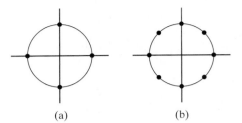

(a)   (b)

Figure 10.38 Signal space diagram for binary CPFSK with (a) $h = 1/2$ and (b) $h = 1/4$.

$\theta = 0, \pm\pi/4, \pm\pi/2, \pm3\pi/4, \pi$ for $h = 1/2$ and $h = 1/4$, respectively, represent the terminal phase states previously shown in the state diagram.

Minimum-Shift Keying (MSK).   MSK is a special form of binary CPFSK in which the modulation index $h = 1/2$. Thus, the phase of the carrier for the MSK signal is

$$\theta(t; \boldsymbol{a}) = \frac{\pi}{2} \sum_{k=-\infty}^{n-1} a_k + \pi a_n q(t - nT_b)$$

(10.4.52)

$$= \theta_n + \frac{\pi}{2}\left(\frac{t - nT_b}{T_b}\right)a_n, \quad nT_b \leq 1 \leq (n+1)T_b,$$

which follows from Equation (10.4.46). The corresponding carrier-modulated signal is

$$u(t) = \sqrt{\frac{2\mathscr{E}_b}{T_b}} \cos[2\pi f_c t + \theta_n + \pi(t - nT_b)a_n/2T_b]$$

(10.4.53)

$$= \sqrt{\frac{2\mathscr{E}_b}{T_b}} \cos\left[2\pi\left(f_c + \frac{1}{4T_b}a_n\right)t - \frac{n\pi}{2}a_n + \theta_n\right].$$

The expression in Equation (10.4.53) indicates that the MSK (binary CPFSK) signal is basically a sinusoid consisting of one of two possible frequencies in the interval $nT_b \leq 1 \leq (n+1)T_b$, namely,

$$f_1 = f_c - \frac{1}{4T_b}$$

(10.4.54)

$$f_2 = f_c + \frac{1}{4T_b}.$$

Hence, the two sinusoidal signals may be expressed as

$$u_i(t) = \sqrt{\frac{2\mathscr{E}_b}{T_b}} \cos\left[2\pi f_i t + \theta_n + \frac{n\pi}{2}(-1)^{i-1}\right], \quad i = 1, 2.$$

(10.4.55)

The frequency separation is $\Delta f = f_2 - f_1 = 1/2T_b$. Recall that this is the minimum frequency separation for orthogonality of the two sinusoids, provided the signals are detected coherently. This explains why binary CPFSK with $h = 1/2$ is called *minimum-shift keying* (MSK). Note that the phase of the carrier in the nth signaling interval is the phase state of the signal that results in phase continuity between adjacent intervals.

It is interesting to demonstrate that MSK is also a form of four-phase PSK. To prove this point, we begin with a four-phase PSK signal, which has the form

$$
u(t) = \sqrt{\frac{2\mathscr{E}_b}{T_b}} \left\{ \left[\sum_{n=-\infty}^{\infty} a_{2n} g_T(t - 2nT_b) \right] \cos 2\pi f_c t \right.
$$

$$
\left. + \left[\sum_{n=-\infty}^{\infty} a_{2n+1} g_T(t - 2nT_b - T_b) \right] \sin 2\pi f_c t \right\},
$$

(10.4.56)

where $g_T(t)$ is a sinusoidal pulse defined as

$$
g_T(t) = \begin{cases} \sin \dfrac{\pi t}{2T_b}, & 0 \le t \le 2T_b \\ \\ 0, & \text{otherwise} \end{cases}
$$

(10.4.57)

and illustrated in Figure 10.39. First, we observe that the four-phase PSK signal consists of two quadrature carriers, $\cos 2\pi f_c t$ and $\sin 2\pi f_c t$, which are amplitude modulated at a rate of one bit per $2T_b$ interval. The even-numbered information bits $\{a_{2n}\}$ are transmitted by modulating the cosine carrier, while the odd-numbered information bits $\{a_{2n+1}\}$ are transmitted by amplitude modulating the sine carrier. Note that the modulation of the two quadrature carriers is staggered in time by T_b and that the transmission rate for each carrier is $1/2T_b$. This type of four-phase modulation is called *offset quadrature* PSK (SQPSK) or *staggered quadrature* PSK (SQPSK).

Figure 10.40 illustrates the SQPSK signal in terms of the two staggered quadrature-modulated binary PSK signals. The corresponding sum of the two quadrature signals is a constant-amplitude, continuous-phase FSK signal, as shown in Figure 10.40.

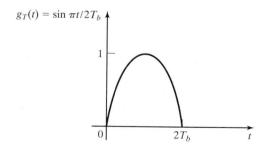

$g_T(t) = \sin \pi t/2T_b$

1

0       $2T_b$ t

Figure 10.39 Sinusoidal pulse shape.

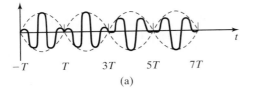

(a)

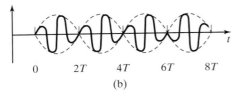

(b)

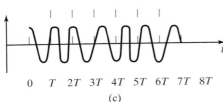

(c)

Figure 10.40   Representation of an MSK signal as a form of two staggered binary PSK signals, each with a sinusoidal envelope. (a) In-phase signal component, (b) quadrature signal component, and (c) MSK signal (a + b).

It is also interesting to compare the waveforms for MSK with the waveforms for staggered QPSK, in which the pulse $g_T(t)$ is rectangular for $0 \le t \le 2T_b$, and with the waveforms conventional quadrature PSK (QPSK), in which the baseband pulse is rectangular in the interval $0 \le t \le 2T_b$. We emphasize that all three of these modulation methods result in identical data rates. The MSK signal is phase continuous. The SQPSK signal with a rectangular baseband pulse is basically two binary PSK signals for which the phase transitions are staggered in time by T_b seconds. Consequently, this signal contains phase jumps of $\pm 90°$ that may occur as often as every T_b seconds. On the other hand, in conventional QPSK with constant envelope, one or both of the information symbols may cause phase transitions as often as every $2T_b$ seconds. These phase jumps may be $\pm 180°$ or $\pm 90°$. An illustration of these three types of four-phase PSK signals is shown in Figure 10.42.

From this description, it is clear that CPFSK is a modulation method with memory. The memory results from the phase continuity of the transmitted carrier phase from one symbol to the next. As a consequence of the continuous phase characteristics, the power spectra of CPFSK signals are narrower than the corresponding FSK signals in which the phase is allowed to change abruptly at the beginning of each symbol.

Continuous-Phase Modulation (CPM).   When the phase of the carrier is expressed in the form of Equation (10.4.46), CPFSK becomes a special case of a general class of continuous-phase modulated (CPM) signals in which the carrier

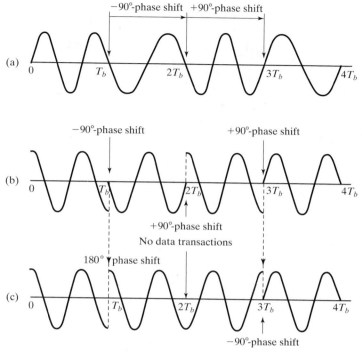

Figure 10.41 Signal waveforms for (a) MSK, (b) offset QPSK (rectangular pulse), and (c) conventional QPSK (rectangular pulse). (From Gronemeyer and McBride; ©1976 IEEE.)

phase is

$$\theta(t; \boldsymbol{a}) = 2\pi \sum_{k=-\infty}^{n} a_k hq(t - kT), \quad nT \le t \le (n+1)T, \tag{10.4.58}$$

where $\{a_k\}$ is the sequence of M-ary information symbols with possible values $\pm 1, \pm 3, \dots, \pm(M-1)$, and $q(t)$ is some arbitrary normalized waveform. Recall that for CPFSK, $q(t) = t/2T$ for $0 \le t \le T$, $q(t) = 0$ for $t < 0$, and $q(t) = 1/2$ for $t > T$.

The waveform $q(t)$ is the integral of a pulse $g_T(t)$ of arbitrary shape, i.e.,

$$q(t) = \int_0^t g_T(\tau) d\tau. \tag{10.4.59}$$

If $g_T(t) = 0$ for $t > T$, the CPM signal is called a *full-response CPM signal*. If the signal pulse $g_T(t)$ is nonzero for $t > T$, the modulated signal is called *partial-response CPM*. In Figure 10.42, we illustrate several pulse shapes for $g_T(t)$ and the corresponding $q(t)$. It is apparent that there is an infinite number of CPM signals that can be obtained by selecting different pulse shapes for $g_T(t)$ and by varying the modulation index h and the number of symbols M.

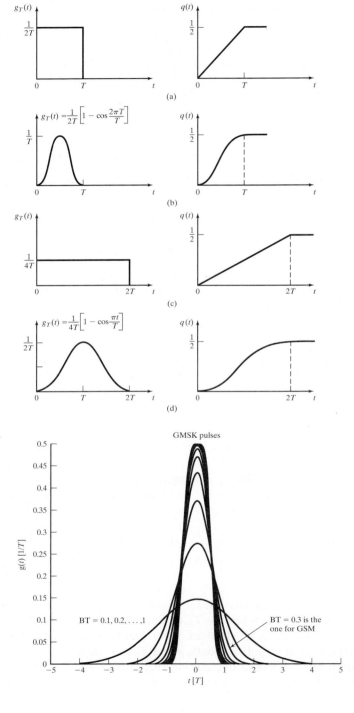

Figure 10.42   Pulse shapes for (a, b) full-response and (c, d) partial response CPM, (e) GMSK pulse shape.

TABLE 10.2 SOME COMMONLY USED CPM PULSE SHAPES

LREC

$$g_T(t) = \begin{cases} \frac{1}{2LT} & (0 \le t \le LT) \\ \\ 0, & \text{otherwise} \end{cases}$$

LRC

$$g_T(t) = \begin{cases} \frac{1}{2LT}\left(1 - \cos\frac{2\pi t}{LT}\right), & 0 \le t \le LT \\ \\ 0, & \text{otherwise} \end{cases}$$

GMSK

$$g_T(t) = \left\{ Q\left[2\pi B\left(t - \frac{T}{2}\right) / (\ln 2)^{1/2}\right] - Q\left[2\pi B\left(t + \frac{T}{2}\right) / (\ln 2)^{1/2}\right]\right\}$$

$$Q(t) = \frac{1}{2\pi}\int_t^\infty e^{-x^2/2}\, dx$$

The primary reason for extending the duration of the pulse $g(t)$ beyond the time interval $0 \le t \le T$ is to further reduce the bandwidth of the transmitted signal. We note that, when the duration of the pulse $g_T(t)$ extends over the time interval $0 \le t \le LT$, where $L > 1$, additional memory is introduced in the CPM signals; hence, the number of phase states increases.

Three popular pulse shapes are given in Table 10.2. LREC denotes a rectangular pulse of duration LT, where L is a positive integer. In this case, $L = 1$ results in a CPFSK signal with the pulse, as shown in Figure 10.42(a). The LREC pulse for $L = 2$ is shown in Figure 10.42(c). LRC denotes a raised cosine pulse of duration LT. The LRC pulses corresponding to $L = 1$ and $L = 2$ are shown in Figures 10.42(b) and (d), respectively.

The last pulse given in Table 10.2 is a Gaussian minimum-shift keying (GMSK) pulse with the bandwidth parameter B, which represents the -3 dB bandwidth of the Gaussian pulse. Figure 10.42(e) illustrates a set of GMSK pulses with time-bandwidth products BT ranging from 0.1 to 1. We observe that the pulse duration increases as the bandwidth of the pulse decreases, as expected. In practical applications, the pulse is usually truncated to some specified fixed duration. GMSK with $BT = 0.3$ is used in the European digital cellular communication system, called GSM. From Figure 10.42(e), we observe that when $BT = 0.3$, the GMSK pulse may be truncated at $|t| = 1.5T$ with a relatively small error incurred for $t > 1.5T$.

In general, CPM signals cannot be represented by discrete points in signal space as in the case of PAM, PSK, and QAM, because the phase of the carrier is time-variant. As we observed in the case of CPFSK, we may employ phase trajectories (or phase trees) to illustrate the signal phase as a function of time. Alternatively, we may use a state trellis or a state diagram to illustrate the terminal phase states and the phase transitions from one state to another. Finally, as in the case of CPFSK, we may represent a CPM signal in signal space by a circle, where the points on the circle correspond to the combined amplitude and phase of the carrier as a function of time. The terminal phase states are usually identified as discrete points on the circle. For example, Figure 10.43 illustrates the signal-space diagrams for binary CPM

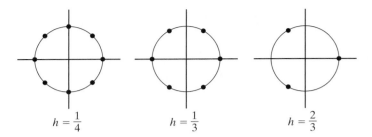

$$h = \frac{1}{4} \qquad\qquad h = \frac{1}{3} \qquad\qquad h = \frac{2}{3}$$

Figure 10.43 Signal-space diagram for CPM.

with $h = 1/4, 1/3$, and $2/3$. Note that the length of the phase trajectory (lengths of the arc) between two terminal phase states increases with an increase in h. An increase in h also results in an increase in the signal bandwidth.

10.5 COMPARISON OF MODULATION METHODS

The digital modulation methods described in this chapter and Chapter 8 can be compared in a number of ways. For example, we can compare them on the basis of the SNR required to achieve a specified probability of error. However, such a comparison would not be very meaningful unless it were made on the basis of some constraint, such as a fixed data rate of transmission.

Suppose that the bit rate R_b is fixed. Consider the channel bandwidth required to transmit the various signals. If we employ M-ary PAM, where $M = 2^k$, the channel bandwidth required to transmit the signal is simply the bandwidth of the signal pulse $g_T(t)$, which depends on its detailed characteristics. For our purposes, we assume that $g_T(t)$ is a pulse of duration T and that its bandwidth W is approximately $1/2T$, where T is the symbol interval. In one symbol interval, we can transmit k information bits, so $T = k/R_b$ seconds. Hence, the channel bandwidth required to transmit the M-ary PAM signal is

$$W = R_b/2k = R_b/2\log_2 M \text{ Hz.} \qquad (10.5.1)$$

If the PAM signal is transmitted at bandpass as a double-sideband suppressed carrier signal, the required channel bandwidth is twice that for the baseband channel. However, the bandwidth of the bandpass PAM signal can be reduced by a factor of two by transmitting only one of the sidebands (either the upper or the lower sideband of the bandpass signal). Thus, the required channel bandwidth of the single-sideband bandpass PAM signal is exactly the same as the bandwidth of the baseband signal.

In the case of QAM, the channel bandwidth is (approximately) $W = 1/T$, but since the information is carried on two quadrature carriers, it follows that $T = 2k/R_b$, where k is the number of information bits/carrier. Hence,

$$\begin{aligned} W &= R_b/2k = R_b/2\log_2 M_{\text{PAM}} \\ &= R_b/\log_2 M_{\text{QAM}}, \end{aligned} \qquad (10.5.2)$$

where the number of signals for M-ary QAM, denoted as M_{QAM}, is equal to the square of the number M_{PAM} of PAM signals.

For M-ary phase modulation, the channel bandwidth required to transmit the multiphase signals is $W = 1/T$, where $T = k/R_b$. Hence,

$$W = R_b/k = R_b/\log_2 M. \tag{10.5.3}$$

Note that PAM, QAM, and PSK signals have the characteristic that, for a fixed bit rate R_b, the channel bandwidth decreases as the number of signals M increases. This means that, with increasing M, the system becomes more bandwidth efficient. On the other hand, examination of Figures 8.45, 10.16, and 10.26 shows that in all these systems, at a given $\mathcal{E}_b/N_0$, increasing M increases the error probability and thus deteriorates the performance. In other words, in these systems, increasing M increases the bandwidth efficiency and decreases the power efficiency. This is a direct consequence of the fact that, in these systems, the dimensionality of the signal space N is one (for PAM) or two (for PSK and QAM) and is independent of M.

Orthogonal signals have completely different bandwidth requirements. For example, if we employ PPM signals, the symbol interval T is subdivided into M subintervals of duration T/M and pulses of width T/M are transmitted in the corresponding subintervals. Consequently, the channel bandwidth required to transmit the PPM signals is

$$W = M/2T = M/2(k/R_b) = M R_b/2\log_2 M \text{ Hz}. \tag{10.5.4}$$

An identical result is obtained if the M orthogonal signals are constructed as M-ary FSK with a minimum frequency separation of $1/2T$ for orthogonality. Biorthogonal and simplex signals result in similar relationships as PPM (orthogonal). In the case of biorthogonal signals, the required bandwidth is one-half of that for orthogonal signals. From the bandwidth relation for orthogonal signals, we can see that for a fixed R_b, increasing M increases the bandwidth proportional to $M/(2\log_2 M)$. This shows that, in this case, increasing M decreases the bandwidth efficiency of the system. On the other hand, examination of Figure 8.46 and Figure 10.32 shows that in these systems, for a fixed $\mathcal{E}_b/N_0$, increasing M improves the performance, and thus the power efficiency, of the system. It is also interesting to note that in orthogonal, biorthogonal, and simplex signaling schemes, the dimensionality of the signal space is not fixed and increases with an increasing M.

From the foregoing discussion, it is clear that the characteristics of PAM, PSK, and QAM are completely different from the characteristics of orthogonal, biorthogonal, and simplex schemes. Therefore, their applications are also quite different.

A compact and meaningful comparison of these modulation methods is based on the normalized data rate (also called the spectral bit rate) R_b/W (bits per second per hertz of bandwidth) versus the SNR/bit ($\mathcal{E}_b/N_0$) required to achieve a given error probability. For PAM, QAM, PSK, and orthogonal signals, we have

$$\text{PAM:} \quad \frac{R_b}{W} = 2 \log_2 M_{\text{PAM}} \tag{10.5.5}$$

$$\text{QAM:} \quad \frac{R_b}{W} = \log_2 M_{\text{QAM}} \tag{10.5.6}$$

$$\text{PSK:} \quad \frac{R_b}{W} = \log_2 M_{\text{PSK}} \tag{10.5.7}$$

$$\text{orthogonal:} \quad \frac{R_b}{W} = \frac{2 \log_2 M}{M}. \tag{10.5.8}$$

Figure 10.44 illustrates the graph of R_b/W (measure of bandwidth efficiency) versus $(\mathcal{E}_b/N_0)$ (measure of power efficiency) for PAM, QAM, PSK, and orthogonal signals

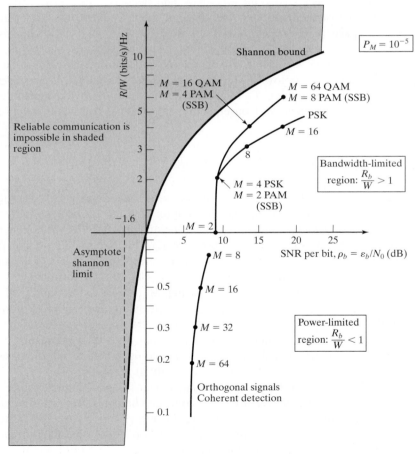

Figure 10.44   Comparison of several modulation methods at 10^{-5} symbol rate.

TABLE 10.3   QAM SIGNAL CONSTELLATIONS

Number of signal points M	Increase in average power (dB) relative to $M = 2$
4	3
8	6.7
16	10.0
32	13.2
64	16.2
128	19.2

for the case in which the symbol error probability is $P_M = 10^{-5}$. As discussed earlier in the case of PAM, QAM, and PSK, increasing the number of signal points M results in a higher bit-rate-to-bandwidth ratio R_b/W. However, the cost of achieving the higher data rate is an increase in the SNR/bit. Consequently, M-ary PAM, QAM, and PSK are appropriate for communication channels that are bandwidth limited, where we desire a bit-rate-to-bandwidth ratio $R_b/W > 1$, and where there is sufficiently high SNR to support multiple signal amplitudes and phases. Telephone channels are examples of such bandlimited channels. The curve denoted by "Shannon limit" illustrates the boundary between the region where reliable communication is possible and the region in which reliable communication is impossible (the shaded area). This curve is given by the relation

$$\frac{\mathscr{E}_b}{N_0} = \frac{2^{R_b/W} - 1}{R_b/W}.$$ (10.5.9)

This relation is derived in Equation (12.6.5).

We have already observed that the cost of doubling the number of phases (increasing the number of bits/symbol by one bit) in PSK approaches 6 dB (a factor of 4) in additional transmitted power for large M. A similar comparison for QAM indicates that the increase in transmitted power is approximately 3 dB per additional bit/symbol. Table 10.3 gives the factor $10 \log_2 (M-1)/3$, which represents the increase in average power required to maintain a given level of performance for QAM as the number of signal points in the rectangular constellation increases. Thus, we observe that QAM (and PAM) is preferable to PSK for large signal constellation sizes.

In contrast to PAM, QAM, and PSK, M-ary orthogonal signals yield a bit-rate-to-bandwidth ratio of $R_b/W \le 1$. As M increases, R_b/W decreases due to an increase in the required channel bandwidth. However, the SNR/bit required to achieve a given error probability (in this case, $P_M = 10^{-5}$) decreases as M increases. Consequently, M-ary orthogonal signals, as well as biorthogonal and simplex signals, are appropriate for power-limited channels that have a sufficiently large bandwidth to accommodate a large number of signals. In this case, as $M \to \infty$, the error probability can be made as small as desired, provided that $\mathscr{E}_b/N_0 > 0.693 \approx (-1.6 \, \text{dB})$. This is the minimum SNR/bit required to achieve reliable transmission in the limit as the channel bandwidth $W \to \infty$ and the corresponding bit-rate-to-bandwidth ratio $R_b/W \to 0$.

10.6 SYMBOL SYNCHRONIZATION FOR CARRIER-MODULATED SIGNALS

The symbol-timing synchronization methods described in Section 8.6 for baseband signals also apply to bandpass signals. Because any carrier-modulated signal can be converted to a baseband signal by a simple frequency translation, symbol timing can be recovered from the received signal after the frequency conversion to baseband.

For QAM signals, the spectral-line methods described in Section 8.6.4 have proven to be particularly suitable for timing recovery. Figure 10.45 illustrates a spectral-line method that is based on filtering out a signal component at the frequency $1/2T$ and squaring the filter output to generate a sinusoidal signal at the desired symbol rate $1/T$. Because the demodulation of the QAM signal is accomplished as previously described [by multiplication of the input signal with the two quadrature-carrier signals $\psi_1(t)$ and $\psi_2(t)$], the in-phase and quadrature signal components at the outputs of the two correlators are used as the inputs to the two bandpass filters tuned to $1/2T$. The two filter outputs are squared (rectified), summed, and then filtered by a narrowband filter tuned to the clock frequency $1/T$. Thus, we generate a sinusoidal signal that is the appropriate clock signal for sampling the outputs of the correlators to recover the information.

In many modern communication systems, the received signal is processed (demodulated) digitally after it has been sampled at the Nyquist rate or faster. In these cases, the symbol timing and carrier phase are recovered by signal-processing operations performed on the signal samples. Thus, a PLL for carrier recovery is implemented as a digital PLL, and the clock recovery loop of a type described in this section is also implemented as a digital loop. References on timing recovery methods based on sampled signals are given in Section 10.7.

10.7 FURTHER READING

In this chapter, we have focused on the transmission of digital information via carrier modulation. We have described conventional carrier modulation methods that

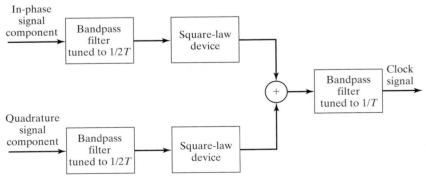

Figure 10.45   Block diagram of a timing recovery method for QAM.

include pulse-amplitude modulation, phase modulation (including phase-shift key-ing and differential phase-shift keying), quadrature-amplitude modulation, frequency modulation, and continuous-phase modulation. The demodulation and detection of these carrier-modulated signals were described for an AWGN channel, and their performance characteristics were evaluated.

More extensive treatments of carrier-modulation techniques may be found in more advanced textbooks on digital communications, such as the book by proakis (2001). Extensive use of MATLAB and Simulink in the simulation of communication systems can be found in the book by Proakis, Salehi, and Bauch (2004).

Timing recovery methods based on sampled signals have been described in a paper by Mueller and Muller (1976). The books by Mengali and D'Andrea (1997) and Meyr et al. (1997) provide comprehensive treatments of synchronization techniques for digital receivers.

CPFSK and CPM are especially suitable for wireless digital communications because these digital modulations are bandwidth efficient and have a constant enve-lope. Therefore, high efficiency nonlinear power amplifiers can be used in the trans-mission of the signals. CPFSK and CPM have been treated extensively in the techni-cal journals and in textbooks. A thorough treatment of CPM can be found in the book by Anderson et al. (1986). The journal papers by Aulin and Sundberg (1981, 1982, and 1984) and by Aulin et al. (1981) provide a detailed analysis of the performance characteristics of CPM. The tutorial paper by Sundberg (1986) gives a very readable overview of CPM, its demodulation, and its performance characteristics. This paper also contains a comprehensive list of references.

PROBLEMS

10.1 A binary PAM signal is generated by exciting a raised-cosine rolloff filter with a 50% rolloff factor and then DSB-SC amplitude modulating it on a sinusoidal carrier, as illustrated in Figure P-10.1. The bit rate is 2400 bps.

1. Determine the spectrum of the modulated binary PAM signal and sketch it.

2. Draw the block diagram illustrating the optimum demodulator/detector for the received signal, which is equal to the transmitted signal plus additive white Gaussian noise.

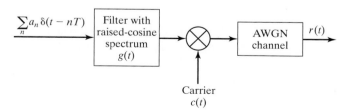

Figure P-10.1

10.2 Consider the signal

$$u(t) = \begin{cases} \frac{A}{T}t \cos 2\pi f_c t, & 0 \le t \le T \\ 0, & \text{otherwise.} \end{cases}$$

1. Determine the impulse response of the matched filter for the preceding signal.

2. Determine the output of the matched filter at $t = T$.

3. Suppose the signal $u(t)$ is passed through a correlator that correlates the input $u(t)$ with $u(t)$. Determine the value of the correlator output at $t = T$. Compare your result with that in Part (2).

10.3 A carrier component is transmitted on the quadrature carrier in a communication system that transmits information via binary PSK. Hence, the received signal has the form

$$v(t) = \pm\sqrt{2P_s} \cos(2\pi f_c t + \phi) + \sqrt{2P_c} \sin(2\pi f_c t + \phi) + n(t),$$

where ϕ is the carrier phase and $n(t)$ is AWGN. The unmodulated carrier component is used as a pilot signal at the receiver to estimate the carrier phase.

1. Sketch a block diagram of the receiver; include the carrier phase estimator.

2. Mathematically illustrate the operations involved in the estimation of the carrier phase ϕ.

3. Express the probability of error for the detection of the binary PSK signal as a function of the total transmitted power $P_T = P_s + P_c$. What is the loss in performance due to the allocation of a portion of the transmitted power to the pilot signal? Evaluate the loss for $P_c/P_T = 0.1$.

10.4 In the demodulation of a binary PSK signal received in white Gaussian noise, a phase-locked loop is used to estimate the carrier phase ϕ.

1. Determine the effect of a phase error $\phi - \widehat{\phi}$ on the probability of error.

2. What is the loss in SNR if the phase error $\phi - \widehat{\phi} = 45°$?

10.5 An ideal voice-band telephone line channel has a bandpass frequency response characteristic spanning the frequency range 600–3000 Hz.

1. Design an $M = 4$ PSK (quadrature PSK or QPSK) system for transmitting data at a rate of 2400 bits/sec and a carrier frequency $f_c = 1800$. For spectral shaping, use a raised-cosine frequency response characteristic. Sketch a block diagram of the system and describe its functional operation.

2. Repeat Part 1 if the bit rate is $R = 4800$ bits/sec.

10.6 Consider the four-phase and eight-phase signal constellations shown in Figure P-10.6. Determine the radii r_1 and r_2 of the circles so that the distance between two adjacent points in the two constellations is d. From this result, determine the additional transmitted energy required in the 8-PSK signal to achieve the same error probability as the four-phase signal at high SNR, where the probability of error is determined by errors in selecting adjacent points.

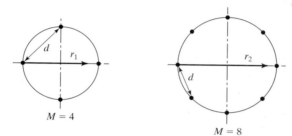

Figure P-10.6

10.7 A voice-band telephone channel passes the frequencies in the band from 300–3300 Hz. We want to design a modem that transmits at a symbol rate of 2400 symbols/sec, and our objective is to achieve 9600 bits/sec. Select an appropriate QAM signal constellation, carrier frequency, and the rolloff factor of a pulse with a raised-cosine spectrum that utilizes the entire frequency band. Sketch the spectrum of the transmitted signal pulse and indicate the important frequencies.

10.8 A 4 kHz bandpass channel will transmit data at a rate of 9600 bits/sec. If $N_0/2 = 10^{-10}$ W/Hz is the spectral density of the additive zero-mean Gaussian noise in the channel, design a QAM modulation and determine the average power that achieves a bit error probability of 10^{-6}. Use a signal pulse with a raised-cosine spectrum having a rolloff factor of at least 50%.

10.9 Consider the two eight-point QAM signal constellation shown in Figure P-10.9. The minimum distance between adjacent points is 2A. Determine the average transmitted power for each constellation assuming that the signal points are equally probable. Which constellation is more power efficient?

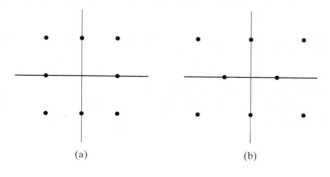

(a)   (b)   **Figure P-10.9**

10.10 The 16-QAM signal constellation shown in Figure P-10.10 is an international standard for telephone line modems (called V.29). Determine the optimum decision boundaries for the detector; assume that the SNR is sufficiently high so that errors only occur between adjacent points.

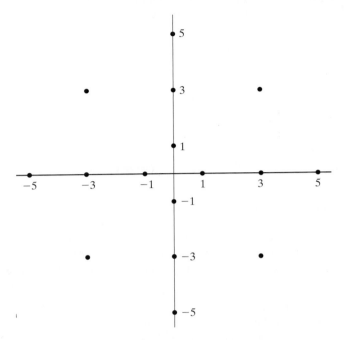

Figure P-10.10

10.11 Specify a Gray code for the 16-QAM V.29 signal constellation shown in Problem 10.10.

10.12 Consider the octal signal point constellations in Figure P-10.12.

   1. The nearest neighbor signal points in the 8-QAM signal constellation are separated in distance by A units. Determine the radii a and b of the inner and outer circles.

   2. The adjacent signal points in the 8-PSK are separated by a distance of A units. Determine the radius r of the circle.

   3. Determine the average transmitter powers for the two signal constellations, and compare the two powers. What is the relative power advantage of one constellation over the other? (Assume that all signal points are equally probable.)

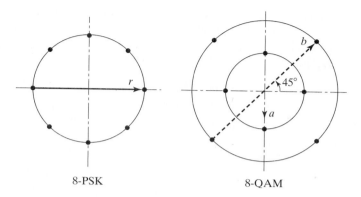

8-PSK   8-QAM   **Figure P-10.12**

10.13 Consider a digital communication system that transmits information via QAM over a voice-band telephone channel at a rate 2400 symbols per second. The additive noise is assumed to be white and Gaussian.

1. Determine the $\mathscr{E}_{\text{bav}}/N_0$ required to achieve an error probability of 10^{-5} at 4800 bps.

2. Repeat Part 1 for a bit rate of 9600 bps.

3. Repeat Part 1 for a bit rate of 19,200 bps.

4. What conclusions do you reach from these results?

10.14 Consider the eight-point QAM signal constellation shown in Figure P-10.12.

1. Assign three data bits to each point of the signal constellation so that the nearest (adjacent) points differ in only one bit position.

2. Determine the symbol rate if the desired bit rate is 90 Mbps.

3. Compare the SNR required for the eight-point QAM modulation with that required for an eight-point PSK modulation having the same error probability.

4. Which signal constellation (eight-point QAM or eight-point PSK) is more immune to phase errors? Explain the reason for your answer.

10.15 In Section 10.4.1, we showed that the minimum frequency separation for the orthogonality of binary FSK signals with coherent detection is $\Delta f = \frac{1}{2T}$. However, a lower error probability is possible with coherent detection of FSK if Δf is increased beyond $\frac{1}{2T}$. Show that the minimum value of the correlation occurs at $\Delta f = \frac{0.715}{T}$; determine the probability of error for this choice of Δf.

10.16 The lowpass equivalent signal waveforms for three signal sets are shown in Figure P-10.16. Each set may be used to transmit one of four equally probable messages over an additive white Gaussian noise channel with a noise power spectral density $\frac{N_0}{2}$.

1. Classify the signal waveforms in Set I, Set II, and Set III. In other words, state the category or class to which each signal set belongs.

2. What is the *average* transmitted energy for each signal set?

3. For signal Set I, specify the average probability of error if the signals are detected coherently.

4. For signal Set II, give a union bound on the probability of a symbol error if the detection is performed (i) coherently and (ii) noncoherently.

5. Is it possible to use noncoherent detection on signal Set III? Explain.

6. Which signal set or signal sets would you select if you wanted to achieve a bit-rate-to-bandwidth ($\frac{R}{W}$) ratio of at least 2? Explain your answer.

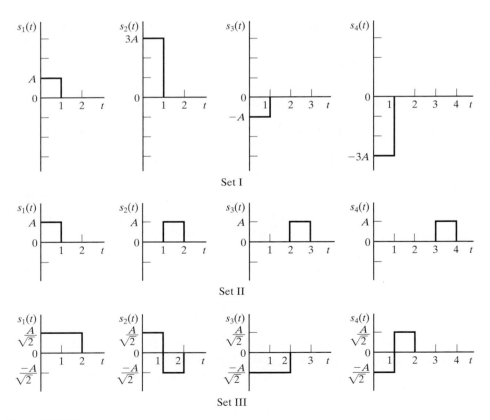

Set I

Set II

Set III

Figure P-10.16

10.17 Consider the phase-coherent demodulator for M-ary FSK signals as shown in Figure 10.28.

1. Assume that the signal

$$u_0(t) = \sqrt{\frac{2\mathcal{E}_s}{T}} \cos 2\pi f_c t, \quad 0 \le t \le T$$

was transmitted; determine the output of the $M - 1$ correlators at $t = T$ that corresponds to the signals $u_m(t)$, $m = 1, 2, \dots, M - 1$, when $\hat{\phi}_m \ne \phi_m$.

2. Show that the minimum frequency separation required for the signal orthogonality at the demodulator when $\hat{\phi}_m \ne \phi_m$ is $\Delta f = \frac{1}{T}$.

10.18 In the demodulation and noncoherent detection of M-ary FSK signals, as illustrated in Figure 10.29, show that the $2M$ noise samples that are given in Equations (10.4.8) and (10.4.9) are zero-mean, mutually independent Gaussian random variables with an equal variance $\sigma^2 = \frac{N_0}{2}$.

10.19 In on–off keying of a carrier modulated signal, the two possible signals are

$$s_0(t) = 0, \qquad\qquad 0 \le t \le T_b$$
$$s_1(t) = \sqrt{\frac{2\mathcal{E}_b}{T_b}} \cos 2\pi f_c t, \quad 0 \le t \le T_b.$$

The corresponding received signals are

$$r(t) = n(t), \qquad\qquad\qquad 0 \le t \le T_b$$
$$r(t) = \sqrt{\frac{2\mathcal{E}_b}{T_b}} \cos(2\pi f_c t + \phi) + n(t), \quad 0 \le t \le T_b,$$

where ϕ is the carrier phase and $n(t)$ is AWGN.

1. Sketch a block diagram of the receiver (demodulator and detector) that employs noncoherent (envelope) detection.

2. Determine the probability density functions for the two possible decision variables at the detector corresponding to the two possible received signals.

3. Derive the probability of error for the detector.

10.20 Digital information is to be transmitted by carrier modulation through an additive Gaussian noise channel with a bandwidth of 100 kHz and $N_0 = 10^{-10}$ W/Hz. Determine the maximum rate that can be transmitted through the channel for four-phase PSK, binary FSK, and four-frequency orthogonal FSK that is detected noncoherently.

10.21 Determine the bit rate that can be transmitted through a 4 kHz voice-band telephone (bandpass) channel if we use the following modulation methods: (1) binary PSK, (2) four-phase PSK, (3) eight-point QAM, (4) binary orthogonal FSK with noncoherent detection, (5) orthogonal four-FSK with noncoherent detection, and (6) orthogonal eight-FSK with noncoherent detection. For Parts 1–3, assume that the transmitter pulse shape has a raised-cosine spectrum with a 50% rolloff.

10.22 (Carrierless QAM or PSK Modem)

Consider the transmission of a QAM or M-ary PSK ($M \geq 4$) signal at a carrier frequency f_c, where the carrier is comparable to the bandwidth of the baseband signal. The bandpass signal may be represented as

$$s(t) = \text{Re}\left[\sum_n a_n g(t - nT)e^{j2\pi f_c t}\right].$$

1. Show that $s(t)$ can be expressed as

$$s(t) = \text{Re}\left[\sum_n a'_n Q(t - nT)\right],$$

where $Q(t)$ is defined as

$$Q(t) = q(t) + j\hat{q}(t)$$

$$q(t) = g(t)\cos 2\pi f_c t$$

$$\hat{q}(t) = g(t)\sin 2\pi f_c t$$

and a'_n is a phase-rotated symbol, i.e., $a'_n = a_n e^{j2\pi f_c nT}$.

2. Using filters with responses $q(t)$ and $\hat{q}(t)$, sketch the block diagram of the modulator and demodulator implementation that does not require the mixer to translate the signal to bandpass at the modulator and to baseband at the demodulator.

10.23 (Carrierless amplitude or phase (CAP) modulation)

In some practical applications in wireline data transmission, the bandwidth of the signal to be transmitted is comparable to the carrier frequency. In such systems, it is possible to eliminate the step of mixing the baseband signal with the carrier component. Instead, the bandpass signal can be directly synthesized by embedding the carrier component in the realization of the shaping filter. Thus, the modem is realized, as shown in the block diagram in Figure P-10.23, where the shaping filters have the impulse responses

$$q(t) = g(t)\cos 2\pi f_c t$$

$$\hat{q}(t) = g(t)\sin 2\pi f_c t$$

and $g(t)$ is a pulse that has a square-root raised cosine spectral characteristic.

1. Show that

$$\int_{-\infty}^{\infty} q(t)\hat{q}(t)\, dt = 0$$

and that the system can be used to transmit two dimensional signals, e.g., PSK and QAM.

2. Under what conditions is this CAP modem identical to the carrierless QAM/PSK modem treated in Problem 10.22?

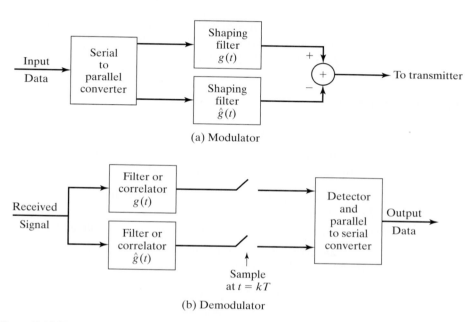

(a) Modulator

(b) Demodulator

Figure P-10.23

10.24 In an MSK signal, the initial state for the phase is either 0 or π radians. Determine the terminal phase state for the following four pairs of input data: (a) 00, (b) 01, (c) 10, and (d) 11.

10.25 A continuous-phase FSK signal with $h = 1/2$ is represented as

$$s(t) = \pm\sqrt{\frac{2\mathcal{E}_b}{T_b}}\cos\left(\frac{\pi t}{2T_b}\right)\cos 2\pi f_c t \pm \sqrt{\frac{2\mathcal{E}_b}{T_b}}\sin\left(\frac{\pi t}{2T_b}\right)\sin 2\pi f_c t, \\ 0 \le t \le 2T_b$$

where the $\pm$ signs depend on the information bits transmitted.

1. Show that this signal has a constant amplitude.

2. Sketch a block diagram of the modulator for synthesizing the signal.

3. Sketch a block diagram of the demodulator and detector for recovering the information.

10.26 Sketch the phase-tree, the state trellis, and the state diagram for partial-response CPM with $h = \frac{1}{2}$ and

$$
g(t) = \begin{cases} \frac{1}{4T}, & 0 \le t \le 2T. \\ 0, & \text{otherwise} \end{cases}
$$

10.27 Determine the number of terminal phase states in the state trellis diagram for (a) a full response binary CPFSK with either $h = \frac{2}{3}$ or $\frac{3}{4}$ and (b) a partial response $L = 3$ binary CPFSK with either $h = \frac{2}{3}$ or $\frac{3}{4}$.

COMPUTER PROBLEMS

10.1 Generation of PSK Signal Waveforms

The objective of this problem is to generate constant envelope PSK signal waveforms described mathematically by the expression

$$
u_m(t) = \sqrt{\frac{2\mathcal{E}_s}{T}} \cos\left(2\pi f_c t + \frac{2\pi m}{M}\right), \quad \begin{array}{l} m = 0, 1, 2, \dots, M - 1. \\ 0 \le t \le T \end{array}
$$

For convenience, the signal amplitude may be normalized to unity.

Generate and plot the PSK signal waveforms for the case in which $f_c = 6/T$ and $M = 8$ over the time interval $0 \le t \le T$.

10.2 Simulation of $M = 4$ PSK Modulation and Detection

The objective of this problem is to estimate the probability of error for a communication system that employs $M = 4$ PSK modulation. The model for the system to be simulated is shown in Figure CP-10.2.

As shown, we simulate the generation of the random vector $y = s_m + n$, which is the output of the signal correlator and the input to the detector. We begin by generating two-bit symbols that are mapped into the corresponding four-phase signal points. To accomplish this task, we use a random number generator that generates a uniform random number in the range $(0, 1)$. This range is subdivided into four equal intervals, $(0, 0.25)$, $(0.25, 0.50)$, $(0.50, 0.75)$, and $(0.75, 1.0)$, where the subintervals correspond to the pairs of information bits 00, 01, 11, and 10, respectively. These pairs of bits are used to select the signal phase vectors s_m, $m = 1, 2, 3, 4$.

The additive noise components n_c and n_s are statistically independent zero-mean Gaussian random variables with an equal variance σ^2. For convenience, we may normalize the signal energy $\mathcal{E}_s$ to unity and control the SNR by scaling the noise variance σ^2.

The detector observes the received signal vector $\boldsymbol{y}$ in each symbol interval and computes the projection (dot product) of $\boldsymbol{y}$ into the four possible signal vectors $\boldsymbol{s}_m$. Its decision is based on selecting the signal point corresponding to the largest projection. The output decisions from the detector are compared with the transmitted symbols, and symbol errors and bit errors are counted.

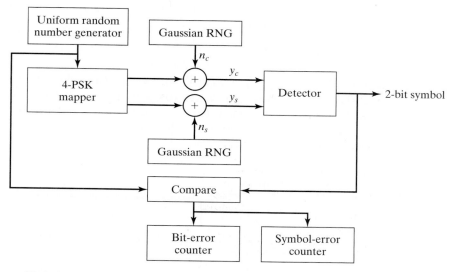

Figure CP-10.2

Perform the simulation of the four-phase system just described for 10,000 symbols (20,000 bits) for the range of SNR $0 \leq 10 \log_{10} \mathcal{E}_b/N_0 \leq 8\, dB$, and plot the estimated probability of a bit error and probability of a symbol error. Also, plot the theoretical bit error probability for $M = 4$ PSK and compare the simulated performance to the theoretical error probability. Comment on the results.

10.3 Simulation of $M = 4$ Differential PSK Modulation and Detection

The objective of this problem is to estimate the probability of error for a communication system that employs $M = 4$ differential PSK (DPSK) modulation. The model for the system to be simulated is shown in Figure CP-10.3.

The $M = 4$ DPSK mapper changes the phase of the transmitted signal relative to the phase of the previous transmission as follows: (a) by 0 radians if the bits to be transmitted are 00, (b) by $\pi/2$ radians if the bits to be transmitted are 01, (c) by π radians if the bits to be transmitted are 11, and (d) by $3\pi/2$ radians if the bits to be transmitted are 10. Two Gaussian random noise generators (RNG) are used to generate the noise components (n_c, n_s), which are uncorrelated, are zero mean, and have the variance σ^2. Hence, the received signal-plus-noise vector is

$$y = \left[\cos \frac{\pi m}{2} + n_c \quad \sin \frac{\pi m}{2} + n_s \right]$$

$$= [y_c \ y_s].$$

The differential detector basically computes the phase difference between two successive received signal vectors y_k and Y_{k-1}. Mathematically, this computation can be performed as

$$y_k y_{k-1}^* = (y_{ck} + jy_{sk})(y_{ck-1} - jy_{sk-1})$$
$$= y_r + j\ y_i,$$

and $\theta_k = \tan^{-1}(y_i/y_r)$ is the phase difference. The value of θ_k is compared with the possible phase differences ($0°$, $90°$, $180°$, and $270°$), and a decision is made in favor of the phase that is closest to θ_k. The detected phase is then mapped into the pair of information bits. The error counter counts the symbol errors in the detected sequence.

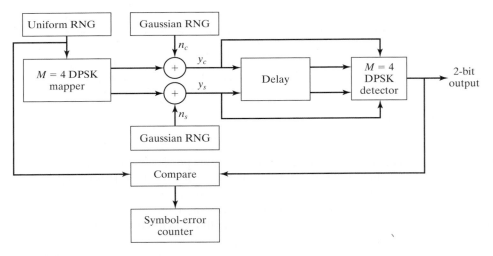

Figure CP-10.3

Perform the simulation of the $M = 4$ DPSK system as previously described for 10,000 symbols (20,000 bits) for the range of SNR $0 \le 10 \log_{10} \mathscr{E}_b/N_0 \le 10$ dB, and plot the estimated probability of a symbol error. Also, plot the theoretical symbol error probability for $M = 4$ PSK and compare the simulated performance for $M = 4$ DPSK to the theoretical error probability for $M = 4$ phase coherent PSK. Comment on the results.

10.4 Simulation of $M = 16$ QAM Modulation and Detection

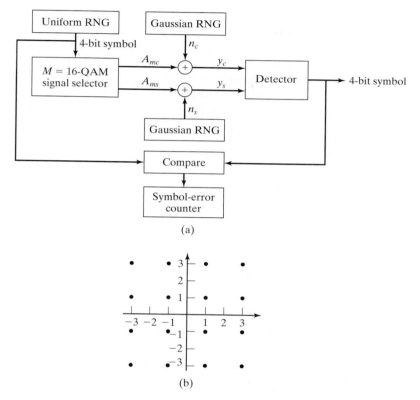

Figure CP-10.4

The objective of this problem is to estimate the probability of error for a communication system that employs $M = 16$ quadrature-amplitude modulation (QAM). The model of the system to be simulated is shown in Figure CP-10.4(a). The uniform random number generator (RNG) is used to generate the sequence of information symbols corresponding to the 16 possible 4-bit combinations of b_1, b_2, b_3 and b_4. The information symbols are mapped into the corresponding signal points, as shown in Figure CP-10.4(b), which have the coordinates $[A_{mc}, A_{ms}]$. Two Gaussian RNG's are used to generate the uncorrelated, zero-mean noise components $[n_c, n_s]$, each of which has a variance σ^2. Consequently, the received signal-plus-noise vector at the input to the detector is

$$y = [A_{mc} + n_c \quad A_{ms} + n_s].$$

The detector computes the Euclidean distance between y and each of the 16 possible transmitted signal vectors; then it selects the signal point that is closest to the received vector y. The error counter counts the symbol errors in the detected sequence.

Perform the simulation of the $M = 16$ QAM system as previously described for 10,000 symbols (40,000 bits) for the range SNR $0 \leq 10 \log_{10} \mathcal{E}_{bav}/N_0 < 13$ dB, and plot the estimated probability of a symbol error. Also, plot the theoretical symbol error probability for $M = 16$ QAM and compare the simulated performance to the theoretical error probability. Comment on the results.

10.5 Demodulation of Binary FSK Signals

The objective of this problem is to digitally implement a correlation-type demodulator in a digital communication system that employs binary FSK signal waveforms, which are given as

$$u_1(t) = \cos 2\pi f_1 t, \quad 0 \leq t \leq T_b$$
$$u_2(t) = \cos 2\pi f_2 t, \quad 0 \leq t \leq T_b,$$

where $f_1 = 1000/T_b$ and $f_2 = f_1 + 1/T_b$. The channel is assumed to impart a phase shift $\phi = 45°$ on each of the transmitted signals. Consequently, the received signal in the absence of noise is

$$r(t) = \cos(2\pi f_i t + \pi/4), \quad i = 1, 2 \ldots, 0 \leq t \leq T_b.$$

The correlation-type demodulator for $u_1(t)$ and $u_2(t)$ is to be implemented digitally.

Suppose we sample the received signal $r(t)$ at a rate $F_s = 5000/T_b$ within the bit interval $0 \leq t \leq T_b$. Thus, the received signal $r(t)$ is represented by the 5000 samples $\{r(n/F_s)\}$. The correlation demodulator multiplies $\{r(n/F_s)\}$ by the sampled versions of $u_1(t) = \cos 2\pi f_1 t$, $v_1(t) = \sin 2\pi f_1 t$, $u_2(t) = \cos 2\pi f_2 t$, and $v_2(t) = \sin 2\pi f_2 t$. Thus, the correlator outputs are

$$y_{1c}(k) = \sum_{n=0}^{k} r\left(\frac{n}{F_s}\right) u_1\left(\frac{n}{F_s}\right), \quad k = 1, 2, \ldots, 5000;$$

$$y_{1s}(k) = \sum_{n=0}^{k} r\left(\frac{n}{F_s}\right) v_1\left(\frac{n}{F_s}\right), \quad k = 1, 2, \ldots, 5000;$$

$$y_{2c}(k) = \sum_{n=0}^{k} r\left(\frac{n}{F_s}\right) u_2\left(\frac{n}{F_s}\right), \quad k = 1, 2, \ldots, 5000;$$

$$y_{2s}(k) = \sum_{n=0}^{k} r\left(\frac{n}{F_s}\right) v_2\left(\frac{n}{F_s}\right), \quad k = 1, 2, \ldots, 5000.$$

The detector is a square-law detector that computes the two decision variables

$$y_1 = y_{1c}^2(5000) + y_{1s}^2(5000)$$
$$y_2 = y_{2c}^2(5000) + y_{2s}^2(5000)$$

and selects the information bit corresponding to the larger decision variable.

Write a program that implements the correlation demodulator for the binary FSK signal processing. Assuming that $r(t) = \cos 2\pi f_1 t$, plot the four sequences $\{y_{1c}(k), y_{1s}(k), y_{2c}(k), y_{2s}(k)\}$ for $0 \le k \le 5000$. Repeat the computation when $r(t) = \cos 2\pi f_2 t$. What are the values of y_1 and y_2 in each of the two experiments?

10.6 Simulation of Binary FSK Modulation and Detection

The objective of this problem is to estimate the probability of error for a communication system that employs binary FSK modulation. The binary FSK waveforms are

$$u_1(t) = \cos 2\pi f_1 t, \qquad\qquad 0 \le t \le T_b$$
$$u_2(t) = \cos 2\pi \left(f_1 + \frac{1}{T_b} \right) t, \ \ 0 \le t \le T_b.$$

The block diagram of the binary FSK system to be simulated is shown in Figure CP-10.6.

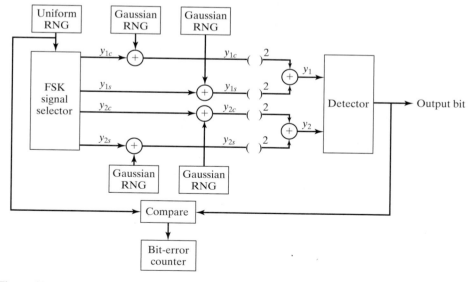

Figure CP-10.6

Since the signals are orthogonal, when $u_1(t)$ is transmitted, the first demodulator output is

$$y_{1c} = \sqrt{\mathcal{E}_b}\cos\phi + n_{1c}$$
$$y_{1s} = \sqrt{\mathcal{E}_b}\sin\phi + n_{1s}$$

and the second demodulator output is

$$y_{2c} = n_{2c}$$
$$y_{2s} = n_{2s},$$

where n_{1c}, n_{1s}, n_{2c}, and n_{2s} are mutual statistically independent, zero-mean Gaussian random variables with equal variance σ^2 and ϕ represents the channel phase shift. The phase shift ϕ may be set to zero for convenience. The square-law detector computes

$$y_1 = y_{1c}^2 + y_{1s}^2$$
$$y_2 = y_{2c}^2 + y_{2s}^2$$

and selects the information bit corresponding to the larger of these two decision variables. An error counter measures the error rate by comparing the transmitted sequence to the output of the detector.

Perform the simulation of the binary FSK system as described for 10,000 bits for the range of the $0 \le 10\log_{10}\mathcal{E}_b/N_0 \le 12$ dB and plot the estimated probability of error. Also plot the theoretical bit error probability of binary FSK and compare the simulated performance to the theoretical error probability. Comment on the results.

11

Selected Topics in Digital Communications

In Chapters 8, 9, and 10, we introduced the basic digital modulation and demodulation techniques and derived their performance in terms of the average probability of error in additive white Gaussian noise channels. In this chapter, we provide additional breadth in our coverage of digital communications by treating several important topics.

The first topic focuses on digital communication through fading multipath channels. This topic is especially important in the design and performance analysis of digital cellular communication systems in particular, and non-line-of-sight radio communication systems in general. The second topic is orthogonal frequency division multiplexed (OFDM) signals and, more generally, multicarrier modulation. OFDM is now widely used in wireless LAN modems and in modems for data transmission through digital subcarrier loops (DSL), which provide high-speed internet access. The third topic is spread spectrum signals and their practical applications in the design of communication systems that are robust to high levels of interference. The fourth topic examined consists of two digital cellular communication systems. The first is based on time division multiple access (TDMA), and the second is based on code division multiple access (CDMA). The last topic concerns digital repeaters in wireline communication systems and link budget analysis in radio communication systems.

11.1 DIGITAL TRANSMISSION IN FADING MULTIPATH CHANNELS

In Chapter 10, our treatment of digital modulation and demodulation methods focused on a time-invariant linear filter channel model with additive white Gaussian noise. Such a channel model is appropriate for characterizing physical channels that are relatively static, e.g., wireline channels. In contrast, wireless communication channels, such as radio and underwater acoustic channels, have transmission characteristics that are time-varying, so a more complex model is needed to describe their behavior.

Physical channels with time-varying transmission characteristics may be characterized as time-varying linear filters. Such linear filters are described by a time-varying impulse response $h(\tau; t)$, where $h(\tau; t)$ is the response of the channel at time t due to an impulse applied at time $t - \tau$. Thus, τ denotes the "age" (elapsed time) variable. The time-varying linear filter model of the channel with additive noise was previously shown in Figure 1.10. The corresponding frequency response of the time-varying channel is given by the Fourier transform of $h(\tau; t)$ with respect to the τ-variable and is denoted as $H(f; t)$. Examples of these types of wireless communication channels include signal transmission via ionospheric propagation and mobile cellular transmission.

Signal Transmission via Ionospheric Propagation in the HF Band. In Chapter 1, we noted that sky-wave propagation, as illustrated in Figure 1.6, results when transmitted signals (in the HF frequency band) are bent or refracted by the ionosphere (which consists of several layers or charged particles ranging in altitude from 30–250 miles above the surface of the earth). As a consequence, the signal arrives at the receiver via different propagation paths at different delays. These signal components are called *multipath components*. The signal multipath components generally have different carrier-phase offsets; hence, they may add destructively at times, resulting in a phenomenon called *signal fading*. To characterize such channel behavior, a time-varying impulse response model is appropriate.

Mobile Cellular Transmission. In mobile cellular radio transmission, the signal transmitted from the base station to the mobile receiver is usually reflected from surrounding buildings, hills, or other obstructions. As a consequence, multiple propagation paths arrive at the receiver at different delays. Hence, the received signal has characteristics similar to those for ionospheric propagation. The same is true of transmission from the mobile station to the base station. Moreover, the traveling speed of the mobile station (i.e. automobile, train) results in frequency offsets, called *Doppler shifts*, of the various frequency components of the signal.

11.1.1 Channel Models for Time-Variant Multipath Channels

There are basically two distinct characteristics of the radio channels we just described. First, the transmitted signal arrives at the receiver via multiple propagation paths, each of which has an associated propagation time delay. Second, the time variations can occur either in the structure of the medium, such as the ionosphere, or the motion and position of the transmitter and receiver relative to the terrain between them. As a result of such time variations, the channel's response to any signal transmitted through it will change with time. In other words, the impulse response of the channel varies with time. In general, the time variations in the received signal appear to be unpredictable to the user of the channel. This leads us to characterize the time-variant multipath channel statistically.

To obtain a statistical description of the channel consider the transmission of an unmodulated carrier

$$s(t) = A \cos 2\pi f_c t. \tag{11.1.1}$$

The received signal in the absence of noise may be expressed as

$$x(t) = A \sum_N \alpha_n(t) \cos[2\pi f_c(t - \tau_n(t)]$$

$$= A\text{Re} \left[\sum_N \alpha_n(t) e^{-j2\pi f_c \tau_n(t)} e^{j2\pi f_c t} \right]$$

$$= \text{Re} \left[c(t) e^{j2\pi f_c t} \right], \tag{11.1.2}$$

where $\alpha_n(t)$ is the time-variant attenuation factor associated with the nth propagation path and $\tau_n(t)$ is the corresponding propagation delay. The complex-valued signal

$$c(t) = \sum_n \alpha_n(t) e^{-j2\pi f_c \tau_n(t)}$$

$$= \sum_n \alpha_n(t) e^{-j\phi_n(t)} \tag{11.1.3}$$

represents the response of the channel to the complex exponential $\exp(j2\pi f_c t)$. Although the input to the channel is a monochromatic signal, i.e., a signal at a single frequency, the output of the channel consists of a signal that contains many different frequency components. These new components are generated as a result of the time variations in the channel response. The r.m.s. (root-mean-square) spectral width of $c(t)$ is called the *Doppler frequency spread* of the channel and is denoted as B_d. This quantity is a measure of how rapidly the signal $c(t)$ is changing with time. If $c(t)$ changes slowly, the Doppler frequency spread is relatively small; if $c(t)$ changes rapidly, the Doppler frequency spread is large.

We may view the received complex-valued signal $c(t)$ in Equation (11.1.3) as the sum of a number of vectors (phasors), each of which has a time-variant amplitude $\alpha_n(t)$ and phase $\phi_n(t)$. In general, it takes large dynamic changes in the physical medium to cause a large change in $\{\alpha_n(t)\}$. On the other hand, the phases $\{\phi_n(t)\}$ will change by 2π radians whenever $\{\tau_n(t)\}$ changes by $1/f_c$. But $1/f_c$ is usually a small number; hence, the phases $\{\phi_n(t)\}$ change by 2π or more radians with relatively small changes of the medium characteristics. We also expect the delays $\{\tau_n(t)\}$ associated with the different signal paths to change at different rates and in an unpredictable (random) manner. This implies that the complex-valued signal $c(t)$ in Equation (11.1.3) can be modeled as a random process. When there are a large number of signal propagation paths, $c(t)$ can be modeled as a zero-mean complex-valued Gaussian random process.

The multipath propagation model for the channel, embodied in the received signal $x(t)$ or, equivalently, $c(t)$ given by Equation (11.1.3), results in signal fading. The fading phenomenon is primarily a result of the time-variant phase factors $\{\phi_n(t)\}$. At times, the complex-valued vectors in $c(t)$ add destructively to reduce the power level of the received signal. At other times, the vectors in $c(t)$ add constructively, to produce a large signal value. The amplitude variations in the received signal due to the time-variant multipath propagation in the channel are usually called *signal fading*.

The time span between the first and the last arriving multipath components in a transmitted signal is called the *multipath* (time) *spread* of the channel. We denote this channel parameter as T_m. A related parameter is the reciprocal of the multipath spread, which provides a measure of the bandwidth over which frequency components of the transmitted signal will be affected similarly by the channel. Thus, we define the channel parameter

$$B_{cb} = \frac{1}{T_m} \tag{11.1.4}$$

and call it the *coherence bandwidth* of the channel. For example, all frequency components of a transmitted signal that fall within the coherence bandwidth B_{cb} will fade simultaneously. If the transmitted signal has a bandwidth $W < B_{cb}$, the channel is called *frequency nonselective*. Thus, at any instant in time, all frequency components of the transmitted signal fade simultaneously. On the other hand, if the transmitted signal has a bandwidth $W > B_{cb}$, the frequencies in the signal separated by an amount greater than B_{cb} will be affected differently by the channel. Hence, at any instant, some frequency components in the transmitted signal may fade, while other frequency components may not. In such a case, the channel is said to be *frequency selective.*

Another related channel parameter is the reciprocal of the Doppler spread, which measures the time interval over which the channel response will change very little. Thus, we define the parameter

$$T_{ct} = \frac{1}{B_d} \tag{11.1.5}$$

and call it the *coherence time* of the channel. For example, a signal that is transmitted at two different time instants, separated in time by an amount less that T_{ct}, will be affected similarly by the channel. Hence, if the signal that is transmitted in the first time instant is highly attenuated by the channel, the signal transmitted in the second time instant will also be highly attenuated. On the other hand, if the time interval between the signal transmissions is much greater than the coherence time T_{ct}, the channel will likely affect the two signal transmissions differently.

Table 11.1 lists the multipath spread and the Doppler spread of several time-varying multipath radio channels.

TABLE 11.1   MULTIPATH SPREAD AND DOPPLER SPREAD OF FADING MULTIPLE CHANNELS

Type of channel	Multipath duration(s)	Doppler spread (Hz)
Shortwave ionospheric propagation (HF)	10^{-3}–10^{-2}	10^{-1}–1
Ionospheric propagation under disturbed auroral conditions	10^{-3}–10^{-2}	10–100
Wireless indoor LANs at 5 GHz	10^{-7}	10^3
Tropospheric scatter	10^{-6}	10
Mobile cellular (UHF)	10^{-5}	100

Example 11.1.1

A shortwave ionospheric radio channel is characterized by a multipath spread of $T_m = 5$ milliseconds and a Doppler spread of $B_d = 0.1$ Hz. Determine the coherence bandwidth and the coherence time of the channel.

Solution   The coherence bandwidth of the channel is

$$B_{cb} = \frac{1}{T_m} = 200 \text{ Hz}.$$

The coherence time of the channel is

$$T_{ct} = \frac{1}{B_d} = 10 \text{ sec.} \qquad \blacksquare$$

Frequency Nonselective Rayleigh Fading Channel.   Consider the transmission of a signal $s(t)$ over a linear time-varying channel with the frequency response $H(f; t)$. If $S(f)$ denotes the spectrum of the transmitted signal, the received signal is given in the frequency domain as $H(f; t)S(f) \equiv R(f)$ and in the time domain as the inverse Fourier transform of $R(f)$, i.e.,

$$r(t) = \int_{-\infty}^{\infty} H(f; t)S(f)e^{j2\pi ft} \, df. \tag{11.1.6}$$

Now, suppose that the bandwidth W of the transmitted signal $s(t)$ satisfies the condition $W \ll B_{cb}$, so that the channel is frequency nonselective. This condition implies that, over the bandwidth range $(-W, W)$ occupied by the transmitted signal, the frequency response is constant in the frequency variable f and may be denoted as

$$H(f; t)|_{f=0} = H(0; t) \equiv c(t). \tag{11.1.7}$$

Therefore, Equation (11.1.6) simplifies to

$$r(t) = c(t) \int_{-\infty}^{\infty} S(f)e^{j2\pi ft} \, df$$

$$= c(t)s(t). \tag{11.1.8}$$

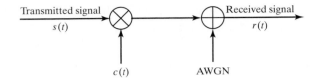

Figure 11.1 Model of a frequency-nonselective time-varying channel with AWGN.

Consequently, in a frequency nonselective channel, the channel distorts the transmitted signal in a multiplicative manner, as illustrated in Figure 11.1.

Another view of the frequency nonselective channel is obtained when the signal $s(t)$ with bandwidth W has a time duration $T \approx 1/W$; since $W \ll B_{cb} = 1/T_m$, it follows that $T \gg T_m$. In this case, the time dispersion due to the channel multipath is much smaller than the time duration T of the transmitted signal. Hence, the channel multipath components, whose amplitudes and phases are given by Equation (11.1.3), are not resolvable. Thus, they are seen only as a single disturbance that multiplies the transmitted signal $s(t)$ as shown in Figure 11.1, and causes fading.

Since the $c(t)$ given in Equation (11.1.3) is complex valued, we may express $c(t)$ as

$$c(t) = c_r(t) + jc_i(t), \tag{11.1.9}$$

where $c_r(t)$ and $c_i(t)$ represent real-valued Gaussian random processes. We assume that $c_r(t)$ and $c_i(t)$ are stationary and statistically independent.

We may also express $c(t)$ as

$$c(t) = \alpha(t)e^{j\phi(t)}, \tag{11.1.10}$$

where

$$\alpha(t) = \sqrt{c_r^2(t) + c_i^2(t)}$$

$$\phi(t) = \tan^{-1} \frac{c_i(t)}{c_r(t)}. \tag{11.1.11}$$

In this representation, if $c_r(t)$ and $c_i(t)$ are Gaussian with zero-mean values, the amplitude $\alpha(t)$ is characterized statistically by the Rayleigh probability distribution and $\phi(t)$ is uniformly distributed over the interval $(0, 2\pi)$. As a consequence, the channel is called a *Rayleigh fading channel*. The Rayleigh fading signal amplitude is described by the PDF

$$f(\alpha) = \frac{\alpha}{\sigma^2} e^{-\alpha^2/2\sigma^2}, \ \alpha \geq 0, \tag{11.1.12}$$

and $f(\alpha) = 0$ for $\alpha < 0$. The parameter $\sigma^2 = E\left(c_r^2\right) = E\left(c_i^2\right)$.

The frequency nonselective channel model illustrated in Figure 11.1 applies when the transmitted signal bandwidth W satisfies the condition $W \ll B_{cb}$. A further simplification occurs when the coherence time T_{ct} of the channel is much larger than the signal time duration T, i.e., $T_{ct} \gg T$. In this case, the channel characteristic $c(t)$ may be treated as a constant over the signal duration T and may be expressed as

$$
\begin{aligned}
c(t) &= \alpha(t)e^{j\phi(t)} \quad , 0 \le t \le T \\
&= \alpha e^{j\phi} \quad\quad , 0 \le t \le T
\end{aligned}
\tag{11.1.13}
$$

We call such a channel a *slowly fading, frequency nonselective channel*.

Example 11.1.2

Consider the radio channel in Example 11.1.1. The signal $s(t)$ transmitted over the channel has a bandwidth $W = 50$ Hz and a time duration of $T \approx 1/W = 20$ milliseconds. Is this a frequency nonselective channel? Is the channel slowly fading?

Solution   Since the bandwidth $W \ll B_{cb} = 200$ Hz, the channel is frequency nonselective. Furthermore, since $T \ll T_{ct} = 10$ seconds, the channel is also slowly fading. ∎

11.1.2 Performance of Binary Modulation in Frequency Nonselective Rayleigh Fading Channels

In this section, we determine the probability of error at the receiver of a binary digital communication system that transmits information through a Rayleigh fading channel. The signal bandwidth W is assumed to be much smaller than the coherence bandwidth B_{cb} of the channel, so that the frequency nonselective channel model that is shown in Figure 11.1 applies, with $c(t) = \alpha(t)e^{j\phi(t)}$. We assume that the time variations in $c(t)$ are very slow compared with the bit interval, so that within the time interval $0 \le t \le T, c(t)$ is a constant, i.e., $c = \alpha e^{j\phi}$, where α is Rayleigh distributed with PDF

$$
f(\alpha) = \begin{cases} \frac{\alpha}{\sigma^2} e^{-\alpha^2/2\sigma^2}, & \alpha \ge 0 \\ \\ 0, & \text{otherwise} \end{cases}
\tag{11.1.14}
$$

and ϕ is uniformly distributed over the interval $(0, 2\pi)$.

Now, suppose that binary antipodal signals, (e.g., binary PSK), are used to transmit the information through the channel. Hence, the two possible signals are

$$
u_m(t) = \sqrt{\frac{2\mathcal{E}_b}{T}} \cos(2\pi f_c t + m\pi) + n(t), \qquad m = 0, 1.
\tag{11.1.15}
$$

The received signal in the interval $0 \le t \le T$ is

$$
r(t) = \alpha\sqrt{\frac{2\mathcal{E}_b}{T}} \cos(2\pi f_c t + m\pi + \phi) + n(t),
\tag{11.1.16}
$$

where ϕ is the carrier-phase offset. Now, assume that ϕ is perfectly estimated at the demodulator, which cross correlates $r(t)$ with

$$
\psi(t) = \sqrt{\frac{2}{T}} \cos(2\pi f_c t + \phi), \qquad 0 \le t \le T.
\tag{11.1.17}
$$

Hence, the input to the detector at the sampling instant is

$$y = \alpha\sqrt{\mathcal{E}_b}\cos m\pi + n, \qquad m = 0, 1. \tag{11.1.18}$$

For a fixed value of α, the probability of error is the familiar form

$$P_2(\alpha) = Q\left(\sqrt{\frac{2\alpha^2\mathcal{E}_b}{N_0}}\right). \tag{11.1.19}$$

$P_2(\alpha)$ is the conditional probability of error for a given value of the channel attenuation α. To determine the probability of error averaged over all possible values of α, we compute

$$P_2 = \int_0^\infty P_2(\alpha)f(\alpha)d\alpha, \tag{11.1.20}$$

where $f(\alpha)$ is the Rayleigh PDF given by Equation (11.1.14). This integral has the simple closed-form expression

$$P_2 = \frac{1}{2}\left[1 - \sqrt{\frac{\overline{\rho}_b}{1 + \overline{\rho}_b}}\right], \tag{11.1.21}$$

where, by definition,

$$\overline{\rho}_b = \frac{\mathcal{E}_b}{N_0}E(\alpha^2). \tag{11.1.22}$$

Hence, $\overline{\rho}_b$ is the average received SNR/bit and $E(\alpha^2) = 2\sigma^2$.

If the binary signals are orthogonal, as in orthogonal FSK, where the two possible transmitted signals are

$$u_m(t) = \sqrt{\frac{2\mathcal{E}_b}{T}}\cos\left[2\pi\left(f_c + \frac{m}{2T}\right)t\right], \qquad m = 0, 1, \tag{11.1.23}$$

then the received signal is

$$r(t) = \alpha\sqrt{\frac{2\mathcal{E}_b}{T}}\cos\left[2\pi\left(f_c + \frac{m}{2T}\right)t + \phi\right] + n(t). \tag{11.1.24}$$

In this case, the received signal is cross correlated with the two signals

$$\psi_1(t) = \sqrt{\tfrac{2}{T}}\cos(2\pi f_c t + \phi)$$

and

$$\psi_2(t) = \sqrt{\tfrac{2}{T}}\cos\left[2\pi\left(f_c + \tfrac{1}{2T}\right)t + \phi\right]. \tag{11.1.25}$$

If $m = 0$, for example, then the two correlator outputs are

$$r_1 = \alpha \sqrt{\mathcal{E}_b} + n_1$$

and

$$r_2 = n_2,$$

(11.1.26)

where n_1 and n_2 are the additive noise components at the outputs of the two correlators. Hence, the probability of error is simply the probability that $r_2 > r_1$. Since the signals are orthogonal, the probability of error for a fixed value of α has the familiar form

$$P_2(\alpha) = Q\left(\sqrt{\frac{\alpha^2 \mathcal{E}_b}{N_0}}\right).$$

(11.1.27)

As in the case of antipodal signals, the average probability of error over all values of α is determined by evaluating the integral in Equation (11.1.20). Thus, we obtain

$$P_2 = \frac{1}{2}\left[1 - \sqrt{\frac{\overline{\rho}_b}{2 + \overline{\rho}_b}}\right],$$

(11.1.28)

where $\overline{\rho}_b$ is the average SNR/bit defined by Equation (11.1.22).

Figure 11.2 illustrates the average probability of error for binary antipodal and orthogonal signals. The striking aspects of these graphs is the slow decay of the probability of error as a function of SNR. In fact, for large values of $\overline{\rho}_b$, the probability of error for binary signals is

$$P_2 \approx \frac{1}{4\overline{\rho}_b}, \quad \text{antipodal signals}$$

and

$$P_2 \approx \frac{1}{2\overline{\rho}_b}, \quad \text{orthogonal signals.}$$

(11.1.29)

Hence, the probability of error in both cases decreases only inversely as the SNR. This is in contrast to the exponential decrease in the case of the AWGN channel. We also note that the difference in SNR between antipodal signals (binary PSK) and orthogonal signals (binary FSK) is 3 dB.

Two other types of binary signal modulation are DPSK and noncoherent FSK. For completeness, we state that the average probability of error for these signals is

$$P_2 = \frac{1}{2(1+\overline{\rho}_b)}, \quad \text{DPSK;}$$

(11.1.30)

$$P_2 = \frac{1}{2+\overline{\rho}_b}, \quad \text{noncoherent FSK.}$$

(11.1.31)

11.1.3 Performance Improvement through Signal Diversity

The basic problem in digital communication through a fading channel is that a large number of errors occur when the channel attenuation is large, i.e., when the channel is in a deep fade. If we can supply the receiver with two or more replicas of the same

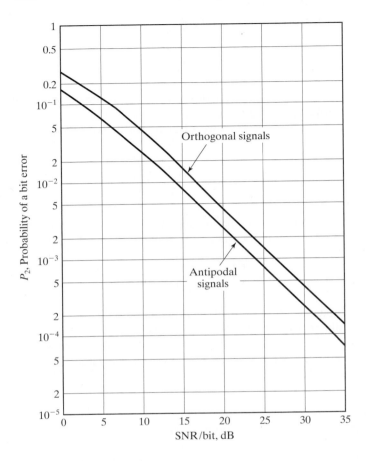

Figure 11.2 Performance of binary signaling in a Rayleigh fading channel.

information signal transmitted through independently fading channels, we considerably reduce the probability that all the transmitted signals will fade simultaneously. If p is the probability that any one signal will fade below some critical value, then p^D is the probability that all D independently fading replicas of the same signal will fade below that critical value. There are several ways that we can provide the receiver with D independently fading replicas of the same information-bearing signal.

One method is to transmit the same information on D carrier frequencies, where the separation between successive carriers equals or exceeds the coherence bandwidth B_{cb} of the channel. This method is called *frequency diversity*.

A second method is to transmit the same information in D different time slots, where the time separation between successive time slots equals or exceeds the coherence time T_{ct} of the channel. This method is called *time diversity*.

Another common method is to use multiple receiving antennas with only one transmitting antenna. The receiving antennas must be spaced sufficiently far apart so that the multipath components in the signal have significantly different propagation

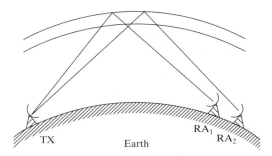

Figure 11.3 Illustration of diversity reception using two receiving antennas.

paths, as illustrated in Figure 11.3. Usually, a separation of a few wavelengths is required between a pair of receiving antennas to obtain signals that fade independently.

Other diversity transmission and reception techniques are angle-of-arrival diversity and polarization diversity.

Given that the information is transmitted to the receiver via D independently fading channels, there are several ways that the receiver may extract the transmitted information from the received signal. The simplest method is for the receiver to monitor the received power level in the D received signals and to select the strongest signal for demodulation and detection. In general, this approach results in frequent switching from one signal to another. A slight modification that leads to a simpler implementation is to use a signal for demodulation and detection as long as the received power level in that signal is above a preset threshold. When the signal falls below the threshold, a switch is made to the channel that has the largest received power level. This method of signal selection is called *selection diversity*.

For better performance, we may use more complex methods for combining the independently fading received signals, as illustrated in Figure 11.4. One method that is appropriate for coherent demodulation and detection requires that the receiver estimate and correct for the different phase offsets on each of the D received signals after demodulation. Then, the phase-corrected signals at the outputs of the D demodulators are summed and fed to the detector. This type of signal combining is called *equal-gain combining*.

In addition, we can estimate the received signal power level for each of the D received signals. Then, we can weight the phase-corrected demodulator outputs in direct proportion to the received signal strength (the square root of the power level) and feed them to the detector. This combiner is called a *maximal-ratio combiner*. On the other hand, we can use orthogonal signals for transmitting the information through D independently fading channels, and the receiver may employ noncoherent demodulation. In such a case, the outputs from the D demodulators may be squared, summed, and fed to the detector. This combiner is called a *square-law combiner*.

Each of these combining methods leads to performance characteristics that result in a probability of error that behaves as $K_D/\overline{\rho}^D$, where K_D is a constant that depends on D and $\overline{\rho}$ is the average SNR/diversity channel. Thus, we achieve an exponential decrease in the error probability. Without providing a detailed derivation, we simply state that, for antipodal signals with maximal ratio combining, the

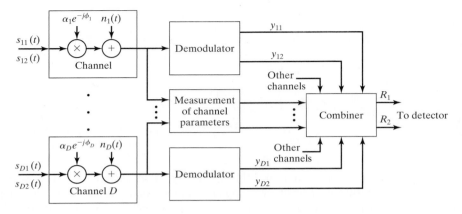

Equal gain combiner: $R_m = \sum_{k=1}^{D} y_{km}\, e^{j\phi_k},\, m = 1, 2$

Maximal ratio combiner: $R_m = \sum_{k=1}^{D} \alpha_k y_{km}\, e^{j\phi_k},\, m = 1, 2$

Square-law combiner: $R_m = \sum_{k=1}^{D} |y_{km}|^2,\, m = 1, 2$

Figure 11.4   Model of binary digital communication system with D-order diversity.

probability of error has the general form

$$P_2 \approx \frac{K_D}{(4\bar{\rho})^D}, \quad \bar{\rho} \gg 1, \tag{11.1.32}$$

where

$$K_D = \frac{(2D-1)!}{D!(D-1)!}. \tag{11.1.33}$$

For binary orthogonal signals with square-law combining, the probability of error has the asymptotic form

$$P_2 = \frac{K_D}{\bar{\rho}^D}, \quad \bar{\rho} \gg 1. \tag{11.1.34}$$

Finally, for binary DPSK with equal gain combining, the probability of error has the asymptotic form

$$P_2 \approx \frac{K_D}{(2\bar{\rho})^D}, \quad \bar{\rho} \gg 1. \tag{11.1.35}$$

These error probabilities are plotted in Figure 11.5 for $D = 1, 2, 4$ as a function of the SNR/bit $\bar{\rho}_b = D\bar{\rho}$. It is apparent that a large reduction in the SNR/bit is achieved with $D = 2$ (dual diversity) compared with no diversity. A further reduction

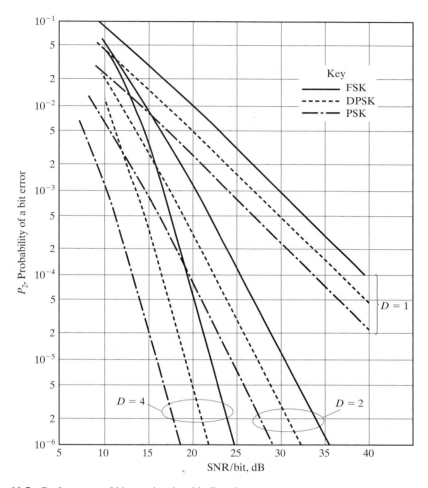

Figure 11.5 Performance of binary signals with diversity.

in SNR is achieved by increasing the order of diversity to $D = 4$, although the additional gain from $D = 2$ to $D = 4$ is smaller than the gain from $D = 1$ to $D = 2$. Beyond $D = 4$, the additional reduction in SNR is significantly smaller.

11.1.4 Frequency Selective Channels and the RAKE Demodulator

The frequency nonselective slowly fading channel model that we previously described applies to many physical radio channels used for digital communications. However, there are communication systems in which the transmitted signal bandwidth $W \gg B_{cb}$, so that the channel is frequency selective. In such a case, a more complex channel model called a *tapped delay line channel model* is employed. The demodulator for a frequency selective channel is also more complex.

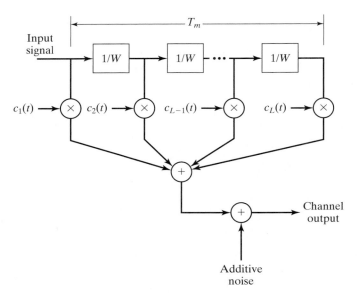

Figure 11.6 Model for time-variant multipath channel.

Tapped delay line channel model.   A general model for a time-variant multipath channel is illustrated in Figure 11.6. The channel model consists of a tapped delay line with uniformly-spaced taps. The tap spacing between adjacent taps is $1/W$, where W is the bandwidth of the signal transmitted through the channel. Hence, $1/W$ is the time resolution that can be achieved by transmitting a signal of bandwidth W. The tap coefficients, denoted as $\{c_n(t) \equiv a_n(t)e^{j\phi_n(t)}\}$ are usually modeled as complex-valued, Gaussian random processes which are mutually uncorrelated. The length of the delay line corresponds to the amount of time dispersion in the multipath channel, which is the *multipath spread*. The multipath spread may be expressed as $T_m = L/W$, where L represents the maximum number of possible multipath signal components.

Example 11.1.3

Determine an appropriate channel model for two-path ionospheric propagation, where the relative time delay between the two received signal paths is one millisecond and the transmitted signal bandwidth W is 10 kHz.

Solution   A 10 kHz signal can provide a time resolution of $1/W = 0.1$ milliseconds. Since the relative time delay between the two received signal paths is one millisecond, the tapped delay line model consists of 10 taps. In this case, only the first tap and the last tap have nonzero, time-varying coefficients, denoted as $c_1(t)$ and $c_2(t)$, as shown in Figure 11.7. Because $c_1(t)$ and $c_2(t)$ represent the signal response of a large number of ionized particles from two different regions of the ionosphere, $c_1(t)$ and $c_2(t)$ are complex-valued, uncorrelated Gaussian random processes. The rate of variation of the tap coefficients determines the value of the Doppler spread for each path.   ∎

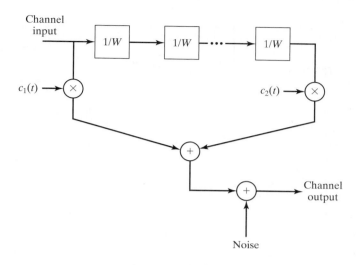

Channel
input

$c_1(t) \rightarrow$

$c_2(t) \rightarrow$

Channel
output

Noise

Figure 11.7 Channel model for two-path channel in Example 11.1.3.

Next we focus on the demodulation of a signal transmitted through a frequency selective channel. The digital information is transmitted at a symbol rate $1/T$ by modulating a single carrier frequency. We assume that the signal duration T satisfies the condition $T \ll T_{ct}$. Consequently, the channel characteristics change very slowly in time so that the channel is slowly fading, but it is frequency selective because $W \gg B_{cb}$. Furthermore, we assume that $T \gg T_m$ so that ISI is negligible. That is, the overlap in time of two successive symbols due to the time dispersion (multipath spread T_m) in the channel is very small relative to the symbol duration T.

Since W is the bandwidth of the bandpass signal, the bandwidth occupancy of the equivalent lowpass signal is $W/2$. Hence, we employ a band-limited lowpass signal $s(t)$. Using the channel model for the frequency-selective channel shown in Figure 11.6, we may express the received signal as

$$r(t) = \sum_{n=1}^{L} c_n(t)s(t - n/W) + n(t), \tag{11.1.36}$$

where $n(t)$ represents the AWGN. Therefore, the frequency-selective channel provides the receiver with up to L replicas of the transmitted signal, where each signal component is multiplied by a corresponding channel tap weight $c_n(t)$, $n = 1, 2, \ldots, L$. Based on the slow fading assumption, the channel coefficients are considered constant over the duration of one or more symbol intervals.

Since there are up to L replicas of the transmitted signal $s(t)$ in $r(t)$, a receiver that optimally processes the signal will achieve the same performance as a communication system with diversity equal to the number of received (resolvable) signal components.

Consider binary signaling over the channel. We have two equal energy signals $s_1(t)$ and $s_2(t)$, which are either antipodal or orthogonal, with a time duration $T \gg T_m$. Since the ISI is negligible, the optimum receiver consists of two correlators or two

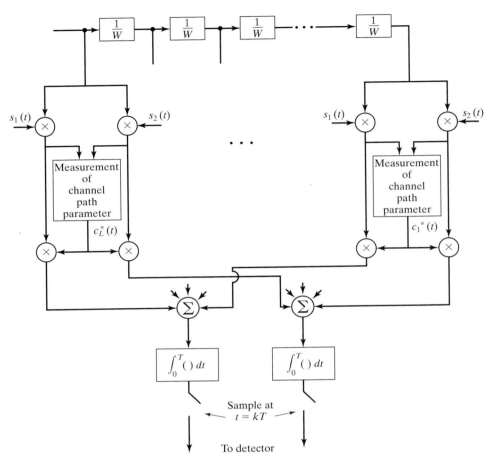

Figure 11.8 RAKE demodulator for a signal transmitted through a frequency selective channel.

matched filters that are matched to the received signals. Now we use the correlator structure that is illustrated in Figure 11.8. The received signal passes through a tapped delay-line filter with a tap spacing of $1/W$, as in the channel model. The number of taps must match the total number of resolvable signal components. At each tap, the signal is multiplied with each of the two possible transmitted signals $s_1(t)$ and $s_2(t)$; then, each multiplier output is phase corrected and weighted by multiplication with $c_n^*(t), n = 1, 2, \ldots, L$. Then, the corresponding phase-aligned and weighted signal components are integrated over the duration of the symbol interval T, and the two integrator outputs are sampled periodically every T seconds. Their outputs are sent to the detector. Thus, we have cross correlated the received signal with each of the two possible transmitted signals at all possible delays introduced by the channel. Note that the multiplication of the signal at each tap with the corresponding tap coefficient $c_n^*(t)$ results in weighting the signal components by the corresponding signal strengths. Hence, the combining of the phase-corrected and weighted signal components corresponds to maximal ratio combining.

To perform maximal ratio combining, we must estimate the channel tap coefficients $c_n(t)$ from the received signal. Since these coefficients are time varying, it is necessary for the estimator to be adaptive; that is, it must be able to track the time variations (see Problem 11.7).

The demodulator structure shown in Figure 11.8 is called a *RAKE demodulator*. Because this demodulator has equally spaced taps with tap coefficients that essentially collect all the signal components in the received signal, its operation has been likened to that of an ordinary garden rake.

Assuming that there are L signal components in the received signal, with corresponding signal strengths that are distinct and Rayleigh distributed, the probability of error for binary signals is well approximated as

$$P_2 = K_L \prod_{k=1}^{L} \frac{1}{\left[2\overline{\rho}_k(1 - \gamma_r)\right]}, \qquad (11.1.37)$$

where $\overline{\rho}_k$ is the average SNR for the kth signal component; that is,

$$\overline{\rho}_k = \frac{\mathscr{E}_b}{N_0} E\left(\alpha_k^2\right), \qquad (11.1.38)$$

in which $\alpha_k = |c_k|$ is the amplitude of the kth tap coefficient, $\gamma_r = -1$ for antipodal signals, $\gamma_r = 0$ for orthogonal signals, and K_L is the constant defined in Equation (11.1.33). If all the signal components have the same strength, the error probability in Equation (11.1.37) reduces to that given by Equation (11.1.32) with $D = L$.

This analysis of the performance of binary modulation focused on Rayleigh fading signal statistics. In general, the Rayleigh distribution is suitable for modeling the signal fading that occurs in ionospheric propagation and in mobile cellular systems. However, there are other statistical models that have been used for a variety of fading multipath channels. The most common statistical models are the Nakagami distribution and the Rician distribution. Section 11.6 gives references to these statistical models.

11.2 MULTICARRIER MODULATION AND OFDM

In Section 9.4, we considered digital transmission through nonideal channels, and we observed that such channels cause intersymbol interference when the reciprocal of the symbol rate is significantly smaller than the time dispersion (duration of the impulse response) of the nonideal channel. In such a case, the receiver employs a channel equalizer to compensate for the channel distortion. If the channel is a bandpass channel with a specified bandwidth, the information-bearing signal may be generated at the baseband and then translated in frequency to the passband of the channel. Thus, the information-bearing signal is transmitted on a single carrier. We also observed that intersymbol interference usually results in some performance degradation, even in the case where the optimum detector is used to recover the information symbols at the receiver.

We can use an alternative approach to the design of a bandwidth-efficient communication system in the presence of channel distortion. Here, we subdivide the available channel bandwidth into a number of equal-bandwidth subchannels, where the bandwidth of each subchannel is sufficiently narrow so that the frequency response characteristics of the subchannels are nearly ideal. Such a subdivision is illustrated in Figure 11.9. Thus, we create $K = W/\Delta f$ subchannels, where different information symbols can be transmitted simultaneously in K subchannels. Consequently, the data is transmitted by frequency-division multiplexing (FDM).

With each subchannel, we associate a carrier

$$x_k(t) = \cos 2\pi f_k t, \quad k = 0, 1, \ldots, K - 1, \tag{11.2.1}$$

where f_k is the mid-frequency in the kth subchannel. By selecting the symbol rate $1/T$ on each of the subchannels to be equal to the separation Δf of the adjacent subcarriers, the subcarriers are orthogonal over the symbol interval T, independently of the relative phase relationship between subcarriers. That is,

$$\int_0^T \cos(2\pi f_k t + \phi_k) \cos(2\pi f_j t + \phi_j) dt = 0, \tag{11.2.2}$$

where $f_k - f_j = n/T, n = 1, 2, \ldots$, independently of the values of the phases ϕ_k and ϕ_j. In this case, we have orthogonal frequency-division multiplexing (OFDM).

Multicarrier modulation (OFDM) has been used in both wireline and radio channels. In particular, it is used in digital subscriber loop (DSL) modems that are used to provide high-speed internet access to homes and businesses. OFDM is also used in wireless local area networks (LANs) that are used in homes and offices for wireless access to the internet.

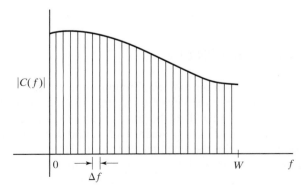

Figure 11.9 Subdivision of the channel bandwidth W into narrowband subchannels of equal width Δf.

11.2.1 Modulation and Demodulation in an OFDM System

In an OFDM system with K subchannels, the subcarrier frequencies are $\{\cos 2\pi f_k t,$ $0 \le k \le K - 1\}$, where adjacent subcarrier frequencies are separated by $\Delta f = 1/T$, i.e., $f_{k+1} - f_k = \Delta f = 1/T$ and T is the symbol interval. The symbol rate $1/T$ is reduced by a factor of K relative to the symbol rate on a single carrier system that employs the entire bandwidth W and transmits data at the same rate as OFDM. Hence, the symbol interval in the OFDM system is $T = KT_s$, where T_s is the symbol interval in the single-carrier system. By selecting K to be sufficiently large, the symbol interval T can be made significantly larger than the time duration of the channel-time dispersion. Thus, intersymbol interference can be made arbitrarily small through the selection of K. In other words, each subchannel appears to have a fixed frequency response $C(f_k)$, $k = 0, 1, \ldots, K - 1$.

Suppose that each subcarrier is modulated with M-ary QAM. Then, the signal on the kth subcarrier may be expressed as [see Equation (10.3.1), where $g_T(t) = \sqrt{2/T}$, $0 \le t \le T$]

$$
\begin{aligned}
u_k(t) &= \sqrt{\frac{2}{T}} A_{kc} \cos 2\pi f_k t + \sqrt{\frac{2}{T}} A_{ks} \sin 2\pi f_k t \\
&= \operatorname{Re}\left[\sqrt{\frac{2}{T}} A_k e^{j\theta_k} e^{j2\pi f_k t} \right] \\
&= \operatorname{Re}\left[\sqrt{\frac{2}{T}} X_k e^{j2\pi f_k t} \right],
\end{aligned}
\tag{11.2.3}
$$

where $X_k = A_k e^{j\theta_k}$ is the signal point from the QAM signal constellation that is transmitted on the kth subcarrier, $A_k = \sqrt{A_{kc}^2 + A_{ks}^2}$, and $\theta_k = \tan^{-1}(A_{ks}/A_{kc})$. The energy per symbol $\mathcal{E}_s$ has been absorbed into $\{X_k\}$.

When the number of subchannels is large, so that the subchannels are sufficiently narrowband, each subchannel can be characterized by a fixed frequency response $C(f_k)$, $k = 0, 1, \ldots, K - 1$. In general, $C(f_k)$ is complex-valued and may be expressed as

$$
C(f_k) = C_k = |C_k| e^{j\phi_k}.
\tag{11.2.4}
$$

Hence, the received signal on the kth subchannel is

$$
\begin{aligned}
r_k(t) &= \sqrt{\frac{2}{T}} |C_k| A_{kc} \cos(2\pi f_k t + \phi_k) + \sqrt{\frac{2}{T}} |C_k| A_{ks} \sin(2\pi f_k t + \phi_k) + n_k(t) \\
&= \operatorname{Re}\left[\sqrt{\frac{2}{T}} C_k X_k e^{j2\pi f_k t} \right] + n_k(t),
\end{aligned}
\tag{11.2.5}
$$

where $n_k(t)$ represents the additive noise in the kth subchannel. We assume that $n_k(t)$ is zero-mean Gaussian and spectrally flat across the bandwidth of the kth subchannel.

We also assume that the channel parameters $|C_k|$ and ϕ_k are known at the receiver. (These parameters are usually estimated by initially transmitting the unmodulated carrier $\cos 2\pi f_k t$ and observing the received signal $|C_k| \cos(2\pi f_k t + \phi_k.)$

The demodulation of the received signal in the kth subchannel may be accomplished by cross correlating $r_k(t)$ with the two basis functions, based on knowledge of the carrier phase $\{\phi_k\}$ at the receiver,

$$\psi_1(t) = \sqrt{\frac{2}{T}} \cos (2\pi f_k t + \phi_k), \quad 0 \le t \le T$$

$$\psi_2(t) = \sqrt{\frac{2}{T}} \sin (2\pi f_k t + \phi_k), \quad 0 \le t \le T \tag{11.2.6}$$

and sampling the output of the cross correlators at $t = T$. Thus, we obtain the received signal vector

$$y_k = (|C_k|A_{kc} + \eta_{kr}, \quad |C_k|A_{ks} + \eta_{ki}), \tag{11.2.7}$$

which can also be expressed as the complex number

$$Y_k = |C_k|X_k + \eta_k, \tag{11.2.8}$$

where $\eta_k = \eta_{kr} + j\eta_{ki}$ represents the additive noise.

The scaling of the transmitted symbol by the channel gain $|C_k|$ can be removed by dividing Y_k by $|C_k|$. Thus, we obtain

$$Y'_k = Y_k/|C_k| = X_k + \eta'_k, \tag{11.2.9}$$

where $\eta'_k = \eta_k/|C_k|$. The normalized variable Y'_k is passed to the detector, which computes the distance metrics between Y'_k and each of the possible signal points in the QAM signal constellation and selects the signal point resulting in the smallest distance.

From this description, it is clear that two cross correlators or two matched filters are required to demodulate the received signal in each subchannel. Therefore, if the OFDM signal consists of K subchannels, the implementation of the OFDM demodulator requires a parallel bank of $2K$ cross correlators or $2K$ matched filters. Furthermore, the modulation process for generating the OFDM signal can also be viewed as exciting a bank of $2K$ parallel filters with symbols taken from an M-ary QAM signal constellation.

The bank of $2K$ parallel filters that generates the modulated signal at the transmitter and demodulates the received signal is equivalent to the computation of the discrete Fourier transform (DFT) and its inverse. Since an efficient computation of the DFT is the fast Fourier transform (FFT) algorithm, a more efficient implementation of the modulation and demodulation processes when K is large, e.g., $K > 20$, is by means of the FFT algorithm. In the next section, we describe the implementation of the modulator and demodulator in an OFDM system that uses the FFT algorithm to compute the DFT.

Since the signals transmitted on the K subchannels of the OFDM system are synchronized, the received signals on any pair of subchannels are orthogonal over the interval $0 \le t \le T$. If the subchannel gains $|C_k|$, $0 \le k \le K - 1$ are sufficiently different across the channel bandwidth, subchannels that yield a higher SNR due to a lower attenuation can be modulated to carry more bits per symbol than subcarriers that yield a lower SNR (high attenuation). Consequently, QAM with different constellation sizes can be used on the different subchannels of an OFDM system. This assignment of different constellation sizes to different subchannels is generally done in practice, e.g., in DSL modems.

11.2.2 An OFDM System Implemented via the FFT Algorithm

In this section, we describe an OFDM system that uses QAM for data transmission on each of the subcarriers and the FFT algorithm in the implementation of the modulator and demodulator.

The basic block diagram of the OFDM system is illustrated in Figure 11.10. A serial-to-parallel buffer subdivides the information sequence into frames of B_f bits. The B_f bits in each frame are parsed into K groups, where the ith group is assigned b_i bits. Hence,

$$\sum_{i=0}^{K-1} b_i = B_f. \tag{11.2.10}$$

We may view the multicarrier modulator as generating K independent QAM subchannels, where the symbol rate for each subchannel is $1/T$ and the signal in each subchannel has a distinct QAM constellation. Hence, the number of signal points for the ith subchannel is $M_i = 2^{b_i}$. We will denote the complex-valued signal points corresponding to the information signals on the K subchannels by

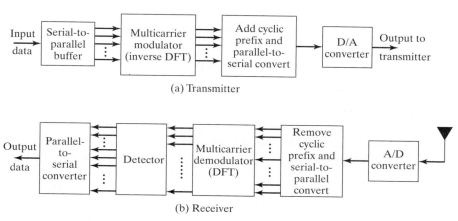

(a) Transmitter

(b) Receiver

Figure 11.10   Block diagram of a multicarrier OFDM digital communication system.

$X_k, k = 0, 1, \ldots, K - 1$. These information symbols $\{X_k\}$ represent the values of the discrete Fourier transform (DFT) of a multicarrier OFDM signal $x(t)$, where the modulation on each subcarrier is QAM. Since $x(t)$ must be a real-valued signal, its N-point DFT $\{X_k\}$ must satisfy the symmetry property $X_{N-k} = X_k^*$. Therefore, we create $N = 2K$ symbols from K information symbols by defining

$$X_{N-k} = X_k^*, \quad k = 1, 2, \ldots, K - 1 \tag{11.2.11}$$

$$X_0' = \mathrm{Re}(X_0) \tag{11.2.12}$$

$$X_K = \mathrm{Im}(X_0). \tag{11.2.13}$$

Note that the information symbol X_0 is split into two parts, both of which are real. If we denote the new sequence of symbols as $\{X_k', k = 0, 1, \ldots, N - 1\}$, the N-point inverse DFT (IDFT) yields the real-valued sequence

$$x_n = \frac{1}{\sqrt{N}} \sum_{k=0}^{N-1} X_k' \, e^{j2\pi nk/N}, \quad n = 0, 1, \ldots, N - 1, \tag{11.2.14}$$

where $1/\sqrt{N}$ is simply a scale factor. The sequence $\{x_n, 0 \leq n \leq N-1\}$ corresponds to samples of the multicarrier OFDM signal $x(t)$ which consists of K subcarriers and may be expressed as

$$x(t) = \frac{1}{\sqrt{N}} \sum_{i=0}^{N-1} X_k' \, e^{j2\pi kt/T}, \quad 0 \leq t \leq T, \tag{11.2.15}$$

where T is the signal duration and $x_n = x(nT/N), n = 0, 1, \ldots, N - 1$. The subcarrier frequencies are $f_k = k/T, k = 0, 1, \ldots, K - 1$. The signal samples $\{x_n\}$ generated by computing the IDFT are passed through a digital-to-analog (D/A) converter, whose output, ideally, is the OFDM signal waveform $x(t)$.

With $x(t)$ as the input to the channel, the channel output at the receiver may be expressed as

$$r(t) = x(t) * c(t) + n(t), \tag{11.2.16}$$

where $c(t)$ is the impulse response of the channel and $*$ denotes convolution. Since the bandwidth Δf of each subchannel is selected to be very small relative to the overall channel bandwidth $W = K\Delta f$, the symbol duration $T = 1/\Delta f$ is larger than the duration of the channel impulse response. To be specific, suppose that the channel impulse response spans $m + 1$ signal samples, where $m \ll N$. A simple way to completely avoid intersymbol interference (ISI) is to insert a time guard of duration mT/N between the transmission of successive data blocks. This allows the response of the channel to die out before the next block of K symbols is transmitted.

An alternative method of avoiding ISI is to append a so-called *cyclic prefix* to each block of N signal samples $\{x_n, 0 \leq n \leq N - 1\}$. The cyclic prefix for the block of samples contains the samples $x_{N-m}, x_{N-m+1}, \ldots, x_{N-1}$. These samples are appended to the beginning of the block, thus creating a signal sequence of length $N + m$ samples, which may be indexed from $n = -m$ to $n = N - 1$, where the first m samples constitute the cyclic prefix. Then, if the sample values of the channel response are $\{c_n, 0 \le n \le m\}$, the convolution of $\{c_n\}$ with $\{x_n, -m \le n \le N - 1\}$ produces the received signal $\{r_n\}$. Since the ISI in any pair of successive signal transmission blocks affects the first m signal samples, we discard the first m samples of $\{r_n\}$ and demodulate the signal based on the received signal samples $\{r_n, 0 \le n \le N - 1\}$.

If we view the channel characteristics in the frequency domain, the channel frequency response at the subcarrier frequencies $f_k = k/T$ is

$$C_k = C\left(\frac{2\pi k}{N}\right) = \sum_{n=0}^{m} c_n e^{-j2\pi nk/N}, \quad k = 0, 1, \ldots, N - 1. \tag{11.2.17}$$

Since the ISI is eliminated through either the cyclic prefix or the time guard band, the demodulated sequence of symbols may be expressed as

$$\hat{X}_k = C_k X'_k + \eta_k, \quad k = 0, 1, \ldots, N - 1, \tag{11.2.18}$$

where $\{\hat{X}_k\}$ is the output of the N-point DFT computed by the demodulator and $\{\eta_k\}$ is the additive noise corrupting the signal.

As illustrated in Figure 11.10, the received signal is demodulated by computing the DFT of the received signal after it has passed through an analog-to-digital (A/D) converter. As in the case of the OFDM modulator, the DFT computation at the demodulator is performed efficiently with the use of the FFT algorithm.

To recover the information symbols from the values of the computed DFT, it is necessary to estimate and compensate for the channel factors $\{C_k\}$. The channel measurement can be accomplished by initially transmitting either a known modulated sequence on each of the subcarriers or simply by transmitting the unmodulated subcarriers. If the channel characteristics vary slowly with time, the time variations can be tracked by using the decisions at the output of the detector in a decision-directed manner. Thus, the multicarrier OFDM system can be made to operate adaptively. The transmission rate on each subcarrier can be optimized by properly allocating the average transmitted power and the number of bits that are transmitted by each subcarrier. The SNR per subchannel may be defined as

$$\mathrm{SNR}_k = \frac{T P_k |C_k|^2}{\sigma_{nk}^2}, \tag{11.2.19}$$

where T is the symbol duration, P_k is the average transmitted power allocated to the kth subchannel, $|C_k|^2$ is the squared magnitude of the frequency response of the kth subchannel, and σ_{nk}^2 is the corresponding noise variance. In subchannels with high SNR, we transmit more bits/symbol by using a larger QAM constellation than we would use with subchannels with low SNR. Thus, the bit rate on each subchannel can be optimized so that the error-rate performance among the subchannels is equalized to satisfy the desired specifications.

Multicarrier OFDM using QAM modulation on each of these subcarriers has been implemented for a variety of applications, including high-speed transmission over telephone lines, such as digital subcarrier lines (DSL). This type of multicarrier OFDM modulation has also been called *discrete-multitone (DMT) modulation*. Multicarrier OFDM is also used in digital audio broadcasting in Europe and other parts of the world, as well as in wireless LANs.

11.2.3 Spectral Characteristics of OFDM Signals

Although the signals transmitted on the subcarriers of an OFDM system are mutually orthogonal in the time domain, i.e.,

$$\int_0^T u_k(t)u_j(t)\,dt = 0 \quad k \neq j, \tag{11.2.20}$$

where $u_k(t)$ is defined in Equation (11.2.3), these signals have significant overlap in the frequency domain. This can be observed by computing the Fourier transform of the signal

$$u_k(t) = \text{Re}\left[\sqrt{\frac{2}{T}}X_k e^{j2\pi f_k t}\right]$$

$$= \sqrt{\frac{2}{T}}A_k \cos\left(2\pi f_k t + \theta_k\right), \quad 0 \leq t \leq T \tag{11.2.21}$$

for several values of k. Figure 11.11 illustrates the magnitude spectrum $|U_k(f)|$ for three adjacent subcarriers. Note the large spectral overlap of the main lobes. Also note that the first sidelobe in the spectrum is only 13 dB down from the main lobe. Hence, there is a significant amount of spectral overlap among the signals transmitted on different subcarriers. Nevertheless, these signals are orthogonal when transmitted synchronously in time.

The large spectral overlap of the OFDM signals has various ramifications when the communication channel is a fading channel. Recall that signal fading, which is caused by time-varying multipath propagation, results in Doppler spreading of the transmitted signal. The multipath propagation of the signal components in the OFDM signal destroys the orthogonality among the subcarriers; when combined with the Doppler spreading, it results in intersubchannel interference (ICI). This ICI produces a significant degradation in the performance (error probability) of the OFDM system. Consequently, OFDM may not be as robust as a single carrier system in radio communications where the receiving terminal is moving at high speed. On the other hand, ICI is not a serious problem in OFDM systems in which the receiving terminal is moving at a low speed, e.g., pedestrian speed. This is the case, for example, in wireless LANs that employ OFDM signals with large ($M = 64$) QAM signal constellations.

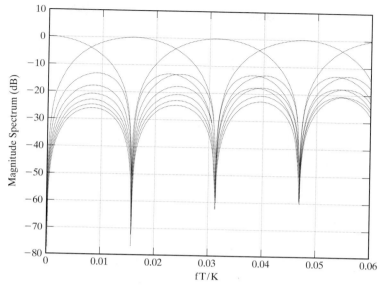

Figure 11.11   An example of the magnitude of the frequency response of adjacent subchannel filters in OFDM system for $f \in \left(0, 0.06 \frac{K}{T}\right)$ and $K = 64$. From Cherubini, et al. (2002) IEEE.

Another type of multicarrier modulation is much more robust than OFDM in the presence of ICI resulting from a large Doppler spread due to high terminal speeds; this is called *filtered multitone (FMT) modulation*. An FMT signal is also generated and demodulated through a bank of parallel filters. However, the filters in FMT are designed to have negligible spectral overlap and extremely sharp frequency rolloff characteristics. For example, Figure 11.12 illustrates the frequency response characteristics in an FMT system. Note that the filter sidelobes are at least 70 dB below the main lobe and the spectral overlap between adjacent filters is negligible. Such filter characteristics provide significant immunity against ICI that may be encountered in highly mobile radio communication environments. Another advantage of FMT is that the signals transmitted on the different subcarriers need not be synchronous.

The large immunity against ICI provided by FMT comes at a price in bandwidth efficiency. In papers that describe the design of FMT systems, Cherubini, et al. (2000, 2002) demonstrate that the increase in bandwidth compared with a conventional OFDM system can be made relatively modest, of the order of 10% to 20%. Therefore, FMT provides an alternative to conventional OFDM in high mobility applications.

11.2.4 Peak-to-Average Power Ratio in OFDM Systems

A major problem with the multicarrier modulation in general and OFDM systems in particular is the high peak-to-average power ratio (PAR) that is inherent in the transmitted signal. Large signal peaks occur in the transmitted signal when the signals in

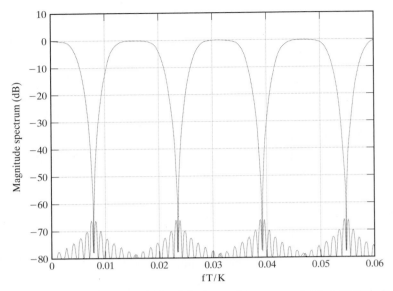

Figure 11.12   An example of the magnitude of the frequency response of adjacent subchannel filters in an FMT system for $f \in \left(0, 0.06\frac{K}{T}\right)$ and design parameters $K = 64$. From Cherubini, et al. (2002) IEEE.

the K subchannels add constructively in phase. Such large signal peaks may saturate the power amplifier at the transmitter, thus, causing intermodulation distortion in the transmitted signal. Intermodulation distortion can be reduced and often avoided by reducing the power in the transmitted signal and, thus, operating the power amplifier at the transmitter in the linear range. Such a power reduction or "power back-off" results in inefficient operation of the OFDM system. For example, if the PAR is 10 dB, the power back-off may be as much as 10 dB to avoid intermodulation distortion.

Various methods have been devised to reduce the PAR in multicarrier systems. One of the simplest methods is to insert different phase shifts in each of the subcarriers. These phase shifts can be selected pseudorandomly, or by means of some algorithm, to reduce the PAR. For example, we may have a small set of stored pseudorandomly selected phase shifts that can be used when the PAR in the modulated subcarriers is large. The information on which set of pseudorandom phase shifts is used in any signal interval can be transmitted to the receiver on one of the K subcarriers. Alternatively, a single set of pseudorandom phase shifts may be employed, where this set is found via computer simulation to reduce the PAR to an acceptable level over the ensemble of possible transmitted data symbols on the K subcarriers.

Another method that can reduce the PAR is to modulate a small subset of the subcarriers with dummy symbols, which are selected to reduce the PAR. Since the dummy symbols do not have to be constrained to take amplitude and phase values from a specified signal constellation, their design is very flexible. The subcarriers carrying the dummy symbols may be distributed across the frequency band. Since modulating subcarriers in this manner results in a lower throughput in data rate, we want to employ only a small percentage of the total subcarriers.

Because of its practical importance, the problem of PAR reduction in multicarrier communication systems has been thoroughly investigated and other methods have been devised. The interested reader may refer to the literature cited in Section 11.6.

11.2.5 Applications of OFDM

OFDM is used in a variety of digital communication systems, including digital audio broadcasting (DAB), digital video broadcasting (DVB), high-speed transmission over telephone lines such as digital subscriber lines (DSL), and wireless local area networks (LAN's). In this section, we briefly describe three of these applications.

Digital Subscriber Lines. As a first application of OFDM, consider high-speed digital transmission over wirelines that connect a telephone subscriber's premises to a telephone central office. These wireline channels typically consist of unshielded twisted-pair wire and are commonly called the *subscriber local loop*. The desire to provide high-speed internet access to homes and businesses over the telephone subscriber loop has resulted in the development of a standard for digital transmission based on OFDM with QAM as the basic modulation method on each of the subcarriers.

The usable bandwidth of a twisted-pair subscriber loop wire is primarily limited by the distance between the subscriber and the central telephone office, i.e., the length of the wire, and by crosstalk interference from other lines in the same cable. For example, a 3 km twisted-pair wireline may have a usable bandwidth of approximately 1.2 MHz. Since the need for high-speed digital transmission is usually in the direction from the central office to the subscriber (the downlink) and the bandwidth is relatively small, the major part of the bandwidth is allocated to the downlink. Consequently, the digital transmission on the subscriber loop is asymmetric, and this transmission mode is called *ADSL (Asymmetric Digital Subscriber Line)*.

In the ADSL standard, the downlink and the uplink maximum data rates are specified as 6.8 Mbps and 640 kbps, respectively, for subscriber lines of approximately 12,000 feet in length, and 1.544 Mbps and 176 kbps, respectively, for subscriber lines of approximately 18,000 feet in length. The low part of the frequency band (0–25 kHz) is reserved for the telephone voice transmission, which requires a nominal bandwidth of 4 kHz. Hence, the frequency band of the subscriber line is separated into two frequency bands via two filters (lowpass and highpass) that have cutoff frequencies of 25 kHz. Thus, the low end frequency for digital transmission is 25 kHz. The ADSL standard specifies that the frequency range of 25 kHz to 1.1 MHz must be subdivided into 256 parallel OFDM subchannels. Hence, the size of the DFT and IDFT in the system implementation shown in Figure 11.10 is $N = 512$. A sampling rate $F_s = 2.208$ MHz is specified, so that the high-end frequency in the signal spectrum is $F_s/2 = 1.104$ MHz. The frequency spacing between two adjacent subcarriers is $\Delta f = 1.104 \times 10^6/256 = 4.3125$ kHz. The channel time dispersion is suppressed by using a cyclic prefix of $N/16 = 32$ samples.

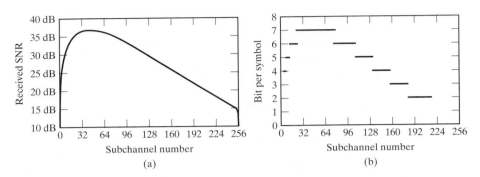

Figure 11.13   Example of a DSL frequency response and bit allocation on the OFDM subchannels.

By measuring the signal-to-noise ratio (SNR) for each subchannel at the receiver and communicating this information to the transmitter via the uplink, the transmitter can select the QAM constellation size in bits/symbol to achieve a desired error probability in each subchannel. The ADSL standard specifies a minimum bit load of 2 bits per subchannel, which corresponds to QPSK modulation. If a subchannel cannot support QPSK at the desired error probability, no information is transmitted over that subchannel. As an example, Figure 11.13 illustrates the received SNR as measured by the receiver for each subchannel and the corresponding number of bits/symbol selected from a QAM signal constellation. Note that the SNR in subchannels 220–256 is too low to support QPSK modulation; hence, no data are transmitted on these subchannels. ADSL channel characteristics and the design of OFDM modems based on the ADSL standard are treated in detail in the books by Bingham (2000) and Starr, et al. (1999). The use of OFDM with variable size QAM signal constellations for each of the subcarriers is often called *discrete multitone (DMT) modulation*.

Wireless LANs.   Wireless local area network (LAN) standards have been developed over the past few years by working groups within the IEEE (Institute of Electrical and Electronics Engineers) and other international standards organizations. These standards make it possible to provide high-speed wireless access to the internet. We will focus on the IEEE 802.11a standard, which is based on OFDM. A typical configuration is illustrated in Figure 11.14, where access point (AP) terminals communicate with several users (laptop computers).

The United States's Federal Communication Commission (FCC) has allocated a 300 MHz spectrum in the 5.2 GHz frequency band to provide wireless LAN services based on the 802.11a standard. A nominal channel bandwidth of 20 MHz is subdivided into 52 subchannels with subcarrier frequency spacing of 312.5 kHz. The OFDM symbol duration is 4 μsec, and the length of the cyclic prefix (guard interval) is 0.8 μsec. By using 48 subchannels to carry data via either BPSK, QPSK, 16-QAM, or 64-QAM, we can achieve (uncoded) data rates of 12–72 Mbps. With channel coding that is usually used to correct transmission errors, as discussed in Chapter 13, the achievable data rates in 802.11a are 6, 9, 12, 18, 24, 36, 48, and

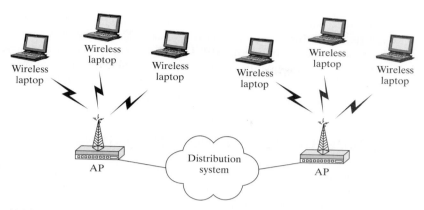

Figure 11.14   A wireless LAN configuration.

54 Mbps. In the remaining four subchannels, pilot tones are transmitted to measure and correct frequency offsets that may occur in the received signal due to terminal mobility.

The multiple access scheme employed in 802.11a is called *carrier sense multiple access with collision avoidance (CSMA/CA)*. In simple terms, a terminal, before starting transmission, senses to see if the channel is available for signal transmission. If no other signal is detected to be above a set threshold, a packet of the signal is sent. After successful reception, the recipient sends back an acknowledgement. After receiving the acknowledgement, the user waits for a certain randomly selected time interval before sending another packet. This standard operates in the 5.2-GHz frequency band; the IEEE 802.11g standard, which is related, operates in the 2.4-GHz frequency band, where OFDM with QAM modulation is employed.

Digital Audio Broadcasting.   Digital audio broadcasting (DAB) systems may be terrestrial, satellite based, or a combination of the two. Eventually, these systems will replace the analog AM and FM broadcasting systems currently in use. Various standards organizations have established standards for DAB in many countries. We shall focus on the European Eureka-147 standard established by the ITU (International Telecommunications Union) and ETSI (European Telecommunication Standard Institute).

The Eureka-147 DAB is a standard that operates in four different modes, where each mode is tailored to a specific frequency band and corresponding application. OFDM is used in all modes. The modulation used on each subchannel is differential QPSK. For example, mode 1 is employed in terrestrial broadcasting in the VHF frequency band. In this mode, there are 1536 subcarriers with a subchannel spacing of 1 kHz. The symbol duration on each subcarrier is 1 msec and the frame duration is 96 msec. The duration of the cyclic prefix is 246 μsec. Another mode is designed for satellite transmission at frequency bands up to 3 GHz. For this mode, there are 192 subchannels, with an adjacent subchannel separation of 8 kHz. The symbol duration is 125 µsec and the frame duration is 24 msec. The cyclic prefix duration is 31 µsec.

To conserve bandwidth, MPEG audio compression is employed in Eureka-147DAB. The audio quality achieved is comparable to CD quality.

For more information on digital audio broadcasting, refer to the book by Hoeg and Lauterback (2001) and the paper by Layer (2001).

11.3 SPREAD-SPECTRUM COMMUNICATION SYSTEMS

In our treatment of signal design for digital communication over an AWGN channel, the major objective has been the efficient utilization of transmitter power and channel bandwidth. As shown in Chapter 13, channel coding allows us to reduce the transmitter power by increasing the transmitted signal bandwidth through code redundancy; this allows us to trade off transmitter power with channel bandwidth. To be specific, let R denote the information (bit) rate at the input to the transmitter and let W denote the channel bandwidth. We also define the ratio W/R as the bandwidth expansion factor, denoted as B_e, of the channel coded transmitted signal. The factor B_e represents the amount of redundancy introduced through channel coding. Thus, by increasing B_e, we can reduce the power in the transmitted signal that is required to achieve a specific level of performance. In most practical communication systems, B_e is in the range of $2 \leq B_e \leq 5$. This is the basic methodology for the design of digital communication systems for AWGN channels.

In practice, other factors influence the design of an efficient digital communication system. For example, in multiple-access communication when two or more transmitters use the same common channel to transmit information, the interference created by the users of the channel limits the performance achieved by the system. The existence of such interference must be considered in the design of a reliable digital communication system.

Even in this complex design problem, the basic system design parameters are transmitter power and channel bandwidth. To overcome the degradation in performance caused by interference, we may further increase the bandwidth of the transmitted signal so that the bandwidth expansion factor $B_e = W/R$ is much greater than unity. This is one characteristic of a *spread-spectrum signal*. A second important characteristic is that the information signal at the modulator is spread in bandwidth by means of a code that is independent of the information sequence. This code has the property of being *pseudorandom*, i.e., it appears random to receivers other than the intended receiver, which uses the knowledge of the code to demodulate the signal. This second characteristic distinguishes a spread-spectrum communication system from the conventional communication system that expands the transmitted signal bandwidth by means of channel code redundancy, as described in Chapter 13. However, channel coding is an important element in the design of an efficient spread-spectrum communication system.

Spread-spectrum signals for digital communications were originally developed and used for military communications either (1) to provide resistance to jamming (antijam protection), or (2) to hide the signal by transmitting it at low power, which made it difficult for an unintended listener to detect its presence in noise (low probability of intercept). However, spread-spectrum signals now provide reliable communications in a variety of civilian applications, including digital cellular communications, cordless telephones, and interoffice wireless communications.

In this section, we present the basic characteristics of spread-spectrum signals and assess their performance in terms of probability of error. We concentrate our discussion on two methods for spreading the signal bandwidth, namely, by direct sequence modulation and by frequency hopping. Both methods require the use of pseudorandom code sequences whose generation is also described. Several applications of spread-spectrum signals are presented.

11.3.1 Model of a Spread-Spectrum Digital Communication System

The basic elements of a spread-spectrum digital communication system are illustrated in Figure 11.15. The channel encoder and decoder and the modulator and demodulator are the basic elements of a conventional digital communication system. In addition, a spread-spectrum system employs two identical pseudorandom sequence generators, one that interfaces with the modulator at the transmitting end and one that interfaces with the demodulator at the receiving end. These two generators produce a pseudorandom or pseudonoise (PN) binary-valued sequence, which is used to spread the transmitted signal at the modulator and to despread the received signal at the demodulator.

Time synchronization of the PN sequence generated at the receiver with the PN sequence contained in the received signal is required to properly despread the received spread-spectrum signal. In a practical system, synchronization is established prior to the transmission of information; this is achieved by transmitting a fixed PN bit pattern, which is designed so that the receiver will detect it with high probability in the presence of interference. After time synchronization of the PN sequence generators is established, the transmission of information commences. In the data mode, the communication system usually tracks the timing of the incoming received signal and keeps the PN sequence generator in synchronism.

Interference is introduced in the transmission of the spread-spectrum signal through the channel. The characteristics of the interference depend to a large extent on its origin. The interference may be generally categorized as either broadband

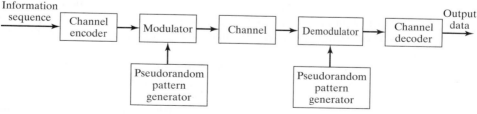

Figure 11.15   Model of a spread-spectrum digital communication system.

or narrowband (partial band) relative to the bandwidth of the information-bearing signal and either continuous in time or discontinuous (pulsed) in time. For example, an interfering signal may consist of a high-power sinusoid in the bandwidth occupied by the information-bearing signal. Such a signal is narrowband. As a second example, the interference generated by other users in a multiple-access channel depends on the type of spread-spectrum signals that are used to transmit information. If all users employ broadband signals, the interference may be characterized as an equivalent broadband noise. If the users employ frequency hopping to generate spread-spectrum signals, the interference from other users may be characterized as narrowband.

Our discussion will focus on the performance of spread-spectrum signals for digital communication in the presence of narrowband and broadband interference. Two types of digital modulation are considered, namely, PSK and FSK. PSK modulation is appropriate for applications where phase coherence between the transmitted signal and the received signal can be maintained over a time interval that spans several symbol (or bit) intervals. On the other hand, FSK modulation is appropriate in applications where phase coherence of the carrier cannot be maintained due to time variations in the transmission characteristics of the communications channel. For example, this may be the case in a communications link between two high-speed aircraft or between a high-speed aircraft and a ground-based terminal.

The PN sequence generated at the modulator is used in conjunction with the PSK modulation to shift the phase of the PSK signal pseudorandomly at a rate that is an integer multiple of the bit rate. The resulting modulated signal is called a *direct sequence (DS) spread-spectrum signal*. When used in conjunction with binary or M-ary ($M > 2$) FSK, the PN sequence is used to select the frequency of the transmitted signal pseudorandomly. The resulting signal is called a *frequency-hopped (FH) spread-spectrum signal*. Although other types of spread-spectrum signals can be generated, our treatment will emphasize DS and FH spread-spectrum communication systems, which are generally used in practice.

11.3.2 Direct Sequence Spread-Spectrum Systems

Consider the transmission of a binary information sequence by means of binary PSK. The information rate is R bits per second, and the bit interval is $T_b = 1/R$ seconds. The available channel bandwidth is B_c Hz, where $B_c \gg R$. At the modulator, the bandwidth of the information signal is expanded to $W = B_c$ Hz by shifting the phase of the carrier pseudorandomly at a rate of W times per second according to the pattern of the PN generator. The basic method for accomplishing the spreading is shown in Figure 11.16.

The information-bearing baseband signal is denoted as $v(t)$ and is expressed as

$$v(t) = \sum_{n=-\infty}^{\infty} a_n g_T(t - nT_b), \qquad (11.3.1)$$

where $\{a_n = \pm 1, -\infty < n < \infty\}$ and $g_T(t)$ is a rectangular pulse of duration T_b. This signal is multiplied by the signal from the PN sequence generator, which may

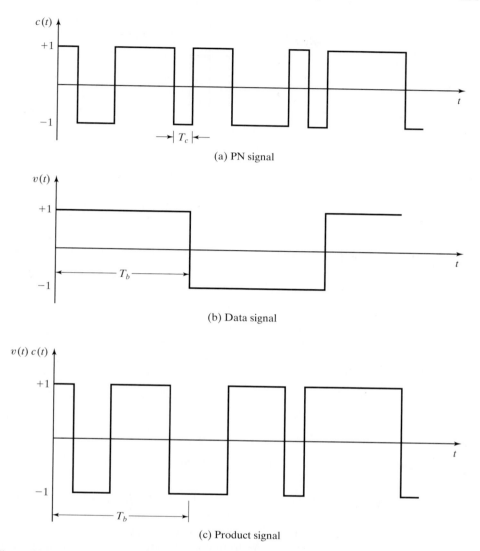

(a) PN signal

(b) Data signal

(c) Product signal

Figure 11.16 Generation of a DS spread-spectrum signal.

be expressed as

$$c(t) = \sum_{n=-\infty}^{\infty} c_n p(t - nT_c), \qquad (11.3.2)$$

where $\{c_n\}$ represents the binary PN code sequence of ± 1's and $p(t)$ is a rectangular pulse of duration T_c, as illustrated in Figure 11.16. This multiplication operation serves to spread the bandwidth of the information-bearing signal (whose bandwidth

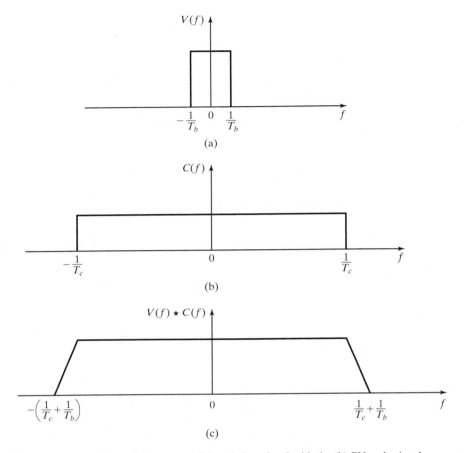

Figure 11.17 Convolution of the spectra of the (a) data signal with the (b) PN code signal.

is approximately R Hz) into the wider bandwidth occupied by PN generator signal $c(t)$ (whose bandwidth is approximately $1/T_c$). The spectrum spreading is illustrated in Figure 11.17, which uses simple rectangular spectra to show the convolution of the two spectra, the narrow spectrum corresponding to the information-bearing signal and the wide spectrum corresponding to the signal from the PN generator.

The product signal $v(t)c(t)$ amplitude modulates the carrier $A_c \cos 2\pi f_c t$ and generates the DSB-SC signal

$$u(t) = A_c v(t)c(t) \cos 2\pi f_c t. \tag{11.3.3}$$

Since $v(t)c(t) = \pm 1$ for any t, it follows that the carrier modulated transmitted signal may also be expressed as

$$u(t) = A_c \cos[2\pi f_c t + \theta(t)], \tag{11.3.4}$$

where $\theta(t) = 0$ when $v(t)c(t) = 1$ and $\theta(t) = \pi$ when $v(t)c(t) = -1$. Therefore, the transmitted signal is a binary PSK signal.

The rectangular pulse $p(t)$ is usually called a *chip*, and its time duration T_c is called the *chip interval*. The reciprocal $1/T_c$ is called the *chip rate* and corresponds (approximately) to the bandwidth W of the transmitted signal. In practical spread-spectrum systems, the ratio of the bit interval T_b to the chip interval T_c is usually selected to be an integer. We denote this ratio as

$$L_c = \frac{T_b}{T_c}. \qquad (11.3.5)$$

Hence, L_c is the number of chips of the PN code sequence per information bit. Another interpretation is that L_c represents the number of possible 180° phase transitions in the transmitted signal during the bit interval T_b.

The demodulation of the signal is illustrated in Figure 11.18. The received signal is first multiplied by a replica of the waveform $c(t)$ generated by the PN code sequence generator at the receiver, which is synchronized to the PN code in the received signal. This operation is called (spectrum) *despreading*, since the effect of multiplication by $c(t)$ at the receiver is to undo the spreading operation at the transmitter. Thus, we have

$$A_c v(t)c^2(t) \cos 2\pi f_c t = A_c v(t) \cos 2\pi f_c t, \qquad (11.3.6)$$

since $c^2(t) = 1$ for all t. The resulting signal $A_c v(t) \cos 2\pi f_c t$ occupies a bandwidth (approximately) of R Hz, which is the bandwidth of the information-bearing signal. Therefore, the demodulator for the despread signal is simply the conventional cross correlator or matched filter that was described in Chapter 8. Since the demodulator has a bandwidth that is identical to the bandwidth of the despread signal, the only additive noise that corrupts the signal at the demodulator is the noise that falls within the information bandwidth of the received signal.

Effect of Despreading on a Narrowband Interference.   It is interesting to investigate the effect of an interfering signal on the demodulation of the desired information-bearing signal. Suppose that the received signal is

$$r(t) = A_c v(t)c(t) \cos 2\pi f_c t + i(t), \qquad (11.3.7)$$

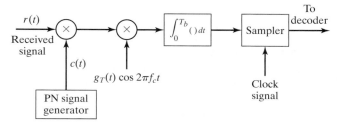

Figure 11.18   Demodulation of DS spread-spectrum signal.

where $i(t)$ denotes the interference. The despreading operation at the receiver yields

$$r(t)c(t) = A_c v(t) \cos 2\pi f_c t + i(t)c(t). \qquad (11.3.8)$$

The effect of multiplying the interference $i(t)$ with $c(t)$ is to spread the bandwidth of $i(t)$ to W Hz.

As an example, consider the sinusoidal interfering signal

$$i(t) = A_I \cos 2\pi f_I t, \qquad (11.3.9)$$

where f_I is a frequency within the bandwidth of the transmitted signal. Its multiplication with $c(t)$ results in a wideband interference with power spectral density $I_0 = P_I/W$, where $P_I = A_I^2/2$ is the average power of the interference. Since the desired signal is demodulated by a matched filter (or correlator) that has a bandwidth R, the total power in the interference at the output of the demodulator is

$$I_0 R = P_I R / W = \frac{P_I}{W/R} = \frac{P_I}{T_b/T_c} = \frac{P_I}{L_c}. \qquad (11.3.10)$$

Therefore, the power in the interfering signal is reduced by an amount equal to the bandwidth expansion factor W/R. The factor $W/R = T_b/T_c = L_c$ is called the *processing gain* of the spread-spectrum system. The reduction in interference power is the basic reason for using spread-spectrum signals to transmit digital information over channels with interference.

In summary, the PN code sequence is used at the transmitter to spread the information-bearing signal into a wide bandwidth for transmission over the channel. By multiplying the received signal with a synchronized replica of the PN code signal, the desired signal is despread back to a narrow bandwidth while any interference signals are spread over a wide bandwidth. The net effect is a reduction in the interference power by the factor W/R, which is the processing gain of the spread-spectrum system.

The PN code sequence $\{c_n\}$ is assumed to be known only to the intended receiver. Any other receiver that does not have knowledge of the PN code sequence cannot demodulate the signal. Consequently, the use of a PN code sequence provides a degree of privacy (or security) that is not possible with conventional modulation. The primary cost for this security and performance gain against interference is an increase in channel bandwidth utilization and in the complexity of the communication system.

Probability of Error.   To derive the probability of error for a direct sequence spread-spectrum system, we assume that the information is transmitted via binary PSK. Within the bit interval $0 \leq t \leq T_b$, the transmitted signal is

$$s(t) = a_0 g_T(t)c(t) \cos 2\pi f_c t, \quad 0 \leq t \leq T_b, \qquad (11.3.11)$$

where $a_o = \pm 1$ is the information symbol, the pulse $g_T(t)$ is defined as

$$
g_T(t) = \begin{cases} \sqrt{\frac{2\mathcal{E}_b}{T_b}}, & 0 \le t \le T_b \\ \\ 0, & \text{otherwise} \end{cases}, \tag{11.3.12}
$$

and $c(t)$ is the output of the PN code generator which, over a bit interval, is expressed as

$$
c(t) = \sum_{n=0}^{L_c-1} c_n p(t - nT_c), \tag{11.3.13}
$$

where L_c is the number of chips per bit, T_c is the chip interval, and $\{c_n\}$ denotes the PN code sequence. The code chip sequence $\{c_n\}$ is uncorrelated (white), i.e.,

$$
E[c_n c_m] = E[c_n]E[c_m], \quad \text{for } n \ne m, \tag{11.3.14}
$$

and each chip is $+1$ or -1 with equal probability. These conditions imply that $E[c_n] = 0$ and $E[c_n^2] = 1$.

The received signal is assumed to be corrupted by an additive interfering signal $i(t)$. Hence,

$$
r(t) = a_o g_T(t - t_d)c(t - t_d)\cos(2\pi f_c t + \phi) + i(t), \tag{11.3.15}
$$

where t_d represents the propagation delay through the channel and ϕ represents the carrier phase shift. Since the received signal $r(t)$ is the output of an ideal bandpass filter in the front end of the receiver, the interference $i(t)$ is also a bandpass signal and may be represented as

$$
i(t) = i_c(t)\cos 2\pi f_c t - i_s(t)\sin 2\pi f_c t, \tag{11.3.16}
$$

where $i_c(t)$ and $i_s(t)$ are the two quadrature components.

Assuming that the receiver is perfectly synchronized to the received signal, we may set $t_d = 0$ for convenience. In addition, the carrier phase is assumed to be perfectly estimated by a phase-locked loop. Then, the signal $r(t)$ is demodulated by first despreading through multiplication by $c(t)$ and then through cross correlating with $g_T(t)\cos(2\pi f_c t + \phi)$, as shown in Figure 11.19. At the sampling instant $t = T_b$, the output of the correlator is

$$
y(T_b) = \mathcal{E}_b + y_i(T_b), \tag{11.3.17}
$$

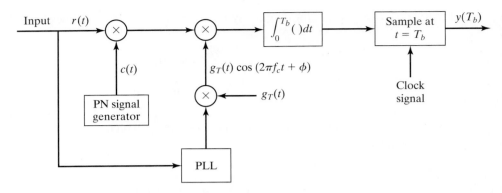

Figure 11.19 DS spread-spectrum signal demodulator.

where $y_i(T_b)$ represents the interference component, which has the form

$$y_i(T_b) = \int_0^{T_b} c(t)i(t)g_T(t)\cos(2\pi f_c t + \phi)dt$$

$$= \sum_{n=0}^{L_c-1} c_n \int_0^{T_b} p(t - nT_c)i(t)g_T(t)\cos(2\pi f_c t + \phi)dt$$

$$= \sqrt{\frac{2\mathcal{E}_b}{T_b}} \sum_{n=0}^{L_c-1} c_n v_n, \tag{11.3.18}$$

where, by definition,

$$v_n = \int_{nT_c}^{(n+1)T_c} i(t)\cos(2\pi f_c t + \phi)dt. \tag{11.3.19}$$

The probability of error depends on the statistical characteristics of the interference component. Clearly, its mean value is

$$E[y_i(T_b)] = 0. \tag{11.3.20}$$

Its variance is

$$E[y_i^2(T_b)] = \frac{2\mathcal{E}_b}{T_b} \sum_{n=0}^{L_c-1} \sum_{m=0}^{L_c-1} E[c_n c_m]E[v_n v_m].$$

But $E[c_n c_m] = \delta_{mn}$. Therefore,

$$E[y_i^2(T_b)] = \frac{2\mathcal{E}_b}{T_b} \sum_{n=0}^{L_c-1} E[v_n^2]$$

$$= \frac{2\mathcal{E}_b}{T_b} L_c E[v^2], \tag{11.3.21}$$

where $v = v_n$, as given by Equation (11.3.19). To determine the variance of v, we must postulate the form of the interference.

First, we will assume that the interference is sinusoidal. Specifically, we assume that the interference is at the carrier frequency and has the form

$$i(t) = \sqrt{2P_I} \, \cos(2\pi f_c t + \Theta_I),$$

(11.3.22)

where P_I is the average power and Θ_I is the phase of the interference, which we assume to be random and uniformly distributed over the interval $(0, 2\pi)$. If we substitute for $i(t)$ in Equation (11.3.19), we obtain

$$v_n = \int_{nT_c}^{(n+1)T_c} \sqrt{2P_I} \, \cos(2\pi f_c t + \Theta_I) \cos(2\pi f_c t + \phi) dt$$

$$= \frac{1}{2} \sqrt{2P_I} \int_{nT_c}^{(n+1)T_c} \cos(\Theta_I - \phi) dt = \frac{T_c}{2} \sqrt{2P_I} \cos(\Theta_I - \phi).$$

(11.3.23)

Since Θ_I is a random variable, v_n is also random. Its mean value is zero, i.e.,

$$E[v_n] = \frac{T_c}{2} \sqrt{2P_I} \int_0^{2\pi} \frac{1}{2\pi} \cos(\Theta_I - \phi) d\Theta_I = 0.$$

(11.3.24)

Its mean square value is

$$E[v_n^2] = \frac{T_c^2 P_I}{2} \frac{1}{2\pi} \int_0^{2\pi} \cos^2(\Theta_I - \phi) d\Theta_I$$

$$= \frac{T_c^2 P_I}{4}.$$

(11.3.25)

We may now substitute for $E[v^2]$ into Equation (11.3.21). Thus, we obtain

$$E\left[y_i^2(T_b)\right] = \frac{\mathscr{E}_b P_I T_c}{2}.$$

(11.3.26)

The ratio of $\{E[y(T_b)]\}^2$ to $E\left[y_i^2(T_b)\right]$ is the signal-to-noise ratio at the detector. In this case, we have

$$(\text{SNR})_D = \frac{\mathscr{E}_b^2}{\mathscr{E}_b P_I T_c / 2} = \frac{2\mathscr{E}_b}{P_I T_c}.$$

(11.3.27)

To see the effect of the spread-spectrum signal, we express the transmitted energy $\mathscr{E}_b$ as

$$\mathscr{E}_b = P_S T_b,$$

(11.3.28)

where P_S is the average signal power. Then, if we substitute for $\mathscr{E}_b$ in Equation (11.3.27), we obtain

$$(\text{SNR})_D = \frac{2P_S T_b}{P_I T_c} = \frac{2P_S}{P_I / L_c},$$

(11.3.29)

where $L_c = T_b/T_c$ is the processing gain. Therefore, the spread- spectrum signal has reduced the power of the interference by the factor L_c.

Another interpretation of the effect of the spread-spectrum signal on the sinusoidal interference is obtained if we express $P_I T_c$ in Equation (11.3.29) as a power spectral density. Since $T_c \simeq 1/W$, we have

$$P_I T_c = P_I/W \equiv I_0, \tag{11.3.30}$$

where I_0 is the power spectral density of an equivalent interference in a bandwidth W. In effect, the spread-spectrum signal has spread the sinusoidal interference over the wide bandwidth W, creating an equivalent spectrally flat noise with the power spectral density I_0. Hence,

$$(\text{SNR})_D = \frac{2\mathcal{E}_b}{I_0}. \tag{11.3.31}$$

The probability of error for a direct sequence spread-spectrum system with binary PSK modulation is easily obtained from the SNR at the detector, if we make an assumption on the probability distribution of the sample $y_i(T_b)$. From Equation (11.3.18), we note that $y_i(T_b)$ consists of a sum of L_c uncorrelated random variables $\{c_n v_n, \quad 0 \leq n \leq L_c - 1\}$, all of which are identically distributed. Since the processing gain L_c is usually large in any practical system, we may use the Central Limit Theorem to justify a Gaussian probability distribution for $y_i(T)$. Under this assumption, the probability of error for the sinusoidal interference is

$$P_2 = Q\left(\sqrt{\frac{2\mathcal{E}_b}{I_0}}\right), \tag{11.3.32}$$

where I_0 is the power spectral density of an equivalent broadband interference. A similar expression for the error probability is obtained when the interference $i(t)$ is a zero-mean broadband random process with a constant power spectral density I_0 over the bandwidth W of the spread-spectrum signal.

Example 11.3.1

The SNR required at the detector to achieve reliable communication in a DS spread-spectrum communication system is 13 dB. If the interference-to-signal power at the receiver is 20 dB, determine the processing gain required to achieve reliable communication.

Solution　We are given $(P_I/P_S)_{\text{dB}} = 20$ dB or, equivalently, $P_I/P_S = 100$. We are also given $(\text{SNR})_D = 13$ dB, or equivalently, $(\text{SNR})_D = 20$. The relation in Equation (11.3.29) may be used to solve for L_c. Thus,

$$L_c = \frac{1}{2}\left(\frac{P_I}{P_S}\right)(\text{SNR})_D = 1000.$$

Therefore, the processing gain required is 1000 or, equivalently, 30 dB.　　■

The Interference Margin.   We may express $\frac{\mathscr{E}_b}{I_0}$ in the Q-function in Equation (11.3.32) as

$$\frac{\mathscr{E}_b}{I_0} = \frac{P_S T_b}{P_I/W} = \frac{P_S/R}{P_I/W} = \frac{W/R}{P_I/P_S}. \tag{11.3.33}$$

Also, suppose we specify a required $\mathscr{E}_b/I_0$ to achieve a desired level of performance. Then, using a logarithmic scale, we may express Equation (11.3.33) as

$$10 \log \frac{P_I}{P_S} = 10 \log \frac{W}{R} - 10 \log \left(\frac{\mathscr{E}_b}{I_0} \right)$$

$$\left(\frac{P_I}{P_S} \right)_{\text{dB}} = \left(\frac{W}{R} \right)_{\text{dB}} - \left(\frac{\mathscr{E}_b}{I_0} \right)_{\text{dB}}. \tag{11.3.34}$$

The ratio $(P_I/P_S)_{\text{dB}}$ is called the *interference margin*. This is the relative power advantage that an interference may have without disrupting the communication system.

Example 11.3.2

Suppose we require an $(\mathscr{E}_b/I_0)_{\text{dB}} = 10$ dB to achieve reliable communication. What is the processing gain that is necessary to provide an interference margin of 20 dB?

Solution   Clearly, if $W/R = 1000$, then $(W/R)_{\text{dB}} = 30$ dB and the interference margin is $(P_J/P_S)_{\text{dB}} = 20$ dB. This means that the average interference power at the receiver may be 100 times the power P_S of the desired signal and we can still maintain reliable communication.   ∎

Performance of Coded Spread-Spectrum Signals.   As shown in Chapter 13, when the transmitted information is coded by a binary linear (block or convolutional) code, the SNR at the output of a soft-decision decoder, at large SNR, is increased by the coding gain, defined as

$$\text{coding gain} = R_c d_{\text{min}}^H, \tag{11.3.35}$$

where R_c is the code rate and d_{min}^H is the minimum Hamming distance of the code. Therefore, the effect of coding is to increase the interference margin by the coding gain. Thus, Equation (11.3.34) may be modified as

$$\left(\frac{P_I}{P_S} \right)_{\text{dB}} = \left(\frac{W}{R} \right)_{\text{dB}} + (CG)_{\text{dB}} - \left(\frac{\mathscr{E}_b}{I_0} \right)_{\text{dB}}, \tag{11.3.36}$$

where $(CG)_{\text{dB}}$ denotes the coding gain in dB.

11.3.3 Some Applications of DS Spread-Spectrum Signals

In this section, we briefly describe the use of DS spread-spectrum signals in four applications. First, we consider an application in which the signal is transmitted at very low power, so that a listener trying to detect the presence of the signal would encounter great difficulty in doing so?. A second application is multiple access radio communications. A third application involves the use of a DS spread-spectrum signal to resolve the multipath in a time-dispersive radio channel. The fourth application is the use of DS spread-spectrum signals in wireless LAN's.

Low-Detectability Signal Transmission.   In this application, the information-bearing signal is transmitted at a very low power level relative to the background channel noise and thermal noise that is generated in the front end of a receiver. If the DS spread-spectrum signal occupies a bandwidth W and the power spectral density of the additive noise is N_0 W/Hz, the average noise power in the bandwidth W is $P_N = W N_0$.

The average received signal power at the intended receiver is P_R. If we wish to hide the presence of the signal from receivers that are in the vicinity of the intended receiver, the signal is transmitted at a power level such that $P_R/P_N \ll 1$. The intended receiver can recover the weak information-bearing signal from the background noise with the aid of the processing gain and the coding gain. However, any other receiver that has no knowledge of the PN code sequence is unable to take advantage of the processing gain and the coding gain. Consequently, the presence of the information-bearing signal is difficult to detect. We say that the transmitted signal has a *low probability of being intercepted* (LPI), and it is called an *LPI signal*.

The probability of error given in Section 11.3.2 also applies to the demodulation and decoding of LPI signals at the intended receiver.

Example 11.3.3

A DS spread-spectrum signal is designed so that the power ratio P_R/P_N at the intended receiver is 10^{-2}. If the desired $\mathscr{E}_b/N_0 = 10$ for acceptable performance, determine the minimum value of the processing gain.

Solution   We may write $\mathscr{E}_b/N_0$ as

$$\frac{\mathscr{E}_b}{N_0} = \frac{P_R T_b}{N_0} = \frac{P_R L_c T_c}{N_0} = \left(\frac{P_R}{W N_0}\right) L_c = \left(\frac{P_R}{P_N}\right) L_c.$$

Since $\mathscr{E}_b/N_0 = 10$ and $P_R/P_N = 10^{-2}$, it follows that the necessary processing gain is $L_c = 1000$.   ∎

Code Division Multiple Access.   The enhancement in performance obtained from a DS spread-spectrum signal through the processing gain and the coding gain can enable many DS spread-spectrum signals to occupy the same channel bandwidth provided that each signal has its own pseudorandom (signature) sequence. Thus, it is possible to have several users transmit messages simultaneously over the same channel bandwidth. This type of digital communication, in which each transmitter–receiver user pair has its own distinct signature code for transmitting over a common channel bandwidth, is called *code division multiple access* (CDMA).

In the demodulation of each DS spread-spectrum signal, the signals from the other simultaneous users of the channel appear as additive interference. The level of interference varies as a function of the number of users of the channel at any given time. A major advantage of CDMA is that a large number of users can be accommodated if each user transmits messages for a short period of time. In such a multiple access system, it is relatively easy to add new users or to decrease the number of users without reconfiguring the system.

Next, we determine the number of simultaneous signals that can be accommodated in a CDMA system. For simplicity, we assume that all signals have identical average powers. In many practical systems, the received signal power level from each user is monitored at a central station and power control is exercised over all simultaneous users via a control channel that instructs the users on whether to increase or decrease their power level. With such power control, if there are N_u simultaneous users, the desired signal-to-noise interference power ratio at a given receiver is

$$\frac{P_S}{P_N} = \frac{P_S}{(N_u - 1)P_S} = \frac{1}{N_u - 1}. \tag{11.3.37}$$

From this relation, we can determine the number of users that can be accommodated simultaneously. Example 11.3.4 illustrates the computation.

Example 11.3.4

Suppose that the desired level of performance for a user in a CDMA system is an error probability of 10^{-6}, which is achieved when $\mathscr{E}_b/I_0 = 20$ (13 dB). Determine the maximum number of simultaneous users that can be accommodated in a CDMA system if the bandwidth-to-bit-rate ratio is 1000 and the coding gain is $R_c d_{\min}^H = 4$ (6 dB).

Solution   From the relationships given in Equations (11.3.36) and (11.3.37), we have

$$\frac{\mathscr{E}_b}{I_0} = \frac{W/R}{N_u - 1} \, R_c d_{\min}^H = 20.$$

If we solve for N_u, we obtain

$$N_u = \frac{W/R}{20} \, R_c d_{\min}^H + 1.$$

For $W/R = 1000$ and $R_c d_{\min}^H = 4$, we obtain the result that $N_u = 201$.   ∎

In determining the maximum number of simultaneous users of the channel, we implicitly assumed that the pseudorandom code sequences used by the various users are uncorrelated and that the interference from other users adds on a power basis only. However, orthogonality of the pseudorandom sequences among N_u users is generally difficult to achieve, especially if N_u is large. In fact, the design of a large set of pseudorandom sequences with good correlation properties is an important problem that has received considerable attention in the technical literature. We shall briefly treat this problem in Section 11.3.4.

CDMA is a viable method for providing digital cellular telephone service to mobile users. In Section 11.4.2, we describe the basic characteristics of the North American digital cellular system that employs CDMA.

Communication Over Channels with Multipath. In Section 11.1, we described the characteristics of fading multipath channels and the design of signals for effective communication through such channels. One example of a fading multipath channel is ionospheric propagation in the HF frequency band (3–30 MHz), where the ionospheric layers serve as signal reflectors. Another example occurs in mobile radio communication systems, where the multipath propagation is due to reflection from buildings, trees, and other obstacles located between the transmitter and the receiver.

Our discussion on signal design in Section 11.1.4 focused on frequency selective channels, where the signal bandwidth W is larger than the coherence bandwidth B_{cb} of the channel. If $W > B_{cb}$, we may consider two approaches to signal design. One approach is to use OFDM. Thus, we subdivide the available bandwidth W into N subchannels such that the bandwidth per channel $\frac{W}{N} < B_{cb}$. In this way, each subchannel is frequency non-selective and the signals in each subchannel satisfy the condition that the symbol interval $T \gg T_m$, where T_m is the multipath spread of the channel. Thus, intersymbol interference is avoided. A second approach is to design the signal to utilize the entire signal bandwidth W and transmit it on a single carrier. In this case, the channel is frequency selective and the multipath components with differential delays of $\frac{1}{W}$ or greater become resolvable.

DS spread spectrum is a particularly effective way to generate a wideband signal for resolving multipath signal components. By separating the multipath components, we may also reduce the effects of fading. For example, in LOS communication systems where there is a direct path and a secondary propagation path resulting from signals reflecting from buildings and surrounding terrain, the demodulator at the receiver may synchronize to the direct signal component and ignore the existence of the multipath component. In such a case, the multipath component becomes a form of interference (ISI) on the demodulation of subsequent transmitted signals.

ISI can be avoided if we are willing to reduce the symbol rate $\frac{1}{T}$ such that $T \gg T_m$. In this case, we employ a DS spread-spectrum signal with a bandwidth W to resolve the multipath. Thus, the channel is frequency selective and the appropriate channel model is the tapped-delay-line model with time-varying coefficients, as shown in Figure 11.6. The optimum demodulator for this channel is a filter matched to the tapped-delay channel model called the RAKE demodulator, as described in Section 11.1.4.

Wireless LANs. Spread-spectrum signals have been used in the IEEE wireless LAN standards 802.11 and 802.11b, which operate in the 2.4 GHz ISM (industrial, scientific, and medical) unlicensed frequency band. The available bandwidth is subdivided into 14 overlapping 22 MHz channels, although not all channels are used in all countries. In the US, only channels 1 through 11 are used.

In the 802.11 standard, an 11-chip Barker sequence is modulated and transmitted at a chip rate of 11 MHz, i.e., the chip duration is 0.909 μsec. The 11-chip Barker sequence is $\{1, -1, 1, 1, -1, 1, 1, 1, -1, -1, -1\}$. As described in the next section, this sequence is desirable because its autocorrelation has sidelobes of less than or equal to 1, compared with the peak autocorrelation value of 11. The Barker sequence is modulated either with BPSK or QPSK. When BPSK is used with 11 chips per bit, a data rate of 1 Mbps is achieved. When QPSK modulation is used with 11 chips per symbol (2 bits), a data rate of 2 Mbps is achieved.

Direct sequence spread spectrum is also used in the higher speed (second generation) IEEE 802.11b wireless LAN standard, which operates in the same 2.4 GHz ISM band. In 802.11b, the 11-MHz chip rate is maintained, but the Barker sequence is replaced by a set of 8-chip waveform sequences, called *Complementary Code Shift Keying* (CCK), which can be viewed as direct sequence spread spectrum modulation with multiple spreading sequences. The use of CCK modulation results in a data rate of 11 Mbps.

11.3.4 Generation of PN Sequences

A pseudorandom or pseudonoise (PN) sequence is a code sequence of 1's and 0's whose autocorrelation has properties similar to those of white noise. In this section, we briefly describe the construction of some PN sequences and their autocorrelation and cross correlation properties.

The most widely known binary PN code sequences are the maximum-length shift-register sequences. A maximum-length shift-register sequence, or m-sequence for short, has the length $L = 2^m - 1$ bits and is generated by an m-stage shift register with linear feedback, as illustrated in Figure 11.20. The sequence is periodic with period L. Each period contains 2^{m-1} ones and $2^{m-1} - 1$ zeros. Table 11.2 lists shift register connections for generating maximum-length sequences.

In DS spread spectrum applications, the binary sequence with elements $\{0, 1\}$ is mapped into a corresponding binary sequence with elements $\{-1, 1\}$. We shall call the equivalent sequence $\{c_n\}$ with elements $\{-1, 1\}$ a *bipolar sequence*.

An important characteristic of a periodic PN sequence is its autocorrelation function, which is usually defined in terms of the bipolar sequences $\{c_n\}$ as

$$R_c(m) = \sum_{n=1}^{L} c_n c_{n+m}, \quad 0 \leq m \leq L - 1, \tag{11.3.38}$$

where L is the period of the sequence. Since the sequence $\{c_n\}$ is periodic with period L, the autocorrelation sequence $\{R_c(m)\}$ is also periodic with period L.

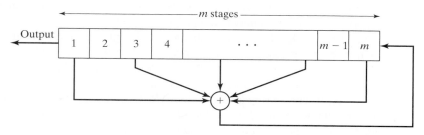

Figure 11.20   General *m*-stage shift register with linear feedback.

TABLE 11.2   SHIFT-REGISTER CONNECTIONS FOR GENERATING MAXIMUM-LENGTH SEQUENCES

m	Stages connected to modulo-2-adder	m	Stages connected to modulo-2-adder	m	Stages connected to modulo-2 adder
2	1, 2	13	1, 10, 11, 13	24	1, 18, 23, 24
3	1, 3	14	1, 5, 9, 14	25	1, 23
4	1, 4	15	1, 15	26	1, 21, 25, 26
5	1, 4	16	1, 5, 14, 16	27	1, 23, 26, 27
6	1, 6	17	1, 15	28	1, 26
7	1, 7	18	1, 12	29	1, 28
8	1, 5, 6, 7	19	1, 15, 18, 19	30	1, 8, 29, 30
9	1, 6	20	1, 18	31	1, 29
10	1, 8	21	1, 20	32	1, 11, 31, 32
11	1, 10	22	1, 22	33	1, 21
12	1, 7, 9, 12	23	1, 19	34	1, 8, 33, 34

Ideally, a PN sequence should have an autocorrelation function that has correlation properties similar to white noise. That is, the ideal autocorrelation sequence for $\{c_n\}$ is $R_c(0) = L$ and $R_c(m) = 0$ for $1 \leq m \leq L-1$. In the case of m-sequences, the autocorrelation sequence is

$$
R_c(m) = \begin{cases} L, & m = 0 \\ -1, & 1 \leq m \leq L - 1. \end{cases} \tag{11.3.39}
$$

For long m-sequences, the size of the off-peak values of $R_c(m)$ relative to the peak value $R_c(0)$, i.e., the ratio $R_c(m)/R_c(0) = -1/L$, is small and, from a practical viewpoint, inconsequential. Therefore, m-sequences are very close to ideal PN sequences when viewed in terms of their autocorrelation function.

In some applications, the cross correlation properties of PN sequences are as important as the autocorrelation properties. For example, in CDMA, each user is assigned a particular PN sequence. Ideally, the PN sequences among users should be mutually uncorrelated so that the level of interference experienced by one user from the transmissions of other users adds on a power basis. However, in practice, the PN sequences of different users exhibit some correlation.

To be specific, consider the class of m-sequences. We know that the periodic cross correlation function between a pair of m-sequences of the same period can have relatively large peaks. Table 11.3 lists the peak magnitude $R_{\max}$ for the periodic cross correlation between pairs of m-sequences for $3 \leq m \leq 12$. It also lists the number of m-sequences of length $L = 2^m - 1$ for $3 \leq m \leq 12$. We observe that the number of m-sequences of length L increases rapidly with m. We also observe that, for most sequences, the peak magnitude $R_{\max}$ of the cross correlation function is a large percentage of the peak value of the autocorrelation function. Consequently, m-sequences are not suitable for CDMA communication systems. Although it is possible to select a small subset of m-sequences that have a relatively lower cross

TABLE 11.3 PEAK CROSS CORRELATIONS OF M-SEQUENCES AND GOLD SEQUENCES

m	$L = 2^{m-1}$	Number	m sequences Peak cross correlation R_{max}	$R_{max}/R(0)$	Gold R_{max}	sequences $R_{min}/R(0)$
3	7	2	5	0.71	5	0.71
4	15	2	9	0.60	9	0.60
5	31	6	11	0.35	9	0.29
6	63	6	23	0.36	17	0.27
7	127	18	41	0.32	17	0.13
8	255	16	95	0.37	33	0.13
9	511	48	113	0.22	33	0.06
10	1023	60	383	0.37	65	0.06
11	2047	176	287	0.14	65	0.03
12	4095	144	1407	0.34	129	0.03

correlation peak value than R_{max}, the number of sequences in the set is usually too small for CDMA applications.

Methods for generating PN sequences with better periodic cross correlation properties than m-sequences have been developed by Gold (1967, 1968) and by Kasami (1966). Gold sequences are constructed by taking a pair of specially selected m-sequences, called *preferred m-sequences*, and forming the modulo-2 sum of the two sequences for each of L cyclically shifted versions of one sequence relative to the other sequence. Thus, L Gold sequences are generated as illustrated in Figure 11.21. For m odd, the maximum value of the cross correlation function between any pair of Gold sequences is $R_{max} = \sqrt{2L}$. For m even, $R_{max} = \sqrt{L}$.

Kasami (1966) described a method for constructing PN sequences by decimating an m-sequence. In Kasami's method of construction, every $2^{m/2} + 1$ bit of

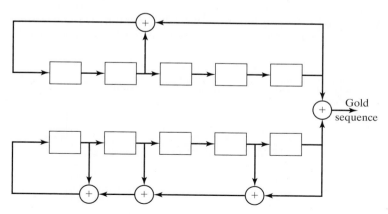

Figure 11.21   Generation of Gold sequences of length 31.

an m-sequence is selected. This method of construction yields a smaller set of PN sequences compared with Gold sequences, but their maximum cross correlation value is $R_{max} = \sqrt{L}$.

It is interesting to compare the peak value of the cross correlation function for Gold sequences and for Kasami sequences with a known lower bound for the maximum cross correlation between any pair of binary sequences of length L. Given a set of N sequences of period L, a lower bound on their maximum cross correlation is

$$R_{max} \geq L\sqrt{\frac{N-1}{NL-1}}, \tag{11.3.40}$$

which, for large values of L and N, is well approximated as $R_{max} \geq \sqrt{L}$. Hence, Kasami sequences satisfy the lower bound, and they are optimal. On the other hand, Gold sequences with m odd have an $R_{max} = \sqrt{2L}$. Hence, they are slightly suboptimal.

11.3.5 Frequency-Hopped Spread Spectrum

In frequency-hopped (FH) spread spectrum, the available channel bandwidth W is subdivided into a large number of non-overlapping frequency slots. In any signaling interval, the transmitted signal occupies one or more of the available frequency slots. The selection of the frequency slot (s) in each signal interval is made pseudorandomly according to the output from a PN generator.

A block diagram of the transmitter and receiver for an FH spread spectrum system is shown in Figure 11.22. The modulation is either binary or M-ary FSK (MFSK). For example, if binary FSK is employed, the modulator selects one of two frequencies, such as f_0 or f_1, corresponding to the transmission of a 0 or a 1. The resulting binary FSK signal is translated in frequency by an amount determined by the output sequence from a PN generator, which is used to select a frequency f_c that is synthesized by the frequency synthesizer. This frequency is mixed with the output of the FSK modulator and the resultant signal is transmitted over the channel. For example, by taking m bits from the PN generator, we may specify $2^m - 1$ possible carrier frequencies. Figure 11.23 illustrates an FH signal pattern.

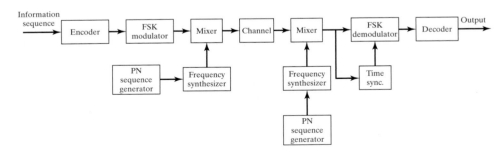

Figure 11.22   Block diagram of an FH spread-spectrum system.

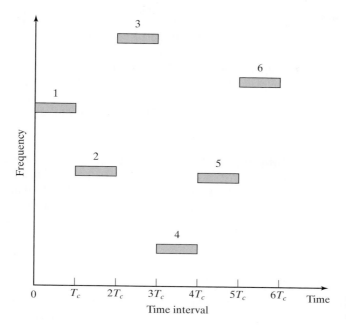

Figure 11.23 An example of an FH pattern.

At the receiver, there is an identical PN sequence generator, which is synchronized with the received signal and is used to control the output of the frequency synthesizer. Thus, the pseudorandom frequency translation introduced at the transmitter is removed at the demodulator by mixing the synthesizer output with the received signal. The resultant signal is then demodulated via an FSK demodulator. A signal for maintaining synchronism of the PN sequence generator with the FH received signal is usually extracted from the received signal.

Binary PSK modulation generally yields better performance than binary FSK. However, it is difficult to maintain phase coherence in the synthesis of the frequencies used in the hopping pattern and, also, in the propagation of the signal over the channel as the signal is hopped from one frequency to another over a wide bandwidth. Consequently, FSK modulation with noncoherent demodulation is usually employed in FH spread-spectrum systems.

The frequency-hopping rate, denoted as R_h, may be either equal to the symbol rate, lower than the symbol rate, or higher than the symbol rate. If R_h is equal to or lower than the symbol rate, the FH system is called a *slow hopping* system. If R_h is higher than the symbol rate, i.e., there are multiple hops per symbol, the FH system is called a *fast hopping* system. However, there is a penalty incurred in subdividing an information symbol into several frequency-hopped elements, because the energy from these separate elements is combined noncoherently.

FH spread-spectrum signals may be used in CDMA where many users share a common bandwidth. In some cases, an FH signal is preferred because of the stringent synchronization requirements inherent in DS spread spectrum signals. Specifically, in a DS system, timing and synchronization must be established to within a fraction of a chip interval $T_c = 1/W$. On the other hand, in an FH system, the chip interval T_c is the time spent in transmitting a signal in a particular frequency slot of bandwidth $B \ll W$. But this interval is approximately $1/B$, which is much larger than $1/W$. Hence, the timing requirements in an FH system are not as stringent as in a DS system.

Next, we shall evaluate the performance of FH spread spectrum systems under the condition that the system is slow hopping.

Slow Frequency-Hopping Systems.   Consider a slow frequency-hopping system in which the hop rate $R_h = 1$ hop per bit. We assume that the interference on the channel is broadband and is characterized as AWGN with power spectral density I_0. Under these conditions, the probability of error for the detection of noncoherently demodulated binary FSK is

$$P_2 = \frac{1}{2}e^{-\rho_b/2}, \tag{11.3.41}$$

where $\rho_b = \mathcal{E}_b/I_0$ is the SNR per bit.

As in the case of a DS spread-spectrum system, we observe that $\mathcal{E}_b$, the energy per bit, can be expressed as $\mathcal{E}_b = P_S T_b = P_S/R$, where P_S is the average transmitted power and R is the bit rate. Similarly, $I_0 = P_I/W$, where P_I is the average power of the broadband interference and W is the available channel bandwidth. Therefore, the SNR ρ_b can be expressed as

$$\rho_b = \frac{\mathcal{E}_b}{I_0} = \frac{W/R}{P_I/P_S}, \tag{11.3.42}$$

where W/R is the processing gain and P_I/P_S is the interference margin for the FH spread-spectrum signal. Note that the relationship in Equation (11.3.42) for the FH spread-spectrum signal is identical to that given by Equation (11.3.33) for the DS spread-spectrum signal. Therefore, frequency hopping provides basically the same benefits as direct sequence spreading.

Applications of FH Spread Spectrum.   FH spread spectrum is a viable alternative to DS spread spectrum for protection against narrowband and broadband interference that is encountered in CDMA. In CDMA systems based on frequency hopping, each transmitter–receiver pair is assigned its own pseudorandom frequency hopping pattern. Aside from this distinguishing feature, the transmitters and receivers of all users may be identical, i.e., they have identical encoders, decoders, modulators, and demodulators.

CDMA systems based on FH spread-spectrum signals are particularly attractive for mobile (i.e., land, air, and sea) users because timing (synchronization) requirements are not as stringent as in a DS spread-spectrum system. In addition, new frequency synthesis techniques and associated hardware make it possible to frequency-hop over bandwidths that are significantly larger, by one or more orders of magnitude, than those currently possible with DS spread spectrum signals. Consequently, larger processing gains are possible by FH, which more than offset the loss in performance inherent in noncoherent detection of the FSK-type signals.

11.4 DIGITAL CELLULAR COMMUNICATION SYSTEMS

In this section, we present an overview of two types of current digital cellular communication systems. One is the GSM (Global System for Mobile Communication) system that is widely used in Europe and other parts of the world. The second is the CDMA system, based on Interim Standard 95 (IS-95), which is widely used in North America and some countries in the Far East.

There are three basic methods that are currently used to provide channel access to multiple users in a communication network. One simple method is to subdivide the available channel bandwidth into a number, say, K, of frequency nonoverlapping subchannels and to assign a subchannel to each user upon request. This method is called *frequency-division multiple access* (FDMA). It is commonly used in wireline channels to accommodate multiple users for voice and data transmission. FDMA is also used in the first generation of cellular communication systems in which analog FM is employed, as described in Section 4.5.

A second method for providing access to multiple users is to subdivide a time interval, called a *frame*, into K nonoverlapping subintervals, each of duration T_f/K, where T_f is the frame duration. Then, each user is assigned to a particular subinterval, or time slot, within a frame. This multiple access method is called *time-division multiple access* (TDMA), and it is used in the GSM system described in Section 11.4.1.

The third method for providing multiple access is CDMA, which was described in Section 11.3.3. In CDMA, each user is assigned a unique spreading sequence, such as, a Gold sequence or a Kasami sequence. All users are allowed to transmit simultaneously over the same channel. In FDMA and TDMA, the signals of the multiple users are nonoverlapping in either frequency or time; however, in CDMA, the transmitted signals completely overlap in both time and frequency. By assigning unique code sequences to multiple users, each user is able to separate its desired signal (via cross correlation) from the other user signals, which appear as interference. As previously indicated, the digital cellular system based on the IS-95 standard employs CDMA.

11.4.1 The GSM System

GSM was developed in Europe to provide a common digital cellular communication system that would serve all of Europe. It is now widely used in many parts of the world. The GSM system employs the frequency band 890–915 MHz for signal transmission from mobile transmitters to base station receivers (uplink or reverse link) and the frequency band 935–960 MHz for transmission from the base stations to the mobile receivers (downlink or forward link). The two 25-MHz frequency

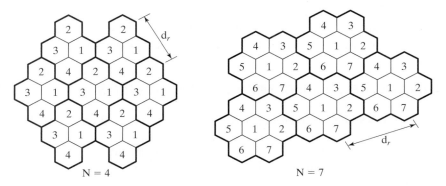

Figure 11.24 Frequency reuse in a cellular system with a reuse factor $N = 4$ or $N = 7$.

bands are each subdivided into 125 channels, where each channel has a bandwidth of 200 kHz. There are two methods to reduce transmission interference from adjacent cells. First, different sets of frequencies are assigned to adjacent base stations. Second, frequencies are reused according to some design plan, such as the frequency plans shown in Figure 11.24, where the frequency reuse factor is either $N = 4$ or $N = 7$. Larger values of N increase the distance d_r between two base stations using the same set of frequencies, thus, reducing co-channel interference. On the other hand, a large value of N reduces the spectral efficiency of the cellular system, since fewer frequencies are assigned to each cell. The maximum radius for a cell is 35 km.

Each 200-kHz frequency band accommodates eight users by creating eight TDMA nonoverlapping time slots, as shown in Figure 11.25. The eight time slots constitute a frame of duration 4.615 msec, and each time slot has a time duration of 576.875 µsec. The information data from the users is transmitted in bursts at a rate of 270.833 kbps. Figure 11.26 illustrates the basic GSM frame structure, where 26 frames form a multiframe and 51 multiframes form a superframe. The framing hierarchy facilitates synchronization and network control. To reduce the effect of fading and interference and provide signal diversity, the carrier frequency is hopped at the frame rate of (nominally) 217 hops/sec.

The functional block diagram of the transmitter and receiver in the GSM system is shown in Figure 11.27. The speech coder is based on a type of linear predictive

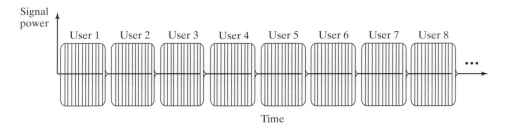

Figure 11.25 TDMA frame in GSM.

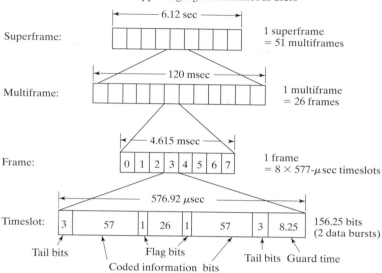

Figure 11.26   GSM frame structure.

coding (LPC) called residual pulse-excited (RPE) linear predictive coding (RPE-LPC). RPE-LPC delivers 260 bits in each 20-msec time interval—hence, a bit rate of 13 kbps. The most significant bits are encoded by a rate 1/2, constraint length $L = 5$ convolutional encoder, and the coded and uncoded bits are block interleaved to produce data at a rate of 22.8 kbps. Thus, the 260 information bits are transformed into 456 coded bits in each 20-msec time interval. The coded bit stream is encrypted. Then, it is organized for burst transmission in time slots that carry 114 coded bits and some overhead bits, as shown in Figure 11.26, including a sequence of 26 bits that measure the channel characteristics in each time slot. Therefore, the 456 coded bits are transmitted in four consecutive bursts, where each burst contains 114 coded bits and occupies one time slot.

To transmit the bits in each time slot, we use GMSK modulation with $BT = 0.3$. This signal pulse is illustrated in Figure 10.42(e). The output of the GMSK modulator is translated to the desired carrier frequency, which is hopped to a different frequency in each frame.

At the receiver, the received signal is dehopped and translated to baseband; this creates in-phase (I) and quadrature (Q) signal components, which are sampled and buffered. The 26 known transmitted bits measure the channel characteristics; thus, they specify the matched filter to the channel corrupted signal. The data bits are passed through the matched filter and the matched filter output is processed by a channel equalizer, which may be realized either as a decision-feedback equalizer (DFE) or a maximum-likelihood sequence detector that is efficiently implemented via the Viterbi algorithm. The bits in a burst at the output of the equalizer are

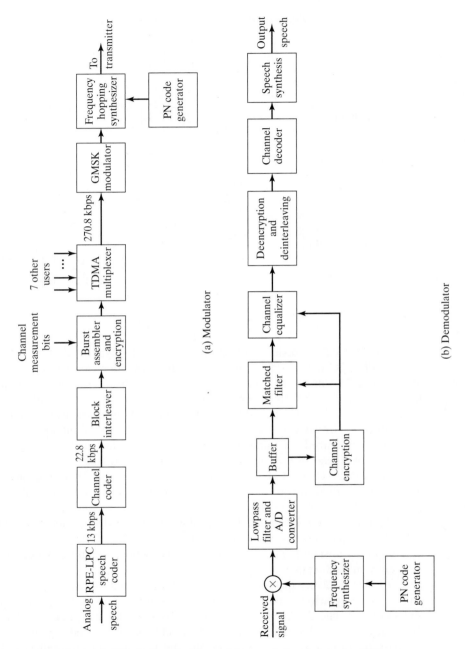

Figure 11.27   Functional block diagram of a modulator and demodulator for GSM.

TABLE 11.4   SUMMARY OF PARAMETERS IN A GSM SYSTEM

System Parameter	Specification
Uplink frequency band	890–915 MHz
Downlink frequency band	935–960 MHz
Number of carriers/band	125
Bandwidth/carrier	200 kHz
Multiple access method	TDMA
Number of users/carrier	8
Data rate/carrier	270.8 kbps
Speech coding rate	13 kHz
Speech encoder	RPE-LPC
Coded speech rate	22.8 kbps
Modulation	GMSK with $BT = 0.30$
Demodulation	Matched filter + equalizer
Interleaver	Block
Frequency hopping rate	217 hops/sec

deassembled, deencrypted, deinterleaved and passed to the channel decoder. The decoded bits are used to synthesize the speech signal that was encoded via RPE-LPC.

In addition to the channels that transmit the digitized speech signals, there are other channels that handle various control and synchronization functions, such as paging, frequency correction, synchronization, and requests for access to send messages. These control channels are additional time slots that contain known sequences of bits for performing the control functions.

Table 11.4 provides a summary of the basic parameters in the GSM system.

11.4.2 CDMA System Based on IS-95

As described in Section 11.3, the enhancement in performance obtained from a DS spread-spectrum signal through the processing gain and coding gain can enable many DS spread-spectrum signals to simultaneously occupy the same channel bandwidth provided that each signal has its own distinct pseudorandom sequence. Direct sequence CDMA has been adopted as one multiple-access method for digital cellular voice communications in North America. This first generation digital cellular (CDMA) communication system was developed by Qualcomm, and it has been standardized and designated as IS-95 by the Telecommunications Industry Association (TIA) for use in the 800 MHz and the 1900 MHz frequency bands. A major advantage of CDMA over other multiple access methods is that the entire frequency band is available at each base station, i.e., the frequency reuse factor $N = 1$.

The nominal bandwidth used for transmission from a base station to the mobile receivers (forward link) is 1.25 MHz. A separate channel, also with a bandwidth of 1.25 MHz, is used for signal transmission from mobile receivers to a base station (reverse link). The signals transmitted in both the forward and the reverse links are

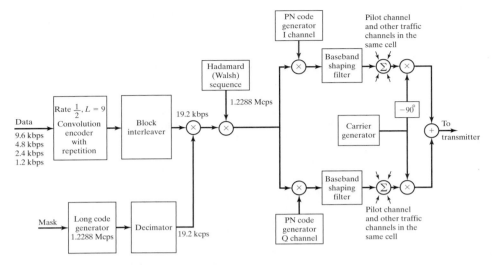

Figure 11.28   Block diagram of a IS-95 forward link.

DS spread spectrum signal and they have a chip rate of 1.2288×10^6 chips per second (1.2288 Mchips/s).

Forward Link.   A block diagram of the modulator for the signals transmitted from a base station to the mobile receivers is shown in Figure 11.28. The speech coder is a code-excited linear predictive (CELP) coder that generates data at the variable rates of 9600, 4800, 2400, and 1200 bits/s, where the data rate is a function of the user's speech activity in frame intervals of 20 ms. The data from the speech coder is encoded by a rate 1/2, constraint length $L = 9$ convolutional code. For lower speech activity, where the data rates are 4800, 2400, or 1200 bits/s, the output symbols from the convolutional encoder are repeated either twice, four times, or eight times to maintain a constant bit rate of 9600 bits/s. At the lower speech activity rates, the transmitter power is reduced by either 3, 6, or 9 dB, so that the transmitted energy per bit remains constant for all speech rates. Thus, a lower speech activity results in a lower transmitter power and, hence, a lower level of interference to other users.

The encoded bits for each frame are passed through a block interleaver, which overcomes the effects of burst errors that may occur in transmission through the channel. The data bits at the output of the block interleaver, which occur at a rate of 19.2 kbits/s, are scrambled by multiplication with the output of a long code (period $N = 2^{42} - 1$) generator running at the chip rate of 1.2288 Mchips/s, but whose output is decimated by a factor of 64 to 19.2 kchips/s. The long code uniquely identifies a call of a mobile station on the forward and reverse links.

Each channel user is assigned a Hadamard (also called a Walsh) sequence of length 64. There are 64 orthogonal Hadamard sequences assigned to each base station; thus, there are 64 channels available. One Hadamard sequence (the all-zero sequence) is used to transmit a pilot signal, which serves as a means for measuring the channel characteristics, including the signal strength and the carrier phase offset. These parameters are used at the receiver to perform phase coherent demodulation. Another Hadamard sequence is used to provide time synchronization. One channel, and possibly more if necessary, is used for paging. That leaves up to 61 channels for allocation to different users.

Each user, using the assigned Hadamard sequence, multiplies the data sequence by the assigned Hadamard sequence. Thus, each encoded data bit is multiplied by the Hadamard sequence of length 64. The resulting binary sequence is now spread by multiplication with two PN sequences of length $N = 2^{15}$; this creates in-phase (I) and quadrature (Q) signal components. Thus, the binary data signal is converted to a four-phase signal and both the I and Q components are filtered by baseband spectral-shaping filters. Different base stations are identified by different offsets of these PN sequences. The signals for all 64 channels are transmitted synchronously so that, in the absence of channel multipath distortion, other signals received at any mobile receiver do not interfere because of the orthogonality of the Hadamard sequences.

At the receiver, a RAKE demodulator resolves the major multipath signal components. These components are then phase-aligned and weighted according to their signal strength, using the estimates of phase and signal strength derived from the pilot signal. These components are combined and passed to the Viterbi soft-decision decoder.

Reverse link.   The reverse link modulator from a mobile transmitter to a base station is different from the forward link modulator. A block diagram of the modulator is shown in Figure 11.29. An important consideration in the design of the modulator is that signals transmitted from the various mobile transmitters to the base station are asynchronous; hence, there is significantly more interference among users. In addition, the mobile transmitters are usually battery operated; consequently, these transmissions are power limited. To compensate for these major limitations, a rate $1/3$, $K = 9$ convolutional code is used in the reverse link. This code has essentially the same coding gain in an AWGN channel as the rate $1/2$ code used in the forward link. However, it has a much higher coding gain in a fading channel. Again, for lower speech activity, output bits from the convolutional encoder are repeated either two, four, or eight times. However, the coded bit rate is 28.8 kbits/s.

For each 20-ms frame, the 576 encoded bits are block-interleaved and passed to the modulator. The data are modulated using an $M = 64$ orthogonal signal set of Hadamard sequences each of length 64. Thus, a 6-bit block of data is mapped into one of the 64 Hadamard sequences. The result is a bit (or chip) rate of 307.2 kbits/s at the output of the modulator. Note that 64-ary orthogonal modulation at an error probability of 10^{-6} requires approximately 3.5 dB less SNR per bit than binary antipodal signaling in an AWGN channel.

To reduce interference to other users, the time position of the transmitted code symbol repetitions is randomized; thus, at the lower speech activity, consecutive bursts are not evenly spaced in time. Following the randomizer, the signal is spread by the output of the long code PN generator, which is running at a rate of

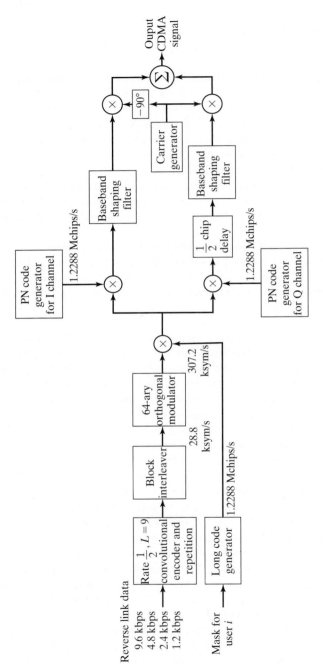

Figure 11.29 Block diagram of the IS-95 reverse link

1.2288 Mchips/s. Hence, there are only four PN chips for every bit of the Hadamard sequence from the modulator, so the processing gain in the reverse link is very small. The resulting 1.2288-Mchips/s binary sequences of length $N = 2^{15}$, whose rate is also 1.2288 Mchips/s, create I and Q signals (a QPSK signal) that are filtered by baseband spectral shaping filters and then passed to quadrature mixers. The Q-channel signal is delayed in time by one-half PN chip relative to the I-channel signal prior to the baseband filter. In effect, the signal at the output of the two baseband filters is an offset QPSK signal.

Although the chips are transmitted as an offset QPSK signal, the demodulator employs noncoherent demodulation of the $M = 64$ orthogonal Hadamard waveforms to recover the encoded data bits. A computationally efficient (fast) Hadamard transform reduces the computational complexity in the demodulation process. The output of the demodulator is then fed to the Viterbi detector, whose output synthesizes the speech signal.

Table 11.5 provides a summary of the basic parameters in the IS-95 system.

11.4.3 Third Generation Cellular Communication Systems

The first generation cellular radio systems employed analog signal transmission, as described in Section 4.5. The GSM and IS-95 digital cellular systems previously described are usually called *second generation cellular radio systems*. They are

TABLE 11.5   SUMMARY OF PARAMETERS IN THE IS-95 SYSTEM

System Parameter	Specification
Uplink frequency band	824–849 MHz
Downlink frequency band	869–894 MHz
Number of carriers/band	20
Bandwidth/carrier	1.25 MHz
Multiple access method	CDMA
Number of users/carrier	60
Chip rate	1.2288 Mcps
Speech coder	Variable rate CELP
Speech rate	9600, 4800, 2400, 1200 bps
Interleaver	Block
Channel encoder	$R = 1/2$, $L = 9$ (D)
	$R = 1/3$, $L = 9$ (U)
Modulation	BPSK with QPSK spreading (D)
	64-ary orthogonal with QPSK spreading (U)
Demodulation	RAKE matched filter with maximal-ratio combining
Signature sequences	Hadamard (Walsh) of length 64
PN sequence	$N = 2^{42} - 1$ (Long code)
	$N = 2^{15}$ (spreading codes)

designed primarily for voice transmission using digital modulation for transmission. These systems have the capability of providing low-rate data transmission, i.e., 64 kbps.

To accommodate the need for multimedia applications, such as internet access, image transmission, and video conferencing, higher data rate cellular systems are currently being developed. These new systems are *third generation cellular radio systems* and have the potential to provide data rates up to 5 Mbps.

The evolution of IS-95 is called CDMA-2000. The channel bandwidth is increased to 3.75 MHz, rather than the 1.25 MHz bandwidth used in IS-95. The corresponding chip rate is 3.684 Mcps. QPSK modulation is used in the downlink to achieve a peak data rate of up to 2.4 Mbps. BPSK is used in the uplink. Higher data rates can be achieved by transmitting on multiple channels.

The evolution of GSM is called UMTS or wideband CDMA (WCDMA). A nominal channel bandwidth of 5 MHz supports a chip rate of 3.84 Mcps. The modulation for both uplink and downlink is BPSK. WCDMA offers symmetrical data rates of 384 kbps at a mobile terminal speed of 120 km/h, 128 kbps at a mobile terminal speed of up to 300 km/h, and a data rate of 2 Mbps at pedestrian speed.

11.5 PERFORMANCE ANALYSIS FOR WIRELINE AND RADIO COMMUNICATION CHANNELS

In the transmission of digital signals through an AWGN channel, the performance of the communication system, measured in terms of the probability of error, depends on the received SNR, $\mathscr{E}_b/N_0$, where $\mathscr{E}_b$ is the transmitted energy/bit and $N_0/2$ is the power spectral density of the additive noise. Hence, the additive noise ultimately limits the performance of the communication system.

In addition to the additive noise, another factor that affects the performance of a communication system is channel attenuation. As indicated in Section 6.5, all physical channels, including wireline and radio channels, are lossy. Hence, the signal is attenuated as it travels through the channel. The simple mathematical model for the attenuation previously shown in Figure 6.17 may also be used for the purpose of digital communication. Consequently, if the transmitted signal is $s(t)$, the received signal is

$$r(t) = \alpha s(t) + n(t), \tag{11.5.1}$$

where α is the attenuation factor. Then, if the energy in the transmitted signal is $\mathscr{E}_b$, the energy in the received signal is $\alpha^2 \mathscr{E}_b$. Consequently, the received signal has an SNR $\alpha^2 \mathscr{E}_b/N_0$. As in the case of analog communication systems, the effect of signal attenuation in digital communication systems is to reduce the energy in the received signal; thus, it will render the communication system more vulnerable to additive noise.

Recall that, in analog communication systems, amplifiers called repeaters are used to periodically boost the signal strength in transmission through the channel. However, each amplifier also boosts the noise in the system. In contrast, digital communication systems allow us to detect and regenerate a clean (noise-free) signal in a transmission channel. Such devices, called *regenerative repeaters*, are frequently used in wireline and fiber optic communication channels.

11.5.1 Regenerative Repeaters

The front end of each regenerative repeater consists of a demodulator/detector that demodulates and detects the transmitted digital information sequence sent by the preceding repeater. Once detected, the sequence is passed to the transmitter side of the repeater, which maps the sequence into signal waveforms that are transmitted to the next repeater. This type of repeater is called a regenerative repeater.

Since a noise-free signal is regenerated at each repeater, the additive noise does not accumulate. However, when errors occur in the detector of a repeater, the errors are propagated forward to the following repeaters in the channel. To evaluate the effect of errors on the performance of the overall system, suppose that the modulation is binary PAM, so that the probability of a bit error for one hop (signal transmission from one repeater to the next repeater in the chain) is

$$P_2 = Q\left(\sqrt{\frac{2\mathscr{E}_b}{N_0}}\right).$$

Since errors occur with low probability, we may ignore the probability that any one bit will be detected incorrectly more than once during the transmission through a channel with K repeaters. Consequently, the number of errors will increase linearly with the number of regenerative repeaters in the channel, and thus the overall probability of error may be approximated as

$$P_2 \approx KQ\left(\sqrt{\frac{2\mathscr{E}_b}{N_0}}\right). \tag{11.5.2}$$

In contrast, the use of K analog repeaters in the channel reduces the received SNR by K; hence, the bit-error probability is

$$P_2 \approx Q\left(\sqrt{\frac{2\mathscr{E}_b}{KN_0}}\right). \tag{11.5.3}$$

Clearly, for the same probability of error performance, the use of regenerative repeaters results in a significant savings in transmitter power over analog repeaters. Hence, in digital communication systems, regenerative repeaters are preferable. However, in wireline telephone channels that are used to transmit both analog and digital signals, analog repeaters are generally employed.

Example 11.5.1

A binary digital communication system transmits data over a wireline channel of length 1000 km. Repeaters are used every 10 km to offset the effect of channel attenuation.

Determine the $\mathscr{E}_b/N_0$ that is required to achieve a probability of a bit error of 10^{-5} if (1) analog repeaters are employed and (2) regenerative repeaters are employed.

Solution   The number of repeaters used in the system is $K = 100$. If regenerative repeaters are used, the $\mathscr{E}_b/N_0$ obtained from Equation (11.5.2) is

$$10^{-5} = 100 Q\left(\sqrt{\frac{2\mathscr{E}_b}{N_0}}\right)$$

$$10^{-7} = Q\left(\sqrt{\frac{2\mathscr{E}_b}{N_0}}\right),$$

which yields approximately 11.3 dB. If analog repeaters are used, the $\mathscr{E}_b/N_0$ obtained from Equation (11.5.3) is

$$10^{-5} = Q\left(\sqrt{\frac{2\mathscr{E}_b}{100N_0}}\right),$$

which yields an $\mathscr{E}_b/N_0$ of 29.6 dB. Hence, the difference on the required SNR is about 18.3 dB, or approximately 70 times the transmitter power of the digital communication system.   ∎

11.5.2 Link Budget Analysis for Radio Channels

In the design of radio communication systems that transmit over LOS microwave satellite channels, we must also consider the effect of the antenna characteristics on determining the SNR at the receiver that is required to achieve a given level of performance. The system design procedure is described next.

Suppose that a transmitting antenna radiates isotropically in free space at a power level P_T Watts, as shown in Figure 11.30. The power density at a distance d from the antenna is $P_T/4\pi d^2$ W/m². If the transmitting antenna has directivity in a particular direction, the power density in that direction is increased by a factor

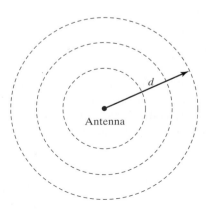

Figure 11.30   Antenna that radiates isotropically in free space.

called the antenna gain G_T. Then, the power density at a distance d is $P_T G_T / 4\pi d^2$ W/m^2. The product $P_T G_T$ is usually called the *effective isotropically* radiated power (EIRP), which is basically the radiated power relative to an isotropic antenna for which $G_T = 1$.

A receiving antenna pointed in the direction of the radiated power gathers a portion of the power that is proportional to its cross-sectional area. Hence, the received power extracted by the antenna is expressed as

$$P_R = \frac{P_T G_T A_R}{4\pi d^2}, \tag{11.5.4}$$

where A_R is the effective area of the antenna. The basic relationship between the antenna gain G_R and its effective area, obtained from basic electromagnetic theory, is

$$A_R = \frac{G_R \lambda^2}{4\pi}, \, m^2 \tag{11.5.5}$$

where λ is the wavelength of the transmitted signal.

If we substitute for A_R from Equation (11.5.5) into Equation (11.5.4), we obtain the expression for the received power:

$$P_R = \frac{P_T G_T G_R}{(4\pi d/\lambda)^2}. \tag{11.5.6}$$

The factor $(4\pi d/\lambda)^2 = \mathscr{L}_s$ is the free-space path loss. Other losses that may be encountered in the transmission of the signal, such as atmospheric losses, are tracked by introducing an additional loss factor $\mathscr{L}_a$. Therefore, the received power may be expressed as

$$P_R = \frac{P_T G_T G_R}{\mathscr{L}_s \mathscr{L}_a}, \tag{11.5.7}$$

or equivalently,

$$P_R|_{\text{dBW}} = P_T|_{\text{dBW}} + G_T|_{\text{dB}} + G_R|_{\text{dB}} - \mathscr{L}_s|_{\text{dB}} - \mathscr{L}_a|_{\text{dB}}. \tag{11.5.8}$$

The effective area for an antenna generally depends on the wavelength λ of the radiated power and the physical dimension of the antenna. For example, a parabolic (dish) antenna of diameter D has an effective area

$$A_R = \frac{\pi D^2}{4}\eta, \tag{11.5.9}$$

where $\pi D^2/4$ is the physical area and η is the *illumination efficiency factor*, which is typically in the range $0.5 \leq \eta \leq 0.6$. Hence, the antenna gain for a parabolic antenna of diameter D is

$$G_R = \eta\left(\frac{\pi D}{\lambda}\right)^2, \quad \text{for a parabolic antenna.} \tag{11.5.10}$$

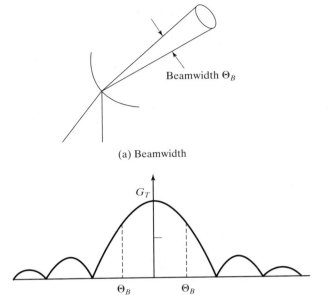

(a) Beamwidth

(b) Antenna pattern

Figure 11.31   A narrow beam antenna and its radiating pattern.

As a second example, a horn antenna of physical area A has an efficiency factor of 0.8, an effective area of $A_R = 0.8A$, and a gain of

$$G_R = \frac{10A}{\lambda^2}, \qquad \text{for a horn antenna.} \tag{11.5.11}$$

Another parameter that is related to the gain (directivity) of an antenna is its beamwidth, which is denoted as Θ_B and illustrated in Figure 11.31. Usually, the beamwidth is measured as the -3 dB width of the antenna pattern. For example, the -3 dB beamwidth of a parabolic antenna is approximately

$$\Theta_B \approx 70\lambda/D \text{ deg,}$$

so that G_T is inversely proportional to Θ_B^2. Hence, a decrease of the beamwidth by a factor of two, which is obtained by doubling the diameter, increases the antenna gain by a factor of four (6 dB).

Example 11.5.2

A satellite in geosynchronous orbit (36,000 km above the earth's surface) radiates 100 W of power (20 dBW). The transmitting antenna has a gain of 18 dB, so that the EIRP = 38 dBW. The earth station employs a 3-meter parabolic antenna, and the downlink is transmitting at a frequency of 4 GHz. Determine the received power.

Solution   The wavelength $\lambda = 0.075$ m. Hence, the free-space path loss is

$$\mathscr{L}_s|_{\text{dB}} = 20 \log \left(\frac{4\pi d}{\lambda} \right) = 195.6 \text{ dB}.$$

Assuming that $\eta = 0.5$, the antenna gain is 39 dB. Since no other losses are assumed, it follows that

$$P_R|_{\text{dBW}} = 20 + 18 + 39 - 195.6$$
$$= -118.6 \text{ dBW},$$

or equivalently,

$$P_R = 1.38 \times 10^{-12} W.$$

We may carry the computation one step further by relating the $\mathscr{E}_b/N_0$ required to achieve a specified level of performance to P_R. Since

$$\frac{\mathscr{E}_b}{N_0} = \frac{T_b P_R}{N_0} = \frac{1}{R_b} \frac{P_R}{N_0}, \tag{11.5.12}$$

it follows that

$$\frac{P_R}{N_0} = R_b \left(\frac{\mathscr{E}_b}{N_0} \right)_{\text{req}}, \tag{11.5.13}$$

where $(\mathscr{E}_b/N_0)_{\text{req}}$ is the required SNR/bit to achieve the desired performance. The relation in Equation (11.5.13) allows us to determine the bit rate R_b. We have

$$10 \log_{10} R_b = \left(\frac{P_R}{N_0} \right)_{\text{dB}} - 10 \log_{10} \left(\frac{\mathscr{E}_b}{N_0} \right)_{\text{req}}. \tag{11.5.14}$$

■

Example 11.5.3

If $(\mathscr{E}_b/N_0)_{\text{req}} = 10$ dB, determine the bit rate for the satellite communication system in Example 11.5.2. Assume that the front end of the receiver has a noise temperature of 300 K, which is typical for a receiver in the 4-GHz range.

Solution   Since $T_0 = 290$ K and $T_e = 10$ K, it follows that

$$N_0 = kT = 4.1 \times 10^{-21} W/\text{Hz},$$

or equivalently, -203.9 dBW/Hz. Then,

$$\left(\frac{P_R}{N_0} \right)_{\text{dB}} = -118.6 + 203.9.$$

Therefore, from Equation (11.5.14), we obtain

$$10 \log_{10} R_b = 85.3 - 10$$
$$= 75.3,$$

or equivalently,

$$R_b = 33.9 \times 10^6 \text{ bps}.$$

We conclude that this satellite channel can support a bit rate of 33.9 Mbps.   ■

11.6 FURTHER READING

We started this chapter with a brief treatment of digital transmission in fading multipath channels. Such channels are encountered in non-LOS radio communication systems, e.g., cellular radio systems. Our treatment of this topic was very brief and focused primarily on the Rayleigh fading channel model. For the most part, this is due to the wide acceptance of this model for describing the fading effects on many radio channels and to its mathematical tractability. Although other statistical models, such as the Rician and Nakagami fading models, may be more appropriate for characterizing fading on some real channels, the general approach for providing reliable communications via signal diversity is still applicable.

More extensive and advanced treatments on fading channels are found in books by Schwartz et al. (1966) and Proakis (2001). The pioneering work on the characterization of fading multipath channels and on signal and receiver design for reliable digital communication over such channels was done by Price (1954, 1956). This early work was followed by significant contributions from Price and Green (1958, 1960), Kailath (1960, 1961), and Green (1962). Diversity transmission and diversity-combining techniques under a variety of channel conditions have been treated in the papers by Pierce (1958), Brennan (1959), Turin (1961, 1962), Pierce and Stein (1960), Barrow (1963), Bello and Nelin (1962a, 1962b, 1963), Price (1962a, 1962b), and Lindsey (1964).

There is a large amount of literature on multicarrier digital communication systems. One of the earliest systems, described by Doeltz et al. (1957), is called Kineplex; it was used for digital transmission in the high-frequency radio band. Other early work on multicarrier system design is described in the papers by Chang (1966) and Saltzberg (1967). The use of DFT for the modulation and demodulation of multicarrier OFDM systems was proposed by Weinstein and Ebert (1971). More recent references on applications of OFDM in practical systems are the papers by Chow et al. (1995) and Bingham (1990). The book by Bahai and Saltzberg (1999) provides a comprehensive treatment of OFDM.

The problem of PAR reduction in multicarrier systems has been investigated by many people. Refer to the papers by Boyd (1986), Popovic (1991), Jones et al. (1994), Wilkinson and Jones (1995), Wulich (1996), Tellado and Cioffi (1998), and Tarokh and Jafarkhani (2000).

Our third topic was a brief, introductory treatment of spread-spectrum modulation and demodulation and the performance evaluation of this type of signaling in the presence of different types of interference. Historically, the primary application of spread-spectrum modulation has been in the design of secure digital communications systems for the military. However, in the last two decades, we have seen a trend toward commercial use of spread-spectrum modulation, especially in mobile cellular communications, in multiple access communications via satellites, and in interoffice radio communications.

A historical account on the development of spread-spectrum communications covering the period 1920–1960 is given in a paper by Scholtz (1982). Tutorial treatments of spread-spectrum modulation that deal with basic concepts are found in papers by Scholtz (1977) and Pickholtz et al. (1982). These papers also contain a large number of references to the previous work. Two tutorial papers by Viterbi (1979, 1985) contain a basic analysis of the performance characteristics of DS and FH spread-spectrum signals.

Comprehensive treatments of various aspects concerning the analysis and design of spread-spectrum signals and systems, including synchronization techniques, are found in books by Simon et al. (1985), Ziemer and Peterson (1985), and Holmes (1982). Also, special issues of the *IEEE Transactions on Communications* (August 1977, May 1982) are devoted to spread-spectrum communications. These special issues contain a collection of papers devoted to a variety of topics, including multiple-access techniques, synchronization techniques, and performance analysis with various types of channel interference. A number of important papers that have been published in IEEE journals have been reprinted in book form by the IEEE press. [See Dixon, ed. (1976) and Cook et al. (1983).]

In this chapter, we also presented two examples of digital cellular communication systems designed for voice transmission. One system, called GSM, is based on TDMA to accommodate multiple users, while the second system employs CDMA. There are many textbooks that provide in-depth treatment of the GSM and IS-95 systems. For the interested reader, we cite the books by Rappaport (1996), Viterbi (1995), and Garg, et al. (1997).

The link budget analysis described in the last section of this chapter applies generally to free space line-of-sight (LOS) channels and is especially relevant to the design of LOS microwave radio communications and satellite communication systems. Radio propagation in non-LOS terrestrial channels is significantly more variable due to terrain characteristics and, consequently, signal attenuation is generally greater than in free space. For example, the modeling of path losses in cellular radio communications is treated in the book by Rappaport (1996).

PROBLEMS

11.1 In the transmission and reception of signals to and from moving vehicles, the transmitted signal frequency is shifted in direct proportion to the speed of the vehicle. The so-called *Doppler frequency shift* imparted to a signal that is received in a vehicle traveling at a velocity v in m/s relative to a (fixed) transmitter is given by the formula

$$f_D = \pm \frac{v}{\lambda},$$

where λ is the wavelength and the sign depends on the direction (moving toward or moving away) that the vehicle is traveling relative to the transmitter. Suppose that a vehicle is traveling at a speed of 100 km/hr relative to a base station in a mobile cellular communication system. The signal is a narrowband signal transmitted at a carrier frequency of 1 GHz.

1. Determine the Doppler frequency shift.

2. What is the bandwidth of a Doppler frequency tracking loop if the loop is designed to track Doppler frequency shifts for vehicles traveling at speeds up to 100 km/hr?

3. Suppose the transmitted signal bandwidth is 1 MHz and is centered at 1 GHz. Determine the Doppler frequency spread between the upper and lower frequencies in the signal.

11.2 A multipath fading channel has a multipath spread of $T_m = 1$ sec and a Doppler spread $B_d = 0.01$ Hz. The total channel bandwidth at bandpass available for signal transmission is $W = 5$ Hz. To reduce the effect of intersymbol interference, the signal designer selects a pulse duration of $T = 10$ sec.

1. Determine the coherence bandwidth and the coherence time.

2. Is the channel frequency selective? Explain.

3. Is the channel fading slowly or rapidly? Explain.

4. Suppose that the channel is used to transmit binary data via (antipodal) coherently detected PSK in a frequency diversity mode. Explain how you would use the available channel bandwidth to obtain frequency diversity, and determine how much diversity is available.

5. For the case in Part 4, what is the approximate SNR required/diversity to achieve an error probability of 10^{-6}?

11.3 The probability of error for binary DPSK and binary FSK with noncoherent detection in an AWGN channel is

$$P_2 = \frac{1}{2} e^{-c\rho_b},$$

where $c = 1$ for DPSK, $c = \frac{1}{2}$ for FSK, and $\rho_b = \frac{\alpha^2 \mathcal{E}_b}{N_0}$, in which α is the attenuation factor. By averaging P_2 over the Rayleigh distributed variable α, as indicated by Equation (11.1.20), verify the expressions for the probability of error for DPSK and FSK in a Rayleigh fading channel.

11.4 A communication system employs dual antenna diversity and binary orthogonal FSK modulation. The received signals are

$$r_1(t) = \alpha_1 s(t) + n_1(t)$$

and

$$r_2(t) = \alpha_2 s(t) + n_2(t),$$

where α_1 and α_2 are statistically i.i.d. Rayleigh random variables, and $n_1(t)$ and $n_2(t)$ are statistically independent, zero-mean white Gaussian random processes with the power spectral density $N_0/2$ W/Hz. The two signals are demodulated, squared, and then combined (summed) prior to detection.

1. Sketch the functional block diagram of the entire receiver, including the demodulator, the combiner, and the detector.

2. Plot the probability of error for the detector and compare the result with the case of no diversity.

11.5 A binary communication system transmits the same information on two diversity channels. The two received signals are

$$r_1 = \pm\sqrt{\mathcal{E}_b} + n_1$$
$$r_2 = \pm\sqrt{\mathcal{E}_b} + n_2,$$

where $E(n_1) = E(n_2) = 0$, $E(n_1^2) = \sigma_1^2$, $E(n_2^2) = \sigma_2^2$, and n_1 and n_2 are uncorrelated Gaussian variables. The detector bases its decision on the linear combination of r_1 and r_2, i.e.,

$$r = r_1 + kr_2.$$

1. Determine the value of k that minimizes the probability of error.

2. Plot the probability of error for $\sigma_1^2 = 1$, $\sigma_2^2 = 3$ and either $k = 1$ or k is the optimum value found in Part 1. Compare the results.

11.6 Suppose the binary antipodal signals $\pm s(t)$ are transmitted over a fading channel and the received signal is

$$r(t) = \pm\alpha s(t) + n(t), \quad 0 \le t \le T,$$

where $n(t)$ is zero-mean white Gaussian noise with autocorrelation function $\frac{N_0}{2}\delta(\tau)$. The energy in the transmitted signal is $\mathcal{E} = \int_0^T |s(t)|^2\,dt$. The channel gain α is specified by the PDF $p(\alpha) = 0.1\delta(\alpha) + 0.9\delta(\alpha - 2)$.

1. Determine the average probability of error P_2 for the demodulator that employs a filter matched to $s(t)$.

2. What value does P_2 approach as $\frac{\mathcal{E}}{N_0}$ approaches infinity?

3. Suppose the same signal is transmitted over two statistically *independently fading* channels with gains α_1 and α_2, where

$$p(a_k) = 0.1\delta(\alpha_k) + 0.9\delta(\alpha_k - 2), \quad k = 1, 2.$$

The noises on the two channels are statistically independent and identically distributed. The demodulator employs a matched filter for each channel and simply adds the two filter outputs to form the decision variable. Determine the average P_2.

4. For the case in Part 3, what value does P_2 approach as $\frac{\mathcal{E}}{N_0}$ approaches infinity?

11.7 Consider the frequency nonselective channel model shown in Figure 11.1. Binary orthogonal signals $u_1(t)$ and $u_2(t)$ are used to transmit information over this channel. The symbol duration $T_b \ll T_{ct}$, so that the channel is assumed to be constant over a time interval (time window) of NT_b, where N is a positive integer. The outputs of the matched filters at the sampling instants are either

$$y_{1k} = c\sqrt{\mathscr{E}_b} + n_{1k}$$

or

$$y_{2k} = n_{2k}$$

when u_{1k} is transmitted or

$$y_{1k} = n_{1k}$$

$$y_{2k} = c\sqrt{\mathscr{E}_b} + n_{2k}$$

when $u_2(t)$ is transmitted. $c = \alpha e^{j\phi}$ is the complex-valued channel coefficient, which is assumed to be constant over the time window $0 \le t \le NT_b$.

To detect the signal coherently over the time window, the channel coefficient c is estimated by averaging the matched filter outputs for $1 \le k \le N$. Thus, the estimate is

$$\hat{c} = \frac{1}{N\sqrt{\mathscr{E}_b}} \sum_{k=1}^{N} (y_{1k} + y_{2k})$$

$$= \frac{1}{N\sqrt{\mathscr{E}_b}} \sum_{k=1}^{N} \left[c\sqrt{\mathscr{E}_b} + n_{1k} + n_{2k} \right].$$

Assuming that the noise terms are statistically independent, zero-mean Gaussian random variables with equal variance $\sigma^2 = N_0/2$, (a) show that the mean value of $\hat{c}$ is the actual channel coefficient c, and (b) show that the variance of the estimate decreases as N is increased. (c) If σ_c^2 is defined as the variance of the estimate and $1/\sigma_c^2$ is defined as the SNR of the estimate, determine the expression for $1/\sigma_c^2$ and thus show that the SNR increases linearly with N and with $\mathscr{E}_b$.

11.8 Show that the sequence $\{x_n, \ 0 \le n \le N-1\}$ given by Equation (11.2.14) corresponds to the samples of $x(t)$ given by Equation (11.2.15). Also, prove that $x(t)$ given by Equation (11.2.15) is a real-valued signal.

11.9 Show that the IDFT of a sequence $\{X_k, \ 0 \le k \le N-1\}$ can be computed by passing the sequence $\{X_k\}$ through a parallel bank of N linear discrete-time filters with system functions

$$H_n(z) = \frac{1}{1 - e^{j2\pi n/N} z^{-1}}, \quad n = 0, 1, \ldots, N-1,$$

where the filter outputs are sampled at $n = N$.

11.10 Demonstrate that a DS spread spectrum signal without coding provides no improvement in performance against additive white Gaussian noise.

11.11 A total of 30 equal-power users are to share a common communication channel by CDMA. Each user transmits information at a rate of 10 kbps via DS spread spectrum and binary PSK. Determine the minimum chip rate needed to obtain a bit error probability of 10^{-5}. Additive noise at the receiver may be ignored in this computation.

11.12 A CDMA system is designed based on DS spread spectrum with a processing gain of 1000 and binary PSK modulation. Determine the number of users, if each user has equal power and the desired level of performance is an error probability of 10^{-6}. Repeat the computation if the processing gain is changed to 500.

11.13 A DS spread spectrum system transmits at a rate of 1000 bps in the presence of a tone interference. The interference power is 20 dB greater than the desired signal, and the required $\mathcal{E}_b/I_0$ to achieve satisfactory performance is 10 dB. Determine the spreading bandwidth required to meet the specifications.

11.14 A DS spread spectrum system is used to resolve the multipath signal component in a two-path radio signal propagation scenario. If the path length of the secondary path is 300 meters longer than that of the direct path, determine the minimum chip rate necessary to resolve the multipath component.

11.15 A CDMA system consists of 15 equal-power users who transmit information at a rate of 10,000 bps each. They use a DS spread spectrum signal operating at a chip rate of 1 MHz. The modulation is binary PSK.

1. Determine the $\mathcal{E}_b/I_0$, where I_0 is the spectral density of the combined interference.

2. What is the processing gain?

3. How much should the processing gain be increased to allow for doubling the number of users without affecting the output SNR?

11.16 An FH binary orthogonal FSK system employs an $m = 15$ stage linear feedback shift register that generates a maximal length sequence. Each state of the shift register selects one of N nonoverlapping frequency bands in the hopping pattern. The bit rate is 100 bits/sec and the hop rate is once per bit. The demodulator employs noncoherent detection.

1. Determine the hopping bandwidth for this channel.

2. What is the processing gain?

3. What is the probability of error in the presence of AWGN?

11.17 A DS binary PSK spread spectrum system has a processing gain of 500. What is the interference margin against a continuous tone interference if the desired error probability is 10^{-5}?

11.18 Suppose that $\{c_{1i}\}$ and $\{c_{2i}\}$ are two binary $(0, 1)$ periodic sequences with periods L_1 and L_2, respectively. Determine the period of the sequence obtained by forming the modulo 2 sum of $\{c_{1i}\}$ and $\{c_{2i}\}$.

11.19 An $m = 10$ maximum-length shift register generates the pseudorandom sequence in a DS spread spectrum system. The chip duration is $T_c = 1\mu$ sec, and the bit duration is $T_b = LT_c$, where L is the length (period) of the m-sequence.

   1. Determine the processing gain of the system in dB.
   2. Determine the interference margin if the required $\mathscr{E}_b/I_0 = 10$ and the interference is a tone interference with an average power P_{av}.

11.20 Figure P-11.20 illustrates the average power multipath delay profile in cellular communication for (a) suburban and urban areas and (b) hilly terrain area. In a GSM system, the bit rate is 270.8 kbps; in an IS-95 system (forward link), the bit rate is 19.2 kbps. Determine the number of bits affected by intersymbol interference in the transmission of the signal through the channels (a) and (b) for each cellular system.

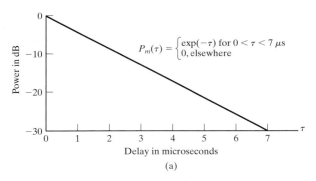

(a)

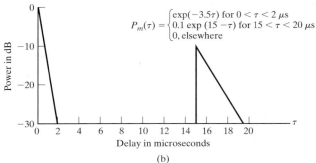

(b)

Figure P-11.20

11.21 For the multipath delay problem shown in Figure P-11.20, determine the number of taps in a RAKE demodulator for the IS-95 forward link that would be needed to span the multipath for channels (a) and (b). Of the total number of RAKE taps, how many will contain signal components and how many will have no signal for channel (b)?

11.22 A widely used model for the Doppler power spectrum of a mobile radio channel is the so-called *Jakes' model*, given as

$$\mathscr{S}(f) = \begin{cases} \frac{1}{\pi f_m} \frac{1}{\sqrt{1-(f/f_m)^2}}, & |f| \leq f_m \\ 0, & \text{otherwise} \end{cases},$$

where $f_m = vf_0/c$ is the maximum Doppler frequency, v is the vehicle speed in m/s, f_0 is the carrier frequency, and c is the speed of light (3×10^8 m/s). Determine f_m for an automobile traveling at 100 km/hr and for a train traveling at 200 km/hr. Plot $\mathscr{S}(f)$ for the two vehicles for a cellular communication system with a carrier frequency of 900 MHz.

11.23 A wireline channel of length 1000 km is used to transmit data via binary PAM. Regenerative repeaters are spaced 50 km apart along the system. Each segment of the channel has an ideal (constant) frequency response over the frequency band $0 \leq f \leq 1200$ and an attenuation of 1 dB/km. The channel noise is AWGN.

1. What is the highest bit rate that can be transmitted without ISI?
2. Determine the required $\mathscr{E}_b/N_0$ to achieve a bit error of $P_2 = 10^{-7}$ for each repeater.
3. Determine the transmitted power at each repeater to achieve the desired $\mathscr{E}_b/N_0$, where $N_0 = 4.1 \times 10^{-21}$ W/Hz.

11.24 Consider a transmission line channel that employs $n - 1$ regenerative repeaters plus the terminal receiver in the transmission of binary information. We assume that the probability of error at the detector of each receiver is p and that errors among repeaters are statistically independent.

1. Show that the binary error probability at the terminal receiver is

$$P_n = \frac{1}{2} \left[1 - (1 - 2p)^n \right].$$

2. If $p = 10^{-6}$ and $n = 100$, determine an approximate value of P_n.

11.25 A digital communication system consists of a transmission line with 100 digital (regenerative) repeaters. Binary antipodal signals are used for transmitting the information. If the overall end-to-end error probability is 10^{-6}, determine the probability of error for each repeater and the required $\mathscr{E}_b/N_0$ to achieve this performance in AWGN.

11.26 A radio transmitter has a power output of $P_T = 1$ Watt at a frequency of 10^9 Hz (1 GHz). The transmitting and receiving antennas are parabolic dishes with a diameter $D = 3$ meters.

  1. Determine the antenna gains.

  2. Determine the EIRP for the transmitter.

  3. The distance (free space) between the transmitting and receiving antennas is 20 km. Determine the signal power at the output of the receiving antenna in dBm.

11.27 A radio communication system transmits at a power level of 0.1 Watt at 1 GHz. The transmitting and receiving antennas are parabolic, and each has a diameter of 1 meter. The receiver is located 30 km from the transmitter.

  1. Determine the gains of the transmitting and receiving antennas.

  2. Determine the EIRP of the transmitted signal.

  3. Determine the signal power from the receiving antenna.

11.28 A satellite in synchronous orbit is used to communicate with an earth station at a distance of 4×10^7 meters. The satellite has an antenna with a gain of 15 dB and a transmitter power of 3 Watts. The earth station uses a 10-m parabolic antenna with an efficiency of 0.6. The frequency band is at $f = 10$ GHz. Determine the received power level at the output of the receiver antenna.

11.29 A spacecraft located 10^8 m from the earth is sending data at a rate of R bps. The frequency band is centered at 2 GHz and the transmitted power is 10 Watts. The earth station uses a parabolic antenna 50 m in diameter, and the spacecraft has an antenna with a gain of 10 dB. The noise temperature of the receiver front end is $T = 300°$ K.

  1. Determine the received power level.

  2. If the desired $\mathscr{E}_b/N_0 = 10$ dB, determine the maximum bit rate that the spacecraft can transmit.

11.30 Consider the front end of the receiver that is shown in the block diagram in Figure P-11.30. The received signal power at the input to the first amplifier is -113 dBm, and the received noise power spectral density is -175 dBm/Hz. The bandpass filter has a bandwidth of 10 MHz, and gains and noise figures are as shown. Determine the signal-to-noise ratio P_s/P_n at the input to the demodulator.

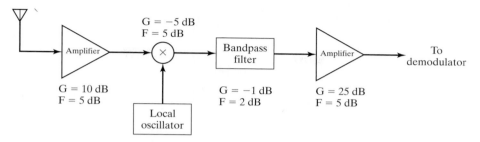

Figure P-11.30

11.31 A satellite in geosynchronous orbit is used as a regenerative repeater in a digital communication system. Consider the satellite-to-earth link in which the satellite antenna has a gain of 6 dB and the earth station antenna has a gain of 50 dB. The downlink is operated at a center frequency of 4 GHz, and the signal bandwidth is 1 MHz. If the required $(\mathcal{E}_b/N_0)$ for reliable communication is 15 dB, determine the transmitted power for the satellite downlink. Assume that $N_0 = 4.1 \times 10^{-21}$ W/Hz.

11.32 One of the Mariner spacecrafts that traveled to the planet Mercury sent its data to earth over a distance of 1.6×10^{11} m. The transmitting antenna had a gain of 27 dB and operated at a frequency $f = 2.3$ GHz. The transmitter power was 17 Watts. The earth station employed a parabolic antenna with a 64-meter diameter and an efficiency of 0.55. The receiver had an effective noise temperature of $T_e = 15°$ K. If the desired SNR/bit $(\mathcal{E}_b/N_0)$ was 6 dB, determine the data rate that could have been supported by the communication link.

11.33 In Example 11.3.3, suppose that the DS spread spectrum signal is transmitted via radio to a receiver at a distance of 2000 km. The transmitter antenna has a gain of 20 dB, while the receiver antenna is omnidirectional. The carrier frequency is 3 MHz, the available channel bandwidth $W = 10^5$ Hz, and the receiver has a noise temperature of 300° K. Determine the required transmitter power and the bit rate of the DS spread spectrum system.

COMPUTER PROBLEMS

11.1 Simulation of A Two-Path Rayleigh Fading Channel

Figure CP-11.1 illustrates a two-path channel model where the time delay between the two paths is T_d. The tap weights $c_1(n)$ and $c_2(n)$ are statistically independent, zero-mean, complex-valued Gaussian random variables. Hence, the channel is a Rayleigh fading channel. The additive noise is a zero-mean,

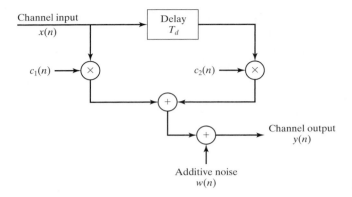

Figure CP-11.1

complex-valued white Gaussian noise sequence $w(n) = w_r(n) + jw_i(n)$, where the real and imaginary components are statistically independent. Write a MAT-LAB program that simulates the channel model with the following conditions:

1. The tap weight sequences $c_1(n)$ and $c_2(n)$ are the outputs of two identical lowpass filters described by the difference equation

$$c_k(n) = 0.9c_k(n-1) + z_k(n) \quad k = 1, 2,$$

where $z_k(n) = z_{kr}(n) + jz_{ki}(n)$ and the real and imaginary components of $z_k(n)$ are statistically independent, zero-mean, and unit variance white Gaussian noise sequences.

2. The additive white Gaussian noise sequence has zero mean and variance σ_w^2 (a variable).

3. The time delay T_d is a positive integer, which is also a variable.

Simulate the appropriate channel model for a two-path ionospheric propagation channel in which the relative time delay between the two received signal paths is 1 msec and the transmitted signal bandwidth is $W = 10$ kHz. Note that a 10 kHz signal provides a time resolution of $1/W = 0.1$ msec, so that the time delay $T_d = 10$ signal samples corresponds to the delay of 1 msec.

Generate and plot $|c_1(n)|$ and $|c_2(n)|$ separately for $1 \le n \le 1000$. Compute and plot the channel output $|y(n)|$ when the input sequence $x(n) = 1$ for $1 \le n \le 1000$, for each value of $\sigma_w^2 = 0, 0.5, 1$.

Repeat this simulation when the signal bandwidth is reduced to 5 kHz.

11.2 Generation of an OFDM signal

Using the 16-point QAM signal constellation shown in Figure 10.20(a), select pseudorandomly each of the information symbols $X_1, X_2, \ldots, X_9$. With $T = 100$ seconds, generate the transmitted signal waveform $x(t)$ given by Equation (11.2.15) for $t = 0, 1, \ldots, 100$ and plot it. Then compute the IDFT values x_n for $n = 0, 1, \ldots, N-1$, by using Equation (11.2.14). Demonstrate that $x(t)$, evaluated at Tn/N, $n = 0, 1, \ldots, N - 1$, corresponds to the IDFT values. Finally, using the IDFT values $\{x_n, 0 \leq n \leq N - 1\}$, compute the DFT, defined as

$$X_k = \frac{1}{\sqrt{N}} \sum_{n=0}^{N-1} x_n e^{-j2\pi k \frac{n}{N}} \quad k = 0, 1, \ldots, N - 1; \tag{11.6.1}$$

thus, demonstrate that the information symbols $\{X_k, 1 \leq k \leq 9\}$ are recovered from the samples of $x(t)$, where $t = nT/N$, $0 \leq n \leq N - 1$.

11.3 Simulation of Direct Sequence Spread Spectrum Signals

The objective of this problem is to demonstrate the effectiveness of a DS spread spectrum signal in suppressing sinusoidal interference via Monte Carlo simulation. The block diagram of the system to be simulated is illustrated in Figure CP-11.3.

A uniform random number generator (RNG) generates a sequence of binary information symbols (± 1). Each information bit is repeated L_c times, where L_c corresponds to the number of PN chips per information bit. The resulting sequence, which contains L_c repetitions per bit, is multiplied by a PN sequence $c(n)$ generated by another uniform RNG. To this product sequence, we add white Gaussian noise with variance $\sigma^2 = N_0/2$ and sinusoidal interference of the form

$$i(n) = A \sin \omega_0 n,$$

where $0 < \omega_0 < \pi$ and the amplitude of the sinusoid is selected to satisfy $A < L_c$. The demodulator performs the cross correlation with the PN sequence and sums (integrates) the blocks of L_c signal samples that constitute each information bit. The output of the summer is fed to the detector, which compares

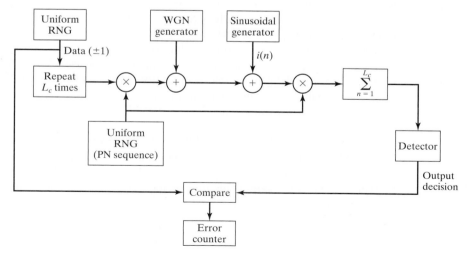

Figure CP-11.3

this signal with the threshold of zero and decides whether the transmitted bit is $+1$ or -1. The error counter tallies the number of errors made by the detector. Plot the measured error probability as a function of the SNR for three different values of the amplitude $A = 0$, $A = 3$, and $A = 10$ of the sinusoidal interference with $L_c = 20$. The SNR is defined as $\mathcal{E}_b/N_0$ and can be varied by setting $\mathcal{E}_b = 1$ and scaling the variance σ^2 of the additive Gaussian noise.

11.4 Simulation of Synchronous CDMA System

Write a MATLAB program that performs a Monte Carlo simulation of four time-synchronous CDMA users when each user employs a distinct Gold sequence of length $L = 31$. The four Gold sequences are as follows:

1	0	0	1	1	0	0	0	0	0	0	1	0	0	1	1	1	1	0	0	1	1	0	0	1	1	1	1	0	1	1
1	0	0	0	1	0	1	0	1	0	1	0	0	0	1	1	0	0	1	0	1	0	1	0	0	0	0	0	1	1	0
1	0	1	0	1	1	1	1	1	1	0	0	0	0	1	0	1	1	1	0	0	1	1	1	1	1	1	1	1	0	0
1	1	1	0	0	1	0	1	0	0	0	0	0	0	0	1	0	1	1	1	1	1	0	0	0	0	0	1	0	0	0

The users employ the binary (± 1) modulation of their representative Gold sequences with the number of chips per bit equal to 31. The receiver for each user correlates the composite CDMA received signal, which is corrupted by additive white Gaussian noise (added on a chip-by-chip basis) with their respective sequence. Using Monte Carlo simulation with $N = 10000$ information bits, estimate and plot the probability of error for each user as a function of SNR.

11.5 Simulation of Asynchronous CDMA system

Repeat Problem CP-11.4 when the four users transmit asynchronously. For example, simulate the case when the four-user CDMA signals are offset in time by one chip relative to one of the other signals. That is, the CDMA signal of user 2 is delayed by one chip relative to the first user, the CDMA signal of user 3 is delayed by one chip relative to the signal of user 2, and the CDMA signal of user 4 is delayed by one chip relative to the signal of user 3. Compare the error probability obtained with asynchronous transmission to that obtained with synchronous transmission.

11.6 Generation of a Maximal Length Shift Register Sequence

Write a MATLAB program that implements an $m = 12$-stage maximum-length shift register and generate three periods of the sequence. Compute and graph the periodic autocorrelation function of the equivalent bipolar sequence given by Equation (11.3.38).

11.7 Simulation of Frequency Hopped FSK

An FH binary orthogonal FSK system employs an $m = 7$-state shift register to generate a periodic maximum-length sequence of length $L = 127$. Each stage of the shift register selects one of $N = 127$ nonoverlapping frequency bands in the hopping pattern. Write a MATLAB program that simulates the selection of the center frequency and the generation of the two frequencies in each of the $N = 127$ frequency bands. Show the frequency selection pattern for the first 10 bit intervals.

12

An Introduction to Information Theory

In Chapter 1, we saw that the essential parts of any communication system are the *information source*, the *communication channel*, and the *destination*. We also saw that these three components of a communication channel are usually the *given* parts of a communication system, i.e., the communication engineer usually does not have much control over them. We also saw that two systems, i.e., the *transmitter* and the *receiver*, connect the information source to the channel and the channel to the destination, respectively. These two systems are completely designed by the communication engineer; therefore, they are under the full control of the designer. The role of these systems is to match the output of their preceding systems (the information source in the case of the transmitter, and the channel in the case of the receiver) to the system that comes after them (the channel in the case of the transmitter, and the destination in the case of the receiver). Another goal in the design of the transmitter and the receiver is to make sure that the signal will be resistant to channel impairments during the transmission. This results in different design techniques and methodologies, depending on the properties of the channel. For instance, in designing signals for transmission over a bandlimited channel, the problems of distortion and intersymbol interference must be considered in the design of the transmitted signal. When many transmitters and receivers share the same communication medium, as in the case of wireless communication, the problem of interference among different users should be addressed as well.

In Chapters 3, 4, 6, 7, 8, 9, 10, and 11, we studied different communication situations and signal design techniques. In Chapters 3, 4, and 6, we studied analog communication systems and the signal-to-noise ratio that each system can provide at the output. In Chapters 7, 8, 9, 10 and 11, we showed how analog signals can be transformed to digital signals and how digital signals can be transmitted. None of these chapters studied the fundamental limits on communications. Even in Chapters 8, 9, 10 and 11, where optimal receivers for a given signal set were designed, we did not design the optimal signal set; thus, there was no discussion of overall communication system optimality.

In this chapter, we study communication systems from another point of view. We study fundamental limits on the representation and transmission of information. In other words, we try to determine the limits of communication. We will answer questions such as the following: What is the highest rate at which information can be reliably transmitted over a communication channel? What is the lowest rate at which information can be compressed and still be retrievable with small or no error? What is the complexity of such optimal systems? These questions belong to a branch of communication theory called *information theory*, a field founded in 1948 by the pioneering work of Claude E. Shannon.

Shannon's fundamental contributions can be classified into two main categories. The first category consists of fundamental limits that apply to information sources. In this category, mathematical models for information sources are developed, and a method for measuring the information content of a source is introduced. Then the fundamental question of source coding is addressed and answered, i.e., What is the minimum rate at which a source can be compressed and still be recoverable from the compressed version? This is the essence of Shannon's source-coding theorem.

The second category of fundamental limits concerns the transmission of information over noisy channels. The fundamental question is: What is the maximum rate at which information can be transmitted reliably over a noisy channel? This question is addressed in the well-known *noisy channel coding theorem*.

12.1 MODELING INFORMATION SOURCES

Communication systems are designed to transmit information. In any communication system, an information source produces the information. The purpose of the communication system is to transmit the output of the source to the destination. In radio broadcasting, for instance, the information source is either a speech source or a music source. In TV broadcasting, the information source is a video source, and the output is a moving image. In fax transmission, the information source produces a still image. In communication between computers, either binary data or ASCII characters (encoded as a sequence of binary digits) are transmitted; therefore, the source can be modeled as a binary or ASCII source. In the storage of binary data on a computer disk, the source is again a binary source.

To study various information sources, we require a mathematical model to represent information sources and to measure their information content. Hartley, Nyquist, and Shannon were the pioneers in defining quantitative measures for information. In this section, we investigate the mathematical modeling of information sources and define a measure of information. In the next few sections, we will see how the output of an information source can be made more compact for easier transmission or storage. The process of representing the output of an information source with a small number of bits is called *data compression*.

The intuitive and common notion of information refers to any new knowledge about something. We can obtain information via hearing, seeing, or other means of perception. The information source, therefore, produces outputs that may interest the receiver of the information (who does not know these outputs in advance). The role of the communication system designer is to make sure that this information is transmitted to the receiver correctly. Since the output of the information source is a time varying unpredictable function (if it is predictable, there is no need to transmit it), it can be modeled as a random process. In previous chapters, we have already seen that the existence of noise in communication channels causes stochastic dependence between the input and output of the channel. Therefore, the communication system designer creates a system that transmits the output of a random process (information source) to a destination via a random medium (channel) and ensures low distortion.

Information sources can be modeled by random processes, and the properties of the random process depend on the nature of the information source. Some basic characteristics of an information source include bandwidth, range of amplitude variations, power content, and statistical properties (e.g., pdf of the amplitude, stationarity, and power spectral density). For example, when we model speech signals, our result is a random process that has all its power in a frequency band of approximately 300–3400 Hz. Therefore, the power spectral density of the speech signal also occupies this band of frequencies. A typical power spectral density for the speech signal is shown in Figure 12.1. Video signals are restored from a still or moving image; therefore, the bandwidth depends on the required resolution. For TV transmission, depending on the system employed (NTSC, PAL, or SECAM), this band is typically between 0–4.5 MHz or 0–6.5 MHz.

There is one common element in each of these processes: They are bandlimited processes and can be sampled at the Nyquist rate or faster and can be reconstructed from the sampled values. Therefore, it makes sense to limit this chapter to discrete-time random processes because all information sources of interest can be modeled by such a process, after they are sampled. The mathematical model for an information source is shown in Figure 12.2. Here, the source is modeled by a discrete-time random process $\{X_i\}_{i=-\infty}^{\infty}$. The random variables X_i are defined over an alphabet that can be either discrete (e.g., for binary data) or continuous (e.g., for sampled speech). The statistical properties of the discrete-time random process depend on the nature of the information source.

Here, we will only study simple models for information sources. The study of more complicated models is mathematically demanding and beyond the scope of

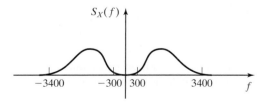

Figure 12.1 Typical power spectrum of a speech signal.

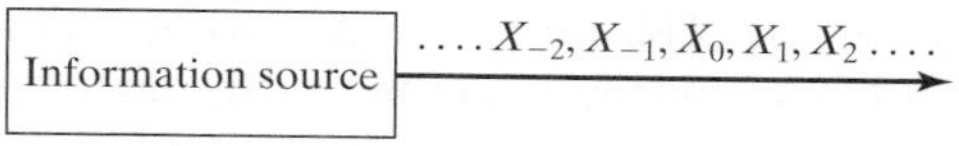

Information source $\xrightarrow{\quad \dots X_{-2}, X_{-1}, X_0, X_1, X_2 \dots \quad}$

Figure 12.2   Mathematical model for a discrete-time information source.

this book. However, even simple models enable us to precisely define a measure of information and bounds on the compression and transmission of information.

The simplest model for the information source that we study is the *discrete memoryless source* (DMS). A DMS is a discrete-time, discrete-amplitude random process in which all X_i's are generated independently and with the same distribution. Therefore, a DMS generates a sequence of i.i.d. (independent and identically distributed) random variables that take values in a discrete set.

Let $\mathcal{A} = \{a_1, a_2, \dots, a_N\}$ denote the set in which the random variable X takes its values, and let the probability mass function for the discrete random variable X be denoted by $p_i = P(X = a_i)$ for $i = 1, 2, \dots, N$. A full description of the DMS is given by the set $\mathcal{A}$, called the **alphabet**, and the probabilities $\{p_i\}_{i=1}^{N}$.

Example 12.1.1

An information source is described by the alphabet $\mathcal{A} = \{0, 1\}$ and $p(X_i = 1) = 1 - P(X_i = 0) = p$. This is an example of a discrete memoryless source, and it generates a sequence of zeros and ones; therefore, it is a *binary* source. In the special case where $p = 0.5$, the source is called a *binary symmetric source*, or BSS, for short. ∎

12.1.1 Measure of Information

To provide a quantitative measure of information, we start with the basic model of an information source and define its information content to satisfy certain intuitive properties. Assume that the source that we are considering is a discrete source. As we have already seen, this is not really a restriction because continuous sources can be sampled and turned into discrete sources. Let the outputs of this source be revealed to an interested party. Let a_1 be the most likely output and a_N be the least likely output. For example, we could imagine the source to represent both the weather conditions and air pollution in a certain city (in the northern hemisphere) during July. In this case, $\mathcal{A}$ represents various combinations of different weather conditions and pollution (such as hot and polluted, hot and lightly polluted, cold and highly polluted, cold and mildly polluted, very cold and lightly polluted, etc.). The question is, Which output conveys more information, a_1 or a_N (the most probable or the least probable output)? Intuitively, revealing a_N (very cold and lightly polluted, in the previous example) reveals the most information. It follows that a rational measure of information for an output of an information source should be a decreasing function of the probability of that output. A second intuitive property of a measure of information is that a small change in the probability of a certain output should not drastically change the information delivered by that output. In other words, the information measure should be a continuous function of the probability of the source output.

Now assume that the information about output a_j can be subdivided into two independent parts called a_{j1} and a_{j2}, i.e., $X_j = (X_{j1}, X_{j2}), a_j = \{a_{j1}, a_{j2}\}$ and

$P(X = a_j) = P(X_{j1} = a_{j1})P(X_{j2} = a_{j2})$. This can happen if we assume that the temperature and pollution were almost independent; therefore, each source output can be subdivided into two independent components. Since the components are independent, revealing the information about one component (temperature) does not provide any information about the other component (pollution); so, intuitively, the amount of information provided by revealing a_j is the sum of the two information components obtained by revealing a_{j1} and a_{j2}. From the preceding discussion, we can conclude that the amount of information revealed about an output a_j with probability p_j must satisfy the following conditions:

1. The information content of output a_j depends only on the probability of a_j and not on the value of a_j. We denote this function by $I(p_j)$, and we call it *self information*.
2. Self information is a continuous function of p_j, i.e., $I(\cdot)$ is a continuous function.
3. Self information is a decreasing function of its argument.
4. If $p_j = p_{j1}p_{j2}$, then $I(p_j) = I(p_{j1}) + I(p_{j2})$.

It can be shown that the only function that satisfies all these properties is the logarithmic function, i.e., $I(x) = -\log(x)$. The base of the logarithm is not important and determines the unit by which the information is measured. If the base is 2, the information is expressed in *bits per source symbol* or *bits per sample*. Note that we can use the relation $\log_2 x = \frac{\log_{10} x}{\log_{10} 2}$ to find logarithms in base 2.

Now that the information revealed about each source output a_i is defined as the self information of that output, which is given by $-\log(p_i)$, we can define the information content of the source as the weighted average of the self information of all source outputs. This is justified by the fact that various source outputs appear with their corresponding probabilities. Therefore, the information revealed by an unidentified source output is the weighted average of the self information of the various source outputs, i.e., $\sum_{i=1}^{N} p_i I(p_i) = \sum_{i=1}^{N} -p_i \log p_i$. The information content of the information source is known as the *entropy* of the source and is denoted by $H(X)$.

Definition 12.1.1.   The entropy of a discrete random variable X is a function of its PMF and is defined by

$$H(X) = -\sum_{i=1}^{N} p_i \log p_i = \sum_{i=1}^{N} p_i \log \left(\frac{1}{p_i} \right), \qquad (12.1.1)$$

where $0 \log 0 = 0$.   ∎

Note that there is a slight abuse of notation here. We would expect $H(X)$ to denote a function of the random variable X; hence, it would be a random variable itself. However, $H(X)$ is a function of the PMF of the random variable X and is a number.

Example 12.1.2

For a binary memoryless source with probabilities p and $1 - p$, we have

$$H(X) = -p \log p - (1 - p) \log(1 - p). \tag{12.1.2}$$

This function, denoted by $H_b(p)$, is known as the *binary entropy function*, and a plot of it is given in Figure 12.3. ∎

From the plot of the binary entropy function, we can see that the entropy of the binary memoryless source is zero when either $p = 0$ or $p = 1$. These two cases correspond to the time when the source generates all zeros or all ones. In both cases, the source is deterministic and completely predictable. On the other hand, entropy is maximized, with a maximum equal to 1, when the two source outputs are equiprobable and each has probability $\frac{1}{2}$. This is the case where the output is least predictable; therefore, its entropy is maximized. This supports the intuitive development of entropy as a measure of the information content of a source (or measure of uncertainty of it).

Example 12.1.3

A source with the bandwidth 4000 Hz is sampled at the Nyquist rate. Assuming that the resulting sequence can be approximately modeled by a DMS with alphabet $\mathscr{A} = \{-2, -1, 0, 1, 2\}$ and with corresponding probabilities $\{\frac{1}{2}, \frac{1}{4}, \frac{1}{8}, \frac{1}{16}, \frac{1}{16}\}$, determine the rate of the source in bits per second.

Solution We have

$$H(X) = \frac{1}{2} \log 2 + \frac{1}{4} \log 4 + \frac{1}{8} \log 8 + 2 \times \frac{1}{16} \log 16 = \frac{15}{8} \text{ bits/sample.}$$

Since we have $2 \times 4000 = 8000$ samples/sec, the source produces information at a rate of $8000 \times \frac{15}{8} = 15,000$ bits/sec. ∎

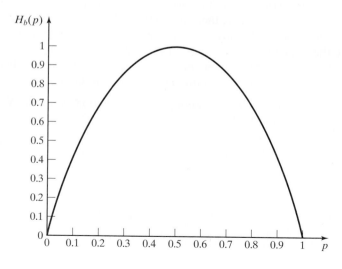

Figure 12.3 The binary entropy function.

Example 12.1.4

A discrete memoryless source has an alphabet of size N and the source outputs are equiprobable (each having a probability of $\frac{1}{N}$). Find the entropy of this source.

Solution   We have

$$H(X) = -\sum_{i=1}^{N} \frac{1}{N} \log \frac{1}{N}$$

$$= \log N.$$ ∎

12.1.2 Joint and Conditional Entropy

When dealing with two or more random variables, we can introduce joint and conditional entropies in exactly the same way that joint and conditional probabilities are introduced. These concepts are especially important when dealing with sources with memory.

Definition 12.1.2.   The *joint entropy* of two discrete random variables (X, Y) is defined by

$$H(X, Y) = -\sum_{x,y} p(x, y) \log p(x, y). \tag{12.1.3}$$

For the case of n random variables $\mathbf{X} = (X_1, X_2, \ldots, X_n)$, we have

$$H(\mathbf{X}) = -\sum_{x_1, x_2, \ldots, x_n} p(x_1, x_2, \ldots, x_n) \log p(x_1, x_2, \ldots, x_n). \tag{12.1.4}$$

∎

As seen, the joint entropy is simply the entropy of a vector-valued random variable.

The conditional entropy of the random variable X, given the random variable Y, can be defined by noting that if $Y = y$, then the PMF of the random variable X will be $p(x|y)$ and the corresponding entropy is $H(X|Y = y) = -\sum_x p(x|y) \log p(x|y)$, which is intuitively the amount of uncertainty in X when we know that $Y = y$. The weighted average of these quantities over all y is the uncertainty in X when Y is known. This quantity is known as the conditional entropy and defined next.

Definition 12.1.3.   The *conditional entropy* of the random variable X, given the random variable Y, is defined by

$$H(X|Y) = -\sum_{x,y} p(x, y) \log p(x|y). \tag{12.1.5}$$

In general, we have

$$H(X_n|X_1, \ldots, X_{n-1}) = -\sum_{x_1, \ldots, x_n} p(x_1, \ldots, x_n) \log p(x_n|x_1, \ldots, x_{n-1}). \tag{12.1.6}$$

∎

Example 12.1.5

Using the chain rule for PMF's, namely, $p(x, y) = p(y)p(x|y)$, as given in Chapter 5, show that $H(X, Y) = H(Y) + H(X|Y)$.

Solution   From the definition of the joint entropy of two random variables, we have

$$H(X, Y) = -\sum_{x,y} p(x, y) \log p(x, y)$$

$$= -\sum_{x,y} p(x, y) \log[p(y)p(x|y)]$$

$$= -\sum_{x,y} p(x, y) \log p(y) - \sum_{x,y} p(x, y) \log p(x|y)$$

$$= -\sum_{y} p(y) \log p(y) - \sum_{x,y} p(x, y) \log p(x|y)$$

$$= H(Y) + H(X|Y), \tag{12.1.7}$$

where, in the last step, we have used

$$\sum_{x} p(x, y) = p(y). \tag{12.1.8}$$

∎

This relation states that the information content of the pair (X, Y) is equal to the information content of Y plus the information content of X after Y is known. Equivalently, it states that the same information is transferred by either revealing the pair (X, Y) or by first revealing Y and then revealing the remaining information in X. This relation can be generalized to the case of n random variables to show the chain rule for entropies, as follows:

$$H(\mathbf{X}) = H(X_1) + H(X_2|X_1) + \cdots + H(X_n|X_1, X_2, \ldots, X_{n-1}). \tag{12.1.9}$$

In the case where the random variables $(X_1, X_2, \ldots, X_n)$ are independent, this relation reduces to

$$H(\mathbf{X}) = \sum_{i=1}^{n} H(X_i). \tag{12.1.10}$$

If the random variable X_n denotes the output of a discrete (not necessarily memoryless) source at time n, then $H(X_2|X_1)$ denotes the new information provided by the source output X_2 to someone who already knows the source output X_1. Similarly, $H(X_n|X_1, X_2, \ldots, X_{n-1})$ denotes the new information in X_n for an observer who has observed $(X_1, X_2, \ldots, X_{n-1})$. The limit of this conditional entropy as n tends to infinity is known as the *entropy rate* of the random process.

Definition 12.1.4.   The *entropy rate* of a stationary discrete-time random process is defined by

$$H = \lim_{n \to \infty} H(X_n|X_1, X_2, \ldots, X_{n-1}). \tag{12.1.11}$$

∎

Stationarity ensures the existence of the limit. It can be proved that an alternative definition of the entropy rate for sources with memory is given by

$$H = \lim_{n \to \infty} \frac{1}{n} H(X_1, X_2, \dots, X_n). \qquad (12.1.12)$$

Entropy rate plays the role of entropy for sources with memory. It is basically a measure of the uncertainty per output symbol of the source.

Example 12.1.6

Two binary random variables X and Y are distributed according to the joint PMF given by

$$P(X = 0, Y = 1) = \frac{1}{4};$$

$$P(X = 1, Y = 1) = \frac{1}{2};$$

$$P(X = 0, Y = 1) = \frac{1}{4}.$$

Determine $H(X, Y)$, $H(X)$, H(Y), $H(X|Y)$, and $H(Y|X)$.

Solution   We have $P(X = 1, Y = 0) = 0$ and $H(X, Y) = -\frac{1}{4} \log \frac{1}{4} - \frac{1}{4} \log \frac{1}{4} - \frac{1}{2} \log \frac{1}{2} = \frac{3}{2}$. Next, we have

$$P(X = 1) = P(X = 1, Y = 0) + P(X = 1, Y = 1) = \frac{1}{2} \Rightarrow P(X = 0) = \frac{1}{2}$$

and

$$P(Y = 1) = P(X = 1, Y = 1) + P(X = 0, Y = 1) = \frac{3}{4} \Rightarrow P(X = 0) = \frac{1}{4}.$$

Thus, we conclude that

$$H(X) = -\frac{1}{2} \log \frac{1}{2} - \frac{1}{2} \log \frac{1}{2} = 1$$

and

$$H(Y) = -\frac{3}{4} \log \frac{3}{4} - \frac{1}{4} \log \frac{1}{4} = 0.8113.$$

Now since

$$H(X, Y) = H(X) + H(Y|X),$$

we have,

$$H(Y|X) = 1.5 - 1 = 0.5$$

and

$$H(X|Y) = H(X, Y) - H(Y) = 1.5 - 0.8113 = 0.6887. \qquad \blacksquare$$

12.1.3 Mutual Information

For discrete random variables, $H(X|Y)$ denotes the entropy (or uncertainty) of the random variable X after the random variable Y is known. Therefore, if the entropy of the random variable X is $H(X)$, then $H(X) - H(X|Y)$ denotes the amount of uncertainty of X that has been removed by revealing random variable Y. In other words, $H(X) - H(X|Y)$ is the amount of information provided by the random variable Y about random variable X. This quantity plays an important role in both source and channel coding and is called the *mutual information* between two random variables.

Definition 12.1.5.   The *mutual information* between two discrete random variables X and Y is denoted by $I(X; Y)$ and is defined by

$$I(X; Y) = H(X) - H(X|Y). \tag{12.1.13}$$

■

Using the expression for entropy and Equation (12.1.5) for conditional entropy, we can determine the following relation for the mutual information between two random variables X and Y:

$$I(X; Y) = \sum_{x \in \mathcal{X}} \sum_{y \in \mathcal{Y}} p(x, y) \log \frac{p(x|y)}{p(x)}$$

$$= \sum_{x \in \mathcal{X}} \sum_{y \in \mathcal{Y}} p(x, y) \log \frac{p(x, y)}{p(x)p(y)}. \tag{12.1.14}$$

Example 12.1.7

Let X and Y be binary random variables with $P(X = 0, Y = 0) = \frac{1}{3}$, $P(X = 1, Y = 0) = \frac{1}{3}$, and $P(X = 0, Y = 1) = \frac{1}{3}$. Find $I(X; Y)$ in this case.

Solution   We have, $P(X = 0) = P(Y = 0) = \frac{2}{3}$; therefore, $H(X) = H(Y) = H_b(\frac{2}{3}) = 0.919$. On the other hand, the (X, Y) pair is a random vector uniformly distributed on three values: $(0, 0)$, $(1, 0)$, and $(0, 1)$. Therefore, $H(X, Y) = \log 3 = 1.585$. From this, we have $H(X|Y) = H(X, Y) - H(Y) = 1.585 - 0.919 = 0.666$ and $I(X; Y) = H(X) - H(X|Y) = 0.919 - 0.666 = 0.253$.   ■

Figure 12.4 represents the relation among entropy, conditional entropy, and mutual information quantities.

12.1.4 Differential Entropy

So far, we have defined entropy and mutual information for discrete sources. If we are dealing with a discrete-time continuous-alphabet source whose outputs are real numbers, no quantity exists that has the intuitive meaning of entropy. In the continuous case, another quantity that resembles entropy, called *differential entropy*, is defined. However, it does not have the intuitive interpretation of entropy as the

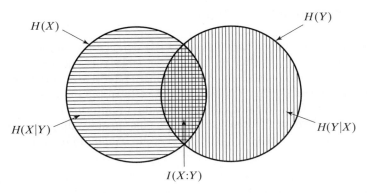

Figure 12.4   Entropy, conditional entropy, and mutual information quantities.

uncertainty in the source output. In fact, to reconstruct the output of a continuous source reliably, an infinite number of bits per source output are required because any output of the source is a real number and the binary expansion of a real number has infinitely many bits.

Definition 12.1.6.   The *differential entropy* of a continuous random variable X with the probability density function $f_X(x)$ is denoted by $h(X)$ and defined by

$$h(X) = -\int_{-\infty}^{\infty} f_X(x) \log f_X(x) dx, \tag{12.1.15}$$

where $0 \log 0 = 0$.   ■

Example 12.1.8

Determine the differential entropy of a random variable X uniformly distributed on $[0, a]$.

Solution   From the definition of differential entropy,

$$h(X) = -\int_0^a \frac{1}{a} \log \frac{1}{a} dx = \log a.$$

Clearly, for $a < 1$, we have $h(X) < 0$, which is in contrast to the nonnegativity of the entropy of discrete sources. Also, for $a = 1$, $h(X) = 0$ without X being deterministic. This is again in contrast to the entropy properties of discrete sources.   ■

Example 12.1.9

Determine the differential entropy of a zero-mean Gaussian random variable with variance σ^2.

Solution   The PDF is $f(x) = \frac{1}{\sqrt{2\pi\sigma^2}}e^{-\frac{x^2}{2\sigma^2}}$. Therefore, using natural logarithms, we find the differential entropy to be

$$h(X) = -\int_{-\infty}^{\infty} \ln\left(\frac{1}{\sqrt{2\pi\sigma^2}}\right) f(x)dx - \int_{-\infty}^{\infty} \ln\left(e^{-\frac{x^2}{2\sigma^2}}\right) f(x)dx$$

$$= \ln\left(\sqrt{2\pi\sigma^2}\right) + \frac{\sigma^2}{2\sigma^2}$$

$$= \frac{1}{2}\ln\left(2\pi e\sigma^2\right),\tag{12.1.16}$$

where $\int_{-\infty}^{\infty} f(x)dx = 1$ and $\int_{-\infty}^{\infty} x^2 f(x)dx = \sigma^2$. Changing the base of the logarithms to 2, we have

$$h(X) = \frac{1}{2}\log_2(2\pi e\sigma^2) \text{ bits.}\tag{12.1.17}$$

■

Extensions of the definition of differential entropy to joint random variables and conditional differential entropy are straightforward. For two random variables, we have

$$h(X, Y) = -\int_{-\infty}^{\infty}\int_{-\infty}^{\infty} f(x, y)\log f(x, y)dx\, dy\tag{12.1.18}$$

and

$$h(X|Y) = h(X, Y) - h(Y).\tag{12.1.19}$$

The mutual information between two continuous random variables X and Y is defined similarly to the discrete case as

$$I(X; Y) = h(Y) - h(Y|X) = h(X) - h(X|Y).\tag{12.1.20}$$

Although differential entropy does not have the intuitive interpretation of discrete source entropy, the mutual information of continuous random variables has the same interpretation as discrete random variables, i.e., the information provided by one random variable about the other random variable.

12.2 THE SOURCE CODING THEOREM

The source coding theorem is one of the three fundamental theorems introduced by Shannon (1948a, 1948b). The source coding theorem establishes a fundamental limit on the rate at which the output of an information source can be compressed without causing a large error probability. We have already seen that the entropy of an information source is a measure of the uncertainty or, equivalently, the information content of the source. Therefore, it is natural that the entropy of the source plays a major role in the statement of the source coding theorem.

The entropy of an information source has a very intuitive meaning. Assume that we are observing outputs of length n of a DMS where n is very large. Then, according to the law of large numbers (see Chapter 5), there is a high probability (that goes to 1 as $n \to \infty$) that letter a_1 is repeated approximately np_1 times, letter a_2 is repeated approximately np_2 times, ..., and letter a_N is repeated approximately np_N times. This means that when n is large enough, with a probability approaching 1, every sequence from the source has the same composition and therefore the same probability. The sequences $\mathbf{x}$ that have this structure are called *typical sequences*. Using the fact that the source is memoryless, the probability of a typical sequence is given by

$$P(\mathbf{X} = \mathbf{x}) \approx \prod_{i=1}^{N} p_i^{np_i}$$

$$= \prod_{i=1}^{N} 2^{np_i \log p_i}$$

$$= 2^{n \sum_{i=1}^{N} p_i \log p_i}$$

$$= 2^{-nH(X)}.$$

(In this chapter, all logarithms are in base 2 and all entropies are in bits.) This means that for large n, almost all the output sequences of length n of the source are equally probable with probability $\approx 2^{-nH(X)}$. These are called typical sequences. On the other hand, the probability of the set of nontypical sequences is negligible.

Since the probability of the typical sequences is almost 1 and each typical sequence has a probability of almost $2^{-nH(X)}$, the total number of typical sequences is almost $2^{nH(X)}$. Therefore, although a source of alphabet size N can produce N^n sequences of length n, the "effective" number of outputs is $2^{nH(X)}$. By "effective number of outputs," we mean that almost nothing is lost by neglecting the other outputs and the probability of having lost anything goes to zero as n goes to infinity. Figure 12.5 gives a schematic diagram of this property. This is a very important result. It tells us that, for all practical purposes, it is enough to consider the set of typical sequences rather than the set of all possible outputs of the source. The error introduced in ignoring nontypical sequences can be made smaller than any given $\epsilon > 0$ by choosing an n that is large enough. This is the essence of *data compression*, the practice of representing the output of the source with a smaller number of sequences than the number of the outputs that the source really produces.

From this result, and since ϵ is an arbitrary positive number, we can only represent the typical source outputs without introducing considerable error. Since the total number of typical sequences is roughly $2^{nH(X)}$, we need $nH(X)$ bits to represent them. However, these bits are used to represent source outputs of length n. Therefore, on the average, any source output requires $H(X)$ bits for an essentially

Set of typical sequences with $\approx 2^{nH(X)}$ elements

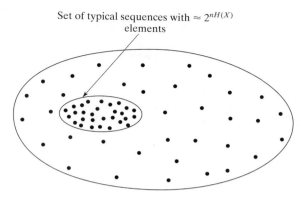

Figure 12.5   The set of typical and nontypical sequences.

error-free representation. This once again justifies the notion of entropy as the amount of information per source output.

So far, we have assumed that the source is discrete and memoryless; therefore, it can be represented by an i.i.d. random variable. Such a source can only be compressed if its PMF is not uniform. For X uniformly distributed, as shown in Example 12.1.4, we have $H(X) = \log N$; therefore, $2^{nH(X)} = 2^{n \log N} = N^n$. This means that the "effective" number of source outputs of length n is equal to the total number of source outputs and no compression is possible.

We have not considered the case where the source has memory. For a source with memory, the outputs of the source are not independent; therefore, previous outputs reveal some information about the future ones. This means that the rate at which fresh information is produced decreases as more and more source outputs are revealed. A classic example of such a case is printed English text, which shows a lot of dependency between letters and words (e.g., a "q" is almost always followed by a "u", a single letter between two spaces is either an "I" or "a", etc.). The entropy per letter for a large text of English is roughly the limit of $H(X_n | X_1, X_2, \ldots, X_{n-1})$ as n becomes large (the entropy rate defined in Section 12.1.2). In general for stationary sources, the entropy rate has the same significance as the entropy for the case of memoryless sources and defines the number of "effective" source outputs for any n that is large enough, i.e., 2^{nH} where H is the entropy rate.

Studies with statistical models of printed English show that the entropy rate converges rather quickly; for $n = 10$, we are very close to the limit. These studies show that for $n = 1$, i.e., a memoryless source model, we have $H(X) = 4.03$ bits/letter. As the memory increases, the size of the space over which conditional probabilities are computed increases rapidly, and it is not easy to find the conditional probabilities required to compute the entropy rate. Some methods for estimating these conditional probabilities have been proposed in the literature and, based on these methods, the entropy of English is estimated to be around 1.3 bits/letter. (In these studies, only the 26 letters of the English alphabet and the space mark have been considered.)

So far, we have given an informal description of the source coding theorem and justified it. A formal statement of the theorem, without proof, is given next. The interested reader is referred to the references at the end of this chapter for a proof.

Theorem 12.2.1. [**Source Coding Theorem**] A source with entropy (or entropy rate) H can be encoded with an arbitrarily small error probability at any rate R (bits/source output) as long as $R > H$. Conversely, if $R < H$, the error probability will be bounded away from zero, independent of the complexity of the encoder and the decoder employed.   ∎

This theorem, first proved by Shannon (1948a), only gives necessary and sufficient conditions for the existence of source codes. It does not provide any algorithm for the design of codes that achieve the performance predicted by this theorem. In the next section, we present two algorithms for the compression of information sources. One is due to Huffman (1952), and the second is due to Lempel and Ziv (1977, 1978).

12.3 SOURCE CODING ALGORITHMS

In the preceding section, we observed that H, the entropy of a source, gives a sharp bound on the rate at which a source can be compressed for reliable reconstruction. This means that at rates above entropy, it is possible to design a code with an error probability as small as desired, whereas at rates below entropy, such a code does not exist. This important result, however, does not provide specific algorithms to design codes approaching this bound. In this section, we will introduce two algorithms to design codes that are close to the entropy bound. These coding methods are the Huffman coding algorithm and the Lempel–Ziv source coding algorithm.

12.3.1 The Huffman Source Coding Algorithm

In Huffman coding, fixed length blocks of the source output are mapped to variable length binary blocks. This is called *fixed-to-variable length* coding. The idea is to map the more frequently occurring fixed length sequences to shorter binary sequences and the less frequently occurring sequences to longer binary sequences, thus achieving good compression ratios. In variable length coding, synchronization is a problem. This means that there should be only one way to parse the binary received sequence into code words. The next example clarifies this point.

Example 12.3.1

Assume that the possible outputs of an information source are $\{a_1, a_2, a_3, a_4, a_5\}$ with the corresponding probabilities $\{\frac{1}{2}, \frac{1}{4}, \frac{1}{4}, \frac{1}{16}, \frac{1}{16}\}$. Consider the following four codes for this source:

Letter	Probability	Codewords			
		Code 1	Code 2	Code 3	Code 4
a_1	$p_1 = \frac{1}{2}$	1	1	0	00
a_2	$p_2 = \frac{1}{4}$	01	10	10	01
a_3	$p_3 = \frac{1}{8}$	001	100	110	10
a_4	$p_4 = \frac{1}{16}$	0001	1000	1110	11
a_5	$p_5 = \frac{1}{16}$	00001	10000	1111	110

In the first code, each codeword ends with a 1. Therefore, as soon as the decoder observes a 1, it knows that the codeword has ended and a new codeword will start. This means that the code is a *self-synchronizing code*. In the second code, each codeword starts with a 1. Therefore, upon observing a 1, the decoder knows that a new codeword has started; hence, the previous bit was the last bit of the previous codeword. This code is again self-synchronizing, but it is not as desirable as the first code. The reason is that, in this code, we have to wait to receive the first bit of the next codeword to recognize that a new codeword has started. However, in Code 1, we recognize the last bit without having to receive the first bit of the next codeword. Both Codes 1 and 2 are therefore *uniquely decodable*. However, only Code 1 is *instantaneous*. Codes 1 and 3 have the convenient property that no codeword is the prefix of another codeword. They satisfy the *prefix condition*. We can prove that a necessary and sufficient condition for a code to be both uniquely decodable *and* instantaneous is that it satisfy the prefix condition. This means that both Codes 1 and 3 are uniquely decodable and instantaneous. However, Code 3 has the advantage of having a smaller average codeword length. In fact, for Code 1, the average codeword length is

$$E[L] = 1 \times \frac{1}{2} + 2 \times \frac{1}{4} + 3 \times \frac{1}{8} + 4 \times \frac{1}{16} + 5 \times \frac{1}{16} = 31/16,$$

and for Code 3, the average codeword length is

$$E[L] = 1 \times \frac{1}{2} + 2 \times \frac{1}{4} + 3 \times \frac{1}{8} + 4 \times \frac{1}{16} + 4 \times \frac{1}{16} = 30/16.$$

Code 4 has a major disadvantage. This code is *not* uniquely decodable. For example, the sequence 110110 can be decoded in two ways, as $a_5 a_5$ or as $a_4 a_2 a_3$. Codes that are not uniquely decodable are not desirable and should be avoided in practice. From this discussion, we can see that the most desirable of the four codes is Code 3, which is uniquely decodable, instantaneous, and has the least average codeword length. This code is an example of a *Huffman code*, which we will discuss, shortly. ∎

As already mentioned, the idea in Huffman coding is to choose codeword lengths such that more probable sequences have shorter codewords. If we can map each source output of probability p_i to a codeword of length approximately $\log \frac{1}{p_i}$ and, at the same time, ensure unique decodability, then we can achieve an average codeword length of approximately $\sum_i p_i \log \frac{1}{p_i}$, which is equal to $H(X)$. Huffman codes are uniquely decodable instantaneous codes with a minimum average codeword length. In this sense, they are optimal. By optimal, we mean that among all codes that satisfy the prefix condition (and hence are uniquely decodable and instantaneous), Huffman codes have the minimum average codeword length. Next, we present the algorithm for the design of the Huffman code. From the algorithm, it is obvious that the resulting code satisfies the prefix condition. The proof of the optimality is omitted; for more information refer to the references at the end of this chapter.

Huffman Coding Algorithm.

1. Sort source outputs in decreasing order of their probabilities.
2. Merge the two least probable outputs into a single output whose probability is the sum of the corresponding probabilities.

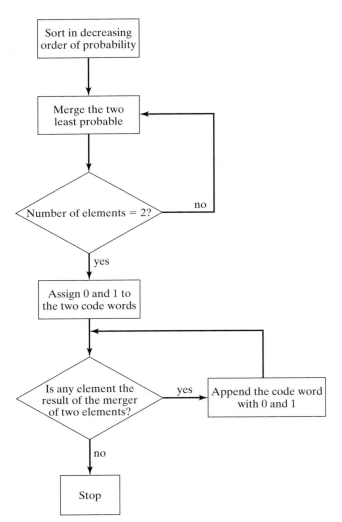

Figure 12.6   Huffman coding algorithm.

3. If the number of remaining outputs is 2, then go to Step 4; otherwise go to Step 2.

4. Arbitrarily assign 0 and 1 as codewords for the two remaining outputs.

5. If an output is the result of the merger of two outputs in a preceding step, append the current codeword with a 0 and a 1 to obtain the codeword for the preceding outputs; then, repeat Step 5. If no output is preceded by another output in a preceding step, then stop.

Figure 12.6 shows a flow chart of this algorithm.

Example 12.3.2

Design a Huffman code for the source given in the preceding Example 12.3.1.

Solution   The following tree diagram summarizes the design steps for code construction and the resulting codewords:

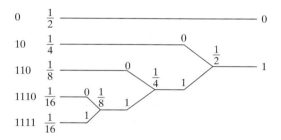

We can show that the average length of a Huffman code, defined by

$$\overline{R} = \sum_{x \in \mathcal{X}} p(x)l(x), \tag{12.3.1}$$

where $l(x)$ is the length of the codeword corresponding to the source output x, satisfies the following inequality:

$$H(X) \leq \overline{R} < H(X) + 1. \tag{12.3.2}$$

From this relation, it is obvious that the *efficiency* of a Huffman code, defined as

$$\eta = \frac{H(X)}{\overline{R}},$$

is always less than or equal to 1.

If the Huffman code is designed for sequences of source letters of length n (the nth extension of the source), we would have

$$H(X^n) \leq \overline{R_n} < H(X^n) + 1,$$

where $\overline{R_n}$ denotes the average codeword length for the extended source sequence; therefore, $\overline{R} = \frac{1}{n}\overline{R_n}$. If our source is memoryless, we also have $H(X^n) = nH(X)$. Substituting these into the preceding equation and dividing by n, we have

$$H(X) \leq \overline{R} < H(X) + \frac{1}{n}. \tag{12.3.3}$$

Therefore, for any n that is large enough, $\overline{R}$ can be made as close to $H(X)$ as desired. It is also obvious that, for discrete sources with memory, $\overline{R}$ approaches the entropy rate of the source as defined in Equation (12.1.12).

Example 12.3.3

A discrete memoryless source with equiprobable outputs and alphabet $\mathcal{A} = \{a_1, a_2, a_3\}$ has the following Huffman code:

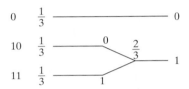

The entropy of the source is $H(X) = \log 3 = 1.585$ and $\overline{R} = \frac{5}{3} = 1.667$. If we use sequences of two letters, we will have the source

$$\mathcal{A}^2 = \{(a_1, a_1), (a_1, a_2), (a_1, a_3), \ldots, (a_3, a_2), (a_3, a_3)\}$$

with the probability vector $\boldsymbol{p}^{(2)} = \{\frac{1}{9}, \frac{1}{9}, \frac{1}{9}, \frac{1}{9}, \frac{1}{9}, \frac{1}{9}, \frac{1}{9}, \frac{1}{9}, \frac{1}{9}\}$. A Huffman code for this source is designed in the following tree diagram:

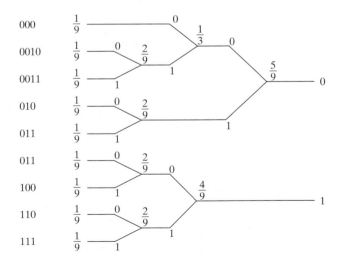

Here, the average codeword length is $\overline{R_2} = 3.222$ bits per pair of source outputs or 1.611 bits per each source output. Comparing this with the previous example, we see that the average length is closer to the entropy of the source. In other words, working with the second extension of the source (i.e., two letters at a time) has improved the efficiency of the coding. If we use the third extension of the source, the efficiency will improve further. It is also obvious that, although the efficiency of the Huffman code improves by increasing n, the complexity of the code design and the encoding/decoding process increases as well.
                                                                                                    ■

12.3.2 The Lempel–Ziv Source Coding Algorithm

We have already seen that Huffman codes are optimal in the sense that, for a given source, they provide a prefix code with minimum average block length. Nevertheless, there are two major problems in implementing Huffman codes. One problem is that the design of Huffman codes depend strongly on the source probabilities (statistics). The source statistics have to be known in advance to design a Huffman code. If we can only observe the source outputs, then we have to do the coding in two passes. In the first pass, we estimate the statistics of the source (which, in the case of sources with memory and in cases where we want to apply Huffman coding to the extension of the source, becomes quite time consuming); in the second pass, coding is done. The other problem with Huffman codes is that if the code is designed for source blocks of length 1, it only employs variations in the frequency of the source outputs and not the source memory. If we want to employ the source memory as well, we have to design the code for blocks of length 2 or more; this exponentially increases the complexity of the algorithm. For instance, encoding ASCII characters with a block length of 1 requires a tree with 256 terminal nodes, but if a block length of two is desired, the size of the tree and the complexity of coding becomes much higher. In certain applications, such as storage in magnetic or optical media, where high transfer rates are desirable, the complexity and speed of Huffman coding becomes a bottleneck.

The Lempel–Ziv algorithm belongs to the class of *universal source coding algorithms*, i.e., algorithms that are independent of the source statistics. This algorithm is a *variable-to-fixed length* coding scheme. This means that any sequence of source outputs is uniquely parsed into phrases of varying length and these phrases are encoded using codewords of equal length. Parsing is done by identifying phrases of the smallest length that have not appeared so far. To this end, the parser observes the source output. As long as the new source output sequence after the last phrase coincides with one of the existing phrases, no new phrase is introduced and another letter from the source is considered. As soon as the new output sequence is different from the previous phrases, it is recognized as a new phrase and encoded. The encoding scheme is simple. The new phrase is the concatenation of a previous phrase and a new source output. To encode it, *the binary expansion of the lexicographic ordering of the previous phrase and the new bit are concatenated*. For example, assume that we want to parse and encode the following sequence:

0100001100001010000010100000110000010100000010.

Parsing the sequence by the rules previously explained results in the following phrases:

$$0, 1, 00, 001, 10, 000, 101, 0000, 01, 010, 00001, 100, 0001, 0100, 0010,$$

Clearly, all the phrases are different and each phrase is one of the previous phrases concatenated with a new source output. The number of phrases is 15. This means that, for each phrase, we need 4 bits, plus an extra bit to represent the new source output. The preceding sequence is encoded by

$$0000\ 0, 0000\ 1, 0001\ 0, 0011\ 1, 0010\ 0, 0011\ 0, 0101\ 1, 0110\ 0, 0001$$

$$1, 1001\ 0, 1000\ 1, 0101\ 0, 0110\ 1, 1010\ 0, 0100\ 0$$

Table 12.1 summarizes this procedure.

This representation can hardly be called a data compression scheme because a sequence of length 44 has been mapped into a sequence of length 75. However, as the length of the original sequence is increased, the compression role of this algorithm becomes more apparent. We can prove that for a stationary and ergodic source, as the length of the sequence increases, the number of bits in the compressed sequence approaches $nH(X)$, where $H(X)$ is the entropy rate of the source. The decompression of the encoded sequence is straightforward and can be done very easily.

One problem with the Lempel–Ziv algorithm is how the number of phrases should be chosen. Here we have chosen 15 phrases, which leaves us 4 bits to represent

TABLE 12.1   SUMMARY OF LEMPEL–ZIV EXAMPLE

	Dictionary Location	Dictionary Contents	Codeword	
1	0001	0	0000	0
2	0010	1	0000	1
3	0011	00	0001	0
4	0100	001	0011	1
5	0101	10	0010	0
6	0110	000	0011	0
7	0111	101	0101	1
8	1000	0000	0110	0
9	1001	01	0001	1
10	1010	010	1001	0
11	1011	00001	1000	1
12	1100	100	0101	0
13	1101	0001	0110	1
14	1110	0100	1010	0
15	1111	0010	0100	0

each phrase. In general, any fixed number of phrases will eventually become too small and overflow would occur. For example, if we were to continue coding this source for additional input letters, we could not add the new phrases to our dictionary because we have assigned four bits for representation of the elements of the dictionary and we already have 15 phrases in the dictionary. To solve this problem, the encoder and decoder must purge their dictionaries of elements which are not useful anymore and substitute new elements for them. The purging method should, of course, be a method on which the encoder and the decoder have agreed.

The Lempel–Ziv algorithm is widely used in practice to compress computer files. The "compress" and "uncompress" utilities under the UNIX© operating system and other compression programs (zip, gzip, etc.) are implementations of various versions of this algorithm.

12.4 MODELING OF COMMUNICATION CHANNELS

The mathematical model for an information source together with a quantitative measure of information, as well as bounds on compression of an information source that are given by the source entropy, were presented in the preceding sections of this chapter. In this section, we study the other important component of a communication system, i.e., the communication channel. We also introduce the concept of coding for the protection of messages against channel errors.

As defined in Chapter 1, a communication channel is any medium over which information can be transmitted or in which information can be stored. Coaxial cable, ionospheric propagation, free space, fiber optic cables, and magnetic and optical disks are examples of communication channels. What is common among these is that they accept signals at their inputs and deliver signals at their outputs at a later time (storage case) or at another location (transmission case). Therefore, each communication channel is characterized by a relation between its input and output. In this sense, a communication channel is a *system*.

There are many factors that cause the output of a communication channel to be different from its input. These factors include attenuation, nonlinearities, bandwidth limitations, multipath propagation, and noise. All these factors contribute to a complex input–output relation in a communication channel. Due to the presence of fading and noise, the input–output relation in a communication channel is generally a stochastic relation.

Channels encountered in practice are generally waveform channels that accept (continuous-time) waveforms as their inputs and produce waveforms as their outputs. Because the bandwidth of any practical channel is limited, by using the sampling theorem a waveform channel becomes equivalent to a discrete-time channel. In a discrete-time channel, both the input and the output are discrete-time signals.

In a discrete-time channel, if the values that the input and output variables can take are finite, or countably infinite, the channel is called a *discrete channel*. An example of a discrete channel is a binary-input binary-output channel. In general, a discrete channel is defined by $\mathcal{X}$, the input alphabet, $\mathcal{Y}$, the output alphabet, and

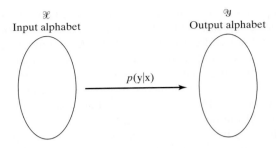

Figure 12.7 A discrete channel.

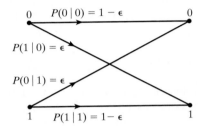

Figure 12.8 The binary-symmetric channel.

$p(\mathbf{y}|\mathbf{x})$, the conditional PMF of the output sequence given the input sequence. A schematic representation of a discrete channel is given in Figure 12.7. In general, the output y_i does not only depend on the input at the same time x_i, but also on the previous inputs (channels with ISI, see Chapter 9) or even on previous and future inputs (in storage channels). Therefore, a channel can have *memory*. However, if a discrete channel does not have memory, it is called a *discrete-memoryless channel* or DMC. For such a channel, for any $\mathbf{y} \in \mathcal{Y}^n$ and $\mathbf{x} \in \mathcal{X}^n$, we have

$$P(\mathbf{y}|\mathbf{x}) = \prod_{i=1}^{n} P(y_i|x_i). \qquad (12.4.1)$$

All channel models that we will discuss in this chapter are memoryless.

A special case of a discrete-memoryless channel is the *binary-symmetric channel* or BSC. Figure 12.8 shows a binary-symmetric channel. In a binary-symmetric channel, $\epsilon = P(0|1) = P(1|0)$ is called the *crossover probability*.

Example 12.4.1

Assume that we are dealing with an additive white Gaussian noise channel with binary antipodal signaling (for instance, a BPSK modulation system). We have already seen in Chapters 8 and 10 that, in such a channel, the error probability of a 1 being detected as zero or a zero being detected as 1 is given by

$$\epsilon = P(1|0) = P(0|1) = Q\left(\sqrt{\frac{2\mathcal{E}_b}{N_0}}\right), \qquad (12.4.2)$$

where N_0 is the noise power spectral density and $\mathscr{E}_b$ denotes the energy content of each of the antipodal signals representing 0 and 1. This discrete channel is an example of a binary symmetric channel.

Example 12.4.2

In an AWGN channel with binary antipodal signaling, the input is either $\sqrt{\mathscr{E}_b}$ or $-\sqrt{\mathscr{E}_b}$. The output is the sum of the input and the Gaussian noise. Therefore, for this binary-input, continuous-output channel, we have $\mathscr{X} = \{\pm\sqrt{\mathscr{E}_b}\}$, $\mathscr{Y} = \mathbb{R}$, and

$$f(y|x) = \frac{1}{\sqrt{2\pi\sigma^2}} e^{-\frac{(y-x)^2}{2\sigma^2}},$$

where σ^2 denotes the variance of the noise. This an example of a discrete input–continuous output channel. ∎

The most important continuous alphabet channel is the discrete-time, additive white Gaussian noise channel with an input power constraint. In this channel, both $\mathscr{X}$ and $\mathscr{Y}$ are the set of real numbers, and the input–output relation is given by

$$Y = X + Z, \tag{12.4.3}$$

where Z denotes the channel noise, which is assumed to be Gaussian with mean equal to 0 and variance equal to P_N. We further assume that inputs to this channel satisfy some power constraint. For example, for large n, input blocks of length n satisfy

$$\frac{1}{n}\sum_{i=1}^{n} x_i^2 \le P, \tag{12.4.4}$$

where P is some fixed power constraint. This channel model is shown in Figure 12.9.

12.5 CHANNEL CAPACITY

We have already seen that $H(X)$ defines a fundamental limit on the rate at which a discrete source can be encoded without errors in its reconstruction. A similar "fundamental limit" exists also for information transmission over communication channels.

Of course, the main objective when transmitting information over any communication channel is *reliability*, which is measured by the probability of correct

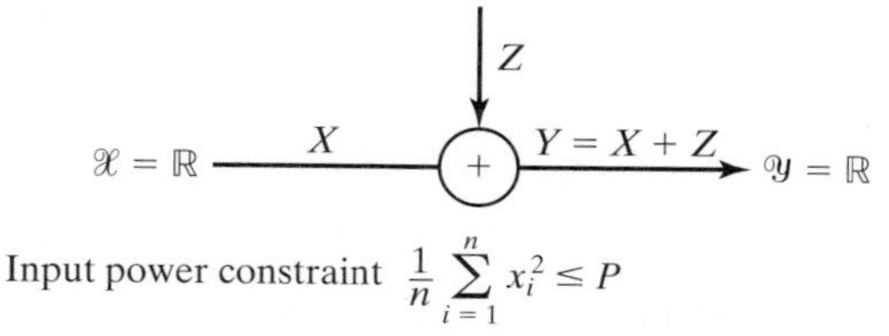

Input power constraint $\frac{1}{n}\sum_{i=1}^{n} x_i^2 \le P$

Figure 12.9 Additive white Gaussian noise channel with power constraint.

reception at the receiver. Due to the presence of noise, at first glance it seems that the error probability is always bounded away from zero by some positive number. However, a fundamental result of information theory is that reliable transmission (that is, transmission with error probability less any given value) is possible even over noisy channels as long as the transmission rate is less than some number, called the *channel capacity*. This remarkable result, first shown by Shannon (1948), is known as the *noisy channel coding theorem*. What the noisy channel coding theorem says is that *the basic limitation that noise causes in a communication channel is not on the reliability of communication, but on the speed of communication*.

Figure 12.10 shows a discrete memoryless channel with four inputs and outputs. If the receiver receives an *a*, it does not know whether an *a* or a *d* was transmitted; if it receives a *b*, it does not know whether an *a* or a *b* was transmitted, etc. Therefore, there always exists a possibility of error. But if the transmitter and the receiver agree that the transmitter only uses the letters *a* and *c*, then there exists no ambiguity. In this case, if the receiver receives an *a* or a *b*, it knows that an *a* was transmitted; if it receives a *c* or a *d*, it knows that a *c* was transmitted. This means that the two symbols *a* and *c* can be transmitted over this channel with no error, i.e., we are able to avoid errors by using only a subset of the possible inputs to the channel. Admittedly, using a smaller subset of the possible inputs reduces the number of possible inputs, but this is the price that must be paid for reliable communication. This is the essence of the noisy channel coding theorem, i.e., using only those inputs whose corresponding possible outputs are disjoint, and thus do not cause ambiguity as to what message was transmitted. The chosen inputs should be "far apart" such that their "images" under the channel operation are non-overlapping (or have negligible overlap).

Looking at the binary symmetric channel and trying to apply this approach, we observe that there is no way that we can have nonoverlapping outputs. In fact, this is the case with most channels. To use the results of this argument for the binary symmetric channel, we have to apply it not to the channel itself, but to the *extension*

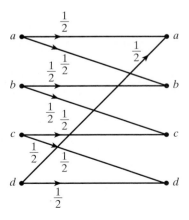

Figure 12.10   An example of a discrete channel.

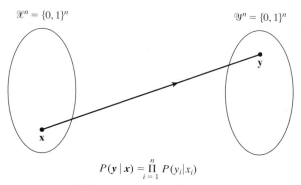

$$P(\boldsymbol{y} \mid \boldsymbol{x}) = \prod_{i=1}^{n} P(y_i|x_i)$$

$\boldsymbol{x}$: A binary sequence of length n.
$\boldsymbol{y}$: A binary sequence of length n.

Figure 12.11   The nth extension of a binary symmetric channel.

channel.[1] The nth extension of a channel with input and output alphabets $\mathscr{X}$ and $\mathscr{Y}$ and conditional probabilities $P(\boldsymbol{y}|\boldsymbol{x})$ is a channel with input and output alphabets $\mathscr{X}^n$ and $\mathscr{Y}^n$ and conditional probability $P(\boldsymbol{y}|\boldsymbol{x}) = \prod_{i=1}^{n} P(y_i|x_i)$. The nth extension of a binary symmetric channel takes binary blocks of length n as its input and its output. This channel is shown in Figure 12.11. By the law of large numbers, discussed in Chapter 5, if n is large enough and a binary sequence of length n is transmitted over the channel, the output will disagree with the input with high probability at $n\epsilon$ positions. The number of possible sequences that disagree with a sequence of length n at $n\epsilon$ positions is given by

$$\binom{n}{n\epsilon} = \frac{n!}{(n - n\epsilon)!(n\epsilon)!}.$$

Using Stirling's approximation $n! \approx n^n e^{-n}\sqrt{2\pi n}$, we obtain

$$\binom{n}{n\epsilon} \approx 2^{nH_b(\epsilon)}, \tag{12.5.1}$$

where $H_b(\epsilon) = -\epsilon \log_2 \epsilon - (1 - \epsilon)\log_2(1 - \epsilon)$ is the binary entropy function as defined in Equation (12.1.2). This means that, for any input block, there are roughly $2^{nH_b(\epsilon)}$ highly probable corresponding outputs. On the other hand, the total number of highly probable output sequences (i.e., typical output sequences) is roughly $2^{nH(Y)}$. The maximum number of input sequences that produce almost nonoverlapping output sequences, therefore, is at most equal to

$$M = \frac{2^{nH(Y)}}{2^{nH_b(\epsilon)}} = 2^{n(H(Y)-H_b(\epsilon))}. \tag{12.5.2}$$

---

[1] Recall that, in Huffman coding, we also applied the coding algorithm to the nth extension of the source to achieve efficient compression.

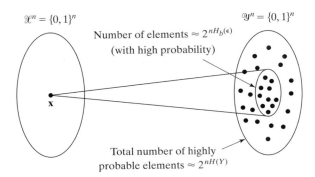

$\mathcal{X}^n = \{0, 1\}^n$

Number of elements $\approx 2^{nH_b(\epsilon)}$
(with high probability)

$\mathcal{Y}^n = \{0, 1\}^n$

x

Total number of highly
probable elements $\approx 2^{nH(Y)}$

Figure 12.12   Schematic representation of a BSC.

This means that in n uses of this channel, we can transmit at most $\log_2 M = n(H(Y) - H_b(\epsilon))$ bits and the transmission rate per channel use is

$$R = \frac{\log_2 M}{n} = H(Y) - H_b(\epsilon). \tag{12.5.3}$$

Figure 12.12 gives a schematic representation of this case.

In the relation $R = H(Y) - H_b(\epsilon)$, ϵ depends on the channel and we cannot control it. However, the probability distribution of the random variable Y depends both on the input distribution $P(x)$ and the channel properties characterized by ϵ. To maximize the transmission rate over the channel, we have to choose a $P(x)$ that maximizes $H(Y)$. If X is a uniformly distributed random variable, such that $p(X = 0) = P(X = 1) = 0.5$, then $H(Y)$ will be maximized at one, therefore we obtain

$$R = 1 - H_b(\epsilon). \tag{12.5.4}$$

We can prove that this rate is the maximum rate at which reliable transmission over the BSC is possible. By *reliable transmission*, we mean that the error probability can tend to zero as the block length n tends to infinity. A plot of the channel capacity in this case is given in Figure 12.13. It is interesting to note that the cases $\epsilon = 0$ and $\epsilon = 1$ both result in $C = 1$. This means that a channel that always flips the input is as good as the channel that transmits the input with no errors. The worst case, of course, happens when the channel flips the input with probability 1/2.

The maximum rate at which we can communicate over a discrete-memoryless channel and still make the error probability approach zero as the code block length increases is called the *channel capacity* and is denoted by C. The noisy channel coding theorem stated next gives the capacity of a general discrete memoryless channel.

Theorem 12.5.1.   [**Noisy Channel Coding Theorem**] The capacity of a discrete memoryless channel is given by

$$C = \max_{P(x)} I(X; Y), \tag{12.5.5}$$

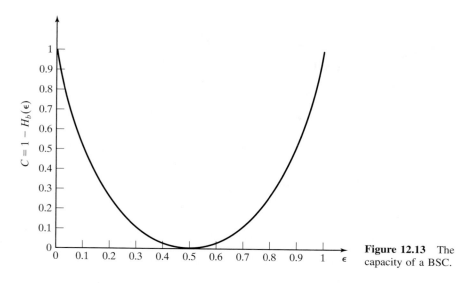

Figure 12.13 The capacity of a BSC.

where $I(X; Y)$ is the mutual information between X and Y, the channel input and output, as defined in Equations (12.1.13) and (12.1.14). If the transmission rate R is less than C, then reliable communication at rate R is possible, and if $R > C$, then reliable communication at rate R is impossible. ∎

In this theorem, R is $\frac{\log_2 M}{n}$, where M is the number of messages transmitted over the nth extension of the channel, and "reliable transmission" refers to transmission wherein the error probability can be made arbitrarily close to zero by increasing n. Both rate R and capacity C are measured in bits per transmission or bits per channel use.

This theorem is one of the most important results in information theory and gives a fundamental limit on the possibility of reliable communication over a noisy channel. According to this theorem, regardless of all other properties, any communication channel is characterized by a number called capacity, which determines how much information can be transmitted over it. Therefore, to compare two channels from an information transmission point of view, it is enough to compare their capacities.

Example 12.5.1

Find the capacity of the channel shown in Figure 12.14.

Solution   We have to find the input distribution that maximizes $I(X; Y)$. We have

$$I(X; Y) = H(Y) - H(Y|X).$$

But

$$H(Y|X) = P(X = a)H(Y|X = a) + P(X = b)H(Y|X = b)$$
$$+ P(X = c)H(Y|X = c).$$

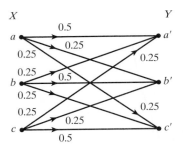

Figure 12.14   The DMC of Example 12.5.1.

From the channel input–output relation, we see that for all three cases ($X = a$, $X = b$, and $X = c$), Y is a ternary random variable with probabilities 0.25, 0.25, and 0.5. Therefore,

$$H(Y|X = a) = H(Y|X = b) = H(Y|X = c) = 1.5.$$

Then, because $P(X = a) + P(X = b) + P(X = c) = 1$, it follows that

$$H(Y|X) = 1.5$$

and

$$I(X; Y) = H(Y) - 1.5.$$

To maximize $I(X; Y)$, it remains to maximize $H(Y)$, which is maximized when Y is an equiprobable random variable. But it is not clear if there exists an input distribution that results in a uniform distribution on the output. However, in this special case (due to the symmetry of the channel), a uniform input distribution results in a uniform output distribution, and for this distribution,

$$H(Y) = \log 3 = 1.585.$$

This means that the capacity of this channel is given by

$$C = 1.585 - 1.5 = .085 \qquad \text{bits/transmission.}$$

12.5.1 Gaussian Channel Capacity

A discrete-time Gaussian channel with input power constraint is characterized by the input–output relation

$$Y = X + Z, \tag{12.5.6}$$

where Z is a zero-mean Gaussian random variable with variance P_N, and for n large enough, an input power constraint of the form

$$\frac{1}{n} \sum_{i=1}^{n} x_i^2 \leq P \tag{12.5.7}$$

applies to any input sequence of length n. For blocks of length n at the input, the output, and the noise, we have

$$y = x + z. \tag{12.5.8}$$

If n is large, by the law of large numbers, we have

$$\frac{1}{n} \sum_{i=1}^{n} z_i^2 = \frac{1}{n} \sum_{i=1}^{n} |y_i - x_i|^2 \to P_N \tag{12.5.9}$$

or

$$|y - x|^2 \to n P_N. \tag{12.5.10}$$

This means that, with probability approaching 1 (as n increases), y will be located in an n-dimensional sphere (hypersphere) that has a radius $\sqrt{n P_N}$ and is centered at x. On the other hand, due to the power constraint of P on the input and the independence of the input and noise, the output power is the sum of the input power and the noise power, i.e.,

$$\frac{1}{n} \sum_{i=1}^{n} y_i^2 \le P + P_N \tag{12.5.11}$$

or

$$|y|^2 \le n(P + P_N). \tag{12.5.12}$$

This implies that the output sequences (again, asymptotically and with high probability) will be inside an n-dimensional hypersphere of radius $\sqrt{n(P + P_N)}$ and centered at the origin. Figure 12.15 shows the sequences in the output space.

The question now is, How many x sequences can we transmit over this channel such that the hyperspheres corresponding to these sequences do not overlap in the output space? Obviously, if this condition is satisfied, then the input sequences can be decoded reliably. An equivalent question is as follows: How many hyperspheres of radius $\sqrt{n P_N}$ can we pack in a hypersphere of radius $\sqrt{n(P_N + P)}$? The answer is roughly the ratio of the volumes of the two hyperspheres. Note that, for an ordinary three-dimensional sphere, the volume is $\frac{4}{3}\pi R^3$; for a two-dimensional case (circle), the "volume" (i.e., the area) is πR^2. We can show that, in general, the volume of an n-dimensional hypersphere is proportional to R^n. If we denote the volume of an n-dimensional hypersphere by

$$V_n = K_n R^n, \tag{12.5.13}$$

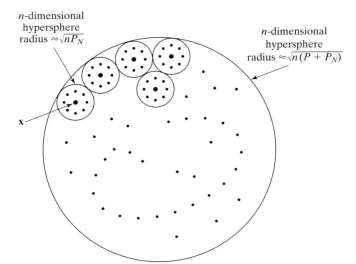

n-dimensional
hypersphere
radius $\approx \sqrt{nP_N}$

n-dimensional
hypersphere
radius $\approx \sqrt{n(P + P_N)}$

x

Figure 12.15 The output
sequences of a Gaussian
channel with power
constraint.

where R denotes the radius and K_n is independent of R, we see that the number of messages that can be reliably transmitted over this channel is equal to

$$M = \frac{K_n(n(P_N + P))^{\frac{n}{2}}}{K_n(nP_N)^{\frac{n}{2}}}$$

$$= \left(\frac{P_N + P}{P_N}\right)^{\frac{n}{2}}$$

$$= \left(1 + \frac{P}{P_N}\right)^{\frac{n}{2}}. \qquad (12.5.14)$$

Therefore, the capacity of a discrete-time, additive white Gaussian noise channel with input power constraint P is given by

$$C = \frac{1}{n} \log M$$

$$= \frac{1}{n} \cdot \frac{n}{2} \log\left(1 + \frac{P}{P_N}\right)$$

$$= \frac{1}{2} \log\left(1 + \frac{P}{P_N}\right). \qquad (12.5.15)$$

When dealing with a continuous-time, bandlimited, additive white Gaussian noise channel with noise power spectral density $\frac{N_0}{2}$, input power constraint P, and bandwidth W, we can sample at the Nyquist rate and obtain a discrete time channel.

The power per sample will be P and the noise power per sample will be

$$P_N = \int_{-W}^{+W} \frac{N_0}{2} df = W N_0.$$

Substituting these results in Equation (12.5.15), we obtain

$$C = \frac{1}{2} \log \left(1 + \frac{P}{N_0 W} \right) \qquad \text{bits/transmission.} \qquad (12.5.16)$$

If we multiply this result by the number of transmissions per second, which is $2W$ by the Nyquist sampling theorem, we obtain the channel capacity in bits/sec:

$$C = W \log \left(1 + \frac{P}{N_0 W} \right) \qquad \text{bits/sec.} \qquad (12.5.17)$$

This is the celebrated Shannon's formula for the capacity of a bandlimited additive white Gaussian noise channel.

Example 12.5.2

Find the capacity of a telephone channel with bandwidth $W = 3000$ Hz, and signal-to-noise ratio of 39 dB.

Solution   The signal-to-noise ratio of 39 dB is equivalent to 7943. Using Shannon's relation, we have

$$C = 3000 \log(1 + 7943) \approx 38{,}867 \quad \text{bits/sec.}$$

12.6 BOUNDS ON COMMUNICATION

From the previous section, the capacity of a bandlimited additive white Gaussian noise channel is given by

$$C = W \log \left(1 + \frac{P}{N_0 W} \right).$$

From this result, the basic factors that determine the channel capacity are the channel bandwidth W, the noise power spectrum N_0, and the signal power P. There exists a trade-off between P and W in the sense that one can compensate for the other. Increasing the input signal power obviously increases the channel capacity, because when we have more power to spend, we can choose a larger number of input levels that are far apart. Therefore we can send more information bits per transmission. However, the increase in capacity as a function of power is logarithmic and slow. This is because if we are transmitting with a certain number of input levels that are Δ apart to allow a certain level of immunity against noise, and we want to increase the number of input levels, we have to introduce new levels with amplitudes higher than the existing levels, and this requires a lot more power. This fact notwithstanding,

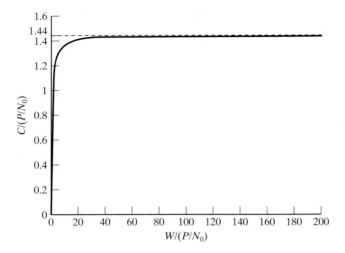

Figure 12.16 Plot of channel capacity versus bandwidth.

the capacity of the channel can be increased to any value by increasing the input power.

The effect of the channel bandwidth, however, is quite different. Increasing W has two contrasting effects. On one hand, on a higher bandwidth channel, we can transmit more samples per second and, therefore, increase the transmission rate. On the other hand, a higher bandwidth means higher input noise power to the receiver, and this reduces its performance. This is seen from the two W's that appear in the relation that describes the channel capacity. To see the effect of increasing the bandwidth, we let the bandwidth W tend to infinity. Then, using L'Hôpital's rule, we obtain

$$\lim_{W \to \infty} W \log(1 + \frac{P}{N_0 W}) = \frac{P}{N_0} \log e = 1.44 \frac{P}{N_0}. \tag{12.6.1}$$

This means that, contrary to the power case, by increasing the bandwidth alone, we cannot increase the capacity to any desired value. Figure 12.16 shows C plotted versus W.

In any practical communication system, we must have $R < C$. If an AWGN channel is employed, we have

$$R < W \log \left(1 + \frac{P}{N_0 W}\right). \tag{12.6.2}$$

By dividing both sides by W and defining $r = \frac{R}{W}$, the *spectral bit rate* (or bandwidth efficiency), we obtain

$$r < \log \left(1 + \frac{P}{N_0 W}\right). \tag{12.6.3}$$

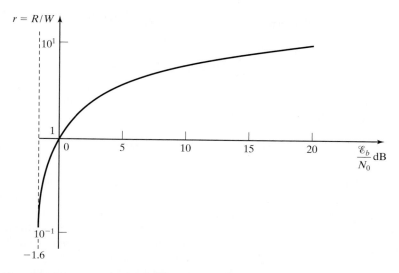

Figure 12.17   Spectral bit rate versus SNR/bit in an optimal system.

If $\mathcal{E}_b$ is the energy per bit, then $\mathcal{E}_b = \frac{P}{R}$. By substituting in the previous relation, we obtain

$$r < \log\left(1 + r\frac{\mathcal{E}_b}{N_0}\right),\tag{12.6.4}$$

or equivalently,

$$\frac{\mathcal{E}_b}{N_0} > \frac{2^r - 1}{r}.\tag{12.6.5}$$

This relation is plotted in Figure 12.17. This equation gives the relation between two important parameters in a communication system. These parameters are r, the spectral bit rate, which is a measure of the bandwidth efficiency of a communication system, and $\frac{\mathcal{E}_b}{N_0}$, the SNR/bit, which is a measure of the power efficiency of a system. With a higher value of r, the system is more bandwidth efficient. With a lower value of $\frac{\mathcal{E}_b}{N_0}$ to achieve a certain error probability, the system is more power efficient.

The curve defined by

$$r = \log\left(1 + r\frac{\mathcal{E}_b}{N_0}\right)\tag{12.6.6}$$

divides the plane into two regions. In one region (below the curve), reliable communication is possible; in another region (above the curve), reliable communication is not possible. The performance of any communication system can be denoted by a point in this plane; the closer the point is to this curve, the closer the performance of the system is to that of an optimal system. From this curve, we see that as r tends to 0,

$$\frac{\mathscr{E}_b}{N_0} = \ln 2 = 0.693 \approx -1.6 \quad \text{dB} \tag{12.6.7}$$

is an absolute minimum for reliable communication. In other words, for reliable communication, we must have

$$\frac{\mathscr{E}_b}{N_0} > 0.693. \tag{12.6.8}$$

In Figure 12.17, when $r \ll 1$, we are dealing with a case where bandwidth is large and the main concern is limitation on power. This case is usually referred to as the *power limited case*. Signaling schemes, with high dimensionality, such as orthogonal, biorthogonal, and simplex schemes can be used in these cases. The case where $r \gg 1$ happens when the bandwidth of the channel is small; therefore, it is referred to as the *bandwidth limited case*. Low-dimensional signaling schemes with crowded constellations (PAM, QAM, and PSK) are implemented in these cases.

In Section 12.2, we introduced the fundamental limit in coding of information sources. This fundamental limit is expressed in terms of the entropy of the source. Entropy gives a lower bound on the rate of the codes that are capable of reproducing the source with no error. If we want to transmit a source U reliably via a channel with capacity C, we require that

$$H(U) < C. \tag{12.6.9}$$

This relation defines the fundamental limit on the transmission of information. Note that we have assumed that for each source output, one transmission over the channel is possible.

Example 12.6.1

A binary source with $P(X = 0) = \frac{1}{4}$ is to be transmitted over a binary symmetric channel with a crossover probability of ϵ. Determine the range of ϵ for reliable communication of the source. Assume that the channel can be used once per source output.

Solution   We have

$$H(U) = -\frac{1}{4} \log_2 \frac{1}{4} - \frac{3}{4} \log_2 \frac{3}{4} = 0.8113.$$

For reliable communication, we need $H(U) < C$. The channel capacity for a BSC is given by $C = 1 - H_b(\epsilon)$. Therefore, we must have

$$0.8113 < 1 - H_b(\epsilon)$$

or

$$H_b(\epsilon) < 0.1887.$$

This is a nonlinear equation and should be solved by numerical methods. This results in $H_b(0.0288) \approx 0.1887$ and two acceptable regions for ϵ, namely, $0 \leq \epsilon < 0.0288$ and $0.9712 < \epsilon \leq 1$. ∎

12.7 FURTHER READING

The source coding theorem and the noisy channel coding theorem, which play a central role in information theory, and the concepts of entropy and channel capacity were first proposed and proved by Shannon (1948). For a detailed discussion of these theorems and their variations, refer to standard books on information theory, such as Gallager (1968), Blahut (1987), and Cover and Thomas (1991).

PROBLEMS

12.1 A discrete memoryless source has an alphabet $\{a_1, a_2, a_3, a_4, a_5, a_6\}$ with corresponding probabilities $\{0.1, 0.2, 0.3, 0.05, 0.15, 0.2\}$. Find the entropy of this source. Compare this entropy with the entropy of a uniformly distributed source with the same alphabet.

12.2 Let random variable X be the output of a discrete memoryless source that is uniformly distributed with size N. Find the entropy of it.

12.3 Show that $H(X) \geq 0$ with equality holding if and only if X is deterministic.

12.4 Let X be a geometrically distributed random variable; that is,

$$P(X = k) = p(1 - p)^{k-1} \qquad k = 1, 2, 3, \dots .$$

1. Find the entropy of X.

2. Knowing that $X > K$, where K is a positive integer, what is the entropy of X?

12.5 Let $Y = g(X)$, where g denotes a deterministic function. Show that, in general, $H(Y) \leq H(X)$. When does equality hold?

12.6 An information source can be modeled as a bandlimited process with a bandwidth of 6000 Hz. This process is sampled at a rate higher than the Nyquist rate to provide a guard band of 2000 Hz. We observe that the resulting samples take values in the set $\mathcal{A} = \{-4, -3, -1, 2, 4, 7\}$ with probabilities $0.2, 0.1, 0.15, 0.05, 0.3, 0.2$. What is the entropy of the discrete time source in bits per output (sample)? What is the information generated by this source in bits per second?

12.7 Let X denote a random variable distributed on the set $\mathcal{A} = \{a_1, a_2, \ldots, a_N\}$ with corresponding probabilities $\{p_1, p_2, \ldots, p_N\}$. Let Y be another random variable defined on the same set, but distributed uniformly. Show that

$$H(X) \leq H(Y)$$

with equality if and only if X is also uniformly distributed. (Hint: First prove the inequality $\ln x \leq x - 1$ with equality only for $x = 1$, then apply this inequality to $\sum_{n=1}^{N} p_n \ln \left(\frac{\frac{1}{N}}{p_n} \right)$.)

12.8 A random variable X is distributed on the set of all positive integers $1, 2, 3, \ldots$ with corresponding probabilities $p_1, p_2, p_3, \ldots$. We further know that the expected value of this random variable is given to be m, i.e.,

$$\sum_{i=1}^{\infty} i p_i = m.$$

Show that, among all random variables that satisfy the preceding condition, the geometric random variable that is defined by

$$p_i = \frac{1}{m} \left(1 - \frac{1}{m} \right)^{i-1} \qquad i = 1, 2, 3, \ldots$$

has the highest entropy. (Hint: Define two distributions on the source, the first one being the foregoing geometric distribution and the second one an arbitrary distribution denoted by q_i. Then apply the approach of Problem 12.7.)

12.9 Two binary random variables, X and Y, are distributed according to the joint distribution $P(X = Y = 0) = P(X = 0, Y = 1) = P(X = Y = 1) = \frac{1}{3}$. Compute $H(X)$, $H(Y)$, $H(X|Y)$, $H(Y|X)$, and $H(X, Y)$.

12.10 Show that if $Y = g(X)$ where g denotes a deterministic function, then $H(Y|X) = 0$.

12.11 A memoryless source has the alphabet $\mathcal{A} = \{-5, -3, -1, 0, 1, 3, 5\}$ with corresponding probabilities $\{0.05, 0.1, 0.1, 0.15, 0.05, 0.25, 0.3\}$.

    1. Find the entropy of the source.

    2. Assume that the source is quantized according to the quantization rule

$$\begin{cases} q(-5) = q(-3) = -4, \\ q(-1) = q(0) = q(1) = 0 \\ q(3) = q(5) = 4 \end{cases} \; .$$

    Find the entropy of the quantized source.

12.12 Using both definitions of the entropy rate of a random process given in Equations (12.1.11) and (12.1.12), prove that for a DMS, the entropy rate and the entropy are equal.

12.13 A Markov process is a process with one step memory, i.e., a process such that

$$p(x_n | x_{n-1}, x_{n-2}, x_{n-3}, \ldots) = p(x_n | x_{n-1})$$

for all n. Show that, for a stationary Markov process, the entropy rate is given by $H(X_n | X_{n-1})$.

12.14 Show that

$$H(X|Y) = \sum_y p(y) H(X | Y = y).$$

12.15 Let X and Y denote two jointly distributed discrete valued random variables.

    1. Show that

$$H(X) = -\sum_{x,y} p(x, y) \log p(x)$$

    and

$$H(Y) = -\sum_{x,y} p(x, y) \log p(y).$$

    2. Use this result to show that

$$H(X, Y) \le H(X) + H(Y).$$

    When does the equality hold? (Hint: Consider the two distributions $p(x, y)$ and $p(x)p(y)$ on the product set $\mathcal{X} \times \mathcal{Y}$, and apply the inequality proved in Problem 12.7 to $\sum_{x,y} p(x, y) \log \frac{p(x)p(y)}{p(x,y)}$.)

12.16 Use the result of Problem 12.15 to show that

$$H(X|Y) \le H(X)$$

with equality if and only if X and Y are independent.

12.17 Show that, in general,

$$H(X_1, X_2, \ldots, X_n) \le \sum_{i=1}^n H(X_i).$$

When does the equality hold?

12.18 Assume that a BSS generates a sequence of n outputs.

1. What is the probability that this sequence consists of all zeros?

2. What is the probability that this sequence consists of all ones?

3. What is the probability that, in this sequence, the first k symbols are ones and the next $n - k$ symbols are zeros?

4. What is the probability that this sequence has k ones and $n - k$ zeros?

5. How would your answers change if, instead of a BSS, we were dealing with a general binary DMS with $P(X_i = 1) = p$.

12.19 Give an estimate of the number of binary sequences of length 10,000 with 3000 zeros and 7000 ones.

12.20 A memoryless ternary source with output alphabet a_1, a_2, and a_3 and corresponding probabilities 0.2, 0.3, and 0.5 produces sequences of length 1000.

1. What is the approximate number of typical sequences in the source output?

2. What is the ratio of typical sequences to nontypical sequences?

3. What is the probability of a typical sequence?

4. What is the number of bits required to represent all output sequences?

5. What is the number of bits required to represent only the typical output sequences?

6. What is the most probable sequence and what is its probability?

7. Is the most probable sequence a typical sequence?

12.21 A source has an alphabet $\{a_1, a_2, a_3, a_4\}$ with corresponding probabilities $\{0.1, 0.2, 0.3, 0.4\}$.

1. Find the entropy of the source.

2. What is the minimum required average codeword length to represent this source for error-free reconstruction?

3. Design a Huffman code for the source and compare the average length of the Huffman code with the entropy of the source.

4. Design a Huffman code for the second extension of the source (take two letters at a time). What is the average codeword length? What is the average number of required binary letters per each source output letter?

5. Which is a more efficient coding scheme: the Huffman coding of the original source or the Huffman coding of the second extension of the source?

12.22 Design a Huffman code for a source with n output letters and corresponding probabilities $\{\frac{1}{2}, \frac{1}{4}, \frac{1}{8}, \ldots, \frac{1}{2^{n-1}}, \frac{1}{2^{n-1}}\}$. Show that the average codeword length for such a source is equal to the source entropy.

12.23 Show that $\{01, 100, 101, 1110, 1111, 0011, 0001\}$ cannot be a Huffman code for *any* source probability distribution.

12.24 Design a *ternary* Huffman code, using 0, 1, and 2 as letters, for a source with output alphabet probabilities given by $\{0.05, 0.1, 0.15, 0.17, 0.18, 0.22, 0.13\}$. What is the resulting average codeword length? Compare the average codeword length with the entropy of the source. (In what base would you compute the logarithms in the expression for the entropy for a meaningful comparison?)

12.25 Design a ternary Huffman code for a source with output alphabet probabilities given by $\{0.05, 0.1, 0.15, 0.17, 0.13, 0.4\}$. (Hint: You can add a dummy source output with zero probability.)

12.26 A discrete memoryless source X has the alphabet $\{-5, -3, -1, 0, 1, 2, 3\}$ with the corresponding probabilities $\{0.08, 0.2, 0.12, 0.15, 0.03, 0.02, 0.4\}$.

1. Design a Huffman code for this source and find $\overline{R}$, the average codeword length of the Huffman code.

2. Determine the entropy of the source and, using the result of Part 1, determine the efficiency of the Huffman code you designed. The efficiency is defined as $\eta = \frac{H(X)}{\overline{R}}$.

3. Now quantize the source using the following quantization rule:

$$\hat{X} = \begin{cases} -2, & X = -5, -3 \\ 0, & X = -1, 0, 1 \\ 2, & X = 2, 3. \end{cases}$$

   What is the absolute minimum required rate (bits per symbol) for perfect reconstruction of $\hat{X}$?

4. Looking at sequences of length 10,000 at the output of the quantized source (i.e., $\hat{X}$), what is the possible number of sequences? What is the number of typical sequences?

5. If a Huffman code is designed for the second extension of the original source before quantization (i.e., taking two letters at a time), what can you say about the average codeword length per individual source symbol? Find the tightest possible bounds on the average codeword length.

12.27 A discrete memoryless information source is described by the alphabet $\mathscr{X} = \{x_1, x_2, x_3, x_4, x_5, x_6\}$ with probabilities $\{1/32, 1/8, 1/2, 1/16, 1/32, 1/4\}$, respectively.

1. Design a Huffman code for this source and determine the average codeword length of the Huffman code.

2. Can you improve the Huffman code by encoding the second extension of this source (in other words, using two letters at a time and designing the Huffman code for that source)? Why?

3. Is there any way to improve the performance of the Huffman code designed in Part 1? (By *improving the performance*, we mean designing a code with a lower average codeword length.)

12.28 A discrete memoryless information source has the alphabet $\{a_1, a_2, a_3, a_4, a_5\}$ with the corresponding probabilities $\{0.1, 0.2, 0.05, 0.3, 0.35\}$.

1. Can this source be compressed at a rate of 2 bits per source symbol such that lossless reconstruction of it is possible?

2. Now consider all sequences of length 1000 that this source can generate. How many of these sequences are possible? Write your answer in exponential form.

3. Approximately how many of the sequences in Part 2 are typical sequences? Write your answer in exponential form.

4. Now assume that you want to merge two letters of the source into one new letter b, say, $b = \{a_i, a_j\}$, such that the resulting source (which now has 4 outputs instead of 5) can be compressed at a rate of 1.5 bits per symbol and can be recovered with no loss. Which two letters would you merge into the new letter b?

12.29 A discrete memoryless source with output alphabet $\{a_i\}_{i=1}^{7}$ and corresponding probabilities $\{0.11, 0.18, 0.1, 0.2, 0.25, 0.05, 0.11\}$ is to be transmitted with no errors to a destination.

1. What is the minimum rate required for transmission of this source?

2. Design a Huffman code for this source. What is the average codeword length of the Huffman code? How is this average codeword length related to the answer in Part 1?

3. A new source is obtained by grouping the outputs of this source as follows: $b_1 = \{a_1, a_2\}$, $b_2 = \{a_3, a_4\}$, $b_3 = \{a_5, a_6\}$, $b_4 = \{a_7\}$. Answer the question in Part 1 for this new source.

12.30 Find the Lempel–Ziv source code for the binary source sequence

0001001000000110000100000001000000101000010000001101000000001100.

Recover the original sequence from the Lempel–Ziv source code. (Hint: You require two passes of the binary sequence to decide on the size of the dictionary.)

12.31 Using the definition of $H(X)$ and $H(X|Y)$, show that

$$I(X;Y) = \sum_{x,y} p(x,y) \log \frac{p(x,y)}{p(x)p(y)}.$$

Now, by using the approach of Problem 12.7, show that $I(X;Y) \geq 0$ with equality if and only if X and Y are independent.

12.32 Show that

1. $I(X;Y) \leq \min\{H(X), H(Y)\}$.

2. If $|\mathcal{X}|$ and $|\mathcal{Y}|$ represent the size of sets $\mathcal{X}$ and $\mathcal{Y}$ respectively, then $I(X;Y) \leq \min\{\log|\mathcal{X}|, \log|\mathcal{Y}|\}$.

12.33 Show that $I(X;Y) = H(X)+H(Y)-H(X,Y) = H(Y)-H(Y|X) = I(Y;X)$.

12.34 Let X denote a binary random variable with $P(X=0) = 1 - P(X=1) = p$, and let Y be a binary random variable that depends on X through $P(Y=1|X=0) = P(Y=0|X=1) = \epsilon$.

1. Find $H(X)$, $H(Y)$, $H(Y|X)$, $H(X,Y)$, $H(X|Y)$, and $I(X;Y)$.

2. For a fixed ϵ, which p maximizes $I(X;Y)$?

3. For a fixed p, which ϵ minimizes $I(X;Y)$?

12.35 Show that

$$I(X;YZW) = I(X;Y) + I(X:Z|Y) + I(X;W|ZY).$$

Can you interpret this relation?

12.36 Let X, Y, and Z be three discrete random variables.

1. Show that if $p(x,y,z) = p(z)p(x|z)p(y|x)$, we have

$$I(X;Y|Z) \leq I(X;Y).$$

2. Show that if $p(x,y,z) = p(x)p(z)p(y|x,z)$, then

$$I(X;Y) \leq I(X;Y|Z).$$

3. In each case, give an example where strict inequality holds.

12.37 Find the capacity of the channel shown in Figure P-12.37.

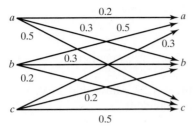

Figure P-12.37

12.38 The channel shown in Figure P-12.38 is known as the *binary erasure channel*. Find the capacity of this channel and plot it as a function of ϵ.

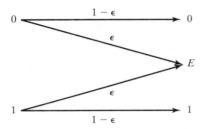

Figure P-12.38

12.39 Find the capacity of the cascade connection of n binary symmetric channels with the same crossover probability ϵ. What is the capacity when the number of channels goes to infinity?

12.40 Using Stirling's approximation $n! \approx n^n e^{-n} \sqrt{2\pi n}$, show that

$$\binom{n}{n\epsilon} \approx 2^{n H_b(\epsilon)}.$$

12.41 Show that the capacity of a binary-input, continuous-output AWGN channel with inputs $\pm A$ and noise variance σ^2 (see Example 12.4.2) is given by

$$C = \frac{1}{2} f\left(\frac{A}{\sigma}\right) + \frac{1}{2} f\left(-\frac{A}{\sigma}\right),$$

where

$$f(x) = \int_{-\infty}^{\infty} \frac{1}{\sqrt{2\pi}} e^{-(u-x)^2/2} \log_2 \frac{2}{1 + e^{-2xu}} \, du.$$

12.42 The matrix whose elements are the transition probabilities of a channel, i.e. $p(y_i|x_j)$'s, is called the channel probability transition matrix. A channel is called *symmetric* if all rows of the channel probability transition matrix are permutations of each other, and all its columns are also permutations of each other. Show that in a symmetric channel, the input probability distribution that achieves capacity is a uniform distribution. What is the capacity of this channel?

12.43 Channels 1, 2 and 3 are shown in Figure P-12.43.

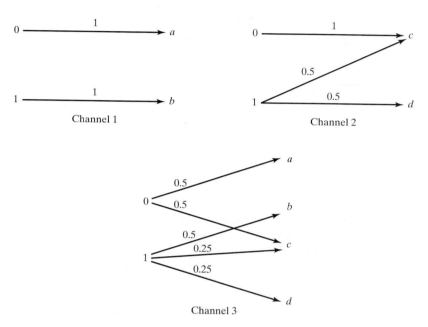

Figure P-12.43

1. Find the capacity of channel 1. What input distribution achieves capacity?

2. Find the capacity of channel 2. What input distribution achieves capacity?

3. Let C denote the capacity of the third channel, and let C_1 and C_2 represent the capacities of the first and second channel. Which of the following relations holds true and why?

    (a) $C < \frac{1}{2}(C_1 + C_2)$.

    (b) $C = \frac{1}{2}(C_1 + C_2)$.

    (c) $C > \frac{1}{2}(C_1 + C_2)$.

12.44 Let C denote the capacity of a discrete memoryless channel with input alphabet $\mathcal{X} = \{x_1, x_2, \ldots, x_N\}$ and output alphabet $\mathcal{Y} = \{y_1, y_2, \ldots, y_M\}$. Show that $C \leq \min\{\log M, \log N\}$.

12.45 The channel C is (known as the Z channel) shown in Figure P-12.45.

1. Find the input probability distribution that achieves capacity.

2. What is the input distribution and capacity for the special cases $\epsilon = 0$, $\epsilon = 1$, and $\epsilon = 0.5$?

3. Show that if n such channels are cascaded, the resulting channel will be equivalent to a Z channel with $\epsilon_1 = \epsilon^n$.

4. What is the capacity of the equivalent Z channel when $n \to \infty$?

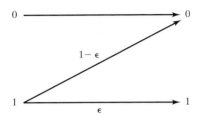

Figure P-12.45

12.46 Find the capacity of the Channels A and B, as shown in Figure P-12.46. What is the capacity of the cascade channel AB? (Hint: Look carefully at the channels, and avoid lengthy math.)

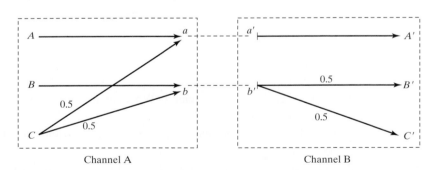

Channel A   Channel B

Figure P-12.46

12.47 Find the capacity of an additive white Gaussian noise channel with a bandwidth of 1 MHz, power of 10 Watts, and noise power spectral density of $\frac{N_0}{2} = 10^{-9}$ Watts/Hz.

12.48 Channel C_1 is an additive white Gaussian noise channel with a bandwidth of W, transmitter power of P, and noise power spectral density of $\frac{N_0}{2}$. Channel C_2 is an additive Gaussian noise channel with the same bandwidth and power as channel C_1 but with noise power spectral density $S_n(f)$. We assume that the total noise power for both channels is the same; that is,

$$\int_{-W}^{W} S_n(f)\,df = \int_{-W}^{W} \frac{N_0}{2}\,df = N_0 W.$$

Which channel do you think has a larger capacity? Give an intuitive reasoning.

12.49 For the channel shown in Figure P-12.49, find the channel capacity and the input distribution that achieves capacity.

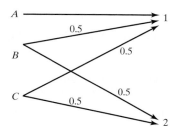

Figure P-12.49

12.50 Consider two discrete memoryless channels represented by $(\mathcal{X}_1, p(y_1|x_1), \mathcal{Y}_1)$ and $(\mathcal{X}_2, p(y_2|x_2), \mathcal{Y}_2)$ with corresponding capacities C_1 and C_2. A new channel is defined by $(\mathcal{X}_1 \times \mathcal{X}_2, p(y_1|x_1)p(y_2|x_2), \mathcal{Y}_1 \times \mathcal{Y}_2)$. This channel is the "product channel" and models the case where $x_1 \in \mathcal{X}_1$ and $x_2 \in \mathcal{X}_2$ are simultaneously transmitted over the two channels with no interference. Prove that the capacity of this new channel is the sum of C_1 and C_2.

12.51 Let $(\mathcal{X}_1, p(y_1|x_1), \mathcal{Y}_1)$ and $(\mathcal{X}_2, p(y_2|x_2), \mathcal{Y}_2)$ represent two discrete memoryless communication channels with inputs $\mathcal{X}_i$, outputs $\mathcal{Y}_i$ and conditional probabilities $p(y_i|x_i)$. Further assume that $\mathcal{X}_1 \cap \mathcal{X}_2 = \emptyset$ and $\mathcal{Y}_1 \cap \mathcal{Y}_2 = \emptyset$. We define the *sum* of these channels as a new channel with the input alphabet $\mathcal{X}_1 \cup \mathcal{X}_2$, output alphabet $\mathcal{Y}_1 \cup \mathcal{Y}_2$, and conditional probability $p(y_i|x_i)$ where i denotes the index of $\mathcal{X}$ to which the input to the channel belongs. This models a communication situation where we have two channels in parallel, and at each transmission interval, we can use one and only one of the channels, the input and output alphabets are disjoint, and therefore at the receiver there is no ambiguity regarding which channel was being used.

1. Show that the capacity of the *sum* channel satisfies $2^C = 2^{C_1} + 2^{C_2}$, where C_1 and C_2 are the capacities of each channel.

2. Using the result of Part 1, show that if $C_1 = C_2 = 0$, we still have $C = 1$. In other words, show that we are able to transmit one bit per transmission using two channels with zero capacity. How do you interpret this result?

3. Find the capacity of the channel shown in Figure P-12.51.

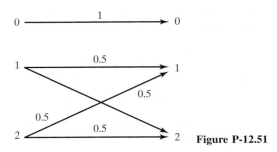

Figure P-12.51

12.52 X is a binary memoryless source with $P(X = 0) = 0.3$. This source output is transmitted over a binary symmetric channel with crossover probability $\epsilon = 0.1$.

1. Assume that the source is directly connected to the channel, i.e., no coding is employed. What is the error probability at the destination?

2. For what values of ϵ is reliable transmission possible (with coding, of course)?

COMPUTER PROBLEMS

12.1 Huffman Coding

The objective of this problem is to design a Huffman code using MATLAB. The discrete memoryless source output is generated from an alphabet $\mathscr{X} = \{x_1, x_2, \ldots, x_9\}$ with corresponding probabilities

$$p = \{0.2, 0.15, 0.13, 0.12, 0.1, 0.09, 0.08, 0.07, 0.06\}.$$

1. Design a Huffman code and sketch the corresponding code tree. Specify the codewords for the nine symbols in the alphabet.

2. Determine the average codeword length of the Huffman code.

3. Determine the entropy of the source and compare it with the average codeword length of the Huffman code.

12.2 Huffman Coding

The objective of this problem is to design Huffman codes using MATLAB. The discrete memoryless source output is generated from the alphabet $\mathscr{X} = \{x_1, x_2, \ldots, x_6\}$ with the corresponding probabilities

$$p = \{0.1, 0.3, 0.05, 0.09, 0.21, 0.25\}.$$

1. Determine the entropy of the source.

2. Design a Huffman code and sketch the corresponding code tree. Specify the codewords for the six symbols in the alphabet.

3. Determine the efficiency of the Huffman code designed in Part 2.

4. Design a Huffman code for the source sequences of length 2 and sketch the codeword tree. Specify the codewords for the symbols of length 2 and determine the efficiency of the code. Compare the efficiency for length 2 sequences with that for length 1.

12.3 Huffman Coding

A discrete memoryless source is generated from the alphabet $\mathcal{X} = \{x_1, x_2, \ldots, x_9\}$ with corresponding probabilities

$$p = \left\{ \frac{1}{2}, \frac{1}{4}, \frac{1}{8}, \frac{1}{16}, \frac{1}{32}, \frac{1}{64}, \frac{1}{128}, \frac{1}{256}, \frac{1}{256}, \right\}.$$

1. Design a Huffman code and sketch the corresponding code tree. Specify the codewords for the nine symbols in the alphabet.

2. Determine the average codeword length of the Huffman code, the entropy of the source, and the efficiency of the Huffman code.

3. Under what conditions is the efficiency of the Huffman code equal to 1?

12.4 Huffman Code for Printed English

The probabilities of the letters of the alphabet occurring in printed English are given in the following table:

Letter	Probability	Letter	Probability	Letter	Probability
A	0.0642	B	0.0127	C	0.0218
D	0.0317	E	0.1031	F	0.0208
G	0.0152	H	0.0467	I	0.0575
J	0.0008	K	0.0049	L	0.0321
M	0.0198	N	0.0574	O	0.0632
P	0.0152	Q	0.0008	R	0.0484
S	0.0514	T	0.0796	U	0.0228
V	0.0083	W	0.0175	X	0.0013
Y	0.0164	Z	0.0005	Space	0.1859

1. Determine the entropy of printed English.

2. Design a Huffman code for printed English.

3. Determine the average codeword length and the efficiency of the Huffman code.

12.5 Capacity of A Bandlimited Additive White Gaussian Noise Channel

The objective of this problem is to compute the capacity of a bandlimited additive white Gaussian noise channel using MATLAB. The capacity of this channel is given by the formula

$$C = W \log_2 \left(1 + \frac{P}{N_0 W} \right),$$

where W is the bandwidth of the channel, P is the average transmitted power, and $N_0/2$ is the power spectral density of the additive white Gaussian noise.

1. Plot the capacity of the channel whose bandwidth $W = 3000$ Hz as a function of P/N_0, for values of P/N_0 between -20 and 30 dB.

2. Plot the capacity of the channel as a function of W when $P/N_0 = 25$ dB. What is the limiting value of the channel capacity as $W \to \infty$?

12.6 Capacity of A Bandlimited Additive White Gaussian Noise Channel

The capacity of an AWGN channel, given as

$$C = W \log_2 \left(1 + \frac{P}{N_0 W} \right)$$

can also be expressed in terms of $\mathcal{E}_b/N_0$, which is the SNR/bit that determines the probability of error. Since the average power $P = \mathcal{E}_b/T_b = C\mathcal{E}_b$, where $1/T_b = C$ is the rate in bits per second, this channel capacity formula may be expressed as

$$\frac{C}{W} = \log_2 \left(1 + \frac{\mathcal{E}_b}{N_0} \frac{C}{W} \right).$$

If we solve for $\mathcal{E}_b/N_0$ as a function of C/W, we obtain

$$\frac{\mathcal{E}_b}{N_0} = \frac{2^{\frac{C}{W}} - 1}{\frac{C}{W}}.$$

1. Plot the normalized capacity C/W as a function of $\mathcal{E}_b/N_0$ for the range of values $0.1 \le C/W \le 10$. It may be convenient to plot the graph as a function of the SNR/bit in dB ($10 \log_{10} \mathcal{E}_b/N_0$).

2. Determine the SNR/bit in the limit as $C/W \to 0$.

13

Coding for Reliable Communications

In Chapter 12, we introduced fundamental bounds on source coding (data compression) and data transmission through a noisy channel. We saw that the fundamental bound for error-free source compression is given by the entropy (or entropy rate, for a source with memory) and the fundamental limit for reliable communication is given by the channel capacity. We also introduced some algorithms for source coding that achieved the bound predicted by theory. In this chapter, we present methods for channel coding in pursuit of achieving the bounds set by theory in Shannon's noisy channel coding theorem, i.e., the channel capacity. It turns out that achieving channel capacity is more difficult than designing good source codes.

13.1 THE PROMISE OF CODING

We begin this section with an example that shows how coding can help us achieve lower error probabilities in digital communications.

Example 13.1.1

In a digital communication system, the transmitter power is P and the rate of the source is R. The system employs an $M = 4$ PSK signaling (QPSK), where a pair of information bits are mapped into any of the four signals shown in the constellation depicted in Figure 13.1. We can readily see that $\mathcal{E}_b = \frac{P}{R}$, and the minimum Euclidean distance between any two signals in this constellation is given by

$$d_{\min}^2 = 4\mathcal{E}_b = 4\frac{P}{R}. \tag{13.1.1}$$

Now assume that instead of transmitting a QPSK signal (which is two dimensional), *three* orthonormal signals are employed to transmit the same *two* bits. For example, we can assume that the orthonormal signals are given by $\psi(t)$, $\psi(t - T)$, and $\psi(t - 2T)$,

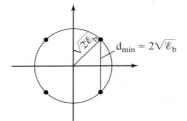

Figure 13.1 Signal constellation for a 4PSK scheme.

where $\psi(t)$ is equal to zero outside the interval $[0, T]$ and

$$\int_0^T \psi^2(t)\, dt = 1. \tag{13.1.2}$$

This is obviously a set of orthonormal signals with a dimensionality of 3. Using these orthonormal basis signals, we construct the following four signals:

$$s_1(t) = \sqrt{\mathscr{E}}\left(+\psi(t) + \psi(t-T) + \psi(t-2T)\right); \tag{13.1.3}$$

$$s_2(t) = \sqrt{\mathscr{E}}\left(+\psi(t) - \psi(t-T) - \psi(t-2T)\right); \tag{13.1.4}$$

$$s_3(t) = \sqrt{\mathscr{E}}\left(-\psi(t) - \psi(t-T) + \psi(t-2T)\right); \tag{13.1.5}$$

$$s_4(t) = \sqrt{\mathscr{E}}\left(-\psi(t) + \psi(t-T) - \psi(t-2T)\right), \tag{13.1.6}$$

or equivalently, in vector notation,

$$s_1 = \sqrt{\mathscr{E}}\,(+1, +1, +1); \tag{13.1.7}$$

$$s_2 = \sqrt{\mathscr{E}}\,(+1, -1, -1); \tag{13.1.8}$$

$$s_3 = \sqrt{\mathscr{E}}\,(-1, -1, +1); \tag{13.1.9}$$

$$s_4 = \sqrt{\mathscr{E}}\,(-1, +1, -1). \tag{13.1.10}$$

The corresponding constellation is shown in the three-dimensional space in Figure 13.2. Obviously, with this choice of codewords, each codeword differs from any other codeword at two components. Therefore, the Euclidean distance between any two signals is given by

$$d_{ij}^2 = |s_i - s_j|^2 = 8\mathscr{E} \quad \text{for } i \neq j. \tag{13.1.11}$$

The energy $\mathscr{E}$ is easily related to the energy per bit $\mathscr{E}_b$. Since two bits are transmitted per waveform,

$$2\mathscr{E}_b = 3\mathscr{E}; \tag{13.1.12}$$

hence,

$$\mathscr{E} = \frac{2}{3}\mathscr{E}_b = \frac{2}{3}\frac{P}{R}. \tag{13.1.13}$$

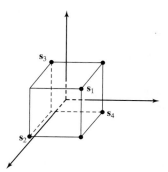

Figure 13.2   Codewords on the vertices of a cube.

Therefore, by substituting the result in Equation (13.1.13) into Equation (13.1.11), we obtain

$$d_{ij}^2 = \frac{16}{3}\frac{P}{R} \quad \text{for } i \neq j. \tag{13.1.14}$$

Comparing this with the minimum distance in the 4-PSK signal, we observe that the minimum-distance squared has increased by a factor of

$$\frac{d_{ij}^2}{d_{4\text{PSK}}^2} = \frac{\frac{16}{3}\frac{P}{R}}{4\frac{P}{R}} = \frac{4}{3}. \tag{13.1.15}$$

Because the error probability is a decreasing function of the minimum Euclidean distance, we have reduced the error probability by employing this new scheme. In fact, we can say that the resulting reduction in error probability is equivalent to the reduction in error probability achieved by an increase in power by a factor of $\frac{4}{3}$. This, in turn, is equivalent to 1.25 dB power gain. This power gain, of course, has not been obtained for free. We see that with this signaling scheme in a time duration of $\frac{2}{R}$, which is the time duration to transmit two bits, we have to transmit three signals. Therefore, the width of these signals is reduced by a factor of $\frac{2}{3}$, and the bandwidth required to transmit them is increased by a factor of $\frac{3}{2}$. A second problem with this scheme is that obviously it is more elaborate and requires a more complex decoding scheme. ■

The foregoing example basically describes what a coding scheme does. Coding results in a lower error probability (which is equivalent to a higher effective SNR) at the price of increasing the bandwidth[1] and the complexity of the system. In the preceding exercise, we have increased the number of dimensions from 2 to 3 (from QPSK to a three-dimensional signaling). This is equivalent to the following mapping:

$$(+1, +1) \rightarrow (+1, +1, +1)$$

$$(+1, -1) \rightarrow (+1, -1, -1)$$

$$(-1, -1) \rightarrow (-1, -1, +1)$$

$$(-1, +1) \rightarrow (-1, +1, -1).$$

[1]There exist coding-modulation schemes that increase the Euclidean distance between codewords, but do not increase the bandwidth.

As seen from this mapping, the role of coding has been to add a parity-check bit to the two information bits. The parity is added in such a way that the number of $+1$'s in the resulting codeword is always an odd number (or, equivalently, the number of -1's is an even number).

In a general signaling scheme with coded waveforms, sequences of length $k = RT$ of the source outputs are mapped into sequences of length n of the form

$$s_i = \sqrt{\mathscr{E}} \underbrace{(\pm 1, \pm 1, \cdots, \pm 1)}_{n}. \tag{13.1.16}$$

These points are located on the vertices of a hypercube of edge length $2\sqrt{\mathscr{E}}$. The ratio

$$R_c = \frac{k}{n} \tag{13.1.17}$$

is defined to be the *code rate*. There exist a total of 2^n vertices of an n-dimensional hypercube, of which we have to choose $M = 2^k$ as codewords. Obviously, we have to select these 2^k vertices so that they are as far apart from each other as possible. This makes the Euclidean distance between them large and, thus, reduces the error probability. In Example 13.1.1, we have $k = 2$ and $n = 3$. We chose $2^k = 4$ points from the possible $2^3 = 8$ vertices of the three-dimensional cube so that they were maximally apart. The rate of the resulting code is $R_c = 2/3$.

Assume that we have chosen 2^k vertices of the hypercube as the codewords and each codeword differs from another codeword in at least $d_{\min}^{\mathrm{H}}$ components. This parameter is called the *minimum Hamming distance of the code* and will be defined more precisely in Section 13.2. The relation between Euclidean distance and Hamming distance is very simple. If the sequences s_i and s_j differ in d_{ij}^{H} locations, then their Euclidean distance d_{ij}^{E} is related to d_{ij}^{H} by

$$\left(d_{ij}^{\mathrm{E}}\right)^2 = \sum_{\substack{1 \le l \le n \\ l:s_i^l \ne s_j^l}} \left(\pm 2\sqrt{\mathscr{E}}\right)^2 = 4 d_{ij}^{\mathrm{H}} \mathscr{E}. \tag{13.1.18}$$

This means that the minimum Euclidean distance can be expressed in terms of the minimum Hamming distance as

$$(d_{\min}^{\mathrm{E}})^2 = 4 d_{\min}^{\mathrm{H}} \mathscr{E}. \tag{13.1.19}$$

Now if we assume that s_i is transmitted and use the union bound (see Section 8.5.3), the probability of a codeword error is upperbounded as

$$P_{M_i} \le M Q \left(\sqrt{\frac{4 d_{\min}^{\mathrm{H}} \mathscr{E}}{2 N_0}} \right)$$

$$\le \frac{M}{2} e^{-d_{\min}^{\mathrm{H}} \mathscr{E}/N_0}, \tag{13.1.20}$$

where in the last step we have used the bound on the Q function introduced in Chapter 5. Noting that the energy content of each codeword is $n\mathscr{E}$ and has to be equal to PT, we have

$$\mathscr{E} = \frac{PT}{n} = \frac{RT}{n}\mathscr{E}_b = \frac{k}{n}\mathscr{E}_b = R_c\mathscr{E}_b, \tag{13.1.21}$$

where we have used the relation $\mathscr{E}_b = \frac{P}{R}$. Hence,

$$P_M \leq \frac{M}{2}e^{-d_{\min}^{\mathrm{H}}R_c\mathscr{E}_b/N_0}. \tag{13.1.22}$$

(The index i has been deleted because the bound is independent of i.) If no coding were employed—that is, if we used all the vertices of a k-dimensional hypercube rather than 2^k vertices of an n-dimensional hypercube—we would have the following union bound on the codeword error probability:

$$P_M \leq M Q\left(\sqrt{\frac{2\mathscr{E}_b}{N_0}}\right)$$

$$\leq \frac{M}{2}e^{-\mathscr{E}_b/N_0}. \tag{13.1.23}$$

Comparing the two bounds, we conclude that coding has resulted in a power gain equivalent to

$$G_{\text{coding}} = d_{\min}^{\mathrm{H}}R_c, \tag{13.1.24}$$

which is called the *asymptotic coding gain*, or simply, the *coding gain*. As seen here, the coding gain is a function of two main code parameters, the minimum Hamming distance and the code rate. Note that, in general, $R_c < 1$ and $d_{\min}^{\mathrm{H}} \geq 1$; therefore, the coding gain can be greater or less than 1.[2] It turns out that there exist many codes that can provide good coding gains. The relation defining the coding gain once again emphasizes that, for a given n and k, the best code is the code that can provide the highest minimum Hamming distance.

To study the bandwidth requirements of coding, we observe that when no coding is used, the width of the pulses employed to transmit one bit is given by

$$T_b = \frac{1}{R}. \tag{13.1.25}$$

After using coding, in the same time duration that k pulses were to be transmitted, we must now transmit n pulses. This means that the duration of each pulse is reduced by a factor of $\frac{k}{n} = R_c$. Therefore, the *bandwidth expansion ratio* is given by

$$B = \frac{W_{\text{coding}}}{W_{\text{no coding}}} = \frac{1}{R_c} = \frac{n}{k}. \tag{13.1.26}$$

---

[2] Although we can even have $d_{\min}^{\mathrm{H}} = 0$ in a very bad code design, we will ignore such cases.

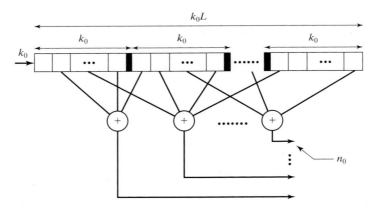

Figure 13.3   A convolutional encoder.

We can prove that in an AWGN channel, there exists a sequence of codes with parameters (n_i, k_i) with fixed rate ($\frac{k_i}{n_i} = R_c$ independent of i) satisfying

$$R_c < \frac{1}{2} \log \left(1 + \frac{P}{N_0 W} \right),$$ (13.1.27)

where $\frac{1}{2} \log \left(1 + \frac{P}{N_0 W} \right)$ is the capacity of the channel in bits per transmission,[3] for which the error probability goes to zero as n_i becomes larger and larger.

In this chapter we study two major types of codes, *block codes* and *convolutional codes*. Block codes are the codes that we have already described. In a block code, the information sequence is broken into blocks of length k and each block is mapped into channel input blocks of length n. This mapping is independent from the previous blocks, i.e., there is no memory from one block to another block. In convolutional codes, we use a shift register of length $k_0 L$, as shown in Figure 13.3. The information bits enter the shift register k_0 bits at a time; then n_0 bits that are linear combinations of various shift register bits, are transmitted over the channel. These n_0 bits depend not only on the recent k_0 bits that just entered the shift register, but also on the $(L - 1)k_0$ previous contents of the shift register that constitute its *state*. The quantity

$$m = L$$ (13.1.28)

is defined as the *constraint length* of the convolutional code, and the number of states of the convolutional code is equal to $2^{(L-1)k_0}$. The rate of a convolutional code is defined as

$$R_c = \frac{k_0}{n_0}.$$ (13.1.29)

---

[3]Recall that the capacity of this channel in bits per second is $W \log \left(1 + \frac{P}{N_0 W} \right)$.

The main difference between block codes and convolutional codes is the existence of memory in convolutional codes.

13.2 LINEAR BLOCK CODES

An (n, k) block code is a collection of $M = 2^k$ binary sequences, each of length n, called *codewords*. A code $\mathscr{C}$ consists of M codewords c_i for $1 \le i \le 2^k$; that is,

$$\mathscr{C} = \{c_1, c_2, \cdots, c_M\},$$

where each c_i is a sequence of length n with components equal to 0 or 1. The collection of the codewords is called the *codebook* or, simply, *the code*.

Definition 13.2.1.   A block code is *linear* if the modulo-2 sum of any two codewords is also a codeword. This requires that if c_i and c_j are codewords, then $c_i \oplus c_j$ must also be a codeword, where $\oplus$ denotes componentwise modulo-2 addition.   ∎

With this definition, we can readily see that a linear block code is a k-dimensional subspace of an n-dimensional space. It is also obvious that the all-zero sequence $\mathbf{0}$ is a codeword of any linear block code, since it can be written as $c_i \oplus c_i$ for any codeword c_i. We further assume that if the information sequence x_1 (of length k) is mapped into the codeword c_1 (of length n) and the information sequence x_2 is mapped into c_2, then $x_1 \oplus x_2$ is mapped into $c_1 \oplus c_2$.

Example 13.2.1

A $(5, 2)$ code is defined by the codewords

$$\mathscr{C}_1 = \{00000, 10100, 01111, 11011\},$$

where the mapping of information bits to codewords is as follows:

$$00 \rightarrow 00000$$
$$01 \rightarrow 01111$$
$$10 \rightarrow 10100$$
$$11 \rightarrow 11011.$$

It is easy to verify that this code is linear. However, the code defined by

$$\mathscr{C}_2 = \{00000, 11100, 01111, 11011\}$$

is not linear because the sum of the second and the third codewords is not a codeword.

∎

Now we will define some of the basic parameters that characterize a code.

Definition 13.2.2.   The *Hamming distance* between two codewords c_i and c_j is the number of components at which the two codewords differ, i.e., the number of components where one codeword is 1 and the other one is 0. The Hamming distance between two codewords is denoted by $d(c_i, c_j)$.[4]   ∎

Definition 13.2.3.   The *Hamming weight*, or simply the *weight* of a codeword c_i, is the number of 1's in the codeword and is denoted by $w(c_i)$.   ∎

Definition 13.2.4.   The minimum distance of a code is the minimum Hamming distance between any two different codewords, i.e.,

$$d_{\min} = \min_{\substack{c_i, c_j \\ i \neq j}} d(c_i, c_j). \tag{13.2.1}$$

∎

Definition 13.2.5.   The minimum weight of a code is the minimum of the weights of the codewords except the all-zero codeword:

$$w_{\min} = \min_{c_i \neq 0} w(c_i). \tag{13.2.2}$$

∎

Theorem 13.2.1.   In any linear code, $d_{\min} = w_{\min}$.

Proof. If c is a codeword, then $w(c) = d(c, 0)$. Also, if c_i and c_j are codewords, so is $c = c_i \oplus c_j$ and, moreover, $d(c_i, c_j) = w(c)$. This implies that in a linear code corresponding to any weight of a codeword, there exists a Hamming distance between two codewords, and corresponding to any Hamming distance, there exists a weight of a codeword. In particular, it shows that $d_{\min} = w_{\min}$.

∎

Generator and Parity Check Matrices.   In an (n, k) linear block code, let the codeword corresponding to the information sequences[5] $e_1 = (1000 \ldots 0)$, $e_2 = (0100 \ldots 0)$, $e_3 = (0010 \ldots 0)$, $\ldots$, $e_k = (0000 \ldots 1)$ be denoted by $g_1, g_2, g_3, \ldots, g_k$, respectively, where each of the g_i sequences is a binary sequence of length n. Now, any information sequence $x = (x_1, x_2, x_3, \ldots, x_k)$ can be written as

$$x = \sum_{i=1}^{k} x_i e_i; \tag{13.2.3}$$

therefore, the corresponding codeword will be

$$c = \sum_{i=1}^{k} x_i g_i. \tag{13.2.4}$$

---

[4]Hamming distance is denoted by d and Euclidean distance is denoted by d^{E}.
[5]In this chapter, all vectors are represented as row vectors.

If we define the *generator matrix* for this code as

$$
G \stackrel{\text{def}}{=}
\begin{bmatrix}
g_1 \\
g_2 \\
\vdots \\
g_k
\end{bmatrix}
=
\begin{bmatrix}
g_{11} & g_{12} & \cdots & g_{1n} \\
g_{21} & g_{22} & \cdots & g_{2n} \\
\vdots & \vdots & \ddots & \vdots \\
g_{k1} & g_{k2} & \cdots & g_{kn}
\end{bmatrix},
\tag{13.2.5}
$$

then, from Equation (13.2.3), we can write

$$
c = xG,
\tag{13.2.6}
$$

where x is a $1 \times k$ row vector and G is the $k \times n$ generator matrix. This shows that any linear combination of the rows of the generator matrix is a codeword. The generator matrix for any linear block code is a $k \times n$ matrix of rank k (because the dimension of the subspace is k by definition). The generator matrix of a code completely describes the code. When the generator matrix is given, the structure of an encoder is quite simple.

Example 13.2.2

Determine the generator matrix for the first code given in Example 13.2.1.

Solution   We have to find the code words corresponding to the information sequences (10) and (01). These are (10100) and (01111), respectively. Therefore,

$$
G = \begin{bmatrix} 10100 \\ 01111 \end{bmatrix}.
\tag{13.2.7}
$$

For the information sequence (x_1, x_2), the codeword is given by

$$
(c_1, c_2, c_3, c_4, c_5) = (x_1, x_2)G
\tag{13.2.8}
$$

or

$$
c_1 = x_1
$$
$$
c_2 = x_2
$$
$$
c_3 = x_1 \oplus x_2
$$
$$
c_4 = x_2
$$
$$
c_5 = x_2.
$$

■

The preceding code has the property that the codeword corresponding to each information sequence starts with a replica of the information sequence itself followed by some extra bits. Such a code is called a *systematic code* and the extra bits following the information sequence in the codeword are called the *parity check bits*. A necessary and sufficient condition for a code to be systematic is that the generator matrix be in the form

$$
G = \begin{bmatrix} I_k & \big| & P \end{bmatrix},
\tag{13.2.9}
$$

where I_k denotes a $k \times k$ identity matrix and P is a $k \times (n - k)$ binary matrix. In a systematic code, we have

$$c_i = \begin{cases} x_i, & 1 \leq i \leq k \\ \sum_{j=1}^{k} p_{ji} x_j, & k+1 \leq i \leq n, \end{cases} \tag{13.2.10}$$

where all summations are modulo 2.

By definition, a linear block code $\mathscr{C}$ is a k-dimensional linear subspace of the n-dimensional space. From linear algebra, we know that if we take all sequences of length n that are orthogonal to all vectors of this k-dimensional linear subspace, the result will be an $(n - k)$-dimensional linear subspace called the *orthogonal complement* of the k-dimensional subspace. This $(n - k)$-dimensional subspace naturally defines an $(n, n - k)$ linear code, which is known as the *dual* of the original (n, k) code $\mathscr{C}$. The dual code is denoted by $\mathscr{C}^\top$. Obviously, the codewords of the original code $\mathscr{C}$ and the dual code $\mathscr{C}^\top$ are orthogonal to each other. In particular, if we denote the generator matrix of the dual code by H, which is an $(n - k) \times n$ matrix, then any codeword of the original code is orthogonal to all rows of H, i.e.,

$$cH^t = 0 \qquad \text{for all } c \in \mathscr{C}. \tag{13.2.11}$$

The matrix H, which is the generator matrix of the dual code $\mathscr{C}^\top$, is called the *parity check matrix* of the original code $\mathscr{C}$. Since all rows of the generator matrix are codewords, we conclude that

$$GH^t = 0. \tag{13.2.12}$$

In the special case of a systematic code, where

$$G = \begin{bmatrix} I_k & | & P \end{bmatrix}, \tag{13.2.13}$$

the parity check matrix has the following form:

$$H = \begin{bmatrix} P^t & | & I_{n-k} \end{bmatrix}. \tag{13.2.14}$$

Example 13.2.3

Find the parity check matrix for the code given in Example 13.2.1.

Solution   Here,

$$G = \begin{bmatrix} 10100 \\ 01111 \end{bmatrix}$$

$$I = \begin{bmatrix} 10 \\ 01 \end{bmatrix}$$

$$P = \begin{bmatrix} 100 \\ 111 \end{bmatrix}.$$

We conclude that

$$\boldsymbol{P}^t = \begin{bmatrix} 11 \\ 01 \\ 01 \end{bmatrix};$$

therefore,

$$\boldsymbol{H} = \begin{bmatrix} 1 & 1 & 1 & 0 & 0 \\ 0 & 1 & 0 & 1 & 0 \\ 0 & 1 & 0 & 0 & 1 \end{bmatrix}.$$

The parity-check equations, given by $\boldsymbol{c}\boldsymbol{H}^t = \boldsymbol{0}$, are

$$c_1 \oplus c_2 \oplus c_3 = 0$$

$$c_2 \oplus c_4 = 0$$

$$c_2 \oplus c_5 = 0.$$

∎

Hamming Codes.   Hamming codes are a class of linear block codes with $n = 2^m - 1$, $k = 2^m - m - 1$ and, regardless of the value of m, have a minimum distance of $d_{\min} = 3$ for some integer $m \geq 2$. This means that for $m = 3$, we have a $(7, 4)$ Hamming code, and for $m = 4$, we have a $(15, 11)$ Hamming code. As we will see later, with this minimum distance, these codes are capable of providing error correction capabilities for single errors. The parity check matrix for these codes has a very simple structure. It consists of all binary sequences of length m except the all-zero sequence. The rate of these codes is given by

$$R_c = \frac{2^m - m - 1}{2^m - 1}, \tag{13.2.15}$$

which is close to 1 for large values of m. Therefore, Hamming codes are high rate codes with a relatively small minimum distance ($d_{\min} = 3$). We will see later that the minimum distance of a code is closely related to its error correcting capabilities. Therefore, Hamming codes have limited error correcting capability.

Example 13.2.4

Find the parity check matrix and the generator matrix of a $(7, 4)$ Hamming code in the systematic form.

Solution   In this case, $m = 3$; therefore, $\boldsymbol{H}$ consists of all binary sequences of length 3 except the all-zero sequence. We generate the parity check matrix in the systematic form as

$$\boldsymbol{H} = \begin{bmatrix} 1 & 0 & 1 & 1 & 1 & 0 & 0 \\ 1 & 1 & 0 & 1 & 0 & 1 & 0 \\ 0 & 1 & 1 & 1 & 0 & 0 & 1 \end{bmatrix} = \begin{bmatrix} \boldsymbol{P}^t \,|\, \boldsymbol{I}_k \end{bmatrix},$$

and the generator matrix is

$$
G = \begin{bmatrix} 1 & 0 & 0 & 0 & 1 & 1 & 0 \\ 0 & 1 & 0 & 0 & 0 & 1 & 1 \\ 0 & 0 & 1 & 0 & 1 & 0 & 1 \\ 0 & 0 & 0 & 1 & 1 & 1 & 1 \end{bmatrix} = \begin{bmatrix} I_k \mid P \end{bmatrix}.
$$

If the information sequence $x = (x_1, x_2, x_3, x_4)$ is encoded by this code, the resulting codeword will be $c = xG$, whose components are given by

$$c_1 = x_1$$

$$c_2 = x_2$$

$$c_3 = x_3$$

$$c_4 = x_4$$

$$c_5 = x_1 \oplus x_3 \oplus x_4$$

$$c_6 = x_1 \oplus x_2 \oplus x_4$$

$$c_7 = x_2 \oplus x_3 \oplus x_4.$$

The parity check equations are obtained from $cH^t = 0$ and are given by

$$c_1 \oplus c_3 \oplus c_4 \oplus c_5 = 0$$

$$c_1 \oplus c_2 \oplus c_4 \oplus c_6 = 0$$

$$c_2 \oplus c_3 \oplus c_4 \oplus c_7 = 0.$$

∎

13.2.1 Decoding and Performance of Linear Block Codes

The purpose of using coding in communication systems is to increase the Euclidean distance between the transmitted signals and, hence, to reduce the error probability at a given transmitted power. This was shown by an example in the previous section. Referring to Figure 13.2, we see that this goal is achieved by choosing the codewords to be as far apart on the vertices of the cube as possible. This means that a good measure for comparing the performance of two codes is the Hamming distance between codewords. Keeping track of all distances between any two codewords is difficult, and in many cases, impossible. Therefore, the comparison between various codes is usually done based on the minimum distance of the code, which, for linear codes, is equal to the minimum weight. It follows that for a given n and k, a code with a larger d_{min} (or w_{min}) performs better than a code with a smaller minimum distance.

Soft-Decision Decoding.   In Chapters 8 and 9, we have seen that the optimum signal-detection scheme on an additive white Gaussian noise channel is detection based on minimizing the Euclidean distance between the received signal and the transmitted signal. This means that after receiving the output of the channel and passing it through the matched filters, we choose one of the message signals that is closest to the received signal in the Euclidean distance sense. In using coded waveforms, the situation is the same. Assuming we are employing binary PSK for transmission of the coded message, a codeword $c_i = (c_{i1}, c_{i2}, \dots, c_{in})$ is mapped into the sequence $s_i(t) = \sum_{k=1}^{n} \psi_{ik}(t - (k-1)T)$, where

$$\psi_{ik}(t) = \begin{cases} \psi(t), & c_{ik} = 1 \\ -\psi(t), & c_{ik} = 0 \end{cases} \tag{13.2.16}$$

and $\psi(t)$ is a signal of duration T and energy $\mathscr{E}$, which is equal to zero outside the interval $[0, T]$. Now the Euclidean distance between two arbitrary signal waveforms is given by Equation (13.1.18)

$$\left(d_{ij}^{\mathrm{E}}\right)^2 = 4d_{ij}^{\mathrm{H}}\mathscr{E}. \tag{13.2.17}$$

This gives a simple relation between the Euclidean and the Hamming distance when a binary PSK signaling scheme (or any antipodal signaling scheme) is employed. Now, using the general relation

$$P_2 = Q\left(\frac{d^{\mathrm{E}}}{\sqrt{2N_0}}\right), \tag{13.2.18}$$

we obtain

$$p(\text{codeword } c_j \text{ received}|\text{codeword } c_i \text{ sent}) = Q\left(\sqrt{\frac{2d_{ij}\mathscr{E}}{N_0}}\right). \tag{13.2.19}$$

Since $d_{ij} \geq d_{\min}$, and since $Q(x)$ is a decreasing function of x, we conclude that

$$p(\text{codeword } c_j \text{ received}|\text{codeword } c_i \text{ sent}) \leq Q\left(\sqrt{\frac{2d_{\min}\mathscr{E}}{N_0}}\right). \tag{13.2.20}$$

Now using the union bound (see Section 8.5.3), we obtain

$$p(\text{error}|\text{codeword } c_i \text{ sent}) \leq (M-1)Q\left(\sqrt{\frac{2d_{\min}\mathscr{E}}{N_0}}\right) \tag{13.2.21}$$

and assuming equiprobable messages, we finally conclude that

$$P_M \leq (M-1)Q\left(\sqrt{\frac{2d_{\min}\mathscr{E}}{N_0}}\right). \tag{13.2.22}$$

There exists a simple relationship between the energy per code element, denoted as $\mathcal{E}$, and the energy per information bit, denoted as $\mathcal{E}_b$. Since a codeword has n elements, the total transmitted energy is $n\mathcal{E}$. But a codeword with n elements carries k information bits. Hence, $k\mathcal{E}_b = n\mathcal{E}$, and it follows that

$$\mathcal{E} = \frac{k}{n}\mathcal{E}_b = R_c\mathcal{E}_b. \tag{13.2.23}$$

From Equations (13.2.22) and (13.2.23), we conclude that

$$P_M \leq (M-1)Q\left(\sqrt{\frac{2d_{\min}R_c\mathcal{E}_b}{N_0}}\right). \tag{13.2.24}$$

Equations (13.2.22) and (13.2.24) are bounds on the codeword error probability of a coded communication system when optimal demodulation is employed. By optimal demodulation, we mean passing the received signal $r(t)$ through a bank of matched filters to obtain the received vector $\mathbf{y}$, and then finding the closest point in the constellation to $\mathbf{y}$ in the Euclidean distance sense. This type of decoding that involves finding the minimum Euclidean distance is called *soft-decision decoding*, and requires real number computation.

Example 13.2.5

Compare the performance of an uncoded data transmission system with the performance of a coded system using the $(7, 4)$ Hamming code given in Example 13.2.4 when applied to the transmission of a binary source with the rate $R = 10^4$ bits/sec. The channel is assumed to be an additive white Gaussian noise channel, the received power is 1 μW and the noise power spectral density is $\frac{N_0}{2} = 10^{-11}$. The modulation scheme for the elements of any codeword is binary PSK.

Solution

1. If no coding is employed, we have

$$P_2 = Q\left(\sqrt{\frac{2\mathcal{E}_b}{N_0}}\right) = Q\left(\sqrt{\frac{2P}{RN_0}}\right). \tag{13.2.25}$$

But $\frac{2P}{RN_0} = \frac{10^{-6}}{10^4 \times 10^{-11}} = 10$; therefore,

$$P_2 = Q(\sqrt{10}) = Q(3.16) \approx 7.86 \times 10^{-4}. \tag{13.2.26}$$

The error probability for four bits will be

$$P_{\text{Error in 4 bits}} = 1 - (1 - p_b)^4 \approx 3.1 \times 10^{-3}. \tag{13.2.27}$$

2. If coding is employed, we have $d_{\min} = 3$ and

$$\frac{\mathcal{E}}{N_0} = R_c\frac{\mathcal{E}_b}{N_0} = R_c\frac{P}{RN_0} = \frac{4}{7} \times 5 = \frac{20}{7}.$$

Therefore, the *message* error probability is given by

$$P_M \leq (M-1)Q\left(\sqrt{\frac{2d_{\min}\mathscr{E}}{N_0}}\right)$$

$$= 15Q\left(\sqrt{3 \times \frac{40}{7}}\right)$$

$$= 15Q(4.14) \approx 2.6 \times 10^{-4}.$$

We see that using this simple code decreases the error probability by a factor of 12. Of course, the price that has been paid is an increase in the bandwidth required for the transmission of the messages. This bandwidth expansion ratio is given by

$$\frac{W_{\text{coded}}}{W_{\text{uncoded}}} = \frac{1}{R_c} = \frac{7}{4} = 1.75.$$ ∎

Hard-Decision Decoding.   A simpler and more frequently used decoding scheme is to make hard binary decisions on the components of the received vector y, and then to find the codeword that is closest to it in the Hamming distance sense. The next example clarifies the distinction between soft and hard decisions.

Example 13.2.6

A (3,1) code consists of the two codewords 000 and 111. The codewords are transmitted using binary PSK modulation with $\mathscr{E} = 1$. The received vector (the sampled outputs of the matched filters) is $y = (.5, .5, -3)$. If soft decision is employed, we have to compare the Euclidean distance between y and the two constellation points $(1, 1, 1)$ and $(-1, -1, -1)$ and choose the smaller one. We have $(d^E(y, (1, 1, 1)))^2 = .5^2 + .5^2 + 4^2 = 16.5$ and $(d^E(y, (-1, -1, -1)))^2 = 1.5^2 + 1.5^2 + (-2)^2 = 8.5$; therefore, a soft-decision decoder would decode y as $(-1, -1, -1)$ or equivalently $(0, 0, 0)$. However, if hard-decision decoding is employed, y is first component-wise detected as 1 or 0. This requires a comparison of the components of y with the zero threshold. The resulting vector $\hat{c}$ is therefore $\hat{c} = (1, 1, 0)$. Now, we have to compare $\hat{c}$ with the $(1, 1, 1)$ and $(0, 0, 0)$ and find the closer one in the Hamming distance sense. The result is, of course, $(1, 1, 1)$. As seen in this example, the results of soft-decision decoding and hard-decision decoding can be quite different. Of course, soft-decision decoding is the optimal detection method and achieves a lower probability of error.   ∎

There are three basic steps involved in hard-decision decoding. First, we perform demodulation by passing the received $r(t)$ through the matched filters and sampling the output to obtain the y vector. This is an n-dimensional vector whose components are real numbers. Second, we compare the components of y with the threshold (usually zero) and quantize each component to one of the two levels (usually 0 and 1) to obtain the estimate $\hat{c}$ of the transmitted codeword, which is an n-dimensional vector with binary components. Finally, we perform decoding by finding the codeword that is closest to $\hat{c}$ in the *Hamming distance sense*. In this section, we present a systematic approach to hard-decision decoding.

$$c_1 \qquad c_2 \qquad c_3 \qquad c_M$$
$$e_1 \qquad e_1 \oplus c_2 \qquad e_1 \oplus c_3 \qquad e_1 \oplus c_M$$
$$e_2 \qquad e_2 \oplus c_2 \qquad e_2 \oplus c_3 \qquad e_2 \oplus c_M$$
$$\vdots \qquad\qquad \vdots \qquad\qquad \vdots \qquad\qquad \vdots$$
$$e_2^{(n-k)}-1 \quad e_2^{(n-k)}-1 \oplus c_2 \quad e_2^{(n-k)}-1 \oplus c_3 \quad e_2^{(n-k)}-1 \oplus c_M$$

Figure 13.4 The standard array.

First, we will define the notion of a *standard array*. Let the code words of the code in question be denoted by $c_1, c_2, \ldots, c_M$, where each of the code words is of length n and $M = 2^k$, and let c_1 denote the all-zero code word. A standard array is a $2^{n-k} \times 2^k$ array whose elements are binary sequences of length n and is generated by writing all the code words in a row starting with the all-zero code word. This constitutes the first row of the standard array. To write the second row, we look among all binary sequences of length n that are not in the first row of the array (i.e., are not codewords). Choose one of these code words that has the minimum weight and call it e_1. Write it under[6] c_1 and write $e_1 \oplus c_i$ under c_i for $2 \le i \le M$. The third row of the array is completed in a similar way. From the binary n-tuples that *have not been used in the first two rows,* we choose one with minimum weight and call it e_2. Then, the elements of the third row become $c_i \oplus e_2$. This process is continued until no binary n-tuples remain to start a new row. Figure 13.4 shows the standard array generated as explained. Each row of the standard array is called a *coset* and the first element of each coset (e_l, in general) is called the *coset leader.*

The standard array has several important properties.

Theorem 13.2.2. All elements of the standard array are distinct.

Proof. Assume two elements of the standard array are equal. This can happen in two ways.

1. The two equal elements belong to the same coset. In this case, we have $e_l \oplus c_i = e_l \oplus c_j$, from which we conclude $c_i = c_j$, which is impossible.
2. The two equal elements belong to two different cosets. Here, we have $e_e \oplus c_i = e_k \oplus c_j$ for $l \ne k$, which means $e_l = e_k \oplus (c_i \oplus c_j)$. By linearity of the code, $c_i \oplus c_j$ is also a code word; let us call it c_m. Therefore, $e_l = e_k \oplus c_m$; hence, e_l and e_k belong to the same coset, which is impossible since, by assumption, $k \ne l$.

∎

From this theorem, we conclude that the standard array contains exactly 2^{n-k} rows.

Theorem 13.2.3. If z_1 and z_2 are elements of the same coset, we have $z_1 H^t = z_2 H^t$.

---

[6]Note that $c_1 \oplus e_1 = e_1$, since $c_1 = (0, 0, \ldots, 0)$.

Problem.   It is enough to note that since z_1 and z_2 are in the same coset, $z_1 = e_l \oplus c_i$ and $z_2 = e_l \oplus c_j$ for some $1 \leq i, j \leq M$. Therefore,

$$z_1 H^t = (e_l \oplus c_i) H^t = e_l H^t + 0 = (e_l \oplus c_j) H^t = z_2 H^t.$$

■

From this theorem, we conclude that each coset of the standard array can be uniquely identified by the product $e_l H^t$, where e_l denotes the coset leader. In general, for any binary sequence z of length n, we define the *syndrome s* as

$$s = zH^t. \tag{13.2.28}$$

If $z = e_l \oplus c_i$, i.e., z belongs to the $(l+1)$st coset, then, obviously, $s = e_l H^t$. The syndrome is a binary sequence of length $n - k$, and corresponding to each coset, there exists a unique syndrome. Obviously, the syndrome corresponding to the first coset, which consists of the code words, is $s = 0$.

Example 13.2.7

Construct the standard array for the $(5, 2)$ code with the code words 00000, 10100, 01111, 11011. Also determine the syndromes corresponding to each coset.

Solution   The generator matrix of the code is

$$G = \begin{bmatrix} 1 & 0 & 1 & 0 & 0 \\ 0 & 1 & 1 & 1 & 1 \end{bmatrix}$$

and the parity check matrix corresponding to G is

$$H = \begin{bmatrix} 1 & 1 & 1 & 0 & 0 \\ 0 & 1 & 0 & 1 & 0 \\ 0 & 1 & 0 & 0 & 1 \end{bmatrix}$$

Using this construction, we obtain the standard array

00000	01111	10100	11011	syndrome = 000
10000	11111	00100	01011	syndrome = 100
01000	00111	11100	10011	syndrome = 111
00010	01101	10110	11001	syndrome = 010
00001	01110	10101	11010	syndrome = 001
11000	10111	01100	00011	syndrome = 011
10010	11101	00110	01001	syndrome = 110
10001	11110	00101	01010	syndrome = 101.

■

Assuming that the received vector y has been component-wise compared with a threshold and the resulting binary vector is $\hat{c}$, we must find the code word which is at minimum Hamming distance from $\hat{c}$. First, we find in which coset $\hat{c}$ is located. To do this, we find the syndrome of $\hat{c}$ by calculating $s = \hat{c} H^t$. After finding s, we refer to the standard array and find the coset corresponding to s. Assume that the coset leader corresponding to this coset is e_l. Because $\hat{c}$ belongs to this coset, it is of the form $e_l \oplus c_i$ for some $1 \le i \le M$. The Hamming distance of $\hat{c}$ from any code word c_j is, therefore,

$$d(\hat{c}, c_j) = w(\hat{c} \oplus c_j) = w(e_l \oplus c_i \oplus c_j). \tag{13.2.29}$$

Because the code is linear, $c_i \oplus c_j = c_k$ for some $1 \le k \le M$. This means that,

$$d(\hat{c}, c_j) = w(c_k \oplus e_l), \tag{13.2.30}$$

but $c_k \oplus e_l$ belongs to the same coset that $\hat{c}$ belongs to. Therefore, to minimize $d(\hat{c}, c_j)$, we have to find the minimum weight element in the coset to which $\hat{c}$ belongs. By construction of the standard array, this element is the coset leader, i.e., we choose $c_k = 0$; therefore, $c_j = c_i$. This means that $\hat{c}$ is decoded into c_i by finding

$$c_i = \hat{c} \oplus e_l. \tag{13.2.31}$$

Therefore, the procedure for hard-decision decoding can be summarized as follows:

1. Find y, the vector representation of the received signal.
2. Compare each component of y to the optimal threshold (usually 0) and make a binary decision on it to obtain the binary vector $\hat{c}$.
3. Find $s = \hat{c} H^t$, the syndrome of $\hat{c}$.
4. Find the coset corresponding to s by using the standard array.
5. Find the coset leader e and decode $\hat{c}$ as $c = \hat{c} \oplus e$.

In this decoding scheme, the difference between the vector $\hat{c}$ and the decoded vector c is e, so the binary n-tuple e is frequently referred to as the *error pattern*. This means that *the coset leaders constitute the set of all correctable error patterns*.

   To obtain error bounds in hard-decision decoding, we note that, since a decision is made on each individual bit, the error probability for each bit for antipodal signaling is

$$P_2 = Q\left(\sqrt{\frac{2\mathscr{E}_b}{N_0}}\right), \tag{13.2.32}$$

and for orthogonal signaling is

$$P_2 = Q\left(\sqrt{\frac{\mathscr{E}_b}{N_0}}\right). \tag{13.2.33}$$

The channel between the input code word c and the output of the hard limiter $\hat{c}$ is a binary-input binary-output channel that can be modeled by a binary-symmetric channel with crossover probability P_2. Because the code is linear, the distance between any two code words c_i and c_j is equal to the distance between the all-zero code word 0 and the code word $c_i \oplus c_j = c_k$. Thus, without loss of generality, we can assume that the all-zero code word is transmitted. If $\mathbf{0}$ is transmitted, the error probability, by the union bound, cannot exceed $(M - 1)$ times the probability of decoding the code word that is closest to $\mathbf{0}$ in the Hamming distance sense. For this code word, denoted by $\mathbf{c}$, which is at distance $d_{\min}$ from $\mathbf{0}$, we have

$$
P(\mathbf{c} \mid \mathbf{0} \text{ sent}) \leq
\begin{cases}
\sum_{i=\frac{d_{\min}+1}{2}}^{d_{\min}} \binom{d_{\min}}{i} P_2^i (1 - P_2)^{d_{\min}-i}, & d_{\min} \text{ odd} \\[2ex]
\sum_{i=\frac{d_{\min}}{2}+1}^{d_{\min}} \binom{d_{\min}}{i} P_2^i (1 - P_2)^{d_{\min}-i} \\[1ex]
\quad + \frac{1}{2} \binom{d_{\min}}{\frac{d_{\min}}{2}} P_2^{\frac{d_{\min}}{2}} (1 - P_2)^{\frac{d_{\min}}{2}}, & d_{\min} \text{ even,}
\end{cases}
$$

or, in general,

$$
P(\mathbf{c} \mid \mathbf{0} \text{ sent}) \leq \sum_{i=[\frac{d_{\min}+1}{2}]}^{d_{\min}} \binom{d_{\min}}{i} P_2^i (1 - P_2)^{d_{\min}-i}. \tag{13.2.34}
$$

Therefore,

$$
P_M \leq (M - 1) \sum_{i=[\frac{d_{\min}+1}{2}]}^{d_{\min}} \binom{d_{\min}}{i} P_2^i (1 - P_2)^{d_{\min}-i}. \tag{13.2.35}
$$

This gives an upper bound on the error probability of a linear block code using hard-decision decoding. A simpler error bound for hard-decision decoding is given by the inequality

$$
P_M \leq (M - 1) \left[P_2 (1 - P_2) \right]^{\frac{d_{\min}}{2}}. \tag{13.2.36}
$$

Equation (13.2.36) is derived in Problem 13.21.

As seen in both soft-decision and hard-decision decoding, $d_{\min}$ plays a major role in bounding the error probability. This means that, for a given (n, k), it is desirable to have codes with large $d_{\min}$.

We can show that the difference between the performance of soft- and hard-decision decoding is roughly 2 dB for an additive white Gaussian noise channel. That is, the error probability of a soft-decision decoding scheme is comparable to the error probability of a hard-decision scheme whose power is 2-dB higher than the soft-decision scheme. We can also show that if, instead of quantization of each component of $\mathbf{y}$ to two levels, an 8-level quantizer (three bits/component) is employed, the performance difference with soft decision (infinite precision) reduces to 0.1 dB. This multilevel quantization scheme, which is a compromise between soft (infinite precision) and hard decision, is also referred to as soft-decision decoding in the literature.

Example 13.2.8

If hard-decision decoding is employed in Example 13.2.5, how will the results change?

Solution   In this example, $P_2 = Q\left(\sqrt{\frac{40}{7}}\right) = Q(2.39) = 0.0084$ and $d_{min} = 3$. Therefore,

$$P_{16} \leq \binom{7}{2} P_2^2 (1 - P_2)^5 + \binom{7}{3} P_2^3 (1 - P_2)^4 + \cdots + P_2^7$$

$$\approx 21 P_2^2 \approx 1.5 \times 10^{-3}.$$

Thus, coding has decreased the error probability by a factor of 2 (it was 12 in the soft-decision case). If, instead of the exact error probability, we used the bound in Equation (13.2.36), we would find that

$$P_{16} \leq (2^k - 1)\,[P_2(1 - P_2)]^{3/2} = 0.0114 = 11.4 \times 10^{-3}. \tag{13.2.37}$$

■

13.2.2 Some Important Linear Block Codes

In general, hard-decision decoding of linear block codes using the standard array and syndrome decoding is too complex to be used in practice for long block lengths. In order to make decoding of linear block codes practical, special classes of linear block codes for which low complexity decoding algorithms exist have been designed and decoding algorithms for these special classes have been devised.

One of the most widely used subclasses of linear block codes is *cyclic codes*. Cyclic codes are linear block codes with the additional property that a cyclic shift of any codeword is itself a codeword. It turns out that cyclic codes have a rich and nice structure if expressed in terms of polynomials. Hard-decision decoding of cyclic codes is simpler than hard-decision decoding for the general class of linear block codes and can be implemented using simple shift-register circuits.

A subclass of cyclic codes called *BCH codes* (Bose, Chaudhuri, and Hocquenghem) are particularly interesting and have been used extensively. BCH codes can be designed for correction of any given number of errors. In many books on coding theory, there are extensive tables for the design of BCH codes that can correct a certain number of errors. There exists an elegant and fast decoding algorithm, called the *Berlekamp–Massey* decoding algorithm, for hard-decision decoding of the BCH codes.

Reed–Solomon codes are a subset of *nonbinary* BCH codes; therefore, they belong to the family of cyclic codes. Unlike other codes that we have studied in this chapter, Reed–Solomon codes are nonbinary codes, i.e., in a codeword $c = (c_1, c_2, \ldots, c_n)$, the elements c_i are not binary 0 or 1 but each c_i is itself a sequence of length k of 0's and 1's. Therefore, the size of the alphabet in Reed–Solomon codes is $q = 2^k$, and the code is a q-ary code. Reed–Solomon codes are particularly useful in correction of bursts of errors as studied in Section 13.2.4. Fading channels and storage channels are examples of channels in which errors tend to occur in bursts. Reed–Solomon codes are widely used in data, music, and video storage on CDs and DVDs.

Reed–Solomon codes can also be concatenated with a binary code to provide higher levels of error protection. The binary code used in concatenation with the Reed–Solomon codes could be either a block code or a convolutional code. The binary encoder and decoder are located right before the modulator and after the demodulator, respectively, and are called the *inner* encoder–decoder pair. We will discuss concatenated codes in Section 13.4.

The detailed structure of cyclic, BCH, and Reed–Solomon codes requires considerable knowledge of finite-field theory and is treated in many standard textbooks on coding theory. The interested reader can refer to the books cited at the end of this chapter.

13.2.3 Error Detection versus Error Correction

Let $\mathscr{C}$ be a linear block code with minimum distance $d_{\min}$. Then if c is transmitted and hard-decision decoding is employed, any codeword will be decoded correctly if the received $\hat{c}$ is closer to c than any other codeword. This situation is shown in Figure 13.5. As shown, around each codeword there is a "Hamming sphere" of radius e_c, where e_c denotes the number of correctable errors. As long as these spheres are disjoint, the code is capable of correcting e_c errors. A little thinking shows that the condition for nonoverlapping spheres is

$$\begin{cases} 2e_c + 1 = d_{\min} & d_{\min} \text{ odd} \\ 2e_c + 2 = d_{\min} & d_{\min} \text{ even,} \end{cases}$$

or

$$e_c = \begin{cases} \frac{d_{\min}-1}{2} & d_{\min} \text{ odd} \\ \frac{d_{\min}-2}{2} & d_{\min} \text{ even,} \end{cases} \tag{13.2.38}$$

which can be summarized as

$$e_c = \left[\frac{d_{\min} - 1}{2} \right]. \tag{13.2.39}$$

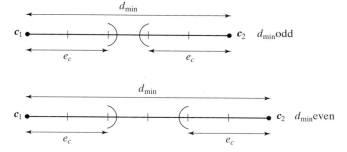

Figure 13.5 Relation between e_c and $d_{\min}$.

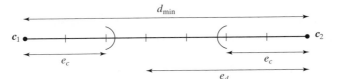

Figure 13.6 Relation between e_c, e_d, and $d_{\min}$.

In some cases, we are interested in decoding procedures that can *detect* errors rather than correct them. For example, in a communication system where a feedback link is available from the receiver to the transmitter, it might be desirable to detect whether an error has occurred and if so, to ask the transmitter via the feedback channel to retransmit the message. If we denote the error detection capability of a code by e_d, then obviously, in the absence of error correction, $e_d = d_{\min} - 1$ because if $d_{\min} - 1$ or fewer errors occur, the transmitted codeword will be converted to a noncodeword sequence and an error is detected. If both error correction and error detection are desirable, then there is naturally a trade-off between these two conditions. Figure 13.6 demonstrates this. From this picture, we see that

$$e_c + e_d = d_{\min} - 1, \tag{13.2.40}$$

with the extra condition $e_c \leq e_d$.

13.2.4 Burst-Error-Correcting Codes

Most of the linear block codes are designed for correcting *random errors*, i.e., errors that occur independently from the location of other channel errors. Certain channel models, including the additive white Gaussian noise channel, can be well modeled as channels with random errors. In some other physical channels, however, the assumption of independently generated errors is not a valid assumption. One such example is a fading channel, as discussed in Section 11.1. In such a channel, if the channel is in deep fade, a large number of errors occur in sequence, i.e., the errors have a *bursty nature*. Obviously, in this channel, the probability of error at a certain location or time depends on whether or not its adjacent bits are received correctly. Another example of a channel with bursty errors is a compact disc. Any physical damage to a compact disc, such as a scratch, damages a sequence of bits; therefore, the errors tend to occur in *bursts*. Of course, any random error-correcting code can be used to correct bursts of errors as long as the number of errors is less than half of the minimum distance of the code. But the knowledge of the bursty nature of errors makes it possible to design more efficient coding schemes. Two particular codes that are designed to be used for burst-error correction are Fire codes and Burton codes. Refer to Lin and Costello (2005) for a discussion of these codes.

An effective method for the correction of error bursts is to *interleave* the coded data such that the location of errors looks random and is distributed over many code words rather than a few code words. In this way, the number of errors that occur in each block is low and can be corrected by using a random error-correcting code.

Encoder
Interleaver
Modulator
Channel
Demodulator
Deinterleaver
Decoder

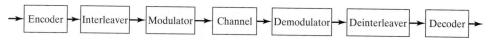

Figure 13.7 Block diagram of a system that employs interleaving/deinterleaving for burst-error correction.

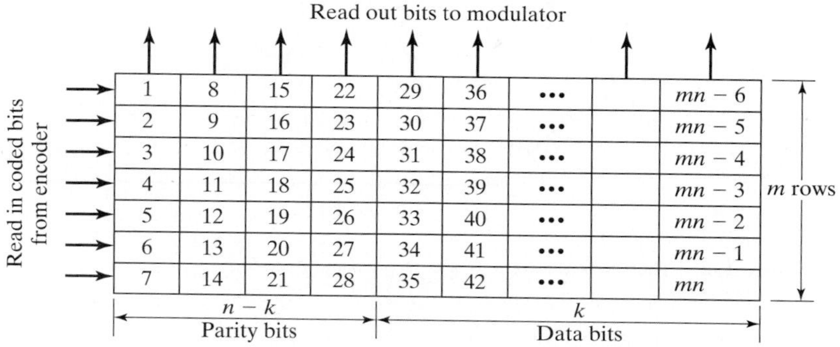
Read out bits to modulator
Read in coded bits from encoder
1 8 15 22 29 36 ••• mn − 6
2 9 16 23 30 37 ••• mn − 5
3 10 17 24 31 38 ••• mn − 4
4 11 18 25 32 39 ••• mn − 3
5 12 19 26 33 40 ••• mn − 2
6 13 20 27 34 41 ••• mn − 1
7 14 21 28 35 42 ••• mn
m rows
n − k
Parity bits
k
Data bits

Figure 13.8 A block interleaver for coded data.

At the receiver, a *deinterleaver* is employed to undo the effect of the interleaver. A block diagram of a coding system employing interleaving/deinterleaving is shown in Figure 13.7.

An *interleaver of depth m* reads m code words of length n each and arranges them in a block with m rows and n columns. Then, this block is read by column and the output is sent to the digital modulator. At the receiver, the output of the detector is supplied to the deinterleaver, which generates the same $m \times n$ block structure. It then reads by row and sends the output to the channel decoder. This is shown in Figure 13.8.

Assume that $m = 8$ and the code in use is a $(15, 11)$ Hamming code capable of correcting one error/code word. Then, the block generated by the interleaver is an 8×15 block containing 120 binary symbols. Obviously, any burst of errors of length 8 or less will result in a maximum of one error/code word; therefore, it can be corrected. If interleaving/deinterleaving was not employed, an error burst of length 8 could possibly result in erroneous detection in 2 code words (or error in detection of up to 22 information bits).

13.3 CONVOLUTIONAL CODES

Convolutional codes are different from block codes because of the existence of memory in the encoding scheme. In block codes, each block of k input bits is mapped into a block of length n of output bits by a rule defined by the code (for example, by $\boldsymbol{G}$) and regardless of the previous inputs to the encoder. The rate of

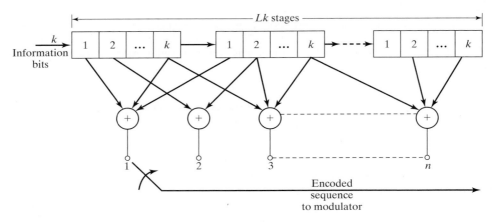

Figure 13.9   The block diagram of a convolutional encoder.

such a code is given by

$$R_c = \frac{k}{n}. \tag{13.3.1}$$

In convolutional codes, each block of k bits is again mapped into a block of n bits to be transmitted over the channel, but these n bits are not only determined by the present k information bits, but also by the previous information bits. This dependence on the previous information bits causes the encoder to be a finite state machine.

To be more specific, the block diagram of a convolutional encoder is given in Figure 13.9. The convolutional encoder consists of a shift register with kL stages where L is called the *constraint length* of the code. At each instant of time, k-information bits enter the shift register and the contents of the last k stages of the shift register are dropped. After the k bits have entered the shift register, n-linear combinations of the contents of the shift register, as shown in the figure, are computed and used to generate the encoded waveform. From this coding procedure, it is obvious that the n-encoder outputs not only depend on the most recent k bits that have entered the encoder, but also on the $(L-1)k$ contents of the first $(L-1)k$ stages of the shift register before the k bits arrived. Therefore, the shift register is a finite state machine with $2^{(L-1)k}$ states. For each k-input bits, we have n-output bits, so the rate of this code is simply

$$R_c = \frac{k}{n}. \tag{13.3.2}$$

Example 13.3.1

A convolutional encoder is shown in Figure 13.10. In this encoder, $k = 1$, $n = 2$, and $L = 3$. Therefore, the rate of the code is $\frac{1}{2}$ and the number of states is $2^{(L-1)k} = 4$. One way to describe such a code (other than drawing the encoder) is to specify how

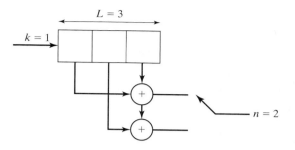

Figure 13.10   A rate $\frac{1}{2}$ convolutional encoder.

the two output bits of the encoder depend on the contents of the shift register. This is usually done by specifying n vectors $\boldsymbol{g}_1, \boldsymbol{g}_2, \ldots, \boldsymbol{g}_n$, known as *generator sequences* of the convolutional code. The ith, $1 \leq i \leq 2^{kL}$, component of $\boldsymbol{g}_j$, $1 \leq j \leq n$, is 1 if the ith stage of the shift register is connected to the combiner corresponding to the jth bit in the output and 0 otherwise. In this example, the generator sequences are given by

$$\boldsymbol{g}_1 = [1 \quad 0 \quad 1]$$
$$\boldsymbol{g}_2 = [1 \quad 1 \quad 1].$$

∎

13.3.1 Basic Properties of Convolutional Codes

Because a convolutional encoder has finite memory, it can easily be represented by a *state-transition diagram*. In the state-transition diagram, each state of the convolutional encoder is represented by a box, and transitions between states are denoted by lines connecting these boxes. On each line, both the input(s) causing that transition and the corresponding output(s) are specified. The number of lines emerging from each state is, therefore, equal to the number of possible inputs to the encoder at that state, which is equal to 2^k. The number of lines merging at each state is equal to the number of states from which a transition is possible to this state. This is equal to the number of possible combinations of bits that leave the encoder as the k bits enter the encoder. This, again, is equal to 2^k. Figure 13.11 shows the state transition diagram for the convolutional code of Figure 13.10.

A second, and more popular method, to describe convolutional codes is to specify their *trellis diagram*. The trellis diagram is a way to show the transition between various states as the time evolves. The trellis diagram is obtained by specifying all states on a vertical axis and repeating this vertical axis along the time axis. Then, each transition from a state to another state is denoted by a line connecting the two states on two adjacent vertical axes. In a sense, the trellis diagram is nothing but a repetition of the state-transition diagram along the time axes. As was the case with the state-transition diagram, we again have 2^k branches of the trellis leaving each state and 2^k branches merging at each state. In the case where $k = 1$, it is common to denote the branch corresponding to a 0 input by a bold line and the branch corresponding to a 1 input to the encoder by a dashed line. Figure 13.12 shows the trellis diagram for the code described by the encoder of Figure 13.10.

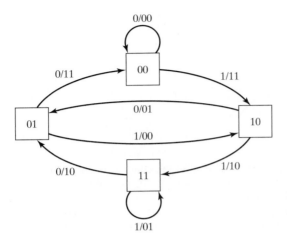

Figure 13.11 State transition diagram for the encoder of Figure 13.10. Encoder input and outputs are shown on transition lines separated by "/"

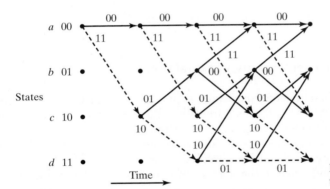

Figure 13.12 Trellis diagram for the encoder of Figure 13.10.

Encoding. The encoding procedure in a convolutional code is very simple. We assume that the encoder, before the first information bit enters it, is loaded with zeros (the all-zero state). The information bits enter the encoder k bits at a time and the corresponding n-output bits are transmitted over the channel. This procedure continues until the last group of k bits is loaded into the encoder and the corresponding n-output bits are sent over the channel. We will assume, for simplicity, that after the last set of k bits, another set of $k(L-1)$ zeros enters the encoder and the corresponding n outputs are transmitted over the channel. This returns the encoder to the all-zero state and makes it ready for the next transmission.

Example 13.3.2

In the convolutional code shown in Figure 13.10, what is the encoded sequence corresponding to the information sequence $x = (1101011)$?

Solution It is enough to note that the encoder is in state 0 before transmission, and after transmission of the last information bit two 0 bits are transmitted. This means that the transmitted sequence is $x_1 = (110101100)$. Using this transmission sequence, we have the code word $c = (111010000100101011)$.

■

The Transfer Function.   For every convolutional code, the transfer function gives information about the various paths through the trellis that start from the all-zero state and return to this state for the first time. According to the coding convention previously described, any code word of a convolutional encoder corresponds to a path through the trellis that starts from the all-zero state and returns to the all-zero state. As we will see in Section 13.3.4, the transfer function of a convolutional code plays a major role in bounding the error probability of the code. To obtain the transfer function of a convolutional code, we split the all-zero state into two states, one denoting the starting state and one denoting the first return to the all-zero state. All the other states are denoted as in-between states. Corresponding to each branch connecting two states, a function of the form $D^\alpha N^\beta J$ is defined where α denotes the number of ones in the output bit sequence for that branch and β is the number of ones in the corresponding input sequence for that branch. The *transfer function* of the convolutional code is, then, the transfer function of the flow graph between the starting all-zero state and the final all-zero state, and will be a function of the tree parameters D, N, and J and denoted by $T(D, N, J)$. Each element of $T(D, N, J)$ corresponds to a path through the trellis starting from the all-zero state and ending at the all-zero state. The exponent of J indicates the number of branches spanned by that path, the exponent of D shows the number of ones in the code word corresponding to that path (or equivalently the Hamming weight of the code word), and finally, the exponent of N indicates the number of ones in the input information sequence. Since $T(D, N, J)$ indicates the properties of all paths through the trellis starting from the all-zero path and returning to it *for the first time,* then, in deriving it, any self loop at the all-zero state is ignored. To obtain the transfer function of the convolutional code, we can use all rules that can be used to obtain the transfer function of a flow graph.

Example 13.3.3

Find the transfer function of the convolutional code of Figure 13.10.

Solution   Figure 13.13 shows the diagram used to find the transfer function of this code. The code has a total of four states denoted by the contents of the first two stages

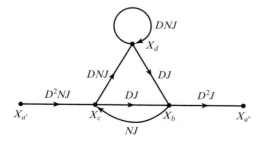

Figure 13.13   Flow graph for finding the transfer function.

of the shift register. We denote these states by the following letters:

$$00 \rightarrow a$$
$$01 \rightarrow b$$
$$10 \rightarrow c$$
$$11 \rightarrow d.$$

As seen in the figure, state a is split into states a' and a'' denoting the starting and returning state. Using the flow graph relations, we can write

$$X_c = X_{a'} D^2 N J + N J X_b$$
$$X_b = D J X_d + D J X_c$$
$$X_d = D N J X_c + D N J X_d$$
$$X_{a''} = D^2 J X_b.$$

Eliminating X_b, X_c, and X_d results in

$$T(D, N, J) = \frac{X_{a''}}{X_{a'}} = \frac{D^5 N J^3}{1 - D N J - D N J^2}. \tag{13.3.3}$$

Now, expanding $T(D, N, J)$ in a polynomial form, we obtain

$$T(D, N, J) = D^5 N J^3 + D^6 N^2 J^4 + D^6 N^2 J^5 + D^7 N^3 J^5 + \cdots. \tag{13.3.4}$$

The term $D^5 N J^3$ indicates that there exists a path through the trellis starting from the all-zero state and returning to the all-zero state for the first time, which spans three branches, corresponding to an input-information sequence containing one 1 (and, therefore, two 0's), and the code word for this path has Hamming weight equal to 5. This path is indicated with bold lines in the Figure 13.14. This path is somewhat similar to the minimum-weight code word in block codes. In fact, this path corresponds to the code word that is at "minimum distance" from the all-zero code word. This minimum distance, which is equal to the minimum power of D in the expansion of $T(D, N, J)$,

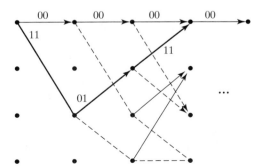

Figure 13.14 The path corresponding to $D^5 N J^3$ in code represented in Figure 13.10.

is called the *free distance of the code* and denoted by d_{free}. The free distance of this code is equal to 5. The general form of the transfer function is, therefore,

$$T(D, N, J) = \sum_{d=d_{\text{free}}}^{\infty} a_d D^d N^{f(d)} J^{g(d)}. \tag{13.3.5}$$

■

A shorter form of transfer function, which only provides information about the weight of the code words, can be obtained from $T(D, N, J)$ by setting $N = J = 1$. This shorter form will be denoted by

$$T_1(D) = \sum_{d=d_{\text{free}}}^{\infty} a_d D^d, \tag{13.3.6}$$

and will later be used in deriving bounds on the error probabilities of the convolutional codes.

Example 13.3.4

For the code of Figure 13.10, we have

$$T_1(D) = \left.\frac{D^5 N J^3}{1 - DNJ - DNJ^2}\right|_{N=J=1}$$

$$= \frac{D^5}{1 - 2D}$$

$$= D^5 + 2D^6 + 4D^7 + \cdots$$

$$= \sum_{i=0}^{\infty} 2^i D^{5+i}.$$

■

Catastrophic Convolutional Codes.   A convolutional code maps a (usually long) sequence of input information bits into a codeword to be transmitted over the channel. The purpose of coding is to provide higher levels of protection against channel noise. Obviously, a code that maps information sequences that are far apart into codewords that are not far apart is not a good code since these two codewords can be mistaken rather easily and the result would be a large number of bit errors in the information stream. A limiting case of this undesirable property happens when two different information sequences in infinitely many positions are mapped into codewords that differ only in a finite number of positions. In such a case, since the codewords differ in a finite number of bits, there always exists the probability that they will be erroneously decoded. This in turn results in an infinite number of errors in detecting the input information sequence. Codes that exhibit this property are called *catastrophic codes* and should be avoided.

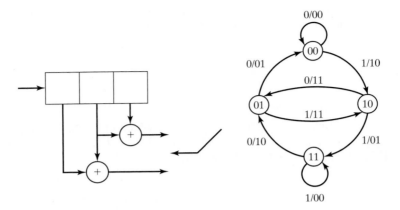

Figure 13.15 Encoder and the state-transition diagram for a catastrophic code.

As an example of a catastrophic code consider the (2, 1) code described by

$$g_1 = [1 \ 1 \ 0]$$

$$g_2 = [0 \ 1 \ 1].$$

The encoder and the state-transition diagram for this code are given in Figure 13.15. As seen in this diagram, there is a self loop in state "11" that corresponds to a "1" input to the encoder and the corresponding output consists of all zeros. Therefore, if an input information stream consists of all ones, the corresponding output will be

$$c = (10010000 \ldots 0001001).$$

If we compare this codeword with the codeword corresponding to the all-zero information sequence

$$c_0 = (0000 \ldots 000),$$

we observe that although the two information sequences are different in a large number of positions, the corresponding output sequences are quite close (the Hamming distance being only 4); therefore, they can be mistaken very easily. The existence of such a self loop (corresponding to k inputs which are not all zeros and for which the n output bits are all zeros), shows that a code is catastrophic and should therefore be avoided.

13.3.2 Maximum Likelihood Decoding of Convolutional Codes — The Viterbi Algorithm

In our discussion of various decoding schemes for block codes, we saw that there exists the possibility of soft- and hard- decision decoding. In soft-decision decoding, y, the vector denoting the outputs of the matched filters, is compared with the various signal points in the constellation of the coded modulation system and the one closest to it in Euclidean distance is chosen. In hard-decision decoding, y is first turned into a binary sequence $\hat{c}$ by making decisions on individual components of y; then the codeword, which is closest to $\hat{c}$ in Hamming distance, is chosen. We see that in both approaches, a fundamental task is *to find a path through the trellis that is at minimum distance from a given sequence*. This fundamental problem arises in many areas of communications and other disciplines of electrical engineering. Particularly, the same problem is encountered in maximum-likelihood sequence estimation when transmitting over bandlimited channels with intersymbol interference (Section 9.3.4), speech recognition, and some pattern classification schemes, etc. All these problems are essentially the same and can be titled as *optimal trellis searching algorithms*. The well-known Viterbi algorithm provides a satisfactory solution to all these problems.

In hard-decision decoding of convolutional codes, we choose a path through the trellis whose codeword, denoted by c, is at minimum Hamming distance from the quantized received sequence $\hat{c}$. In hard-decision decoding, the channel is binary memoryless (the fact that the channel is memoryless follows from the fact that the channel noise is assumed to be white). Because the desired path starts from the all-zero state and returns back to the all-zero state, we assume that this path spans a total of m branches, and since each branch corresponds to n bits of the encoder output, the total number of bits in c (and also in $\hat{c}$) is mn. We denote the sequence of bits corresponding to the ith branch by c_i and $\hat{c}_i$, respectively, where $1 \leq i \leq m$ and each c_i and $\hat{c}_i$ is of length n. The Hamming distance between c and $\hat{c}$ is therefore

$$d(c, \hat{c}) = \sum_{i=1}^{m} d(c_i, \hat{c}_i). \tag{13.3.7}$$

In soft-decision decoding, we have a similar situation, but with three differences:

1. Instead of $\hat{c}$, we are dealing directly with the vector y, which is the vector output of the optimal (matched filter type or correlators type) digital demodulator.

2. Instead of the binary sequence c, we are dealing with the corresponding sequence c' with

$$c'_{ij} = \begin{cases} \sqrt{\mathscr{E}}, & c_{ij} = 1 \\ -\sqrt{\mathscr{E}}, & c_{ij} = 0 \end{cases} \quad \text{for } 1 \leq i \leq m \text{ and } 1 \leq j \leq n.$$

3. Instead of Hamming distance, we are using Euclidean distance. This is a consequence of the fact that the channel under study is an additive white Gaussian noise channel.

Thus, we have

$$d_E^2(c', y) = \sum_{i=1}^{m} d_E^2(c'_i, y_i). \tag{13.3.8}$$

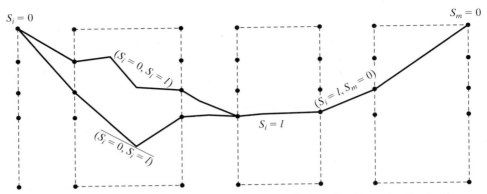

Figure 13.16   Comparison of the optimal path ($S_1 = 0$, $S_i = l$, $S_m = 0$) with a suboptimal path consisting of the concatenation of $\overline{(S_1 = 0, S_i = l)}$ and ($S_i = l$, $S_m = 0$).

From Equations (13.3.7) and (13.3.8), we see that the generic form of the problem we have to solve is as follows: Given a sequence a, find a path through the trellis, start at the all-zero state $S_1 = 0$ and end at the all-zero state $S_m = 0$, such that some distance measure between a and a sequence b corresponding to that path is minimized.[7]. The important fact that makes this problem easy to solve is that the distance between a and b in both cases of interest can be written as the sum of the distances corresponding to individual branches of the path. This is easily observed from Equations (13.3.7) and (13.3.8).

Assume that the general problem is formulated as minimizing a metric μ in the form

$$\mu(a, b) = \sum_{i=1}^{m} \mu(a_i, b_i),$$

where, for soft- and hard-decision decoding, μ represents the Euclidean distance and the Hamming distance, respectively. First, we observe that if the path ($S_1 = 0$, $S_i = l$, $S_m = 0$), $1 \leq i \leq m$, is the optimal path starting from the all-zero state, terminating at the all-zero state, and passing through state l at time i, then the metric contribution of *part of this path* ($S_1 = 0$, $S_i = l$) is lower than the metric contribution of any other path connecting $S_1 = 0$ to $S_i = l$ and denoted by $\overline{(S_1 = 0, S_i = l)}$. Otherwise, the path consisting of the concatenation of $\overline{(S_1 = 0, S_i = l)}$ and ($S_i = l$, $S_m = 0$) would be the optimal path. This is shown in Figure 13.16.

This shows that at each state l at time i, *only one path* connecting the all-zero state to state l should be saved. This path is the path corresponding to the lowest metric connecting $S_1 = 0$ to $S_i = l$, i.e., the path at the minimum Hamming, or Euclidean, distance from the received sequence. The path of minimum metric connecting $S_1 = 0$ to $S_i = l$ is called the *survivor path*, or simply the *survivor*, at $S_i = l$.

----

[7]The problem can also be formulated as a maximization problem. For example, instead of minimizing the Euclidean distance, we could maximize the correlation.

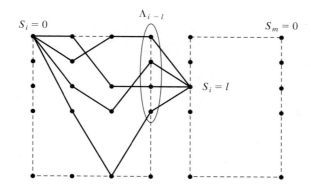

Figure 13.17   Finding a new survivor from old survivors.

Let Λ_{i-1} denote the set of states at time $i - 1$ that are connected with a branch to $S_i = l$. If the path from $S_1 = 0$ to $S_i = l$ is a survivor, then this path must be the concatenation of a survivor path at $S_{i-1} = \lambda$, for some $\lambda \in \Lambda_{i-1}$, and the branch connecting $S_{i-1} = \lambda$ to $S_i = l$. Therefore, in order to find the survivor path at state $S_i = l$, it is sufficient to have the survivors (and their metrics) for all $S_{i-1} = \lambda$, $\lambda \in \Lambda_{i-1}$, then append them to the branches connecting the elements of Λ_{i-1} to $S_i = l$, find the metric of the resulting paths from $S_1 = 0$ to $S_i = l$, and pick the one with the minimum metric. This will be the new survivor path at $S_i = l$. This process is started at $S_1 = 0$ and is finished at $S_m = 0$. The final survivor at $S_m = 0$ represents the optimal path and the best (maximum likelihood) match to the received sequence. This process is shown in Figure 13.17.

At each step, the new survivor metric is

$$\mu(S_1 = 0, S_i = l) = \min_{\lambda \in \Lambda_{i-1}} \{\mu(S_1 = 0, S_{i-1} = \lambda) + \mu(S_{i-1} = \lambda, S_i = l)\}. \quad (13.3.9)$$

Having found the survivor metric, the new survivor at $S_i = l$ is the path ($S_1 = 0, S_{i-1} = \lambda, S_i = l$) for the λ that minimizes the survivor metric in Equation (13.3.9).

This procedure can be summarized in the *Viterbi algorithm*:

1. Parse the received sequence into m subsequences, each of length n.
2. Draw a trellis of depth m for the code under study. For the last $L - 1$ stages of the trellis, draw only paths corresponding to the all-zero input sequences. (This is done because we know that the input sequence has been padded with $k(L - 1)$ zeros.)
3. Set $i = 1$ and set the metric of the initial all-zero state equal to zero.
4. Find the distance from the ith subsequence of the received sequence to all branches connecting the ith stage states to the $(i + 1)^{\text{st}}$ stage states of the trellis.
5. Add these distances to the metrics of the ith stage states to obtain the metric candidates for the $(i+1)^{\text{st}}$ stage states. For each state of the $(i+1)^{\text{st}}$ stage, there are 2^k candidate metrics, each corresponding to one branch ending at that state.

6. For each state at the $(i+1)^{\text{st}}$ stage, choose the minimum of the metric candidates; then, label the branch corresponding to this minimum value as the *survivor* and assign the minimum of the metric candidates as the metrics of the $(i+1)^{\text{st}}$ stage states.

7. If $i=m$, go to Step 8. Otherwise, increase i by 1 and go to Step 4.

8. Starting with the all-zero state at the final stage, go back through the trellis along the survivors to reach the initial all-zero state. This path is the optimal path and the input bit sequence corresponding to it is the maximum likelihood decoded information sequence. To obtain the most likely input bit sequence, remove the last $k(L-1)$ zeros from this sequence.

As seen from this algorithm, the decoding delay and the amount of memory required for decoding a long information sequence is unacceptable. The decoding cannot be started until the whole sequence (which, in the case of convolutional codes, can be very long) is received, and all surviving paths have to be stored. In practice, a suboptimal solution that does not cause these problems is desirable. One such approach, called *path memory truncation*, is that the decoder at each stage only searches δ stages back in the trellis instead of searching back to the start of the trellis. With this approach, at the $(\delta+1)^{\text{th}}$ stage, the decoder makes a decision on the input bits corresponding to the first stage of the trellis (the first k bits) and future received bits do not change this decision. This means that the decoding delay will be $k\delta$ bits, and it is required only to keep the surviving paths corresponding to the last δ stages. Computer simulations have shown that if $\delta \approx 5L$, the degradation in performance due to path memory truncation is negligible.

Example 13.3.5

Assume that, in hard-decision decoding, the quantized received sequence is

$$\hat{c} = (01101111010001).$$

The convolutional code is given in Figure 13.10. Find the maximum likelihood information sequence and the number of errors.

Solution The code is a $(2, 1)$ code with $L = 3$. The length of the received sequence $\hat{c}$ is 14. This means that $m = 7$ and we have to draw a trellis of depth 7. Also note that because the input information sequence is padded with $k(L-1) = 2$ zeros, for the final two stages of the trellis, we will only draw the branches corresponding to all-zero inputs. This also means that the actual length of the input sequence is 5, which, after padding with two zeros, has increased to 7. The trellis diagram for this case is shown in Figure 13.18. The parsed received sequence $\hat{c}$ is also shown in this figure. In drawing the trellis in the last two stages, we have considered only the zero inputs to the encoder. (In the final two stages, there are no dashed lines corresponding to 1 inputs.) Now the metric of the initial all-zero state is set to zero and the metrics of the next stage are computed. In this step, there is only one branch entering each state; therefore, there is no comparison, and the metrics (which are the Hamming distances between that part of the received sequence and the branches of the trellis) are added to the metric of the previous state. In the next stage, there is

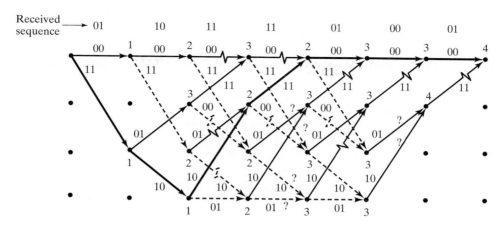

Figure 13.18   The trellis diagram for the Viterbi decoding of the sequence (01101111010001).

no comparison either. In the third stage, we actually have two branches entering each state. This means that a comparison has to be made, and survivors are to be chosen. From the two branches that enter each state, the one that corresponds to the least total accumulated metric remains as a survivor and the other branches are deleted (marked with $\approx$ on the graph). If, at any stage, two paths result in the same metric, each one of them can be a survivor. Such cases have been marked by a "?" in the trellis diagram. The procedure is continued to the final all-zero state of the trellis. Starting from that state, we move along the surviving paths to the initial all-zero state. This path, which is denoted by a heavy path through the trellis, is the optimal path. The input bit sequence corresponding to this path is 1100000 (where the last two zeros are not information bits, but are added to return the encoder to the all-zero state). Therefore, the information sequence is 11000. The corresponding codeword for the selected path is 11101011000000, which is at Hamming distance 4 from the received sequence. No other path through the trellis is at a Hamming distance less than 4 from the received $\hat{c}$.

■

For soft-decision decoding, a similar procedure is followed with squared Euclidean distance substituted for Hamming distance.

13.3.3 Other Decoding Algorithms for Convolutional Codes

The Viterbi algorithm provides maximum-likelihood decoding for convolutional codes. However, as we have already seen, the complexity of the algorithm is proportional to the number of states in the trellis diagram. This means that the complexity of the algorithm increases exponentially with the constraint length of the convolutional code. Therefore, the Viterbi algorithm can be applied only to codes with low constraint lengths. For higher constraint-length codes, other suboptimal decoding schemes have been proposed. These include the sequential decoding of Wozencraft (1957), the Fano algorithm (1963), the stack algorithm [Zigangirov (1966) and

Jelinek (1969)], the feedback-decoding algorithm [Heller (1975)], and majority logic decoding [Massey (1963)].

13.3.4 Bounds on the Error Probability of Convolutional Codes

Finding bounds on the error performance of convolutional codes is different from the method used to find error bounds for block codes because here we are dealing with sequences of very large length and, since the free distance of these codes is usually small, some errors will eventually occur. The number of errors is a random variable that depends both on the channel characteristics (SNR in soft-decision decoding and crossover probability in hard-decision decoding) and the length of the input sequence. The longer the input sequence, the higher the probability of making errors. Therefore, it makes sense to normalize the number of bit errors to the length of the input sequence. A measure that is usually adopted for comparing the performance of convolutional codes is the expected number of bits received in error/input bit.

To find a bound on the average number of bits in error for each input bit, we first derive a bound on the average number of bits in error for each input sequence of length k. To determine this, assume that the all-zero sequence is transmitted[8] and, up to stage l in the decoding, there has been no error. Now k-information bits enter the encoder and result in moving to the next stage in the trellis. We are interested in finding a bound on the expected number of errors that can occur due to this input block of length k. Because we are assuming that there has been no error up to stage l, then the all-zero path through the trellis has the minimum metric up to this stage. Now, moving to the next stage, which is stage $(l + 1)$th, it is possible that another path through the trellis will have a metric less than the all-zero path and, therefore, cause errors. If this happens, we must have a path through the trellis that merges with the all-zero path, for the first time, at the $(l + 1)$st stage and has a metric less than the all-zero path. Such an event is called the *first-error event* and the corresponding probability is called *the first-error-event probability*. This situation is depicted in Figure 13.19.

Our first step would be bounding the first-error-event probability. Let $P_2(d)$ denote the probability that a path through the trellis, which is at Hamming distance d from the all-zero path, is the survivor at the $(l + 1)$st stage. Denoting the number of paths of weight d by a_d, we can bound the first-error-event probability by

$$P_e \leq \sum_{d=d_{\text{free}}}^{\infty} a_d P_2(d), \tag{13.3.10}$$

where, on the right-hand side, we have included all paths through the trellis that merge with the all-zero path at the $(l + 1)$st stage. The value of $P_2(d)$ depends on whether soft- or hard-decision decoding is employed.

---

[8]Because of the linearity of convolutional codes we can, without loss of generality, make this assumption.

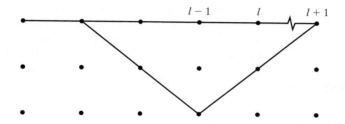

Figure 13.19 The path corresponding to the first-error event.

For soft-decision decoding, if antipodal signaling (binary PSK) is used, we have

$$P_2(d) = Q\left(\frac{d^E}{\sqrt{2N_0}}\right)$$

$$= Q\left(\sqrt{\frac{2\mathcal{E}d}{N_0}}\right)$$

$$= Q\left(\sqrt{2R_c d\frac{\mathcal{E}_b}{N_0}}\right); \tag{13.3.11}$$

therefore,

$$P_e \leq \sum_{d=d_{\text{free}}}^{\infty} a_d Q\left(\sqrt{2R_c d\frac{\mathcal{E}_b}{N_0}}\right). \tag{13.3.12}$$

Using the upper bound on the Q function, we have

$$Q\left(\sqrt{2R_c d\frac{\mathcal{E}_b}{N_0}}\right) \leq \frac{1}{2}e^{-R_c d\mathcal{E}_b/N_0}. \tag{13.3.13}$$

Now, noting that

$$e^{-R_c d\mathcal{E}_b/N_0} = D^d\big|_{D=e^{-R_c \mathcal{E}_b/N_0}}, \tag{13.3.14}$$

we finally obtain

$$P_e \leq \frac{1}{2}\sum_{d=d_{\text{free}}}^{\infty} a_d D^d\bigg|_{D=e^{-R_c \mathcal{E}_b/N_0}} = \frac{1}{2}T_1(D)\big|_{D=e^{-R_c \mathcal{E}_b/N_0}}. \tag{13.3.15}$$

This is a bound on the first-error-event probability. To find a bound on the average number of bits in error for k-input bits, $\overline{P}_b(k)$, we note that each path through the trellis causes a certain number of input bits to be decoded erroneously. For a general $D^d N^{f(d)} J^{g(d)}$ in the expansion of $T(D, N, J)$, there are $f(d)$ nonzero-input bits. This means that the average number of input bits in error can be obtained by multiplying the probability of choosing each path by the total number of input errors that would result if that path were chosen. Hence, the average number of bits in error, in the soft-decision case, can be bounded by

$$\overline{P}_b(k) \leq \sum_{d=d_{\text{free}}}^{\infty} a_d f(d) Q\left(\sqrt{2 R_c d \frac{\mathscr{E}_b}{N_0}}\right)$$

$$\leq \frac{1}{2} \sum_{d=d_{\text{free}}}^{\infty} a_d f(d) e^{-R_c d \mathscr{E}_b/N_0}. \tag{13.3.16}$$

If we define

$$T_2(D, N) = T(D, N, J)|_{J=1}$$

$$= \sum_{d=d_{\text{free}}}^{\infty} a_d D^d N^{f(d)}, \tag{13.3.17}$$

we have

$$\frac{\partial T_2(D, N)}{\partial N} = \sum_{d=d_{\text{free}}}^{\infty} a_d f(d) D^d N^{f(d)-1}. \tag{13.3.18}$$

Therefore, using Equations (13.3.16) and (13.3.18), we obtain

$$\overline{P}_b(k) \leq \frac{1}{2} \left. \frac{\partial T_2(D, N)}{\partial N}\right|_{N=1, D=e^{-R_c \frac{\mathscr{E}_b}{N_0}}}. \tag{13.3.19}$$

To obtain the average number of bits in error for each input bit, we have to divide this bound by k. Thus, the final result is

$$\overline{P}_b = \frac{1}{2k} \left. \frac{\partial T_2(D, N)}{\partial N}\right|_{N=1, D=e^{-R_c \frac{\mathscr{E}_b}{N_0}}}. \tag{13.3.20}$$

For high SNRs, the first term corresponding to the minimum distance is the dominant term, and we have the approximation

$$\overline{P}_b \approx \frac{1}{2k} a_{d_{\min}} f(d_{\min}) e^{-R_c d_{\min} \mathscr{E}_b/N_0}. \tag{13.3.21}$$

For hard-decision decoding, the basic procedure follows this derivation. The only difference is the bound on $P_2(d)$. It can be shown that (see Problem 13.21) $P_2(d)$ can be bounded by

$$P_2(d) \leq \left[4p(1-p)\right]^{d/2}, \tag{13.3.22}$$

where p is the crossover probability of the binary symmetric channel. Using this result, it is straightforward to show that in hard-decision decoding, the probability of error is upperbounded as

$$\overline{P}_b \leq \frac{1}{k} \left. \frac{\partial T_2(D, N)}{\partial N} \right|_{N=1, D=\sqrt{4p(1-p)}}. \tag{13.3.23}$$

A comparison of hard-decision decoding and soft-decision decoding for convolutional codes shows that here, as in the case for linear block codes, soft-decision decoding outperforms hard-decision decoding by a margin of roughly 2 dB in additive white Gaussian noise channels.

Convolutional Codes with Good Distance Properties.   From this analysis, it is obvious that d_{free} plays a major role in the performance of convolutional codes. For a given n and k, the free distance of a convolutional code depends on the constraint length of the code. Searching for convolutional codes with good distance properties has been extensively carried out in the literature. Tables 13.1 and 13.2 summarize the result of computer simulations carried out for rate $\frac{1}{2}$ and rate $\frac{1}{3}$ convolutional codes. In these tables, for each constraint length, the convolutional code that achieves the highest free distance is tabulated. For this code, the generators $\mathbf{g}_i$ are given in octal form. The resulting free distance of the code is also given in these tables.

13.4 GOOD CODES BASED ON COMBINATION OF SIMPLE CODES

As we have seen in the preceding sections, the performance of block and convolutional codes depends on the distance properties of the code and, in particular, the

TABLE 13.1   RATE $\frac{1}{2}$ MAXIMUM FREE DISTANCE CODES

Constraint length L	Generators in octal		d_{free}
3	5	7	5
4	15	17	6
5	23	35	7
6	53	75	8
7	133	171	10
8	247	371	10
9	561	753	12
10	1167	1545	12
11	2335	3661	14
12	4335	5723	15
13	10533	17661	16
14	21675	27123	16

Odenwalder (1970) and Larsen (1973).

TABLE 13.2   RATE $\frac{1}{3}$ MAXIMUM FREE DISTANCE CODES

Constraint length L	Generators in octal			d_{free}
3	5	7	7	8
4	13	15	17	10
5	25	33	37	12
6	47	53	75	13
7	133	145	175	15
8	225	331	367	16
9	557	663	711	18
10	1117	1365	1633	20
11	2353	2671	3175	22
12	4767	5723	6265	24
13	10533	10675	17661	24
14	21645	35661	37133	26

Odenwalder (1970) and Larsen (1973).

minimum distance in block codes and the free distance in convolutional codes. In order to design block codes with a given rate and with high minimum distance, we must increase n, the block length of the code. Increasing n increases the complexity of the decoding. In most decoding algorithms, the complexity of the decoding increases exponentially with increasing the block length of the code.

For convolutional codes, increasing the free distance at a given rate requires increasing the constraint length of the code. But increasing the constraint length of the code increases the number of states in the code trellis, which in turn increases the decoding complexity. Again, the decoding complexity increases exponentially with the constraint length of the convolutional code.

Various methods have been proposed to increase the effective block length of the code while keeping the complexity tractable. Most of these methods are based on combining simple codes to generate more complex codes. The decoding of the resulting code is performed by using methods for decoding the simple component codes. The resulting decoding is a suboptimal decoding scheme, which usually performs satisfactorily. Here, we discuss three widely used methods for combining simple codes to generate more complex codes. These techniques generate *product codes, concatenated codes,* and *turbo codes.*

13.4.1 Product Codes

The structure of product codes (or array codes) is very similar to a crossword puzzle. Product codes are generated by using two linear block codes arranged in a matrix form. Two linear block codes, one with parameters n_1, k_1, and $d_{\min 1}$ and another with

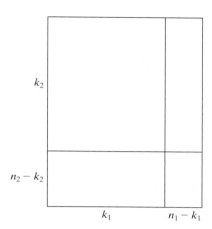

Figure 13.20   The structure of a product code.

parameters n_2, k_2, and $d_{\min 2}$, are used in a matrix of the form shown in Figure 13.20. The resulting code is an $(n_1 n_2, k_1 k_2)$ linear block code. We can show that the minimum distance of the resulting code is the product of the minimum distances of the component codes, i.e., $d_{\min} = d_{\min 1} d_{\min 2}$, and is capable of correcting $\lfloor \frac{d_{\min 1} d_{\min 2} - 1}{2} \rfloor$, using optimal hard-decision decoding. But we can also decode using the properties of the component codes, as we would solve a crossword puzzle. Using the row codes, we can make the best guess for the bit values; then, using the column codes, we can improve these guesses. This process, which can be repeated in an iterative fashion, improving the quality of the guess in each step, is known as *iterative decoding* and is very similar to the way a crossword puzzle is solved. To employ this decoding procedure, we need decoding schemes for the row and column codes that are capable of providing *guesses* about each individual bit. In other words, decoding schemes with soft outputs (usually, the likelihood values) are desirable. In Section 13.4.3, we will describe such decoding procedures in our discussion of turbo codes.

13.4.2 Concatenated Codes

A concatenated code consists of two codes, an inner code and an outer code, connected serially, as shown in Figure 13.21. The inner code is a binary block or convolutional code and the outer code is typically a Reed–Solomon code. If the inner code is an (n, k) code, the combination of the inner encoder, digital modulator, waveform channel, digital demodulator, and the inner decoder can be considered as a channel whose input and output are binary blocks of length k, or equivalently, elements of a q-ary alphabet where $q = 2^k$. Now the Reed–Solomon code (the outer code) can be used on this q-ary input, q-ary output channel to provide further error protection. It can easily be seen that if the rates of the inner and the outer codes are r_c and R_c, respectively, the rate of the concatenated code will be

$$R_{cc} = R_c r_c. \tag{13.4.1}$$

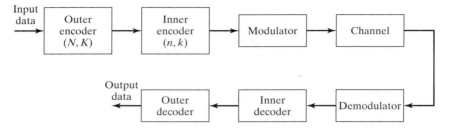

Figure 13.21   Block diagram of a communication system with concatenated coding.

Also, the minimum distance of the concatenated code is the product of the minimum distances of the inner and the outer codes. In concatenated codes, the performance of the inner code has a major impact on the overall performance of the code. That is why convolutional codes with soft decoding using the Viterbi algorithm are commonly employed for the inner code.

13.4.3 Turbo Codes

Turbo codes are a special class of concatenated codes where there exists an interleaver between two parallel or serial encoders. The existence of the interleaver results in very large code word lengths with excellent performance, particularly at low SNRs. Using these codes, it is possible to get as close as a fraction of a dB to the Shannon limit at low SNRs.

The structure of a turbo encoder is shown in Figure 13.22. The turbo encoder consists of two constituent codes separated by an interleaver of length N. The constituent codes are usually *recursive systematic convolutional codes (RSCC)* of rate $\frac{1}{2}$, and the same code is usually employed as the two constituent codes. Recursive convolutional codes are different from the nonrecursive convolutional codes by the existence of feedback in their shift-register realization. Therefore, unlike the nonrecursive convolutional codes, which are realized as finite impulse response (FIR) digital filters, recursive convolutional codes are infinite impulse response (IIR) filters. An example of a recursive systematic convolutional encoder is shown in Figure 13.23.

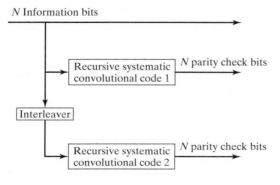

Figure 13.22   Block diagram of a turbo encoder.

Information bits

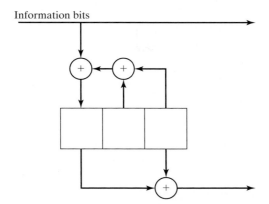

Figure 13.23   A recursive systematic convolutional encoder.

In the encoder structure shown in Figure 13.22, N information bits enter the first encoder. The same information bits are interleaved and applied to the second encoder. Because the encoders are systematic, each encoder generates the N-information bits applied to its input followed by N parity check bits. After the encoding, the N information bits and the $2N$ parity check bits of the two encoders, i.e., a total of $3N$ bits, are transmitted over the channel. Therefore, the overall rate is $R = N/3N = 1/3$. If higher rates are desirable, the parity check bits are punctured, i.e., only some of parity check bits, according to a regular pattern, are transmitted.

The interleaver in turbo codes is usually very long, in the order of thousands of bits. Pseudorandom interleavers perform well, although some improvement in the performance of the code can be obtained by a clever choice of the interleaver. This improvement is more noticeable at short interleaver lengths. Also note that unlike nonrecursive convolutional codes in which a sequence of zeros padded to the message sequence guarantees that the encoder returns to the all-zero state, returning a component code to the all-zero state requires padding the information sequence with a particular nonzero sequence. Due to the existence of the interleaver, it is, in most cases, impossible to return both codes to the all-zero state.

Since turbo codes have two constituent-code components, an iterative algorithm is appropriate for their decoding, as illustrated in Figure 13.24. Any decoding method that yields the likelihood of the bits as its output can be used in the iterative decoding scheme. One such decoding scheme is the *maximum a posteriori Probability (MAP)* decoding method of *Bahl, Cocke, Jelinek, and Raviv (BCJR)*, as described in Bahl et al. (1974). Another popular method with lower complexity (and degraded performance) is the soft-output Viterbi algorithm (SOVA) of Hagenauer and Hoher (1989). Using either method, the likelihoods of different bits are computed and passed to the second decoder. The second decoder computes the likelihood ratios and passes them to the first decoder; this process is repeated until the likelihoods suggest high probability of correct decoding for each bit. At this point, the final decision is made. In the iterative decoding procedure shown in Figure 13.24, y_s, y_{1p}, and y_{2p} represent the received data corresponding to the systematic bits, parity check bits for

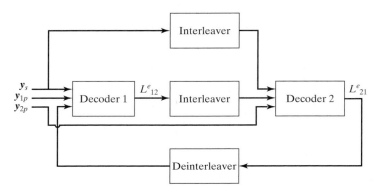

Figure 13.24 The iterative decoding scheme for turbo codes.

encoder 1 and parity check bits for encoder 2, respectively, L_{12}^e and L_{21}^e represent the information passed between the two decoders.

13.4.4 The BCJR Algorithm

The BCJR (Bahl, Jelinek, Coke, and Raviv) algorithm is a trellis-based decoding algorithm that decodes *each information bit* in a convolutional code based on the MAP criteria. This is in contrast to the Viterbi algorithm that makes a maximum-likelihood decision on a *sequence* of information bits based on the maximum-likelihood criterion. As a by-product, the BCJR algorithm generates the probabilities, or likelihoods, for the transmitted bits that can be used in the iterative decoding of the turbo codes.

Assume that the code is described by a trellis shown in Figure 13.25, where we are assuming that the depth of the trellis for the received sequence is N stages and we are considering transition between stages $i - 1$ and i. The number of states is denoted by M and the total received sequence is $\boldsymbol{y}_{[1,N]}$. Figure 13.25 is drawn for $M = 4$ and only two of the transitions between the states at times $i - 1$ and i are shown. For $1 \le i \le N$, denote the received sequence up to time i by $\boldsymbol{y}_{[1,i]}$ and from time $i + 1$ to time N by $\boldsymbol{y}_{[i,N]}$. In other words, $\boldsymbol{y}_{[1,i]} = (\boldsymbol{y}_1, \boldsymbol{y}_2, \dots, \boldsymbol{y}_i)$ and $\boldsymbol{y}_{[i,N]} = (\boldsymbol{y}_i, \boldsymbol{y}_{i+1}, \dots, \boldsymbol{y}_N)$. Let random variable $S(i)$ denote the state of the trellis at time i. Because there are a total of M possible states, we assume that $S(i)$ takes its values in the set of integers from 0 to $M - 1$, where 0 corresponds to the all-zero state of the trellis.

For $0 \le s, s' \le M - 1$ and $1 \le i \le N - 1$, we define the probabilities

$$\alpha_i(s) = P\left(S(i) = s, \boldsymbol{y}_{[1,i]}\right) \tag{13.4.2}$$

$$\beta_i(s) = P\left(\boldsymbol{y}_{[i+1,N]}|S(i) = s\right) \tag{13.4.3}$$

$$\gamma_i(s, s') = P\left(S(i) = s', \boldsymbol{y}_i|S(i - 1) = s\right) \tag{13.4.4}$$

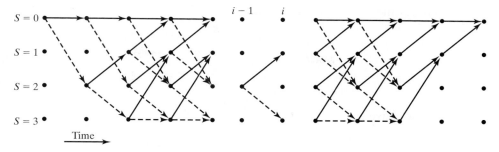

Figure 13.25   The trellis for the BCJR algorithm.

with the boundary conditions

$$\alpha_0(s) = \begin{cases} 1 & s = 0 \\ 0 & s = 1, 2, \ldots, M - 1 \end{cases} ; \qquad (13.4.5)$$

$$\beta_N(s) = \begin{cases} 1 & s = 0 \\ 0 & s = 1, 2, \ldots, M - 1 \end{cases} . \qquad (13.4.6)$$

The boundary conditions state that we start and end in the all-zero state.

Having received $y_{[1,N]}$, the MAP detector declares $\hat{u}_i = 1$ if the likelihood ratio $\Lambda(u_i)$ satisfies

$$\Lambda(u_i) = \frac{P\left(u_i = 1 | y_{[1,N]}\right)}{P\left(u_i = 0 | y_{[1,N]}\right)} = \frac{P\left(u_i = 1, y_{[1,N]}\right)}{P\left(u_i = 0, y_{[1,N]}\right)} > 1. \qquad (13.4.7)$$

But

$$P\left(u_i = 1, y_{[1,N]}\right) = \sum_{(s,s') \in \Sigma^1} P\left(S(i-1) = s, S(i) = s', y_{[1,N]}\right) \qquad (13.4.8)$$

$$P\left(u_i = 0, y_{[1,N]}\right) = \sum_{(s,s') \in \Sigma^0} P\left(S(i-1) = s, S(i) = s', y_{[1,N]}\right), \qquad (13.4.9)$$

where Σ^1 denotes the set of all (s, s') pair of states such that $u_i = 1$ causes a transition from $S(i - 1) = s$ to $S(i) = s'$. Similarly, Σ^0 denotes the set of all-state pairs (s, s') such that a transition from $S(i - 1) = s$ to $S(i) = s'$ corresponds to an input $u_i = 0$. For instance, in Figure 13.25, $(2, 1) \in \Sigma^0$ and $(2, 3) \in \Sigma^1$.

Substituting Equations (13.4.8) and (13.4.9) in Equation (13.4.7) and defining

$$\pi_i(s, s', y) = P\left(S(i-1) = s, S(i) = s', y_{[1,N]}\right) \qquad (13.4.10)$$

reduces the likelihood ratio $\Lambda(u_i)$ to

$$\Lambda(u_i) = \frac{\displaystyle\sum_{(s,s')\in\Sigma^1} \pi_i(s, s', \mathbf{y})}{\displaystyle\sum_{(s,s')\in\Sigma^0} \pi_i(s, s', \mathbf{y})}. \tag{13.4.11}$$

But we have

$$\begin{aligned}
\pi_i(s, s', \mathbf{y}) &= P\left(S(i-1) = s, S(i) = s', \mathbf{y}_{[1,i-1]}, \mathbf{y}_i, \mathbf{y}_{[i+1,N]}\right) \\
&= P\left(S(i-1) = s, S(i) = s', \mathbf{y}_{[1,i-1]}, \mathbf{y}_i\right) \\
&\quad \times P\left(\mathbf{y}_{[i+1,N]}|S(i-1) = s, S(i) = s', \mathbf{y}_{[1,i-1]}, \mathbf{y}_i\right) \\
&= P\left(S(i-1) = s, \mathbf{y}_{[1,i-1]}\right) \times P\left(S(i) = s', \mathbf{y}_i|S(i-1) = s, \mathbf{y}_{[1,i-1]}\right) \\
&\quad \times P\left(\mathbf{y}_{[i+1,N]}|S(i-1) = s, S(i) = s', \mathbf{y}_{[1,i-1]}, \mathbf{y}_i\right) \\
&\overset{(a)}{=} P\left(S(i-1) = s, \mathbf{y}_{[1,i-1]}\right) P\left(S(i) = s', \mathbf{y}_i|S(i-1) = s\right) \\
&\quad \times P\left(\mathbf{y}_{[i+1,N]}|S(i) = s'\right) \\
&= \alpha_{i-1}(s)\gamma_i(s, s')\beta_i(s'), \tag{13.4.12}
\end{aligned}$$

where the equality in (a) is because the channel is memoryless.

In Problems 13.22 and 13.23, we prove the following recursive relations for $\alpha_i(s)$ and $\beta_i(s')$

$$\alpha_i(s) = \sum_{s'} \alpha_{i-1}(s')\gamma_i(s', s) \tag{13.4.13}$$

$$\beta_i(s') = \sum_{s} \beta_{i+1}(s)\gamma_{i+1}(s', s). \tag{13.4.14}$$

Equations (13.4.13) and (13.4.14) are called the *forward recursion* and the *backward recursion*, respectively. Assuming that we know the values of $\gamma_i(s', s)$, we can use the forward and the backward recursions with the initial and final conditions given in Equations (13.4.5) and (13.4.6) to find values of $\alpha_i(s)$ and $\beta_i(s')$ for all $1 \le i \le N - 1$ and $0 \le s, s' \le M - 1$. Then we can substitute these values in Equation (13.4.12) to determine the values of $\pi_i(s, s', \mathbf{y})$, which can then be substituted in Equation (13.4.11) to find the likelihood ratios. Having found the likelihood ratios, we can detect u_i's.

It remains to show how the values of $\gamma_i(s', s)$ can be computed. These values depend on the characteristics of the channel. For a memoryless channel, we have

$$\begin{aligned}
\gamma_i(s', s) &= P\left(S(i) = s, \mathbf{y}_i|S(i-1) = s'\right) \\
&= P\left(S(i) = s|S(i-1) = s'\right) P\left(\mathbf{y}_i|S(i) = s, S(i-1) = s'\right) \\
&= P(u_i) P(\mathbf{y}_i|u_i). \tag{13.4.15}
\end{aligned}$$

For instance, in an AWGN channel with BPSK modulation, the transmitted signal is $(2u_i - 1)\sqrt{\mathcal{E}}$; hence,

$$P(y_i | u_i) = \frac{1}{\sqrt{\pi N_0}} e^{-\frac{\left(y_i - (2u_i - 1)\sqrt{\mathcal{E}}\right)^2}{N_0}}. \tag{13.4.16}$$

Therefore, starting with prior probabilities $P(u_i)$, we can use Equations (13.4.15) and (13.4.16) to determine the values of $\gamma_i(s', s)$, which can be used to find the values of α and β. In summary, the BCJR algorithm can be carried out as follows:

1. Initialize the algorithm with the boundary conditions given in Equations (13.4.5) and (13.4.6) (trellis termination condition).
2. Given prior probabilities of u_i (usually assumed equiprobable to be either 0 or 1), the channel characteristics, and the received sequence, compute the γ's using Equations (13.4.15) and (13.4.16).
3. Use the forward recursion given in Equation (13.4.13) to compute the α's.
4. Use the backward recursion given in Equation (13.4.14) to compute the β's.
5. Use Equation (13.4.12) to determine the values of $\pi_i(s, s', \mathbf{y})$.
6. Compute the likelihood ratio of u_i's using Equation (13.4.11) and make the decision.

Note that, in the BCJR algorithm, we have to pass the trellis once in the forward direction (to find α's) and once in the backward direction (to find β's). Therefore, the complexity of this algorithm is roughly twice the complexity of the Viterbi algorithm. Also, in the forward pass, the values of α have to be stored; thus, the storage requirements of the algorithm are quite demanding.

Because the BCJR algorithm provides the likelihood ratios for the transmitted bits, it can be employed in the iterative decoding of turbo codes, as shown in Figure 13.24.

13.4.5 Performance of Turbo Codes

Turbo codes are characterized by excellent performance at low SNR's. The performance of turbo codes improves with increasing the length of the interleaver and the number of iterations. The original turbo code studied by Berrou et al. (1993) used the recursive systematic convolutional encoder shown in Figure 13.26. In this code, the forward connections to the code can be represented by $g_1 = [1\ 0\ 0\ 0\ 1]$ and the feedback connection can be described by $g_2 = [1\ 1\ 1\ 1\ 1]$. These vectors can be represented in octal form as 21 and 37, respectively. It is customary to refer to such a code as a 21/37 recursive systematic convolutional code. This code was employed in a turbo code with a constraint length of $N = 65536$ and puncturing was used to increase its rate to 1/2. The performance of the resulting code using the BCJR decoding algorithm is shown in Figure 13.27. After 18 iterations, the performance of this code is only 0.7 dBs from the Shannon limit.

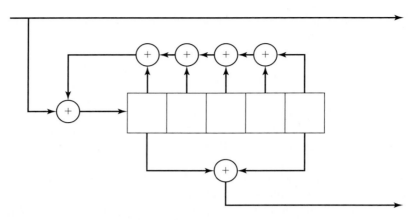

Figure 13.26   The 21/37 recursive systematic convolutional code.

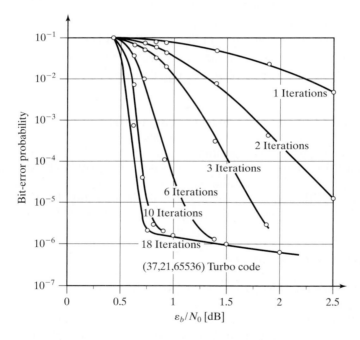

Figure 13.27   The performance plots for the (37,21,65536) turbo code for different number of iterations.

One problem with the turbo codes is the existence of the error floor. As seen in Figure 13.27, the probability of error decreases sharply with increasing $\mathcal{E}_b/N_0$ up to a certain point. After this point, the error probability decreases very slowly. The existence of the error floor is a consequence of the distance properties of turbo codes. Its turns out that, although turbo codes have excellent performance at low signal-to-noise ratios, they have rather poor minimum distance. The reason they can perform well is that, although the distance properties are poor, the number of paths at

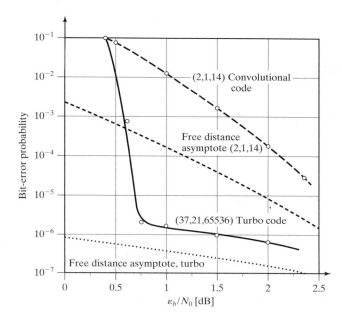

Figure 13.28 Comparison of performance of a turbo code and a convolutional code.

low distance (called the *multiplicity* of that distance) is very low. In ordinary convolutional codes, we can design codes with much better minimum distance, but much higher multiplicity at low distances. Now referring to Equation (13.3.21), we see that for low SNR's the effect of multiplicity ($a_{d_{min}}$) is more important on the performance of the code, whereas at high SNR's, the minimum distance of the code plays a major role; thus, the performance of turbo codes at high SNR's sharply degrades.

Figure 13.28 compares the performance of the 37/21 turbo code with a rate 1/2 convolutional code with constraint length 14, and soft-decision Viterbi decoding. In the same plot, performance bounds (using the union bound) for both codes are also plotted.

13.5 LOW-DENSITY PARITY CHECK CODES

Low-density parity check (LDPC) codes are block codes, closely related to turbo codes with an iterative decoding algorithm, which achieve excellent error performance at rates very close to channel capacity. These codes were first introduced by Gallager in the early 1960s, but, since the decoding algorithm for these codes was not practical with the technology of that time, were soon forgotten. The introduction of turbo codes in early 1990s and their similarity to LDPC codes rekindled the interest in low-density parity check codes and new classes of these codes were introduced and analyzed. In particular, the class of *irregular low-density parity check codes* was introduced, which is capable of achieving rates very close to the channel capacity.

Low-density parity check codes are linear block codes with a sparse parity check matrix H. In other words, the number of ones in each row and column of H is small compared to the size of the matrix. More formally, we can define a low-density parity check code as an (n, k) linear block code with an $(n - k) \times n$ parity check matrix H with the following properties:

1. The number of 1's in each row is ρ.
2. The number of 1's in each column is γ
3. No two columns have more than two rows with 1's in those two row locations.
4. Both ρ and γ are small compared to the dimensions of H.

An example of the parity check matrix for a $(12, 3)$ LDPC code with $\rho = 4$ and $\gamma = 3$ is given in Equation (13.5.1).

$$H = \begin{bmatrix} 0 & 0 & 1 & 0 & 0 & 1 & 1 & 1 & 0 & 0 & 0 & 0 \\ 1 & 1 & 0 & 0 & 1 & 0 & 0 & 0 & 0 & 0 & 0 & 1 \\ 0 & 0 & 0 & 1 & 0 & 0 & 0 & 0 & 1 & 1 & 1 & 0 \\ 0 & 1 & 0 & 0 & 0 & 1 & 1 & 0 & 0 & 1 & 0 & 0 \\ 1 & 0 & 1 & 0 & 0 & 0 & 0 & 1 & 0 & 0 & 1 & 0 \\ 0 & 0 & 0 & 1 & 1 & 0 & 0 & 0 & 1 & 0 & 0 & 1 \\ 1 & 0 & 0 & 1 & 1 & 0 & 1 & 0 & 0 & 0 & 0 & 0 \\ 0 & 0 & 0 & 0 & 0 & 1 & 0 & 1 & 0 & 0 & 1 & 1 \\ 0 & 1 & 1 & 0 & 0 & 0 & 0 & 0 & 1 & 1 & 0 & 0 \end{bmatrix}. \qquad (13.5.1)$$

The density of a LDPC code is denoted by r and is defined as

$$r = \frac{\rho}{n}. \qquad (13.5.2)$$

The density of the code with H given in Equation (13.5.1) is $r = \frac{1}{3}$.

If $c = (c_1, c_2, \ldots, c_n)$ is a codeword, then by Equation (13.2.11)

$$cH^t = 0.$$

This relation yields a total of $n - k$ linear equations, each containing ρ of the c_i's. Each equation states that from the ρ terms appearing in it, an odd number can be 1's. Note that each c_i, for $1 \le i \le n$, appears in exactly γ of these equations. Also, by the third property of LDPC codes, any arbitrary pair of components c_i and c_j, $i \ne j$ appear in not more than one of the $n - k$ equations. For instance, for this code, we have the following parity check equations:

$$c_3 \oplus c_6 \oplus c_7 \oplus c_8 = 0 \qquad (13.5.3)$$

$$c_1 \oplus c_2 \oplus c_5 \oplus c_{12} = 0 \qquad (13.5.4)$$

$$c_4 \oplus c_9 \oplus c_{10} \oplus c_{11} = 0 \tag{13.5.5}$$

$$c_2 \oplus c_6 \oplus c_7 \oplus c_{10} = 0 \tag{13.5.6}$$

$$c_1 \oplus c_3 \oplus c_8 \oplus c_{11} = 0 \tag{13.5.7}$$

$$c_4 \oplus c_5 \oplus c_9 \oplus c_{12} = 0 \tag{13.5.8}$$

$$c_1 \oplus c_4 \oplus c_5 \oplus c_7 = 0 \tag{13.5.9}$$

$$c_6 \oplus c_8 \oplus c_{11} \oplus c_{12} = 0 \tag{13.5.10}$$

$$c_2 \oplus c_3 \oplus c_9 \oplus c_{10} = 0. \tag{13.5.11}$$

As it is seen here, there is a total of $n - k = 9$ equations, each containing $\rho = 4$ variables and each c_i appearing in $\gamma = 3$ of the equations. Also note that any pair of components appears in at most one equation. For instance, the pair c_1 and c_2 appears only in the second equation and the pair c_1 and c_6 does not appear in any of the equations.

The preceding parity check equations can be described in terms of a special graph, called the *Tanner graph*. A Tanner graph consists of n *variable nodes* and $n - k$ *check nodes*. The check nodes correspond to the $n - k$ parity check equations and the variable nodes correspond to the n components of a codeword. If a variable appears in a parity check equation, then an edge connects the corresponding variable node to the corresponding check node. Therefore, each check node is connected to ρ variable nodes and each variable node is connected to γ of the check nodes. The graph shown in Figure 13.29 represents the Tanner graph for the code $\boldsymbol{H}$ given in Equation (13.5.1). In this graph, variable nodes are denoted by circles and check nodes are denoted by squares.

The decoding of low-density parity check codes is usually done by an iterative decoding scheme, which is very similar to the iterative scheme used in decoding turbo codes. This iterative scheme is called the *message passing algorithm*. In this algorithm, each variable node receives the corresponding soft output of the channel (the output of the matched filter); then, based on the received value, it computes the

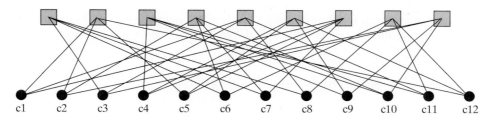

Figure 13.29   The Tanner graph for the LDPC code with $\boldsymbol{H}$ given in Equation (13.5.1).

probability of that node value being 1. Then each node provides the probability of the node value being 1 to all check nodes that are connected to that variable node. Each check node receives the probabilities of different variable nodes connected to that node. Based on the values of these probabilities and the parity equation at that node, which states that an even number of the variables connected to that node can be 1, the check node determines new estimates of the probabilities and sends them back to the variable nodes. These probabilities are updated at variable nodes and the iteration is repeated until a final decision is made at the variable nodes.

These low-density parity check codes are *regular LDPC codes*. *Irregular LDPC codes* are a generalized class of low-density parity check codes in which the number of 1's in rows and columns is variable, but the overall number of 1's in H is low. This added flexibility in the structure of irregular LDPC codes makes it possible to optimize the distribution of 1's in rows and columns, thus improving the performance. In general, irregular LDPC codes perform better than regular LDPC codes. Irregular LDPC codes are among the most effective coding schemes that can achieve performance within a fraction of dB from channel capacity.

13.6 CODING FOR BANDWIDTH-CONSTRAINED CHANNELS

In the two major classes of codes studied so far, i.e., block and convolutional codes, an improvement in the performance of the communication system is achieved by expanding bandwidth. In both cases, the Euclidean distance between the transmitted coded waveforms is increased by use of coding, but at the same time, the bandwidth is increased by a factor of $\frac{n}{k} = \frac{1}{R_c}$. These codes have wide applications in cases where there is enough bandwidth and the communication system designer is not under tight bandwidth constraints. An example of such a case is a deep-space communication system. However, in many practical applications, we are dealing with communication channels with strict bandwidth constraints and the bandwidth expansion due to coding may not be acceptable. For example, in the transmission of digital data over telephone channels (modem design), we are dealing with a channel that has a restricted bandwidth, and the overhead due to coding imposes a major restriction on the transmission rate. In this section, we will discuss an integral coding and modulation scheme called *trellis-coded modulation* that is particularly useful for bandwidth-constrained channels.

13.6.1 Combined Coding and Modulation

Use of block or convolutional codes introduces redundancy that, in turn, causes increased Euclidean distance between the coded waveforms. On the other hand, the dimensionality of the transmitted signal will increase from k-dimensions/transmission to n-dimensions/transmission, if binary PSK modulation is employed. This increase in dimensionality results in an increase in bandwidth since bandwidth and dimensionality are proportional. If we want to reap the benefits of coding and, at the same time, not increase the bandwidth, we have to use a modulation scheme other than binary PSK, i.e., a scheme that is more bandwidth efficient. This means that we have to employ a multilevel/multiphase-modulation scheme to reduce the bandwidth. Of course using a multilevel/multiphase-modulation scheme results in a more "crowded" constellation and, at a constant power level, decreases the minimum Euclidean distance within the constellation. This certainly has a negative effect on the error performance of the overall coding-modulation scheme. But, as we will see next, this reduction of the minimum Euclidean distance within the constellation can be well compensated by the increase in the Hamming distance due to coding such that the overall performance shows considerable improvement.

As an example, assume that, in the coding stage, we want to use a rate $\frac{2}{3}$ code. If the rate of the source is R bits/sec, the number of encoder output binary symbols/sec will be $\frac{3}{2}R$. If we want to use a constellation such that the bandwidth requirement is equal to the bandwidth requirement of the uncoded signal (no bandwidth expansion), we must assign m dimensions for each output binary symbol such that the resulting number of dimensions/sec is equal the number of dimensions/sec of the uncoded data, which is R. Therefore, we must have

$$R = \frac{3}{2}Rm; \qquad (13.6.1)$$

hence,

$$m = \frac{2}{3} \qquad \text{dimension/binary symbol.} \qquad (13.6.2)$$

This means that the constellation should be designed so we have two dimensions for every three binary symbols. But three binary symbols are equivalent to eight points in the constellation; therefore, we can achieve our goal with an eight-point constellation in the two-dimensional space. One such constellation is, of course, an 8-PSK modulation scheme. Therefore, if we use a rate $\frac{2}{3}$ code in conjunction with an 8-PSK modulation scheme, there will be no bandwidth expansion.

Now, examine how much coding gain we can obtain from such a scheme. Assuming that the available power is P with no coding, we have

$$\mathscr{E}_b = \frac{P}{R}; \qquad (13.6.3)$$

therefore, for the minimum Euclidean distance between two sequences, we have

$$d^2 = \frac{4P}{R}. \qquad (13.6.4)$$

If two information bits are mapped into a point in an 8-PSK constellation, the energy of this point is

$$\mathscr{E}_s = \frac{2P}{R}. \qquad (13.6.5)$$

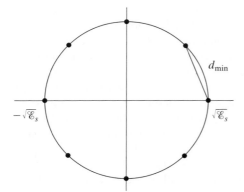

Figure 13.30 The 8-PSK constellation used for bandwidth-efficient coding.

From this, we can derive an expression for the minimum Euclidean distance within the constellation (see Figure 13.30) as

$$d_{\min}^2 = 4\frac{2P}{R}\sin^2\frac{\pi}{8} = 2(2 - \sqrt{2})\frac{P}{R}. \tag{13.6.6}$$

Obviously, the minimum Euclidean distance has been decreased. To see this effect, we derive the loss due to using this constellation:

$$\frac{d_{\min}^2}{d^2} = \frac{2}{2 - \sqrt{2}} = 2 + \sqrt{2} = 3.141 \sim 5.33\,\text{dB}. \tag{13.6.7}$$

This loss has to be compensated by the code. Of course, the rate $\frac{2}{3}$ code employed here should not only compensate for this loss, but should also provide additional gain to justify its use of the overall coding-modulation scheme. We can use any block or convolutional code that can provide the minimum distance required to achieve a certain overall coding gain. For example, if we need an overall coding gain of 3 dB, the code must provide a coding gain of 8.33 dB to compensate for the 5.33-dB loss due to modulation and to provide an extra 3-dB coding gain. A code that can provide such a high coding gain is a very complex (long constraint length) code requiring a sophisticated encoder and decoder. However, by interpreting coding and modulation as a single entity, as shown in Section 13.6.2, we see that a comparable performance can be achieved using a much simpler coding scheme.

13.6.2 Trellis-Coded Modulation

Trellis-coded modulation, or TCM, is a simple method for designing coded-modulation schemes that can achieve good overall performance. This coding-modulation scheme is based on the concept of *mapping by set partitioning* developed by Ungerboeck (1982). Mapping by set partitioning can be used in conjunction with both block and convolutional codes, but due to the existence of a simple optimal soft-decision decoding algorithm for convolutional codes (the Viterbi algorithm), it has been mostly used with convolutional codes. When used with convolutional codes, the resulting coded-modulation scheme is known as *trellis-coded modulation*.[9]

Set Partitioning Principles.   The key point in the partitioning of a constellation is to find subsets of it that are similar and the points inside each partition that are maximally separated. Starting with the original constellation, we partition it into two subsets that are congruent and ensure that the points within each partition are maximally separated. Then, we apply the same principle to each partition and continue. The point at which the partitioning is stopped depends on the code that we are using. This will be discussed shortly.

An example of set partitioning is shown in Figure 13.31. We start with an 8-PSK constellation with power $\mathcal{E}_s$. The minimum distance within this constellation is

$$d_0 = \sqrt{(2 - \sqrt{2})\mathcal{E}_s}. \tag{13.6.8}$$

This constellation is partitioned into two subsets denoted by B_0 and B_1. Note that B_0 and B_1 are congruent. There are many ways that the original 8-PSK constellation can be partitioned into two congruent subsets, but B_0 and B_1 provide the maximum intra-partition distance. This distance is easily seen to be

$$d_1 = \sqrt{2\mathcal{E}_s}. \tag{13.6.9}$$

Now, we further partition B_0 and B_1 to obtain C_0, C_1, C_2, and C_3. The intra partition distance now has increased to

$$d_2 = 2\sqrt{\mathcal{E}_s}. \tag{13.6.10}$$

We can still go one step further to obtain eight partitions, each containing a single point. The corresponding subsets are denoted by D_0 through D_7. Another example of set partitioning applied to a QAM constellation is given in Figure 13.32. This partitioning follows the general rules for set partitioning as just described.

Coded Modulation.   The block diagram of a coded-modulation scheme is shown in Figure 13.33. A block of length k-input bits is broken into two subblocks of lengths k_1 and k_2. The first k_1 bits are applied to an (n_1, k_1) binary encoder. The output of the encoder consists of n_1 bits. These bits are used to choose one of 2^{n_1} partitions in the constellation. This means that the constellation has been partitioned into 2^{n_1} subsets. After the constellation is chosen, the remaining k_2 bits are used to choose one of the points in the chosen constellation. This means that there exist 2^{k_2} points in each partition. Therefore, the partitioning that is used contains 2^{n_1} subsets and each subset contains 2^{k_2} points. This gives us a rule for how large a constellation is required and how many steps in partitioning of this constellation must be taken.

---

[9]To be more precise, the term trellis-coded modulation is used when a general trellis code is employed.

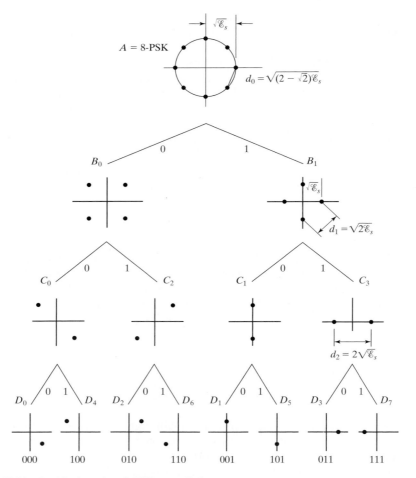

Figure 13.31   Partitioning of an 8-PSK constellation.

Ungerboeck (1982) has shown that, by choosing $n_1 = k_1 + 1$ and $k_2 = 1$ and using simple convolutional codes, we can design coded modulation schemes that achieve an overall coding gain between 3 and 6 dB. One such scheme is shown in Figure 13.34. In this coding scheme, $k_1 = 1$, $n_1 = 2$, and $k_2 = 1$. The constellation contains $2^{n_1 + k_2} = 8$ points, which are partitioned into $2^{n_1} = 4$ subsets, each containing $2^{k_2} = 2$ points. The constellation chosen here is an 8-PSK constellation, and it is partitioned as shown in Figure 13.31. The convolutional code employed can be any rate $\frac{k_1}{n_1} = \frac{1}{2}$ code. The constraint length of this code is a design parameter and can be chosen to provide the desired coding gain. Higher constraint lengths, of course, provide higher coding gains at the price of increased encoder–decoder complexity. In this very simple example, the constraint length has been chosen to be equal to 3. The (one-stage) trellis diagram of this code is also shown in Figure 13.34.

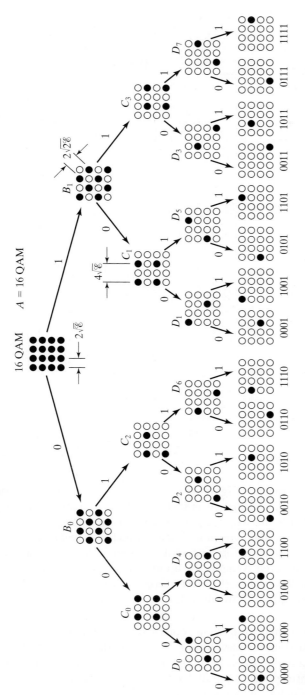

Figure 13.32 Set partitioning of a 16-QAM constellation.

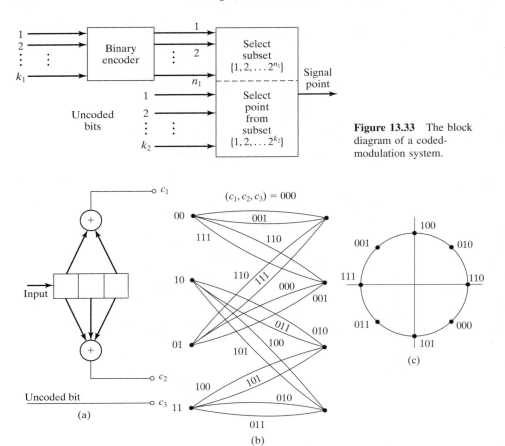

Figure 13.33   The block diagram of a coded-modulation system.

Figure 13.34   A simple TCM scheme.

The trellis diagram shown in Figure 13.34 is the trellis diagram of an ordinary convolutional code. The main difference is that we have two paths connecting two states. The reason for this is the existence of the extra $k_2 = 1$ bit, which chooses a point in each partition. In fact, the two parallel paths connecting two states correspond to a partition, and any single path corresponds to a point in the partition. One final question remains to be answered: What is the optimal assignment of the constellation points to the branches of the code trellis? Extensive computer simulations as well as heuristic reasoning result in the following rules:

1. Parallel transitions, when they occur, correspond to signal points in a single partition at the last stage of partitioning (thus providing the largest Euclidean distance). In this example, $C_0 = \{D_0, D_4\}$, $C_2 = \{D_2, D_6\}$, $C_1 = \{D_1, D_5\}$, and $C_3 = \{D_3, D_7\}$ correspond to parallel transitions. These points are separated by the maximum Euclidean distance of $d_2 = 2\sqrt{\mathscr{E}_s}$.

2. The transitions originating from, or merging into, any state are assigned partitions, in the last stage of partitioning, that have a single parent partition in the preceding stage. In this example, $\{C_0, C_2\}$ and $\{C_1, C_3\}$ are such partitions, having parents B_0 and B_1, respectively. The maximum distance in this case is $d_1 = \sqrt{2\mathscr{E}_s}$.

3. The signal points should occur with equal frequency.

To see how the trellis-coded modulation scheme of Figure 13.34 performs, we have to find the minimum Euclidean distance between two paths originating from a node and merging into another node. This distance, known as the *free Euclidean distance* and denoted by D_{fed}, is an important characteristic of a trellis-coded modulation scheme. One obvious candidate for D_{fed} is the Euclidean distance between two parallel transitions. The Euclidean distance between two parallel transitions is $d_2 = 2\sqrt{\mathscr{E}_s}$. Another candidate path is shown in Figure 13.35. However, the Euclidean distance between these two paths is $d^2 = d_0^2 + 2d_1^2 = 4.58\mathscr{E}_s$. Obviously, this is larger than the distance between two parallel transitions. It is easily verified that for this code, the free Euclidean distance is $D_{\text{fed}} = d_2 = 2\sqrt{\mathscr{E}_s}$. To compare this result with an uncoded scheme, we note that in an uncoded scheme,

$$d_{\text{uncoded}}^2 = 4\mathscr{E}_b = 4\frac{P}{R} \tag{13.6.11}$$

and in the coded scheme,

$$d_{\text{coded}}^2 = 4\mathscr{E}_s = 8\mathscr{E}_b. \tag{13.6.12}$$

Therefore, the coding gain is given by

$$G_{\text{coding}} = \frac{d_{\text{coded}}^2}{d_{\text{uncoded}}^2} = 2 \approx 3 \text{ dB}. \tag{13.6.13}$$

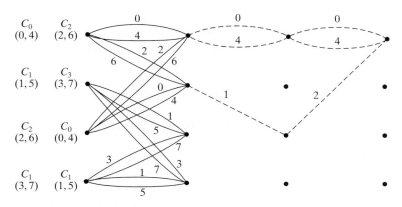

Figure 13.35   Two candidate minimum distance paths.

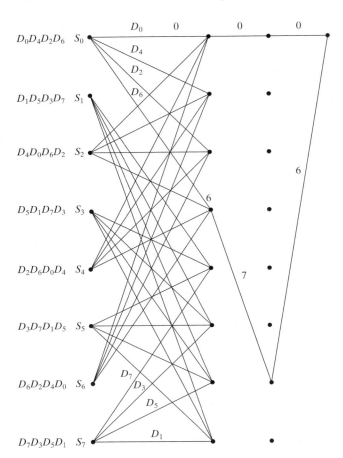

Figure 13.36　An 8-state Ungerboeck encoder.

Thus, this simple coding scheme is capable of achieving a 3-dB coding gain without increasing bandwidth. Of course, the price paid for this better performance is increased complexity in encoding and decoding.

Instead of a 4-state trellis, a trellis with a higher number of states yields higher coding gains. Extensive computer simulations by Ungerboeck indicate that with 8, 16, 32, 64, 128, and 256 states, coding gains in the range of 3.6–5.75 dB can be achieved. The trellis diagram for an 8-state trellis is shown in Figure 13.36.

Decoding of Trellis-Coded Modulation Codes.　The decoding of trellis-coded modulation is performed in two steps. Because each transition in the trellis corresponds to a partition of the signal set, and each partition generally corresponds to a number of signal points, the first step is to find the most likely signal point in each partition. This is accomplished by finding the point in each partition that is closest in Euclidean distance to the received point. This first step in decoding of a trellis-coded modulation scheme is called *subset decoding*. After this step, corresponding to each transition in the trellis, there exists only one point (the most likely one), and only one Euclidean distance (the distance between the received point and this most likely point). The second step of the decoding procedure is to use this Euclidean distance to find a path through the trellis whose total Euclidean distance from the received sequence is minimum. This is done by applying the Viterbi algorithm.

Trellis-coded modulation is now widely used in high-speed telephone line modems. Without coding, high-speed modems achieved data rates up to 14,400 bits/sec with a 64-QAM signal constellation. The added coding gain provided by trellis-coded modulation has made it possible to more than double the speed of transmission.

13.7 CODING AND DIVERSITY FOR FADING CHANNELS: MULTIPLE ANTENNA SYSTEMS AND SPACE-TIME CODES

In Section 11.1.3, we saw that using diversity is an effective technique for combatting fading. The use of two or more antennas at the receiving terminal of a digital communication system is a common method for achieving spatial diversity to mitigate the effects of signal fading. Typically, the receiving antennas must be separated by one or more wavelengths to ensure that the received signal undergoes statistically independent fading. Spatial receiver diversity is especially attractive because the signal diversity is achieved without expanding the signal transmission bandwidth.

Spatial diversity can also be achieved by using multiple antennas at the transmitter terminal of the communication system. For example, suppose we employ two antennas at the transmitter (T1 and T2) and one antenna at the receiver, as shown in Figure 13.37. Two information symbols are transmitted in each symbol interval. In the kth-symbol interval, antenna T1 transmits the signal $s_{1k} g_T(t)$ and antenna T2

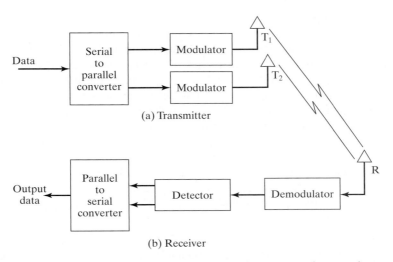

(a) Transmitter

(b) Receiver

Figure 13.37 Communication system employing two transmit antennas and one receive antenna.

simultaneously transmits the signal $s_{2k}g_T(t)$, where $g_T(t)$ is some basic signal pulse and s_{1k} and s_{2k} are symbols selected from some signal constellation, e.g., QAM or PSK. The signal at the receiving antenna may be expressed as

$$r(t) = c_1 s_{1k} g_T(t) + c_2 s_{2k} g_T(t) + n(t), \tag{13.7.1}$$

where the fading channel is assumed to be slowly fading and frequency nonselective with complex valued channel coefficients c_1 and c_2, and $n(t)$ denotes the additive noise. The receiver processes the received signal by passing it through a filter matched to $g_T(t)$. Hence, at the sampling instant, the matched-filter output may be expressed as

$$r_k = c_1 s_{1k} + c_2 s_{2k} + n_k. \tag{13.7.2}$$

In the next symbol interval, T1 transmits the signal $s_{2k}^* g_T(t)$ and T2 transmits $-s_{1k}^* g_T(t)$, where the asterisk denotes the complex conjugate. Hence, the output of the matched filter may be expressed as

$$r_{k+1} = c_1 s_{2k}^* - c_2 s_{1k}^* + n_{k+1}. \tag{13.7.3}$$

We assume that the receiver knows the channel coefficients c_1 and c_2, which are obtained by periodically transmitting pilot signals through the channel and measuring the channel coefficients c_1 and c_2. Thus, it is possible for the receiver to recover s_{1k} and s_{2k} and, thus, achieve dual diversity. The recovery of s_{1k} and s_{2k} from r_k and r_{k+1} is done in Problem 13.28. The scheme for achieving dual transmit diversity is especially suitable when the receiving terminal is a small mobile platform, such as a cell phone, where it may be physically difficult to have more than one receiving antenna on the platform.

In addition to providing signal diversity, multiple transmit antennas can create multiple spatial channels to increase the data transmission rate. In particular, suppose that we employ N transmit antennas and M receive antennas, where each of the N transmit antennas is used to transmit a different symbol in every symbol interval. Figure 13.38 illustrates the block diagram of a digital communication system that employs N transmit antennas and M receive antennas. The system shown in this figure includes an encoder and an interleaver at the transmitter and a deinterleaver and a decoder at the receiver. In the system, all signals from the N transmit antennas are transmitted synchronously through the common channel. Thus, we achieve an N-fold increase in data rate compared with a system that uses a single transmit antenna. Assuming that the channel is frequency nonselective and slowly fading, so that the channel coefficients can be accurately measured, we can achieve Mth-order spatial diversity by using M receive antennas, where $M \geq N$. Additional diversity, equal to the minimum distance of the code, can be achieved by encoding the data, provided that we employ a sufficiently long interleaver to ensure statistically independent channel fading.

Example 13.7.1

Consider a communication system that employs two transmit and two receive antennas. The encoder implements a rate $1/2$ convolutional code with free distance $d_{\text{free}} = 6$. The

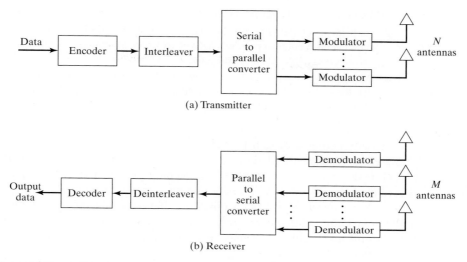

(a) Transmitter

(b) Receiver

Figure 13.38 A digital communication system with multiple transmit and receive antennas.

interleaver is sufficiently long to assure that successive code symbols are separated in time by at least the coherence time of the channel. The channel is slowly fading and frequency nonselective. The decoder employs the Viterbi algorithm to recover the data.

1. Express mathematically the signal that is transmitted through the channel by the two antennas, the signal received at each of the received antennas, and the signal at the output of the demodulator (input to the decoder).
2. What is the order of diversity achieved by the receiver if the decoder performs soft-decision decoding?
3. What is the order of diversity achieved by the receiver if the decoder performs hard-decision decoding?

Solution

1. The signal at the input to the channel at the kth-symbol interval is

$$x(t) = s_{1k}g_T(t) + s_{2k}g_T(t), \qquad 0 \le t \le T.$$

The signal received at the two receive antennas are

and
$$r_1(t) = c_{11}s_{1k}g_T(t) + c_{12}s_{2k}g_T(t) + n_1(t)$$
$$r_2(t) = c_{21}s_{1k}g_T(t) + c_{22}s_{2k}g_T(t) + n_2(t),$$

where the $\{c_{ij}\}$ are the channel coefficients, s_{1k} and s_{2k} are the channel symbols, and $n_1(t)$ and $n_2(t)$ are the additive noise terms at the input to the receiver. The demodulator (matched filter) outputs are

and
$$r_1 = c_{11}s_{1k} + c_{12}s_{2k} + n_1$$
$$r_2 = c_{21}s_{1k} + c_{22}s_{2k} + n_2.$$

These two signals are fed to the decoder.

2. With ideal interleaving, all the coded symbols fade independently so the received signal at each antenna provides a diversity order equal to $d_{\text{free}} = 6$. If the decoder is a soft-decision decoder, the total order of diversity for the two antennas is $2d_{\text{free}} = 12$.

3. With hard-decision decoding, it can be shown with the aid of Equation (13.3.22) that the order of signal diversity from each antenna is $d_{\text{free}}/2$. Hence, the total order of diversity for the two antennas is equal to $d_{\text{free}} = 6$.   ∎

An alternative approach to coding for multiple antenna systems is to employ a new class of codes, called *space-time codes*. A space-time code may be either a block code or a trellis code that is especially designed for multiple transmit and receive antennas to achieve diversity and an increase in data rate.

We consider a trellis code to illustrate the encoding process. A block diagram of the transmitter and the receiver is illustrated in Figure 13.39 for the case of two transmit and receive antennas. We assume that the data is binary and a signal constellation of 2^k signal points is selected. Thus, any k-input data bits are mapped into one of 2^k signal points. As an example, suppose that the modulation is 4-PSK and the encoder is represented by the 4-state trellis shown in Figure 13.40. This trellis generates a rate $\frac{1}{2}$ code, where each input symbol, represented by the possible phase values 0, 1, 2 and 3 is encoded into two output symbols, as specified by the trellis code. Each of the two output symbols is passed to (identical) interleavers. The symbol (phase) sequence at the output of each interleaver modulates a basic pulse shape $g_T(t)$ and the two symbols are transmitted synchronously through the channel. At the receiver, the signals received at the antennas are passed through matched filters, deinterleaved, and fed to the trellis (Viterbi) decoder. The 4-state trellis code can be shown to provide dual diversity and the use of two antennas at the receiver results in fourth-order diversity for the communication system.

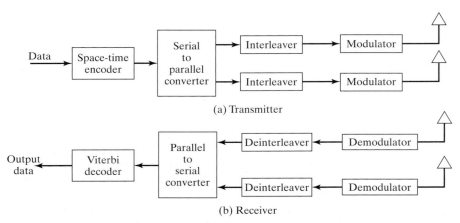

(a) Transmitter

(b) Receiver

Figure 13.39 Block diagram of communication system employing a space-time code for two transmit and two receive antennas.

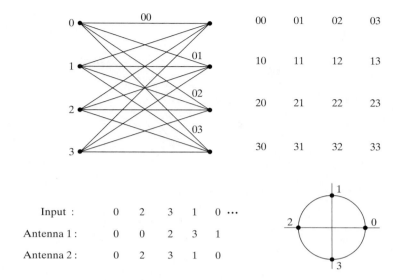

Figure 13.40   4-PSK, 4-state, space-time code.

As a second example, we consider a rate 1/2, 8-state trellis code which is used with 8-PSK modulation, as shown in Figure 13.41. In this case, we transmit 3 bits/symbol, so that the throughput is 50% higher than the rate $\frac{1}{2}$, 4-state trellis code. With two transmit and two receive antennas, the code also results in fourth-order diversity.

For the sake of brevity, we will not discuss methods for designing space-time codes. Refer to the papers by Tarokh, et al. (1998, 1999), which describe the design of trellis codes and block codes.

13.8 PRACTICAL APPLICATIONS OF CODING

In the previous sections, we have seen that coding can be employed to improve the effective SNR and, thus, enhance the performance of the digital communication system. Block and convolutional codes and combinations of them in the form of concatenated and turbo codes as discussed earlier, have been applied to communication situations where bandwidth is not a major concern; thus, some bandwidth expansion due to coding is allowed. On the other hand, in cases where bandwidth is a major concern, as in digital communication over telephone channels, coded modulation can be employed. By using coding, the performance of practical digital communication systems has improved up to 9 dB, depending on the application and the type of the code employed. In this section, we discuss applications of coding to three digital communication cases. These include deep-space communications, telephone-line modems, and coding for compact discs (CDs).

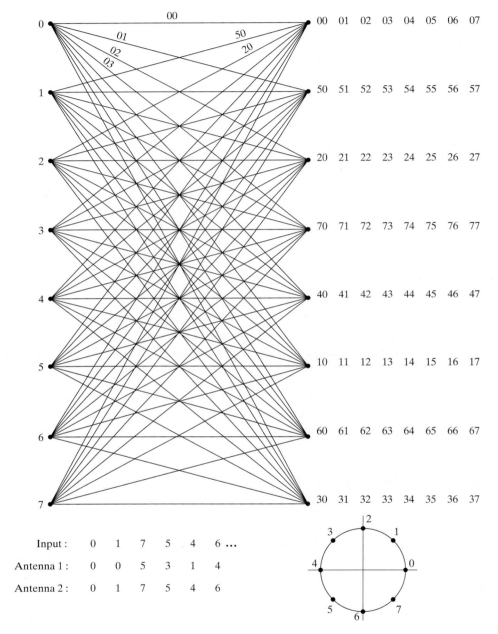

Figure 13.41   8-PSK, 8-state, space-time code.

13.8.1 Coding for Deep-Space Communications

Deep-space communication channels are characterized by very low SNR's and practically no bandwidth limitations. The transmitter power is usually obtained from onboard solar cells and, therefore, is typically limited to 20–30 W. The physical dimensions of the transmitting antenna are also quite limited and, therefore, its gain is also limited. The enormous distance between the transmitter and the receiver and lack of repeaters results in a very low SNR at the receiver. The channel noise can be characterized by a white Gaussian random process. These channels are very well modeled as AWGN channels. Because bandwidth is not a major concern on these channels, both block and convolutional codes can be applied.

In the Viking orbiters and landers mission to Mars, a (32, 6) block code (Reed–Muller code) was employed. It provided a coding gain of approximately 4 dB with respect to an uncoded PSK system at an error rate of 10^{-6}. Later, in the Voyager space mission to outer planets (Mars, Jupiter, and Saturn), convolutional codes with Viterbi decoding were employed. Two codes that were designed at the Jet Propulsion Laboratory (JPL) for that mission were a (2, 1) convolutional code with a constraint length of $L = 7$ with

$$\boldsymbol{g}_1 = [1 \quad 1 \quad 0 \quad 1 \quad 1 \quad 0 \quad 1]$$
$$\boldsymbol{g}_2 = [1 \quad 0 \quad 0 \quad 1 \quad 1 \quad 1 \quad 1]$$

and a (3, 1) convolutional code with $L = 7$ and

$$\boldsymbol{g}_1 = [1 \quad 1 \quad 0 \quad 1 \quad 1 \quad 0 \quad 1]$$
$$\boldsymbol{g}_2 = [1 \quad 0 \quad 0 \quad 1 \quad 1 \quad 1 \quad 1]$$
$$\boldsymbol{g}_3 = [1 \quad 0 \quad 1 \quad 0 \quad 1 \quad 1 \quad 1].$$

The first code has a free distance of $d_{\text{free}} = 10$, and the second code has $d_{\text{free}} = 15$. Both codes were decoded using the Viterbi algorithm and a soft-decoding scheme, in which the output was quantized to $Q = 8$ levels. The first code provides a coding gain of 5.1 dB with respect to an uncoded PSK system operating at an error rate of 10^{-5}. The second code provides a gain of 5.7 dB. Both operate about 4.5 dB above the theoretical limit predicted by Shannon's formula.

In subsequent missions of the Voyager to Uranus, in 1986, the (2, 1) convolutional code with $L = 7$ was used as an inner code in a concatenated coding scheme where a (255, 223) Reed–Solomon code served as the outer code. Viterbi decoding followed by a Reed–Solomon decoder at the Earth terminal provided a total coding gain of 8 dB at an error rate of 10^{-6}. This system operated at a data rate of approximately 30 kbits/sec.

Other decoding algorithms for convolutional codes have also been applied to certain deep-space communication projects. For NASA's Pioneer 9 mission, a (2, 1) convolutional code with a constraint length of $L = 21$ was designed with generator sequences (in octal representation)

$$g_1 = [4 \quad 0 \quad 0 \quad 0 \quad 0 \quad 0 \quad 0]$$
$$g_2 = [7 \quad 1 \quad 5 \quad 4 \quad 7 \quad 3 \quad 7],$$

which employed Fano's algorithm with a soft-decision decoding scheme and eight levels of output quantization. Pioneers 10, 11, and 12 and Helios A and B German solar orbiter missions employed a (2, 1) convolutional code with a constraint length of $L = 32$. The generator sequences for this code (in octal representation) are

$$g_1 = [7 \quad 3 \quad 3 \quad 5 \quad 3 \quad 3 \quad 6 \quad 7 \quad 6 \quad 7 \quad 2]$$
$$g_2 = [5 \quad 3 \quad 3 \quad 5 \quad 3 \quad 3 \quad 6 \quad 7 \quad 6 \quad 7 \quad 2].$$

This code has a free distance of $d_{free} = 23$. For decoding, again, the Fano decoding algorithm with eight-level output quantization was employed. Majority logic decoding has also been used in a number of coding schemes designed for the INTELSAT communication satellites. As an example, an (8, 7) code with $L = 48$ designed to operate at 64 kbits/sec on an INTELSAT satellite was capable of improving the error rate from 10^{-4} to 5×10^{-8}. In the Galileo space mission, a (4, 1, 14) convolutional code was used, resulting in a spectral bit-rate of $\frac{1}{4}$, which at $\mathcal{E}_b/N_0$ of 1.75 dB, achieved an error probability of 10^{-5}. This code performed 2.5 dB from the Shannon limit. Using an outer (255, 223) Reed–Solomon code improves the coding gain by another 0.8 dB, resulting in a concatenated code operating 1.7-dB from the Shannon limit. Turbo codes performing at a spectral-bit rate of 0.5 and operating only 0.7-dB from the Shannon limit, compare quite favorably with all these systems.

13.8.2 Coding for Telephone-Line Modems

Telephone-line channels are characterized by a limited bandwidth, typically between 300–3000 Hz, and a rather high SNR, which is usually 28 dB or more. Therefore, in designing coding schemes for telephone-line channels, we are faced with bandwidth limitation. This is in direct contrast to the deep-space communication channel, which is primarily power limited. This corresponds to the case of $r \gg 1$ in Figure 12.17. Because bandwidth is limited, we have to use low dimensional signaling schemes. Since power is rather abundant, we can employ multilevel modulation schemes. As we have already seen in Section 13.6, trellis-coded modulation is an appropriate scheme to be employed in such a case.

Historically, the first modems on telephone channels (prior to the 1960s) employed frequency-shift keying with asynchronous detection and achieved bit rates in the range of 300–1200 bits/sec. Later, in the early 1960s, the first generation of synchronous modems employing 4-PSK modulation achieved bit rates of up to 2400 bits/sec. Advances in equalization techniques allowed for more sophisticated constellations, which resulted in higher bit rates. These included 8-PSK modems achieving a bit rate of 4800 bits/sec and 16-point QAM modems that increased the bit rate to 9600 bits/sec. In the early 1980s, modems with a bit rate of 14,400 bits/sec were introduced; they employed a 64-point QAM signal constellation. All these improvements were results of advances in equalization and signal-processing techniques and also improvements in the characteristics of telephone lines.

The advent of trellis-coded modulation made it possible to design coded-modulation systems that improved overall system performance without requiring excess bandwidth. Trellis-coded modulation schemes based on variations of the original Ungerboeck's codes and introduced by Wei (1984) were adopted as standard by the CCITT standard committees. These codes are based on linear or nonlinear convolutional codes to guarantee invariance to 180°- or 90°-phase rotations. This is crucial in applications where differential encoding is employed to avoid phase ambiguities when a PLL is employed for carrier-phase estimation at the receiver. These codes achieve a coding gain comparable to Ungerboeck's codes with the same number of states but, at the same time, provide the required phase invariance. In Figure 13.42, we have shown the combination of the differential encoder, the nonlinear convolutional encoder, and the signal mapping for the 8-state trellis-coded modulation system that is adopted in the CCITT V.32 standard.

13.9 FURTHER READING

Papers by Golay (1949), Hamming (1950), Hocquenghem (1959), Bose and Ray-Chaudhuri (1960a, 1960b), and Reed and Solomon (1960) are landmark papers in the development of block codes. Convolutional codes were introduced by Elias (1955) and various methods for their decoding were developed by Wozencraft and Reiffen (1961), Fano (1963), Zigangirov (1966), Viterbi (1967), and Jelinek (1969). Trellis-coded modulation was introduced by Ungerboeck (1982) and later developed by Forney (1988a,b). Product codes were introduced by Elias (1954) and concatenated codes were developed and analyzed by Forney (1966). Berrou, Glavieux, and Thitimajshima (1993) introduced turbo codes. Low-density parity check codes were introduced by Gallager (1962, 1963). Refer to books on coding theory including Berlekamp (1968), Peterson and Weldon (1972), MacWilliams and Sloane (1977), Lin and Costello (2005), Blahut (1983), Wicker (1995), Johannesson and Zigangirov (1999), and Biglieri, et. al. (1991). Refer to the book by Heegard and Wicker (1999) on turbo codes.

PROBLEMS

13.1 In Example 13.2.1, find the minimum distance of the code. Which codeword(s) is (are) minimum weight?

13.2 In Example 13.2.3, verify that all codewords of the original code satisfy

$$cH^t = 0.$$

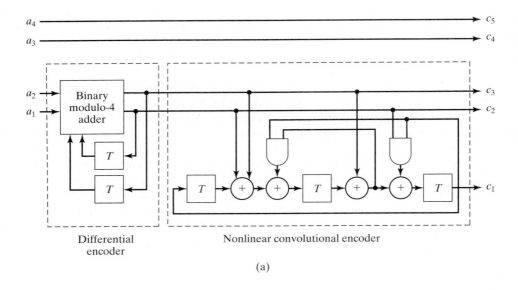

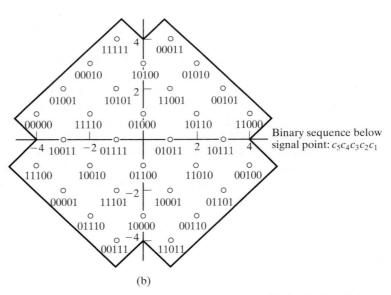

Figure 13.42 (a) Differential encoder, nonlinear convolutional encoder, and (b) signal constellation adopted in the V.32 standard.

13.3 By listing all codewords of the $(7, 4)$ Hamming code, verify that its minimum distance is equal to 3.

13.4 Find the parity check matrix and the generator matrix of a $(15, 11)$ Hamming code in the systematic form.

13.5 Show that the minimum Hamming distance of a linear block code is equal to the minimum number of columns of its parity check matrix that are linearly dependent. From this, conclude that the minimum Hamming distance of a Hamming code is always equal to 3.

13.6 A *simple repetition* code of blocklength n is a simple code consisting of only two codewords: $\underbrace{(0, 0, \ldots, 0)}_{n}$ and $\underbrace{(1, 1, \ldots, 1)}_{n}$. Find the parity check matrix and the generator matrix of this code in the systematic form. What is the rate and the minimum distance of this code?

13.7 The matrix

$$
G = \begin{bmatrix} 1 & 0 & 0 & 1 & 1 & 0 \\ 0 & 1 & 0 & 1 & 0 & 1 \\ 0 & 0 & 1 & 0 & 1 & 1 \end{bmatrix}
$$

is the generator matrix of a $(6, 3)$ linear code. This code is *extended* by adding an overall parity check bit to each codeword so that the Hamming weight of each resulting codeword is even.

1. Find the parity check matrix of the extended code.
2. What is the minimum distance of the extended code?
3. Find the coding gain of the extended code.

13.8 Compare the block error probability of an uncoded system with a system that uses a $(15, 11)$ Hamming code. The transmission rate is $R = 10^4$ bps and the channel is AWGN with a received power of $1 \ \mu$Watt and a noise power spectral density of $\frac{N_0}{2} = 10^{-11}$ Watts/Hz. The modulation scheme is binary PSK and soft-decision decoding is employed. Repeat the comparison when hard-decision decoding is employed.

13.9 Generate the standard array for a $(7,4)$ Hamming code. Use it to decode the received sequence $(1, 1, 1, 0, 1, 0, 0)$.

13.10 In a coded communication system, M messages $1, 2, \ldots, M = 2^k$ are transmitted by M *baseband* signals $x_1(t), x_2(t), \ldots, x_M(t)$, each of duration nT. The general form of $x_i(t)$ is given by

$$
x_i(t) = \sum_{j=0}^{n-1} \psi_{ij}(t - jT),
$$

where $\psi_{ij}(t)$ can be either of the two signals $\psi_1(t)$ or $\psi_2(t)$ where $\psi_1(t) = \psi_2(t) \equiv 0$ for all $t \notin [0, T]$. We further assume that $\psi_1(t)$ and $\psi_2(t)$ have equal energy $\mathscr{E}$ and the channel is ideal (no attenuation) with additive white Gaussian noise of power spectral density $\frac{N_0}{2}$. This means that the received signal is $r(t) = x(t) + n(t)$, where $x(t)$ is one of the $x_i(t)$'s and $n(t)$ represents the noise.

1. With $\psi_1(t) = -\psi_2(t)$, show that N, the dimensionality of the signal space, satisfies $N = n$.

2. Show that, in general, $N \leq 2n$.

3. With $M = 2$, show that for general $\psi_1(t)$ and $\psi_2(t)$,

$$P(\text{Error}|x_1(t) \text{ sent}) \leq \underset{R^N}{\int \cdots \int} \sqrt{f(\boldsymbol{r}|\boldsymbol{x}_1)f(\boldsymbol{r}|\boldsymbol{x}_2)} \, d\boldsymbol{r},$$

where $\boldsymbol{r}$, $\boldsymbol{x}_1$, and $\boldsymbol{x}_2$ are the vector representations of $r(t)$, $x_1(t)$, and $x_2(t)$ in the N-dimensional space.

4. Using the result of Part 3, show that for general M,

$$P(\text{Error}|x_m(t) \text{ sent}) \leq \sum_{\substack{1 \leq m' \leq M \\ m' \neq m}} \underset{R^N}{\int \cdots \int} \sqrt{f(\boldsymbol{r}|\boldsymbol{x}_m)f(\boldsymbol{r}|\boldsymbol{x}_{m'})} \, d\boldsymbol{r}.$$

5. Show that

$$\underset{R^N}{\int \cdots \int} \sqrt{f(\boldsymbol{r}|\boldsymbol{x}_m)f(\boldsymbol{r}|\boldsymbol{x}_{m'})} \, d\boldsymbol{r} = e^{-\frac{|\boldsymbol{x}_m - \boldsymbol{x}_{m'}|^2}{4N_0}},$$

and therefore,

$$P(\text{Error}|x_m(t) \text{ sent}) \leq \sum_{\substack{1 \leq m' \leq M \\ m' \neq m}} e^{-\frac{|\boldsymbol{x}_m - \boldsymbol{x}_{m'}|^2}{4N_0}}.$$

13.11 A convolutional code is described by

$$g_1 = [1\ 0\ 0]$$
$$g_2 = [1\ 0\ 1]$$
$$g_3 = [1\ 1\ 1].$$

1. Draw the encoder corresponding to this code.

2. Draw the state transition diagram for this code.

3. Draw the trellis diagram for this code.

4. Find the transfer function and the free distance of this code.

5. Verify whether this code is catastrophic or not.

13.12 Show that, in the trellis diagram of a convolutional code, 2^k branches enter each state and 2^k branches leave each state.

13.13 The block diagram of a binary convolutional code is shown in Figure P–13.13.

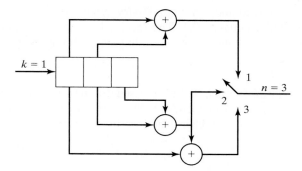

Figure P-13.13

1. Draw the state diagram for the code.

2. Find $T(D)$, the transfer function of the code.

3. What is d_{free}, the minimum free distance of the code?

4. Assume that a message has been encoded by this code and transmitted over a binary symmetric channel with an error probability of $p = 10^{-5}$. If the received sequence is $r = (110, 110, 110, 111, 010, 101, 101)$, use the Viterbi algorithm to find the transmitted bit sequence.

5. Find an upper bound to the bit error probability of the code when the preceding binary symmetric channel is employed. Make any reasonable approximations.

13.14 The block diagram of a (3, 1) convolutional code is shown in Figure P–13.14.

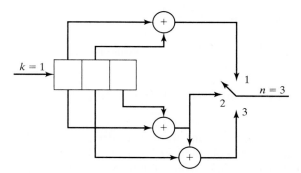

Figure P-13.14

1. Draw the state diagram of the code.

2. Find the transfer function $T(D)$ of the code.

3. Find the minimum free distance (d_{free}) of the code and show the corresponding path (at distance d_{free} from the all-zero codeword) on the trellis.

4. Assume that four information bits (x_1, x_2, x_3, x_4), followed by two zero bits, have been encoded and sent via a binary symmetric channel with a crossover probability equal to 0.1. The received sequence is (111, 111, 111, 111, 111, 111). Use the Viterbi decoding algorithm to find the most likely data sequence.

13.15 The convolutional code of Problem 13.11 is used for transmission over an AWGN channel with hard-decision decoding. The output of the demodulator-detector is (101001011110111 . . .). Using the Viterbi algorithm, find the transmitted sequence.

13.16 Repeat Problem 13.11 for a code with

$$g_1 = [1\ 1\ 0]$$
$$g_2 = [1\ 0\ 1]$$
$$g_3 = [1\ 1\ 1].$$

13.17 Show the paths corresponding to all codewords of weight 6 in Example 13.3.3.

13.18 Consider the convolutional code generated by the encoder shown in Figure P–13.18.

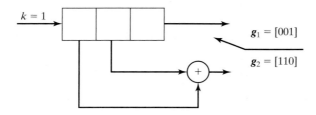

Figure P-13.18

1. Find the transfer function of the code in the form $T(N, D)$.

2. Find d_{free} of the code.

3. If the code is used on a channel using hard-decision Viterbi decoding, assuming the crossover probability of the channel is $p = 10^{-6}$, use the hard-decision bound to find an upper bound on the average bit error probability of the code.

13.19 Let x_1 and x_2 be two codewords of length n with distance d; assume that these two codewords are transmitted via a binary symmetric channel with crossover probability p. Let $P_2(d)$ denote the error probability in the transmission of these two codewords.

1. Show that

$$P_2(d) \leq \sum_{i=1}^{2^n} \sqrt{p(y_i|x_1)p(y_i|x_2)},$$

where the summation is over all binary sequences y_i.

2. From your answer to Part 1, conclude that

$$P_2(d) \leq \left[4p(1-p)\right]^{\frac{d}{2}}.$$

3. Using the result of Part 2, prove Equation (13.2.36).

13.20 The complementary error function $\text{erfc}(x)$ is defined by

$$\text{erfc}(x) = \frac{2}{\sqrt{\pi}} \int_x^\infty e^{-t^2}\, dt.$$

1. Express $Q(x)$ in terms of $\text{erfc}(x)$.
2. Using the inequality

$$\text{erfc}\left(\sqrt{x+y}\right) \leq \text{erfc}\left(\sqrt{x}\right) e^{-y} \qquad x \geq 0, \ y \geq 0,$$

prove that the bound on the average bit error probability of a convolutional code is

$$p_b \leq \frac{1}{2k}\text{erfc}\left(\sqrt{d_\text{free} R_c \gamma_b}\right) e^{d_\text{free} R_c \gamma_b} \left.\frac{\partial T(D, N)}{\partial N}\right|_{N=1, D=e^{-R_c\gamma_b}},$$

where $\gamma_b = \frac{\mathcal{E}_b}{N_0}$ and we assume that soft-decision decoding is employed.

13.21 A product code is designed using an $(n_1, k_1) = (7, 4)$ systematic Hamming code and an $(n_2, k_2) = (6, 2)$ systematic code with generator matrix

$$G = \begin{bmatrix} 1 & 0 & 0 & 1 & 1 & 1 \\ 0 & 1 & 1 & 1 & 1 & 0 \end{bmatrix}$$

as its component codes.

1. Determine the minimum distance of the product code.
2. If the product code is decoded using optimal hard-decision decoding, what is the maximum number of errors it can correct?

3. Consider an information sequence of length $k = k_1 k_2 = 8$ consisting of all 1's. Use an $n_2 \times n_1 = 6 \times 7$ matrix similar to Figure 13.20 to find all the $n_1 n_2 - k_1 k_2 = 34$ parity check bits for this sequence.

4. Assume that, instead of optimal hard-decision decoding, we employ a simple decoding scheme in which hard-decision decoding is first applied to the rows and then to the columns. Show that, using this simple strategy, the code can correct all error patterns of weight 3 but it cannot correct all error patterns of weight 4.

13.22 By writing

$$\alpha_i(s) = \sum_{s'} P\left(S(i-1) = s', S(i) = s, \mathbf{y}_{[1,i]}\right),$$

prove the forward recursion equation for α given in Equation (13.4.13).

13.23 By writing

$$\beta_i(s') = \sum_{s} P\left(S(i+1) = s, \mathbf{y}_{[i+1,N]} | S(i) = s'\right),$$

prove the backward recursion equation for β given in Equation (13.4.14).

13.24 Show that, in a regular low-density parity check code, the rate of the code can be found as

$$R = 1 - \frac{\gamma}{\rho}.$$

13.25 A trellis-coded modulation system uses an 8-ary PAM signal set given by $\{\pm 1, \pm 3, \pm 5, \pm 7\}$ and the 4-state trellis encoder shown in Figure 13.34(a).

1. Using the set partitioning rules, partition the signal set into four subsets.

2. If the channel is an additive white Gaussian noise channel, and at the output of the matched filter the sequence $(-.2, 1.1, 6, 4, -3, -4.8, 3.3)$ is observed, what is the most likely transmitted sequence?

13.26 Using Equations (13.7.2) and (13.7.3), and assuming that the receiver has knowledge of channel coefficients, show that the receiver can recover s_{1k} and s_{2k} by computing the following quantities:

$$c_1^* r_k - c_2 r_{k+1}^* = \left(|c_1|^2 + |c_2|^2\right) s_{1k} + c_1^* n_k - c_2 n_{k+1}^*$$

$$c_2^* r_k + c_1 r_{k+1}^* = \left(|c_1|^2 + |c_2|^2\right) s_{2k} + c_2^* n_k + c_1 n_{k+1}^*.$$

COMPUTER PROBLEMS

13.1 Error Probability in Repetition Code

The crossover probability in a binary symmetric channel is $p = 0.3$. Evaluate and plot the error probability P_e for a simple repetition block code of length n, where n is odd, for $n = 3, 5, \ldots, 41$. Plot a graph of P_e versus n.

13.2 Linear Block Codes

The generator matrix for a (10,4) linear block code is given by

$$G = \begin{bmatrix} 1 & 0 & 0 & 1 & 1 & 1 & 0 & 1 & 1 & 1 \\ 1 & 1 & 1 & 0 & 0 & 0 & 1 & 1 & 1 & 0 \\ 0 & 1 & 1 & 0 & 1 & 1 & 0 & 1 & 0 & 1 \\ 1 & 1 & 0 & 1 & 1 & 1 & 1 & 0 & 0 & 1 \end{bmatrix}.$$

Determine all codewords and the minimum weight of the code.

13.3 Hamming Codes

The objective of this problem is to determine all the codewords of the $(15, 11)$ Hamming code.

1. Determine the parity check matrix of the $(15, 11)$ Hamming code, which has the form

$$H = \left[P^t \big| I_k \right].$$

2. Determine the generator matrix G, which has the form

$$G = \left[I_k \big| P \right].$$

3. Use the generator matric to generate and list all the codewords in the code.

13.4 Performance of Hard-Decision Decoding

The $(15, 11)$ Hamming code is used with antipodal signaling, and hard-decision decoding is used at the receiver to detect the codewords. The error probability for a single bit in any codeword is given by

$$P_2 = Q \left(\sqrt{\frac{2R_c \mathscr{E}_b}{N_0}} \right),$$

where $\mathscr{E}_b / N_0$ is the SNR/bit and R_c is the code rate $(11/15)$. Compute and plot the (upper bound) codeword error probability as a function of $\mathscr{E}_b / N_0$.

13.5 Soft-Decision Decoding

The (15, 11) Hamming code is used with antipodal signaling, and soft-decision decoding is used at the receiver to detect the codewords. Using the upper bound on the error probability

$$P_M = (M-1)Q\left(\sqrt{\frac{2d_{min}R_c\mathscr{E}_b}{N_0}}\right),$$

evaluate and graph P_M as a function of the SNR/bit $\mathscr{E}_b/N_0$.

13.6 Convolutional Encoder

Determine the output sequence of the convolutional encoder shown in Figure CP–13.6 when the information sequence is

1 0 0 1 1 1 0 0 1 1 0 0 0 0 1 1 1.

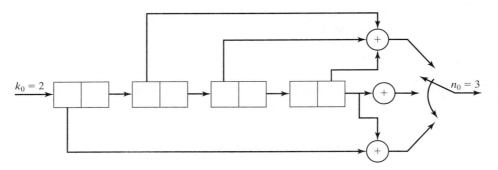

Figure CP-13.6

The encoder is initially in the all-zero state.

13.7 Viterbi Decoding

The encoder shown in Figure CP–13.7 is used to transmit a sequence of 5 information bits followed by two zeros to flush out the decoder. The quantized (hard-decision) received sequence at the input to the Viterbi decoder is

$$y = 0\ 1\ 1\ 0\ 1\ 1\ 1\ 1\ 0\ 1\ 0\ 0\ 0\ 1.$$

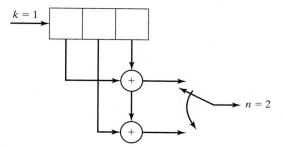

$k = 1$

$n = 2$

Figure CP-13.7

1. Sketch the trellis for this convolutional code and label the branches.
2. Determine the maximum-likelihood 5 bit information sequence at the output of the Viterbi decoder.

References

ADLER, R. L., COPPERSMITH, D., AND HASSNER, M. (1983), "Algorithms for Sliding Block Codes," *IEEE Trans. Inform. Theory,* vol. IT-29, pp. 5–22, January.

ANDERSON, J. B., AULIN, T., AND SUNDBERG, C. W. (1986), *Digital Phase Modulation,* Plenum, New York.

AULIN, T., RYDBECK, N., AND SUNDBERG, C. W. (1981), "Continuous Phase Modulation— Part II: Partial Response Signaling," *IEEE Trans. Commun.,* vol. COM-29, pp. 210–225, March.

AULIN, T., AND SUNDBERG, C. W. (1981), "Continuous Phase Modulation—Part I: Full Response Signaling," *IEEE Trans. Commun.,* vol. COM-29, pp. 196–209, March.

AULIN, T., AND SUNDBERG, C. W. (1982a), "On the Minimum Euclidean Distance for a Class of Signal Space Codes," *IEEE Trans. Inform. Theory,* vol. IT-28, pp. 43–55, January.

AULIN, T., AND SUNDBERG, C. W. (1982b), "Minimum Euclidean Distance and Power Spectrum for a Class of Smoothed Phase Modulation Codes with Constant Envelope," *IEEE Trans. Commun.,* vol. COM-30, pp. 1721–1729, July.

AULIN, T., AND SUNDBERG, C. W. (1984), "CPM—An Efficient Constant Amplitude Modulation Scheme," *Int. J. Satellite Commun.,* vol. 2, pp. 161–186.

BAHAI, A. R. S., AND SALTZBERG, B. R. (1999), *Multi-Carrier Digital Communications—Theory and Applications of OFDM,* Kluwer, Plenum, New York.

BAHL, L. R., COCKE, J., JELINEK, F., AND RAVIV, J. (1974), "Optimal Decoding of Linear Codes for Minimizing Symbol Error Rate," *IEEE Trans. Inform. Theory,* vol. IT-20, pp. 284–287, March.

BARROW, B. (1963), "Diversity Combining of Fading Signals with Unequal Mean Strengths," *IEEE Trans. Commun. Sys.,* vol. CS-11, pp. 73–78, March.

BELLO, P. A., AND NELIN, B. D. (1962a), "Predetection Diversity Combining with Selectively Fading Channels," *IRE Trans. Commun. Sys.,* vol. CS-10, pp. 32–44, March.

BELLO, P. A., AND NELIN, B. D. (1962b), "The Influence of Fading Spectrum on the Binary Error Probabilities of Incoherent and Differentially Coherent Matched Filter Receivers," *IRE Trans. Commun. Sys.,* vol. CS-11, pp. 170–186, June.

831

BELLO, P. A., AND NELIN, B. D. (1963), "The Effect of Frequency Selective Fading on the Binary Error Probabilities of Incoherent and Differentially Coherent Matched Filter Receivers," *IEEE Trans. Commun. Sys.,* vol. CS-11, pp. 170–186, June.

BENEDETTO, S., BIGLIERI, E., AND CASTELLANI, V. (1987), *Digital Transmission Theory,* Prentice-Hall, Englewood Cliffs, N.J.

BERGER, T., AND TUFTS, D. W. (1967), "Optimum Pulse Amplitude Modulation, Part I: Transmitter-Receiver Design and Bounds from Information Theory," *IEEE Trans. Inform. Theory,* vol. IT-13, pp. 196–208.

BERGER, T. (1971), *Rate Distortion Theory: A Mathematical Basis for Data Compression,* Prentice-Hall, Englewood Cliffs, N.J.

BERLEKAMP, E. R. (1968), *Algebraic Coding Theory,* McGraw-Hill, New York.

BERROU, C., GLAVIEUX, A., AND THITIMAJSHIMA, P. (1993), "Near Shannon Limit Error Correcting Coding and Decoding: Turbo Codes," *Proc. IEEE Int. Conf. Commun.,* pp. 1064–1070, May, Geneva, Switzerland.

BIGLIERI, E., DIVSALAR, D., MCLANE, P. J., AND SIMON, M. K. (1991), *Introduction to Trellis-Coded Modulation with Applications,* Macmillan., New York.

BINGHAM, J. A. C. (1990), "Multicarrier Modulation for Data Transmission: An Idea Whose Time Has Come," *IEEE Commun. Mag.,* vol. 28, pp. 5–14, May.

BLAHUT, R. E. (1983), *Theory and Practice of Error Control Codes,* Addison-Wesley, Reading, Mass.

BLAHUT, R. E. (1987), *Principles and Practice of Information Theory,* Addison-Wesley, Reading, Mass.

BLAHUT, R. E. (1990), *Digital Transmission of Information,* Addison-Wesley, Reading, Mass.

BOSE, R. C., AND RAY-CHAUDHURI, D. K. (1960a), "On a Class of Error Correcting Binary Group Codes," *Inform. Control,* vol. 3, pp. 68–79, March.

BOSE, R. C., AND RAY-CHAUDHURI, D. K. (1960b), "Further Results in Error Correcting Binary Group Codes," *Inform. Control,* vol. 3, pp. 279–290, September.

BOYD, S. (1986), "Multitone Signals with Low Crest Factor," *IEEE Trans. Circ. Sys.,* vol. CAS-33, pp. 1018–1022.

BRACEWELL, R. (1965), *The Fourier Transform and Its Applications,* 2nd Ed., McGraw-Hill, New York.

BRENNAN, D. G. (1959), "Linear Diversity Combining Techniques," *Proc. IRE,* vol. 47, pp. 1075–1102, June.

CARLSON, A. B. (1986), *Communication Systems,* 3rd Ed., McGraw-Hill, New York.

CHANG, R. W. (1966), "Synthesis of Bandlimited Orthogonal Signals for Multichannel Data Transmission," *Bell Sys. Tech. J.,* vol-45, pp. 1775–1796, December.

CHERUBINI, G. ELEFTHERIOU, E., AND OLCER, S. (2002), "Filtered Multitone Modulation for Very High-Speed Digital Subscriber Lines", *IEEE J. Select.. Area. Comm.,* vol. 20, issue 5 pp. 1016–1028. June.

CHOW, J. S., CIOFFI, J. M., AND BINGHAM, J. A. C. (1995), "A Practical Discrete Multitone Transceiver Loading Algorithm for Data Transmission over Spectrally Shaped Channels," *IEEE Trans. Commun.,* vol. 43, pp. 357–363, October.

CLARKE, K. K., AND HESS, D. T. (1971), *Communication Circuits: Analysis and Design,* Addison-Wesley, Reading, Mass.

COOK, C. E., ELLERSICK, F. W., MILSTEIN, L. B., AND SCHILLING, D. L. (1983), *Spread Spectrum Communications,* IEEE Press, New York.

COUCH, L. W., II (1993), *Digital and Analog Communication Systems,* 4th Ed., Macmillan, New York.

COVER, T. M., AND THOMAS, J. A. (1991), *Elements of Information Theory,* Wiley-Interscience, New York.

DAVENPORT, W. B., Jr., AND ROOT, W. L. (1987), *Random Signals and Noise,* McGraw-Hill, New York.

DELLER, J. P., PROAKIS, J. G., AND HANSEN, H. L. (2000), *Discrete-Time Processing of Speech Signals,* IEEE Press, Piscataway, N.J.

DIXON, R. C. (1976), *Spread Spectrum Techniques,* IEEE Press, New York.

DOELTZ, M. L., HEALD, E. T., AND MARTIN, D. L. (1957), "Binary Data Transmission Techniques for Linear Systems," *Proc. IRE,* vol-45, pp. 656–661, May.

ELIAS, P. (1954), "Error-Free Coding," *IRE Trans. Inform. Theory,* vol. IT-4, pp. 29–39, September.

ELIAS, P. (1955), "Coding for Noisy Channels," *IRE Conv. Rec.,* vol. 3, pt. 4, pp. 37–46.

FANO, R. M. (1963), "A Heuristic Discussion of Probabilistic Decoding," *IEEE Trans. Inform. Theory,* vol. IT-9, pp. 64–74, April.

FORNEY, G. D., Jr. (1966), *Concatenated Codes,* MIT Press, Cambridge, Mass.

FORNEY, G. D., Jr. (1972), "Maximum-Likelihood Sequence Estimation of Digital Sequences in the Presence of Intersymbol Interference," *IEEE Trans. Inform. Theory,* vol. 18, pp. 363–378, May.

FORNEY, G. D., Jr. (1988a), "Coset Codes I: Introduction and Geometrical Classification," *IEEE Trans. Inform. Theory,* vol. IT-34, pp. 671–680, September.

FORNEY, G. D., Jr. (1988b), "Coset Codes II: Binary Lattices and Related Codes," *IEEE Trans. Inform. Theory,* vol. IT-34, pp. 671–680, September.

FRANASZEK, P. A. (1968), "Sequence-State Coding for Digital Transmission," *Bell Sys. Tech. J.,* vol. 27, p. 143.

FRANASZEK, P. A. (1969), "On Synchronous Variable Length Coding for Discrete Noiseless Channels," *Inform. Control,* vol. 15, pp. 155–164.

FRANASZEK, P. A. (1970), "Sequence-State Methods for Run-Length-Limited Coding," *IBM J. Res. Dev.,* pp. 376–383, July.

FRANKS, L. E. (1969), *Signal Theory,* Prentice-Hall, Englewood Cliffs, N.J.

FRANKS, L. E. (1980), "Carrier and Bit Synchronization in Data Communication—A Tutorial Review," *IEEE Trans. Commun.,* vol. COM-28, pp. 1107–1121, August.

FREIMAN, C. E., AND WYNER, A. D. (1964), "Optimum Block Codes for Noiseless Input Restricted Channels," *Inform. Control,* vol. 7, pp. 398–415.

GABOR, A. (1967), "Adaptive Coding for Self Clocking Recording," *IEEE Trans. Elect. Comput.,* vol. EC-16, p. 866.

GALLAGER, R. G. "Low-Density Parity-Check Codes," *IRE Trans. Inform. Theory,* vol. 8, pp. 21–28. Jan.

GALLAGER, R. G. (1963), *Low-Density Parity-Check Codes,* MIT Press, Cambridge, MA.

GALLAGER, R. G. (1968), *Information Theory and Reliable Communication,* Wiley, New York.

GARDNER, F. M. (1979), *Phaselock Techniques,* Wiley, New York.

GARG, V. K., SMOLIK, K., AND WILKES, J. E. (1997), *Applications of CDMA in Wireless/Personal Communications,* Prentice Hall, Upper Saddle River, NJ.

GERSHO, A., AND GRAY, R. M. (1992), *Vector Quantization and Signal Compression,* Kluwer, Boston.

GERST, I., AND DIAMOND, J. (1961), "The Elimination of Intersymbol Interference by Input Pulse Shaping," *Proc. IRE,* vol. 53, July.

GIBSON, J. D. (1993), *Principles of Digital and Analog Communications,* 2nd Ed., Macmillan, New York.

GIBSON, J. D., BERGER, T., LOOKABAUGH, T., LINDBERG, D., AND BAKER, R. L. (1998), *Digital Compression for Multimedia: Principles and Standards,* Morgan Kaufmann, San Francisco, Calif.

GOLAY, M. J. E. (1949), "Notes on Digital Coding," *Proc. IRE,* vol. 37, p. 657, June.

GOLD, R. (1967), "Maximal Recursive Sequences with 3-Valued Recursive Cross Correlation Functions," *IEEE Trans. Inform. Theory,* vol. IT-14, pp. 154–156, January.

GOLOMB, S. W. (1967), *Shift Register Sequences,* Holden-Day, San Francisco, Calif.

GRAY, R. M., AND DAVISSON, L. D. (1986), *Random Processes: A Mathematical Approach for Engineers,* Prentice-Hall, Englewood Cliffs, N.J.

GREEN, P. E., Jr. (1962), "Radar Astronomy Measurement Techniques," MIT Lincoln Laboratory, Lexington, Mass., Tech. Report No. 282, December.

GRONEMEYER, S. A., AND McBRIDE, A. L. (1976), "MSK and Offset QPSK Modulation," *IEEE Trans. Commun.,* vol. COM-24, pp. 809–820, August.

GUPTA, S. C. (1975), "Phase-Locked Loops," *Proc. IEEE,* vol. 63, pp. 291–306, February.

HAGENAUER, J., AND HOEHER, P. (1989), "A Viterbi Algorithm with Soft Decision Outputs and Its Applications," *Proc. IEEE Globecom Conf.,* pp. 1680–1686, November, Dallas, Tex.

HAMMING, R. W. (1950), "Error Detecting and Error Correcting Codes," *Bell Sys. Tech. J.,* vol. 29, pp. 147–160, April.

HARTLEY, R. V. (1928), "Transmission of Information," *Bell Sys. Tech. J.,* vol. 7, p. 535.

HAYKIN, S. (2000), *Communication Systems,* 4th Ed., Wiley, New York.

HEEGARD, C., AND WICKER, S. B. (1999), *Turbo Coding,* Kluwer, Boston.

HELLER, J. A. (1975), "Feedback Decoding of Convolutional Codes," in *Advances in Communication Systems,* vol. 4, A. J. Viterbi (Ed.), Academic, New York.

HELSTROM, C. W. (1991), *Probability and Stochastic Processes for Engineers,* Macmillan, New York.

HOCQUENGHEM, A. (1959), "Codes Correcteurs d'Erreurs," *Chiffers,* vol. 2, pp. 147–156.

HOEG, W., AND LAUTERBACK, T. (2001), *Digital Audio Broadcasting: Principles and Applications of Digital Radio,* John Wiley and Sons, New York.

HOLMES, J. K. (1982), *Coherent Spread Spectrum Systems,* Wiley-Interscience, New York.

HUFFMAN, D. A. (1952), "A Method for the Construction of Minimum Redundancy Codes," *Proc. IRE,* vol. 40, pp. 1098–1101, September.

INGLE, V. AND PROAKIS J.G.(2000), *Digital Signal Processing Using Matlab,* Brooks-Cole, Pacific Grove, CA.

JACOBY, G. V. (1977), "A New Look-Ahead Code for Increased Data Density," *IEEE Trans. Magnetics,* vol. MAG-13, pp. 1202–1204.

JAYANT, N. S., AND NOLL. P. (1984), *Digital Coding of Waveforms,* Prentice-Hall, Englewood Cliffs, N.J.

JELINEK, F. (1969), "Fast Sequential Decoding Algorithm Using a Stack," *IBM J. Res. Dev.,* vol. 13, pp. 675–685, November.

JOHANNESON, R. AND ZIGANGIROV, K. S. (1999), *Fundamentals of Convolutional Coding,* IEEE Press, New York.

JONES, A. E., WILKINSON, T. A., AND BARTON, S. K. (1994), "Block Coding Schemes for Reduction of Peak-to-Mean Envelope Power Ratio of Multicarrier Transmission Systems," *Elect. Lett.,* vol. 30, pp. 2098–2099, December.

KAILATH, T. (1960), "Correlation Detection of Signals Perturbed by a Random Channel," *IRE Trans. Inform. Theory,* vol. IT-6, pp. 361–366, June.

KAILATH, T. (1961), "Channel Characterization: Time-Variant Dispersive Channels," in *Lectures on Communication Theory,* Chapter 6, E. Baghdadi (Ed.), McGraw-Hill, New York.

KARABED, R., AND SIEGEL, P. H. (1991), "Matched-Spectral Null Codes for Partial-Response Channels," *IEEE Trans. Inform. Theory,* vol. IT-37, pp. 818–855, May.

KASAMI, T. (1966), "Weight Distribution Formula for Some Class of Cyclic Codes," Coordinated Science Laboratory, University of Illinois, Urbana, Ill., Tech. Report No. R-285, April.

KOTELNIKOV, V. A. (1947), "The Theory of Optimum Noise Immunity," Ph.D. Dissertation, Molotov Energy Institute, Moscow. Translated by R. A. Silverman, McGraw-Hill, New York, 1959.

KRETZMER, E. R. (1966), "Generalization of a Technique for Binary Data Communication," *IEEE Trans. Commun. Tech.,* vol. COM-14, pp. 67–68, February.

LARSON, K. J. (1973), "Short Convolutional Codes with Maximal Free Distance for Rates 1/2, 1/3, 1/4," *IEEE Trans. Inform Theory,* vol. IT-19, pp. 371–372, May.

LAYER, D. H. (2001), *Digital Radio Takes to the Road,* IEEE Spectrum, vol. 38, no. 7, pp. 40–46. July.

LENDER, A. (1963), "The Duobinary Technique for High Speed Data Transmission," *AIEE Trans. Commun. Elect.,* vol. 82, pp. 214–218.

LEON-GARCIA, A. (1994), *Probability and Random Processes for Electrical Engineering,* 2nd Ed., Addison-Wesley, Reading Mass.

LIN, S., AND COSTELLO, D. J., Jr. (2005), *Error Control Coding: Fundamentals and Applications,* Prentice-Hall, Englewood Cliffs, N.J.

LINDE, J., BUZO, A., AND GRAY, R. M. (1980), "An Algorithm for Vector Quantizer Design," *IEEE Trans. Commun.,* vol. COM-28, pp. 84–95, January.

LINDSEY, W. C. (1964), "Error Probabilities for Ricean Fading Multichannel Reception of Binary and *N*-ary Signals," *IEEE Trans. Inform. Theory,* vol. IT-10, pp. 339–350, October.

LINDSEY, W. C. (1972), *Synchronization Systems in Communications,* Prentice-Hall, Englewood Cliffs, N.J.

LINDSEY, W. C., AND CHIE, C. M. (1981), "A Survey of Digital Phase-Locked Loops," *Proc. IEEE,* vol. 69, pp. 410–432.

LINDSEY, W. C., AND SIMON, M. K. (1973), *Telecommunication Systems Engineering,* Prentice-Hall, Englewood Cliffs, N.J.

LLOYD, S. P. (1957), "Least Square Quantization in PCM," Reprinted in *IEEE Trans. Inform. Theory,* vol. IT-28, pp. 129–137, March 1982.

LUCKY, R. W. (1965), "Automatic Equalization for Digital Communication," *Bell Sys. Tech. J.,* vol. 45, pp. 255–286, April.

LUCKY, R. W. (1966), "Techniques for Adaptive Equalization for Digital Communication," *Bell Sys. Tech. J.,* vol. 45, pp. 255–286.

LUCKY, R. W., SALZ, J., AND WELDON, E. J., Jr. (1968), *Principles of Data Communication,* McGraw-Hill, New York.

LYON, D. L. (1975a), "Timing Recovery in Synchronous Equalized Data Communication," *IEEE Trans. Commun.,* vol. COM-23, pp. 269–274, February.

LYON, D. L. (1975b), "Envelope-Derived Timing Recovery in QAM and SQAM Systems," *IEEE Trans. Commun.,* vol. COM-23, pp. 1327–1331, November.

MacWILLIAMS, F. J., AND SLOANE, J. J. (1977), *The Theory of Error Correcting Codes,* North Holland, New York.

MARKEL, J. D., AND GRAY, A. H., Jr. (1976), *Linear Prediction of Speech,* Springer-Verlag, New York.

MASSEY, J. L. (1963), *Threshold Decoding,* MIT Press, Cambridge, Mass.

MAX, J. (1960), "Quantizing for Minimum Distortion," *IRE Trans. Inform. Theory,* vol. IT-6, pp. 7–12, March.

McMAHON, M. A. (1984), *The Making of a Profession—A Century of Electrical Engineering in America,* IEEE Press.

MENGALI, U., AND D'ANDREA, A. N. (1997), *Synchronization Techniques for Digital Receivers,* Plenum Press, New York.

MEYR, H., AND ASCHEID, G. (1990), *Synchronization in Digital Communication,* Wiley-Interscience, New York.

MEYR, H., MOENECLAEY, M., AND FECHTEL, S. A. (1997), *Digital Communication Receivers, Synchronization, Channel Estimation, and Signal Processing,* vol. 2, John Wiley and Sons, New York.

MILLMAN, S., Ed. (1984), *A History of Engineering and Science in the Bell System— Communications Sciences (1925–1980),* AT&T Bell Laboratories.

MUELLER, K. H., AND MULLER, M. S. (1976), "Timing Recovery in Digital Synchronous Data Receivers," *IEEE Trans. Commun.,* vol. COM-24, pp. 516–531, May.

NELSON, R. (1995), *Probability, Stochastic Processes and Queueing Theory: The Mathematics of Computer Modeling,* Springer Verlag.

NORTH, D. O. (1943), "An Analysis of the Factors Which Determine Signal/Noise Discrimination in Pulse-Carrier Systems," RCA Tech. Report No. 6, PTR-6C.

NYQUIST, H. (1924), "Certain Factors Affecting Telegraph Speed," *Bell Sys. Tech. J.,* vol. 3, p. 324.

NYQUIST, H. (1928), "Certain Topics in Telegraph Transmission Theory," *AIEE Trans.,* vol. 47, pp. 617–644.

ODENWALDER, J. P. (1970), "Optimal Decoding of Convolutional Codes," Ph.D. Dissertation, Dept. of Systems Sciences, UCLA, Los Angeles.

OLSEN, J. D. (1977), "Nonlinear Binary Sequences with Asymptotically Optimum Periodic Cross Correlation," Ph.D. Dissertation, University of Southern California, Los Angeles.

OMURA, J. (1971), "Optimal Receiver Design for Convolutional Codes and Channels with Memory via Control Theoretical Concepts," *Inform. Sci.,* vol. 3, pp. 243–266.

OPPENHEIM, A. V., WILLSKY, A. S., AND YOUNG, I. T. (1983), *Signals and Systems,* Prentice-Hall, Englewood Cliffs, N.J.

PAPOULIS, A. (1962), *The Fourier Integral and Its Applications,* McGraw-Hill, New York.

PAPOULIS, A. (1991), *Probability, Random Variables, and Stochastic Processes,* 3rd Ed., McGraw-Hill, New York.

PETERSON, W. W., AND WELDON, E. J., Jr. (1972), *Error Correcting Codes,* 2nd Ed., MIT Press, Cambridge, Mass.

PETERSON, R. L., ZIEMER, R. E., AND BORTH, D. E. (1995), *Introduction to Spread Spectrum Communications,* Prentice-Hall, Englewood Cliffs NJ.

PICKHOLTZ, R. L., SCHILLING, D. L., AND MILSTEIN, L. B. (1982), "Theory of Spread Spectrum Communications—A Tutorial," *IEEE Trans. Commun.,* vol. COM-30, pp. 855–884, May.

PIERCE, J. N. (1958), "Theoretical Diversity Improvement in Frequency Shift Keying," *Proc. IRE,* vol. 46, pp. 903–910, May.

PIERCE, J. N., AND STEIN, S. (1960), "Multiple Diversity with Non-Independent Fading," *Proc. IRE,* vol. 48, pp. 89–104, January.

POPOVIC, B. M. (1991), "Synthesis of Power Efficient Multitone Signals with Flat Amplitude Spectrum," *IEEE Trans. Commun.,* vol. 39, pp. 1031–1033, July.

PRICE, R. (1954), "The Detection of Signals Perturbed by Scatter and Noise," *IRE Trans. Inform. Theory,* vol. PGIT-4, pp. 163–170, September.

PRICE, R. (1956), "Optimum Detection of Random Signals in Noise with Application to Scatter Multipath Communication," *IRE Trans. Inform. Theory,* vol. IT-2, pp. 125–135, December.

PRICE, R. (1962a), "Error Probabilities for Adaptive Multichannel Reception of Binary Signals," MIT Lincoln Laboratory, Lexington, Mass., Tech. Report No. 258, July.

PRICE, R. (1962b), "Error Probabilities for Adaptive Multichannel Reception of Binary Signals," *IRE Trans. Inform. Theory,* vol. IT-8, pp. 308–316, September.

PRICE, R., AND GREEN, P. E., Jr. (1958), "A Communication Technique for Multipath Channels," *Proc. IRE,* vol. 46, pp. 555–570, March.

PRICE, R., AND GREEN, P. E., Jr. (1960), "Signal Processing in Radio Astronomy—Communication via Fluctuating Multipath Media," MIT Lincoln Laboratory, Lexington, Mass., Tech. Report No. 2334, October.

PROAKIS, J. G. (2001), *Digital Communications,* 4th Ed., McGraw-Hill, New York.

PROAKIS, J. G. AND MANOLAKIS, D. G. (1996), *Digital Signal Processing: Principles, Algorithms, and Applications,* 3d Ed., Prentice Hall, Upper Saddle River, NJ.

PROAKIS, J. G. AND SALEHI, M. (2002), *Communication Systems Engineering,* 2d Ed., Prentice Hall, Upper Saddle River, NJ.

PROAKIS, J. G., SALEHI, M., AND BAUCH, G. (2004), *Contemporary Communication Systems Using Matlab and Simulink,* 2d Ed., Thomson, Stamford, CT.

RABINER, L. R., AND SCHAFER, R. W. (1979), *Digital Processing of Speech Signals,* Prentice-Hall, Englewood Cliffs, N.J.

RAPPAPORT, T. S. (1996), *Wireless Communications: Principles and Practice,* Prentice Hall, Upper Saddle River, NJ.

REED, I. S., AND SOLOMON, G. (1960), "Polynomial Codes over Certain Finite Fields," *SIAM J.,* vol. 8, pp. 300–304, June.

RYDER, J. D., AND FINK, D. G. (1984), *Engineers and Electronics,* IEEE Press.

SAKRISON, D. J. (1968), *Communication Theory: Transmission of Waveforms and Digital Information,* New York, Wiley.

SALTZBERG, B. R. (1967), "Performance of an Efficient Data Transmission System," *IEEE Trans. Commun.,* vol. COM-15, pp. 805–811, December.

SARWATE, D. V., AND PURSLEY, M. B. (1980), "Crosscorrelation Properties of Pseudorandom and Related Sequences," *Proc. IEEE,* vol. 68, pp. 2399–2419, September.

SCHOLTZ, R. A. (1977), "The Spread Spectrum Concept," *IEEE Trans. Commun.,* vol. COM-25, pp. 748–755, August.

SCHOLTZ, R. A. (1979), "Optimal CDMA Codes," *1979 National Telecommunications Conf. Record,* Washington, D.C., pp. 54.2.1–54.2.4, November.

SCHOLTZ, R. A. (1982), "The Origins of Spread Spectrum," *IEEE Trans. Commun.,* vol. COM-30, pp. 822–854, May.

SCHOUHAMER IMMINK, K. A. (1991), *Coding Techniques for Digital Recorders,* Prentice-Hall, Englewood-Cliffs, N.J.

SCHWARTZ, M., BENNETT, W. R., AND STEIN, S. (1966), *Communication Systems and Techniques,* McGraw-Hill, New York.

SHANMUGAM, K. S. (1979), *Digital and Analog Communication Systems,* Wiley, New York.

SHANNON, C. E. (1948a), "A Mathematical Theory of Communication," *Bell Sys. Tech. J.,* vol. 27, pp. 379–423, July.

SHANNON, C. E. (1948b), "A Mathematical Theory of Communication," *Bell Sys. Tech. J.,* vol. 27, pp. 623–656, October.

SIMON, M. K., OMURA, J. K., SCHOLTZ, R. A., AND LEVITT, B. K. (1985), *Spread Spectrum Communications vol. I, II, III,* Computer Science Press, Rockville, Md.

SMITH, J. W. (1965), "The Joint Optimization of Transmitted Signal and Receiving Filter for Data Transmission Systems," *Bell Sys. Tech. J.,* vol. 44, pp. 1921–1942, December.

STARK, H., AND WOODS, J. W. (1994), *Probability, Random Processes and Estimation Theory for Engineers,* 2nd Ed., Prentice-Hall, Upper Saddle River, N.J.

STARR, T., CIOFFI, J. M., AND SILVERMAN, P.J. (1998), *Understanding Digital Subscriber Line Technology,* Prentice Hall, Upper Saddle River, NJ.

STENBIT, J. P. (1964), "Table of Generators for BCH Codes," *IEEE Trans. Inform. Theory,* vol. IT-10, pp. 390–391, October.

STIFFLER, J. J. (1971), *Theory of Synchronous Communications,* Prentice-Hall, Englewood Cliffs, N.J.

STREMLER, F. G. (1990), *Introduction to Communication Systems,* 3rd Ed., Addison-Wesley, Reading, Mass.

SUNDBERG, C. W. (1986), "Continuous Phase Modulation," *IEEE Commun. Mag.,* vol. 24, pp. 25–38, April.

TANG, D. L., AND BAHL, L. R. (1970), "Block Codes for a Class of Constrained Noiseless Channels," *Inform. Control,* vol. 17, pp. 436–461.

TAROKH, V., SESHADRI, N., AND CALDERBANK, A. R. (1998), "Space-Time Codes for High Data Rate Wireless Communications: Performance Analysis and Code Construction," *IEEE Trans. Inform. Theory,* vol. 44, pp. 744-765, March.

TAROKH, V., JAFARKHANI, H., AND CALDERBANK, A. R. (1999), "Space-Time Block Codes from Orthogonal Designs," *IEEE Trans. Inform. Theory,* vol. IT-45, pp. 1456–1467, July.

TAROKH, V., AND JAFARKHANI, H. (2000), "On the Computation and Reduction of the Peak-to-Average Power Ratio in Multicarrier Communications," *IEEE Trans. Commun.,* vol. 48, pp. 37–44, January.

TAUB, H., AND SCHILLING, D. L. (1986), *Principles of Communication Systems,* 2nd Ed., McGraw-Hill, New York.

TELLADO, J., AND CIOFFI, J. M. (1998), "Efficient Algorithms for Reducing PAR in Multicarrier Systems," *Proc. 1998 IEEE Int. Symp. Inform. Theory,* p. 191, August 16–21, Cambridge, Mass. Also in *Proc. 1998 Globecom,* Nov. 8–12, Sydney, Australia.

TUFTS, D. W. (1965), "Nyquist's Problem—The Joint Optimization of Transmitter and Receiver in Pulse Amplitude Modulation," *Proc. IEEE,* vol. 53, pp. 248–259, March.

TURIN, G. L. (1961), "On Optimal Diversity Reception," *IRE Trans. Inform. Theory,* vol. IT-7, pp. 154–166, March.

TURIN, G. L. (1962), "On Optimal Diversity Reception II," *IRE Trans. Commun. Sys.,* vol. CS-12, pp. 22–31, March.

UNGERBOECK, G. (1974), "Adaptive Maximum Likelihood Receiver for Carrier Modulated Data Transmission Systems," *IEEE Trans. Commun.,* vol. COM-22, pp. 624–636, May.

UNGERBOECK, G. (1982), "Channel Coding with Multilevel/Phase Signals," *IEEE Trans. Inform. Theory,* vol. IT-28, pp. 55–67, January.

VITERBI, A. J. (1966), *Principles of Coherent Communication*, McGraw-Hill, New York.

VITERBI, A. J. (1967), "Error Bounds for Convolutional Codes and an Asymptotically Optimum Decoding Algorithm," *IEEE Trans. Inform. Theory,* vol. IT-13, pp. 260–269, April.

VITERBI, A. J. (1979), "Spread Spectrum Communication—Myths and Realities," *IEEE Commun. Mag.,* vol. 17, pp. 11–18, May.

VITERBI, A. J. (1985), "When Not to Spread Spectrum—A Sequel," *IEEE Commun. Mag.,* vol. 23, pp. 12–17, April.

VITERBI, A. J. (1995), *CDMA, Principles of Spread Spectrum Communications*, Addison-Wesley, Reading, MA.

VITERBI, A. J., AND JACOBS, I. M. (1975), "Advances in Coding and Modulation for Noncoherent Channels Affected by Fading, Partial Band, and Multiple-Access Interference," in *Advances in Communication Systems,* vol. 4, A. J. Viterbi, (Ed.), Academic, New York.

WEI, L. F. (1984), "Rotationally Invariant Convolutional Channel Coding with Expanded Signal Space. Part I: 180°, Part II: Nonlinear Codes," *IEEE J. Selected Areas Commun.,* vol. SAC-2, pp. 659–687. September.

WEINSTEIN, SW. B., AND EBERT, P. M. (1971), "Data Transmission by Frequency Division Multiplexing Using the Discrete Fourier Transform," *IEEE Trans. Commun.,* vol. COM-19, pp. 628–634, October.

WICKER, S. B. (1995), *Error Control Systems for Digital Communication and Storage*, Prentice Hall, Upper Saddle River, NJ.

WIDROW, B. (1966), "Adaptive Filters, I: Fundamentals," Tech. Report No. 6764-6, Stanford Electronic Laboratories, Stanford University, Stanford, Calif., December.

WIENER, N. (1949), *The Extrapolation, Interpolation, and Smoothing of Stationary Time-Series with Engineering Applications,* Wiley, New York. (The original work appeared in an MIT Radiation Laboratory Report, 1942).

WILKINSON, T. A., AND JONES, A. E. (1995), "Minimization of the Peak-to-Mean Envelope Power Ratio of Multicarrier Transmission Systems by Block Coding," *Proc. IEEE Vehicular Tech. Conf.,* pp. 825–829, July.

WONG, E., AND HAJEK, B. (1985), *Stochastic Processes in Engineering Systems,* Springer-Verlag, New York.

WOZENCRAFT, J. M. (1957), "Sequential Decoding for Reliable Communication," *IRE Natl. Conv. Rec.,* vol. 5, pt. 2, pp. 11–25.

WOZENCRAFT, J. M., AND JACOBS, I. M. (1965), *Principles of Communication Engineering,* Wiley, New York.

WOZENCRAFT, J. M., AND REIFFEN, B. (1961), *Sequential Decoding,* MIT Press, Cambridge, Mass.

WULICH, D. (1996), "Reduction of Peak-to-Mean Ratio of Multicarrier Modulation Using Cyclic Coding," *Elect. Lett.,* vol. 32, pp. 432–433, February.

ZIEMER, R. E., AND PETERSON, R. L. (1985), *Digital Communications and Spread Spectrum Systems,* Macmillan, New York.

ZIEMER, R. E., AND TRANTER, W. H. (2002), *Principles of Communications: Systems, Modulation, and Noise,* Houghton Mifflin, Boston.

ZIGANGIROV, K. S. (1966), "Some Sequential Decoding Procedures," *Problemy Peredachi Informatsii,* vol. 2, pp. 13–25.

ZIV, J., AND LEMPEL, A. (1977), "A Universal Algorithm for Sequential Data Compression," *IEEE Trans. Inform. Theory,* vol-IT23, pp. 337–343.

ZIV, J., AND LEMPEL, A. (1978), "Compression of Individual Sequences via Variable Rate Coding," *IEEE Trans. Inform. Theory,* vol. IT-24, pp. 530–536.

Index

F